ALLE ZEIT WACH 1842

B. Platzer · A. Polt · G. Maurer

Thermophysical Properties of Refrigerants

With 148 Figures

Springer-Verlag Berlin Heidelberg NewYork
London Paris Tokyo Hong Kong

Dipl.-Ing. Bernhard Platzer
Lehrstuhl für Technische Thermodynamik
Universität Kaiserslautern
Erwin-Schrödinger-Straße
6750 Kaiserslautern
FR Germany

Dr.-Ing. Axel Polt
Im unteren Grain 8a
6705 Deidesheim
FR Germany

Professor Dr.-Ing. Gerd Maurer
Lehrstuhl für Technische Thermodynamik
Universität Kaiserslautern
Erwin-Schrödinger-Straße
6750 Kaiserslautern
FR Germany

ISBN 3-540-16112-0 Springer-Verlag Berlin Heidelberg NewYork
ISBN 0-387-16112-0 Springer-Verlag NewYork Berlin Heidelberg

Library of Congress Cataloging-in-Publication Data
Platzer, Bernhard
Thermophysical properties of refrigerants / Bernhard Platzer, Axel Polt, Gerd Maurer.
Includes bibliographical references.
ISBN 0-387-16112-0 (U.S.)
1. Refrigerants--Thermal properties.
I. Polt, Axel. II. Maurer, Gerd. III. Title.
TP492.7.P56 1990
621.5'64--dc20 89-21761

Offsetprinting: Color-Druck Dorfi GmbH, Berlin; Bookbinding: Lüderitz & Bauer, Berlin
2161/3020 543210 – Printed on acid-free paper.

PREFACE

Halogenated derivatives of simple hydrocarbons, like chloro- and fluoromethanes and -ethanes, are important chemicals in many industries for example as working media in organic Rankine cycles and refrigerating processes. Although there are ecological reasons against using these substances, it is expected that they cannot be substituted in all cases. But optimizing the processes in which these substances have to be used can also contribute to minimizing the possible damage. This book summarizes the extensive experimental material available on the thermodynamic properties of 16 pure and mixed halogenated hydrocarbons, so that it can easily be used both by process and design engineers.

The project was initiated by the late Professor Dr.-Ing. Eberhard Bender and Prof. Dr.-Ing. Dr.-Ing. E.h. H.D. Baehr. We would like to express our gratitude especially to Prof. Baehr for his continuing interest and to Springer-Verlag for their patience. Besides the authors, quite a large number of students contributed to the completion of the book. Among them we are most obliged to Mr. L. Hoffmann and Mr. F. Strepp for their assistance in preparing tables, diagrams and charts.

We hope that this book will prove to be useful to many colleagues, and we welcome proposals, advice, and comments.

Kaiserslautern, January 1990

B. Platzer
A.W. Polt
G. Maurer

Contents

I Introduction . . . 1

II Fundamentals . . . 2

1 Fundamental Equations . . . 2
2 Ideal Gas State Properties . . . 3
2.1 Ideal Gas Standard Enthalpy and Entropy . . . 3
2.2 Ideal Gas State Heat Capacity . . . 3
3 Equation of State . . . 4
3.1 Selection of an Equation of State . . . 4
3.2 Parameter Estimation . . . 4

III Results . . . 7

Refrigerant R11 (CCl_3F) . . . 7
Experimental Data and Comparison with Compilation . . . 10
Charts . . . 19
References . . . 33
Diagrams . . . 35
Refrigerant R12 (CCl_2F_2) . . . 41
Experimental Data and Comparison with Compilation . . . 43
Charts . . . 54
References . . . 66
Diagrams . . . 69
Refrigerant R13 ($CClF_3$) . . . 75
Experimental Data and Comparison with Compilation . . . 77
Charts . . . 86
References . . . 99
Diagrams . . . 100
Refrigerant R14 (CF_4) . . . 107
Experimental Data and Comparison with Compilation . . . 109
Charts . . . 117
References . . . 131
Diagrams . . . 132
Refrigerant R21 ($CHCl_2F$) . . . 139
Experimental Data and Comparison with Compilation . . . 141
Charts . . . 150
References . . . 165
Diagrams . . . 167
Refrigerant R22 ($CHClF_2$) . . . 173
Experimental Data and Comparison with Compilation . . . 175
Charts . . . 184
References . . . 197
Diagrams . . . 199
Refrigerant R23 (CHF_3) . . . 205
Experimental Data and Comparison with Compilation . . . 207
Charts . . . 216
References . . . 230
Diagrams . . . 231
Refrigerant R12B1 ($CBrClF_2$) . . . 237
Experimental Data and Comparison with Compilation . . . 239
Charts . . . 246
References . . . 256
Diagrams . . . 257

Refrigerant R13B1 ($CBrF_3$) 263
Experimental Data and Comparison with Compilation 265
Charts 273
References 284
Diagrams 285
Refrigerant R113 (CCl_2FCClF_2) 291
Experimental Data and Comparison with Compilation 293
Charts 302
References 316
Diagrams 317
Refrigerant R114 ($CClF_2CClF_2$) 323
Experimental Data and Comparison with Compilation 325
Charts 334
References 346
Diagrams 347
Refrigerant R115 ($CClF_2CF_3$) 353
Experimental Data and Comparison with Compilation 355
Charts 364
References 376
Diagrams 377
Refrigerant RC318 (C_4F_8) 383
Experimental Data and Comparison with Compilation 385
Charts 393
References 406
Diagrams 407
Refrigerant R500 (73.8 wt.% R12/26.2 wt.% R152a) 413
Experimental Data and Comparison with Compilation 415
Charts 423
References 434
Diagrams 435
Refrigerant R502 (48.8 wt.% R22/51.2 wt.% R115) 441
Experimental Data and Comparison with Compilation 443
Charts 452
References 463
Diagrams 465
Refrigerant R503 (59.9 wt.% R13/40.1 wt.% R23) 471
Experimental Data and Comparison with Compilation 473
Charts 480
References 488
Diagrams 489

Nomenclature

$\vec{a}$	parameter vector of the *Bender* equation of state
a_i	element of $\vec{a}$
b_i	parameter in equation (20)
c_i	parameter in equation (21)
c_i^*	parameter in equation (22)
c_p	isobaric heat capacity $(\partial h/\partial T)_p$
c_v	isochoric heat capacity $(\partial u/\partial T)_v$
d_i	parameter in equation (15)
h	enthalpy
IT	isothermal throttling coefficient $(\partial h/\partial p)_T$
JT	Joule-Thomson coefficient $(\partial T/\partial p)_h$
k_p	number of experimental p-ρ-T data points
k_s	number of experimental saturation properties
n	number of data points
p	pressure
r	enthalpy of vaporization
R	gas constant
s	entropy
SS	sum of squares
T	thermodynamic temperature in Kelvin
u	internal energy
v	volume
w	speed of sound
w_{DK}	weighting factor for the vapor pressure in equation (16)
w_ρ	weighting factor for the ratio of fugacity coefficients in equation (16)
Δ	difference
μ	Joule-Thomson coefficient $(\partial T/\partial p)_h$
ρ	density
σ_p^*	total uncertainty of p_{exp}
σ_p	absolute uncertainty of measuring p
σ_ρ	absolute uncertainty of measuring ρ
σ_T	absolute uncertainty of measuring T
τ	abreviation for $1 - T/T_c$
φ	fugacity coefficient

subscripts

0	standard state ($T_0 = 298.15$ K ; $p_0 = 0.1$ MPa)
c	critical point
calc	calculated result
exp	experimental result
gas	vapor phase
k	number of experimental data point
id. gas	ideal gas

superscripts

o	ideal gas state
′	saturated liquid phase
″	saturated vapor phase
s	saturation

I Introduction

Halogenated hydrocarbons are often used as working media in refrigerating units and organic Rankine cycles for heat recovery systems and heat pumps. Reliable information on thermodynamic properties of the working media are essential for designing and simulating such processes. This book is aimed at providing that information for the most important halogenated hydrocarbons. Experimental results for thermodynamic properties such as vapor pressure, saturation densities, pressure-volume-temperature data, heat capacity, etc. are carefully evaluated and used to describe consistently the gaseous and liquid regions. The method applied combines the *Bender* equation of state with an empirical equation for the ideal gas state heat capacity. Results are presented for 16 halogenated hydrocarbons (Table A1) in several ways:

- parameters for the equation are reported,
- tables of thermodynamic properties (p, T, v, h, s) are supplied,
- some results are also summarized graphically ($p, h-$ $T, s-$ and $h, s-$ diagrams).

In order to enhance the applicability of the results computer software is available from the authors on request.

Table A1: Some physico-chemical properties of hydrocarbons

Substance	Molecular weight	Normal boiling point * K	Critical properties p_c MPa	T_c K	ρ_c g/cm^3	Ideal gas state properties ** h_0 kJ/kg	s_0 kJ/kg K
R11	137.380	296.38	4.4026	471.15	0.5590	405.41	1.6937
R12	120.925	242.99	4.1290	384.95	0.5600	372.02	1.6995
R13	104.470	191.49	3.8770	301.88	0.5820	329.62	1.6876
R14	88.010	144.92	3.7500	227.51	0.6257	437.07	2.4363
R21	102.920	281.68	5.1812	451.48	0.5260	460.99	1.9232
R22	86.457	232.06	4.9900	369.33	0.5130	430.68	1.9866
R23	70.019	190.86	4.8162	299.01	0.5200	397.39	2.0675
R12B1	165.370	268.94	4.2500	426.88	0.6732	346.77	1.5417
R13B1	148.930	215.06	3.9611	340.15	0.7448	313.37	1.5232
R113	187.390	320.37	3.4100	487.25	0.5760	374.59	1.5520
R114	170.930	276.40	3.2570	418.83	0.5800	355.45	1.5586
R115	154.480	233.74	3.1600	353.10	0.6135	332.50	1.5554
RC318	200.040	266.86	2.7775	388.38	0.6200	335.50	1.5015
R500	99.300	239.43	4.4256	378.70	0.4970	409.35	1.8637
R502	111.600	227.53	4.0645	355.31	0.5610	373.11	1.7510
R503	87.280	185.06	4.3316	292.40	0.5605	344.60	1.8113

* calculated from the *Bender* equation of state (p^s/MPa = 0.1)

** from this work (p_0/MPa = 0.1 , T_0/K = 298.15)

II Fundamentals

II.1 Fundamental Equations

The thermodynamic behavior of a pure isotropic fluid is completely described by the combined knowledge of

- the ideal gas standard state enthalpy $h_0 = h(T_0, p_0)_{\text{id.gas}}$,
- the ideal gas standard state entropy $s_0 = s(T_0, p_0)_{\text{id.gas}}$,
- the ideal gas state heat capacity $c_p^\circ(T)$ and,
- the equation of state $p = p(\rho, T)$ which describes the gaseous as well as the liquid phase.

For example, applying well-known procedures one finds for

- the internal energy

$$u(T,\rho) = h_0 - RT_0 + \int\limits_{T_0}^{T} (c_p^\circ - R)\mathrm{d}T + \int\limits_{0}^{\rho} \left[T\left(\frac{\partial p}{\partial T}\right)_\rho - p\right]\frac{\mathrm{d}\rho}{\rho^2} , \tag{1}$$

- the enthalpy

$$h = u + pv , \tag{2}$$

- the entropy

$$s(T,\rho) = s_0 + \int\limits_{T_0}^{T} c_p^\circ \frac{\mathrm{d}T}{T} - R\ln\left(\frac{RT\rho}{p_0}\right) + \int\limits_{0}^{\rho}\left[\frac{R}{\rho} - \frac{1}{\rho^2}\left(\frac{\partial p}{\partial T}\right)_\rho\right]\mathrm{d}\rho , \tag{3}$$

- the isochoric heat capacity

$$c_v(T,\rho) = c_p^\circ - R - \int\limits_{0}^{\rho} \frac{T}{\rho^2}\left(\frac{\partial^2 p}{\partial T^2}\right)_\rho \mathrm{d}\rho , \tag{4}$$

- the isobaric heat capacity

$$c_p(T,p) = c_v(T,\rho) + \frac{T}{\rho^2}\frac{\left(\frac{\partial p}{\partial T}\right)_\rho^2}{\left(\frac{\partial p}{\partial \rho}\right)_T} , \tag{5}$$

- the speed of sound

$$w(T,\rho) = \sqrt{\frac{c_p}{c_v}\left(\frac{\partial p}{\partial \rho}\right)_T} , \tag{6}$$

- the differential Joule-Thomson coefficient

$$\mu(T,\rho) = \left(\frac{\partial T}{\partial p}\right)_h = \frac{1}{\rho c_p}\left[\frac{T}{\rho}\frac{\left(\frac{\partial p}{\partial T}\right)_\rho}{\left(\frac{\partial p}{\partial \rho}\right)_T} - 1\right] , \tag{7}$$

- the fugacity coefficient

$$\ln\varphi(T,\rho) = \ln\left(\frac{RT\rho}{p}\right) + \frac{p}{RT\rho} - 1 + \frac{1}{RT}\int\limits_{0}^{\rho}[p(T,\rho) - RT\rho]\frac{\mathrm{d}\rho}{\rho^2} . \tag{8}$$

- Saturation properties are found by calculating the vapor pressure $p^s(T)$ and the coexisting densities $\rho'(T)$ and $\rho''(T)$ by combining the equations

$$\varphi(T,\rho') = \varphi(T,\rho'') \ , \tag{9}$$

$$p^s(T) = p(T,\rho') = p(T,\rho'') \ , \tag{10}$$

and then applying the results for ρ' and ρ'' to calculate the thermodynamic properties along the saturation curve, e.g.,

$$h'(T) = h(\rho',T) \ \text{and} \ h''(T) = h(\rho'',T) \ .$$

II.2 Ideal Gas State Properties

II.2.1 Ideal Gas Standard Enthalpy and Entropy

Instead of adopting the ideal gas standard state enthalpy h_0 and entropy s_0 from other sources, we follow a commonly used approach and assign fixed numbers to the enthalpy and the entropy of the saturated pure liquid at a fixed temperature:

$$h'(T/\mathrm{K} = 273.15)/\mathrm{kJ\ kg^{-1}} = 200 \ , \tag{11}$$

$$s'(T/\mathrm{K} = 273.15)/\mathrm{kJ\ kg^{-1}\ K^{-1}} = 1.0 \ . \tag{12}$$

However, for refrigerant R14 (because of its critical temperature of 227.51 K), the following normalization is made:

$$h'(T/\mathrm{K} = 130)/\mathrm{kJ\ kg^{-1}} = 200 \ ,$$

$$s'(T/\mathrm{K} = 130)/\mathrm{kJ\ kg^{-1}\ K^{-1}} = 1.0 \ .$$

Thus numbers for the ideal gas standard state ($T_0/\mathrm{K} = 298.15$; $p_0/\mathrm{MPa} = 0.1$) enthalpy and entropy from this work do not agree with numbers from other sources, but conversions are easily made as follows:

$$\Delta h = h - h_{\text{ this work}} = h(T_0,p_0)_{\text{id. gas}} - h(T_0,p_0)_{\text{id. gas, this work}} \ , \tag{13}$$

$$\Delta s = s - s_{\text{ this work}} = s(T_0,p_0)_{\text{id. gas}} - s(T_0,p_0)_{\text{id. gas, this work}} \ . \tag{14}$$

Therefore the enthalpy and entropy of the ideal gas in the standard state resulting from the normalization are given for each substance in Table A1.

II.2.2 Ideal Gas State Heat Capacity

The temperature dependence of the specific isobaric heat capacity c_p is described by

$$c_p^{\circ}\ /\mathrm{kJ\ kg^{-1}\ K^{-1}} = \sum_{i=1}^{5} d_i \cdot (T/\mathrm{K})^{i-1}. \tag{15}$$

The parameters d_i were fitted to literature data for the ideal gas state heat capacity. Parameters are reported for each substance.

II.3 Equation of State

II.3.1 Selection of an Equation of State

Many equations have been used to describe the p-ρ-T behavior of a pure fluid. Special equations exist for some substances like water, ammonia and oxygen, as well as more general equations like the *Benedict-Webb-Rubin(BWR)* equation */Bene40/* and its many extensions, for example by *Strobrigde /Stro62/*, *Starling /Star73/* (*BWRS* equation) and *Bender /Bend71/*. For the description of the p-ρ-T behavior of refrigerants it seems reasonable to use only one equation, which we call the standard equation. In selecting such an equation many criteria may be used. Here the two most important ones were

- the capability of describing reliably the gaseous as well as the liquid region including the saturation properties and
- the complexity and form of the equation as well as the number of parameters which have to be fitted to experimental data.

The *Bender* equation of state */Bend71/* (Table A2) seems to be a reasonable compromise.

Table A2: *Bender* equation of state (*Bender*-EOS)

$$p(\rho,T) = \rho T[R + B\rho + C\rho^2 + D\rho^3 + E\rho^4 + F\rho^5 + (G + H\rho^2)\rho^2 \exp(-a_{20}\rho^2)]$$

with temperature functions

$$B = a_{01} - a_{02}/T - a_{03}/T^2 - a_{04}/T^3 - a_{05}/T^4$$
$$C = a_{06} + a_{07}/T + a_{08}/T^2$$
$$D = a_{09} + a_{10}/T$$
$$E = a_{11} + a_{12}/T$$
$$F = a_{13}/T$$
$$G = a_{14}/T^3 + a_{15}/T^4 + a_{16}/T^5$$
$$H = a_{17}/T^3 + a_{18}/T^4 + a_{19}/T^5$$

As an extended *BWR* equation it reduces to the virial equation of state and should therefore be adequate for describing the gas phase reliably. Furthermore, it is proven that this equation can be used to describe the saturation properties and the liquid region reasonably well, especially when the Maxwell criterion is included in the parameter estimation procedure. The *Bender* equation is an analytical equation. Thus it offers the advantage of simple procedures for differentiating and integrating. The price one has to pay is a deficiency in the neighborhood of the critical point. With 20 parameters the *Bender* equation lies between some simpler equations, like the *BWR* and *BWRS* equation, and more sophisticated ones, for example the *Jacobsen* equation. As has been proven for many substances */Polt87/*, the *Bender* equation is suitable for describing experimental results for the thermodynamic properties of pure fluids, mostly within experimental uncertainty. Parameters for this equation are now available for about 53 pure fluids */Polt87/*; furthermore a FORTRAN 77 software package is available from the authors for the direct use of this equation in computer simulations.

II.3.2 Parameter Estimation

The parameters of the *Bender* equation for a pure fluid were determined from experimental results for the p-ρ-T area $(p_{\text{exp}}, \rho_{\text{exp}}, T_{\text{exp}})$, vapor pressure $p^s_{\text{exp}}(T)$ and coexisting densities ρ'_{exp} and ρ''_{exp} by minimization of the sum of squares:

$$\text{SS} = \sum_{k=1}^{k_p} \left[\frac{p_{\text{exp}} - p_{\text{calc}}(\rho_{\text{exp}}, T_{\text{exp}}, \vec{a})}{\sigma_p^*}\right]_k^2 + \sum_{k=1}^{k_s} w_{DK,k} \cdot \left[p_{\text{calc}}(T_{\text{exp}}, \rho''_{\text{exp}}) - p_{\text{calc}}(T_{\text{exp}}, \rho'_{\text{exp}})\right]_k^2$$

$$+ \sum_{k=1}^{k_s} w_{\rho,k} \cdot \left[\ln \frac{\varphi''_{\text{calc}}(T_{\text{exp}}, \rho''_{\text{exp}})}{\varphi'_{\text{calc}}(T_{\text{exp}}, \rho'_{\text{exp}})}\right]_k^2 . \qquad (16)$$

The uncertainty of the experimental results are taken into account by $\sigma_p^*, w_{DK,k}$ and $w_{\rho,k}$.

The value of σ_p^* is calculated from estimated absolute uncertainties of the experimental results for pressure (σ_p), density (σ_ρ) and temperature (σ_T):

$$\sigma_p^* = \sqrt{\sigma_p^2 + \left(\frac{\partial p}{\partial \rho}\right)_T^2 \sigma_\rho^2 + \left(\frac{\partial p}{\partial T}\right)_\rho^2 \sigma_T^2} \ . \tag{17}$$

The differential coefficients $(\partial p/\partial \rho)_T$ and $(\partial p/\partial T)_\rho$ require knowledge of an equation of state. Therefore the parameter fitting follows a multi-step procedure. At first only p-ρ-T data are used to find a first estimate for the parameter vector $\vec{a}$ by minimizing

$$\mathrm{SS} = \sum_{k=1}^{k_p} \left[\frac{p_{\mathrm{exp}} - p_{\mathrm{calc}}(\rho_{\mathrm{exp}}, T_{\mathrm{exp}}, \vec{a})}{p_{\mathrm{exp}}}\right]_k^2 \ . \tag{18}$$

The resulting (preliminary) parameter vector $\vec{a}$ is then used to calculate the differential coefficients. In the next step the complete sum of squares (16) is used to find a new parameter vector $\vec{a}$. By repeating this step for between 2 and 4 times the final parameter vector $\vec{a}$ is found.

The calculation requires weighting factors $w_{DK,k}$ and $w_{\rho,k}$ for the inclusion of saturation properties. Generally the assumption

$$w_{DK,k} = w_{\rho,k} = w_i \tag{19}$$

is used, where index i is to demonstrate that the number for the weighting factor might depend on component i Numbers for w_i vary between about 100 and 180 depending on subjective criteria and on the number and quality of data points.

The procedure described above requires experimental results for densities of coexisting phases which are generally not available. Therefore precorrelated saturation properties were used to determine parameter vector $\vec{a}$.

For the precorrelation the vapor pressure equation by *Wagner /Wagn74/*

$$\ln\left(\frac{p^s}{p_c}\right) = \left(\frac{T_c}{T}\right) \sum_{i=1}^{21} b_i \cdot \tau^{(i-3)/2} \tag{20}$$

and the saturated liquid density equation by *Ewers /Ewer81/*

$$\frac{\rho' - \rho_c}{\rho_c} = \sum_{i=1}^{20} c_i \cdot \tau^{(i-7)/3} \tag{21}$$

with $\tau = 1 - (T/T_c)$ were used.

It has been proved that either 4 or 3 terms (instead of 21 or 20) are sufficient to describe p^s and ρ' within the experimental uncertainty. Which combination of b_i and c_i yields the best representation depends on the fluid.

Besides these so-called "quasi-optimal" equations, "standard" equations were also used. The standard equation for the vapor pressure is (20) with i = 5, 6, 9, 15, that for the saturated liquid density is (21) with i = 8, 9 and 16. Furthermore a standard equation for the dew point density ρ'' was used:

$$\ln\left(\frac{\rho''}{\rho_c}\right) = [c_1^*\tau^{1/3} + c_2^*\tau^{2/3} + c_3^*\tau + c_4^*\tau^{5/3} + c_5^*\tau^4 + c_6^*\tau^9] \ . \tag{22}$$

This equation was applied only when experimental results for the dew point density were available. Otherwise, calculated dew point densities were used in (16) determined from precorrelated vapor pressure data and the preliminary equation of state:

$$\rho'' = \rho_{\mathrm{gas}}(T, p^s, \vec{a}) \ . \tag{23}$$

The precorrelations require critical pressures p_c, critical temperatures T_c and critical densities ρ_c. These properties were carefully selected from published data; in some case numbers for the critical pressure reported in the literature were slightly changed to give a better representation of reliable vapor pressure data (Table A1).

References

[Bend71] BENDER, E.: Die Berechnung von Phasengleichgewichten mit der thermischen Zustandsgleichung - dargestellt an den reinen Fluiden Argon, Stickstoff und ihren Gemischen; Habilitationsschrift, Ruhr-Universität Bochum (1971)

[Bene40] BENEDICT, M.; WEBB, G.B. AND RUBIN, L.C.: An Empirical Equation for Thermodynamic Properties of Light Hydrocarbons and their Mixtures; J. Chem. Phys. 8(1940) 334/345

[Ewer81] EWERS, J.: Eine Methode zur Optimierung der Struktur von Zustandsgleichungen und ihre Anwendung zur Aufstellung einer Fundamentalgleichung für Sauerstoff; Dissertation, Ruhr-Universität Bochum (1981)

[KMR 81] - : Kältemaschinenregeln, 7. Auflage; Verlag C.F. Müller, Karlsruhe (1981)

[Polt87] POLT, A.W.: Zur Beschreibung der thermodynamischen Eigenschaften reiner Fluide mit "Erweiterten BWR-Gleichungen"; Dissertation, Universität Kaiserslautern (1987)

[Star73] STARLING, K.E.: Fluid Thermodynamic Properties for Light Petroleum Systems; Gulf Publishing Company, Houston (1973)

[Stro62] STROBRIDGE, T.R.: The Thermodynamic Properties of Nitrogen from 64 to 300K between 0.1 and 200 atmospheres; NBS Techn. Note No.129 (1962)

[Wagn74] WAGNER, W.: Eine mathematisch statistische Methode zum Aufstellen thermodynamischer Gleichungen - gezeigt am Beispiel der Dampfdruckkurve reiner Fluide; Fortschr.-Ber. VDI-Z. Reihe 3, Nr.39 Düsseldorf (1974)

III Results

Thermodynamic properties for 16 halogenated hydrocarbons (Table A1) are presented in this chapter. The substances include 9 halogenated methane derivatives, 3 derivatives of ethane, octofluorinated cyclobutane and three azeotropic mixtures (R500, R502, R503), which are dealt here in the same way as pure substances. To demonstrate the application of the following results, refrigerant R11 is discussed in detail while for all other substances the results are only given as tables, figures, diagrams and charts.

Refrigerant R11 (CCl_3F)

The critical properties of R11 are given together with those of all other substances in Table A1.

Table R11.1 lists the references for experimental pressure-density-temperature data considered here. For each author the investigated range of state is characterized by minimum and maximum numbers for pressure, density and temperature, furthermore the number of data points is given. Full information on the data sources are given at the end of this chapter. Figure R11.1 shows the distribution of the experimental results used for fitting the parameters of the *Bender* equation in a pressure-temperature and in a temperature-density diagram.

Sources for experimental results for the vapor pressure and the saturated liquid density are cited in Tables R11.2 and R11.3. These results were used to precorrelate the vapor pressure and the saturated liquid density. The four coefficients for the standard vapor pressure equation [i.e. (20) with $i = 5, 6, 9, 15$] are given in Table R11.4; those for the "quasi- optimal" vapor pressure equation are given in Table R11.5. The calculated temperatures at the normal boiling point (i.e. $p^s/\mathrm{MPa} = 0.1$) are

$$\begin{aligned} T/\mathrm{K} &= 296.41 \text{ (quasi} - \text{optimal equation)} , \\ &= 296.40 \text{ (standard eqation)} , \\ &= 296.38 \text{ (equation of state, cf. Chart R11.2 : Saturation properties)}. \end{aligned}$$

The "quasi-optimal" saturated density equation is (21) with $i = 8$, 9 and 15 (Table R11.7), it differs only slightly in its structure from the standard equation ($i = 8$, 9, 16) (Table R11.6). For 296.38 K the calculated saturated liquid phase volumes are

$$\begin{aligned} v'/\mathrm{cm}^3 \cdot \mathrm{g}^{-1} &= 0.6758 \text{ (quasi} - \text{optimal equation)} , \\ &= 0.6759 \text{ (standard equation)} , \\ &= 0.6758 \text{ (equation of state, cf. Chart R11.2 : Saturation properties)}. \end{aligned}$$

The resulting coefficients of the *Bender* equation are given in Table R11.8. Columns with headings "Specific units" and "Molar units" report the coefficients of the *Bender* equation with density in g cm^{-3} and mole dm^{-3} respectively, while pressure and temperature are always in MPa and Kelvin.

The limits for pressure, density and temperature are those of the range of state covered by experimental results. It is recommended that Figs. R11.1, 4 and 5 be used to check that the state of interest is within the range of experimental data.

Table R11.9 gives a summary of the description of p-ρ-T data by the *Bender* equation of state. The gaseous, liquid and critical regions are defined in Fig. A1 .

Deviations are reported between experimental and calculated pressures $p(T = T_{\mathrm{exp}}, \rho = \rho_{\mathrm{exp}})$ and densities $\rho(T = T_{\mathrm{exp}}, p = p_{\mathrm{exp}})$:

$$p \; : \; \Delta/\% = \sqrt{\frac{1}{n}\sum_{i=1}^{n}\left[\frac{p_{\mathrm{exp}} - p_{\mathrm{calc}}(\rho_{\mathrm{exp}}, T_{\mathrm{exp}})}{p_{\mathrm{exp}}}\right]_i^2} \cdot 100 \; , \tag{24}$$

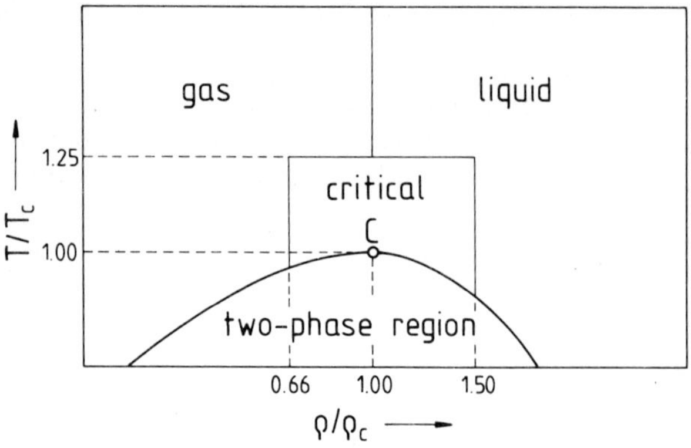

Figure A1. Definitions of gaseous, liquid and critical region

$$\rho \, : \, \Delta/\% = \sqrt{\frac{1}{n} \sum_{i=1}^{n} \left[\frac{\rho_{\text{exp}} - \rho_{\text{calc}}(p_{\text{exp}}, T_{\text{exp}})}{\rho_{\text{exp}}} \right]_i^2} \cdot 100 \, . \tag{25}$$

Figure R11.2 gives a detailed comparison between calculated and measured pressures and densities:

$$\Delta p/\% = \frac{p_{\text{exp}} - p_{\text{calc}}(\rho_{\text{exp}}, T_{\text{exp}})}{p_{\text{exp}}} \cdot 100 \, , \tag{26}$$

$$\Delta \rho/\% = \frac{\rho_{\text{exp}} - \rho_{\text{calc}}(p_{\text{exp}}, T_{\text{exp}})}{\rho_{\text{exp}}} \cdot 100 \, . \tag{27}$$

The comparison with experimental vapor pressure and saturated liquid density data based on the precorrelation as well as on the resulting equation of state is given in Tables R11.2 and R11.3. Additionally Figs. R11.3 and R11.4 show the comparison between experimental results and those from the equation of state. Deviations were calculated corresponding to

$$p^s \, : \, \Delta/\% = \frac{1}{n} \cdot \sum_{i=1}^{n} \left| \frac{p^s_{\text{exp}} - p^s_{\text{calc}}(T_{\text{exp}})}{p^s_{\text{exp}}} \right|_i \cdot 100 \, , \tag{28}$$

$$\rho' \, : \, \Delta/\% = \frac{1}{n} \cdot \sum_{i=1}^{n} \left| \frac{\rho'_{\text{exp}} - \rho'_{\text{calc}}(T_{\text{exp}})}{\rho'_{\text{exp}}} \right|_i \cdot 100 \, , \tag{29}$$

and individual deviations by

$$\Delta/\% = \frac{x_{\text{exp}} - x_{\text{calc}}(T_{\text{exp}})}{x_{\text{exp}}} \cdot 100 \, , \qquad x = \rho', p^s \, . \tag{30}$$

Table R11.10 contains information on the ideal gas state heat capacity c_p^0. The source and temperature region of heat capacity data used to fit parameters of (15) as well as the parameters are reported.

As shown in Fig. R11.5, a comparatively large number of experimental data for the isobaric heat capacity c_p and the velocity of sound w are available in the literature. Experimental results for the isobaric heat capacity c_p are compared with the results of the correlation in Fig. R11.6, indicating that the differences correspond to the scatter of the experimental data.

As an example for a comparison with other correlations of the thermodynamic properties, Fig. R11.7 shows deviations between results taken from *Kältemaschinenregeln* /*KMR* 81/ and this work for vapor pressure, saturated liquid density, dew point density and heat of vaporization.

Charts R11.1 to 3 summarize the results from the *Bender* equation of state and the ideal gas heat capacity equation. Charts R11.1 and 2 are temperature and pressure tables of saturation properties starting at around the lowest temperature for which experimental vapor pressure data is available and ending about 1 K below the critical temperature.

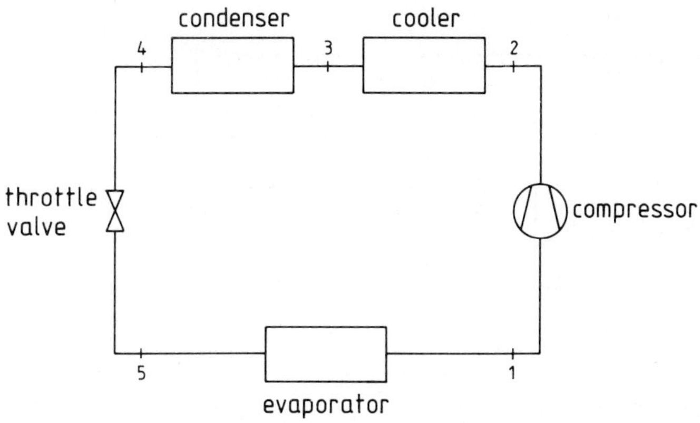

Figure A2. Scheme of an idealized heat pump process

Chart R11.3 gives pressure, temperature, volume and enthalpy data in the liquid and the gaseous region for temperatures covered by experimental vapor pressure and p-ρ-T data up to the highest pressure for which experimental p-ρ-T data were available. Therefore the procedure for calculating thermodynamic properties is not supported in every case by experimental results; in some cases the calculated results represent extrapolations whose reliability is difficult to estimate. Therefore it is recommended that a check be made against Fig. R11.1 (for p-ρ-T data) and Table R11.10 (for ideal gas state heat capacity data) when data from the charts are used.

The results of Charts R11.1 to R11.3 are shown in Diagrams R11. In order to demonstrate the use of the charts and diagrams, an idealized heat pump process is discussed.
Saturated vapor of the refrigerant R11 leaves the evaporator under a pressure of 0.04 MPa (1). In a compressor the pressure is increased isentropically to p_2 = 0.5 MPa (2). At constant pressure the vapor is cooled to saturation temperature (3) and afterwards completely condensed (4). The saturated liquid is throttled adiabatically to the evaporation pressure (5); (5) is in the two phase region because the specific enthalpy before and after the throttling is the same.

From Diagrams R11, temperature, pressure, enthalpy and entropy of states 1 to 5 can be read roughly; precise results can be read from Charts R11.1 to R11.3. Numbers from both procedures are reported in Table R11.0.

Table R11.0: Results for state variables at states (1) to (5) as read from Diagrams R11 and Charts R11.1 to R11.3

State	Pressure p MPa	Temperature T K		Enthalpy h kJ/kg		Entropy s kJ/kg K		Vapor content	
		Diagrams	Charts	Diagrams	Charts	Diagrams	Charts	Diagrams	Charts
1	0.04	~273	272.98	~390	390.50	~1.70	1.6978		
2	0.5	~369	368.84	~440	440.40	~1.70	1.6978		
3	0.5	~351	351.27	~429	428.63	~1.68	1.6760		
4	0.5	~351	351.27	~271	270.84	~1.23	1.2268		
5	0.04	~273	272.98	~271	270.84	~1.26	1.2595	~0.37	0.3724

Table R11.1: List of references for p, ρ, T data

Author	n	p/MPa min	p/MPa max	ρ/g cm^{-3} min	ρ/g cm^{-3} max	T/K min	T/K max
BENNING, A.F. *[Benn 40a]*	34	0.10	2.16	0.0052	0.11	318	477
KREMENEWSKAJA, E.A. *[Krem 75]*	71	0.53	4.17	0.0220	0.29	383	473
RIVKIN, S.L. *[Rivk 71]*	166	0.80	19.76	0.8007	1.57	273	473
RIVKIN, S.L. *[Rivk 72]*	16	3.23	4.88	0.2045	0.79	453	473

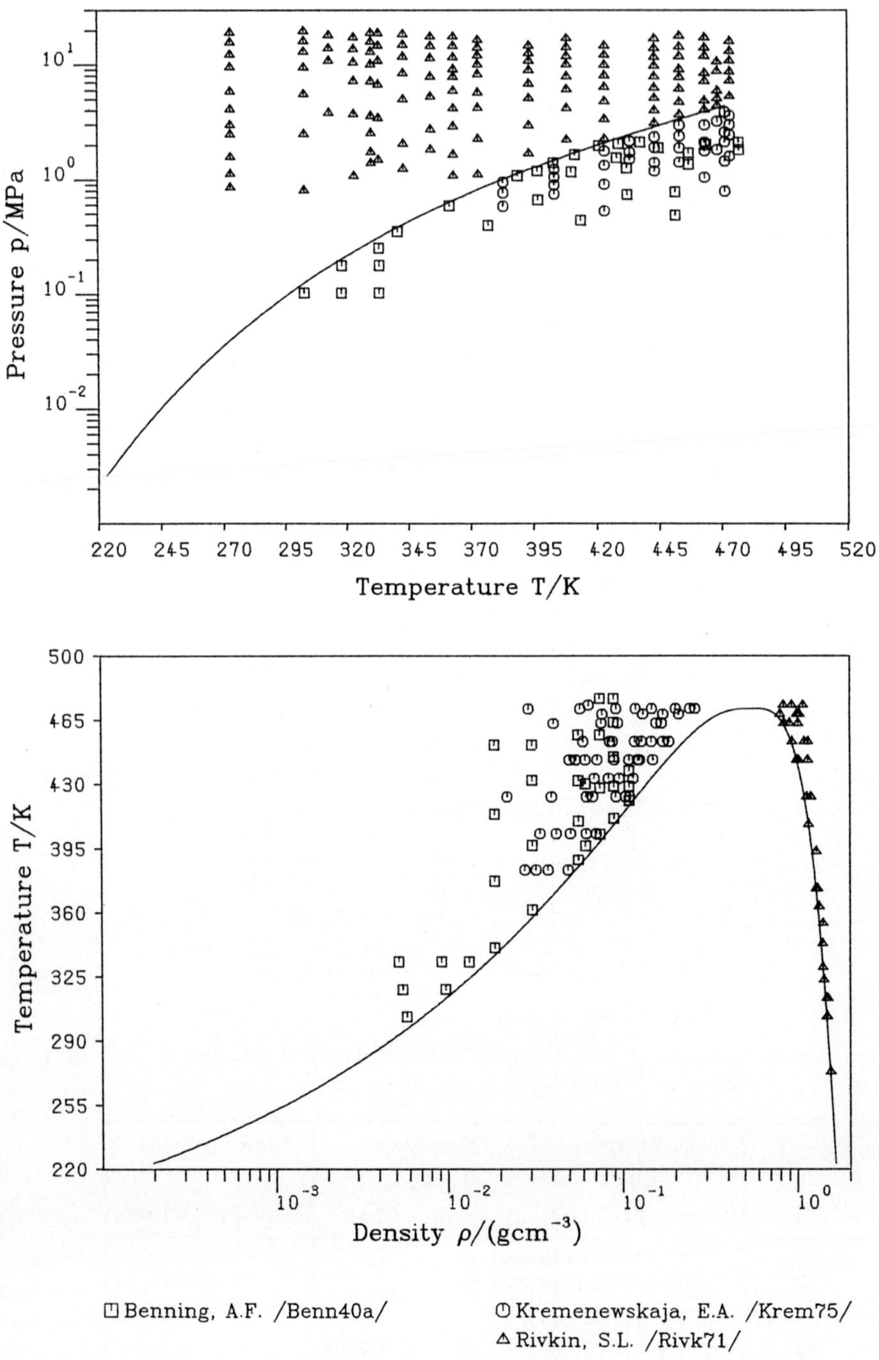

Fig.R11.1: Distribution of p, ρ, T data projected in the p, T and in the T, ρ plane

Table R11.2: Experimental vapor pressure data

Author	Δ / % std	opt	*EOS*
BENNING, A.F. *[Benn 40]*	0.36	0.37	0.44
FERNÁNDEZ-FASSNACHT, E. *[Fern 84]*	1.11	1.07	1.11
JÄGER, H.P. *[Jaeg 69]*	0.20	0.24	0.20
JÄGER, H.P. *[Jaeg 73]*	2.40	2.40	2.40
KREMENEWSKAJA, E.A. *[Krem 75]*	0.14	0.14	0.15
OSBORNE, D.W. *[Osbo 41]*	0.42	0.34	0.43
PAWLOWA, I.A. *[Pawl 55]*	1.90	1.80	2.00
RIEDEL, L. *[Ried 39]*	0.17	0.06	0.17

Table R11.3: Experimental saturated liquid density data

Author	Δ / % std	opt	*EOS*
BENNING, A.F. *[Benn 40b]*	0.04	0.04	0.10
DESREUX, V. *[Desr 32]*	0.02	0.01	0.02
JÄGER, H.P. *[Jaeg 69]*	0.20	0.20	0.20
SWARTS, F. *[Swar 93]*	0.02	0.02	0.02

Table R11.4: Coefficients of the standard vapor pressure equation

Index i	b_i
5	−6.987448 E+00
6	1.581291 E+00
9	−2.785606 E+00
15	−1.680212 E+00

Table R11.5: Coefficients of the quasi-optimal vapor pressure equation

Index i	b_i
5	−7.218636 E+00
6	3.159518 E+00
7	−2.910622 E+00
14	−4.022189 E+00

Table R11.6: Coefficients of the standard saturated liquid density equation

Index i	c_i
8	1.655554 E+00
9	8.528106 E−01
16	3.336934 E−01

Table R11.7: Coefficients of the quasi-optimal saturated liquid density equation

Index i	c_i
8	1.661132 E+00
9	8.380188 E−01
15	2.931265 E−01

Table R11.8: Coefficients of the *Bender-EOS*

Range of validity: one-phase region

$0.10 \leq p/\mathrm{MPa} \leq 19.76$

$0.00519 \leq \rho/\mathrm{g\ cm^{-3}} \leq 1.57210$

$273.15 \leq T/\mathrm{K} \leq 476.79$

	Specific units	Molar units
a_{01}	1.705383618 E−01	3.218615592 E+00
a_{02}	2.068212766 E+02	3.903392636 E+03
a_{03}	−6.739246197 E+04	−1.271915753 E+06
a_{04}	1.978352454 E+07	3.733796893 E+08
a_{05}	−1.246506232 E+09	−2.352564169 E+10
a_{06}	3.038910608 E−02	7.879314966 E−02
a_{07}	−4.681206402 E+01	−1.213747439 E+02
a_{08}	2.221679971 E+04	5.760391966 E+04
a_{09}	−1.051694925 E−02	−3.746138472 E−03
a_{10}	1.286247364 E+02	4.581614515 E+01
a_{11}	1.209342705 E−01	5.917890831 E−03
a_{12}	−2.450623959 E+02	−1.199207222 E+01
a_{13}	9.108007375 E+01	6.123002628 E−01
a_{14}	−1.208838000 E+08	−3.134286122 E+08
a_{15}	1.171918609 E+11	3.038561190 E+11
a_{16}	−2.826547007 E+13	−7.328696698 E+13
a_{17}	3.001247218 E+08	1.468653453 E+07
a_{18}	−3.235004169 E+11	−1.583041882 E+10
a_{19}	8.601099624 E+13	4.208928404 E+12
a_{20}	3.200000000 E+00	6.039444608 E−02

Table R11.9: Deviations between experimental and predicted p, ρ, T data

Region	n	Δ / %	
		p	ρ
Gas	102	0.21	0.27
Critical	3	0.70	0.61
Liquid	162	3.70	0.12
Total	267	2.98	0.21

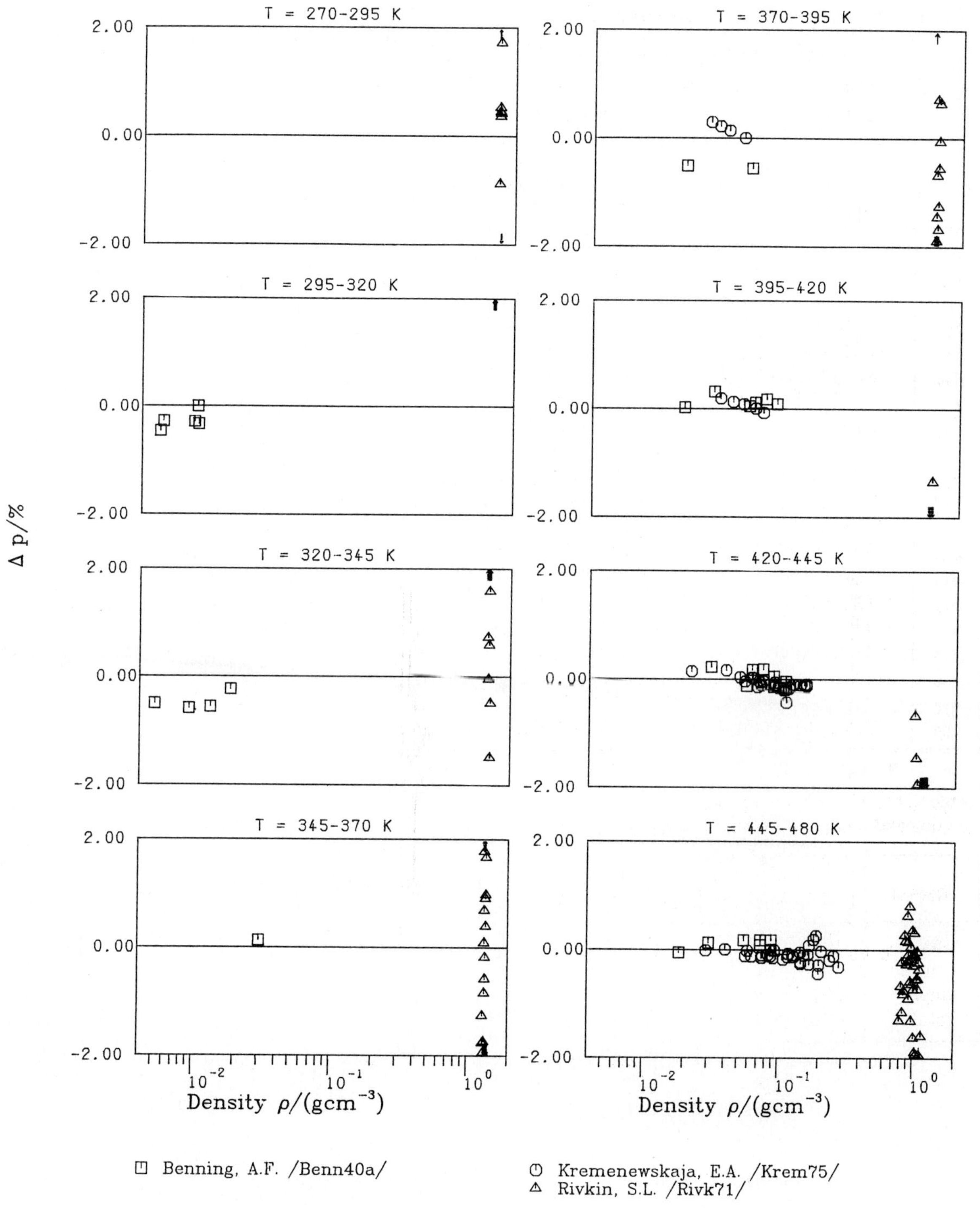

Fig.R11.2: Deviations between experimental and calculated p, ρ, T data

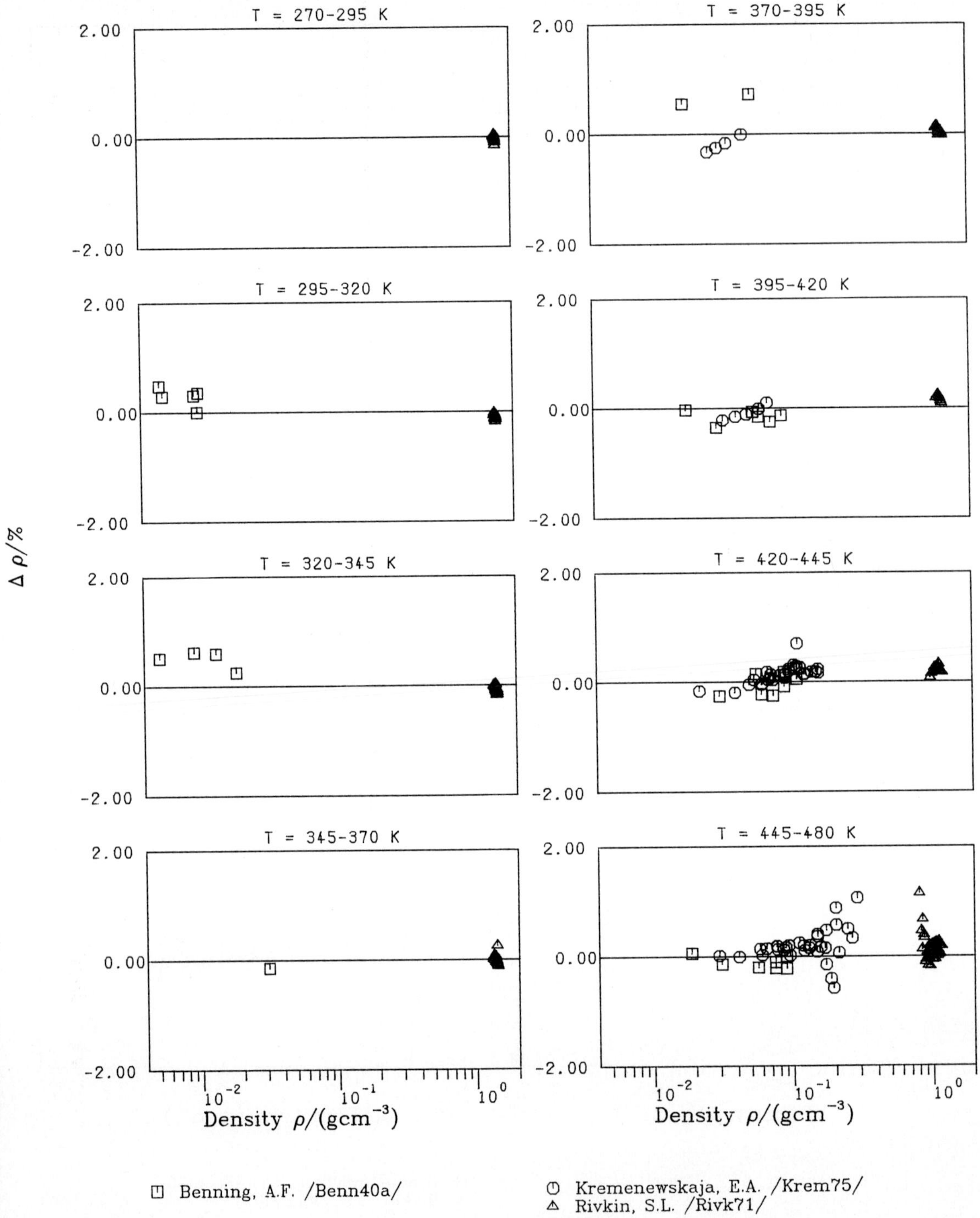

Fig.R11.2: Deviations between experimental and calculated p, ρ, T data *-continued*

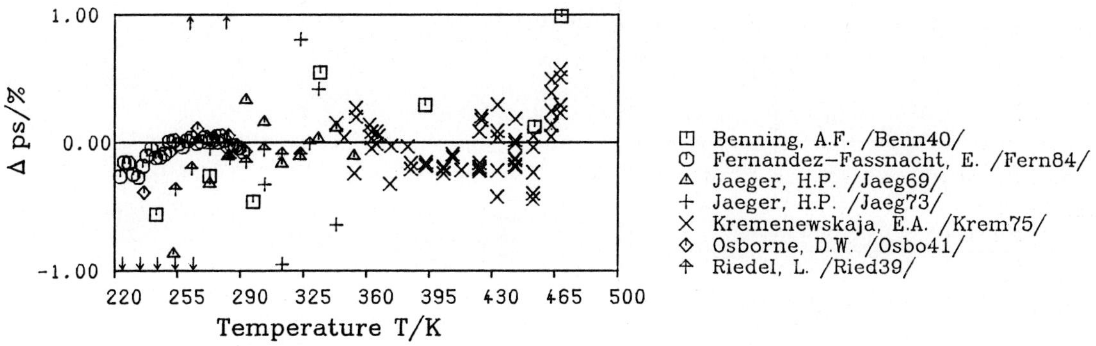

Fig.R11.3: Deviations between experimental and calculated vapor pressures

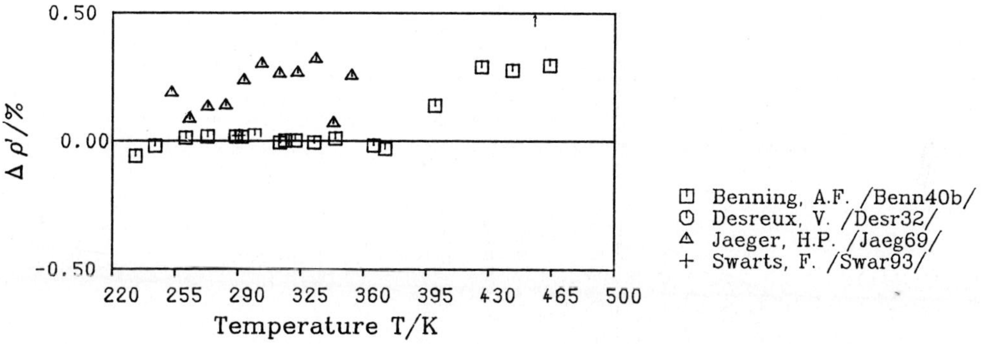

Fig.R11.4: Deviations between experimental and calculated saturated liquid densities

Table R11.10: Information on ideal gas state heat capacity equation

Author	T / K min	T / K max
CHEN, S.S. *[Chen 76]*	100	800

Coefficients of equation (15)	
Index i	d_i
1	8.964088 E−02
2	2.677897 E−03
3	−4.762636 E−06
4	4.222620 E−09
5	−1.499175 E−12

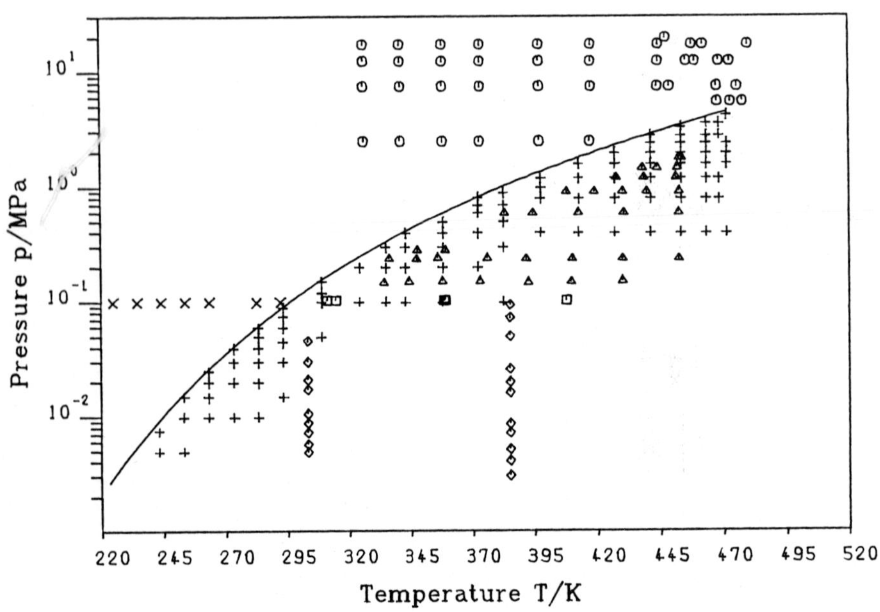

□ cp: Benning, A.F. /Benn40c/
○ cp: Grishkov, A.Y. /Gris71/
△ cp: Gruzdev, V.A. /Gruz75/
+ w: Kolomov, Y.L. /Kolo68/
× w: Meyer, K.J. /Meye69/
◇ w: Redkozubov, V.A. /Redk76/

Fig.R11.5: Distribution of $c_p(p,T)$ and $w(p,T)$ data

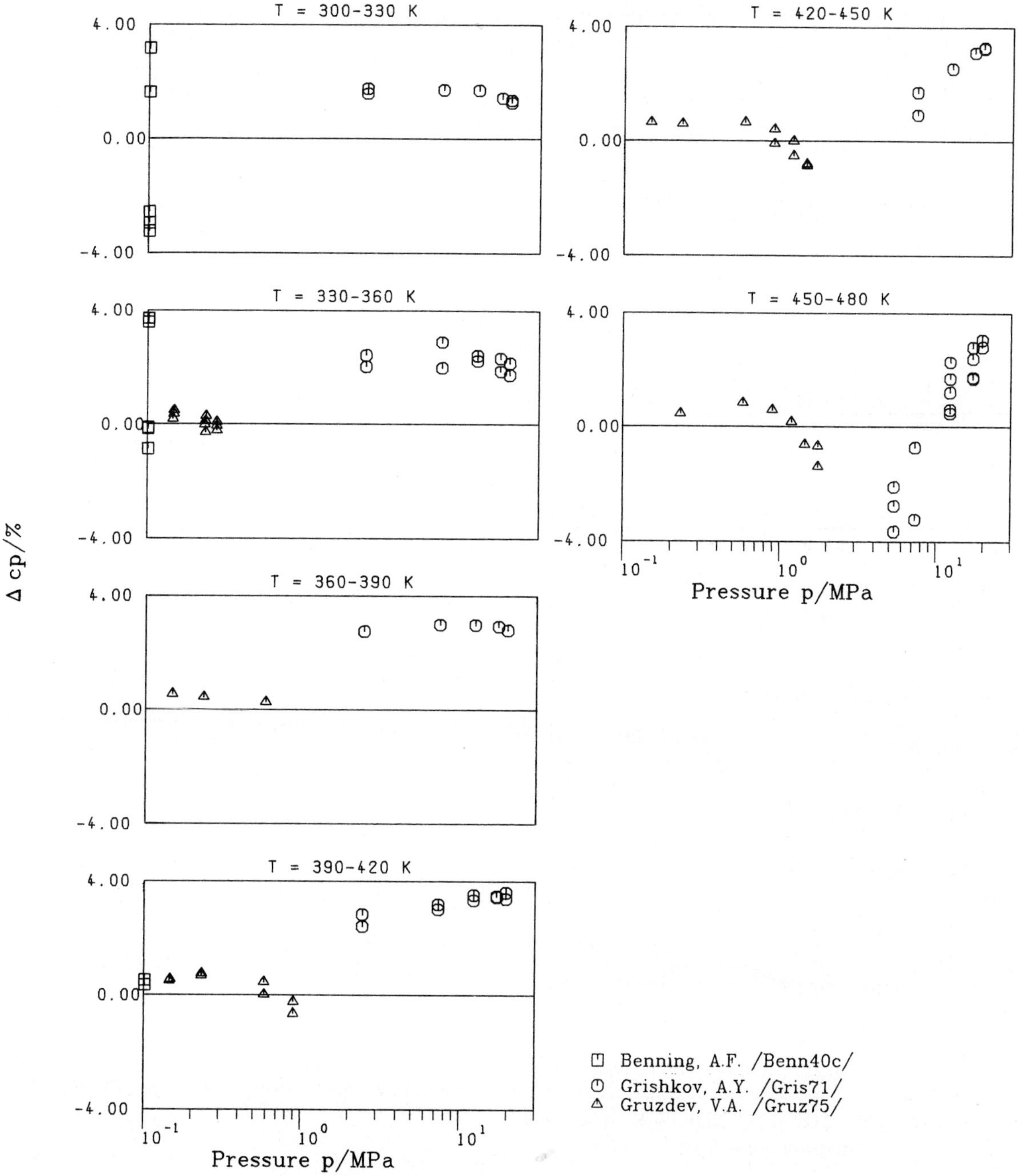

Fig.R11.6: Deviations between experimental and calculated specific isobaric heat capacities c_p

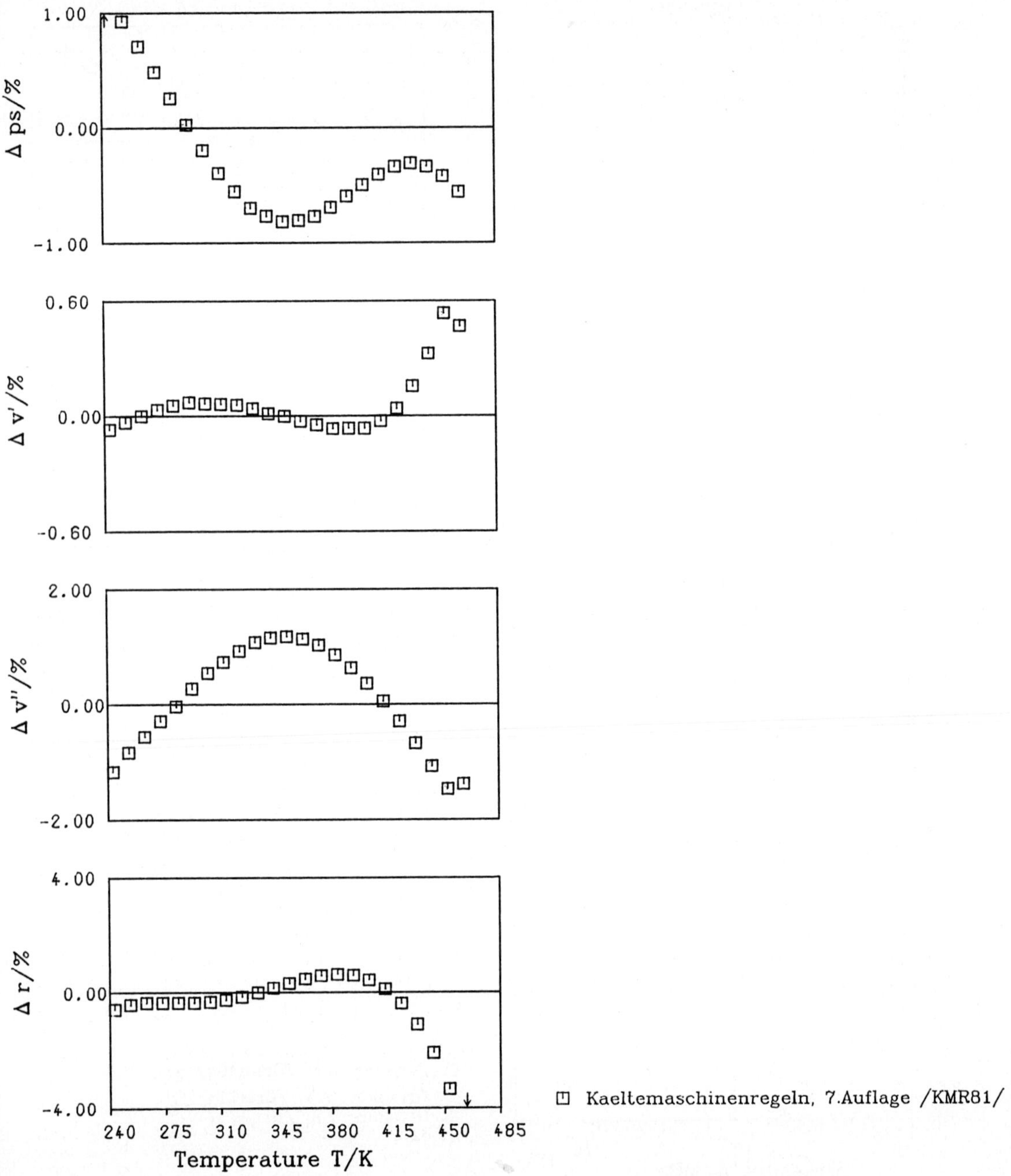

Fig.R11.7: Deviations between saturation properties from the *Kältemaschinenregeln* and those calculated by the *Bender-EOS*

Chart R11.1: Saturation properties

T K	p MPa	v' dm^3/kg	v'' dm^3/kg	h' kJ/kg	h'' kJ/kg	r kJ/kg	s' kJ/kgK	s'' kJ/kgK
225.0	0.00298	0.6096	4553.5	160.00	366.40	206.40	0.8394	1.7567
226.0	0.00319	0.6104	4275.9	160.76	366.89	206.13	0.8427	1.7548
227.0	0.00341	0.6112	4017.9	161.52	367.38	205.85	0.8461	1.7530
228.0	0.00364	0.6120	3777.8	162.29	367.87	205.58	0.8495	1.7511
229.0	0.00389	0.6128	3554.3	163.06	368.36	205.30	0.8529	1.7494
230.0	0.00415	0.6137	3346.0	163.84	368.85	205.01	0.8562	1.7476
231.0	0.00442	0.6145	3151.9	164.62	369.35	204.72	0.8596	1.7459
232.0	0.00471	0.6153	2970.9	165.41	369.84	204.43	0.8630	1.7442
233.0	0.00501	0.6161	2801.9	166.19	370.33	204.14	0.8664	1.7426
234.0	0.00534	0.6169	2644.1	166.98	370.83	203.85	0.8698	1.7409
235.0	0.00567	0.6178	2496.7	167.78	371.33	203.55	0.8732	1.7393
236.0	0.00603	0.6186	2358.8	168.58	371.82	203.25	0.8766	1.7378
237.0	0.00640	0.6194	2229.7	169.38	372.32	202.94	0.8800	1.7363
238.0	0.00680	0.6203	2109.0	170.18	372.82	202.63	0.8834	1.7348
239.0	0.00721	0.6211	1995.8	170.99	373.32	202.32	0.8867	1.7333
240.0	0.00765	0.6219	1889.8	171.80	373.82	202.01	0.8901	1.7318
241.0	0.00810	0.6228	1790.3	172.62	374.32	201.70	0.8935	1.7304
242.0	0.00858	0.6236	1697.0	173.44	374.82	201.38	0.8969	1.7290
243.0	0.00908	0.6245	1609.4	174.26	375.32	201.06	0.9003	1.7277
244.0	0.00961	0.6253	1527.1	175.08	375.82	200.74	0.9037	1.7264
245.0	0.01016	0.6262	1449.7	175.91	376.32	200.41	0.9070	1.7251
246.0	0.01074	0.6271	1377.0	176.73	376.82	200.09	0.9104	1.7238
247.0	0.01135	0.6279	1308.5	177.57	377.32	199.76	0.9138	1.7225
248.0	0.01198	0.6288	1244.1	178.40	377.83	199.43	0.9171	1.7213
249.0	0.01264	0.6297	1183.4	179.24	378.33	199.10	0.9205	1.7201
250.0	0.01333	0.6306	1126.2	180.07	378.84	198.76	0.9239	1.7189
251.0	0.01405	0.6314	1072.2	180.91	379.34	198.43	0.9272	1.7178
252.0	0.01481	0.6323	1021.4	181.76	379.85	198.09	0.9306	1.7166
253.0	0.01559	0.6332	973.35	182.60	380.35	197.75	0.9339	1.7155
254.0	0.01641	0.6341	928.01	183.45	380.86	197.41	0.9373	1.7144
255.0	0.01727	0.6350	885.17	184.30	381.36	197.06	0.9406	1.7134
256.0	0.01816	0.6359	844.68	185.15	381.87	196.72	0.9439	1.7124
257.0	0.01909	0.6368	806.39	186.00	382.38	196.37	0.9472	1.7113
258.0	0.02006	0.6377	770.17	186.86	382.88	196.02	0.9506	1.7103
259.0	0.02107	0.6386	735.88	187.71	383.39	195.68	0.9539	1.7094
260.0	0.02211	0.6395	703.41	188.57	383.90	195.32	0.9572	1.7084
261.0	0.02320	0.6405	672.65	189.43	384.40	194.97	0.9605	1.7075
262.0	0.02434	0.6414	643.50	190.29	384.91	194.62	0.9638	1.7066
263.0	0.02552	0.6423	615.86	191.16	385.42	194.26	0.9671	1.7057
264.0	0.02674	0.6432	589.63	192.02	385.93	193.91	0.9703	1.7048
265.0	0.02801	0.6442	564.74	192.89	386.44	193.55	0.9736	1.7040
266.0	0.02933	0.6451	541.11	193.76	386.94	193.19	0.9769	1.7031
267.0	0.03070	0.6461	518.67	194.63	387.45	192.83	0.9801	1.7023
268.0	0.03212	0.6470	497.34	195.50	387.96	192.46	0.9834	1.7015
269.0	0.03359	0.6480	477.06	196.37	388.47	192.10	0.9866	1.7008
270.0	0.03512	0.6489	457.78	197.24	388.98	191.74	0.9899	1.7000
271.0	0.03670	0.6499	439.43	198.12	389.49	191.37	0.9931	1.6993
272.0	0.03834	0.6508	421.97	198.99	390.00	191.00	0.9963	1.6985
273.0	0.04003	0.6518	405.35	199.87	390.50	190.64	0.9995	1.6978
274.0	0.04179	0.6528	389.51	200.75	391.01	190.27	1.0027	1.6971

Chart R11.1: Saturation properties–*continued*

T K	p MPa	v' dm³/kg	v'' dm³/kg	h' kJ/kg	h'' kJ/kg	r kJ/kg	s' kJ/kgK	s'' kJ/kgK
275.0	0.04360	0.6537	374.42	201.63	391.52	189.90	1.0059	1.6965
276.0	0.04548	0.6547	360.03	202.51	392.03	189.52	1.0091	1.6958
277.0	0.04743	0.6557	346.31	203.39	392.54	189.15	1.0123	1.6952
278.0	0.04943	0.6567	333.22	204.27	393.05	188.78	1.0155	1.6945
279.0	0.05151	0.6577	320.73	205.15	393.56	188.40	1.0186	1.6939
280.0	0.05365	0.6587	308.81	206.04	394.06	188.03	1.0218	1.6933
281.0	0.05587	0.6597	297.42	206.92	394.57	187.65	1.0249	1.6927
282.0	0.05816	0.6607	286.54	207.81	395.08	187.27	1.0281	1.6922
283.0	0.06052	0.6617	276.14	208.69	395.59	186.89	1.0312	1.6916
284.0	0.06295	0.6627	266.20	209.58	396.09	186.51	1.0343	1.6911
285.0	0.06546	0.6638	256.70	210.47	396.60	186.13	1.0375	1.6906
286.0	0.06805	0.6648	247.60	211.36	397.11	185.75	1.0406	1.6900
287.0	0.07072	0.6658	238.90	212.25	397.61	185.37	1.0437	1.6895
288.0	0.07348	0.6669	230.57	213.14	398.12	184.98	1.0468	1.6891
289.0	0.07631	0.6679	222.59	214.03	398.63	184.60	1.0498	1.6886
290.0	0.07923	0.6689	214.95	214.92	399.13	184.21	1.0529	1.6881
291.0	0.08224	0.6700	207.62	215.82	399.64	183.82	1.0560	1.6877
292.0	0.08534	0.6711	200.60	216.71	400.14	183.43	1.0590	1.6872
293.0	0.08853	0.6721	193.87	217.60	400.65	183.04	1.0621	1.6868
294.0	0.09181	0.6732	187.41	218.50	401.15	182.65	1.0651	1.6864
295.0	0.09518	0.6743	181.22	219.40	401.65	182.26	1.0682	1.6860
296.0	0.09865	0.6753	175.27	220.29	402.16	181.86	1.0712	1.6856
297.0	0.10222	0.6764	169.56	221.19	402.66	181.47	1.0742	1.6852
298.0	0.10589	0.6775	164.08	222.09	403.16	181.07	1.0772	1.6849
299.0	0.10966	0.6786	158.82	222.99	403.66	180.68	1.0802	1.6845
300.0	0.11353	0.6797	153.76	223.88	404.16	180.28	1.0832	1.6842
301.0	0.11751	0.6808	148.90	224.78	404.66	179.88	1.0862	1.6838
302.0	0.12159	0.6819	144.22	225.68	405.16	179.48	1.0892	1.6835
303.0	0.12579	0.6831	139.73	226.59	405.66	179.08	1.0922	1.6832
304.0	0.13010	0.6842	135.40	227.49	406.16	178.68	1.0951	1.6829
305.0	0.13452	0.6853	131.24	228.39	406.66	178.27	1.0981	1.6826
306.0	0.13905	0.6865	127.24	229.29	407.16	177.87	1.1010	1.6823
307.0	0.14370	0.6876	123.38	230.19	407.65	177.46	1.1039	1.6820
308.0	0.14847	0.6888	119.67	231.10	408.15	177.05	1.1069	1.6817
309.0	0.15337	0.6899	116.09	232.00	408.65	176.64	1.1098	1.6815
310.0	0.15838	0.6911	112.65	232.91	409.14	176.23	1.1127	1.6812
311.0	0.16352	0.6922	109.32	233.81	409.63	175.82	1.1156	1.6810
312.0	0.16879	0.6934	106.12	234.72	410.13	175.41	1.1185	1.6807
313.0	0.17419	0.6946	103.04	235.63	410.62	174.99	1.1214	1.6805
314.0	0.17972	0.6958	100.06	236.53	411.11	174.58	1.1243	1.6803
315.0	0.18538	0.6970	97.187	237.44	411.60	174.16	1.1272	1.6800
316.0	0.19118	0.6982	94.416	238.35	412.09	173.74	1.1300	1.6798
317.0	0.19711	0.6994	91.742	239.26	412.58	173.32	1.1329	1.6796
318.0	0.20319	0.7006	89.159	240.17	413.07	172.90	1.1357	1.6794
319.0	0.20941	0.7019	86.666	241.08	413.55	172.47	1.1386	1.6793
320.0	0.21577	0.7031	84.258	241.99	414.04	172.05	1.1414	1.6791
321.0	0.22228	0.7044	81.931	242.90	414.53	171.62	1.1442	1.6789
322.0	0.22893	0.7056	79.683	243.81	415.01	171.20	1.1471	1.6787
323.0	0.23574	0.7069	77.510	244.73	415.49	170.77	1.1499	1.6786
324.0	0.24270	0.7081	75.410	245.64	415.97	170.33	1.1527	1.6784

Chart R11.1: Saturation properties—*continued*

T K	p MPa	v' dm³/kg	v'' dm³/kg	h' kJ/kg	h'' kJ/kg	r kJ/kg	s' kJ/kgK	s'' kJ/kgK
325.0	0.24982	0.7094	73.379	246.55	416.46	169.90	1.1555	1.6783
326.0	0.25709	0.7107	71.415	247.47	416.93	169.47	1.1583	1.6781
327.0	0.26452	0.7120	69.514	248.38	417.41	169.03	1.1611	1.6780
328.0	0.27212	0.7133	67.676	249.30	417.89	168.59	1.1639	1.6779
329.0	0.27988	0.7146	65.897	250.22	418.37	168.15	1.1666	1.6777
330.0	0.28780	0.7159	64.174	251.14	418.84	167.71	1.1694	1.6776
331.0	0.29589	0.7173	62.507	252.05	419.32	167.26	1.1722	1.6775
332.0	0.30416	0.7186	60.892	252.97	419.79	166.82	1.1749	1.6774
333.0	0.31259	0.7199	59.328	253.89	420.26	166.37	1.1777	1.6773
334.0	0.32121	0.7213	57.813	254.81	420.73	165.92	1.1804	1.6772
335.0	0.33000	0.7227	56.345	255.73	421.20	165.47	1.1831	1.6771
336.0	0.33897	0.7240	54.922	256.66	421.67	165.01	1.1859	1.6770
337.0	0.34812	0.7254	53.543	257.58	422.14	164.56	1.1886	1.6769
338.0	0.35746	0.7268	52.206	258.50	422.60	164.10	1.1913	1.6768
339.0	0.36698	0.7282	50.909	259.43	423.06	163.64	1.1940	1.6767
340.0	0.37670	0.7296	49.651	260.35	423.53	163.18	1.1967	1.6767
341.0	0.38661	0.7311	48.431	261.28	423.99	162.71	1.1994	1.6766
342.0	0.39671	0.7325	47.248	262.20	424.45	162.24	1.2021	1.6765
343.0	0.40700	0.7339	46.099	263.13	424.91	161.78	1.2048	1.6764
344.0	0.41750	0.7354	44.984	264.06	425.36	161.30	1.2075	1.6764
345.0	0.42820	0.7369	43.901	264.99	425.82	160.83	1.2102	1.6763
346.0	0.43910	0.7383	42.850	265.92	426.27	160.35	1.2128	1.6763
347.0	0.45021	0.7398	41.830	266.85	426.72	159.87	1.2155	1.6762
348.0	0.46153	0.7413	40.838	267.78	427.17	159.39	1.2181	1.6762
349.0	0.47306	0.7428	39.875	268.72	427.62	158.91	1.2208	1.6761
350.0	0.48480	0.7444	38.940	269.65	428.07	158.42	1.2234	1.6761
351.0	0.49676	0.7459	38.030	270.59	428.52	157.93	1.2261	1.6760
352.0	0.50894	0.7475	37.146	271.52	428.96	157.44	1.2287	1.6760
353.0	0.52133	0.7490	36.287	272.46	429.40	156.94	1.2314	1.6760
354.0	0.53396	0.7506	35.451	273.40	429.84	156.44	1.2340	1.6759
355.0	0.54681	0.7522	34.639	274.34	430.28	155.94	1.2366	1.6759
356.0	0.55988	0.7538	33.849	275.28	430.72	155.44	1.2392	1.6759
357.0	0.57319	0.7554	33.080	276.22	431.15	154.93	1.2418	1.6758
358.0	0.58674	0.7570	32.332	277.16	431.58	154.42	1.2444	1.6758
359.0	0.60052	0.7587	31.604	278.11	432.02	153.91	1.2471	1.6758
360.0	0.61454	0.7604	30.896	279.05	432.44	153.39	1.2497	1.6757
361.0	0.62880	0.7620	30.207	280.00	432.87	152.87	1.2522	1.6757
362.0	0.64330	0.7637	29.535	280.95	433.30	152.35	1.2548	1.6757
363.0	0.65805	0.7654	28.882	281.89	433.72	151.83	1.2574	1.6757
364.0	0.67305	0.7671	28.245	282.85	434.14	151.30	1.2600	1.6757
365.0	0.68831	0.7689	27.625	283.80	434.56	150.76	1.2626	1.6756
366.0	0.70381	0.7706	27.021	284.75	434.98	150.23	1.2652	1.6756
367.0	0.71958	0.7724	26.433	285.70	435.39	149.69	1.2677	1.6756
368.0	0.73560	0.7742	25.860	286.66	435.81	149.14	1.2703	1.6756
369.0	0.75189	0.7760	25.301	287.62	436.22	148.60	1.2729	1.6756
370.0	0.76845	0.7778	24.757	288.58	436.62	148.05	1.2754	1.6756
371.0	0.78527	0.7796	24.226	289.54	437.03	147.49	1.2780	1.6755
372.0	0.80236	0.7815	23.708	290.50	437.43	146.93	1.2805	1.6755
373.0	0.81972	0.7834	23.203	291.46	437.84	146.37	1.2831	1.6755
374.0	0.83737	0.7853	22.711	292.43	438.23	145.81	1.2856	1.6755

Chart R11.1: Saturation properties—*continued*

T K	p MPa	v' dm³/kg	v'' dm³/kg	h' kJ/kg	h'' kJ/kg	r kJ/kg	s' kJ/kgK	s'' kJ/kgK
375.0	0.85529	0.7872	22.231	293.40	438.63	145.24	1.2882	1.6755
376.0	0.87349	0.7891	21.762	294.36	439.02	144.66	1.2907	1.6755
377.0	0.89198	0.7910	21.305	295.33	439.42	144.08	1.2933	1.6754
378.0	0.91075	0.7930	20.859	296.31	439.81	143.50	1.2958	1.6754
379.0	0.92981	0.7950	20.424	297.28	440.19	142.91	1.2983	1.6754
380.0	0.94917	0.7970	19.999	298.26	440.58	142.32	1.3009	1.6754
381.0	0.96882	0.7991	19.585	299.23	440.96	141.72	1.3034	1.6754
382.0	0.98877	0.8011	19.180	300.21	441.33	141.12	1.3059	1.6753
383.0	1.00903	0.8032	18.784	301.19	441.71	140.51	1.3084	1.6753
384.0	1.02958	0.8053	18.398	302.18	442.08	139.90	1.3110	1.6753
385.0	1.05044	0.8074	18.021	303.16	442.45	139.29	1.3135	1.6753
386.0	1.07162	0.8096	17.653	304.15	442.82	138.67	1.3160	1.6752
387.0	1.09310	0.8118	17.293	305.14	443.18	138.04	1.3185	1.6752
388.0	1.11490	0.8140	16.941	306.13	443.54	137.41	1.3210	1.6752
389.0	1.13702	0.8162	16.597	307.13	443.90	136.77	1.3235	1.6751
390.0	1.15947	0.8184	16.261	308.12	444.26	136.13	1.3260	1.6751
391.0	1.18223	0.8207	15.933	309.12	444.61	135.49	1.3286	1.6751
392.0	1.20533	0.8230	15.612	310.12	444.96	134.83	1.3311	1.6750
393.0	1.22875	0.8254	15.298	311.13	445.30	134.17	1.3336	1.6750
394.0	1.25251	0.8277	14.991	312.13	445.64	133.51	1.3361	1.6749
395.0	1.27660	0.8301	14.691	313.14	445.98	132.84	1.3386	1.6749
396.0	1.30103	0.8326	14.397	314.15	446.32	132.16	1.3411	1.6748
397.0	1.32581	0.8350	14.110	315.17	446.65	131.48	1.3436	1.6748
398.0	1.35093	0.8375	13.829	316.18	446.97	130.79	1.3461	1.6747
399.0	1.37640	0.8401	13.554	317.20	447.30	130.09	1.3486	1.6747
400.0	1.40222	0.8426	13.285	318.23	447.62	129.39	1.3511	1.6746
401.0	1.42840	0.8452	13.021	319.25	447.93	128.68	1.3536	1.6745
402.0	1.45494	0.8479	12.763	320.28	448.25	127.97	1.3561	1.6744
403.0	1.48183	0.8505	12.511	321.31	448.55	127.24	1.3586	1.6744
404.0	1.50909	0.8533	12.264	322.35	448.86	126.51	1.3611	1.6743
405.0	1.53672	0.8560	12.021	323.38	449.16	125.77	1.3636	1.6742
406.0	1.56472	0.8588	11.784	324.43	449.45	125.03	1.3662	1.6741
407.0	1.59310	0.8616	11.552	325.47	449.75	124.28	1.3687	1.6740
408.0	1.62185	0.8645	11.325	326.52	450.03	123.51	1.3712	1.6739
409.0	1.65098	0.8674	11.102	327.57	450.31	122.75	1.3737	1.6738
410.0	1.68050	0.8704	10.883	328.62	450.59	121.97	1.3762	1.6737
411.0	1.71040	0.8734	10.669	329.68	450.86	121.18	1.3787	1.6736
412.0	1.74070	0.8765	10.459	330.75	451.13	120.39	1.3812	1.6734
413.0	1.77139	0.8796	10.254	331.81	451.40	119.58	1.3838	1.6733
414.0	1.80248	0.8828	10.052	332.88	451.65	118.77	1.3863	1.6732
415.0	1.83396	0.8860	9.8541	333.96	451.91	117.95	1.3888	1.6730
416.0	1.86586	0.8893	9.6602	335.04	452.15	117.11	1.3913	1.6729
417.0	1.89816	0.8926	9.4699	336.12	452.39	116.27	1.3939	1.6727
418.0	1.93088	0.8960	9.2832	337.21	452.63	115.42	1.3964	1.6725
419.0	1.96401	0.8995	9.1001	338.30	452.86	114.56	1.3989	1.6724
420.0	1.99756	0.9030	8.9204	339.40	453.08	113.68	1.4015	1.6722
421.0	2.03153	0.9066	8.7440	340.50	453.30	112.80	1.4040	1.6720
422.0	2.06593	0.9103	8.5709	341.61	453.51	111.90	1.4066	1.6718
423.0	2.10076	0.9140	8.4009	342.73	453.72	110.99	1.4092	1.6716
424.0	2.13603	0.9178	8.2340	343.84	453.92	110.07	1.4117	1.6713

Chart R11.1: Saturation properties—*continued*

T K	p MPa	v' dm^3/kg	v'' dm^3/kg	h' kJ/kg	h'' kJ/kg	r kJ/kg	s' kJ/kgK	s'' kJ/kgK
425.0	2.17173	0.9217	8.0701	344.97	454.11	109.14	1.4143	1.6711
426.0	2.20788	0.9257	7.9091	346.10	454.29	108.20	1.4169	1.6708
427.0	2.24447	0.9298	7.7509	347.24	454.47	107.24	1.4195	1.6706
428.0	2.28151	0.9339	7.5955	348.38	454.64	106.26	1.4220	1.6703
429.0	2.31901	0.9382	7.4428	349.53	454.80	105.28	1.4246	1.6700
430.0	2.35697	0.9425	7.2926	350.68	454.96	104.27	1.4273	1.6698
431.0	2.39539	0.9469	7.1450	351.84	455.10	103.26	1.4299	1.6694
432.0	2.43428	0.9515	6.9999	353.01	455.24	102.22	1.4325	1.6691
433.0	2.47364	0.9561	6.8571	354.19	455.37	101.17	1.4351	1.6688
434.0	2.51347	0.9609	6.7167	355.38	455.48	100.11	1.4378	1.6684
435.0	2.55379	0.9658	6.5785	356.57	455.59	99.02	1.4404	1.6681
436.0	2.59459	0.9708	6.4425	357.77	455.69	97.92	1.4431	1.6677
437.0	2.63588	0.9760	6.3086	358.98	455.78	96.80	1.4458	1.6673
438.0	2.67767	0.9813	6.1768	360.20	455.86	95.66	1.4485	1.6669
439.0	2.71995	0.9867	6.0470	361.43	455.92	94.49	1.4512	1.6664
440.0	2.76274	0.9924	5.9191	362.67	455.98	93.31	1.4539	1.6660
441.0	2.80604	0.9981	5.7930	363.92	456.02	92.10	1.4566	1.6655
442.0	2.84986	1.0041	5.6687	365.18	456.05	90.87	1.4594	1.6650
443.0	2.89420	1.0103	5.5462	366.45	456.06	89.62	1.4622	1.6645
444.0	2.93906	1.0166	5.4253	367.73	456.06	88.33	1.4650	1.6639
445.0	2.98445	1.0232	5.3060	369.03	456.05	87.02	1.4678	1.6633
446.0	3.03038	1.0300	5.1883	370.34	456.02	85.68	1.4706	1.6627
447.0	3.07686	1.0370	5.0720	371.66	455.97	84.31	1.4735	1.6621
448.0	3.12388	1.0443	4.9571	373.00	455.91	82.91	1.4763	1.6614
449.0	3.17146	1.0519	4.8435	374.35	455.83	81.47	1.4792	1.6607
450.0	3.21960	1.0599	4.7312	375.72	455.72	80.00	1.4822	1.6600
451.0	3.26831	1.0681	4.6201	377.11	455.60	78.49	1.4852	1.6592
452.0	3.31759	1.0767	4.5100	378.52	455.45	76.93	1.4882	1.6584
453.0	3.36746	1.0857	4.4010	379.95	455.28	75.33	1.4912	1.6575
454.0	3.41791	1.0951	4.2929	381.40	455.08	73.68	1.4943	1.6566
455.0	3.46897	1.1050	4.1856	382.88	454.86	71.98	1.4974	1.6556
456.0	3.52063	1.1155	4.0790	384.38	454.60	70.23	1.5006	1.6546
457.0	3.57291	1.1265	3.9731	385.91	454.32	68.41	1.5038	1.6535
458.0	3.62581	1.1381	3.8677	387.47	453.99	66.53	1.5071	1.6523
459.0	3.67934	1.1505	3.7626	389.06	453.63	64.57	1.5104	1.6511
460.0	3.73351	1.1637	3.6578	390.69	453.22	62.53	1.5138	1.6498
461.0	3.78834	1.1778	3.5530	392.37	452.76	60.39	1.5173	1.6483
462.0	3.84383	1.1930	3.4480	394.09	452.25	58.16	1.5209	1.6468
463.0	3.89999	1.2094	3.3426	395.87	451.68	55.81	1.5246	1.6452
464.0	3.95685	1.2273	3.2366	397.71	451.04	53.33	1.5284	1.6434
465.0	4.01440	1.2469	3.1295	399.62	450.31	50.69	1.5324	1.6414
466.0	4.07267	1.2685	3.0209	401.61	449.49	47.88	1.5365	1.6393
467.0	4.13167	1.2928	2.9101	403.71	448.55	44.84	1.5408	1.6369
468.0	4.19[illegible]42	1.3205	2.7965	405.94	447.47	41.54	1.5454	1.6342
469.0	4.25194	1.3524	2.6788	408.33	446.21	37.89	1.5504	1.6312
470.0	4.31324	1.3905	2.5551	410.94	444.71	33.77	1.5558	1.6276

Chart R11.2: Saturation properties

p MPa	T K	v' dm^3/kg	v'' dm^3/kg	h' kJ/kg	h'' kJ/kg	r kJ/kg	s' kJ/kgK	s'' kJ/kgK
0.0031	225.57	0.6101	4392.7	160.43	366.68	206.24	0.8413	1.7556
0.0032	226.04	0.6105	4264.1	160.79	366.91	206.11	0.8429	1.7547
0.0033	226.51	0.6108	4143.1	161.15	367.13	205.99	0.8444	1.7539
0.0034	226.96	0.6112	4028.9	161.49	367.36	205.87	0.8460	1.7530
0.0035	227.39	0.6116	3921.1	161.83	367.57	205.74	0.8474	1.7522
0.0036	227.82	0.6119	3819.2	162.16	367.78	205.63	0.8489	1.7515
0.0037	228.24	0.6122	3722.5	162.48	367.99	205.51	0.8503	1.7507
0.0038	228.65	0.6126	3630.8	162.79	368.19	205.39	0.8517	1.7500
0.0039	229.05	0.6129	3543.7	163.10	368.38	205.28	0.8530	1.7493
0.0040	229.44	0.6132	3460.8	163.41	368.58	205.17	0.8544	1.7486
0.0041	229.82	0.6135	3381.8	163.70	368.77	205.06	0.8557	1.7479
0.0042	230.20	0.6138	3306.5	164.00	368.95	204.96	0.8569	1.7473
0.0043	230.57	0.6141	3234.6	164.28	369.13	204.85	0.8582	1.7466
0.0044	230.93	0.6144	3165.8	164.56	369.31	204.75	0.8594	1.7460
0.0045	231.28	0.6147	3100.0	164.84	369.48	204.64	0.8606	1.7454
0.0046	231.63	0.6150	3037.0	165.11	369.66	204.54	0.8618	1.7448
0.0047	231.97	0.6153	2976.6	165.38	369.82	204.44	0.8629	1.7443
0.0048	232.30	0.6155	2918.6	165.64	369.99	204.35	0.8640	1.7437
0.0049	232.63	0.6158	2862.9	165.90	370.15	204.25	0.8652	1.7432
0.0050	232.95	0.6161	2809.4	166.16	370.31	204.16	0.8663	1.7426
0.0051	233.27	0.6163	2757.9	166.41	370.47	204.06	0.8673	1.7421
0.0052	233.58	0.6166	2708.4	166.66	370.62	203.97	0.8684	1.7416
0.0053	233.89	0.6168	2660.6	166.90	370.78	203.88	0.8694	1.7411
0.0054	234.20	0.6171	2614.6	167.14	370.93	203.79	0.8705	1.7406
0.0055	234.49	0.6173	2570.1	167.38	371.07	203.70	0.8715	1.7401
0.0056	234.79	0.6176	2527.3	167.61	371.22	203.61	0.8725	1.7397
0.0057	235.08	0.6178	2485.9	167.84	371.36	203.52	0.8734	1.7392
0.0058	235.36	0.6181	2445.8	168.07	371.51	203.44	0.8744	1.7388
0.0059	235.64	0.6183	2407.1	168.29	371.64	203.35	0.8754	1.7383
0.0060	235.92	0.6185	2369.6	168.51	371.78	203.27	0.8763	1.7379
0.0061	236.19	0.6187	2333.4	168.73	371.92	203.19	0.8772	1.7375
0.0062	236.46	0.6190	2298.2	168.95	372.05	203.11	0.8781	1.7371
0.0063	236.73	0.6192	2264.2	169.16	372.18	203.02	0.8790	1.7367
0.0064	236.99	0.6194	2231.1	169.37	372.31	202.94	0.8799	1.7363
0.0065	237.25	0.6196	2199.1	169.58	372.44	202.86	0.8808	1.7359
0.0066	237.50	0.6198	2168.0	169.78	372.57	202.79	0.8817	1.7355
0.0067	237.76	0.6200	2137.8	169.99	372.70	202.71	0.8825	1.7351
0.0068	238.00	0.6203	2108.4	170.19	372.82	202.63	0.8834	1.7347
0.0069	238.25	0.6205	2079.9	170.39	372.94	202.56	0.8842	1.7344
0.0070	238.49	0.6207	2052.2	170.58	373.06	202.48	0.8850	1.7340
0.0071	238.73	0.6209	2025.2	170.78	373.18	202.41	0.8858	1.7337
0.0072	238.97	0.6211	1999.0	170.97	373.30	202.33	0.8866	1.7333
0.0073	239.21	0.6213	1973.4	171.16	373.42	202.26	0.8874	1.7330
0.0074	239.44	0.6215	1948.5	171.35	373.54	202.19	0.8882	1.7326
0.0076	239.89	0.6218	1900.7	171.72	373.76	202.04	0.8898	1.7320
0.0078	240.34	0.6222	1855.2	172.08	373.99	201.90	0.8913	1.7314
0.0080	240.78	0.6226	1811.9	172.44	374.20	201.77	0.8928	1.7307
0.0082	241.21	0.6230	1770.7	172.79	374.42	201.63	0.8942	1.7301
0.0084	241.62	0.6233	1731.3	173.13	374.63	201.50	0.8956	1.7296
0.0086	242.04	0.6237	1693.8	173.47	374.83	201.37	0.8970	1.7290

Chart R11.2: Saturation properties—*continued*

p MPa	T K	v' dm^3/kg	v'' dm^3/kg	h' kJ/kg	h'' kJ/kg	r kJ/kg	s' kJ/kgK	s'' kJ/kgK
0.0088	242.44	0.6240	1657.9	173.80	375.04	201.24	0.8984	1.7284
0.0090	242.83	0.6243	1623.5	174.12	375.23	201.11	0.8997	1.7279
0.0092	243.22	0.6247	1590.6	174.44	375.43	200.99	0.9010	1.7274
0.0094	243.60	0.6250	1559.1	174.75	375.62	200.87	0.9023	1.7269
0.0096	243.98	0.6253	1528.8	175.06	375.81	200.74	0.9036	1.7264
0.0098	244.35	0.6256	1499.7	175.37	375.99	200.63	0.9048	1.7259
0.0100	244.71	0.6260	1471.7	175.67	376.17	200.51	0.9061	1.7254
0.0102	245.07	0.6263	1444.8	175.96	376.35	200.39	0.9073	1.7250
0.0104	245.42	0.6266	1418.9	176.25	376.53	200.28	0.9084	1.7245
0.0106	245.76	0.6269	1394.0	176.54	376.70	200.17	0.9096	1.7241
0.0108	246.10	0.6272	1369.9	176.82	376.87	200.06	0.9107	1.7236
0.0110	246.43	0.6274	1346.7	177.10	377.04	199.95	0.9119	1.7232
0.0115	247.25	0.6282	1292.1	177.77	377.45	199.68	0.9146	1.7222
0.0120	248.03	0.6288	1241.9	178.43	377.85	199.42	0.9173	1.7213
0.0125	248.79	0.6295	1195.6	179.06	378.23	199.17	0.9198	1.7203
0.0130	249.53	0.6301	1152.8	179.68	378.60	198.92	0.9223	1.7195
0.0135	250.24	0.6308	1113.0	180.27	378.96	198.68	0.9247	1.7186
0.0140	250.93	0.6314	1076.0	180.85	379.30	198.45	0.9270	1.7178
0.0145	251.60	0.6320	1041.4	181.42	379.64	198.22	0.9292	1.7171
0.0150	252.25	0.6325	1009.1	181.97	379.97	198.00	0.9314	1.7164
0.0155	252.88	0.6331	978.76	182.50	380.29	197.79	0.9335	1.7157
0.0160	253.50	0.6337	950.28	183.03	380.60	197.58	0.9356	1.7150
0.0165	254.10	0.6342	923.48	183.54	380.91	197.37	0.9376	1.7143
0.0170	254.69	0.6347	898.19	184.03	381.21	197.17	0.9396	1.7137
0.0175	255.26	0.6352	874.31	184.52	381.50	196.97	0.9415	1.7131
0.0180	255.82	0.6357	851.71	185.00	381.78	196.78	0.9433	1.7125
0.0185	256.37	0.6362	830.29	185.46	382.06	196.59	0.9451	1.7120
0.0190	256.90	0.6367	809.96	185.92	382.33	196.41	0.9469	1.7114
0.0195	257.43	0.6372	790.64	186.37	382.59	196.22	0.9487	1.7109
0.0200	257.94	0.6377	772.25	186.81	382.85	196.05	0.9504	1.7104
0.0210	258.94	0.6386	738.01	187.66	383.36	195.70	0.9537	1.7094
0.0220	259.89	0.6394	706.79	188.48	383.84	195.36	0.9568	1.7085
0.0230	260.82	0.6403	678.19	189.27	384.31	195.04	0.9599	1.7077
0.0240	261.71	0.6411	651.90	190.04	384.76	194.72	0.9628	1.7068
0.0250	262.57	0.6419	627.64	190.78	385.20	194.42	0.9656	1.7061
0.0260	263.40	0.6427	605.18	191.50	385.62	194.12	0.9684	1.7053
0.0270	264.21	0.6434	584.33	192.20	386.03	193.83	0.9710	1.7046
0.0280	264.99	0.6442	564.93	192.88	386.43	193.55	0.9736	1.7040
0.0290	265.75	0.6449	546.81	193.54	386.82	193.28	0.9761	1.7033
0.0300	266.49	0.6456	529.86	194.19	387.20	193.01	0.9785	1.7027
0.0310	267.22	0.6463	513.97	194.81	387.56	192.75	0.9808	1.7022
0.0320	267.92	0.6469	499.04	195.43	387.92	192.49	0.9831	1.7016
0.0330	268.60	0.6476	484.98	196.02	388.27	192.25	0.9853	1.7011
0.0340	269.27	0.6482	471.72	196.61	388.61	192.00	0.9875	1.7005
0.0350	269.93	0.6488	459.19	197.18	388.94	191.76	0.9896	1.7001
0.0360	270.56	0.6494	447.33	197.73	389.27	191.53	0.9917	1.6996
0.0370	271.19	0.6500	436.09	198.28	389.58	191.30	0.9937	1.6991
0.0380	271.80	0.6506	425.43	198.81	389.89	191.08	0.9957	1.6987
0.0390	272.40	0.6512	415.29	199.34	390.20	190.86	0.9976	1.6982
0.0400	272.98	0.6518	405.64	199.85	390.50	190.64	0.9995	1.6978

Chart R11.2: Saturation properties–*continued*

p MPa	T K	v' dm^3/kg	v'' dm^3/kg	h' kJ/kg	h'' kJ/kg	r kJ/kg	s' kJ/kgK	s'' kJ/kgK
0.0410	273.56	0.6523	396.44	200.36	390.79	190.43	1.0013	1.6974
0.0420	274.12	0.6529	387.67	200.85	391.07	190.22	1.0031	1.6971
0.0430	274.67	0.6534	379.29	201.34	391.35	190.02	1.0049	1.6967
0.0440	275.21	0.6540	371.28	201.81	391.63	189.82	1.0066	1.6963
0.0450	275.75	0.6545	363.61	202.28	391.90	189.62	1.0083	1.6960
0.0460	276.27	0.6550	356.26	202.74	392.17	189.42	1.0100	1.6956
0.0470	276.78	0.6555	349.22	203.20	392.43	189.23	1.0116	1.6953
0.0480	277.29	0.6560	342.46	203.64	392.69	189.04	1.0132	1.6950
0.0490	277.79	0.6565	335.96	204.08	392.94	188.86	1.0148	1.6947
0.0500	278.28	0.6570	329.72	204.51	393.19	188.68	1.0163	1.6944
0.0510	278.76	0.6575	323.71	204.94	393.43	188.49	1.0179	1.6941
0.0520	279.23	0.6579	317.92	205.36	393.67	188.32	1.0194	1.6938
0.0530	279.70	0.6584	312.35	205.77	393.91	188.14	1.0208	1.6935
0.0540	280.16	0.6588	306.97	206.18	394.14	187.97	1.0223	1.6932
0.0550	280.61	0.6593	301.78	206.58	394.37	187.80	1.0237	1.6930
0.0560	281.06	0.6598	296.77	206.97	394.60	187.63	1.0251	1.6927
0.0570	281.50	0.6602	291.93	207.36	394.82	187.46	1.0265	1.6925
0.0580	281.93	0.6606	287.25	207.75	395.05	187.30	1.0279	1.6922
0.0590	282.36	0.6611	282.73	208.13	395.26	187.14	1.0292	1.6920
0.0600	282.78	0.6615	278.35	208.50	395.48	186.98	1.0305	1.6917
0.0610	283.20	0.6619	274.10	208.87	395.69	186.82	1.0319	1.6915
0.0620	283.61	0.6623	269.99	209.24	395.90	186.66	1.0331	1.6913
0.0630	284.02	0.6628	266.01	209.60	396.10	186.51	1.0344	1.6911
0.0640	284.42	0.6632	262.14	209.96	396.31	186.35	1.0357	1.6909
0.0650	284.82	0.6636	258.39	210.31	396.51	186.20	1.0369	1.6906
0.0660	285.21	0.6640	254.75	210.66	396.71	186.05	1.0381	1.6904
0.0670	285.60	0.6644	251.22	211.00	396.90	185.90	1.0393	1.6902
0.0680	285.98	0.6648	247.78	211.34	397.10	185.76	1.0405	1.6901
0.0690	286.36	0.6652	244.44	211.68	397.29	185.61	1.0417	1.6899
0.0700	286.73	0.6655	241.19	212.01	397.48	185.47	1.0428	1.6897
0.0710	287.10	0.6659	238.03	212.34	397.67	185.33	1.0440	1.6895
0.0720	287.47	0.6663	234.96	212.67	397.85	185.19	1.0451	1.6893
0.0730	287.83	0.6667	231.96	212.99	398.03	185.05	1.0462	1.6891
0.0740	288.19	0.6670	229.05	213.31	398.22	184.91	1.0473	1.6890
0.0760	288.89	0.6678	223.44	213.93	398.57	184.64	1.0495	1.6886
0.0780	289.58	0.6685	218.10	214.55	398.92	184.37	1.0516	1.6883
0.0800	290.26	0.6692	213.03	215.15	399.26	184.11	1.0537	1.6880
0.0820	290.92	0.6699	208.19	215.75	399.60	183.85	1.0557	1.6877
0.0840	291.57	0.6706	203.57	216.33	399.93	183.60	1.0577	1.6874
0.0860	292.21	0.6713	199.16	216.90	400.25	183.35	1.0597	1.6871
0.0880	292.84	0.6719	194.95	217.46	400.56	183.10	1.0616	1.6869
0.0900	293.45	0.6726	190.91	218.01	400.87	182.86	1.0635	1.6866
0.0920	294.06	0.6732	187.04	218.55	401.18	182.63	1.0653	1.6864
0.0940	294.65	0.6739	183.33	219.09	401.48	182.39	1.0671	1.6861
0.0960	295.24	0.6745	179.77	219.61	401.77	182.16	1.0689	1.6859
0.0980	295.81	0.6751	176.35	220.13	402.06	181.94	1.0706	1.6857
0.1000	296.38	0.6758	173.07	220.63	402.35	181.71	1.0724	1.6855
0.1020	296.94	0.6764	169.90	221.13	402.63	181.49	1.0740	1.6853
0.1040	297.49	0.6770	166.86	221.63	402.90	181.28	1.0757	1.6850
0.1060	298.03	0.6775	163.92	222.11	403.18	181.06	1.0773	1.6849

Chart R11.2: Saturation properties—*continued*

p MPa	T K	v' dm^3/kg	v'' dm^3/kg	h' kJ/kg	h'' kJ/kg	r kJ/kg	s' kJ/kgK	s'' kJ/kgK
0.1080	298.56	0.6781	161.09	222.59	403.44	180.85	1.0789	1.6847
0.1100	299.09	0.6787	158.36	223.07	403.71	180.64	1.0805	1.6845
0.1150	300.37	0.6801	151.92	224.22	404.35	180.13	1.0843	1.6840
0.1200	301.61	0.6815	146.01	225.34	404.97	179.64	1.0880	1.6836
0.1250	302.81	0.6828	140.55	226.42	405.57	179.15	1.0916	1.6832
0.1300	303.98	0.6842	135.50	227.47	406.15	178.68	1.0950	1.6829
0.1350	305.11	0.6854	130.80	228.49	406.71	178.23	1.0984	1.6825
0.1400	306.21	0.6867	126.43	229.48	407.26	177.78	1.1016	1.6822
0.1450	307.27	0.6879	122.35	230.44	407.79	177.35	1.1048	1.6819
0.1500	308.31	0.6891	118.53	231.38	408.31	176.92	1.1078	1.6816
0.1550	309.33	0.6903	114.94	232.30	408.81	176.51	1.1108	1.6814
0.1600	310.32	0.6914	111.58	233.20	409.30	176.10	1.1136	1.6811
0.1650	311.28	0.6926	108.41	234.07	409.77	175.70	1.1164	1.6809
0.1700	312.23	0.6937	105.42	234.92	410.24	175.31	1.1192	1.6807
0.1750	313.15	0.6948	102.59	235.76	410.69	174.93	1.1218	1.6805
0.1800	314.05	0.6959	99.912	236.58	411.14	174.56	1.1244	1.6802
0.1850	314.93	0.6969	97.375	237.38	411.57	174.19	1.1270	1.6801
0.1900	315.80	0.6980	94.966	238.17	411.99	173.82	1.1294	1.6799
0.1950	316.65	0.6990	92.676	238.94	412.41	173.47	1.1319	1.6797
0.2000	317.48	0.7000	90.496	239.69	412.81	173.12	1.1342	1.6795
0.2100	319.09	0.7020	86.436	241.17	413.60	172.43	1.1388	1.6792
0.2200	320.65	0.7039	82.730	242.59	414.36	171.77	1.1433	1.6790
0.2300	322.16	0.7058	79.335	243.96	415.09	171.13	1.1475	1.6787
0.2400	323.61	0.7076	76.211	245.29	415.79	170.50	1.1516	1.6785
0.2500	325.03	0.7094	73.328	246.58	416.47	169.89	1.1556	1.6783
0.2600	326.39	0.7112	70.658	247.83	417.12	169.29	1.1594	1.6781
0.2700	327.72	0.7129	68.178	249.05	417.76	168.71	1.1631	1.6779
0.2800	329.02	0.7146	65.869	250.23	418.38	168.14	1.1667	1.6777
0.2900	330.27	0.7163	63.712	251.39	418.97	167.59	1.1702	1.6776
0.3000	331.50	0.7179	61.694	252.51	419.55	167.04	1.1735	1.6774
0.3100	332.69	0.7195	59.800	253.61	420.12	166.51	1.1768	1.6773
0.3200	333.86	0.7211	58.021	254.68	420.67	165.98	1.1800	1.6772
0.3300	335.00	0.7227	56.344	255.73	421.20	165.47	1.1831	1.6771
0.3400	336.11	0.7242	54.763	256.76	421.72	164.96	1.1862	1.6770
0.3500	337.20	0.7257	53.268	257.77	422.23	164.47	1.1891	1.6769
0.3600	338.27	0.7272	51.853	258.75	422.73	163.98	1.1920	1.6768
0.3700	339.31	0.7287	50.512	259.71	423.21	163.50	1.1949	1.6767
0.3800	340.34	0.7301	49.238	260.66	423.68	163.02	1.1976	1.6766
0.3900	341.34	0.7315	48.027	261.59	424.14	162.55	1.2003	1.6766
0.4000	342.32	0.7330	46.874	262.50	424.60	162.09	1.2030	1.6765
0.4100	343.29	0.7344	45.775	263.40	425.04	161.64	1.2056	1.6764
0.4200	344.24	0.7357	44.726	264.28	425.47	161.19	1.2081	1.6764
0.4300	345.17	0.7371	43.724	265.14	425.89	160.75	1.2106	1.6763
0.4400	346.08	0.7385	42.766	266.00	426.31	160.31	1.2130	1.6763
0.4500	346.98	0.7398	41.849	266.83	426.71	159.88	1.2154	1.6762
0.4600	347.87	0.7411	40.969	267.66	427.11	159.46	1.2178	1.6762
0.4700	348.74	0.7424	40.126	268.47	427.50	159.03	1.2201	1.6761
0.4800	349.59	0.7438	39.317	269.27	427.89	158.62	1.2224	1.6761
0.4900	350.44	0.7450	38.539	270.06	428.26	158.21	1.2246	1.6761
0.5000	351.27	0.7463	37.791	270.84	428.63	157.80	1.2268	1.6760

Chart R11.2: Saturation properties–*continued*

p MPa	T K	v' dm³/kg	v'' dm³/kg	h' kJ/kg	h'' kJ/kg	r kJ/kg	s' kJ/kgK	s'' kJ/kgK
0.5100	352.09	0.7476	37.071	271.60	429.00	157.39	1.2290	1.6760
0.5200	352.89	0.7489	36.378	272.36	429.35	157.00	1.2311	1.6760
0.5300	353.69	0.7501	35.709	273.10	429.70	156.60	1.2332	1.6759
0.5400	354.47	0.7514	35.065	273.84	430.05	156.21	1.2352	1.6759
0.5500	355.25	0.7526	34.443	274.57	430.39	155.82	1.2373	1.6759
0.5600	356.01	0.7538	33.842	275.29	430.72	155.43	1.2392	1.6759
0.5700	356.76	0.7550	33.261	275.99	431.05	155.05	1.2412	1.6758
0.5800	357.50	0.7562	32.700	276.69	431.37	154.68	1.2432	1.6758
0.5900	358.24	0.7574	32.157	277.39	431.69	154.30	1.2451	1.6758
0.6000	358.96	0.7586	31.631	278.07	432.00	153.93	1.2470	1.6758
0.6100	359.68	0.7598	31.122	278.75	432.31	153.56	1.2488	1.6758
0.6200	360.39	0.7610	30.628	279.42	432.61	153.19	1.2507	1.6757
0.6300	361.08	0.7622	30.150	280.08	432.91	152.83	1.2525	1.6757
0.6400	361.77	0.7633	29.685	280.73	433.20	152.47	1.2543	1.6757
0.6500	362.46	0.7645	29.235	281.38	433.49	152.11	1.2560	1.6757
0.6600	363.13	0.7656	28.798	282.02	433.78	151.76	1.2578	1.6757
0.6700	363.80	0.7668	28.373	282.65	434.06	151.40	1.2595	1.6757
0.6800	364.46	0.7679	27.960	283.28	434.33	151.05	1.2612	1.6757
0.6900	365.11	0.7691	27.558	283.90	434.61	150.71	1.2629	1.6756
0.7000	365.76	0.7702	27.168	284.52	434.88	150.36	1.2645	1.6756
0.7100	366.39	0.7713	26.788	285.13	435.14	150.02	1.2662	1.6756
0.7200	367.03	0.7724	26.418	285.73	435.40	149.67	1.2678	1.6756
0.7300	367.65	0.7736	26.058	286.33	435.66	149.33	1.2694	1.6756
0.7400	368.27	0.7747	25.707	286.92	435.92	149.00	1.2710	1.6756
0.7600	369.49	0.7769	25.032	288.09	436.42	148.33	1.2741	1.6756
0.7800	370.69	0.7791	24.389	289.24	436.90	147.67	1.2772	1.6755
0.8000	371.86	0.7812	23.778	290.37	437.38	147.01	1.2802	1.6755
0.8200	373.02	0.7834	23.195	291.48	437.84	146.36	1.2831	1.6755
0.8400	374.15	0.7855	22.639	292.57	438.29	145.72	1.2860	1.6755
0.8600	375.26	0.7877	22.108	293.65	438.73	145.09	1.2888	1.6755
0.8800	376.35	0.7898	21.599	294.71	439.16	144.46	1.2916	1.6755
0.9000	377.43	0.7919	21.113	295.75	439.58	143.83	1.2943	1.6754
0.9200	378.49	0.7940	20.646	296.78	439.99	143.21	1.2970	1.6754
0.9400	379.53	0.7961	20.199	297.80	440.39	142.60	1.2997	1.6754
0.9600	380.55	0.7981	19.769	298.80	440.79	141.99	1.3023	1.6754
0.9800	381.56	0.8002	19.356	299.78	441.17	141.38	1.3048	1.6753
1.0000	382.56	0.8023	18.959	300.76	441.54	140.78	1.3073	1.6753
1.0200	383.54	0.8043	18.576	301.72	441.91	140.19	1.3098	1.6753
1.0400	384.50	0.8064	18.208	302.67	442.27	139.60	1.3122	1.6753
1.0600	385.45	0.8084	17.853	303.61	442.62	139.01	1.3146	1.6753
1.0800	386.39	0.8104	17.510	304.54	442.96	138.42	1.3170	1.6752
1.1000	387.32	0.8125	17.180	305.46	443.30	137.84	1.3193	1.6752
1.1500	389.58	0.8175	16.401	307.71	444.11	136.40	1.3250	1.6751
1.2000	391.77	0.8225	15.685	309.89	444.88	134.98	1.3305	1.6750
1.2500	393.90	0.8275	15.023	312.03	445.61	133.58	1.3358	1.6749
1.3000	395.96	0.8325	14.409	314.11	446.30	132.19	1.3410	1.6748
1.3500	397.96	0.8374	13.839	316.15	446.96	130.81	1.3460	1.6747
1.4000	399.91	0.8424	13.307	318.14	447.59	129.45	1.3509	1.6746
1.4500	401.81	0.8474	12.811	320.09	448.19	128.10	1.3557	1.6745
1.5000	403.67	0.8523	12.345	322.00	448.76	126.76	1.3603	1.6743

Chart R11.2: Saturation properties–*continued*

p MPa	T K	v' dm³/kg	v'' dm³/kg	h' kJ/kg	h'' kJ/kg	r kJ/kg	s' kJ/kgK	s'' kJ/kgK
1.5500	405.48	0.8573	11.908	323.88	449.30	125.42	1.3648	1.6742
1.6000	407.24	0.8623	11.497	325.72	449.82	124.09	1.3693	1.6740
1.6500	408.97	0.8673	11.109	327.53	450.31	122.77	1.3736	1.6738
1.7000	410.65	0.8724	10.743	329.32	450.77	121.45	1.3778	1.6736
1.7500	412.30	0.8774	10.396	331.07	451.21	120.14	1.3820	1.6734
1.8000	413.92	0.8825	10.068	332.80	451.63	118.83	1.3861	1.6732
1.8500	415.50	0.8877	9.7558	334.50	452.03	117.53	1.3901	1.6729
1.9000	417.06	0.8928	9.4592	336.18	452.41	116.22	1.3940	1.6727
1.9500	418.58	0.8980	9.1768	337.84	452.76	114.92	1.3979	1.6724
2.0000	420.07	0.9033	8.9075	339.48	453.10	113.62	1.4017	1.6722
2.1000	422.98	0.9139	8.4046	342.70	453.72	111.01	1.4091	1.6716
2.2000	425.78	0.9248	7.9438	345.85	454.25	108.40	1.4163	1.6709
2.3000	428.49	0.9360	7.5196	348.94	454.72	105.78	1.4233	1.6702
2.4000	431.12	0.9475	7.1276	351.98	455.12	103.13	1.4302	1.6694
2.5000	433.66	0.9593	6.7637	354.98	455.45	100.47	1.4369	1.6686
2.6000	436.13	0.9715	6.4248	357.93	455.70	97.77	1.4434	1.6676
2.7000	438.53	0.9842	6.1078	360.85	455.89	95.04	1.4499	1.6666
2.8000	440.86	0.9973	5.8104	363.74	456.01	92.27	1.4563	1.6656
2.9000	443.13	1.0111	5.5304	366.61	456.06	89.45	1.4625	1.6644
3.0000	445.34	1.0255	5.2658	369.47	456.04	86.57	1.4687	1.6631
3.1000	447.49	1.0406	5.0151	372.32	455.94	83.63	1.4749	1.6618
3.2000	449.59	1.0566	4.7766	375.16	455.77	80.60	1.4810	1.6603
3.3000	451.64	1.0736	4.5490	378.02	455.51	77.49	1.4871	1.6587
3.4000	453.65	1.0917	4.3310	380.88	455.16	74.27	1.4932	1.6569
3.5000	455.60	1.1112	4.1214	383.78	454.71	70.93	1.4993	1.6550
3.6000	457.51	1.1324	3.9189	386.70	454.15	67.45	1.5055	1.6529
3.7000	459.38	1.1554	3.7225	389.68	453.48	63.80	1.5117	1.6506
3.8000	461.21	1.1809	3.5308	392.73	452.66	59.93	1.5181	1.6480
3.9000	463.00	1.2094	3.3426	395.87	451.68	55.81	1.5246	1.6452
4.0000	464.75	1.2418	3.1563	399.13	450.50	51.37	1.5314	1.6419
4.1000	466.46	1.2795	2.9697	402.57	449.07	46.50	1.5385	1.6382
4.2000	468.14	1.3247	2.7800	406.27	447.31	41.04	1.5461	1.6338
4.3000	469.79	1.3817	2.5823	410.36	445.06	34.71	1.5546	1.6284

Chart R11.3: Thermodynamic properties

p/MPa	0.10			0.15			0.20		
T K	v dm³/kg	h kJ/kg	s kJ/kgK	v dm³/kg	h kJ/kg	s kJ/kgK	v dm³/kg	h kJ/kg	s kJ/kgK
230.0	0.6136	163.88	0.8562	0.6136	163.90	0.8561	0.6135	163.92	0.8561
240.0	0.6219	171.84	0.8900	0.6218	171.86	0.8900	0.6218	171.88	0.8900
250.0	0.6305	180.11	0.9238	0.6305	180.13	0.9237	0.6304	180.15	0.9237
260.0	0.6395	188.60	0.9571	0.6394	188.62	0.9571	0.6394	188.64	0.9570
270.0	0.6488	197.27	0.9898	0.6488	197.29	0.9897	0.6488	197.31	0.9897
280.0	0.6586	206.05	1.0217	0.6586	206.07	1.0217	0.6585	206.09	1.0216
290.0	0.6689	214.93	1.0529	0.6689	214.95	1.0528	0.6688	214.97	1.0528
300.0	175.45	404.46	1.6925	0.6797	223.90	1.0832	0.6796	223.91	1.0831
310.0	181.98	410.33	1.7118	119.29	409.32	1.6849	0.6910	232.92	1.1127
320.0	188.45	416.25	1.7306	123.78	415.33	1.7040	91.377	414.36	1.6844
330.0	194.88	422.23	1.7490	128.21	421.39	1.7226	94.828	420.51	1.7033
340.0	201.26	428.26	1.7670	132.60	427.49	1.7409	98.224	426.69	1.7218
350.0	207.61	434.35	1.7846	136.94	433.64	1.7587	101.58	432.90	1.7398
360.0	213.92	440.49	1.8019	141.25	439.83	1.7761	104.89	439.14	1.7574
370.0	220.21	446.68	1.8189	145.53	446.06	1.7932	108.17	445.43	1.7746
380.0	226.49	452.92	1.8355	149.79	452.34	1.8099	111.43	451.75	1.7914
390.0	232.74	459.21	1.8519	154.03	458.67	1.8264	114.67	458.11	1.8080
400.0	238.97	465.55	1.8679	158.26	465.04	1.8425	117.89	464.52	1.8242
410.0	245.19	471.93	1.8837	162.46	471.45	1.8583	121.09	470.96	1.8401
420.0	251.40	478.36	1.8992	166.66	477.90	1.8739	124.28	477.44	1.8557
430.0	257.60	484.83	1.9144	170.84	484.39	1.8892	127.45	483.96	1.8710
440.0	263.78	491.34	1.9294	175.01	490.93	1.9042	130.62	490.51	1.8861
450.0	269.96	497.89	1.9441	179.17	497.50	1.9189	133.77	497.10	1.9009

p/MPa	0.30			0.40			0.50		
230.0	0.6135	163.97	0.8560	0.6134	164.01	0.8559	0.6134	164.05	0.8558
240.0	0.6217	171.93	0.8899	0.6217	171.97	0.8898	0.6216	172.01	0.8897
250.0	0.6304	180.19	0.9236	0.6303	180.23	0.9235	0.6302	180.27	0.9234
260.0	0.6393	188.68	0.9569	0.6392	188.72	0.9568	0.6392	188.76	0.9567
270.0	0.6487	197.35	0.9896	0.6486	197.38	0.9895	0.6485	197.42	0.9894
280.0	0.6584	206.13	1.0215	0.6583	206.17	1.0214	0.6582	206.20	1.0213
290.0	0.6687	215.00	1.0527	0.6686	215.04	1.0526	0.6685	215.08	1.0525
300.0	0.6795	223.95	1.0830	0.6793	223.98	1.0829	0.6792	224.02	1.0828
310.0	0.6909	232.95	1.1125	0.6907	232.99	1.1124	0.6906	233.02	1.1123
320.0	0.7030	242.02	1.1413	0.7028	242.05	1.1412	0.7026	242.08	1.1411
330.0	0.7159	251.14	1.1694	0.7157	251.17	1.1693	0.7155	251.19	1.1691
340.0	63.761	424.97	1.6936	0.7296	260.36	1.1967	0.7293	260.38	1.1965
350.0	66.140	431.33	1.7120	48.341	429.63	1.6910	0.7443	269.65	1.2234
360.0	68.474	437.71	1.7300	50.203	436.17	1.7095	39.182	434.52	1.6926
370.0	70.770	444.11	1.7475	52.019	442.71	1.7274	40.723	441.21	1.7109
380.0	73.036	450.53	1.7646	53.798	449.24	1.7448	42.221	447.88	1.7287
390.0	75.276	456.97	1.7814	55.548	455.78	1.7618	43.684	454.53	1.7460
400.0	77.495	463.45	1.7978	57.273	462.34	1.7784	45.118	461.19	1.7628
410.0	79.696	469.95	1.8138	58.978	468.92	1.7946	46.530	467.85	1.7793
420.0	81.881	476.49	1.8296	60.665	475.52	1.8105	47.921	474.52	1.7953
430.0	84.052	483.06	1.8450	62.338	482.14	1.8261	49.297	481.20	1.8111
440.0	86.212	489.66	1.8602	63.997	488.79	1.8414	50.658	487.91	1.8265
450.0	88.361	496.29	1.8751	65.646	495.47	1.8564	52.008	494.63	1.8416

Chart R11.3: Thermodynamic properties—*continued*

p/MPa	0.60			0.80			1.00		
T K	v dm³/kg	h kJ/kg	s kJ/kgK	v dm³/kg	h kJ/kg	s kJ/kgK	v dm³/kg	h kJ/kg	s kJ/kgK
230.0	0.6133	164.10	0.8558	0.6132	164.18	0.8556	0.6131	164.27	0.8554
240.0	0.6216	172.05	0.8896	0.6214	172.14	0.8895	0.6213	172.22	0.8893
250.0	0.6301	180.31	0.9234	0.6300	180.40	0.9232	0.6299	180.48	0.9230
260.0	0.6391	188.80	0.9566	0.6389	188.88	0.9565	0.6388	188.97	0.9563
270.0	0.6484	197.46	0.9893	0.6482	197.54	0.9891	0.6480	197.62	0.9889
280.0	0.6581	206.24	1.0212	0.6579	206.32	1.0211	0.6577	206.39	1.0209
290.0	0.6683	215.11	1.0524	0.6681	215.19	1.0522	0.6679	215.26	1.0520
300.0	0.6791	224.05	1.0827	0.6788	224.12	1.0825	0.6786	224.19	1.0822
310.0	0.6904	233.05	1.1122	0.6901	233.12	1.1120	0.6898	233.18	1.1117
320.0	0.7025	242.11	1.1409	0.7021	242.17	1.1407	0.7018	242.23	1.1404
330.0	0.7153	251.22	1.1690	0.7149	251.28	1.1687	0.7145	251.33	1.1685
340.0	0.7291	260.40	1.1964	0.7286	260.45	1.1961	0.7282	260.50	1.1958
350.0	0.7441	269.67	1.2233	0.7435	269.71	1.2230	0.7430	269.75	1.2226
360.0	31.777	432.72	1.6778	0.7597	279.08	1.2493	0.7591	279.11	1.2490
370.0	33.150	439.61	1.6967	0.7777	288.58	1.2754	0.7769	288.60	1.2750
380.0	34.470	446.44	1.7149	24.692	443.27	1.6912	0.7968	298.26	1.3008
390.0	35.749	453.22	1.7325	25.763	450.38	1.7097	19.681	447.17	1.6899
400.0	36.995	459.98	1.7496	26.788	457.41	1.7275	20.598	454.56	1.7086
410.0	38.214	466.73	1.7663	27.780	464.38	1.7447	21.468	461.82	1.7265
420.0	39.412	473.48	1.7825	28.743	471.31	1.7614	22.302	468.98	1.7438
430.0	40.592	480.24	1.7984	29.685	478.22	1.7776	23.109	476.08	1.7605
440.0	41.757	487.00	1.8140	30.608	485.12	1.7935	23.893	483.13	1.7767
450.0	42.908	493.77	1.8292	31.516	492.01	1.8090	24.659	490.16	1.7925
p/MPa	2.00			4.00			6.00		
230.0	0.6125	164.69	0.8546	0.6114	165.55	0.8531	0.6103	166.41	0.8515
240.0	0.6207	172.64	0.8885	0.6194	173.48	0.8868	0.6182	174.33	0.8852
250.0	0.6292	180.89	0.9221	0.6278	181.72	0.9204	0.6265	182.55	0.9187
260.0	0.6380	189.37	0.9554	0.6365	190.18	0.9536	0.6350	191.00	0.9518
270.0	0.6472	198.01	0.9880	0.6455	198.80	0.9861	0.6438	199.60	0.9843
280.0	0.6567	206.78	1.0199	0.6548	207.54	1.0179	0.6530	208.32	1.0160
290.0	0.6668	215.63	1.0509	0.6646	216.37	1.0489	0.6626	217.12	1.0469
300.0	0.6773	224.54	1.0812	0.6749	225.26	1.0790	0.6726	225.98	1.0770
310.0	0.6884	233.52	1.1106	0.6856	234.20	1.1083	0.6830	234.89	1.1062
320.0	0.7001	242.54	1.1392	0.6970	243.18	1.1368	0.6940	243.83	1.1346
330.0	0.7126	251.62	1.1672	0.7090	252.20	1.1646	0.7056	252.82	1.1622
340.0	0.7260	260.75	1.1944	0.7218	261.28	1.1917	0.7180	261.84	1.1891
350.0	0.7404	269.96	1.2211	0.7355	270.42	1.2182	0.7311	270.92	1.2154
360.0	0.7560	279.27	1.2473	0.7503	279.63	1.2442	0.7451	280.05	1.2412
370.0	0.7731	288.69	1.2732	0.7663	288.95	1.2697	0.7602	289.27	1.2664
380.0	0.7921	298.27	1.2987	0.7838	298.38	1.2948	0.7766	298.59	1.2913
390.0	0.8135	308.04	1.3241	0.8033	307.97	1.3198	0.7945	308.04	1.3158
400.0	0.8381	318.08	1.3495	0.8251	317.76	1.3445	0.8143	317.64	1.3401
410.0	0.8672	328.48	1.3752	0.8500	327.80	1.3693	0.8365	327.43	1.3643
420.0	0.9030	339.40	1.4015	0.8792	338.19	1.3944	0.8617	337.47	1.3885
430.0	9.5882	462.34	1.6939	0.9145	349.04	1.4199	0.8909	347.83	1.4129
440.0	10.188	470.97	1.7137	0.9594	360.60	1.4465	0.9256	358.61	1.4377
450.0	10.734	479.20	1.7322	1.0219	373.33	1.4751	0.9682	369.96	1.4632

Chart R11.3: Thermodynamic properties—*continued*

p/MPa	8.00			10.00			19.75		
T K	v dm³/kg	h kJ/kg	s kJ/kgK	v dm³/kg	h kJ/kg	s kJ/kgK	v dm³/kg	h kJ/kg	s kJ/kgK
230.0	0.6093	167.28	0.8499	0.6082	168.15	0.8484	0.6036	172.40	0.8412
240.0	0.6171	175.18	0.8836	0.6159	176.04	0.8820	0.6108	180.24	0.8746
250.0	0.6252	183.39	0.9171	0.6239	184.23	0.9154	0.6183	188.37	0.9078
260.0	0.6336	191.82	0.9501	0.6322	192.64	0.9484	0.6260	196.72	0.9405
270.0	0.6422	200.41	0.9825	0.6407	201.22	0.9808	0.6340	205.23	0.9726
280.0	0.6513	209.11	1.0142	0.6496	209.90	1.0124	0.6421	213.84	1.0040
290.0	0.6606	217.89	1.0450	0.6587	218.66	1.0431	0.6505	222.52	1.0344
300.0	0.6704	226.72	1.0749	0.6683	227.47	1.0730	0.6592	231.23	1.0640
310.0	0.6806	235.60	1.1040	0.6782	236.32	1.1020	0.6681	239.97	1.0926
320.0	0.6913	244.51	1.1323	0.6886	245.20	1.1302	0.6774	248.73	1.1204
330.0	0.7025	253.45	1.1598	0.6995	254.10	1.1576	0.6871	257.49	1.1474
340.0	0.7144	262.43	1.1866	0.7110	263.04	1.1842	0.6971	266.27	1.1736
350.0	0.7269	271.45	1.2128	0.7231	272.01	1.2102	0.7075	275.05	1.1990
360.0	0.7403	280.52	1.2383	0.7360	281.02	1.2356	0.7185	283.85	1.2238
370.0	0.7547	289.66	1.2634	0.7497	290.09	1.2605	0.7299	292.67	1.2480
380.0	0.7701	298.88	1.2880	0.7643	299.23	1.2849	0.7419	301.53	1.2716
390.0	0.7869	308.21	1.3122	0.7801	308.45	1.3088	0.7546	310.42	1.2947
400.0	0.8052	317.66	1.3361	0.7972	317.78	1.3324	0.7679	319.36	1.3173
410.0	0.8253	327.26	1.3598	0.8158	327.24	1.3558	0.7820	328.35	1.3395
420.0	0.8478	337.06	1.3835	0.8362	336.84	1.3789	0.7970	337.41	1.3614
430.0	0.8732	347.09	1.4071	0.8589	346.63	1.4020	0.8130	346.55	1.3829
440.0	0.9022	357.41	1.4308	0.8843	356.64	1.4250	0.8301	355.78	1.4041
450.0	0.9362	368.10	1.4548	0.9132	366.91	1.4481	0.8483	365.10	1.4250

References for R11

[Benn40] BENNING, A.F. AND MCHARNESS, R.C.: Thermodynamic Properties of Fluorochloromethanes and -Ethanes; Ind. Eng. Chem. 32(1940)4 497/499

[Benn40a] BENNING, A.F. AND MCHARNESS, R.C.: Thermodynamic Properties of Fluorochloromethanes and -Ethanes; Ind. Eng. Chem. 32(1940)5 698/701

[Benn40b] BENNING, A.F. AND MCHARNESS, R.C.: Thermodynamic Properties of Fluorochloromethanes and -Ethanes; Ind. Eng. Chem. 32(1940)6 814/816

[Benn40c] BENNING, A.F.; MCHARNESS, R.C.; MARKWOOD JR., W.H. AND SMITH, W.J.: Thermodynamic Properties of Fluorochloromethanes and -Ethanes; Ind. Eng. Chem. 32(1940)7 976/980

[Chav81] CHAVEZ, M.; TSUMURA, R. AND DEL RIO, F.: Speed of Sound in Saturated Liquid Trichlorofluoromethane; J. Chem. Eng. Data 26(1981)1 1/2

[Chen76] CHEN, S.S.; WILHOIT, R.C. AND ZWOLINSKI, B.J.: Ideal Gas Thermodynamic Properties of Six Chlorofluoromethanes; J. Phys. Chem. Ref. Data 5(1976)3 571/580

[Desr35] DESREUX, V.: Etude sur le Parachor; Bull. soc. chim. Belg. 44(1935) 249/287

[Doer77] DÖRING, R.: Thermophysical Properties of Ammonia (R717) and of Fluorinated Refrigerants R11 and R113; Proc. Symp. Thermophys. Prop. (1977)7 574/588

[Doer78] DÖRING, R.: Kaltron 11; Kali-Chemie Hannover (1978)

[Dvor74] DVORAK, Z. AND PETRAK, J.: Tepelne vlastnosti chladiva R11; CHT 8 Prag (1974)

[Fern84] FERNÁNDEZ-FASSNACHT, E. AND DEL RIO, F.: The vapor pressure of CCl_3F from 218 to 291 K; J. Chem. Therm. 16(1984) 1003/1006

[Gris71] GRISKOV, A.Y.: Experimental Investigation of the isobaric specific heat of Freon 11 at supercritical parameters of state; Thermophysical Properties of Matter and Substances 4(1971) 33/43

[Gruz75] GRUZDEV, V.A. AND SUMSKAJA, A.I.: Experimental Investigation of the Isobaric Heat Capacity of Freons: R-11, R-12, R-13, R-21, R-22, R-23; Thermophysical Properties of Matter and Substances 8(1975) 108/120

[Jaeg69] JÄGER, H.P.: Thermodynamische Eigenschaften binärer Gemische aus Kältemaschinenölen und Monofluortrichlormethan (R11) (1. Teil); Kältetechnik-Klimatisierung 21(1969)10 302/306

[Jaeg73] JÄGER, H.P.: Dichte- und Dampfdruckmeßwerte von verschiedenen Kältemitteln; Die Kälte (1973)7 276/278

[KMR 81] - : Kältemaschinenregeln, 7. Auflage; Verlag C.F. Müller, Karlsruhe (1981)

[Kolo68] KOLOMOV, Y.L.; SOLOVJEV, A.N. AND SELUDJAKOV, J.P.: Measurement of the speed of sound and calculation of entropy in Freon 11; Zhurnal Prikladnoi Mekhaniki i Tekhnicheskoi Fiziki 9(1968)2 141/145

[Koma67] KOMAROV, S.G.; SOLOVJEV, A.N. AND SELUDJAKOV, J.P: Speed of sound in Freons on the saturation line; Zhurnal Prikladnoi Mekhaniki i Tekhnicheskoi Fiziki 8(1967)5 116/119

[Krem75] KREMENEWSKAJA, E.A. AND RIVKIN, S.L.: Saturation Curve of Freon-11; Thermophysical Properties of Matter and Substances 8(1975) 40/45

[Krem75a] KREMENEWSKAJA, E.A. AND RIVKIN, S.L.: Thermodynamic Properties of Freon-11; Thermophysical Properties of Matter and Substances 8(1975) 46/64

[Lipp78] LIPPOLD, R.: ILKA-Berechnungskatalog: Stoffdaten von Kältemitteln und Kälteölen; VEB Kombinat Luft- und Kältetechnik (1978)

[Meye69] MEYER, K.J.: Über den Zusammenhang zwischen der Schallgeschwindigkeit und der Wärmeleitfähigkeit bei flüssigen Fluor-Chlor-Derivaten des Methans und Äthans; Kältetechnik-Klimatisierung 21(1969)9 270/275

[Nozd76] NOZDREV, V.F. AND REDKOZUBOV, V.A.: Study of the ultraaccustic properties of Freon 11 and Freon 12 along the saturation line in a wide temperature range; Primenenie ul'traakustiki Issledovaniju vescestva (1976) 3/5

[Oguc77] OGUCHI, K.: Physical properties of refrigerants. PVT relationship; Reito 52(1977) 869/889

[Osbo41] OSBORNE, D.W.; GARNER, C.S.; DOESCHER, R.N. AND YOST, D.M.: The Heat Capacity, Entropy, Heats of Fusion and Vaporization and Vapor Pressure of Fluorotrichloromethane; J. Am. Chem. Soc. 63(1941)12 3496/3499

[Pawl55] PAWLOWA, I.A.: Sammelwerk wissenschaftlicher Arbeiten; Cholodilnaja Technika Gostorgidat (1955) ref. in *[Doer78]*

[Redk76] REDKOZUBOV, V.A.: Rate of Propagation of Ultrasonic Waves in Freons 11, 12 and 13; Primenenie ul'traakustiki Issledovaniju vescestva (1976)26 14/19

[Ried39] RIEDEL, L.: Dampfdruckmessungen an Monofluor-Trichlormethan; Zeitschrift für die gesamte Kälteindustrie 46(1939)12 197/198

[Rivk71] RIVKIN, S.L. AND KREMNEVSKAJA, E.A.: Investigation of the Density of Freon-11; Thermophysical Properties of Matter and Substances 4(1971) 5/17

[Rivk72] RIVKIN, S.L. AND KREMNEVSKAJA, E.A.: Investigation of the Thermal Properties of Freon-11 vapor; Akademija Nauk Belorusskoj SSR.Vesci.Seryja fizika-enerhetyčnych navuk 4(1972) 59/64

[Romb67] ROMBUSCH, U.K. AND GIESEN, H.: Dampftafel des Kältemittels R11; Kältetechnik-Klimatisierung 19(1967)9 DKV-Arbeitsblatt 1-86

[Swar93] SWARTS, F.: ref. in *[Benn40b]*

[Varg75] VARGAFTIK, N.B.: Tables on the thermophysical properties of liquids and gases; Hemisphere Publishing Corporation Washington (1975)

R 11. p

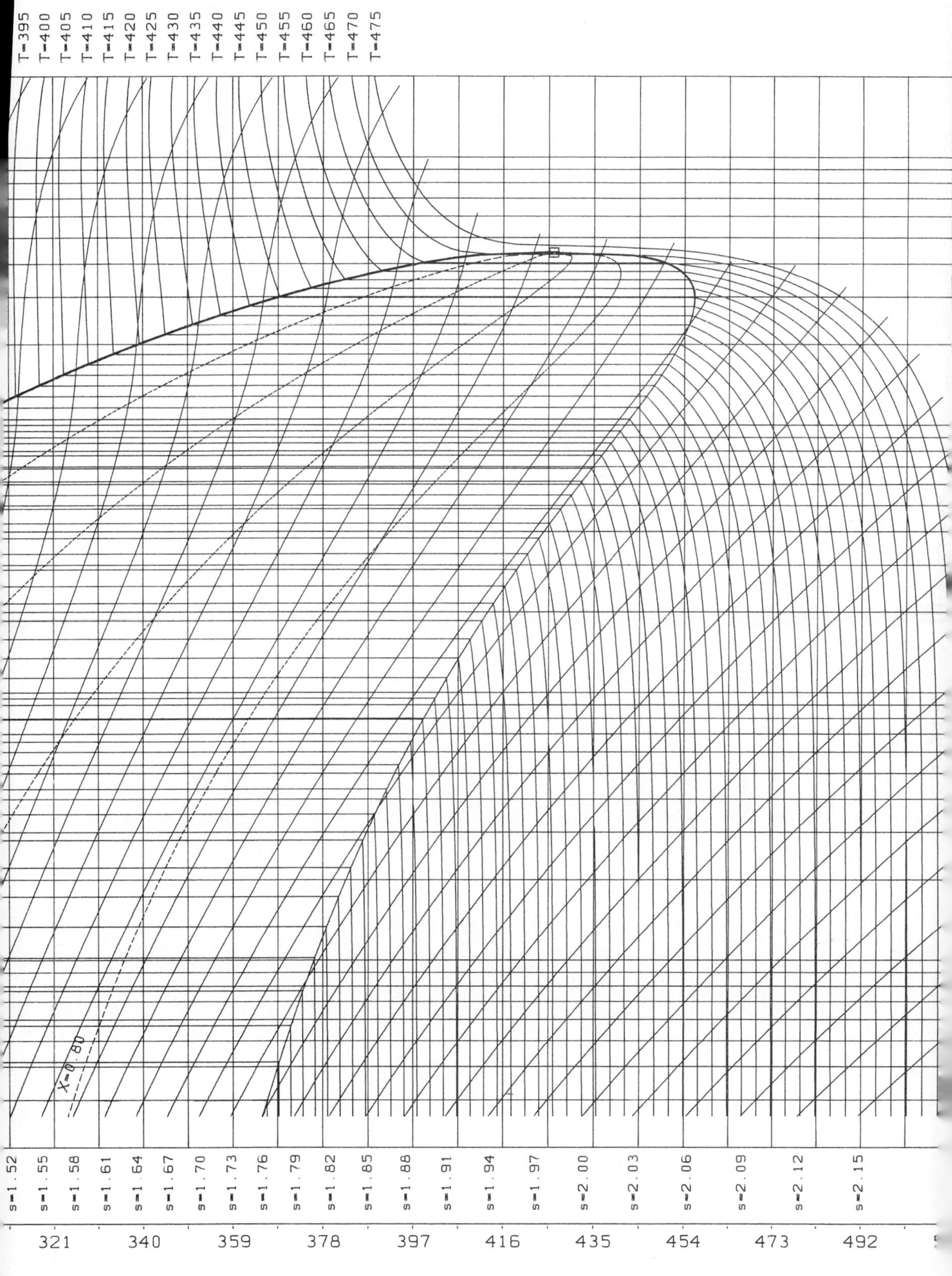

py h/(kJ/kg)

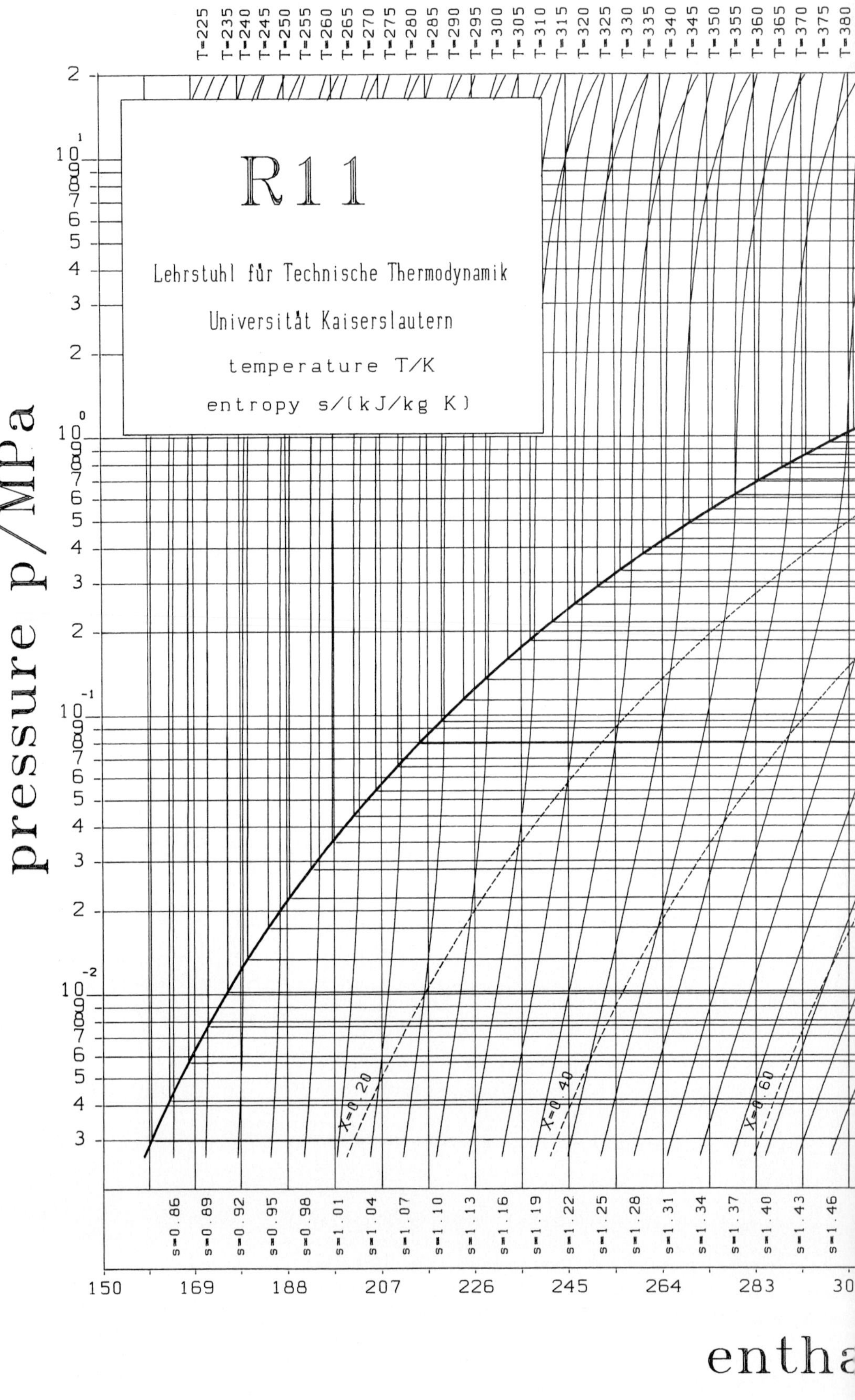

R11
Lehrstuhl für Technische Thermodynamik
Universität Kaiserslautern
temperature T/K
entropy s/(kJ/kg K)
pressure p/MPa
T=225 T=235 T=240 T=245 T=250 T=255 T=260 T=265 T=270 T=275 T=280 T=285 T=290 T=295 T=300 T=305 T=310 T=315 T=320 T=325 T=330 T=335 T=340 T=345 T=350 T=355 T=360 T=365 T=370 T=375 T=380
s=0.86 s=0.89 s=0.92 s=0.95 s=0.98 s=1.01 s=1.04 s=1.07 s=1.10 s=1.13 s=1.16 s=1.19 s=1.22 s=1.25 s=1.28 s=1.31 s=1.34 s=1.37 s=1.40 s=1.43 s=1.46
X=0.20
X=0.40
X=0.60
150 169 188 207 226 245 264 283 30
entha

R 11.Ts

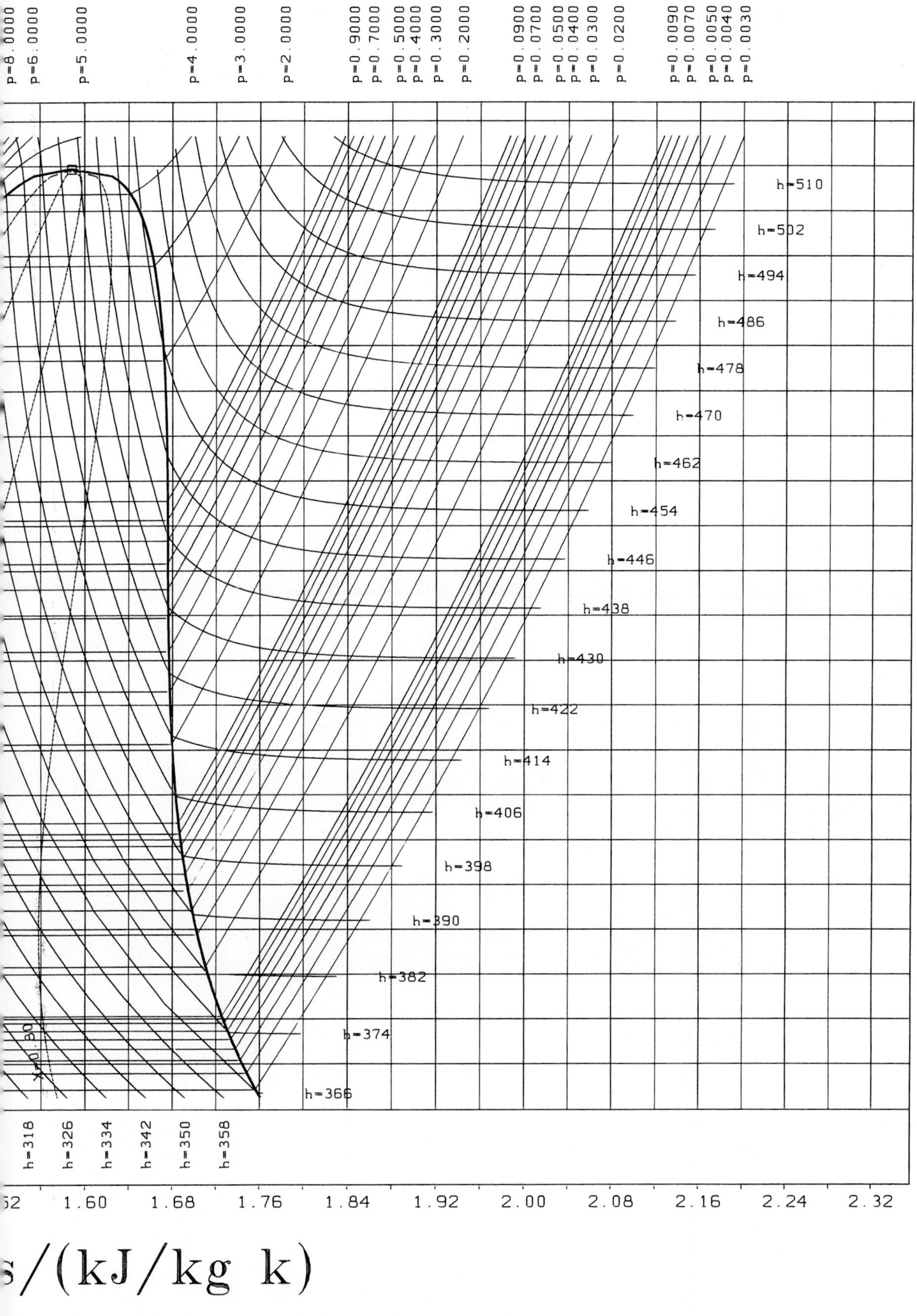

p=8.0000
p=6.0000
p=5.0000
p=4.0000
p=3.0000
p=2.0000
p=0.9000
p=0.7000
p=0.5000
p=0.4000
p=0.3000
p=0.2000
p=0.0900
p=0.0700
p=0.0500
p=0.0400
p=0.0300
p=0.0200
p=0.0090
p=0.0070
p=0.0050
p=0.0040
p=0.0030
h=510
h=502
h=494
h=486
h=478
h=470
h=462
h=454
h=446
h=438
h=430
h=422
h=414
h=406
h=398
h=390
h=382
h=374
h=366
x=0.30
h=318
h=326
h=334
h=342
h=350
h=358
1.60
1.68
1.76
1.84
1.92
2.00
2.08
2.16
2.24
2.32
s/(kJ/kg k)

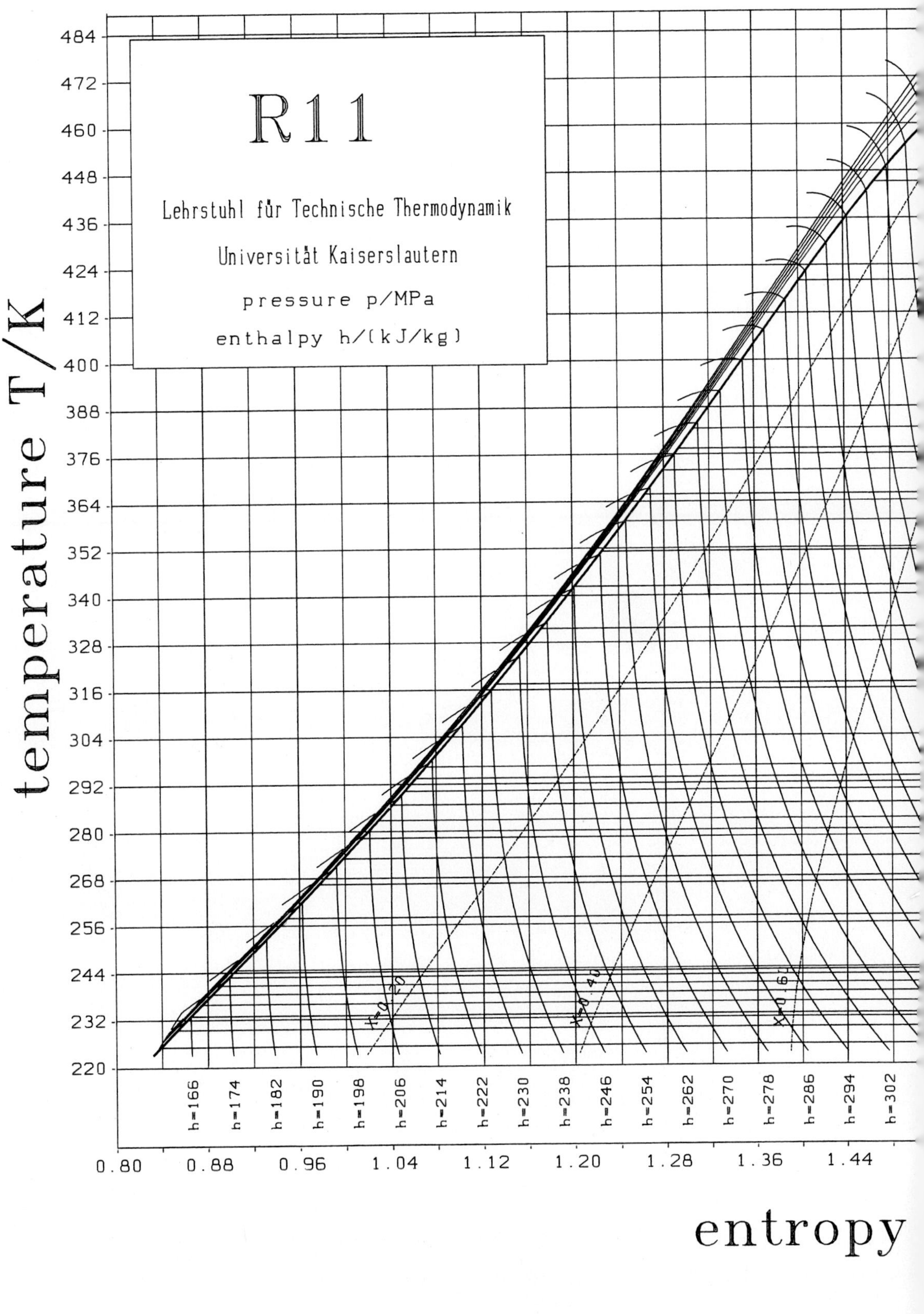

R11
Lehrstuhl für Technische Thermodynamik
Universität Kaiserslautern
pressure p/MPa
enthalpy h/(kJ/kg)
temperature T/K
entropy

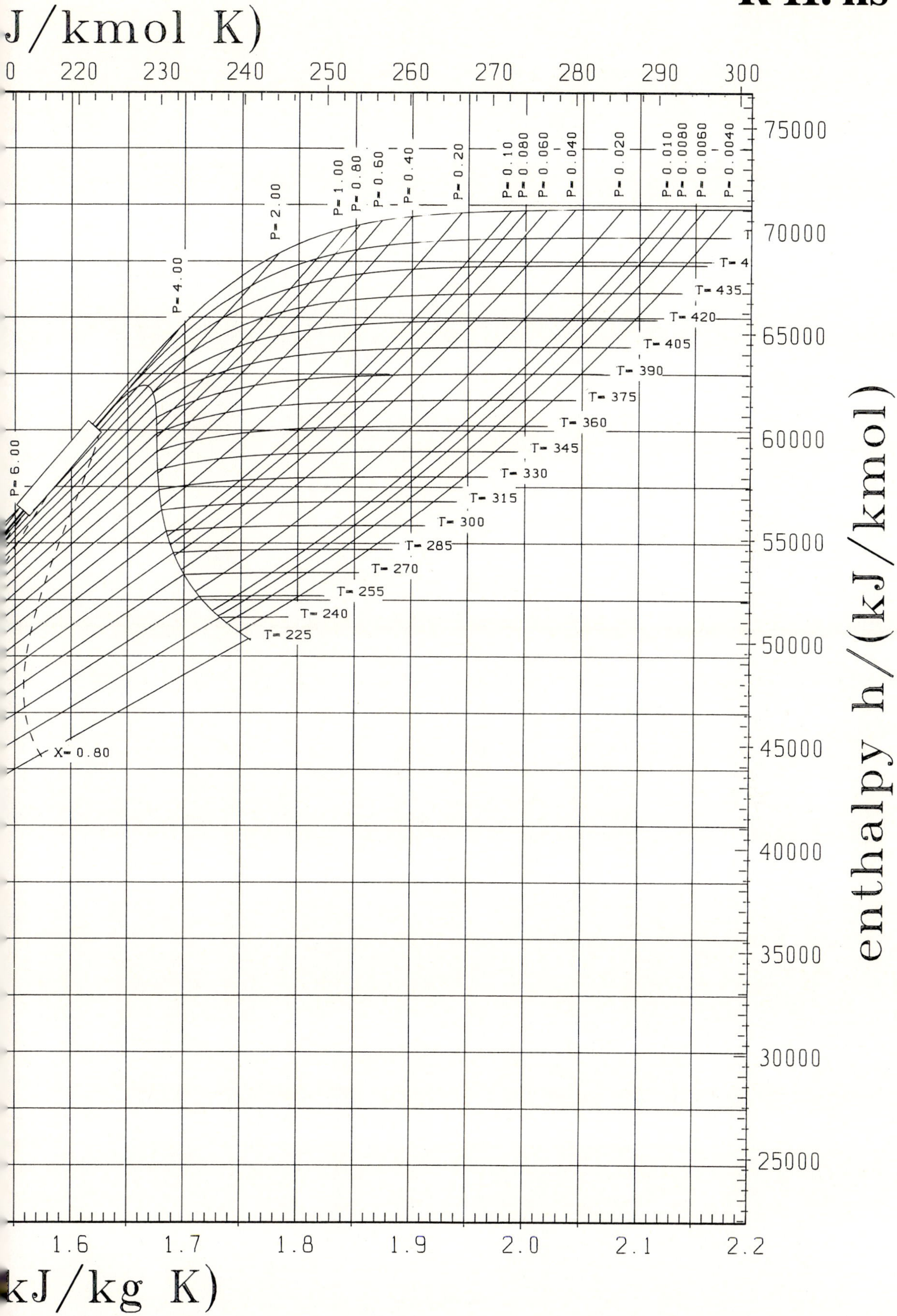
J/kmol K)
0
220
230
240
250
260
270
280
290
300
75000
70000
65000
60000
55000
50000
45000
40000
35000
30000
25000
enthalpy h/(kJ/kmol)
P= 6.00
P= 4.00
P= 2.00
P= 1.00
P= 0.80
P= 0.60
P= 0.40
P= 0.20
P= 0.10
P= 0.080
P= 0.060
P= 0.040
P= 0.020
P= 0.010
P= 0.0080
P= 0.0060
P= 0.0040
T= 4
T= 435
T= 420
T= 405
T= 390
T= 375
T= 360
T= 345
T= 330
T= 315
T= 300
T= 285
T= 270
T= 255
T= 240
T= 225
X= 0.80
1.6
1.7
1.8
1.9
2.0
2.1
2.2
kJ/kg K)

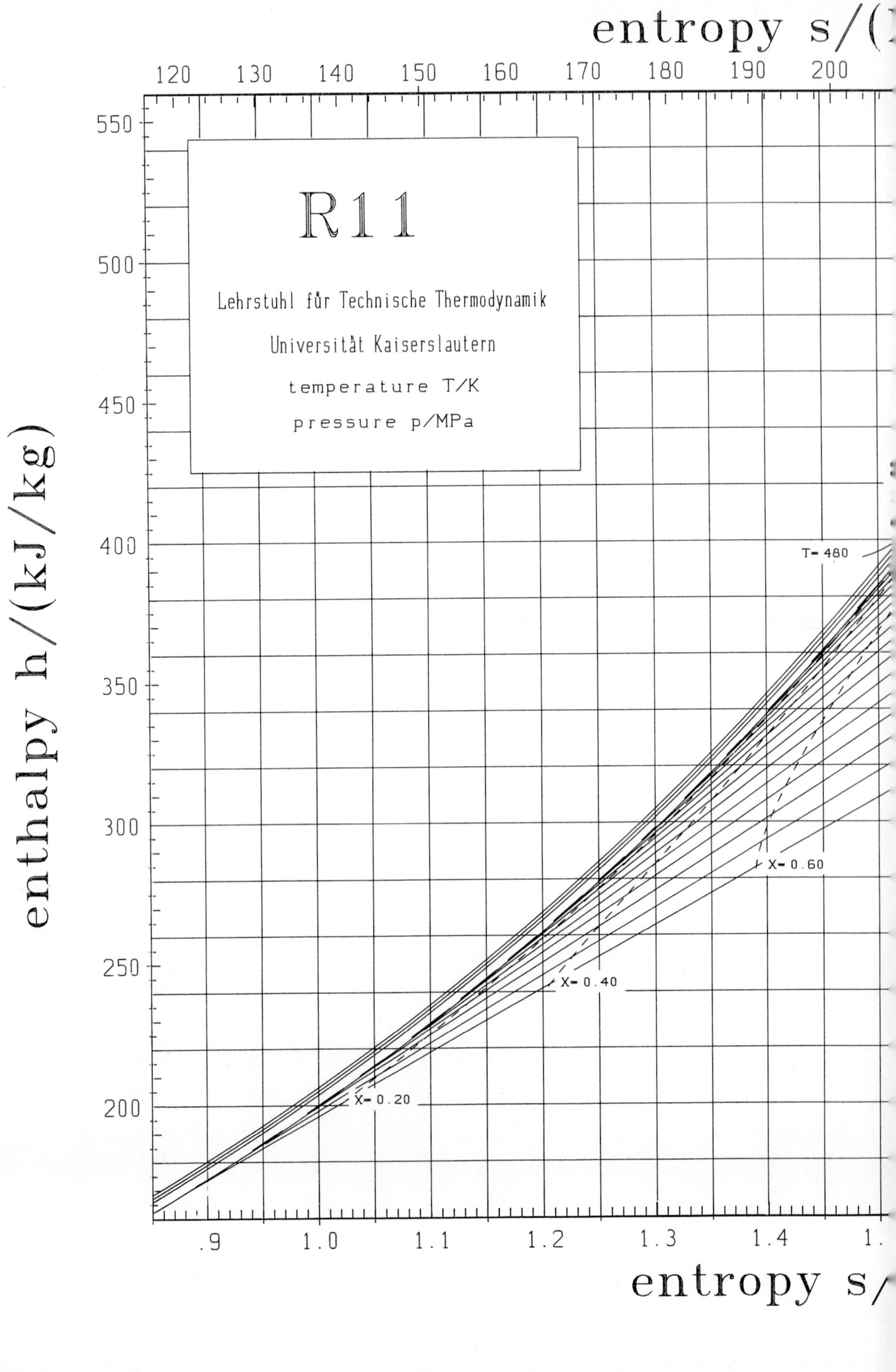
entropy s/(
120
130
140
150
160
170
180
190
200
550
500
450
400
350
300
250
200
enthalpy h/(kJ/kg)
R11
Lehrstuhl für Technische Thermodynamik
Universität Kaiserslautern
temperature T/K
pressure p/MPa
T= 480
X= 0.60
X= 0.40
X= 0.20
.9
1.0
1.1
1.2
1.3
1.4
1.
entropy s/

Refrigerant R12 (CCl_2F_2)

Table R12.1: List of references for p, ρ, T data

Author	n	p/MPa min	p/MPa max	ρ/g cm^{-3} min	ρ/g cm^{-3} max	T/K min	T/K max
BUFFINGTON, R.M. *[Buff 31]*	29	0.55	1.88	0.0289	0.08	303	399
KELLS, L.F. *[Kell 55]*	73	0.86	6.43	0.0439	0.48	326	474
KUBOTA, H. *[Kubo 74]*	40	0.10	1.80	0.0396	0.10	303	348
KUMAGAI, A. *[Kuma 78]*	37	9.65	160.82	1.3127	1.70	253	313
MCHARNESS, R.C. *[McHa 55]*	141	0.84	13.65	0.0463	0.98	310	483
MICHELS, A. *[Mich 66]*	186	0.71	8.17	0.0363	0.70	323	423
OGUCHI, K. *[Oguc 79]*	19	0.88	29.47	1.2399	1.36	293	353
PERELSTEIN, I.I. *[Pere 71]*	71	0.81	8.50	0.0389	0.64	340	471
WATANABE, K. *[Wata 77]*	104	0.11	10.86	0.0049	0.99	273	403

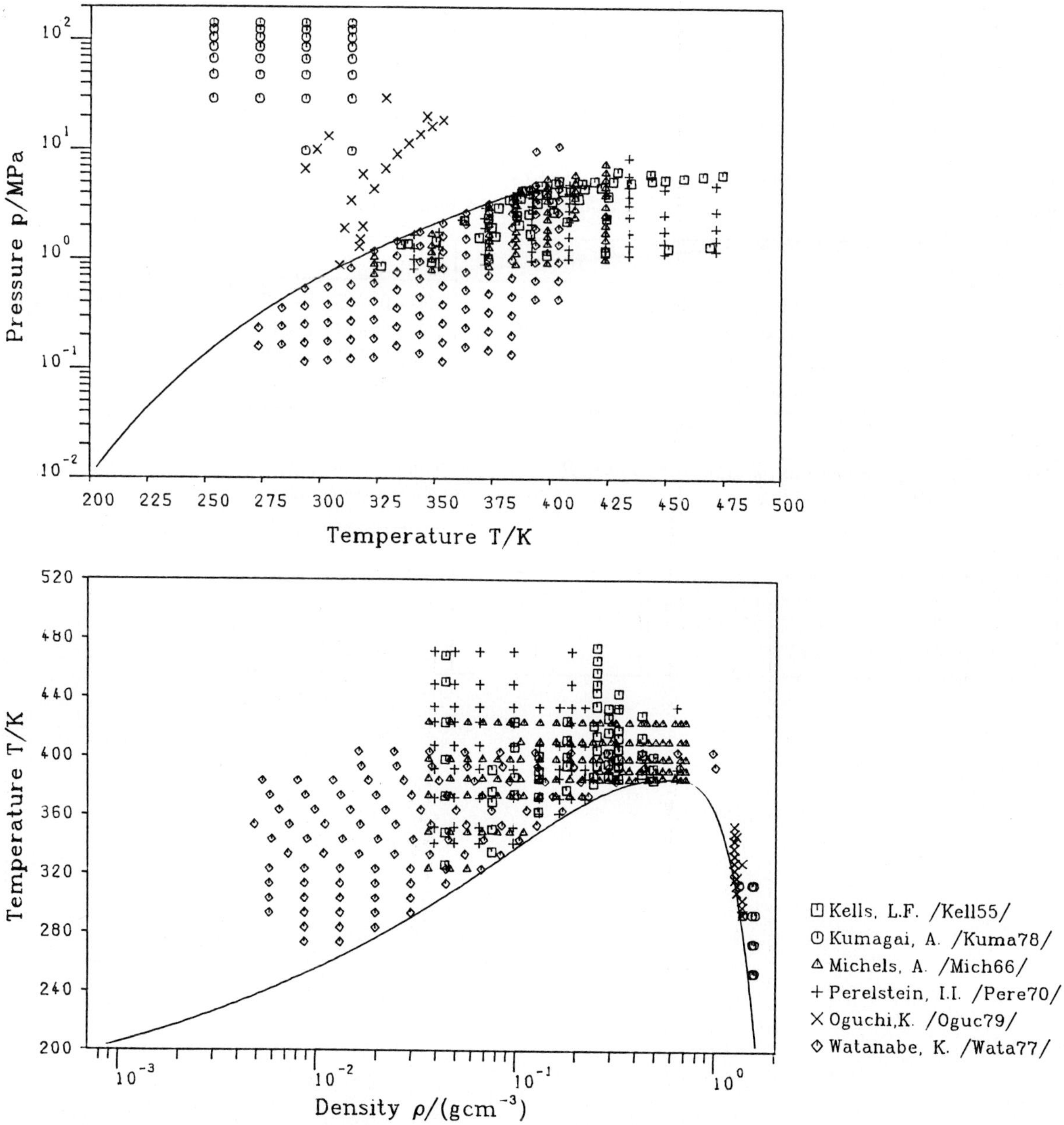

Fig.R12.1: Distribution of p, ρ, T data projected in the p, T and in the T, ρ plane

Table R12.2: Experimental vapor pressure data

Author	Δ / %		
	std	opt	*EOS*
GILKEY, W.K. *[Gilk 31]*	0.17	0.11	0.21
JÄGER, H.P. *[Jaeg 73]*	0.34	0.36	0.35
KELLS, L.F. *[Kell 55]*	0.70	0.68	0.68
MCHARNESS, R.C. *[McHa 55]*	0.37	0.30	0.52
MICHELS, A. *[Mich 66]*	0.16	0.14	0.11
WATANABE, K. *[Wata 77]*	0.05	0.06	0.05

Table R12.3: Experimental saturated liquid density data

Author	Δ / %		
	std	opt	*EOS*
JÄGER, H.P. *[Jaeg 73]*	0.17	0.16	0.15
MCHARNESS, R.C. *[McHa 55]*	0.10	0.10	0.15

Table R12.4: Coefficients of the standard vapor pressure equation

Index i	b_i
5	−6.909344 E+00
6	1.512469 E+00
9	−2.663723 E+00
15	−1.491543 E+00

Table R12.5: Coefficients of the quasi-optimal vapor pressure equation

Index i	b_i
5	−7.072553 E+00
6	2.654128 E+00
7	−2.217211 E+00
12	−2.798711 E+00

Table R12.6: Coefficients of the standard saturated liquid density equation

Index i	c_i
8	1.665115 E+00
9	8.735025 E−01
16	2.672822 E−01

Table R12.7: Coefficients of the quasi-optimal saturated liquid density equation

Index i	c_i
8	1.659521 E+00
9	8.861593 E−01
17	2.957567 E−01

Table R12.8: Coefficients of the *Bender-EOS*

Range of validity: one-phase region

$0.11 \leq p/\text{MPa} \leq 160.82$

$0.00486 \leq \rho/\text{g cm}^{-3} \leq 1.70413$

$253.15 \leq T/\text{K} \leq 473.81$

	Specific units	Molar units
a_{01}	1.763566032 E−01	2.578837147 E+00
a_{02}	1.923446709 E+02	2.812628353 E+03
a_{03}	−6.159685370 E+04	−9.007218986 E+05
a_{04}	1.677564138 E+07	2.453077819 E+08
a_{05}	−1.145276316 E+09	−1.674721022 E+10
a_{06}	−1.011324086 E−01	−1.788292844 E−01
a_{07}	8.797085662 E+00	1.555561225 E+01
a_{08}	1.965561180 E+03	3.475640541 E+03
a_{09}	5.088451389 E−01	1.088052909 E−01
a_{10}	−2.066358831 E+02	−4.418451832 E+01
a_{11}	−2.161867623 E−01	−5.589971295 E−03
a_{12}	1.357733272 E+02	3.510709876 E+00
a_{13}	−3.242385410 E+00	−1.013820837 E−02
a_{14}	3.228988726 E+07	5.709720073 E+07
a_{15}	−8.965654616 E+09	−1.585368748 E+10
a_{16}	1.393324389 E+11	2.463772068 E+11
a_{17}	−3.574632682 E+07	−9.242977632 E+05
a_{18}	2.424361787 E+10	6.268706119 E+08
a_{19}	−1.143925061 E+12	−2.957863000 E+10
a_{20}	2.050000000 E+00	2.997685403 E−02

Table R12.9: Deviations between experimental and predicted p, ρ, T data

Region	n	Δ / %	
		p	ρ
Gas	338	0.22	0.30
Critical	90	0.18	3.24
Liquid	55	2.01	0.09
Total	483	0.72	1.45

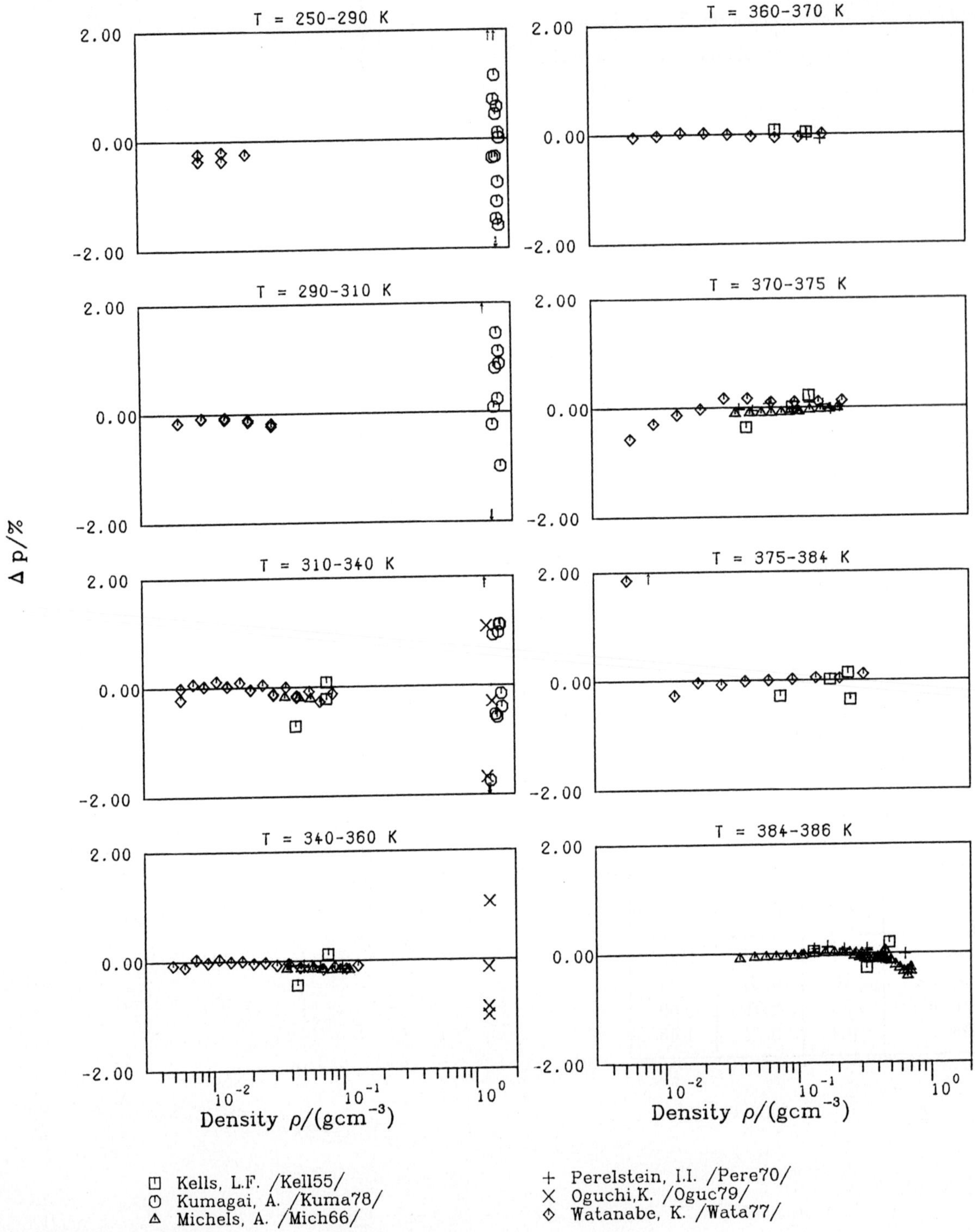

Fig.R12.2: Deviations between experimental and calculated p, ρ, T data

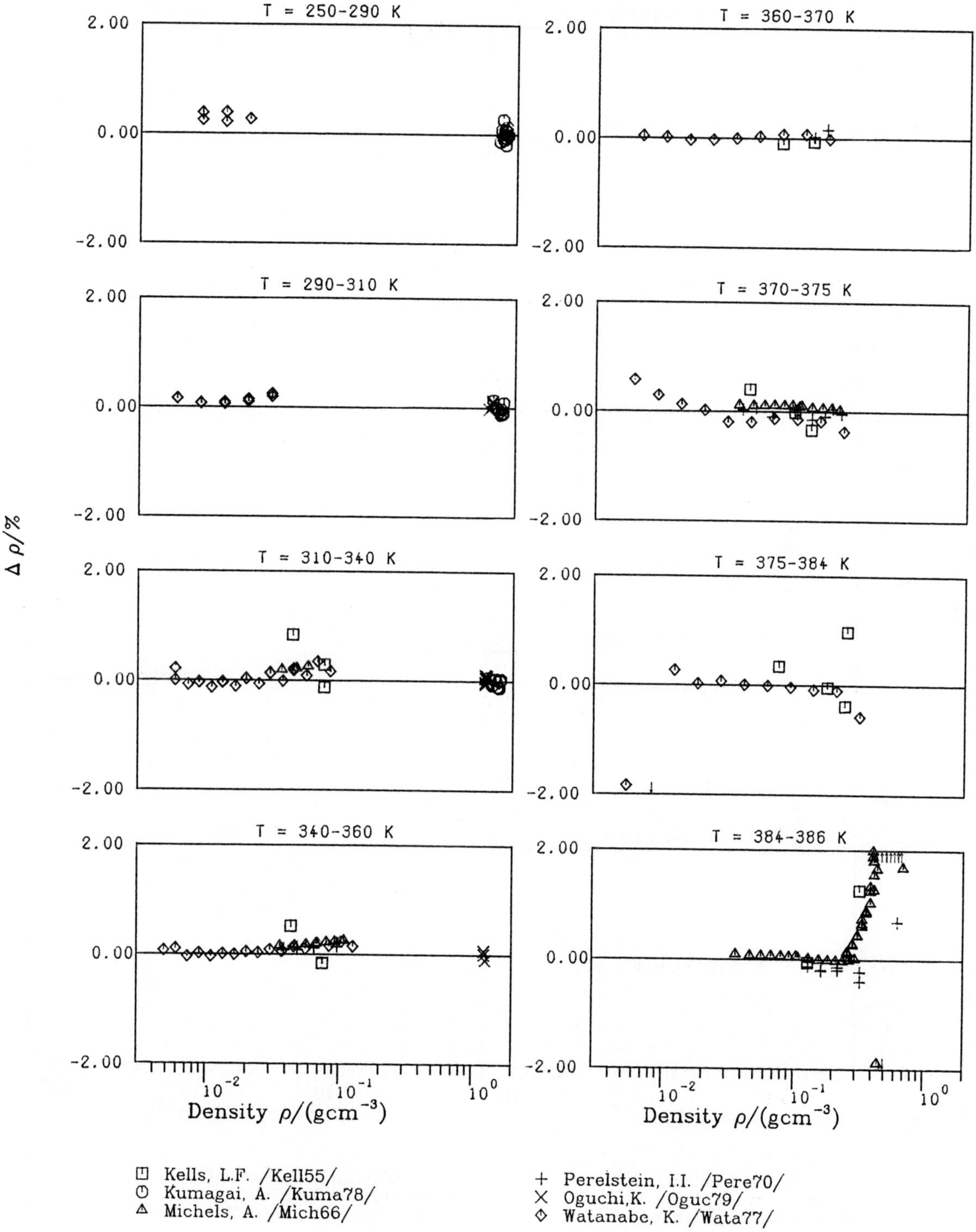

Fig.R12.2: Deviations between experimental and calculated p, ρ, T data *-continued*

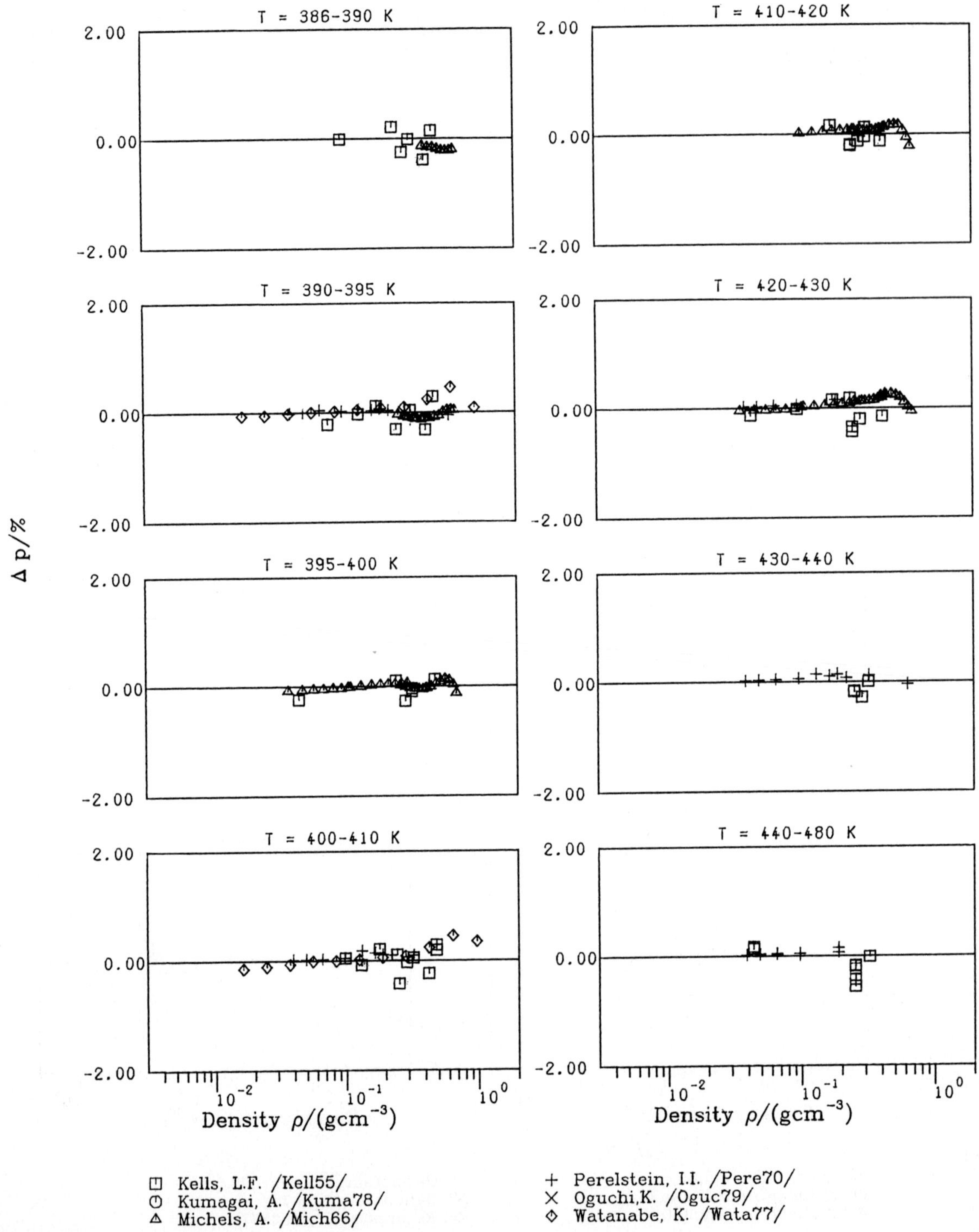

Fig.R12.2: Deviations between experimental and calculated p, ρ, T data *-continued*

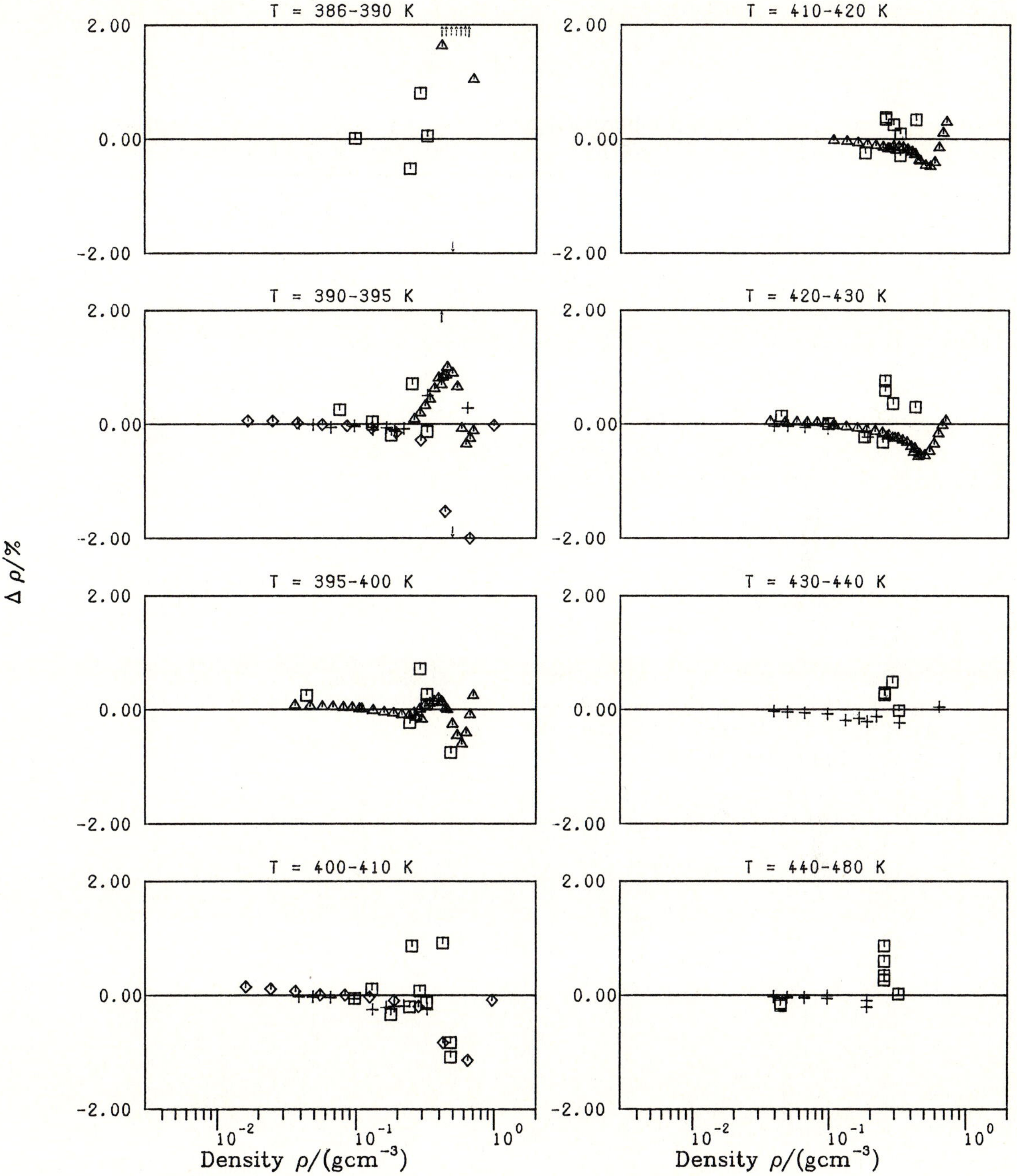

□ Kells, L.F. /Kell55/
○ Kumagai, A. /Kuma78/
△ Michels, A. /Mich66/
+ Perelstein, I.I. /Pere70/
× Oguchi,K. /Oguc79/
◇ Watanabe, K. /Wata77/

Fig.R12.2: Deviations between experimental and calculated p, ρ, T data *-continued*

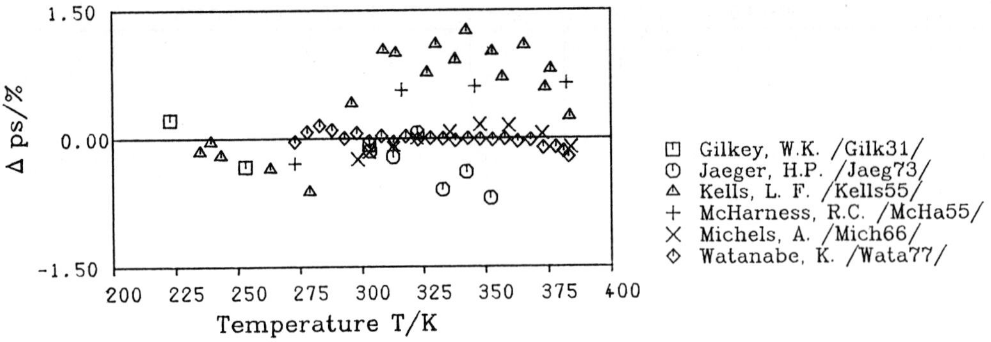

Fig.R12.3: Deviations between experimental and calculated vapor pressures

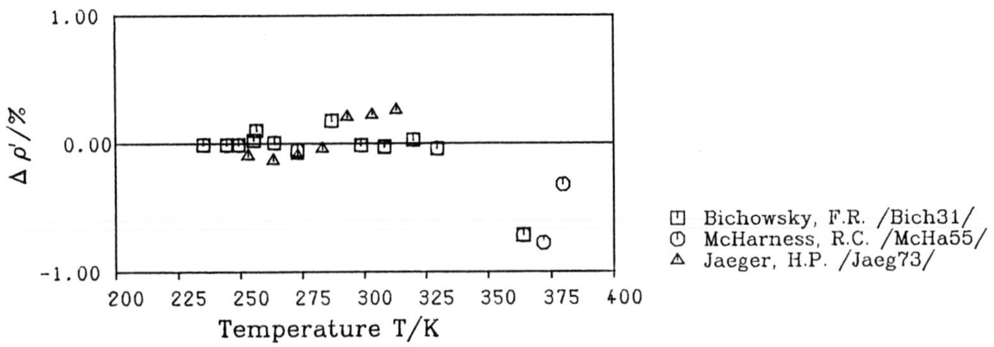

Fig.R12.4: Deviations between experimental and calculated saturated liquid densities

Table R12.10: Information on ideal gas state heat capacity equation

Author	T / K min	T / K max
BARHO, W. *[Barh 66]*	200	550

Coefficients of equation (15)	
Index i	d_i
1	9.042906 E−02
2	2.689024 E−03
3	−4.122018 E−06
4	3.242061 E−09
5	−1.066667 E−12

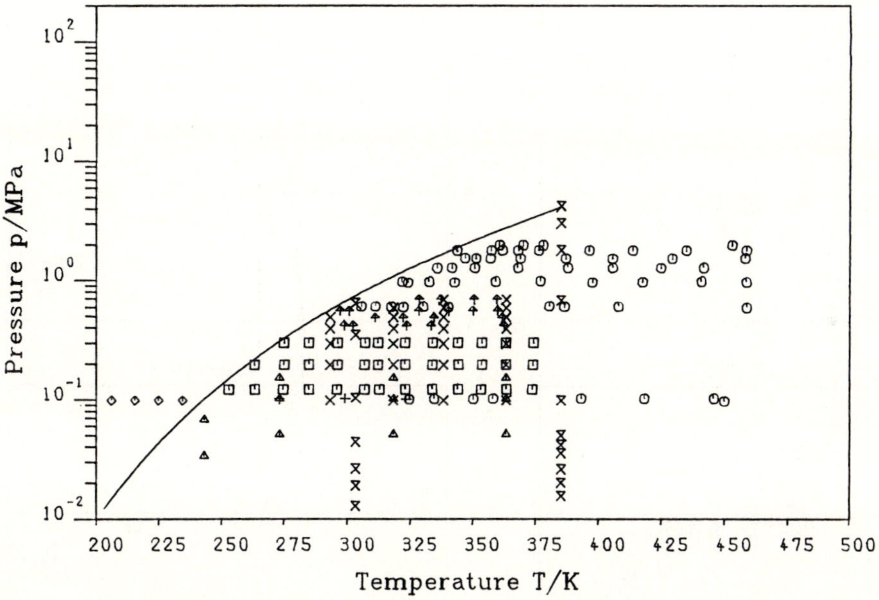

□ cp: Tanishita, I. /Tani72/ ○ cp: Gruzdev, V.A. /Gruz75/ △ cp: Masi, J. /Masi52/ + cp: Buffington, R.M. /Buff31a/ × cp: Ernst, G. /Erns69/ ◇ w: Meyer, K.J. /Meye69/ ↑ w: Woodburn, J. /Wood66/ ⌛ w: Redkozubov, V.A. /Redk76/

Fig.R12.5: Distribution of $c_p(p,T)$ and $w(p,T)$ data

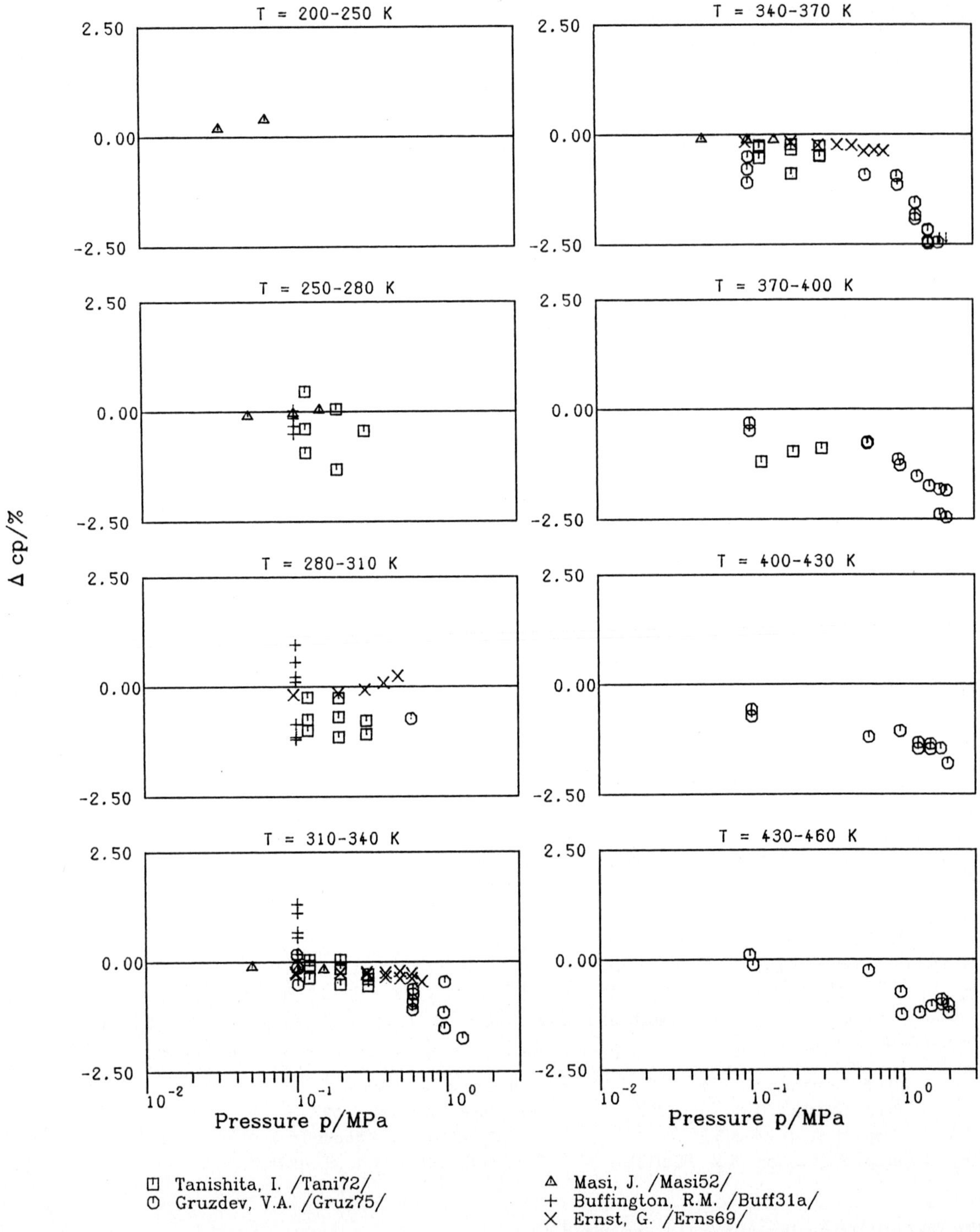

Fig.R12.6: Deviations between experimental and calculated specific isobaric heat capacities c_p

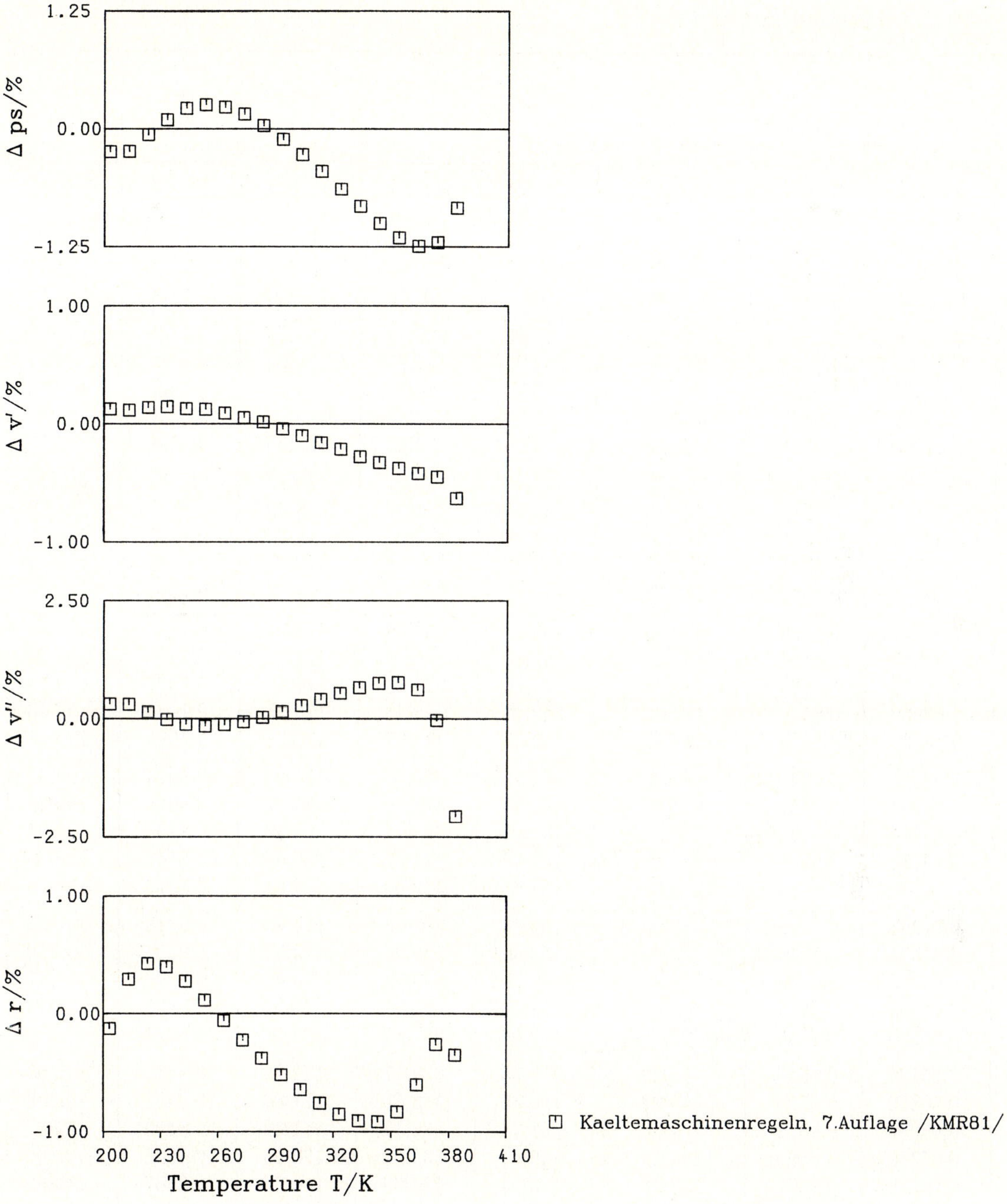

Fig.R12.7: Deviations between saturation properties from the *Kältemaschinenregeln* and those calculated by the *Bender-EOS*

Chart R12.1: Saturation properties

T K	p MPa	v' dm^3/kg	v'' dm^3/kg	h' kJ/kg	h'' kJ/kg	r kJ/kg	s' kJ/kgK	s'' kJ/kgK
200.0	0.00995	0.6224	1373.0	135.21	317.87	182.66	0.7264	1.6397
201.0	0.01063	0.6234	1290.6	135.94	318.34	182.41	0.7300	1.6375
202.0	0.01136	0.6244	1213.9	136.67	318.82	182.14	0.7337	1.6354
203.0	0.01212	0.6254	1142.6	137.42	319.29	181.87	0.7373	1.6333
204.0	0.01293	0.6265	1076.3	138.17	319.76	181.59	0.7410	1.6312
205.0	0.01378	0.6275	1014.4	138.93	320.24	181.30	0.7448	1.6292
206.0	0.01467	0.6286	956.71	139.70	320.71	181.01	0.7485	1.6272
207.0	0.01561	0.6296	902.89	140.48	321.18	180.70	0.7523	1.6252
208.0	0.01661	0.6307	852.64	141.27	321.66	180.39	0.7560	1.6233
209.0	0.01765	0.6317	805.69	142.06	322.13	180.08	0.7598	1.6215
210.0	0.01875	0.6328	761.79	142.86	322.61	179.75	0.7637	1.6196
211.0	0.01990	0.6339	720.73	143.67	323.09	179.42	0.7675	1.6178
212.0	0.02111	0.6350	682.28	144.48	323.56	179.08	0.7713	1.6161
213.0	0.02238	0.6361	646.27	145.30	324.04	178.74	0.7752	1.6144
214.0	0.02371	0.6372	612.52	146.12	324.52	178.39	0.7791	1.6127
215.0	0.02511	0.6383	580.86	146.96	325.00	178.04	0.7829	1.6110
216.0	0.02657	0.6394	551.15	147.79	325.47	177.68	0.7868	1.6094
217.0	0.02810	0.6405	523.25	148.63	325.95	177.32	0.7907	1.6078
218.0	0.02969	0.6417	497.04	149.48	326.43	176.95	0.7946	1.6063
219.0	0.03137	0.6428	472.40	150.33	326.91	176.58	0.7985	1.6047
220.0	0.03311	0.6439	449.22	151.19	327.39	176.20	0.8024	1.6033
221.0	0.03494	0.6451	427.39	152.05	327.86	175.82	0.8062	1.6018
222.0	0.03684	0.6463	406.84	152.91	328.34	175.43	0.8101	1.6004
223.0	0.03883	0.6474	387.48	153.78	328.82	175.04	0.8140	1.5990
224.0	0.04090	0.6486	369.22	154.65	329.30	174.65	0.8179	1.5976
225.0	0.04306	0.6498	352.00	155.53	329.78	174.25	0.8218	1.5963
226.0	0.04531	0.6510	335.74	156.40	330.26	173.85	0.8257	1.5950
227.0	0.04765	0.6522	320.39	157.29	330.74	173.45	0.8296	1.5937
228.0	0.05009	0.6534	305.88	158.17	331.21	173.04	0.8335	1.5924
229.0	0.05263	0.6546	292.17	159.06	331.69	172.63	0.8374	1.5912
230.0	0.05526	0.6558	279.20	159.95	332.17	172.22	0.8412	1.5900
231.0	0.05800	0.6570	266.92	160.85	332.65	171.80	0.8451	1.5888
232.0	0.06085	0.6582	255.30	161.74	333.12	171.38	0.8490	1.5877
233.0	0.06381	0.6595	244.29	162.64	333.60	170.96	0.8528	1.5866
234.0	0.06688	0.6607	233.86	163.54	334.08	170.54	0.8567	1.5855
235.0	0.07006	0.6620	223.96	164.45	334.55	170.11	0.8605	1.5844
236.0	0.07336	0.6632	214.57	165.35	335.03	169.68	0.8644	1.5834
237.0	0.07678	0.6645	205.67	166.26	335.51	169.25	0.8682	1.5823
238.0	0.08033	0.6658	197.21	167.17	335.98	168.81	0.8720	1.5813
239.0	0.08400	0.6671	189.17	168.08	336.46	168.38	0.8758	1.5803
240.0	0.08780	0.6683	181.53	168.99	336.93	167.94	0.8796	1.5794
241.0	0.09174	0.6696	174.27	169.91	337.41	167.50	0.8834	1.5784
242.0	0.09581	0.6710	167.36	170.83	337.88	167.05	0.8872	1.5775
243.0	0.10002	0.6723	160.78	171.75	338.35	166.61	0.8910	1.5766
244.0	0.10438	0.6736	154.52	172.67	338.82	166.16	0.8948	1.5757
245.0	0.10888	0.6749	148.56	173.59	339.30	165.71	0.8985	1.5749
246.0	0.11353	0.6763	142.88	174.51	339.77	165.26	0.9023	1.5740
247.0	0.11833	0.6776	137.46	175.43	340.24	164.80	0.9060	1.5732
248.0	0.12328	0.6790	132.29	176.36	340.71	164.35	0.9097	1.5724
249.0	0.12840	0.6804	127.36	177.29	341.18	163.89	0.9134	1.5716

Chart R12.1: Saturation properties–*continued*

T K	p MPa	v' dm^3/kg	v'' dm^3/kg	h' kJ/kg	h'' kJ/kg	r kJ/kg	s' kJ/kgK	s'' kJ/kgK
250.0	0.13368	0.6817	122.66	178.22	341.64	163.43	0.9172	1.5709
251.0	0.13912	0.6831	118.16	179.15	342.11	162.97	0.9208	1.5701
252.0	0.14473	0.6845	113.87	180.08	342.58	162.50	0.9245	1.5694
253.0	0.15051	0.6859	109.77	181.01	343.04	162.04	0.9282	1.5687
254.0	0.15647	0.6874	105.85	181.94	343.51	161.57	0.9319	1.5680
255.0	0.16261	0.6888	102.10	182.88	343.97	161.10	0.9355	1.5673
256.0	0.16893	0.6902	98.514	183.81	344.44	160.63	0.9392	1.5666
257.0	0.17544	0.6917	95.081	184.75	344.90	160.15	0.9428	1.5660
258.0	0.18214	0.6931	91.794	185.68	345.36	159.68	0.9464	1.5653
259.0	0.18903	0.6946	88.647	186.62	345.82	159.20	0.9500	1.5647
260.0	0.19612	0.6961	85.631	187.56	346.28	158.72	0.9536	1.5641
261.0	0.20341	0.6976	82.740	188.50	346.74	158.23	0.9572	1.5635
262.0	0.21090	0.6991	79.969	189.44	347.19	157.75	0.9608	1.5629
263.0	0.21860	0.7006	77.312	190.38	347.65	157.26	0.9644	1.5623
264.0	0.22651	0.7021	74.762	191.33	348.10	156.78	0.9679	1.5618
265.0	0.23464	0.7037	72.315	192.27	348.56	156.28	0.9715	1.5612
266.0	0.24298	0.7052	69.966	193.22	349.01	155.79	0.9750	1.5607
267.0	0.25155	0.7068	67.710	194.16	349.46	155.30	0.9785	1.5602
268.0	0.26034	0.7084	65.543	195.11	349.91	154.80	0.9821	1.5597
269.0	0.26937	0.7099	63.461	196.06	350.36	154.30	0.9856	1.5592
270.0	0.27863	0.7116	61.459	197.01	350.80	153.80	0.9891	1.5587
271.0	0.28812	0.7132	59.534	197.95	351.25	153.29	0.9925	1.5582
272.0	0.29786	0.7148	57.683	198.91	351.69	152.79	0.9960	1.5577
273.0	0.30785	0.7164	55.902	199.86	352.13	152.28	0.9995	1.5573
274.0	0.31808	0.7181	54.187	200.81	352.58	151.77	1.0029	1.5568
275.0	0.32856	0.7198	52.537	201.76	353.01	151.25	1.0064	1.5564
276.0	0.33931	0.7214	50.948	202.72	353.45	150.73	1.0098	1.5560
277.0	0.35031	0.7231	49.417	203.67	353.89	150.21	1.0133	1.5555
278.0	0.36158	0.7249	47.943	204.63	354.32	149.69	1.0167	1.5551
279.0	0.37312	0.7266	46.521	205.59	354.76	149.17	1.0201	1.5547
280.0	0.38493	0.7283	45.151	206.55	355.19	148.64	1.0235	1.5543
281.0	0.39702	0.7301	43.829	207.51	355.62	148.11	1.0269	1.5539
282.0	0.40939	0.7319	42.555	208.47	356.04	147.57	1.0303	1.5536
283.0	0.42205	0.7337	41.325	209.43	356.47	147.04	1.0336	1.5532
284.0	0.43500	0.7355	40.138	210.40	356.89	146.50	1.0370	1.5528
285.0	0.44823	0.7373	38.992	211.36	357.32	145.95	1.0404	1.5525
286.0	0.46177	0.7391	37.886	212.33	357.74	145.41	1.0437	1.5521
287.0	0.47561	0.7410	36.817	213.30	358.15	144.86	1.0470	1.5518
288.0	0.48975	0.7429	35.785	214.27	358.57	144.30	1.0504	1.5514
289.0	0.50421	0.7448	34.787	215.24	358.98	143.75	1.0537	1.5511
290.0	0.51898	0.7467	33.823	216.21	359.39	143.19	1.0570	1.5508
291.0	0.53406	0.7486	32.891	217.18	359.80	142.62	1.0603	1.5505
292.0	0.54947	0.7506	31.989	218.16	360.21	142.06	1.0636	1.5501
293.0	0.56521	0.7526	31.118	219.13	360.62	141.48	1.0669	1.5498
294.0	0.58128	0.7546	30.274	220.11	361.02	140.91	1.0702	1.5495
295.0	0.59769	0.7566	29.458	221.09	361.42	140.33	1.0735	1.5492
296.0	0.61443	0.7586	28.667	222.07	361.82	139.75	1.0768	1.5489
297.0	0.63152	0.7607	27.902	223.05	362.21	139.16	1.0801	1.5486
298.0	0.64896	0.7628	27.161	224.04	362.60	138.57	1.0833	1.5483
299.0	0.66675	0.7649	26.444	225.02	362.99	137.97	1.0866	1.5480

Chart R12.1: Saturation properties–*continued*

T K	p MPa	v' dm³/kg	v'' dm³/kg	h' kJ/kg	h'' kJ/kg	r kJ/kg	s' kJ/kgK	s'' kJ/kgK
300.0	0.68490	0.7670	25.748	226.01	363.38	137.37	1.0898	1.5477
301.0	0.70341	0.7692	25.075	227.00	363.77	136.76	1.0931	1.5475
302.0	0.72229	0.7714	24.422	228.00	364.15	136.15	1.0963	1.5472
303.0	0.74154	0.7736	23.789	228.99	364.53	135.54	1.0996	1.5469
304.0	0.76116	0.7758	23.175	229.99	364.91	134.92	1.1028	1.5466
305.0	0.78117	0.7781	22.579	230.98	365.28	134.29	1.1060	1.5463
306.0	0.80156	0.7804	22.002	231.98	365.65	133.66	1.1093	1.5461
307.0	0.82234	0.7827	21.441	232.99	366.02	133.03	1.1125	1.5458
308.0	0.84351	0.7850	20.897	233.99	366.38	132.39	1.1157	1.5455
309.0	0.86508	0.7874	20.369	235.00	366.74	131.74	1.1189	1.5452
310.0	0.88706	0.7898	19.857	236.01	367.10	131.09	1.1221	1.5450
311.0	0.90944	0.7923	19.359	237.02	367.45	130.43	1.1253	1.5447
312.0	0.93224	0.7947	18.876	238.04	367.81	129.77	1.1285	1.5444
313.0	0.95545	0.7972	18.406	239.05	368.15	129.10	1.1317	1.5442
314.0	0.97909	0.7998	17.949	240.07	368.50	128.42	1.1349	1.5439
315.0	1.00315	0.8023	17.506	241.09	368.84	127.74	1.1381	1.5436
316.0	1.02764	0.8049	17.075	242.12	369.17	127.05	1.1413	1.5433
317.0	1.05257	0.8076	16.655	243.15	369.51	126.36	1.1445	1.5431
318.0	1.07795	0.8103	16.248	244.18	369.83	125.66	1.1476	1.5428
319.0	1.10377	0.8130	15.851	245.21	370.16	124.95	1.1508	1.5425
320.0	1.13004	0.8158	15.466	246.25	370.48	124.23	1.1540	1.5422
321.0	1.15677	0.8186	15.090	247.29	370.80	123.51	1.1572	1.5419
322.0	1.18396	0.8214	14.725	248.33	371.11	122.77	1.1604	1.5416
323.0	1.21162	0.8243	14.369	249.38	371.42	122.03	1.1635	1.5413
324.0	1.23975	0.8272	14.023	250.43	371.72	121.29	1.1667	1.5410
325.0	1.26836	0.8302	13.686	251.49	372.02	120.53	1.1699	1.5407
326.0	1.29744	0.8332	13.358	252.54	372.31	119.77	1.1731	1.5404
327.0	1.32702	0.8363	13.038	253.60	372.60	118.99	1.1762	1.5401
328.0	1.35709	0.8394	12.726	254.67	372.88	118.21	1.1794	1.5398
329.0	1.38766	0.8426	12.423	255.74	373.16	117.42	1.1826	1.5395
330.0	1.41873	0.8458	12.127	256.81	373.43	116.62	1.1858	1.5391
331.0	1.45032	0.8491	11.838	257.89	373.70	115.81	1.1889	1.5388
332.0	1.48241	0.8525	11.557	258.97	373.96	114.99	1.1921	1.5385
333.0	1.51503	0.8559	11.282	260.06	374.22	114.16	1.1953	1.5381
334.0	1.54817	0.8593	11.015	261.15	374.46	113.32	1.1985	1.5378
335.0	1.58185	0.8629	10.754	262.24	374.71	112.47	1.2017	1.5374
336.0	1.61606	0.8665	10.499	263.34	374.94	111.60	1.2049	1.5370
337.0	1.65082	0.8702	10.250	264.45	375.17	110.73	1.2081	1.5366
338.0	1.68612	0.8739	10.007	265.56	375.40	109.84	1.2113	1.5362
339.0	1.72198	0.8777	9.7697	266.67	375.61	108.94	1.2145	1.5358
340.0	1.75841	0.8816	9.5380	267.79	375.82	108.03	1.2177	1.5354
341.0	1.79540	0.8856	9.3117	268.92	376.02	107.10	1.2209	1.5350
342.0	1.83296	0.8897	9.0905	270.05	376.22	106.17	1.2241	1.5345
343.0	1.87110	0.8938	8.8743	271.19	376.40	105.21	1.2273	1.5341
344.0	1.90983	0.8981	8.6630	272.33	376.58	104.24	1.2305	1.5336
345.0	1.94916	0.9024	8.4564	273.48	376.75	103.26	1.2338	1.5331
346.0	1.98908	0.9069	8.2543	274.64	376.91	102.26	1.2370	1.5326
347.0	2.02961	0.9114	8.0566	275.81	377.05	101.25	1.2403	1.5321
348.0	2.07076	0.9161	7.8631	276.98	377.19	100.22	1.2435	1.5315
349.0	2.11252	0.9209	7.6737	278.16	377.32	99.17	1.2468	1.5310

Chart R12.1: Saturation properties—*continued*

T K	p MPa	v' dm³/kg	v'' dm³/kg	h' kJ/kg	h'' kJ/kg	r kJ/kg	s' kJ/kgK	s'' kJ/kgK
350.0	2.15492	0.9258	7.4882	279.34	377.44	98.10	1.2501	1.5304
351.0	2.19794	0.9308	7.3066	280.54	377.55	97.01	1.2534	1.5298
352.0	2.24162	0.9360	7.1287	281.74	377.64	95.90	1.2567	1.5292
353.0	2.28594	0.9413	6.9543	282.95	377.73	94.78	1.2600	1.5285
354.0	2.33092	0.9468	6.7833	284.17	377.80	93.62	1.2634	1.5278
355.0	2.37657	0.9524	6.6157	285.40	377.85	92.45	1.2667	1.5271
356.0	2.42290	0.9583	6.4512	286.64	377.90	91.25	1.2701	1.5264
357.0	2.46990	0.9643	6.2898	287.90	377.92	90.03	1.2734	1.5256
358.0	2.51761	0.9705	6.1313	289.16	377.94	88.78	1.2768	1.5248
359.0	2.56601	0.9769	5.9756	290.43	377.93	87.50	1.2803	1.5240
360.0	2.61513	0.9835	5.8226	291.72	377.91	86.19	1.2837	1.5231
361.0	2.66496	0.9904	5.6721	293.02	377.87	84.85	1.2872	1.5222
362.0	2.71554	0.9975	5.5241	294.33	377.80	83.47	1.2907	1.5213
363.0	2.76685	1.0049	5.3784	295.66	377.72	82.06	1.2942	1.5203
364.0	2.81892	1.0126	5.2349	297.00	377.61	80.61	1.2977	1.5192
365.0	2.87175	1.0207	5.0935	298.36	377.48	79.12	1.3013	1.5181
366.0	2.92537	1.0291	4.9539	299.74	377.32	77.58	1.3050	1.5169
367.0	2.97977	1.0380	4.8161	301.14	377.13	75.99	1.3086	1.5157
368.0	3.03499	1.0472	4.6800	302.56	376.92	74.36	1.3123	1.5144
369.0	3.09102	1.0570	4.5452	304.00	376.66	72.66	1.3161	1.5130
370.0	3.14789	1.0673	4.4117	305.47	376.37	70.90	1.3199	1.5115
371.0	3.20561	1.0782	4.2793	306.97	376.04	69.07	1.3238	1.5099
372.0	3.26421	1.0899	4.1477	308.50	375.66	67.16	1.3277	1.5083
373.0	3.32369	1.1023	4.0168	310.07	375.23	65.16	1.3318	1.5064
374.0	3.38409	1.1156	3.8861	311.67	374.74	63.06	1.3359	1.5045
375.0	3.44542	1.1301	3.7553	313.32	374.18	60.85	1.3401	1.5024
376.0	3.50772	1.1458	3.6241	315.03	373.54	58.51	1.3445	1.5001
377.0	3.57101	1.1630	3.4919	316.80	372.81	56.01	1.3490	1.4975
378.0	3.63533	1.1822	3.3580	318.65	371.98	53.33	1.3537	1.4947
379.0	3.70070	1.2038	3.2216	320.59	371.01	50.41	1.3586	1.4916
380.0	3.76719	1.2285	3.0813	322.65	369.87	47.21	1.3638	1.4880
381.0	3.83483	1.2575	2.9354	324.88	368.51	43.63	1.3694	1.4839
382.0	3.90371	1.2928	2.7808	327.34	366.85	39.51	1.3756	1.4791
383.0	3.97388	1.3380	2.6124	330.15	364.74	34.59	1.3828	1.4731
384.0	4.06133	1.3684	1.5495	332.30	338.99	6.69	1.3881	1.4055

Chart R12.2: Saturation properties

p MPa	T K	v' dm³/kg	v'' dm³/kg	h' kJ/kg	h'' kJ/kg	r kJ/kg	s' kJ/kgK	s'' kJ/kgK
0.0102	200.37	0.6228	1341.5	135.48	318.05	182.57	0.7277	1.6389
0.0104	200.66	0.6230	1317.5	135.69	318.19	182.49	0.7288	1.6382
0.0106	200.95	0.6233	1294.4	135.90	318.32	182.42	0.7298	1.6376
0.0108	201.24	0.6236	1272.0	136.11	318.45	182.35	0.7309	1.6370
0.0110	201.51	0.6239	1250.5	136.31	318.59	182.27	0.7319	1.6364
0.0115	202.19	0.6246	1199.9	136.81	318.91	182.09	0.7344	1.6350
0.0120	202.85	0.6253	1153.3	137.30	319.22	181.92	0.7368	1.6336
0.0125	203.48	0.6259	1110.3	137.77	319.51	181.74	0.7391	1.6323
0.0130	204.09	0.6266	1070.6	138.24	319.80	181.57	0.7414	1.6310
0.0135	204.68	0.6272	1033.7	138.69	320.08	181.40	0.7436	1.6298
0.0140	205.25	0.6278	999.31	139.13	320.36	181.23	0.7457	1.6287
0.0145	205.81	0.6284	967.24	139.56	320.62	181.06	0.7478	1.6275
0.0150	206.35	0.6289	937.24	139.98	320.88	180.90	0.7498	1.6265
0.0155	206.88	0.6295	909.11	140.39	321.13	180.74	0.7518	1.6255
0.0160	207.39	0.6300	882.68	140.79	321.37	180.58	0.7538	1.6245
0.0165	207.89	0.6306	857.80	141.18	321.61	180.43	0.7556	1.6235
0.0170	208.38	0.6311	834.33	141.57	321.84	180.27	0.7575	1.6226
0.0175	208.86	0.6316	812.16	141.95	322.07	180.12	0.7593	1.6217
0.0180	209.32	0.6321	791.18	142.32	322.29	179.97	0.7611	1.6209
0.0185	209.78	0.6326	771.30	142.68	322.51	179.82	0.7628	1.6200
0.0190	210.22	0.6331	752.43	143.04	322.72	179.68	0.7645	1.6192
0.0195	210.66	0.6335	734.49	143.39	322.92	179.54	0.7662	1.6184
0.0200	211.08	0.6340	717.42	143.73	323.13	179.39	0.7678	1.6177
0.0210	211.91	0.6349	685.63	144.41	323.52	179.12	0.7710	1.6162
0.0220	212.71	0.6358	656.64	145.06	323.90	178.84	0.7741	1.6149
0.0230	213.47	0.6366	630.09	145.69	324.27	178.58	0.7770	1.6136
0.0240	214.21	0.6374	605.67	146.30	324.62	178.32	0.7799	1.6123
0.0250	214.93	0.6382	583.15	146.89	324.96	178.07	0.7826	1.6111
0.0260	215.62	0.6390	562.29	147.47	325.29	177.82	0.7853	1.6100
0.0270	216.29	0.6397	542.93	148.03	325.61	177.58	0.7879	1.6089
0.0280	216.94	0.6405	524.91	148.58	325.92	177.34	0.7904	1.6079
0.0290	217.57	0.6412	508.08	149.12	326.22	177.11	0.7929	1.6069
0.0300	218.19	0.6419	492.34	149.64	326.52	176.88	0.7953	1.6060
0.0310	218.78	0.6426	477.58	150.15	326.81	176.66	0.7976	1.6051
0.0320	219.37	0.6432	463.71	150.65	327.08	176.44	0.7999	1.6042
0.0330	219.94	0.6439	450.65	151.13	327.36	176.22	0.8021	1.6034
0.0340	220.49	0.6445	438.33	151.61	327.62	176.01	0.8043	1.6025
0.0350	221.03	0.6451	426.69	152.08	327.88	175.81	0.8064	1.6018
0.0360	221.56	0.6457	415.68	152.53	328.13	175.60	0.8084	1.6010
0.0370	222.08	0.6463	405.24	152.98	328.38	175.40	0.8105	1.6003
0.0380	222.59	0.6469	395.33	153.42	328.62	175.20	0.8124	1.5996
0.0390	223.08	0.6475	385.91	153.85	328.86	175.01	0.8144	1.5989
0.0400	223.57	0.6481	376.94	154.28	329.09	174.82	0.8163	1.5982
0.0410	224.05	0.6486	368.40	154.69	329.32	174.63	0.8181	1.5976
0.0420	224.51	0.6492	360.25	155.10	329.55	174.45	0.8199	1.5969
0.0430	224.97	0.6497	352.46	155.50	329.77	174.26	0.8217	1.5963
0.0440	225.42	0.6503	345.02	155.90	329.98	174.08	0.8235	1.5957
0.0450	225.86	0.6508	337.90	156.29	330.19	173.91	0.8252	1.5952
0.0460	226.30	0.6513	331.07	156.67	330.40	173.73	0.8269	1.5946
0.0470	226.72	0.6518	324.52	157.04	330.60	173.56	0.8285	1.5940

Chart R12.2: Saturation properties–*continued*

p MPa	T K	v' dm³/kg	v'' dm³/kg	h' kJ/kg	h'' kJ/kg	r kJ/kg	s' kJ/kgK	s'' kJ/kgK
0.0480	227.14	0.6523	318.24	157.41	330.80	173.39	0.8302	1.5935
0.0490	227.56	0.6528	312.21	157.78	331.00	173.22	0.8318	1.5930
0.0500	227.96	0.6533	306.40	158.14	331.20	173.06	0.8333	1.5925
0.0510	228.36	0.6538	300.82	158.49	331.39	172.89	0.8349	1.5920
0.0520	228.76	0.6543	295.44	158.84	331.58	172.73	0.8364	1.5915
0.0530	229.14	0.6547	290.26	159.19	331.76	172.57	0.8379	1.5910
0.0540	229.53	0.6552	285.27	159.53	331.94	172.41	0.8394	1.5906
0.0550	229.90	0.6557	280.44	159.86	332.12	172.26	0.8409	1.5901
0.0560	230.27	0.6561	275.79	160.19	332.30	172.10	0.8423	1.5897
0.0570	230.64	0.6566	271.29	160.52	332.47	171.95	0.8437	1.5893
0.0580	231.00	0.6570	266.94	160.84	332.65	171.80	0.8451	1.5888
0.0590	231.35	0.6574	262.74	161.16	332.82	171.65	0.8465	1.5884
0.0600	231.70	0.6579	258.67	161.48	332.98	171.51	0.8478	1.5880
0.0610	232.05	0.6583	254.72	161.79	333.15	171.36	0.8492	1.5876
0.0620	232.39	0.6587	250.90	162.09	333.31	171.22	0.8505	1.5873
0.0630	232.73	0.6591	247.20	162.40	333.47	171.07	0.8518	1.5869
0.0640	233.06	0.6595	243.61	162.70	333.63	170.93	0.8531	1.5865
0.0650	233.39	0.6600	240.12	162.99	333.79	170.79	0.8544	1.5861
0.0660	233.72	0.6604	236.74	163.29	333.94	170.66	0.8556	1.5858
0.0670	234.04	0.6608	233.45	163.58	334.10	170.52	0.8568	1.5854
0.0680	234.36	0.6612	230.26	163.86	334.25	170.38	0.8581	1.5851
0.0690	234.67	0.6616	227.16	164.15	334.40	170.25	0.8593	1.5848
0.0700	234.98	0.6619	224.14	164.43	334.55	170.12	0.8605	1.5844
0.0710	235.29	0.6623	221.20	164.71	334.69	169.99	0.8616	1.5841
0.0720	235.59	0.6627	218.34	164.98	334.84	169.86	0.8628	1.5838
0.0730	235.89	0.6631	215.56	165.25	334.98	169.73	0.8640	1.5835
0.0740	236.19	0.6635	212.85	165.52	335.12	169.60	0.8651	1.5832
0.0760	236.77	0.6642	207.63	166.05	335.40	169.35	0.8673	1.5826
0.0780	237.35	0.6649	202.68	166.58	335.67	169.10	0.8695	1.5820
0.0800	237.91	0.6657	197.96	167.09	335.94	168.85	0.8717	1.5814
0.0820	238.46	0.6664	193.46	167.59	336.20	168.61	0.8738	1.5809
0.0840	239.00	0.6671	189.17	168.08	336.46	168.38	0.8758	1.5803
0.0860	239.53	0.6677	185.07	168.56	336.71	168.14	0.8778	1.5798
0.0880	240.05	0.6684	181.16	169.04	336.96	167.91	0.8798	1.5793
0.0900	240.56	0.6691	177.40	169.51	337.20	167.69	0.8818	1.5788
0.0920	241.06	0.6697	173.81	169.97	337.44	167.47	0.8837	1.5784
0.0940	241.56	0.6704	170.36	170.42	337.67	167.25	0.8855	1.5779
0.0960	242.05	0.6710	167.05	170.87	337.90	167.03	0.8874	1.5775
0.0980	242.52	0.6716	163.88	171.31	338.13	166.82	0.8892	1.5770
0.1000	242.99	0.6723	160.82	171.74	338.35	166.61	0.8910	1.5766
0.1020	243.46	0.6729	157.88	172.17	338.57	166.40	0.8927	1.5762
0.1040	243.91	0.6735	155.05	172.59	338.78	166.20	0.8944	1.5758
0.1060	244.36	0.6741	152.32	173.00	339.00	165.99	0.8961	1.5754
0.1080	244.81	0.6747	149.68	173.41	339.21	165.80	0.8978	1.5750
0.1100	245.24	0.6753	147.14	173.81	339.41	165.60	0.8994	1.5747
0.1150	246.31	0.6767	141.17	174.80	339.91	165.12	0.9034	1.5738
0.1200	247.34	0.6781	135.67	175.75	340.40	164.65	0.9073	1.5729
0.1250	248.34	0.6795	130.59	176.67	340.87	164.19	0.9110	1.5721
0.1300	249.31	0.6808	125.90	177.57	341.32	163.75	0.9146	1.5714
0.1350	250.25	0.6821	121.53	178.44	341.76	163.31	0.9181	1.5707

Chart R12.2: Saturation properties–*continued*

p MPa	T K	v' dm^3/kg	v'' dm^3/kg	h' kJ/kg	h'' kJ/kg	r kJ/kg	s' kJ/kgK	s'' kJ/kgK
0.1400	251.16	0.6834	117.47	179.29	342.19	162.89	0.9214	1.5700
0.1450	252.05	0.6846	113.67	180.12	342.60	162.48	0.9247	1.5693
0.1500	252.91	0.6858	110.12	180.93	343.00	162.08	0.9279	1.5687
0.1550	253.76	0.6870	106.79	181.71	343.39	161.68	0.9310	1.5681
0.1600	254.58	0.6882	103.66	182.48	343.78	161.30	0.9340	1.5676
0.1650	255.38	0.6893	100.71	183.23	344.15	160.92	0.9369	1.5670
0.1700	256.17	0.6905	97.934	183.97	344.51	160.55	0.9398	1.5665
0.1750	256.93	0.6916	95.306	184.68	344.87	160.18	0.9426	1.5660
0.1800	257.68	0.6927	92.818	185.39	345.21	159.83	0.9453	1.5655
0.1850	258.42	0.6938	90.460	186.08	345.55	159.48	0.9479	1.5651
0.1900	259.14	0.6948	88.221	186.75	345.88	159.13	0.9505	1.5646
0.1950	259.84	0.6959	86.092	187.41	346.21	158.79	0.9531	1.5642
0.2000	260.54	0.6969	84.066	188.06	346.52	158.46	0.9556	1.5638
0.2100	261.88	0.6989	80.291	189.33	347.14	157.81	0.9604	1.5630
0.2200	263.18	0.7009	76.847	190.55	347.73	157.18	0.9650	1.5622
0.2300	264.43	0.7028	73.690	191.74	348.30	156.56	0.9695	1.5615
0.2400	265.65	0.7047	70.787	192.88	348.85	155.97	0.9738	1.5609
0.2500	266.82	0.7065	68.107	193.99	349.38	155.39	0.9779	1.5603
0.2600	267.96	0.7083	65.625	195.07	349.89	154.82	0.9819	1.5597
0.2700	269.07	0.7101	63.320	196.12	350.39	154.27	0.9858	1.5591
0.2800	270.15	0.7118	61.173	197.14	350.87	153.72	0.9896	1.5586
0.2900	271.19	0.7135	59.168	198.14	351.33	153.19	0.9932	1.5581
0.3000	272.22	0.7151	57.292	199.11	351.79	152.68	0.9968	1.5576
0.3100	273.21	0.7168	55.532	200.06	352.23	152.17	1.0002	1.5572
0.3200	274.19	0.7184	53.877	200.99	352.66	151.67	1.0036	1.5567
0.3300	275.14	0.7200	52.319	201.89	353.07	151.18	1.0069	1.5563
0.3400	276.06	0.7215	50.849	202.78	353.48	150.70	1.0100	1.5559
0.3500	276.97	0.7231	49.460	203.65	353.88	150.23	1.0132	1.5556
0.3600	277.86	0.7246	48.144	204.50	354.26	149.76	1.0162	1.5552
0.3700	278.73	0.7261	46.897	205.33	354.64	149.31	1.0192	1.5548
0.3800	279.59	0.7276	45.713	206.15	355.01	148.86	1.0221	1.5545
0.3900	280.42	0.7291	44.587	206.95	355.37	148.42	1.0249	1.5542
0.4000	281.24	0.7305	43.516	207.74	355.72	147.98	1.0277	1.5539
0.4100	282.05	0.7319	42.494	208.52	356.06	147.55	1.0304	1.5536
0.4200	282.84	0.7334	41.519	209.28	356.40	147.12	1.0331	1.5533
0.4300	283.62	0.7348	40.588	210.03	356.73	146.70	1.0357	1.5530
0.4400	284.38	0.7362	39.697	210.76	357.05	146.29	1.0383	1.5527
0.4500	285.13	0.7375	38.844	211.49	357.37	145.88	1.0408	1.5524
0.4600	285.87	0.7389	38.027	212.20	357.68	145.48	1.0433	1.5522
0.4700	286.60	0.7402	37.243	212.91	357.98	145.08	1.0457	1.5519
0.4800	287.31	0.7416	36.490	213.60	358.28	144.68	1.0481	1.5517
0.4900	288.02	0.7429	35.767	214.28	358.58	144.29	1.0504	1.5514
0.5000	288.71	0.7442	35.072	214.96	358.86	143.91	1.0528	1.5512
0.5100	289.39	0.7455	34.403	215.62	359.15	143.53	1.0550	1.5510
0.5200	290.07	0.7468	33.758	216.27	359.42	143.15	1.0573	1.5508
0.5300	290.73	0.7481	33.137	216.92	359.69	142.77	1.0595	1.5505
0.5400	291.39	0.7494	32.538	217.56	359.96	142.40	1.0616	1.5503
0.5500	292.03	0.7507	31.960	218.19	360.22	142.04	1.0638	1.5501
0.5600	292.67	0.7519	31.401	218.81	360.48	141.67	1.0659	1.5499
0.5700	293.30	0.7532	30.861	219.43	360.74	141.31	1.0679	1.5497

Chart R12.2: Saturation properties–*continued*

p MPa	T K	v' dm³/kg	v'' dm³/kg	h' kJ/kg	h'' kJ/kg	r kJ/kg	s' kJ/kgK	s'' kJ/kgK
0.5800	293.92	0.7544	30.340	220.03	360.99	140.95	1.0700	1.5495
0.5900	294.53	0.7557	29.835	220.63	361.23	140.60	1.0720	1.5494
0.6000	295.14	0.7569	29.346	221.23	361.47	140.25	1.0740	1.5492
0.6100	295.74	0.7581	28.872	221.81	361.71	139.90	1.0759	1.5490
0.6200	296.33	0.7593	28.414	222.39	361.95	139.55	1.0779	1.5488
0.6300	296.91	0.7605	27.969	222.97	362.18	139.21	1.0798	1.5486
0.6400	297.49	0.7617	27.537	223.53	362.40	138.87	1.0817	1.5485
0.6500	298.06	0.7629	27.118	224.10	362.63	138.53	1.0835	1.5483
0.6600	298.62	0.7641	26.712	224.65	362.85	138.20	1.0854	1.5481
0.6700	299.18	0.7653	26.317	225.20	363.07	137.86	1.0872	1.5480
0.6800	299.73	0.7665	25.933	225.75	363.28	137.53	1.0890	1.5478
0.6900	300.28	0.7676	25.559	226.29	363.49	137.20	1.0907	1.5477
0.7000	300.82	0.7688	25.196	226.82	363.70	136.87	1.0925	1.5475
0.7100	301.35	0.7700	24.843	227.35	363.90	136.55	1.0942	1.5474
0.7200	301.88	0.7711	24.499	227.88	364.10	136.23	1.0959	1.5472
0.7300	302.40	0.7723	24.164	228.40	364.30	135.91	1.0976	1.5471
0.7400	302.92	0.7734	23.838	228.91	364.50	135.59	1.0993	1.5469
0.7600	303.94	0.7757	23.210	229.93	364.88	134.96	1.1026	1.5466
0.7800	304.94	0.7780	22.613	230.93	365.26	134.33	1.1058	1.5464
0.8000	305.92	0.7802	22.045	231.91	365.62	133.71	1.1090	1.5461
0.8200	306.89	0.7824	21.503	232.88	365.98	133.10	1.1121	1.5458
0.8400	307.84	0.7847	20.986	233.83	366.32	132.49	1.1152	1.5456
0.8600	308.77	0.7869	20.491	234.76	366.66	131.89	1.1181	1.5453
0.8800	309.68	0.7891	20.019	235.69	366.99	131.30	1.1211	1.5451
0.9000	310.58	0.7912	19.566	236.60	367.31	130.71	1.1240	1.5448
0.9200	311.47	0.7934	19.132	237.49	367.62	130.13	1.1268	1.5446
0.9400	312.34	0.7956	18.716	238.38	367.92	129.55	1.1296	1.5443
0.9600	313.19	0.7977	18.316	239.25	368.22	128.97	1.1323	1.5441
0.9800	314.04	0.7999	17.932	240.11	368.51	128.40	1.1350	1.5439
1.0000	314.87	0.8020	17.563	240.96	368.79	127.83	1.1377	1.5436
1.0200	315.69	0.8041	17.207	241.80	369.07	127.27	1.1403	1.5434
1.0400	316.50	0.8063	16.865	242.63	369.34	126.71	1.1429	1.5432
1.0600	317.29	0.8084	16.534	243.45	369.60	126.15	1.1454	1.5430
1.0800	318.08	0.8105	16.216	244.26	369.86	125.60	1.1479	1.5428
1.1000	318.86	0.8126	15.908	245.06	370.11	125.05	1.1504	1.5425
1.1500	320.75	0.8178	15.184	247.03	370.72	123.69	1.1564	1.5420
1.2000	322.58	0.8231	14.517	248.94	371.29	122.34	1.1622	1.5415
1.2500	324.36	0.8283	13.901	250.81	371.83	121.01	1.1678	1.5409
1.3000	326.09	0.8335	13.330	252.64	372.33	119.70	1.1733	1.5404
1.3500	327.77	0.8387	12.799	254.42	372.82	118.40	1.1787	1.5399
1.4000	329.40	0.8439	12.304	256.17	373.27	117.10	1.1838	1.5393
1.4500	330.99	0.8491	11.841	257.88	373.70	115.82	1.1889	1.5388
1.5000	332.54	0.8543	11.407	259.56	374.10	114.54	1.1938	1.5383
1.5500	334.05	0.8595	11.000	261.21	374.48	113.27	1.1987	1.5377
1.6000	335.53	0.8648	10.617	262.83	374.83	112.01	1.2034	1.5372
1.6500	336.98	0.8701	10.256	264.42	375.17	110.75	1.2080	1.5366
1.7000	338.39	0.8754	9.9140	265.99	375.48	109.49	1.2125	1.5361
1.7500	339.77	0.8807	9.5907	267.54	375.78	108.24	1.2169	1.5355
1.8000	341.12	0.8861	9.2841	269.06	376.05	106.99	1.2213	1.5349
1.8500	342.45	0.8915	8.9929	270.56	376.30	105.74	1.2255	1.5343

Chart R12.2: Saturation properties–*continued*

p MPa	T K	v' dm³/kg	v'' dm³/kg	h' kJ/kg	h'' kJ/kg	r kJ/kg	s' kJ/kgK	s'' kJ/kgK
1.9000	343.75	0.8970	8.7159	272.04	376.54	104.49	1.2297	1.5337
1.9500	345.02	0.9025	8.4520	273.51	376.75	103.24	1.2338	1.5331
2.0000	346.27	0.9081	8.2003	274.96	376.95	101.99	1.2379	1.5324
2.1000	348.70	0.9194	7.7297	277.80	377.29	99.48	1.2458	1.5311
2.2000	351.05	0.9311	7.2981	280.59	377.55	96.96	1.2535	1.5297
2.3000	353.31	0.9430	6.9002	283.33	377.75	94.42	1.2611	1.5283
2.4000	355.51	0.9554	6.5318	286.03	377.88	91.85	1.2684	1.5268
2.5000	357.63	0.9682	6.1892	288.69	377.93	89.24	1.2756	1.5251
2.6000	359.69	0.9814	5.8692	291.32	377.92	86.59	1.2827	1.5234
2.7000	361.69	0.9953	5.5691	293.93	377.82	83.90	1.2896	1.5216
2.8000	363.64	1.0098	5.2866	296.51	377.65	81.14	1.2965	1.5196
2.9000	365.53	1.0251	5.0195	299.09	377.40	78.31	1.3032	1.5175
3.0000	367.37	1.0413	4.7658	301.66	377.06	75.40	1.3100	1.5152
3.1000	369.16	1.0586	4.5239	304.24	376.62	72.38	1.3167	1.5128
3.2000	370.90	1.0771	4.2921	306.83	376.07	69.25	1.3234	1.5101
3.3000	372.60	1.0972	4.0686	309.44	375.40	65.96	1.3301	1.5072
3.4000	374.26	1.1193	3.8520	312.10	374.60	62.50	1.3370	1.5040
3.5000	375.88	1.1438	3.6403	314.82	373.62	58.81	1.3439	1.5004
3.6000	377.45	1.1714	3.4315	317.63	372.45	54.82	1.3511	1.4963
3.7000	378.99	1.2035	3.2230	320.57	371.02	50.45	1.3585	1.4916
3.8000	380.49	1.2420	3.0110	323.72	369.24	45.52	1.3665	1.4861
3.9000	381.95	1.2907	2.7893	327.20	366.95	39.75	1.3753	1.4794
4.0000	383.37	1.3585	2.5452	331.32	363.80	32.48	1.3857	1.4704

Chart R12.3: Thermodynamic properties

p/MPa	0.10			0.15			0.20		
T K	v dm³/kg	h kJ/kg	s kJ/kgK	v dm³/kg	h kJ/kg	s kJ/kgK	v dm³/kg	h kJ/kg	s kJ/kgK
250.0	166.04	342.31	1.5927	0.6817	178.22	0.9171	0.6816	178.24	0.9171
260.0	173.42	348.04	1.6152	113.74	347.14	1.5849	0.6961	187.56	0.9536
270.0	180.74	353.87	1.6372	118.79	353.05	1.6072	87.779	352.20	1.5852
280.0	188.01	359.79	1.6587	123.78	359.04	1.6290	91.640	358.27	1.6072
290.0	195.23	365.80	1.6798	128.73	365.11	1.6503	95.448	364.41	1.6288
300.0	202.42	371.90	1.7005	133.63	371.27	1.6711	99.213	370.62	1.6498
310.0	209.57	378.09	1.7208	138.49	377.50	1.6916	102.94	376.90	1.6704
320.0	216.69	384.36	1.7407	143.33	383.81	1.7116	106.64	383.26	1.6906
330.0	223.79	390.71	1.7602	148.14	390.20	1.7313	110.31	389.69	1.7104
340.0	230.87	397.15	1.7794	152.93	396.67	1.7506	113.95	396.19	1.7298
350.0	237.93	403.66	1.7983	157.70	403.21	1.7695	117.58	402.76	1.7489
360.0	244.97	410.25	1.8169	162.45	409.83	1.7882	121.19	409.40	1.7676
370.0	252.00	416.91	1.8351	167.19	416.51	1.8065	124.79	416.11	1.7859
380.0	259.02	423.64	1.8531	171.92	423.27	1.8245	128.37	422.89	1.8040
390.0	266.03	430.44	1.8707	176.64	430.09	1.8422	131.94	429.73	1.8218
400.0	273.02	437.31	1.8881	181.34	436.97	1.8596	135.50	436.63	1.8393
410.0	280.01	444.24	1.9052	186.04	443.92	1.8768	139.05	443.60	1.8565
420.0	286.99	451.23	1.9221	190.73	450.93	1.8937	142.60	450.63	1.8734
430.0	293.97	458.29	1.9387	195.41	458.00	1.9103	146.14	457.71	1.8901
440.0	300.94	465.40	1.9550	200.09	465.13	1.9267	149.67	464.85	1.9065
450.0	307.90	472.57	1.9711	204.76	472.31	1.9428	153.19	472.04	1.9226
460.0	314.86	479.80	1.9870	209.43	479.54	1.9587	156.71	479.29	1.9386
470.0	321.81	487.07	2.0027	214.09	486.83	1.9744	160.23	486.58	1.9543

p/MPa	0.30			0.40			0.50		
250.0	0.6815	178.27	0.9169	0.6813	178.30	0.9168	0.6812	178.34	0.9166
260.0	0.6959	187.59	0.9535	0.6957	187.62	0.9533	0.6955	187.65	0.9532
270.0	0.7115	197.01	0.9890	0.7113	197.04	0.9889	0.7111	197.07	0.9887
280.0	59.429	356.65	1.5752	0.7283	206.55	1.0235	0.7280	206.57	1.0233
290.0	62.119	362.94	1.5973	45.396	361.38	1.5737	35.306	359.72	1.5542
300.0	64.759	369.27	1.6188	47.487	367.86	1.5957	37.082	366.38	1.5767
310.0	67.356	375.67	1.6398	49.529	374.38	1.6170	38.802	373.04	1.5986
320.0	69.918	382.12	1.6602	51.532	380.94	1.6378	40.476	379.71	1.6198
330.0	72.451	388.63	1.6803	53.503	387.54	1.6582	42.115	386.42	1.6404
340.0	74.960	395.21	1.6999	55.446	394.20	1.6780	43.723	393.16	1.6605
350.0	77.447	401.84	1.7191	57.367	400.91	1.6975	45.307	399.94	1.6802
360.0	79.916	408.54	1.7380	59.268	407.67	1.7165	46.870	406.77	1.6994
370.0	82.370	415.31	1.7565	61.153	414.48	1.7352	48.415	413.65	1.7183
380.0	84.810	422.13	1.7747	63.023	421.35	1.7535	49.944	420.57	1.7367
390.0	87.238	429.01	1.7926	64.880	428.28	1.7715	51.461	427.54	1.7548
400.0	89.655	435.95	1.8102	66.727	435.26	1.7892	52.966	434.57	1.7726
410.0	92.063	442.95	1.8275	68.564	442.30	1.8066	54.461	441.64	1.7901
420.0	94.463	450.01	1.8445	70.392	449.39	1.8236	55.947	448.76	1.8072
430.0	96.856	457.12	1.8612	72.213	456.53	1.8405	57.425	455.93	1.8241
440.0	99.243	464.29	1.8777	74.027	463.72	1.8570	58.896	463.15	1.8407
450.0	101.62	471.50	1.8939	75.836	470.97	1.8733	60.361	470.42	1.8571
460.0	104.00	478.77	1.9099	77.638	478.26	1.8893	61.821	477.74	1.8731
470.0	106.37	486.09	1.9256	79.436	485.60	1.9051	63.276	485.10	1.8890

Chart R12.3: Thermodynamic properties–*continued*

p/MPa	0.60			0.80			1.00		
T K	v dm³/kg	h kJ/kg	s kJ/kgK	v dm³/kg	h kJ/kg	s kJ/kgK	v dm³/kg	h kJ/kg	s kJ/kgK
250.0	0.6810	178.37	0.9165	0.6807	178.44	0.9162	0.6804	178.51	0.9160
260.0	0.6953	187.68	0.9530	0.6950	187.75	0.9527	0.6946	187.81	0.9524
270.0	0.7108	197.09	0.9885	0.7104	197.15	0.9882	0.7099	197.20	0.9879
280.0	0.7277	206.60	1.0231	0.7272	206.64	1.0227	0.7267	206.69	1.0224
290.0	0.7464	216.22	1.0569	0.7458	216.26	1.0565	0.7451	216.29	1.0561
300.0	30.105	364.80	1.5604	0.7666	226.02	1.0896	0.7657	226.05	1.0891
310.0	31.620	371.63	1.5827	22.562	368.56	1.5556	0.7892	236.01	1.1218
320.0	33.083	378.44	1.6044	23.783	375.71	1.5783	18.122	372.67	1.5559
330.0	34.506	385.26	1.6253	24.950	382.79	1.6001	19.158	380.11	1.5788
340.0	35.895	392.09	1.6457	26.075	389.85	1.6212	20.141	387.45	1.6007
350.0	37.257	398.96	1.6656	27.167	396.90	1.6416	21.082	394.72	1.6218
360.0	38.596	405.86	1.6851	28.233	403.96	1.6615	21.991	401.97	1.6422
370.0	39.916	412.79	1.7041	29.276	411.03	1.6809	22.874	409.20	1.6620
380.0	41.220	419.77	1.7227	30.302	418.13	1.6998	23.736	416.43	1.6813
390.0	42.510	426.79	1.7409	31.311	425.26	1.7183	24.580	423.68	1.7001
400.0	43.788	433.86	1.7588	32.307	432.42	1.7365	25.410	430.94	1.7185
410.0	45.055	440.97	1.7764	33.292	439.61	1.7542	26.227	438.22	1.7365
420.0	46.314	448.13	1.7936	34.267	446.84	1.7717	27.033	445.53	1.7541
430.0	47.564	455.33	1.8106	35.233	454.11	1.7888	27.830	452.87	1.7713
440.0	48.807	462.58	1.8272	36.192	461.42	1.8056	28.619	460.25	1.7883
450.0	50.044	469.88	1.8436	37.144	468.77	1.8221	29.401	467.65	1.8049
460.0	51.275	477.22	1.8598	38.090	476.16	1.8383	30.176	475.10	1.8213
470.0	52.501	484.60	1.8756	39.031	483.59	1.8543	30.946	482.57	1.8374
p/MPa	2.00			4.00			6.00		
250.0	0.6788	178.85	0.9146	0.6759	179.54	0.9119	0.6731	180.25	0.9094
260.0	0.6928	188.12	0.9510	0.6894	188.77	0.9481	0.6862	189.43	0.9454
270.0	0.7078	197.49	0.9863	0.7038	198.07	0.9833	0.7000	198.69	0.9803
280.0	0.7241	206.93	1.0207	0.7193	207.45	1.0174	0.7149	208.00	1.0142
290.0	0.7420	216.48	1.0542	0.7362	216.90	1.0505	0.7309	217.38	1.0471
300.0	0.7618	226.17	1.0870	0.7547	226.47	1.0830	0.7484	226.84	1.0792
310.0	0.7842	236.03	1.1194	0.7753	236.18	1.1148	0.7676	236.42	1.1106
320.0	0.8100	246.16	1.1515	0.7985	246.09	1.1463	0.7889	246.16	1.1415
330.0	0.8405	256.64	1.1837	0.8252	256.26	1.1775	0.8129	256.11	1.1721
340.0	0.8784	267.65	1.2166	0.8566	266.78	1.2090	0.8403	266.32	1.2026
350.0	8.5024	380.58	1.5429	0.8950	277.81	1.2409	0.8724	276.87	1.2332
360.0	9.2259	389.75	1.5687	0.9448	289.58	1.2741	0.9110	287.89	1.2642
370.0	9.8669	398.38	1.5924	1.0169	302.69	1.3100	0.9593	299.54	1.2961
380.0	10.455	406.70	1.6145	1.1650	319.62	1.3551	1.0240	312.14	1.3297
390.0	11.005	414.81	1.6356	3.5255	384.23	1.5234	1.1217	326.41	1.3668
400.0	11.528	422.79	1.6558	4.1841	398.90	1.5605	1.3125	344.50	1.4126
410.0	12.028	430.67	1.6753	4.6643	410.45	1.5891	1.7893	370.22	1.4760
420.0	12.512	438.50	1.6942	5.0672	420.76	1.6139	2.3444	392.19	1.5290
430.0	12.981	446.29	1.7125	5.4251	430.38	1.6366	2.7679	407.93	1.5661
440.0	13.439	454.06	1.7303	5.7530	439.58	1.6577	3.1092	420.79	1.5957
450.0	13.887	461.81	1.7478	6.0594	448.49	1.6777	3.4020	432.18	1.6213
460.0	14.327	469.56	1.7648	6.3496	457.20	1.6969	3.6635	442.69	1.6444
470.0	14.760	477.31	1.7815	6.6270	465.77	1.7153	3.9032	452.65	1.6658

Chart R12.3: Thermodynamic properties—*continued*

p/MPa	8.00			10.00			15.00		
T K	v dm³/kg	h kJ/kg	s kJ/kgK	v dm³/kg	h kJ/kg	s kJ/kgK	v dm³/kg	h kJ/kg	s kJ/kgK
250.0	0.6705	180.97	0.9069	0.6680	181.70	0.9045	0.6622	183.58	0.8987
260.0	0.6831	190.12	0.9428	0.6803	190.82	0.9402	0.6737	192.62	0.9341
270.0	0.6965	199.32	0.9775	0.6932	199.98	0.9748	0.6858	201.70	0.9684
280.0	0.7108	208.58	1.0112	0.7070	209.19	1.0083	0.6984	210.80	1.0015
290.0	0.7261	217.89	1.0438	0.7217	218.43	1.0407	0.7118	219.92	1.0335
300.0	0.7427	227.27	1.0756	0.7375	227.74	1.0723	0.7261	229.07	1.0645
310.0	0.7607	236.74	1.1067	0.7545	237.12	1.1030	0.7413	238.27	1.0947
320.0	0.7805	246.35	1.1372	0.7731	246.61	1.1332	0.7576	247.54	1.1241
330.0	0.8025	256.12	1.1673	0.7935	256.24	1.1628	0.7752	256.90	1.1529
340.0	0.8272	266.09	1.1970	0.8161	266.04	1.1921	0.7942	266.38	1.1812
350.0	0.8553	276.34	1.2267	0.8414	276.05	1.2211	0.8149	276.00	1.2091
360.0	0.8877	286.91	1.2565	0.8699	286.32	1.2500	0.8376	285.77	1.2366
370.0	0.9262	297.89	1.2866	0.9026	296.89	1.2789	0.8625	295.74	1.2639
380.0	0.9729	309.40	1.3173	0.9406	307.82	1.3081	0.8900	305.90	1.2910
390.0	1.0318	321.61	1.3490	0.9855	319.18	1.3376	0.9206	316.29	1.3180
400.0	1.1103	334.77	1.3823	1.0399	331.05	1.3677	0.9548	326.92	1.3449
410.0	1.2218	349.28	1.4181	1.1070	343.53	1.3985	0.9930	337.78	1.3718
420.0	1.3886	365.50	1.4572	1.1918	356.68	1.4302	1.0360	348.90	1.3985
430.0	1.6213	382.76	1.4978	1.2993	370.50	1.4627	1.0843	360.24	1.4252
440.0	1.8853	399.13	1.5355	1.4321	384.75	1.4954	1.1385	371.77	1.4517
450.0	2.1408	413.67	1.5682	1.5857	398.94	1.5273	1.1988	383.46	1.4780
460.0	2.3753	426.62	1.5966	1.7496	412.57	1.5573	1.2653	395.22	1.5039
470.0	2.5896	438.43	1.6220	1.9144	425.41	1.5849	1.3374	406.97	1.5291
p/MPa	20.00			40.00			80.00		
250.0	0.6570	185.52	0.8932	0.6401	193.64	0.8739	0.6162	210.90	0.8425
260.0	0.6679	194.49	0.9284	0.6491	202.45	0.9084	0.6232	219.55	0.8765
270.0	0.6792	203.50	0.9624	0.6583	211.26	0.9417	0.6303	228.18	0.9090
280.0	0.6910	212.51	0.9952	0.6679	220.05	0.9737	0.6376	236.76	0.9403
290.0	0.7034	221.53	1.0269	0.6778	228.81	1.0044	0.6449	245.28	0.9702
300.0	0.7165	230.57	1.0575	0.6880	237.55	1.0340	0.6524	253.75	0.9989
310.0	0.7303	239.63	1.0872	0.6986	246.28	1.0626	0.6601	262.16	1.0264
320.0	0.7450	248.74	1.1161	0.7096	254.99	1.0903	0.6679	270.52	1.0530
330.0	0.7606	257.91	1.1444	0.7210	263.72	1.1172	0.6760	278.85	1.0786
340.0	0.7774	267.17	1.1720	0.7330	272.47	1.1433	0.6842	287.17	1.1034
350.0	0.7953	276.52	1.1991	0.7454	281.26	1.1688	0.6927	295.48	1.1275
360.0	0.8146	285.99	1.2258	0.7585	290.10	1.1937	0.7013	303.79	1.1510
370.0	0.8353	295.60	1.2521	0.7721	299.00	1.2180	0.7103	312.13	1.1738
380.0	0.8578	305.35	1.2781	0.7864	307.97	1.2420	0.7195	320.50	1.1961
390.0	0.8821	315.25	1.3038	0.8013	317.02	1.2655	0.7289	328.91	1.2180
400.0	0.9085	325.31	1.3293	0.8169	326.15	1.2886	0.7386	337.38	1.2394
410.0	0.9371	335.54	1.3545	0.8332	335.36	1.3113	0.7486	345.89	1.2604
420.0	0.9681	345.92	1.3796	0.8503	344.66	1.3338	0.7589	354.47	1.2811
430.0	1.0018	356.44	1.4043	0.8680	354.05	1.3558	0.7695	363.12	1.3014
440.0	1.0382	367.09	1.4288	0.8865	363.51	1.3776	0.7803	371.83	1.3215
450.0	1.0774	377.85	1.4530	0.9058	373.05	1.3990	0.7914	380.60	1.3412
460.0	1.1195	388.67	1.4768	0.9257	382.65	1.4201	0.8027	389.45	1.3606
470.0	1.1643	399.54	1.5001	0.9464	392.30	1.4409	0.8144	398.36	1.3798

References for R12

[ASHR81] ASHRAE: Thermophysical properties of refrigerants; ASHRAE (1981)

[Baeh65] BAEHR, H.D. AND HICKEN, E. : Die thermodynamischen Eigenschaften von CCl_2F_2 ($R12$) im kältetechnisch wichtigen Zustandsbereich; Kältetechnik 17(1965)5 143/150

[Barh66] BARHO, W.: Berechnung der Molwärme der Fluor-Chlor-Derivate von Methan und Ethan im Zustand idealer Gase; Dissertation Universität Karlsruhe (1966)

[Bich31] BICHOWSKY, F.R. AND GILKEY, W.K.: Thermodynamic properties of dichlorodifluoromethane, a new refrigerant; Ind. Eng. Chem. 23(1931) 366/367

[Buff31] BUFFINGTON, R.M. AND GILKEY, W.K.: Thermodynamic properties of dichlorodifluoromethane, a new refrigerant; Ind. Eng. Chem. 23(1931) 254/256

[Buff31a] BUFFINGTON, R.M. AND FLEISCHER, J.: The thermodynamic properties of dichlordifluoromethane, a new refrigerant; Ind. Eng. Chem. 23(1931) 1290/1294

[DuPo63] - : Thermodynamic properties of Freon-refrigerants, Freon Products Division; E.I. DuPont de Nemours & Co, Wilmington Delaware (1963)

[Erns69] ERNST, G.: Zur universellen Darstellung des Realanteils der spezifischen Wärme von Gasen; Chem. Ing. Techn. 41(1969) 544/551

[Gilk31] GILKEY, W.K.; GERARD, F.W. AND BIXLER, M.E.: Thermodynamic properties of dichlordifluoromethane, a new refrigerant; Ind. Eng. Chem. 23(1931) 364/366

[Gruz75] GRUZDEV, V.A.; SHELUDYAKOV, E.P.; KIRIJANENKO, A.A.; KOLOTOV, J.L.; LAVROV, B.A.; SCHILJAKOV, A.A.; SCHESTOVA, A.I. AND SCHUMSKAJA A.I.: Experimental investigation of the isobaric heat capacity of Freons: R-11, R-12, R-13, R-22, R-23; Thermophysical properties of Matter and Substances 8(1975) 108/120

[Hoech78] HOECHST AG: Physikalische Daten für Frigen; Frigen Informationsdienst (1978)

[Hwan61] HWANG, Y.T.: The constant heat capacities of gaseous tetrafluoromethane, chlorodifluoromethane,dichlorotetrafluoroethane, and chloropentafluoroethane; Ph.D. Diss. University of Michigan (1961)

[Jaeg73] JÄGER, H.P.: Dichte- und Dampfdruckmesswerte von verschiedenen Kältemitteln; Die Kälte (1973)7 276/278

[Kali] KALI-CHEMIE AG: Kaltron 12, thermodynamische Eigenschaften; Kali Chemie Hannover

[Kell55] KELLS, L.F.; ORFEO, S.R. AND MEARS, W.H.: Thermodynamic properties of "Freon-12"; Refrigerating Engineering 63(1955)9 32/44

[KMR 81] Kältemaschinenregeln, 7. Auflage; Verlag C.F. Müller, Karlsruhe (1981)

[Koke71] KOKERNAK, R.P. AND FELDMAN, C.L.: The velocity of sound in liquid $R12$; ASHRAE Journal (1971) 59/62

[Koma67] KOMAROV, S.G. ;SOLOV'EV, A.N. AND SHELUDYAKOV, E.P.: Speed of sound in freons on the saturation line; Zhurnal Prikladnoi Mekhaniki i Tekhnicheskoi Fiziki 8(1967)5 116/119

[Kubo74] KUBOTA, H.; TANAKA, Y. AND MAKITA, T.: Compressibility of fluorochlorocarbons by the Burnett method; 4th International Conference on High Pressure, Kyoto (1974) 538/542

[Kuma78] KUMAGAI, A. AND IWASAKI, H.: Pressure-volume-temperature relationships of several polar liquids; J. Chem. Eng. Data 23(1978)3 193/195

[Masi52] MASI, J.: Thermodynamic properties of gaseous difluorodichloromethane ($R12$); J. Am. Chem. Soc. 74(1952) 4738/4741

[McHa55] MCHARNESS, R.C.; EISEMAN, B.J. AND MARTIN, J.J.: The new thermodynamic properties of "Freon-12"; Refrigerating Engineering 63(1955)9 32/44

[Meye69] MEYER, K.J.: Über den Zusammenhang zwischen der Schallgeschwindigkeit und der Wärmeleitfähigkeit bei flüssigen Fluor-Chlor-Derivaten des Methans und Ethans; Kältetechnik-Klimatisierung 21(1969) 270/275

[Mich66] MICHELS, A.; WASSENAAR, T.; WOLKERS, G.J.; PRINS, CHR. AND L.V.D. KLUNDERT: P-v-T data and thermodynamical properties of Freon-12 (CCl_2F_2) and Freon-13 ($CClF_3$) fluorocarbons at temperatures between 0^0 C and 1500^0 C and at pressures up to 400 atm; J. Chem. Eng. Data 11(1966)10 449/452

[Nozd76] NOZDREV, V.F. AND REDKOZUBOV, V.A.: Study of the ultraacustic properties of Freon 11 and Freon 12 along the saturation line in a wide temperature range; Primenie ultraakustiki Issledovaniju vescestva (1976)62 3/5

[Oguc79] OGUCHI, K.: Experimental study of p-v-T relationship of dichlorodifluoromethane ($R12$) in liquid state; International Congress of Refrigeration, Sept. 1979 Venice; Paper B1, 27

[Pere70] PERELSTEIN, I.I.: Experimental investigation of compressibility of Freon-12; Thermophysical Properties of Matter and Substances 2(1970) 224/234

[Pere71] PERELSTEIN, I.I.: Thermodynamc properties of $R12$ and $R13$; Thermophysical Properties of Matter and Substances 4(1971) 65/95

[Pool72] POOLE, G.R. AND AZIZ, R.A.: Sound velocity in liquid CCl_2F_2 and the law of corresponding states; AIChEJ 18(1972) 430/432

[Pras81] PRASAD, M.: Low-pressure compression factors for $R12$ gas; J. Chem. Eng. Data 26(1981) 361/362

[Redk76] REDKOZUBOV, V.A.: Rate of propagation of ultrasonic waves in freons 11, 12 and 13; Primenie ultraakustiki Issledovaniju vescestva (1976) 14/19

[Ried39] RIEDEL, L.: Bestimmung der spezifischen Wärme von flüssigem Difluordichlormethan; Zeitschrift für die gesamte Kälteindustrie 46(1939) 105/106

[Tani72] TANISHITA, I.; WATANABE, K.; KIJIMA, J.; OGUCHI, K. AND KONDO, H.: Status report on the thermodynamic properties research of fluorocarbon Refrigerants; Second report, Keio University, Tokyo (1972)

[Trie69] TRIEMER, E.: Berechnung eines neuen log p,i-Diagramms für das Kältemittel $R12$; Luft- und Kältetechnik 5(1969) 272/274

[Varg75] VARGAFTIK, N.B.: Tables on the thermophysical properties of liquids and gases; Hemisphere Publishing Corporation Washington (1975)

[Wata77] WATANABE, K.; TANAKA, T. AND OGUCHI, K.: Compressibility and virial coefficients of dichlorodifluoromethane ($R12$) with Burnett apparatus; 7th Symposium on Thermophysical Properties, ASME, New York (1977), 470/479

[Wats75] WATSON, J.T.R.: Thermophysical properties of refrigerant $R12$; Department of Industrie, National Engineering Laboratory, Edingburgh (1975)

[Wood66] WOODBURN, J.; METTREY, M.T. AND HOA, B.L.: Velocity of sound in Refrigerant 12; ASHRAE Journal (1966) 74/77

R 12. ph

T=360
T=365
T=370
T=375
T=380
T=385
T=390
T=395
T=400
T=405
T=410
T=415
T=420
T=425
T=430
T=435
T=440
T=445
T=450
T=455
T=460
T=465
T=470
s=1.60
s=1.63
s=1.66
s=1.69
s=1.72
s=1.75
s=1.78
s=1.81
s=1.84
s=1.87
s=1.90
s=1.93
s=1.96
s=1.99
s=2.02
s=2.05
s=2.08
s=2.11
320
339
358
377
396
415
434
453
472
491

h/(kJ/kg)

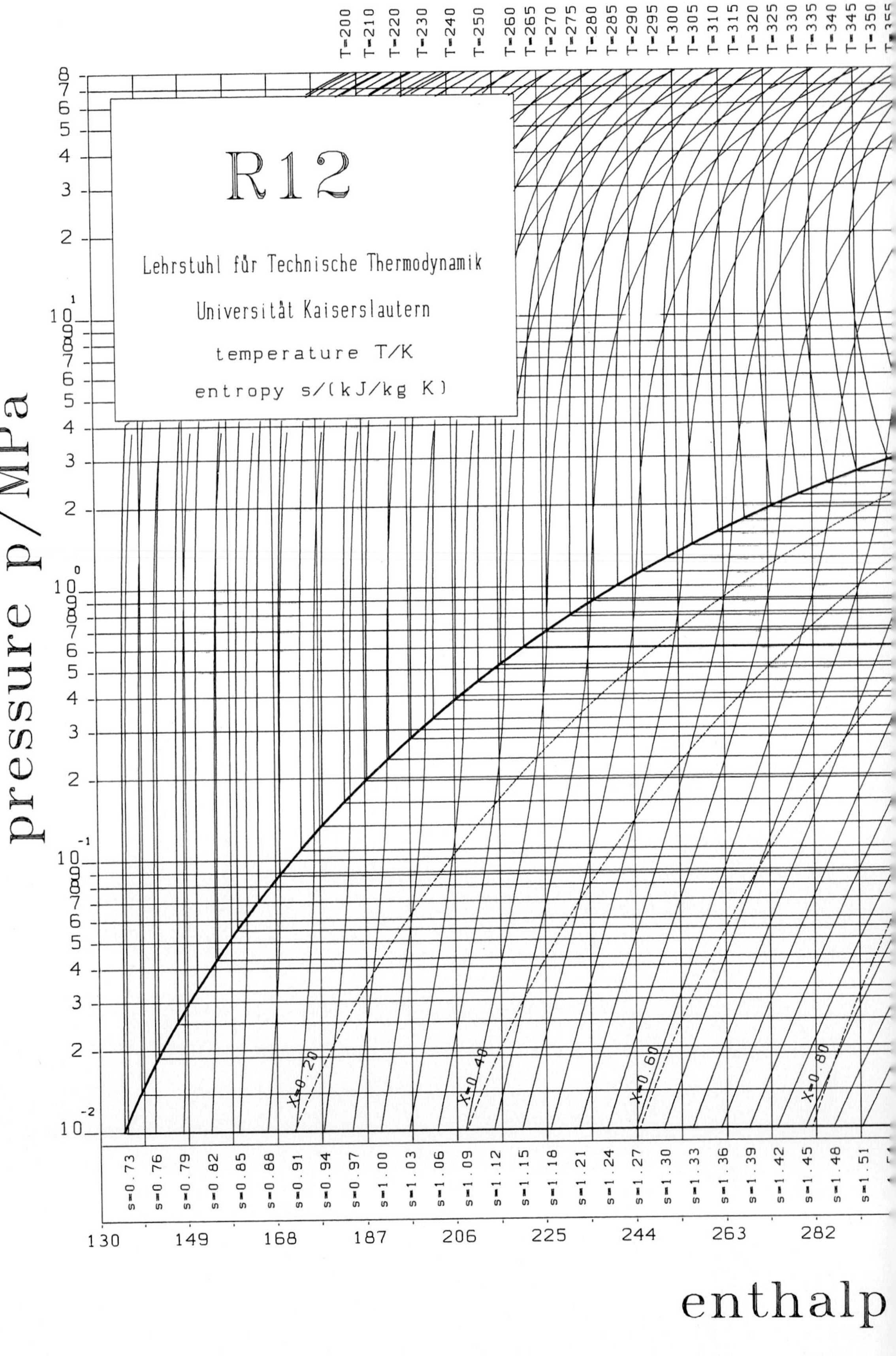
R12
Lehrstuhl für Technische Thermodynamik
Universität Kaiserslautern
temperature T/K
entropy s/(kJ/kg K)
pressure p/MPa
enthalp
T=200
T=210
T=220
T=230
T=240
T=250
T=260
T=265
T=270
T=275
T=280
T=285
T=290
T=295
T=300
T=305
T=310
T=315
T=320
T=325
T=330
T=335
T=340
T=345
T=350
X=0.20
X=0.40
X=0.60
X=0.80
s=0.73
s=0.76
s=0.79
s=0.82
s=0.85
s=0.88
s=0.91
s=0.94
s=0.97
s=1.00
s=1.03
s=1.06
s=1.09
s=1.12
s=1.15
s=1.18
s=1.21
s=1.24
s=1.27
s=1.30
s=1.33
s=1.36
s=1.39
s=1.42
s=1.45
s=1.48
s=1.51
130
149
168
187
206
225
244
263
282

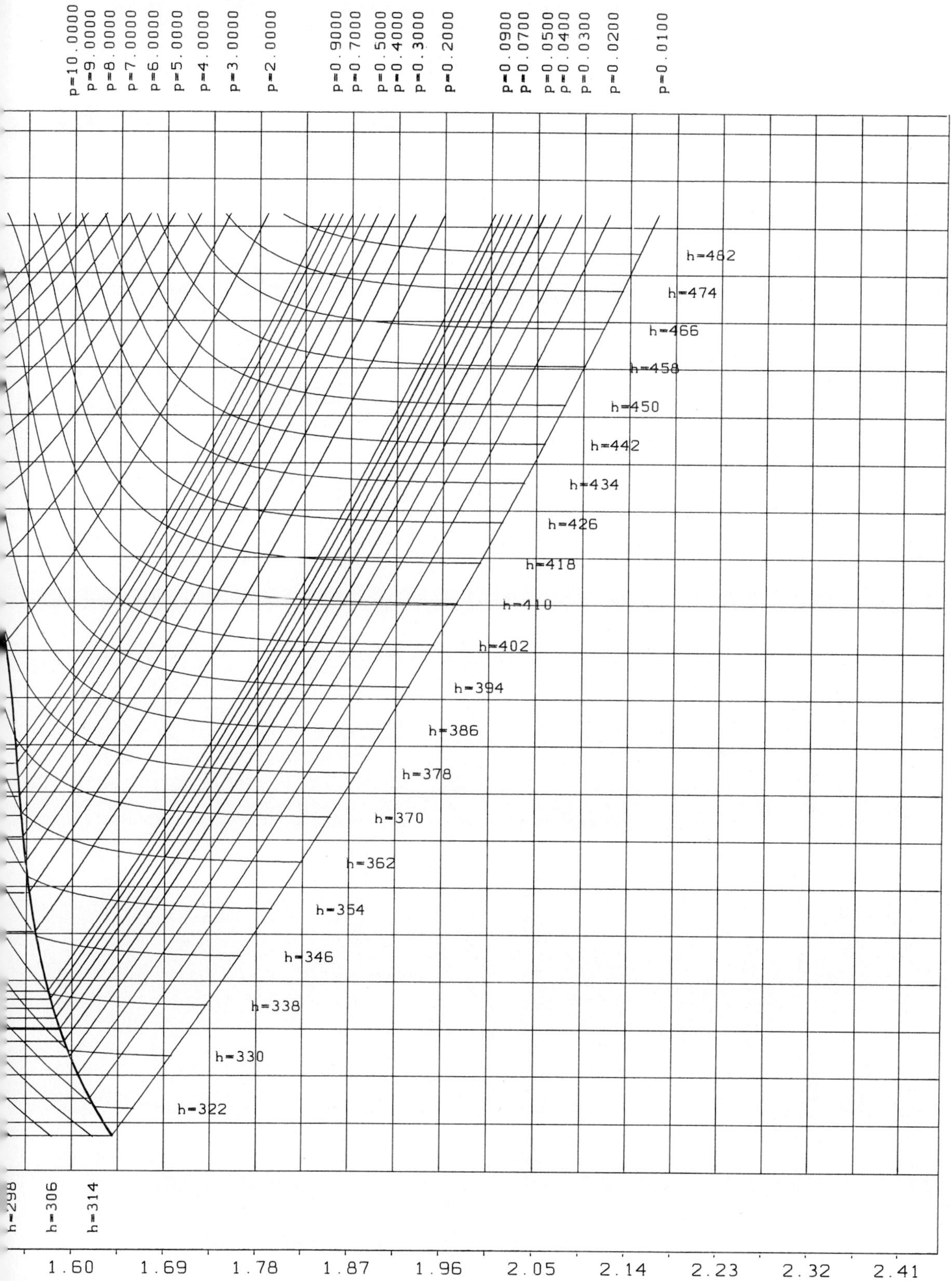

p=10.0000
p=9.0000
p=8.0000
p=7.0000
p=6.0000
p=5.0000
p=4.0000
p=3.0000
p=2.0000
p=0.9000
p=0.7000
p=0.5000
p=0.4000
p=0.3000
p=0.2000
p=0.0900
p=0.0700
p=0.0500
p=0.0400
p=0.0300
p=0.0200
p=0.0100
h=482
h=474
h=466
h=458
h=450
h=442
h=434
h=426
h=418
h=410
h=402
h=394
h=386
h=378
h=370
h=362
h=354
h=346
h=338
h=330
h=322
h=298
h=306
h=314
1.60
1.69
1.78
1.87
1.96
2.05
2.14
2.23
2.32
2.41

/(kJ/kg k)

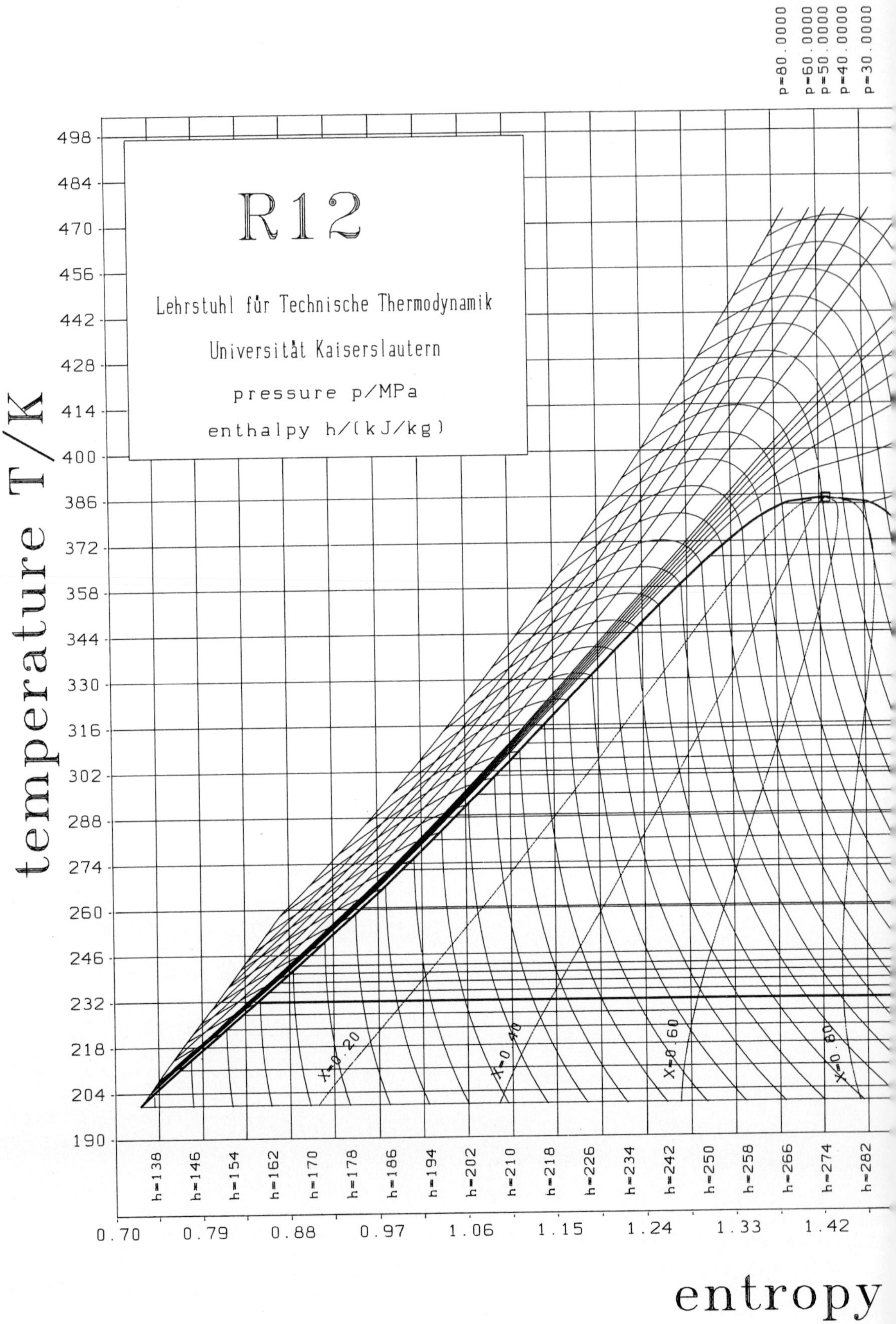
R12
Lehrstuhl für Technische Thermodynamik
Universität Kaiserslautern
pressure p/MPa
enthalpy h/(kJ/kg)
temperature T/K
entropy
p=80.0000
p=60.0000
p=50.0000
p=40.0000
p=30.0000
498
484
470
456
442
428
414
400
386
372
358
344
330
316
302
288
274
260
246
232
218
204
190
X=0.20
X=0.40
X=0.60
X=0.80
h=138
h=146
h=154
h=162
h=170
h=178
h=186
h=194
h=202
h=210
h=218
h=226
h=234
h=242
h=250
h=258
h=266
h=274
h=282
0.70
0.79
0.88
0.97
1.06
1.15
1.24
1.33
1.42

R 12. hs

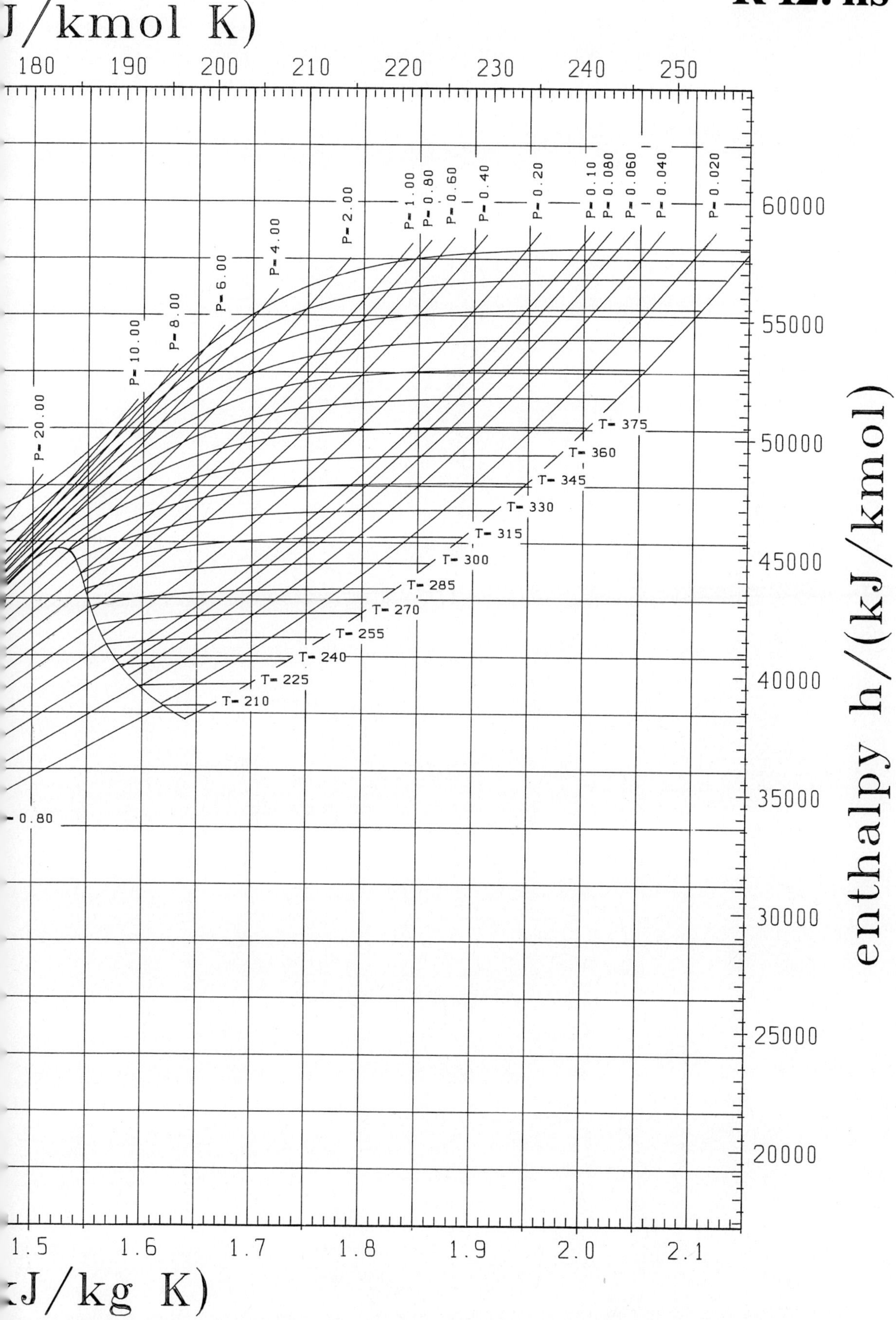
J/kmol K)
180
190
200
210
220
230
240
250
P= 20.00
P= 10.00
P= 8.00
P= 6.00
P= 4.00
P= 2.00
P= 1.00
P= 0.80
P= 0.60
P= 0.40
P= 0.20
P= 0.10
P= 0.080
P= 0.060
P= 0.040
P= 0.020
T= 375
T= 360
T= 345
T= 330
T= 315
T= 300
T= 285
T= 270
T= 255
T= 240
T= 225
T= 210
= 0.80
60000
55000
50000
45000
40000
35000
30000
25000
20000
enthalpy h/(kJ/kmol)
1.5
1.6
1.7
1.8
1.9
2.0
2.1
kJ/kg K)

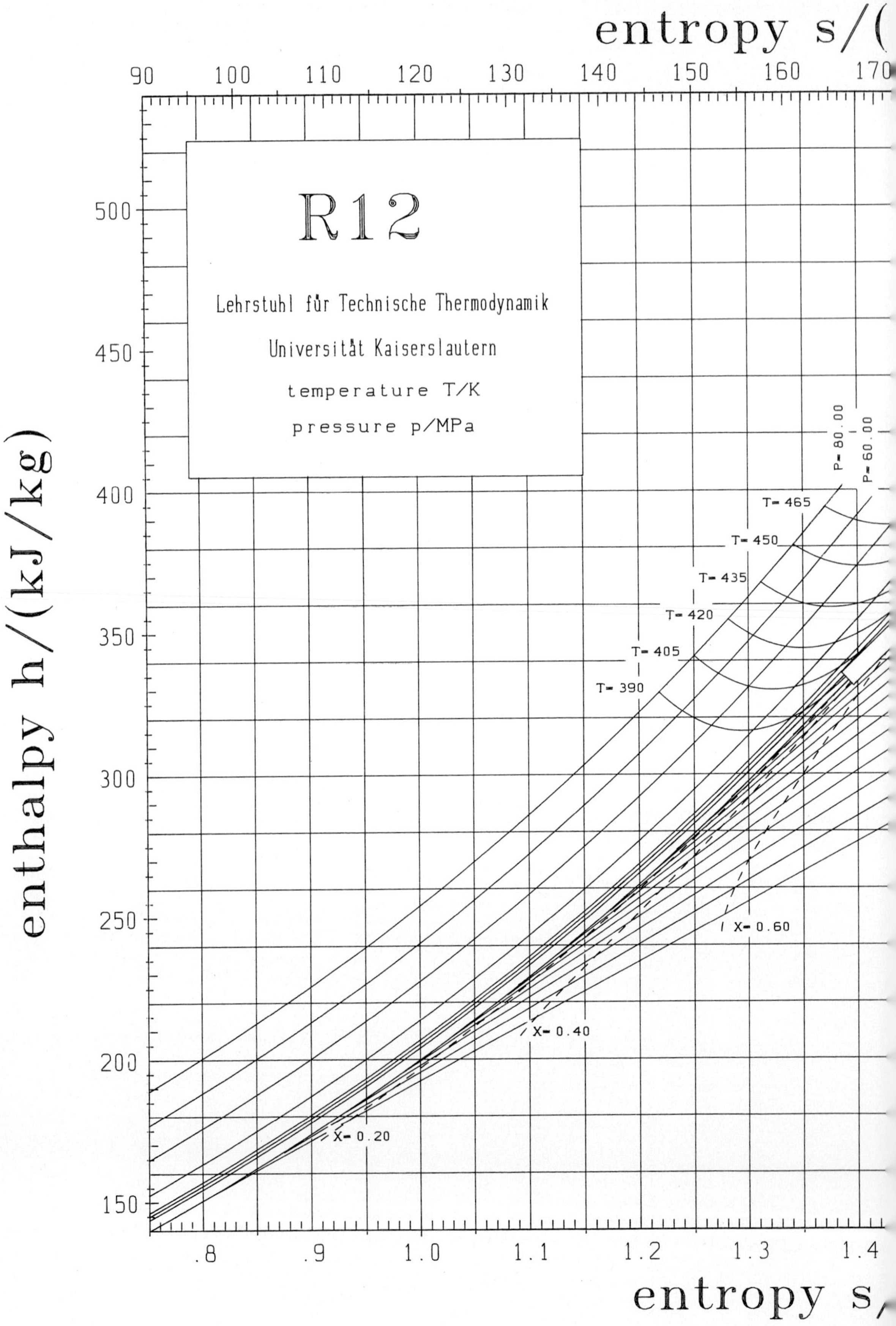
entropy s/(
90
100
110
120
130
140
150
160
170
R12
Lehrstuhl für Technische Thermodynamik
Universität Kaiserslautern
temperature T/K
pressure p/MPa
P= 80.00
P= 60.00
T= 465
T= 450
T= 435
T= 420
T= 405
T= 390
X= 0.60
X= 0.40
X= 0.20
enthalpy h/(kJ/kg)
500
450
400
350
300
250
200
150
.8
.9
1.0
1.1
1.2
1.3
1.4
entropy s/

Refrigerant R13 ($CClF_3$)

Table R13.1: List of references for p, ρ, T data

Author	n	p/MPa min	p/MPa max	ρ/g cm^{-3} min	ρ/g cm^{-3} max	T/K min	T/K max
ALBRIGHT, L.F. *[Albr 52]*	87	0.29	14.04	0.0184	0.71	215	402
GELLER, V.Z. *[Gell 79]*	68	5.00	50.00	0.9909	1.89	98	296
KUBOTA, H. *[Kubo 74]*	60	0.10	3.00	0.0036	0.20	298	348
MICHELS, A. *[Mich 66]*	280	1.52	47.24	0.0900	1.05	273	423
OGUCHI, K. *[Oguc 75]*	204	1.72	10.65	0.1015	1.04	273	372
RASSKAZOV, D. *[Rass 79]*	227	0.47	27.08	0.0167	1.32	248	448
RIEDEL, L. *[Ried 41]*	66	0.10	3.02	0.0045	0.19	194	333

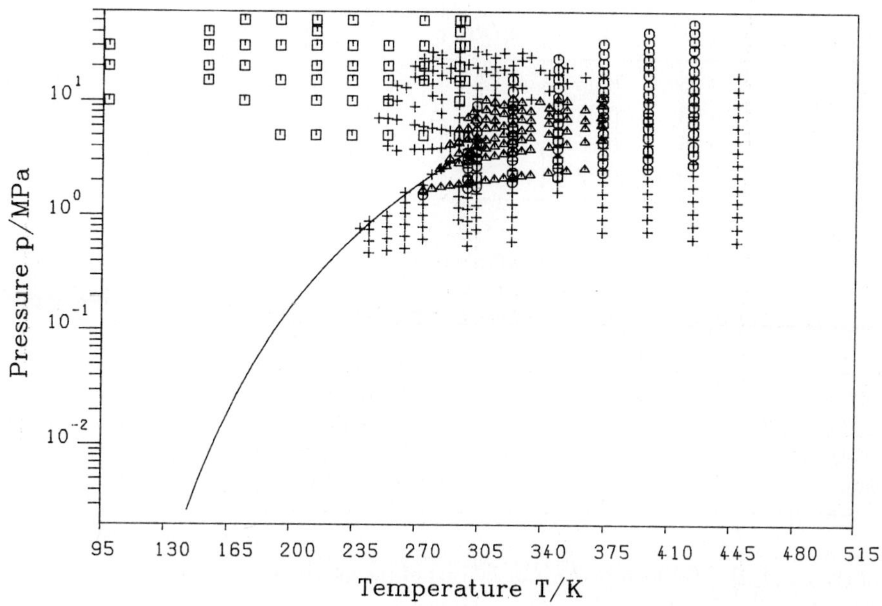

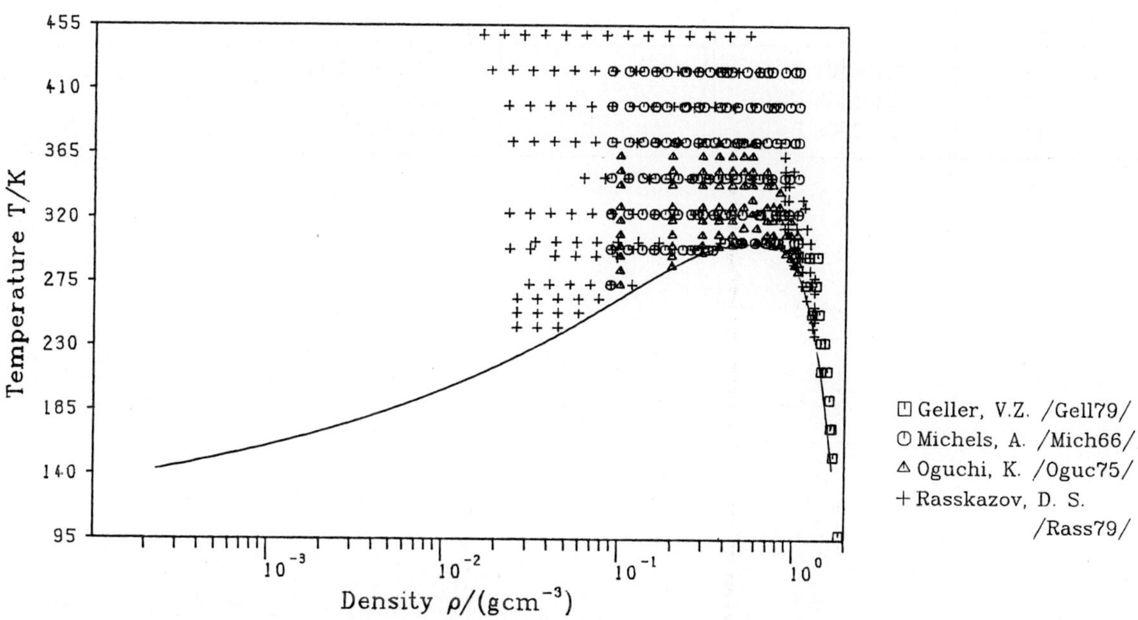

Fig.R13.1: Distribution of p, ρ, T data projected in the p, T and in the T, ρ plane

Table R13.2: Experimental vapor pressure data

Author	Δ / %		
	std	opt	*EOS*
ALBRIGHT, L.F. *[Albr 52]*	0.09	0.07	0.07
JÄGER, H.P. *[Jaeg 73]*	0.84	0.83	0.84
KUBIC, W. *[Kubi 80]*	0.71	0.73	0.70
OGUCHI, K. *[Oguc 75a]*	0.04	0.04	0.11
RIEDEL, L. *[Ried 41]*	0.24	0.21	0.21

Table R13.3: Experimental saturated liquid density data

Author	Δ / %		
	std	opt	*EOS*
ALBRIGHT, L.F. *[Albr 52]*	0.05	0.05	0.07
OGUCHI, K. *[Oguc 75]*	0.08	0.08	0.45
RIEDEL, L. *[Ried 41]*	0.25	0.25	0.23
SHAVANDRIN, A.M. *[Shav 79]*	0.49	0.49	0.70

Table R13.4: Coefficients of the standard vapor pressure equation

Index i	b_i
5	−6.790888 E+00
6	1.228308 E+00
9	−2.115934 E+00
15	−2.084861 E+00

Table R13.5: Coefficients of the quasi-optimal vapor pressure equation

Index i	b_i
5	−6.907036 E+00
6	2.109753 E+00
7	−1.792165 E+00
13	−3.272448 E+00

Table R13.6: Coefficients of the standard saturated liquid density equation

Index i	c_i
8	1.635829 E+00
9	8.405156 E−01
16	3.121800 E−01

Table R13.7: Coefficients of the quasi-optimal saturated liquid density equation

Index i	c_i
8	1.635829 E+00
9	8.405156 E−01
16	3.121800 E−01

Table R13.8: Coefficients of the *Bender-EOS*

Range of validity: one-phase region

$$0.47 \le p/\text{MPa} \le 50.00$$
$$0.01675 \le \rho/\text{g cm}^{-3} \le 1.87810$$
$$98.15 \le T/\text{K} \le 448.15$$

	Specific units	Molar units
a_{01}	1.040849972 E−01	1.135981671 E+00
a_{02}	8.027824713 E+01	8.761552559 E+02
a_{03}	−1.172605387 E+04	−1.279779280 E+05
a_{04}	4.067044392 E+06	4.438764481 E+07
a_{05}	−1.334451787 E+08	−1.456418131 E+09
a_{06}	1.072863360 E−01	1.223261192 E−01
a_{07}	−5.721805774 E+01	−6.523909018 E+01
a_{08}	1.608445758 E+04	1.833923450 E+04
a_{09}	−7.479359657 E−02	−8.909037691 E−03
a_{10}	6.227483781 E+01	7.417866003 E+00
a_{11}	1.488925716 E−01	1.852810505 E−03
a_{12}	−1.269855916 E+02	−1.580201320 E+00
a_{13}	4.534102215 E+01	5.894417244 E−02
a_{14}	−8.496437880 E+06	−9.687498999 E+06
a_{15}	4.742476665 E+09	5.407294044 E+09
a_{16}	−6.837878955 E+11	−7.796437338 E+11
a_{17}	1.499138743 E+06	1.865519537 E+04
a_{18}	−4.939234387 E+09	−6.146354557 E+07
a_{19}	1.533128481 E+12	1.907816169 E+10
a_{20}	2.900000000 E+00	3.165054461 E−02

Table R13.9: Deviations between experimental and predicted p, ρ, T data

Region	n	Δ / %	
		p	ρ
Gas	326	0.16	0.31
Critical	196	0.31	2.30
Liquid	235	1.78	0.19
Total	757	1.02	1.21

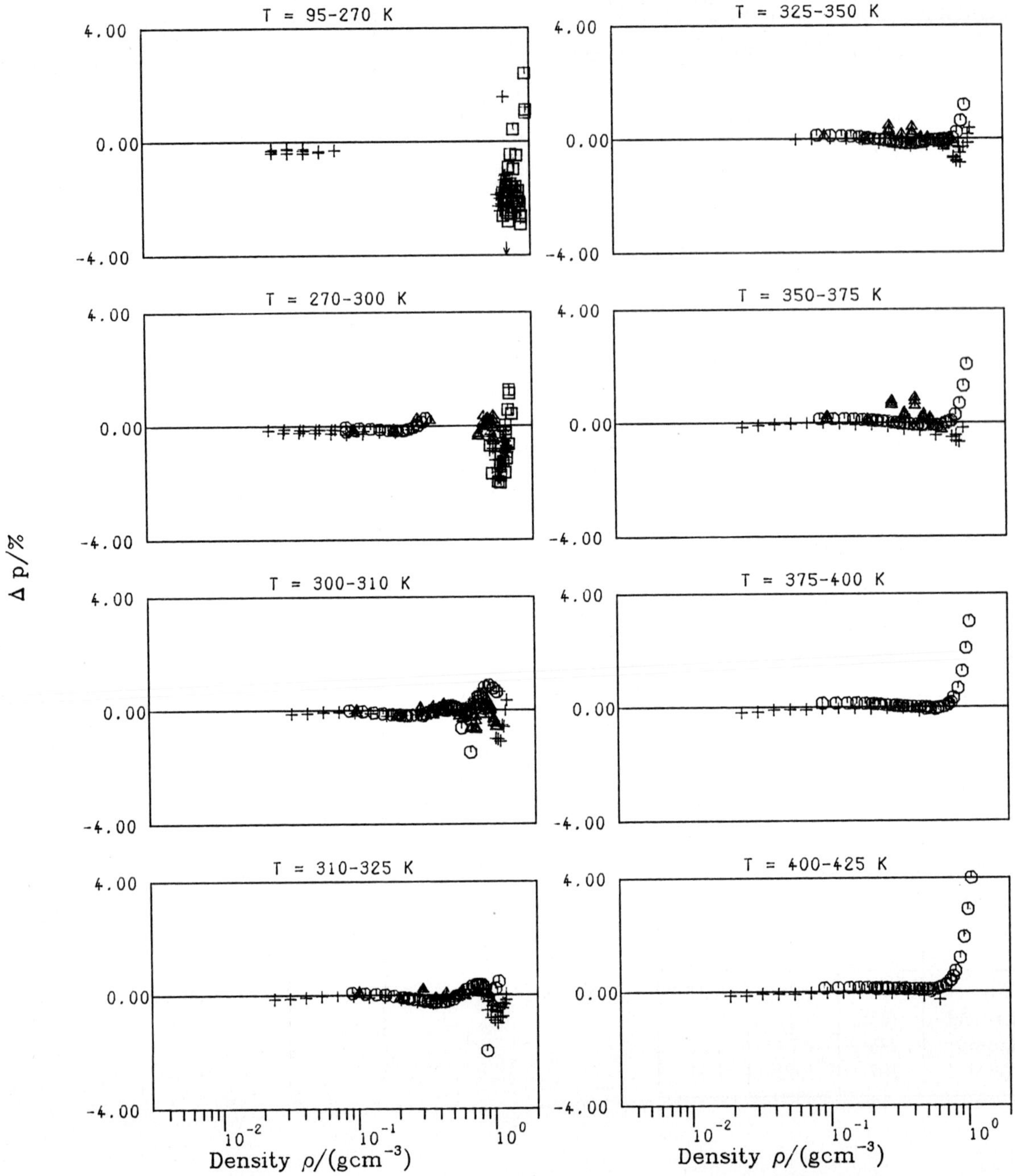

□ Geller, V.Z. /Gell79/
○ Michels, A. /Mich66/
△ Oguchi, K. /Oguc75/
+ Rasskazov, D. S. /Rass79/

Fig.R13.2: Deviations between experimental and calculated p, ρ, T data

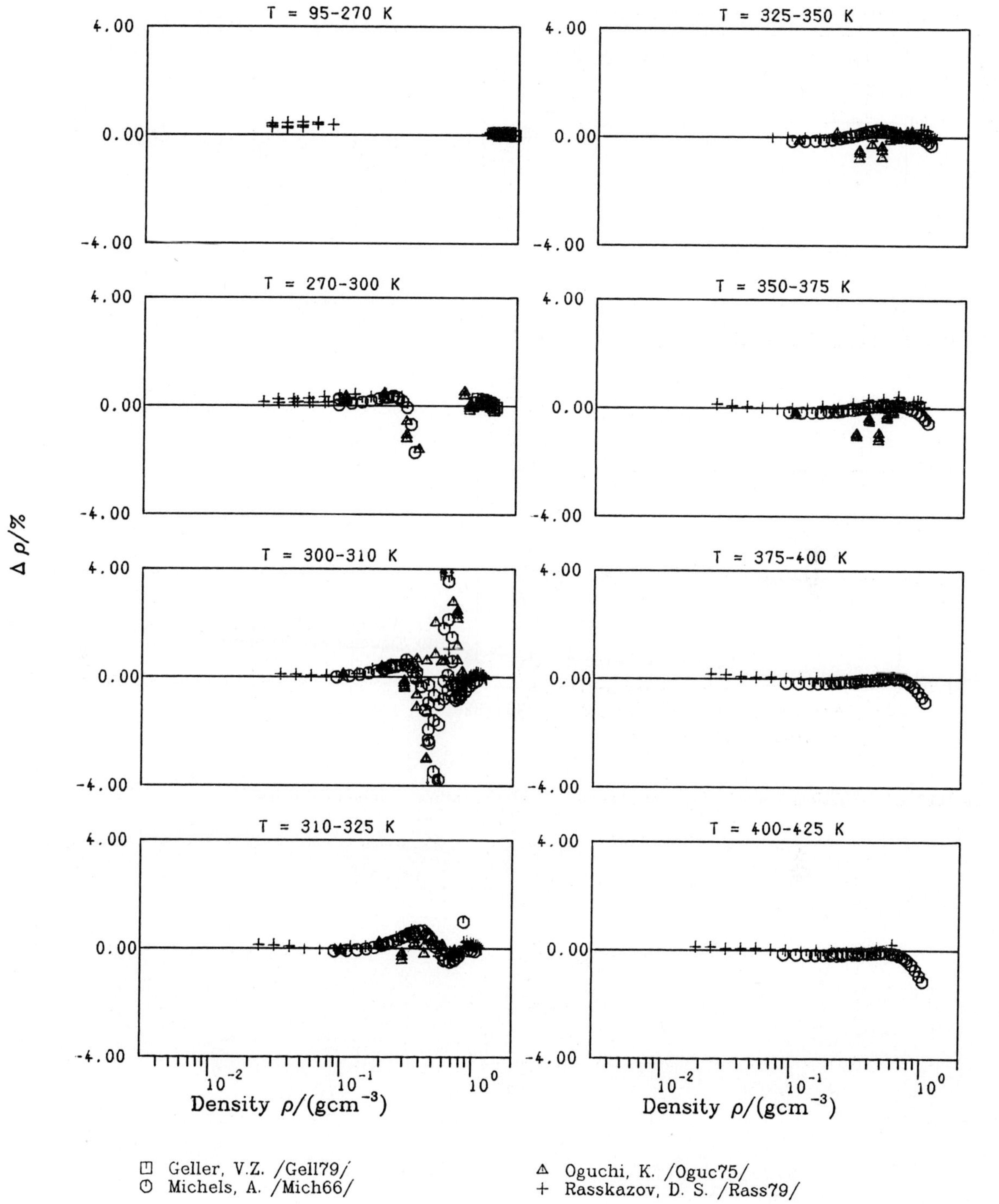

Fig.R13.2: Deviations between experimental and calculated p, ρ, T data *-continued*

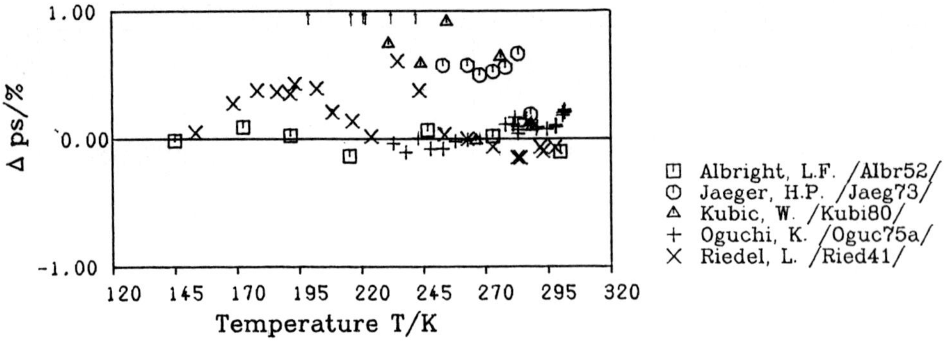

Fig.R13.3: Deviations between experimental and calculated vapor pressures

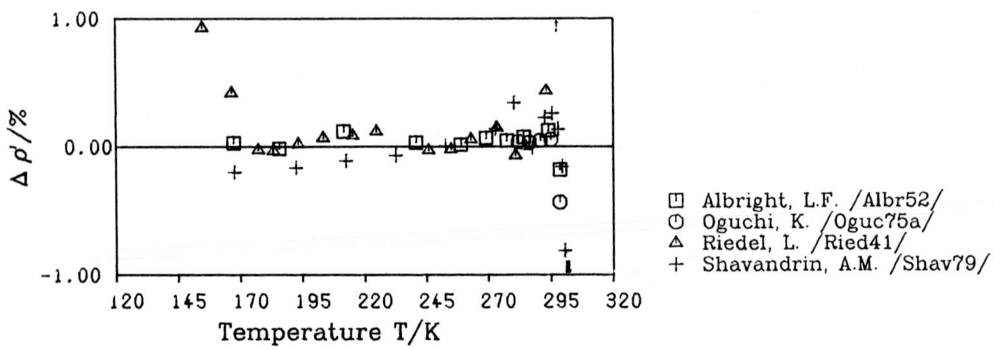

Fig.R13.4: Deviations between experimental and calculated saturated liquid densities

Table R13.10: Information on ideal gas state heat capacity equation

Author	T / K min	T / K max
CHEN, S.S. *[Chen 76]*	100	600

Coefficients of equation (15)	
Index i	d_i
1	1.971309 E−01
2	1.438638 E−03
3	1.746775 E−06
4	−6.830178 E−09
5	5.030396 E−12

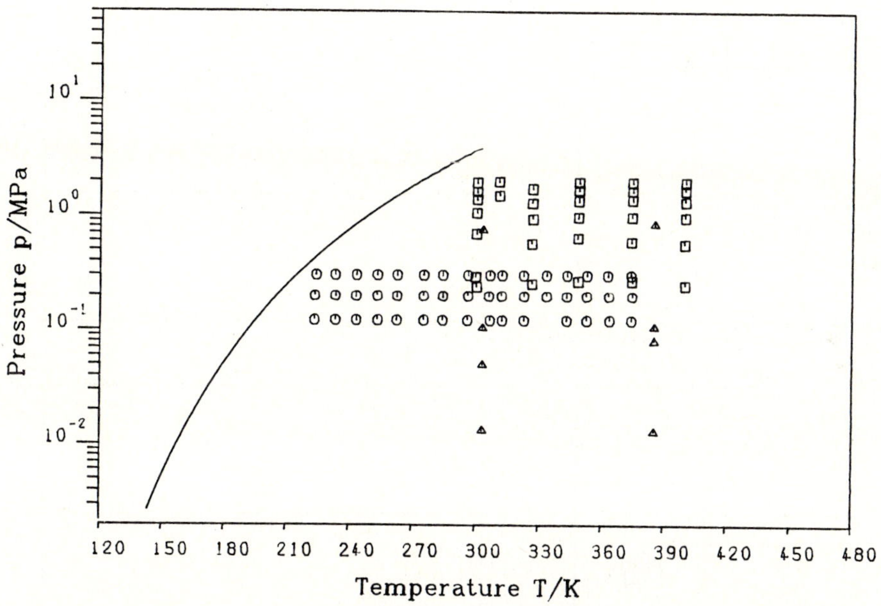

□ cp: Gruzdev, V.A. /Gruz75/ ○ cp: Tanishita, I. /Tani72/ △ w : Redkozubov, D. /Redk76/

Fig.R13.5: Distribution of $c_p(p,T)$ and $w(p,T)$ data

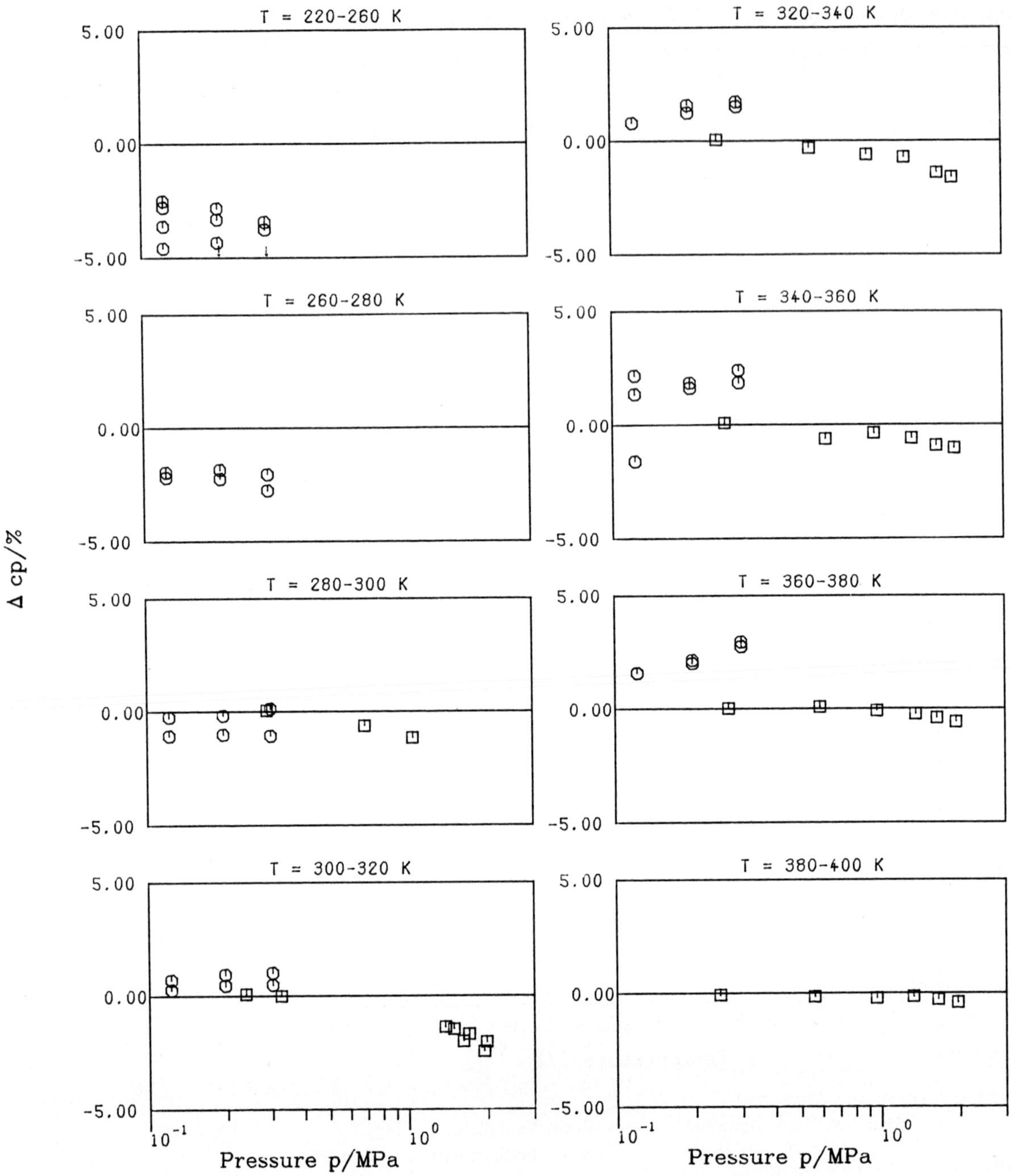

□ Gruzdev, V.A. /Gruz75/ ○ Tanishita, I. /Tani72/

Fig.R13.6: Deviations between experimental and calculated specific isobaric heat capacities c_p

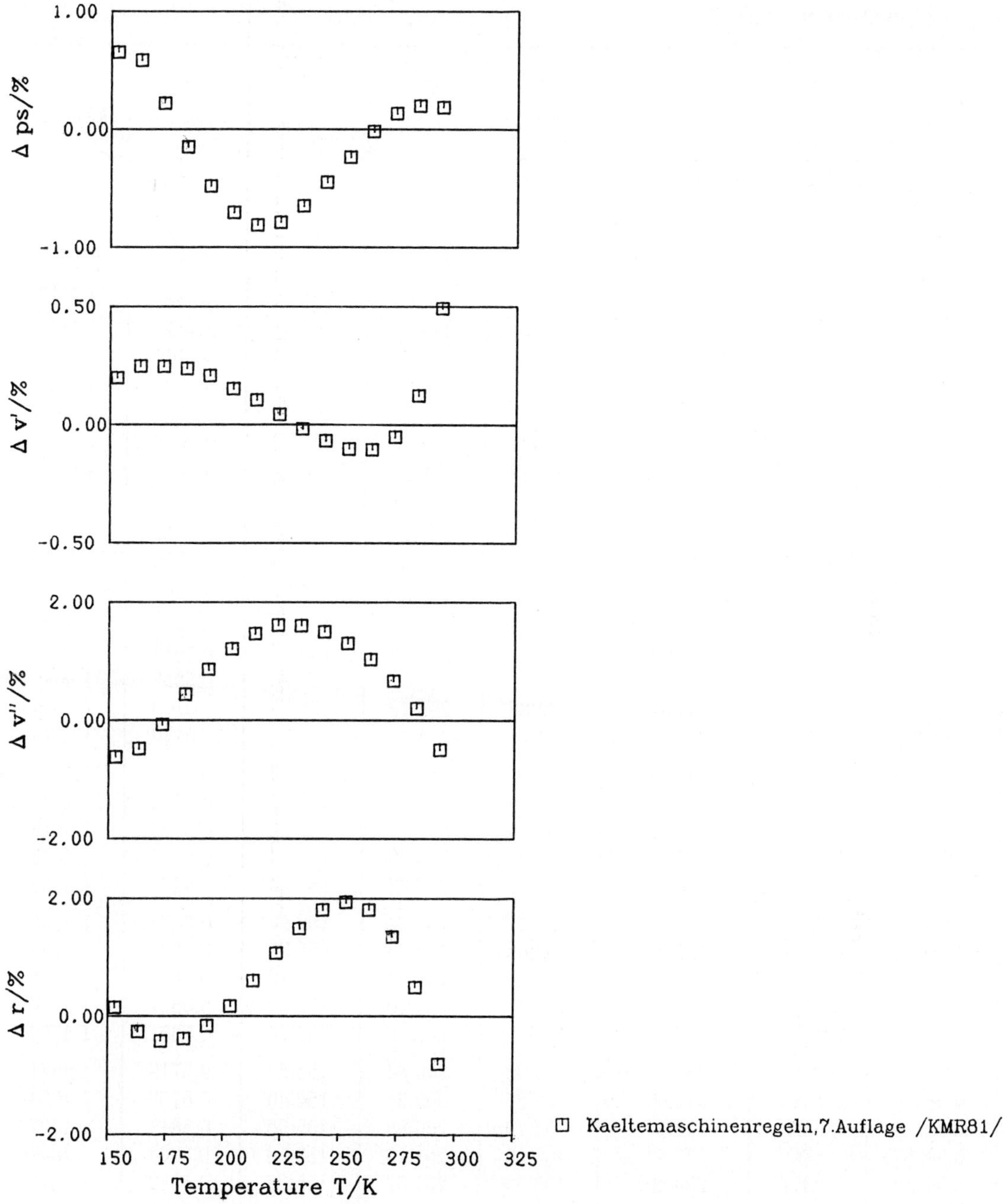

Fig.R13.7: Deviations between saturation properties from the *Kältemaschinenregeln* and those calculated by the *Bender-EOS*

Chart R13.1: Saturation properties

T K	p MPa	v' dm³/kg	v'' dm³/kg	h' kJ/kg	h'' kJ/kg	r kJ/kg	s' kJ/kgK	s'' kJ/kgK
145.0	0.00322	0.5919	3571.0	77.39	247.26	169.87	0.4113	1.5828
146.0	0.00356	0.5931	3249.2	78.28	247.67	169.39	0.4174	1.5777
147.0	0.00394	0.5944	2961.1	79.16	248.09	168.92	0.4235	1.5726
148.0	0.00434	0.5956	2702.6	80.04	248.50	168.46	0.4294	1.5677
149.0	0.00478	0.5969	2470.3	80.92	248.92	168.00	0.4353	1.5628
150.0	0.00525	0.5981	2261.2	81.79	249.33	167.54	0.4411	1.5581
151.0	0.00577	0.5994	2072.8	82.65	249.75	167.10	0.4469	1.5535
152.0	0.00632	0.6006	1902.7	83.52	250.17	166.65	0.4526	1.5490
153.0	0.00692	0.6019	1748.9	84.37	250.59	166.21	0.4582	1.5446
154.0	0.00757	0.6032	1609.6	85.23	251.00	165.77	0.4638	1.5402
155.0	0.00826	0.6045	1483.3	86.08	251.42	165.34	0.4693	1.5360
156.0	0.00901	0.6058	1368.6	86.93	251.84	164.91	0.4748	1.5319
157.0	0.00981	0.6071	1264.4	87.78	252.26	164.48	0.4802	1.5278
158.0	0.01067	0.6084	1169.4	88.63	252.68	164.05	0.4856	1.5239
159.0	0.01158	0.6097	1082.9	89.48	253.10	163.63	0.4909	1.5200
160.0	0.01257	0.6110	1003.9	90.32	253.53	163.20	0.4962	1.5162
161.0	0.01362	0.6123	931.68	91.17	253.95	162.78	0.5015	1.5125
162.0	0.01474	0.6137	865.60	92.01	254.37	162.36	0.5067	1.5089
163.0	0.01594	0.6150	805.07	92.86	254.79	161.93	0.5119	1.5053
164.0	0.01721	0.6164	749.54	93.70	255.21	161.51	0.5170	1.5019
165.0	0.01857	0.6177	698.55	94.54	255.63	161.09	0.5222	1.4985
166.0	0.02001	0.6191	651.67	95.39	256.06	160.67	0.5273	1.4951
167.0	0.02154	0.6205	608.53	96.23	256.48	160.24	0.5323	1.4919
168.0	0.02317	0.6219	568.78	97.08	256.90	159.82	0.5374	1.4887
169.0	0.02489	0.6233	532.12	97.92	257.32	159.40	0.5424	1.4855
170.0	0.02672	0.6247	498.27	98.77	257.74	158.97	0.5474	1.4825
171.0	0.02865	0.6261	466.99	99.62	258.16	158.55	0.5523	1.4795
172.0	0.03070	0.6275	438.06	100.46	258.58	158.12	0.5573	1.4765
173.0	0.03286	0.6289	411.26	101.31	259.00	157.69	0.5622	1.4737
174.0	0.03514	0.6304	386.43	102.16	259.42	157.26	0.5671	1.4709
175.0	0.03755	0.6318	363.39	103.02	259.84	156.83	0.5719	1.4681
176.0	0.04008	0.6333	342.00	103.87	260.26	156.40	0.5768	1.4654
177.0	0.04275	0.6348	322.11	104.72	260.68	155.96	0.5816	1.4627
178.0	0.04557	0.6362	303.62	105.58	261.10	155.52	0.5864	1.4601
179.0	0.04852	0.6377	286.40	106.44	261.52	155.08	0.5912	1.4576
180.0	0.05163	0.6392	270.35	107.29	261.94	154.64	0.5960	1.4551
181.0	0.05489	0.6407	255.39	108.15	262.35	154.20	0.6007	1.4527
182.0	0.05832	0.6423	241.42	109.02	262.77	153.75	0.6055	1.4503
183.0	0.06191	0.6438	228.37	109.88	263.18	153.30	0.6102	1.4479
184.0	0.06568	0.6454	216.18	110.75	263.60	152.85	0.6149	1.4456
185.0	0.06962	0.6469	204.76	111.61	264.01	152.40	0.6196	1.4434
186.0	0.07375	0.6485	194.08	112.48	264.42	151.94	0.6243	1.4411
187.0	0.07807	0.6501	184.07	113.35	264.84	151.48	0.6289	1.4390
188.0	0.08259	0.6517	174.69	114.23	265.25	151.02	0.6336	1.4369
189.0	0.08731	0.6533	165.88	115.10	265.66	150.55	0.6382	1.4348
190.0	0.09223	0.6549	157.61	115.98	266.06	150.09	0.6428	1.4327
191.0	0.09737	0.6565	149.84	116.86	266.47	149.61	0.6474	1.4307
192.0	0.10274	0.6582	142.53	117.74	266.88	149.14	0.6520	1.4287
193.0	0.10833	0.6598	135.66	118.62	267.28	148.66	0.6565	1.4268
194.0	0.11415	0.6615	129.19	119.51	267.69	148.18	0.6611	1.4249

Chart R13.1: Saturation properties—*continued*

T K	p MPa	v' dm³/kg	v'' dm³/kg	h' kJ/kg	h'' kJ/kg	r kJ/kg	s' kJ/kgK	s'' kJ/kgK
195.0	0.12021	0.6632	123.09	120.39	268.09	147.69	0.6656	1.4230
196.0	0.12653	0.6649	117.34	121.28	268.49	147.21	0.6702	1.4212
197.0	0.13309	0.6666	111.92	122.17	268.89	146.71	0.6747	1.4194
198.0	0.13992	0.6684	106.80	123.07	269.29	146.22	0.6792	1.4177
199.0	0.14701	0.6701	101.96	123.97	269.68	145.72	0.6837	1.4159
200.0	0.15438	0.6719	97.397	124.86	270.08	145.21	0.6882	1.4142
201.0	0.16202	0.6736	93.077	125.76	270.47	144.71	0.6926	1.4125
202.0	0.16996	0.6754	88.990	126.67	270.86	144.19	0.6971	1.4109
203.0	0.17819	0.6773	85.121	127.57	271.25	143.68	0.7015	1.4093
204.0	0.18673	0.6791	81.456	128.48	271.64	143.16	0.7060	1.4077
205.0	0.19557	0.6809	77.983	129.39	272.03	142.63	0.7104	1.4062
206.0	0.20473	0.6828	74.690	130.30	272.41	142.11	0.7148	1.4046
207.0	0.21421	0.6847	71.566	131.22	272.79	141.57	0.7192	1.4031
208.0	0.22403	0.6866	68.600	132.14	273.17	141.03	0.7236	1.4016
209.0	0.23419	0.6885	65.784	133.06	273.55	140.49	0.7280	1.4002
210.0	0.24469	0.6905	63.108	133.98	273.93	139.95	0.7323	1.3987
211.0	0.25554	0.6924	60.564	134.91	274.30	139.39	0.7367	1.3973
212.0	0.26676	0.6944	58.145	135.84	274.67	138.84	0.7410	1.3959
213.0	0.27834	0.6964	55.842	136.77	275.04	138.28	0.7454	1.3946
214.0	0.29030	0.6985	53.650	137.70	275.41	137.71	0.7497	1.3932
215.0	0.30265	0.7005	51.563	138.64	275.78	137.14	0.7541	1.3919
216.0	0.31539	0.7026	49.573	139.58	276.14	136.56	0.7584	1.3906
217.0	0.32853	0.7047	47.677	140.52	276.50	135.98	0.7627	1.3893
218.0	0.34207	0.7068	45.868	141.46	276.86	135.39	0.7670	1.3880
219.0	0.35603	0.7089	44.142	142.41	277.21	134.80	0.7713	1.3868
220.0	0.37042	0.7111	42.494	143.36	277.56	134.20	0.7756	1.3856
221.0	0.38524	0.7133	40.921	144.32	277.91	133.60	0.7798	1.3843
222.0	0.40049	0.7155	39.417	145.27	278.26	132.99	0.7841	1.3831
223.0	0.41619	0.7178	37.980	146.23	278.60	132.37	0.7884	1.3820
224.0	0.43235	0.7200	36.606	147.20	278.94	131.75	0.7926	1.3808
225.0	0.44898	0.7224	35.292	148.16	279.28	131.12	0.7969	1.3796
226.0	0.46607	0.7247	34.034	149.13	279.62	130.49	0.8011	1.3785
227.0	0.48365	0.7270	32.830	150.10	279.95	129.84	0.8054	1.3774
228.0	0.50171	0.7294	31.677	151.08	280.28	129.20	0.8096	1.3762
229.0	0.52027	0.7319	30.572	152.06	280.60	128.54	0.8138	1.3751
230.0	0.53934	0.7343	29.513	153.04	280.92	127.88	0.8180	1.3740
231.0	0.55891	0.7368	28.497	154.03	281.24	127.21	0.8223	1.3730
232.0	0.57901	0.7393	27.523	155.02	281.56	126.54	0.8265	1.3719
233.0	0.59964	0.7419	26.589	156.01	281.87	125.86	0.8307	1.3708
234.0	0.62081	0.7445	25.692	157.01	282.18	125.17	0.8349	1.3698
235.0	0.64253	0.7471	24.830	158.01	282.48	124.47	0.8391	1.3687
236.0	0.66480	0.7498	24.003	159.01	282.78	123.77	0.8433	1.3677
237.0	0.68764	0.7525	23.208	160.02	283.07	123.05	0.8475	1.3667
238.0	0.71105	0.7552	22.443	161.03	283.37	122.33	0.8516	1.3656
239.0	0.73504	0.7580	21.709	162.05	283.65	121.60	0.8558	1.3646
240.0	0.75962	0.7609	21.002	163.07	283.94	120.87	0.8600	1.3636
241.0	0.78481	0.7637	20.321	164.09	284.21	120.12	0.8642	1.3626
242.0	0.81060	0.7667	19.667	165.12	284.49	119.37	0.8684	1.3616
243.0	0.83700	0.7696	19.036	166.16	284.76	118.60	0.8725	1.3606
244.0	0.86404	0.7726	18.429	167.19	285.02	117.83	0.8767	1.3596

Chart R13.1: Saturation properties—*continued*

T K	p MPa	v' dm³/kg	v'' dm³/kg	h' kJ/kg	h'' kJ/kg	r kJ/kg	s' kJ/kgK	s'' kJ/kgK
245.0	0.89171	0.7757	17.844	168.23	285.28	117.05	0.8809	1.3586
246.0	0.92002	0.7788	17.280	169.28	285.53	116.25	0.8851	1.3576
247.0	0.94899	0.7820	16.737	170.33	285.78	115.45	0.8892	1.3566
248.0	0.97862	0.7853	16.213	171.39	286.03	114.64	0.8934	1.3557
249.0	1.00892	0.7885	15.707	172.45	286.26	113.82	0.8976	1.3547
250.0	1.03990	0.7919	15.219	173.51	286.50	112.98	0.9017	1.3537
251.0	1.07157	0.7953	14.747	174.58	286.72	112.14	0.9059	1.3527
252.0	1.10394	0.7988	14.292	175.66	286.94	111.28	0.9101	1.3517
253.0	1.13703	0.8023	13.853	176.74	287.15	110.41	0.9143	1.3507
254.0	1.17083	0.8060	13.428	177.83	287.36	109.53	0.9185	1.3497
255.0	1.20536	0.8097	13.018	178.92	287.56	108.64	0.9226	1.3487
256.0	1.24063	0.8134	12.621	180.02	287.76	107.73	0.9268	1.3477
257.0	1.27664	0.8173	12.237	181.13	287.94	106.82	0.9310	1.3467
258.0	1.31342	0.8212	11.866	182.24	288.12	105.88	0.9352	1.3456
259.0	1.35097	0.8252	11.506	183.35	288.29	104.94	0.9394	1.3446
260.0	1.38929	0.8293	11.158	184.48	288.45	103.98	0.9436	1.3435
261.0	1.42841	0.8335	10.821	185.61	288.61	103.00	0.9478	1.3425
262.0	1.46832	0.8378	10.495	186.75	288.75	102.01	0.9521	1.3414
263.0	1.50904	0.8422	10.178	187.89	288.89	101.00	0.9563	1.3403
264.0	1.55059	0.8468	9.8716	189.04	289.02	99.97	0.9605	1.3392
265.0	1.59297	0.8514	9.5742	190.20	289.13	98.93	0.9648	1.3381
266.0	1.63619	0.8562	9.2857	191.37	289.24	97.87	0.9691	1.3370
267.0	1.68027	0.8610	9.0058	192.55	289.34	96.79	0.9733	1.3358
268.0	1.72521	0.8661	8.7340	193.73	289.42	95.69	0.9776	1.3347
269.0	1.77103	0.8712	8.4701	194.93	289.50	94.57	0.9819	1.3335
270.0	1.81775	0.8765	8.2137	196.13	289.56	93.43	0.9862	1.3323
271.0	1.86536	0.8820	7.9645	197.35	289.61	92.26	0.9906	1.3310
272.0	1.91389	0.8876	7.7223	198.57	289.65	91.07	0.9949	1.3298
273.0	1.96335	0.8934	7.4866	199.81	289.67	89.86	0.9993	1.3285
274.0	2.01374	0.8995	7.2573	201.05	289.67	88.62	1.0037	1.3271
275.0	2.06509	0.9057	7.0341	202.31	289.67	87.35	1.0081	1.3258
276.0	2.11741	0.9121	6.8167	203.58	289.64	86.06	1.0125	1.3243
277.0	2.17071	0.9188	6.6048	204.87	289.60	84.73	1.0170	1.3229
278.0	2.22501	0.9257	6.3982	206.17	289.54	83.37	1.0215	1.3214
279.0	2.28032	0.9329	6.1967	207.48	289.45	81.98	1.0260	1.3199
280.0	2.33665	0.9404	6.0000	208.81	289.35	80.54	1.0306	1.3183
281.0	2.39403	0.9482	5.8078	210.16	289.23	79.07	1.0352	1.3166
282.0	2.45246	0.9564	5.6200	211.52	289.08	77.56	1.0399	1.3149
283.0	2.51198	0.9650	5.4363	212.90	288.90	76.00	1.0446	1.3131
284.0	2.57259	0.9740	5.2565	214.31	288.70	74.39	1.0493	1.3112
285.0	2.63431	0.9834	5.0802	215.74	288.46	72.73	1.0541	1.3093
286.0	2.69717	0.9934	4.9074	217.19	288.20	71.01	1.0590	1.3073
287.0	2.76118	1.0039	4.7376	218.67	287.89	69.22	1.0639	1.3051
288.0	2.82637	1.0151	4.5707	220.18	287.55	67.37	1.0690	1.3029
289.0	2.89277	1.0271	4.4064	221.72	287.16	65.43	1.0741	1.3005
290.0	2.96038	1.0399	4.2443	223.30	286.72	63.41	1.0793	1.2980
291.0	3.02925	1.0536	4.0841	224.93	286.22	61.30	1.0846	1.2953
292.0	3.09941	1.0685	3.9255	226.60	285.67	59.07	1.0901	1.2924
293.0	3.17087	1.0847	3.7680	228.32	285.04	56.71	1.0957	1.2893
294.0	3.24369	1.1025	3.6111	230.11	284.32	54.21	1.1016	1.2859

Chart R13.1: Saturation properties–*continued*

T K	p MPa	v' dm^3/kg	v'' dm^3/kg	h' kJ/kg	h'' kJ/kg	r kJ/kg	s' kJ/kgK	s'' kJ/kgK
295.0	3.31789	1.1222	3.4541	231.98	283.51	51.53	1.1076	1.2823
296.0	3.39352	1.1444	3.2963	233.94	282.57	48.64	1.1140	1.2783
297.0	3.47062	1.1696	3.1365	236.01	281.49	45.48	1.1206	1.2738
298.0	3.54926	1.1990	2.9733	238.23	280.23	42.00	1.1278	1.2687
299.0	3.62950	1.2341	2.8044	240.66	278.71	38.06	1.1356	1.2629
300.0	3.71141	1.2778	2.6261	243.37	276.85	33.48	1.1443	1.2559

Chart R13.2: Saturation properties

p MPa	T K	v' dm³/kg	v'' dm³/kg	h' kJ/kg	h'' kJ/kg	r kJ/kg	s' kJ/kgK	s'' kJ/kgK
0.0031	144.63	0.5914	3701.2	77.05	247.10	170.05	0.4090	1.5848
0.0032	144.94	0.5918	3593.0	77.33	247.23	169.90	0.4109	1.5832
0.0033	145.24	0.5922	3491.1	77.60	247.36	169.76	0.4128	1.5816
0.0034	145.53	0.5926	3395.0	77.86	247.48	169.62	0.4146	1.5801
0.0035	145.82	0.5929	3304.3	78.12	247.60	169.48	0.4163	1.5786
0.0036	146.10	0.5933	3218.4	78.37	247.71	169.35	0.4180	1.5771
0.0037	146.38	0.5936	3137.1	78.61	247.83	169.22	0.4197	1.5757
0.0038	146.64	0.5939	3059.9	78.85	247.94	169.09	0.4213	1.5744
0.0039	146.91	0.5942	2986.6	79.08	248.05	168.97	0.4229	1.5731
0.0040	147.16	0.5946	2916.8	79.31	248.15	168.85	0.4245	1.5718
0.0041	147.42	0.5949	2850.3	79.53	248.26	168.73	0.4260	1.5705
0.0042	147.66	0.5952	2786.9	79.75	248.36	168.61	0.4274	1.5693
0.0043	147.90	0.5955	2726.3	79.96	248.46	168.50	0.4289	1.5681
0.0044	148.14	0.5958	2668.4	80.17	248.56	168.39	0.4303	1.5670
0.0045	148.37	0.5961	2613.0	80.37	248.66	168.29	0.4316	1.5658
0.0046	148.60	0.5964	2560.0	80.57	248.75	168.18	0.4330	1.5647
0.0047	148.83	0.5966	2509.1	80.77	248.84	168.08	0.4343	1.5637
0.0048	149.05	0.5969	2460.3	80.96	248.94	167.98	0.4356	1.5626
0.0049	149.26	0.5972	2413.4	81.15	249.03	167.88	0.4369	1.5616
0.0050	149.47	0.5974	2368.4	81.33	249.11	167.78	0.4381	1.5606
0.0051	149.68	0.5977	2325.0	81.51	249.20	167.69	0.4393	1.5596
0.0052	149.89	0.5980	2283.3	81.69	249.29	167.59	0.4405	1.5586
0.0053	150.09	0.5982	2243.1	81.87	249.37	167.50	0.4417	1.5577
0.0054	150.29	0.5985	2204.3	82.04	249.45	167.41	0.4428	1.5567
0.0055	150.49	0.5987	2166.9	82.21	249.54	167.33	0.4440	1.5558
0.0056	150.68	0.5990	2130.8	82.38	249.62	167.24	0.4451	1.5549
0.0057	150.87	0.5992	2095.9	82.54	249.70	167.15	0.4462	1.5541
0.0058	151.06	0.5994	2062.2	82.71	249.77	167.07	0.4472	1.5532
0.0059	151.24	0.5997	2029.6	82.87	249.85	166.99	0.4483	1.5524
0.0060	151.43	0.5999	1998.0	83.02	249.93	166.91	0.4493	1.5515
0.0061	151.61	0.6001	1967.5	83.18	250.00	166.83	0.4503	1.5507
0.0062	151.78	0.6004	1937.9	83.33	250.08	166.75	0.4514	1.5499
0.0063	151.96	0.6006	1909.2	83.48	250.15	166.67	0.4523	1.5491
0.0064	152.13	0.6008	1881.4	83.63	250.22	166.59	0.4533	1.5484
0.0065	152.30	0.6010	1854.4	83.78	250.29	166.52	0.4543	1.5476
0.0066	152.47	0.6012	1828.2	83.92	250.36	166.44	0.4552	1.5469
0.0067	152.64	0.6014	1802.8	84.06	250.43	166.37	0.4562	1.5461
0.0068	152.80	0.6017	1778.1	84.21	250.50	166.30	0.4571	1.5454
0.0069	152.96	0.6019	1754.0	84.34	250.57	166.23	0.4580	1.5447
0.0070	153.13	0.6021	1730.7	84.48	250.64	166.16	0.4589	1.5440
0.0071	153.28	0.6023	1708.0	84.62	250.70	166.09	0.4598	1.5433
0.0072	153.44	0.6025	1685.9	84.75	250.77	166.02	0.4607	1.5426
0.0073	153.60	0.6027	1664.3	84.88	250.83	165.95	0.4615	1.5420
0.0074	153.75	0.6029	1643.4	85.02	250.90	165.88	0.4624	1.5413
0.0076	154.05	0.6032	1603.1	85.27	251.03	165.75	0.4640	1.5400
0.0078	154.34	0.6036	1564.8	85.52	251.15	165.62	0.4657	1.5388
0.0080	154.63	0.6040	1528.3	85.77	251.27	165.50	0.4673	1.5375
0.0082	154.92	0.6044	1493.6	86.01	251.39	165.38	0.4688	1.5364
0.0084	155.19	0.6047	1460.4	86.25	251.50	165.26	0.4703	1.5352
0.0086	155.46	0.6051	1428.8	86.48	251.62	165.14	0.4718	1.5341

Chart R13.2: Saturation properties—*continued*

p MPa	T K	v' dm³/kg	v'' dm³/kg	h' kJ/kg	h'' kJ/kg	r kJ/kg	s' kJ/kgK	s'' kJ/kgK
0.0088	155.73	0.6054	1398.5	86.70	251.73	165.03	0.4733	1.5330
0.0090	155.99	0.6058	1369.6	86.93	251.84	164.91	0.4747	1.5319
0.0092	156.25	0.6061	1341.9	87.15	251.95	164.80	0.4761	1.5309
0.0094	156.50	0.6064	1315.3	87.36	252.05	164.69	0.4775	1.5298
0.0096	156.75	0.6067	1289.8	87.57	252.16	164.59	0.4788	1.5288
0.0098	156.99	0.6071	1265.3	87.78	252.26	164.48	0.4801	1.5279
0.0100	157.23	0.6074	1241.7	87.98	252.36	164.38	0.4814	1.5269
0.0102	157.47	0.6077	1219.0	88.18	252.46	164.28	0.4827	1.5260
0.0104	157.70	0.6080	1197.2	88.38	252.56	164.18	0.4839	1.5251
0.0106	157.93	0.6083	1176.2	88.57	252.65	164.08	0.4852	1.5242
0.0108	158.15	0.6086	1155.9	88.76	252.75	163.99	0.4864	1.5233
0.0110	158.37	0.6089	1136.3	88.95	252.84	163.89	0.4876	1.5224
0.0115	158.91	0.6096	1090.3	89.40	253.07	163.66	0.4904	1.5203
0.0120	159.43	0.6103	1048.0	89.84	253.29	163.44	0.4932	1.5184
0.0125	159.93	0.6109	1009.0	90.27	253.50	163.23	0.4958	1.5165
0.0130	160.42	0.6116	972.88	90.68	253.70	163.02	0.4984	1.5147
0.0135	160.89	0.6122	939.34	91.08	253.90	162.83	0.5009	1.5129
0.0140	161.35	0.6128	908.12	91.46	254.09	162.63	0.5033	1.5112
0.0145	161.79	0.6134	878.99	91.84	254.28	162.44	0.5056	1.5096
0.0150	162.22	0.6140	851.73	92.20	254.46	162.26	0.5078	1.5081
0.0155	162.64	0.6145	826.18	92.55	254.64	162.08	0.5100	1.5066
0.0160	163.05	0.6151	802.16	92.90	254.81	161.91	0.5121	1.5052
0.0165	163.45	0.6156	779.56	93.24	254.98	161.74	0.5142	1.5038
0.0170	163.84	0.6162	758.24	93.56	255.14	161.58	0.5162	1.5024
0.0175	164.22	0.6167	738.10	93.88	255.30	161.42	0.5181	1.5011
0.0180	164.59	0.6172	719.04	94.20	255.46	161.26	0.5200	1.4998
0.0185	164.95	0.6177	700.97	94.50	255.61	161.11	0.5219	1.4986
0.0190	165.31	0.6182	683.83	94.80	255.76	160.96	0.5237	1.4974
0.0195	165.65	0.6186	667.53	95.09	255.91	160.81	0.5255	1.4963
0.0200	165.99	0.6191	652.02	95.38	256.05	160.67	0.5272	1.4951
0.0210	166.65	0.6200	623.14	95.94	256.33	160.39	0.5306	1.4930
0.0220	167.29	0.6209	596.80	96.47	256.60	160.12	0.5338	1.4909
0.0230	167.90	0.6217	572.68	96.99	256.86	159.86	0.5368	1.4890
0.0240	168.49	0.6226	550.49	97.49	257.10	159.61	0.5398	1.4871
0.0250	169.06	0.6233	530.02	97.97	257.35	159.37	0.5427	1.4854
0.0260	169.61	0.6241	511.08	98.44	257.58	159.14	0.5454	1.4837
0.0270	170.15	0.6249	493.48	98.89	257.80	158.91	0.5481	1.4820
0.0280	170.67	0.6256	477.10	99.34	258.02	158.69	0.5507	1.4805
0.0290	171.17	0.6263	461.82	99.76	258.24	158.47	0.5532	1.4790
0.0300	171.66	0.6270	447.51	100.18	258.44	158.26	0.5556	1.4775
0.0310	172.14	0.6277	434.10	100.59	258.64	158.06	0.5580	1.4761
0.0320	172.61	0.6284	421.49	100.98	258.84	157.86	0.5602	1.4748
0.0330	173.06	0.6290	409.63	101.37	259.03	157.66	0.5625	1.4735
0.0340	173.51	0.6297	398.43	101.75	259.22	157.47	0.5647	1.4722
0.0350	173.94	0.6303	387.86	102.11	259.40	157.29	0.5668	1.4710
0.0360	174.36	0.6309	377.85	102.47	259.58	157.10	0.5688	1.4698
0.0370	174.78	0.6315	368.36	102.83	259.75	156.92	0.5709	1.4687
0.0380	175.18	0.6321	359.35	103.17	259.92	156.75	0.5728	1.4676
0.0390	175.58	0.6327	350.79	103.51	260.09	156.58	0.5747	1.4665
0.0400	175.97	0.6332	342.65	103.84	260.25	156.41	0.5766	1.4655

Chart R13.2: Saturation properties–*continued*

p MPa	T K	v' dm³/kg	v'' dm³/kg	h' kJ/kg	h'' kJ/kg	r kJ/kg	s' kJ/kgK	s'' kJ/kgK
0.0410	176.35	0.6338	334.88	104.17	260.41	156.24	0.5785	1.4645
0.0420	176.72	0.6344	327.47	104.49	260.57	156.08	0.5803	1.4635
0.0430	177.09	0.6349	320.40	104.80	260.72	155.92	0.5820	1.4625
0.0440	177.45	0.6354	313.63	105.11	260.87	155.76	0.5838	1.4616
0.0450	177.80	0.6359	307.16	105.41	261.02	155.61	0.5855	1.4607
0.0460	178.15	0.6365	300.95	105.71	261.16	155.46	0.5871	1.4598
0.0470	178.49	0.6370	295.00	106.00	261.31	155.31	0.5888	1.4589
0.0480	178.83	0.6375	289.29	106.29	261.45	155.16	0.5904	1.4580
0.0490	179.16	0.6380	283.80	106.57	261.58	155.01	0.5920	1.4572
0.0500	179.48	0.6385	278.53	106.85	261.72	154.87	0.5935	1.4564
0.0510	179.80	0.6389	273.45	107.12	261.85	154.73	0.5950	1.4556
0.0520	180.12	0.6394	268.57	107.39	261.98	154.59	0.5965	1.4548
0.0530	180.43	0.6399	263.86	107.66	262.11	154.45	0.5980	1.4541
0.0540	180.73	0.6403	259.32	107.92	262.24	154.32	0.5995	1.4533
0.0550	181.03	0.6408	254.93	108.18	262.37	154.18	0.6009	1.4526
0.0560	181.33	0.6412	250.70	108.44	262.49	154.05	0.6023	1.4519
0.0570	181.62	0.6417	246.61	108.69	262.61	153.92	0.6037	1.4512
0.0580	181.91	0.6421	242.66	108.94	262.73	153.79	0.6050	1.4505
0.0590	182.19	0.6426	238.84	109.18	262.85	153.67	0.6064	1.4498
0.0600	182.47	0.6430	235.13	109.43	262.97	153.54	0.6077	1.4492
0.0610	182.75	0.6434	231.55	109.66	263.08	153.42	0.6090	1.4485
0.0620	183.02	0.6438	228.08	109.90	263.19	153.29	0.6103	1.4479
0.0630	183.29	0.6443	224.71	110.13	263.31	153.17	0.6116	1.4472
0.0640	183.56	0.6447	221.45	110.36	263.42	153.05	0.6128	1.4466
0.0650	183.82	0.6451	218.28	110.59	263.52	152.93	0.6141	1.4460
0.0660	184.08	0.6455	215.20	110.82	263.63	152.81	0.6153	1.4454
0.0670	184.34	0.6459	212.21	111.04	263.74	152.70	0.6165	1.4448
0.0680	184.59	0.6463	209.31	111.26	263.84	152.58	0.6177	1.4443
0.0690	184.84	0.6467	206.49	111.48	263.95	152.47	0.6189	1.4437
0.0700	185.09	0.6471	203.74	111.69	264.05	152.36	0.6200	1.4432
0.0710	185.34	0.6474	201.07	111.91	264.15	152.24	0.6212	1.4426
0.0720	185.58	0.6478	198.47	112.12	264.25	152.13	0.6223	1.4421
0.0730	185.82	0.6482	195.94	112.33	264.35	152.02	0.6234	1.4415
0.0740	186.06	0.6486	193.48	112.53	264.45	151.91	0.6245	1.4410
0.0760	186.53	0.6493	188.74	112.94	264.64	151.70	0.6267	1.4400
0.0780	186.98	0.6500	184.23	113.34	264.83	151.49	0.6288	1.4390
0.0800	187.43	0.6508	179.94	113.73	265.01	151.28	0.6309	1.4381
0.0820	187.87	0.6515	175.85	114.11	265.19	151.08	0.6330	1.4371
0.0840	188.30	0.6521	171.95	114.49	265.37	150.88	0.6350	1.4362
0.0860	188.73	0.6528	168.22	114.86	265.54	150.68	0.6369	1.4353
0.0880	189.14	0.6535	164.66	115.23	265.71	150.49	0.6388	1.4345
0.0900	189.55	0.6542	161.25	115.59	265.88	150.30	0.6407	1.4336
0.0920	189.95	0.6548	157.98	115.94	266.05	150.11	0.6426	1.4328
0.0940	190.35	0.6555	154.85	116.28	266.21	149.92	0.6444	1.4320
0.0960	190.74	0.6561	151.84	116.63	266.36	149.74	0.6462	1.4312
0.0980	191.12	0.6567	148.95	116.96	266.52	149.56	0.6479	1.4305
0.1000	191.49	0.6573	146.17	117.29	266.67	149.38	0.6497	1.4297
0.1020	191.87	0.6579	143.49	117.62	266.82	149.20	0.6514	1.4290
0.1040	192.23	0.6585	140.92	117.94	266.97	149.03	0.6530	1.4283
0.1060	192.59	0.6591	138.43	118.26	267.12	148.86	0.6547	1.4276

Chart R13.2: Saturation properties—*continued*

p MPa	T K	v' dm³/kg	v'' dm³/kg	h' kJ/kg	h'' kJ/kg	r kJ/kg	s' kJ/kgK	s'' kJ/kgK
0.1080	192.94	0.6597	136.04	118.57	267.26	148.69	0.6563	1.4269
0.1100	193.29	0.6603	133.73	118.88	267.40	148.52	0.6579	1.4262
0.1150	194.14	0.6617	128.29	119.63	267.74	148.11	0.6617	1.4246
0.1200	194.97	0.6631	123.29	120.36	268.07	147.71	0.6655	1.4231
0.1250	195.76	0.6645	118.68	121.07	268.39	147.32	0.6691	1.4216
0.1300	196.53	0.6658	114.40	121.76	268.70	146.94	0.6726	1.4202
0.1350	197.28	0.6671	110.44	122.43	269.00	146.57	0.6760	1.4189
0.1400	198.01	0.6684	106.74	123.08	269.29	146.21	0.6792	1.4176
0.1450	198.72	0.6696	103.29	123.71	269.57	145.86	0.6824	1.4164
0.1500	199.41	0.6708	100.06	124.33	269.84	145.51	0.6855	1.4152
0.1550	200.08	0.6720	97.029	124.94	270.11	145.17	0.6885	1.4141
0.1600	200.74	0.6732	94.182	125.53	270.37	144.84	0.6915	1.4130
0.1650	201.38	0.6743	91.500	126.11	270.62	144.51	0.6943	1.4119
0.1700	202.00	0.6755	88.971	126.67	270.86	144.19	0.6971	1.4109
0.1750	202.62	0.6766	86.580	127.23	271.10	143.88	0.6998	1.4099
0.1800	203.21	0.6777	84.317	127.77	271.34	143.57	0.7025	1.4090
0.1850	203.80	0.6787	82.171	128.30	271.56	143.26	0.7051	1.4080
0.1900	204.37	0.6798	80.134	128.82	271.78	142.96	0.7076	1.4071
0.1950	204.94	0.6808	78.198	129.33	272.00	142.67	0.7101	1.4063
0.2000	205.49	0.6819	76.354	129.84	272.21	142.38	0.7125	1.4054
0.2100	206.56	0.6839	72.921	130.82	272.62	141.81	0.7173	1.4038
0.2200	207.59	0.6858	69.787	131.76	273.02	141.25	0.7218	1.4022
0.2300	208.59	0.6877	66.916	132.68	273.40	140.71	0.7262	1.4008
0.2400	209.56	0.6896	64.274	133.57	273.76	140.19	0.7304	1.3994
0.2500	210.49	0.6914	61.836	134.44	274.11	139.67	0.7345	1.3980
0.2600	211.40	0.6932	59.578	135.28	274.45	139.17	0.7384	1.3968
0.2700	212.28	0.6950	57.481	136.10	274.78	138.68	0.7423	1.3956
0.2800	213.14	0.6967	55.528	136.90	275.10	138.20	0.7460	1.3944
0.2900	213.97	0.6984	53.704	137.68	275.40	137.72	0.7496	1.3933
0.3000	214.79	0.7001	51.997	138.44	275.70	137.26	0.7531	1.3922
0.3100	215.58	0.7017	50.396	139.18	275.99	136.80	0.7566	1.3911
0.3200	216.35	0.7033	48.890	139.91	276.27	136.36	0.7599	1.3901
0.3300	217.11	0.7049	47.473	140.62	276.54	135.92	0.7632	1.3892
0.3400	217.85	0.7065	46.136	141.32	276.80	135.48	0.7663	1.3882
0.3500	218.57	0.7080	44.871	142.01	277.06	135.05	0.7694	1.3873
0.3600	219.28	0.7095	43.675	142.68	277.31	134.63	0.7725	1.3865
0.3700	219.97	0.7110	42.540	143.34	277.55	134.22	0.7754	1.3856
0.3800	220.65	0.7125	41.463	143.98	277.79	133.81	0.7783	1.3848
0.3900	221.32	0.7140	40.439	144.62	278.02	133.40	0.7812	1.3840
0.4000	221.97	0.7155	39.464	145.24	278.25	133.01	0.7840	1.3832
0.4100	222.61	0.7169	38.534	145.86	278.47	132.61	0.7867	1.3824
0.4200	223.24	0.7183	37.648	146.46	278.69	132.22	0.7894	1.3817
0.4300	223.86	0.7197	36.800	147.06	278.90	131.84	0.7920	1.3810
0.4400	224.46	0.7211	35.990	147.64	279.10	131.46	0.7946	1.3803
0.4500	225.06	0.7225	35.214	148.22	279.30	131.08	0.7971	1.3796
0.4600	225.65	0.7239	34.471	148.79	279.50	130.71	0.7996	1.3789
0.4700	226.23	0.7252	33.757	149.35	279.69	130.34	0.8021	1.3782
0.4800	226.79	0.7266	33.073	149.90	279.88	129.98	0.8045	1.3776
0.4900	227.35	0.7279	32.415	150.45	280.07	129.62	0.8069	1.3770
0.5000	227.91	0.7292	31.782	150.99	280.25	129.26	0.8092	1.3764

Chart R13.2: Saturation properties—*continued*

p MPa	T K	v' dm³/kg	v'' dm³/kg	h' kJ/kg	h'' kJ/kg	r kJ/kg	s' kJ/kgK	s'' kJ/kgK
0.5100	228.45	0.7305	31.174	151.52	280.42	128.90	0.8115	1.3757
0.5200	228.99	0.7318	30.587	152.05	280.60	128.55	0.8138	1.3752
0.5300	229.51	0.7331	30.022	152.56	280.77	128.20	0.8160	1.3746
0.5400	230.03	0.7344	29.477	153.08	280.94	127.86	0.8182	1.3740
0.5500	230.55	0.7357	28.951	153.58	281.10	127.52	0.8204	1.3735
0.5600	231.05	0.7369	28.443	154.08	281.26	127.18	0.8225	1.3729
0.5700	231.55	0.7382	27.952	154.58	281.42	126.84	0.8246	1.3724
0.5800	232.05	0.7394	27.477	155.07	281.57	126.51	0.8267	1.3718
0.5900	232.54	0.7407	27.018	155.55	281.72	126.17	0.8287	1.3713
0.6000	233.02	0.7419	26.573	156.03	281.87	125.85	0.8307	1.3708
0.6100	233.49	0.7432	26.142	156.50	282.02	125.52	0.8327	1.3703
0.6200	233.96	0.7444	25.725	156.97	282.16	125.19	0.8347	1.3698
0.6300	234.43	0.7456	25.320	157.43	282.31	124.87	0.8367	1.3693
0.6400	234.88	0.7468	24.928	157.89	282.44	124.55	0.8386	1.3689
0.6500	235.34	0.7480	24.547	158.35	282.58	124.23	0.8405	1.3684
0.6600	235.79	0.7492	24.177	158.80	282.72	123.92	0.8424	1.3679
0.6700	236.23	0.7504	23.817	159.24	282.85	123.60	0.8442	1.3675
0.6800	236.67	0.7516	23.468	159.69	282.98	123.29	0.8461	1.3670
0.6900	237.10	0.7528	23.128	160.12	283.10	122.98	0.8479	1.3666
0.7000	237.53	0.7539	22.798	160.56	283.23	122.67	0.8497	1.3661
0.7100	237.96	0.7551	22.477	160.99	283.35	122.36	0.8515	1.3657
0.7200	238.38	0.7563	22.164	161.42	283.47	122.06	0.8532	1.3653
0.7300	238.79	0.7574	21.859	161.84	283.59	121.76	0.8550	1.3648
0.7400	239.20	0.7586	21.562	162.26	283.71	121.45	0.8567	1.3644
0.7600	240.02	0.7609	20.991	163.09	283.94	120.85	0.8601	1.3636
0.7800	240.81	0.7632	20.448	163.90	284.16	120.26	0.8634	1.3628
0.8000	241.59	0.7655	19.931	164.70	284.38	119.67	0.8667	1.3620
0.8200	242.36	0.7677	19.438	165.49	284.58	119.09	0.8699	1.3613
0.8400	243.11	0.7700	18.967	166.27	284.79	118.52	0.8730	1.3605
0.8600	243.85	0.7722	18.518	167.04	284.98	117.94	0.8761	1.3598
0.8800	244.58	0.7744	18.087	167.80	285.17	117.38	0.8791	1.3590
0.9000	245.30	0.7766	17.676	168.54	285.36	116.81	0.8821	1.3583
0.9200	246.00	0.7788	17.281	169.28	285.53	116.25	0.8851	1.3576
0.9400	246.69	0.7810	16.902	170.01	285.71	115.70	0.8879	1.3570
0.9600	247.37	0.7832	16.538	170.73	285.87	115.15	0.8908	1.3563
0.9800	248.05	0.7854	16.189	171.44	286.04	114.60	0.8936	1.3556
1.0000	248.71	0.7876	15.853	172.14	286.19	114.06	0.8964	1.3550
1.0200	249.36	0.7897	15.529	172.83	286.35	113.52	0.8991	1.3543
1.0400	250.00	0.7919	15.217	173.52	286.50	112.98	0.9018	1.3537
1.0600	250.64	0.7941	14.916	174.20	286.64	112.45	0.9044	1.3530
1.0800	251.26	0.7962	14.626	174.87	286.78	111.91	0.9070	1.3524
1.1000	251.88	0.7984	14.347	175.53	286.92	111.38	0.9096	1.3518
1.1500	253.39	0.8037	13.687	177.16	287.24	110.07	0.9159	1.3503
1.2000	254.85	0.8091	13.080	178.75	287.53	108.78	0.9220	1.3488
1.2500	256.26	0.8144	12.519	180.31	287.80	107.50	0.9279	1.3474
1.3000	257.64	0.8198	11.999	181.83	288.06	106.22	0.9337	1.3460
1.3500	258.97	0.8251	11.515	183.32	288.29	104.96	0.9393	1.3446
1.4000	260.28	0.8305	11.064	184.79	288.50	103.71	0.9448	1.3432
1.4500	261.54	0.8359	10.642	186.23	288.69	102.46	0.9501	1.3419
1.5000	262.78	0.8413	10.247	187.64	288.86	101.22	0.9554	1.3406

Chart R13.2: Saturation properties–*continued*

p MPa	T K	v' dm^3/kg	v'' dm^3/kg	h' kJ/kg	h'' kJ/kg	r kJ/kg	s' kJ/kgK	s'' kJ/kgK
1.5500	263.99	0.8467	9.8759	189.03	289.02	99.99	0.9605	1.3392
1.6000	265.16	0.8522	9.5263	190.39	289.15	98.76	0.9655	1.3379
1.6500	266.32	0.8577	9.1965	191.74	289.27	97.53	0.9704	1.3366
1.7000	267.44	0.8632	8.8848	193.07	289.38	96.31	0.9752	1.3353
1.7500	268.54	0.8688	8.5897	194.38	289.47	95.08	0.9800	1.3340
1.8000	269.62	0.8745	8.3097	195.68	289.54	93.86	0.9846	1.3327
1.8500	270.68	0.8802	8.0436	196.96	289.60	92.64	0.9892	1.3314
1.9000	271.72	0.8860	7.7905	198.22	289.64	91.41	0.9937	1.3301
1.9500	272.73	0.8919	7.5491	199.47	289.66	90.19	0.9981	1.3288
2.0000	273.73	0.8978	7.3188	200.71	289.67	88.96	1.0025	1.3275
2.1000	275.67	0.9100	6.8880	203.16	289.65	86.49	1.0111	1.3248
2.2000	277.54	0.9225	6.4923	205.57	289.57	84.00	1.0194	1.3221
2.3000	279.35	0.9355	6.1270	207.95	289.42	81.48	1.0276	1.3193
2.4000	281.10	0.9491	5.7883	210.30	289.21	78.92	1.0357	1.3164
2.5000	282.80	0.9632	5.4727	212.63	288.94	76.31	1.0436	1.3135
2.6000	284.45	0.9781	5.1774	214.94	288.60	73.65	1.0515	1.3104
2.7000	286.04	0.9938	4.8997	217.26	288.18	70.93	1.0592	1.3072
2.8000	287.60	1.0106	4.6375	219.57	287.69	68.12	1.0669	1.3038
2.9000	289.11	1.0284	4.3888	221.89	287.11	65.22	1.0746	1.3002
3.0000	290.58	1.0477	4.1516	224.24	286.44	62.20	1.0824	1.2964
3.1000	292.01	1.0686	3.9242	226.61	285.66	59.05	1.0901	1.2924
3.2000	293.40	1.0916	3.7049	229.03	284.76	55.72	1.0981	1.2880
3.3000	294.76	1.1173	3.4917	231.52	283.71	52.19	1.1061	1.2832
3.4000	296.08	1.1464	3.2828	234.11	282.49	48.38	1.1145	1.2779
3.5000	297.38	1.1801	3.0756	236.83	281.04	44.22	1.1233	1.2720
3.6000	298.63	1.2205	2.8669	239.74	279.30	39.56	1.1327	1.2651
3.7000	299.86	1.2711	2.6514	242.97	277.14	34.16	1.1430	1.2570

Chart R13.3: Thermodynamic properties

p/MPa	0.50			0.60			0.80		
T K	v dm³/kg	h kJ/kg	s kJ/kgK	v dm³/kg	h kJ/kg	s kJ/kgK	v dm³/kg	h kJ/kg	s kJ/kgK
150.0	0.5978	81.991	0.4405	0.5977	82.032	0.4404	0.5976	82.114	0.4401
160.0	0.6106	90.518	0.4956	0.6105	90.558	0.4954	0.6104	90.638	0.4952
170.0	0.6242	98.952	0.5467	0.6241	98.990	0.5465	0.6239	99.068	0.5463
180.0	0.6387	107.46	0.5953	0.6386	107.50	0.5952	0.6384	107.57	0.5949
190.0	0.6543	116.12	0.6421	0.6542	116.15	0.6420	0.6539	116.22	0.6416
200.0	0.6713	124.97	0.6875	0.6711	125.00	0.6874	0.6708	125.07	0.6870
210.0	0.6899	134.05	0.7318	0.6897	134.08	0.7316	0.6893	134.14	0.7312
220.0	0.7108	143.39	0.7753	0.7105	143.41	0.7751	0.7100	143.46	0.7746
230.0	32.233	281.60	1.3823	0.7341	153.05	0.8179	0.7334	153.08	0.8174
240.0	34.318	288.03	1.4096	27.860	286.53	1.3905	0.7607	163.07	0.8599
250.0	36.317	294.45	1.4358	29.619	293.14	1.4175	21.180	290.33	1.3863
260.0	38.250	300.88	1.4610	31.304	299.72	1.4433	22.576	297.28	1.4135
270.0	40.133	307.35	1.4855	32.934	306.32	1.4682	23.903	304.16	1.4395
280.0	41.976	313.87	1.5092	34.521	312.95	1.4923	25.177	311.02	1.4644
290.0	43.788	320.47	1.5323	36.073	319.63	1.5157	26.412	317.89	1.4885
300.0	45.573	327.14	1.5550	37.598	326.37	1.5386	27.616	324.79	1.5119
310.0	47.337	333.89	1.5771	39.100	333.19	1.5610	28.794	331.73	1.5347
320.0	49.082	340.73	1.5988	40.583	340.08	1.5828	29.951	338.74	1.5569
330.0	50.812	347.65	1.6201	42.050	347.04	1.6043	31.090	345.80	1.5787
340.0	52.529	354.66	1.6410	43.503	354.09	1.6253	32.215	352.94	1.6000
350.0	54.234	361.75	1.6616	44.944	361.22	1.6460	33.328	360.14	1.6209
360.0	55.930	368.93	1.6818	46.376	368.43	1.6663	34.430	367.42	1.6414
370.0	57.617	376.19	1.7017	47.798	375.72	1.6862	35.523	374.77	1.6615
380.0	59.296	383.53	1.7213	49.213	383.09	1.7059	36.608	382.20	1.6813
390.0	60.969	390.95	1.7405	50.621	390.53	1.7252	37.685	389.69	1.7008
400.0	62.636	398.45	1.7595	52.023	398.05	1.7443	38.757	397.26	1.7199
410.0	64.298	406.02	1.7782	53.420	405.65	1.7630	39.823	404.90	1.7388
420.0	65.955	413.67	1.7967	54.812	413.32	1.7815	40.884	412.61	1.7574
430.0	67.607	421.40	1.8148	56.200	421.06	1.7997	41.940	420.38	1.7756
440.0	69.256	429.19	1.8327	57.584	428.86	1.8177	42.993	428.22	1.7937
450.0	70.902	437.04	1.8504	58.964	436.74	1.8354	44.042	436.12	1.8114

p/MPa	1.00			2.00			4.00		
150.0	0.5974	82.196	0.4399	0.5968	82.607	0.4387	0.5954	83.432	0.4362
160.0	0.6102	90.718	0.4949	0.6094	91.119	0.4936	0.6079	91.924	0.4910
170.0	0.6237	99.145	0.5460	0.6228	99.533	0.5446	0.6209	100.31	0.5419
180.0	0.6381	107.64	0.5946	0.6370	108.01	0.5931	0.6348	108.76	0.5902
190.0	0.6536	116.29	0.6413	0.6523	116.64	0.6397	0.6496	117.35	0.6366
200.0	0.6704	125.13	0.6867	0.6688	125.45	0.6849	0.6656	126.11	0.6815
210.0	0.6889	134.19	0.7309	0.6868	134.48	0.7289	0.6829	135.07	0.7252
220.0	0.7094	143.51	0.7742	0.7068	143.74	0.7720	0.7020	144.24	0.7679
230.0	0.7327	153.11	0.8169	0.7294	153.28	0.8144	0.7232	153.67	0.8098
240.0	0.7597	163.08	0.8593	0.7552	163.15	0.8564	0.7472	163.37	0.8511
250.0	16.024	287.18	1.3589	0.7858	173.44	0.8984	0.7748	173.41	0.8921
260.0	17.279	294.61	1.3881	0.8236	184.28	0.9409	0.8076	183.88	0.9331
270.0	18.443	301.85	1.4154	0.8736	195.99	0.9851	0.8481	194.91	0.9748
280.0	19.542	308.98	1.4413	7.9156	295.99	1.3503	0.9017	206.79	1.0180
290.0	20.594	316.06	1.4662	8.7283	305.11	1.3823	0.9824	220.22	1.0650

Chart R13.3: Thermodynamic properties–*continued*

p/MPa	1.00			2.00			4.00		
T K	v dm³/kg	h kJ/kg	s kJ/kgK	v dm³/kg	h kJ/kg	s kJ/kgK	v dm³/kg	h kJ/kg	s kJ/kgK
300.0	21.610	323.14	1.4902	9.4421	313.62	1.4112	1.1662	238.49	1.1269
310.0	22.598	330.23	1.5134	10.094	321.80	1.4380	3.2148	293.20	1.3068
320.0	23.562	337.36	1.5360	10.704	329.79	1.4634	3.9393	308.17	1.3544
330.0	24.508	344.54	1.5581	11.283	337.66	1.4876	4.4633	319.72	1.3900
340.0	25.438	351.76	1.5797	11.838	345.47	1.5109	4.9028	329.97	1.4206
350.0	26.354	359.05	1.6008	12.374	353.25	1.5335	5.2935	339.54	1.4483
360.0	27.260	366.40	1.6215	12.894	361.03	1.5554	5.6517	348.70	1.4741
370.0	28.155	373.82	1.6418	13.402	368.82	1.5767	5.9866	357.60	1.4985
380.0	29.042	381.30	1.6618	13.900	376.62	1.5975	6.3040	366.34	1.5218
390.0	29.922	388.85	1.6814	14.388	384.46	1.6179	6.6076	374.96	1.5442
400.0	30.796	396.46	1.7007	14.869	392.34	1.6378	6.9000	383.51	1.5659
410.0	31.664	404.14	1.7196	15.343	400.25	1.6574	7.1833	392.01	1.5868
420.0	32.526	411.89	1.7383	15.811	408.21	1.6765	7.4589	400.49	1.6073
430.0	33.385	419.69	1.7567	16.274	416.21	1.6954	7.7280	408.95	1.6272
440.0	34.239	427.57	1.7748	16.733	424.26	1.7139	7.9914	417.41	1.6466
450.0	35.090	435.50	1.7926	17.187	432.35	1.7321	8.2500	425.88	1.6657
p/MPa	6.00			8.00			10.00		
150.0	0.5941	84.260	0.4338	0.5929	85.092	0.4314	0.5917	85.927	0.4291
160.0	0.6064	92.735	0.4885	0.6049	93.550	0.4860	0.6035	94.369	0.4836
170.0	0.6192	101.10	0.5392	0.6175	101.90	0.5366	0.6158	102.69	0.5341
180.0	0.6327	109.52	0.5874	0.6307	110.29	0.5846	0.6288	111.06	0.5819
190.0	0.6471	118.07	0.6336	0.6448	118.81	0.6306	0.6425	119.55	0.6278
200.0	0.6626	126.79	0.6783	0.6598	127.48	0.6751	0.6570	128.18	0.6721
210.0	0.6793	135.68	0.7217	0.6759	136.32	0.7183	0.6727	136.98	0.7150
220.0	0.6975	144.78	0.7640	0.6934	145.35	0.7603	0.6895	145.95	0.7567
230.0	0.7176	154.11	0.8054	0.7125	154.59	0.8013	0.7078	155.11	0.7974
240.0	0.7400	163.68	0.8462	0.7336	164.05	0.8416	0.7278	164.47	0.8372
250.0	0.7654	173.53	0.8864	0.7572	173.75	0.8812	0.7499	174.05	0.8763
260.0	0.7947	183.73	0.9264	0.7839	183.74	0.9203	0.7746	183.87	0.9149
270.0	0.8295	194.34	0.9664	0.8148	194.06	0.9593	0.8026	193.97	0.9530
280.0	0.8723	205.50	1.0070	0.8513	204.77	0.9983	0.8347	204.38	0.9908
290.0	0.9280	217.43	1.0488	0.8956	215.99	1.0376	0.8724	215.16	1.0286
300.0	1.0072	230.59	1.0934	0.9517	227.86	1.0779	0.9174	226.37	1.0666
310.0	1.1401	246.09	1.1442	1.0268	240.63	1.1197	0.9727	238.10	1.1051
320.0	1.4364	266.79	1.2099	1.1339	254.67	1.1643	1.0425	250.46	1.1443
330.0	1.9815	290.44	1.2827	1.2948	270.30	1.2123	1.1326	263.53	1.1846
340.0	2.4782	308.16	1.3357	1.5247	287.04	1.2623	1.2490	277.28	1.2256
350.0	2.8749	321.93	1.3756	1.7977	303.22	1.3092	1.3942	291.43	1.2666
360.0	3.2098	333.80	1.4090	2.0709	317.76	1.3502	1.5622	305.47	1.3062
370.0	3.5054	344.61	1.4387	2.3253	330.75	1.3858	1.7414	318.95	1.3431
380.0	3.7743	354.78	1.4658	2.5590	342.62	1.4175	1.9215	331.67	1.3770
390.0	4.0237	364.53	1.4911	2.7748	353.69	1.4462	2.0965	343.66	1.4082
400.0	4.2583	374.00	1.5151	2.9761	364.22	1.4729	2.2642	355.03	1.4370
410.0	4.4813	383.26	1.5380	3.1657	374.36	1.4979	2.4244	365.91	1.4638
420.0	4.6949	392.39	1.5599	3.3456	384.21	1.5217	2.5775	376.41	1.4892
430.0	4.9007	401.41	1.5812	3.5177	393.85	1.5443	2.7242	386.62	1.5132
440.0	5.1000	410.36	1.6018	3.6831	403.33	1.5661	2.8654	396.60	1.5361
450.0	5.2938	419.26	1.6218	3.8429	412.69	1.5872	3.0017	406.40	1.5582

Chart R13.3: Thermodynamic properties—*continued*

p/MPa	15.00			20.00			50.00		
T K	v dm³/kg	h kJ/kg	s kJ/kgK	v dm³/kg	h kJ/kg	s kJ/kgK	v dm³/kg	h kJ/kg	s kJ/kgK
150.0	0.5887	88.028	0.4234	0.5860	90.145	0.4180	0.5719	103.09	0.3885
160.0	0.6001	96.434	0.4777	0.5969	98.520	0.4720	0.5810	111.34	0.4418
170.0	0.6119	104.72	0.5279	0.6082	106.76	0.5220	0.5903	119.44	0.4909
180.0	0.6242	113.03	0.5754	0.6200	115.03	0.5693	0.5998	127.54	0.5372
190.0	0.6371	121.45	0.6209	0.6323	123.39	0.6145	0.6095	135.69	0.5813
200.0	0.6508	130.00	0.6648	0.6452	131.87	0.6580	0.6195	143.92	0.6235
210.0	0.6653	138.69	0.7072	0.6588	140.48	0.7000	0.6298	152.24	0.6641
220.0	0.6808	147.54	0.7483	0.6731	149.23	0.7407	0.6404	160.65	0.7032
230.0	0.6973	156.54	0.7884	0.6884	158.12	0.7802	0.6514	169.15	0.7410
240.0	0.7152	165.71	0.8274	0.7046	167.14	0.8186	0.6627	177.74	0.7775
250.0	0.7345	175.05	0.8655	0.7220	176.32	0.8560	0.6744	186.42	0.8129
260.0	0.7556	184.58	0.9029	0.7407	185.64	0.8926	0.6865	195.18	0.8473
270.0	0.7787	194.30	0.9396	0.7607	195.12	0.9283	0.6991	204.03	0.8807
280.0	0.8043	204.23	0.9757	0.7824	204.75	0.9634	0.7120	212.97	0.9132
290.0	0.8327	214.37	1.0113	0.8060	214.54	0.9977	0.7254	221.99	0.9449
300.0	0.8646	224.75	1.0464	0.8316	224.50	1.0315	0.7392	231.09	0.9757
310.0	0.9006	235.37	1.0813	0.8595	234.61	1.0647	0.7535	240.26	1.0058
320.0	0.9415	246.24	1.1158	0.8901	244.89	1.0973	0.7682	249.51	1.0352
330.0	0.9882	257.37	1.1500	0.9235	255.32	1.1294	0.7834	258.84	1.0639
340.0	1.0415	268.74	1.1840	0.9600	265.89	1.1609	0.7990	268.23	1.0919
350.0	1.1022	280.32	1.2175	0.9999	276.58	1.1919	0.8151	277.68	1.1193
360.0	1.1707	292.05	1.2506	1.0433	287.39	1.2224	0.8317	287.18	1.1461
370.0	1.2467	303.86	1.2829	1.0901	298.27	1.2522	0.8486	296.75	1.1723
380.0	1.3295	315.66	1.3144	1.1404	309.20	1.2813	0.8660	306.35	1.1979
390.0	1.4176	327.35	1.3448	1.1940	320.15	1.3098	0.8838	316.01	1.2230
400.0	1.5095	338.88	1.3740	1.2503	331.09	1.3375	0.9020	325.70	1.2475
410.0	1.6036	350.20	1.4019	1.3091	341.97	1.3643	0.9205	335.43	1.2715
420.0	1.6985	361.30	1.4287	1.3698	352.80	1.3904	0.9394	345.19	1.2951
430.0	1.7932	372.17	1.4542	1.4320	363.54	1.4157	0.9587	354.98	1.3181
440.0	1.8872	382.83	1.4788	1.4952	374.18	1.4402	0.9782	364.79	1.3406
450.0	1.9799	393.31	1.5023	1.5590	384.72	1.4639	0.9981	374.62	1.3627

References for R13

[Albr52] ALBRIGHT, L.F. AND MARTIN, J.J.: Thermodynamic properties of chlortrifluoromethane; Ind. Eng. Chem. 44(1952)1 188/198

[Chen76] CHEN, S.S.; WILHOIT, R.C. AND ZWOLINSKI, B.J.: Ideal Gas Thermodynamic Properties of Six Chlorofluoromethanes; J. Phys. Chem. Ref. Data 5(1976)3 571/580

[Gell79] GELLER, V.Z. ; PORITSCHANSKIJ, E.G. ; SVETLIUNIJ, P.I. AND ELKIN, JU.G.: Density and EOS of liquid Freons 13, 22, 23 and 152A; Kholod. Tekh. Tekhnol 29(1979) 43/48

[Gruz75] GRUZDEV, V.A.; SHELUDYAKOV, E.P.; KIRIJANENKO, A.A.; KOLOTOV, J.L.; LAVROV, B.A.; SCHILJAKOV, A.A.; SCHESTOVA, A.I. AND SCHUMSKAJA A.I.: Experimental investigation of the isobaric heat capacity of Freons: R-11, R-12, R-13, R-22, R-23; Thermophysical properties of Matter and Substances 8(1975) 108/120

[Hoec78] HOECHST AG: Physikalische Daten für Frigen; Frigen-Informationsdienst (1978)

[Jaeg73] JÄGER, H.P.: Dichte- und Dampfdruckmeßwerte von verschiedenen Kältemitteln; Die Kälte (1973)7 276/278

[KMR 81] - : Kältemaschinenregeln, 7. Auflage; Verlag C.F. Müller, Karlsruhe (1981)

[Kubi80] KUBIC, W.L. AND STEIN, F.P.: An Experimental and Correlative Study of the Vapor-Liquid Equilibria of the Tetrafluoromethane-Chlorotrifluoromethane System; Fluid Phase Equilibria (1980/81)5 289/304

[Kubo74] KUBOTA, H. ; TANAKA, Y. AND MAKITA, T.: Compressibility of Fluorocarbons by the Burnett Method; Review of Physical Chem. of Japan (1974) 538/542

[Mich66] MICHELS, A. ; WASSENAR, T. ; WOLKERS, G.J. ; PRINS, CHR. AND L.V.D. KLUNDERT: P-v-T data and thermodynamical properties of Freon-12 (CCl_2F_2) and Freon-13 ($CClF_3$) fluorocarbons at temperatures between 0^0 C and 1500^0 C and at pressures up to 400 atm; J. Chem. Eng. Data 11(1966)10 449/452

[Oguc75] OGUCHI, K. ; TANISHITA, J. ; WATANABE, K. ; YAMAGUCHI, T. AND SASAYMA, A.: Experimental study of p-v-T properties of fluorocarbon refrigerant R13. I. p-v-T properties; Bull. JSME 18(1975)126 1448/1455

[Oguc75a] OGUCHI, K. ; TANISHITA, J. ; WATANABE, K. ; YAMAGUCHI, T. AND SASAYMA, A.: Experimental study of p-v-T properties of fluorocarbon refrigerant R13. II. Experimental results of the saturation pressure and determination of the critical parameters; Bull. JSME 18(1975)126 1456/1464

[Otto66] OTTO, J. AND THOMAS, W.: Die Füllmenge verflüssigter Gase in Behältern; Teil 4: Trifluormonochlormethan (R13) und Trifluormonobrommethan (R13B1); PTB Mitteilungen (1966)6 526/528

[Pere71] PERELSTEIN, I.I.: Thermodynamic properties of Freon 12 and Freon 13; Thermophysical Properties of Matter and Substances 4(1971) 65/95

[Rass79] RASSKAZOV, D.S. ; PETROV, E.K. AND USCHMAJKIN, Z.R.: Thermal properties of R13; Thermophysical Properties of Matter and Substances 13(1979) 49/56

[Redk76] REDKOZUBOV, V.A.: Rate of Propagation of Ultrasonic Waves in Freons 11, 12 and 13; Primenenie ul'traakustiki Issledovaniju vescestva 26(1976) 14/19

[Ried41] RIEDEL, L.: Bestimmung der thermischen Eigenschaften von Trifluormonochlormethan; Zeitschrift für die gesamte Kälteindustrie 48(1941)1 9/13

[Shav79] SHAVANDRIN, A.M. AND LI, S.A.: Temperature-Density Parameters of Freon-13 on the Saturation Line; Inzhenerno-Fizicheskii Zhurnal 37(1979)5 830/834

[Tani72] TANISHITA, I. ; WATANABE, K. ; KIJIMA, J. ; OGUCHI, K. AND KONDO, H.: Status Report on the Thermodynamic Properties Research of Fluorocarbon Refrigerants; Second report, Keio University, Tokyo (1972)

R 13. ph

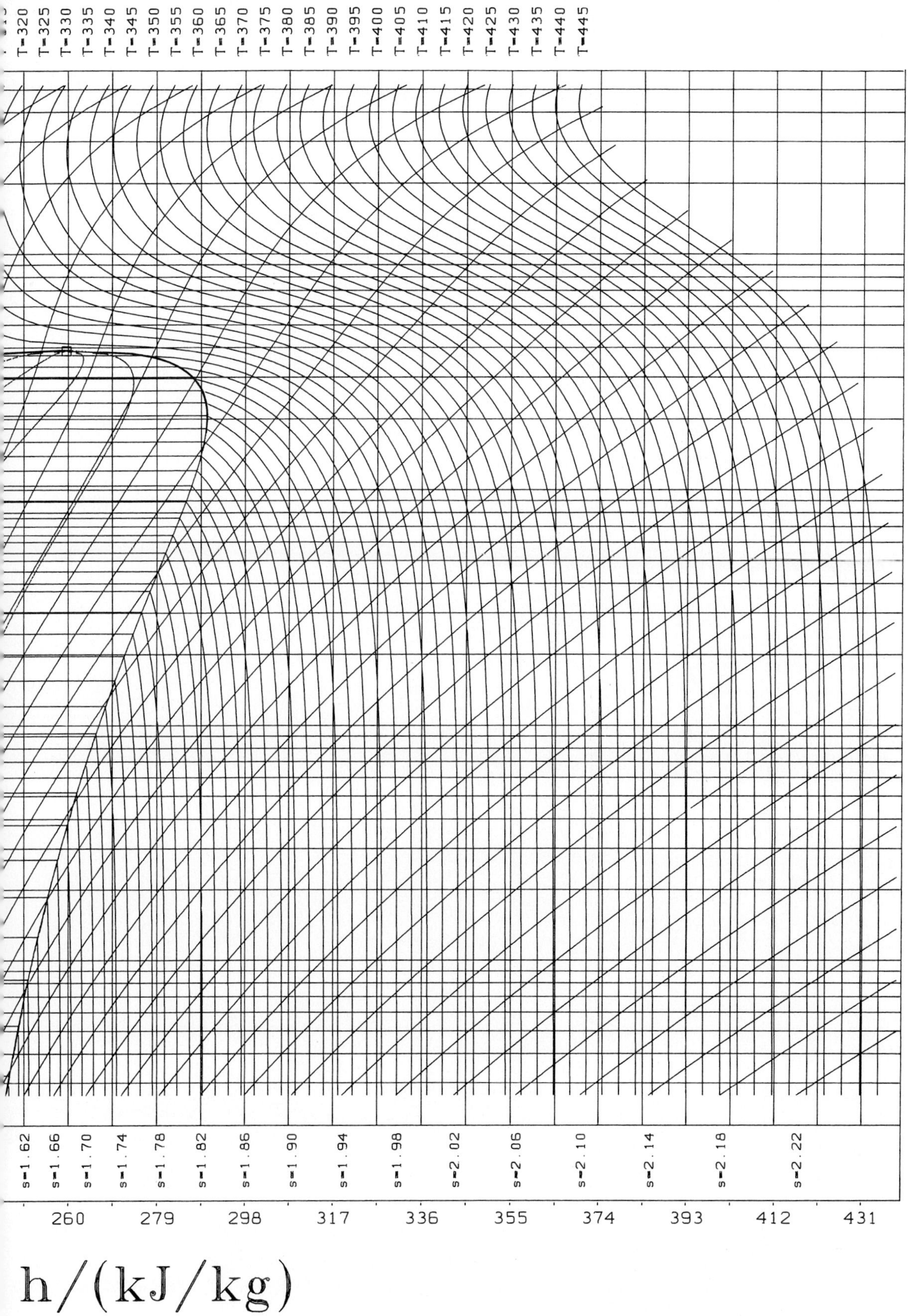

T=320
T=325
T=330
T=335
T=340
T=345
T=350
T=355
T=360
T=365
T=370
T=375
T=380
T=385
T=390
T=395
T=400
T=405
T=410
T=415
T=420
T=425
T=430
T=435
T=440
T=445
s=1.62
s=1.66
s=1.70
s=1.74
s=1.78
s=1.82
s=1.86
s=1.90
s=1.94
s=1.98
s=2.02
s=2.06
s=2.10
s=2.14
s=2.18
s=2.22
260
279
298
317
336
355
374
393
412
431
h/(kJ/kg)

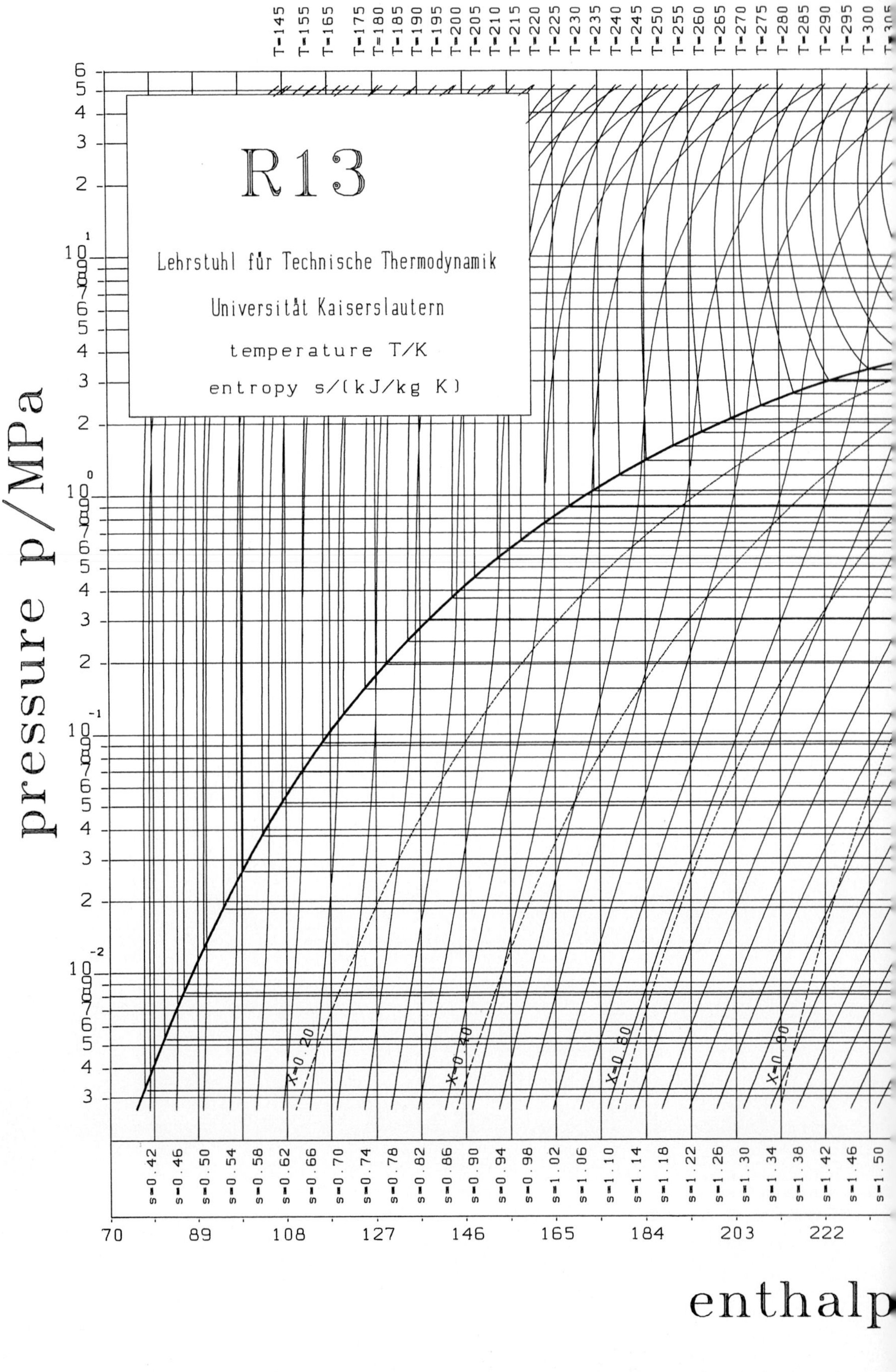

R13
Lehrstuhl für Technische Thermodynamik
Universität Kaiserslautern
temperature T/K
entropy s/(kJ/kg K)
pressure p/MPa
enthalp
T=145
T=155
T=165
T=175
T=180
T=185
T=190
T=195
T=200
T=205
T=210
T=215
T=220
T=225
T=230
T=235
T=240
T=245
T=250
T=255
T=260
T=265
T=270
T=275
T=280
T=285
T=290
T=295
T=300
X=0.20
X=0.40
X=0.60
X=0.80
s=0.42
s=0.46
s=0.50
s=0.54
s=0.58
s=0.62
s=0.66
s=0.70
s=0.74
s=0.78
s=0.82
s=0.86
s=0.90
s=0.94
s=0.98
s=1.02
s=1.06
s=1.10
s=1.14
s=1.18
s=1.22
s=1.26
s=1.30
s=1.34
s=1.38
s=1.42
s=1.46
s=1.50
70
89
108
127
146
165
184
203
222

R 13. Ts

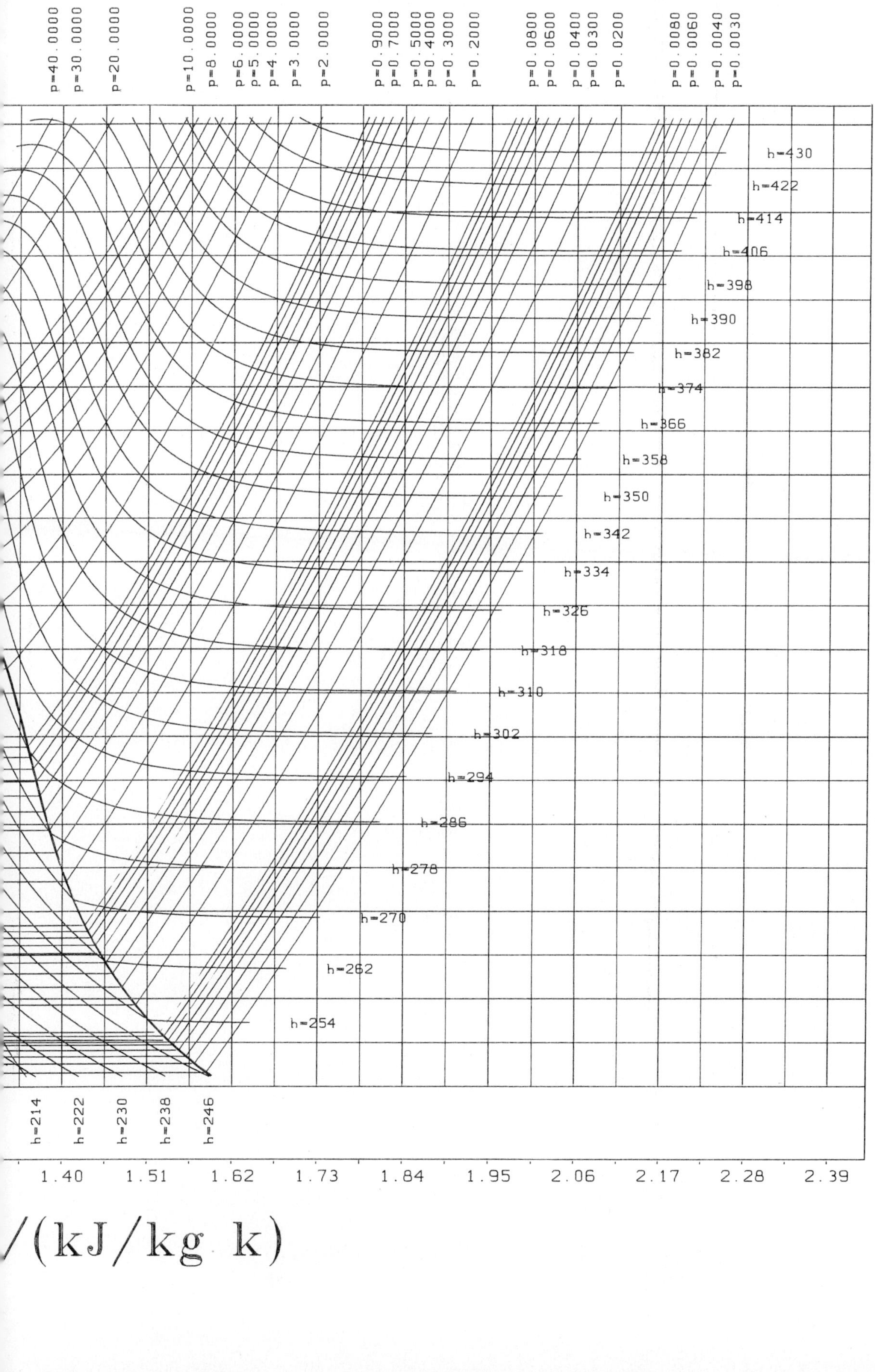

p=40.0000
p=30.0000
p=20.0000
p=10.0000
p=8.0000
p=6.0000
p=5.0000
p=4.0000
p=3.0000
p=2.0000
p=0.9000
p=0.7000
p=0.5000
p=0.4000
p=0.3000
p=0.2000
p=0.0800
p=0.0600
p=0.0400
p=0.0300
p=0.0200
p=0.0080
p=0.0060
p=0.0040
p=0.0030
h=430
h=422
h=414
h=406
h=398
h=390
h=382
h=374
h=366
h=358
h=350
h=342
h=334
h=326
h=318
h=310
h=302
h=294
h=286
h=278
h=270
h=262
h=254
h=214
h=222
h=230
h=238
h=246
1.40
1.51
1.62
1.73
1.84
1.95
2.06
2.17
2.28
2.39
/(kJ/kg k)

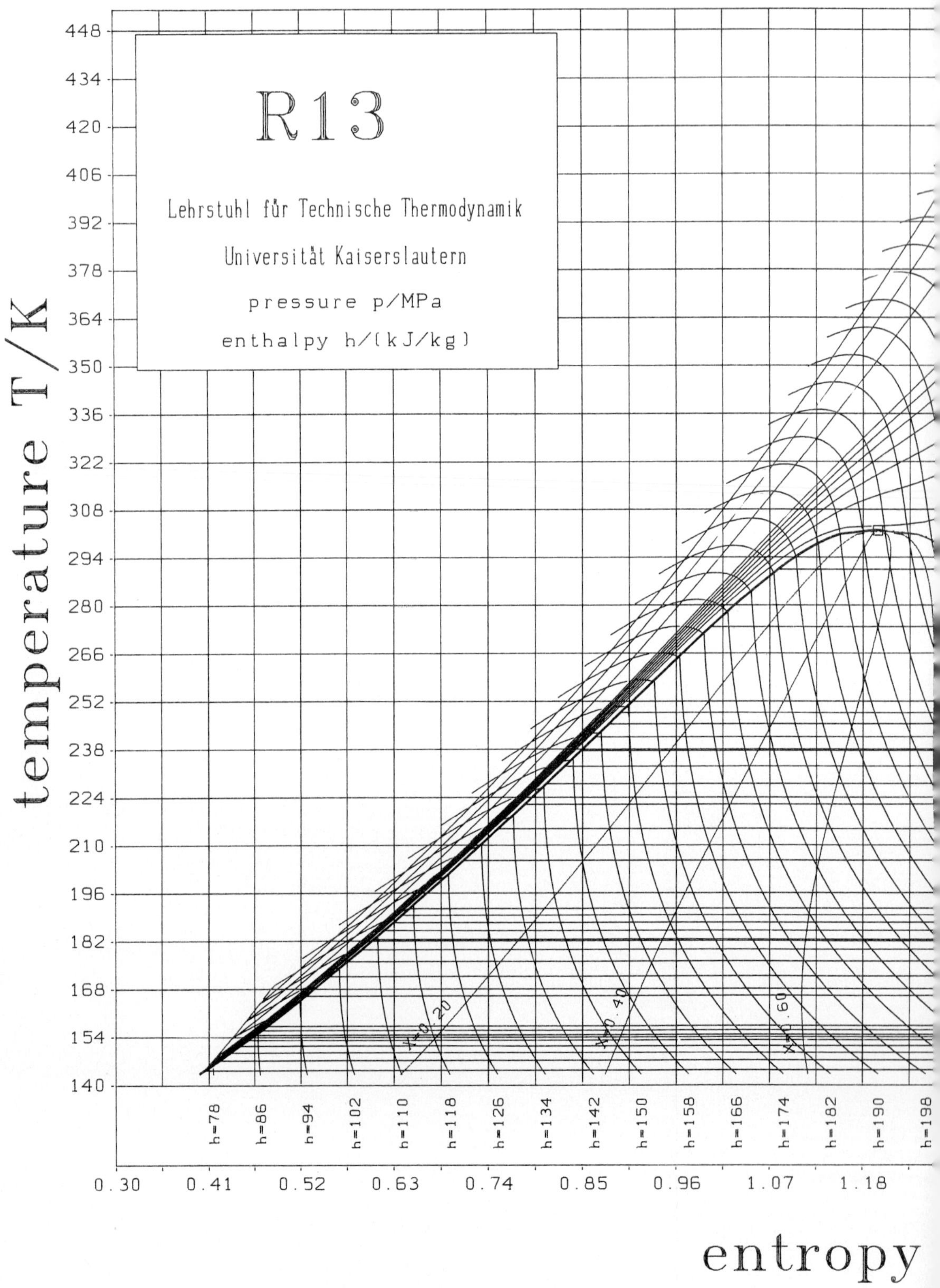

R13
Lehrstuhl für Technische Thermodynamik
Universität Kaiserslautern
pressure p/MPa
enthalpy h/(kJ/kg)
temperature T/K
448
434
420
406
392
378
364
350
336
322
308
294
280
266
252
238
224
210
196
182
168
154
140
x=0.20
x=0.40
x=0.60
h=78
h=86
h=94
h=102
h=110
h=118
h=126
h=134
h=142
h=150
h=158
h=166
h=174
h=182
h=190
h=198
0.30
0.41
0.52
0.63
0.74
0.85
0.96
1.07
1.18
entropy

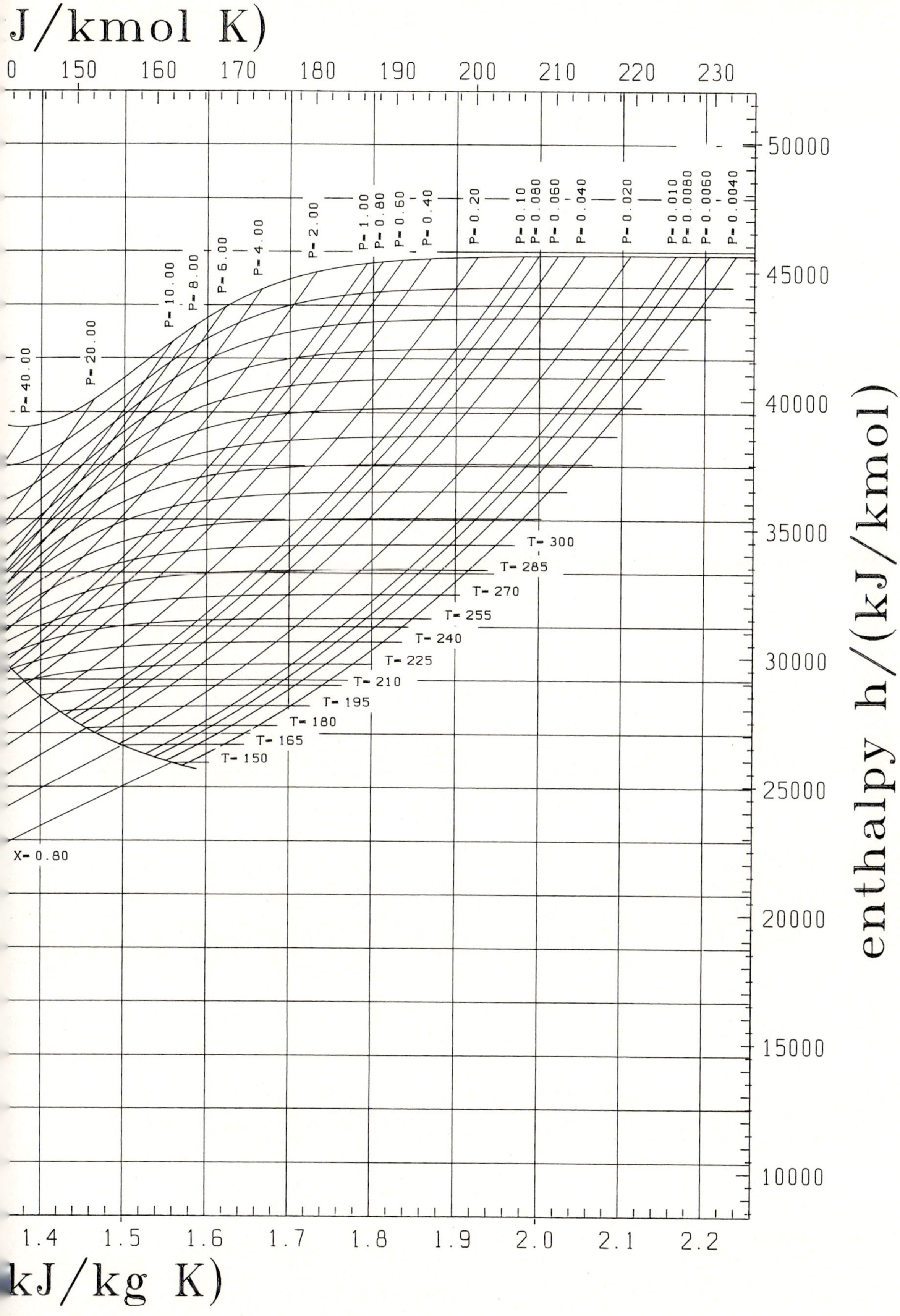
J/kmol K)
0
150
160
170
180
190
200
210
220
230
50000
45000
40000
35000
30000
25000
20000
15000
10000
enthalpy h/(kJ/kmol)
P= 40.00
P= 20.00
P= 10.00
P= 8.00
P= 6.00
P= 4.00
P= 2.00
P= 1.00
P= 0.80
P= 0.60
P= 0.40
P= 0.20
P= 0.10
P= 0.080
P= 0.060
P= 0.040
P= 0.020
P= 0.010
P= 0.0080
P= 0.0060
P= 0.0040
T= 300
T= 285
T= 270
T= 255
T= 240
T= 225
T= 210
T= 195
T= 180
T= 165
T= 150
X= 0.80
1.4
1.5
1.6
1.7
1.8
1.9
2.0
2.1
2.2
kJ/kg K)

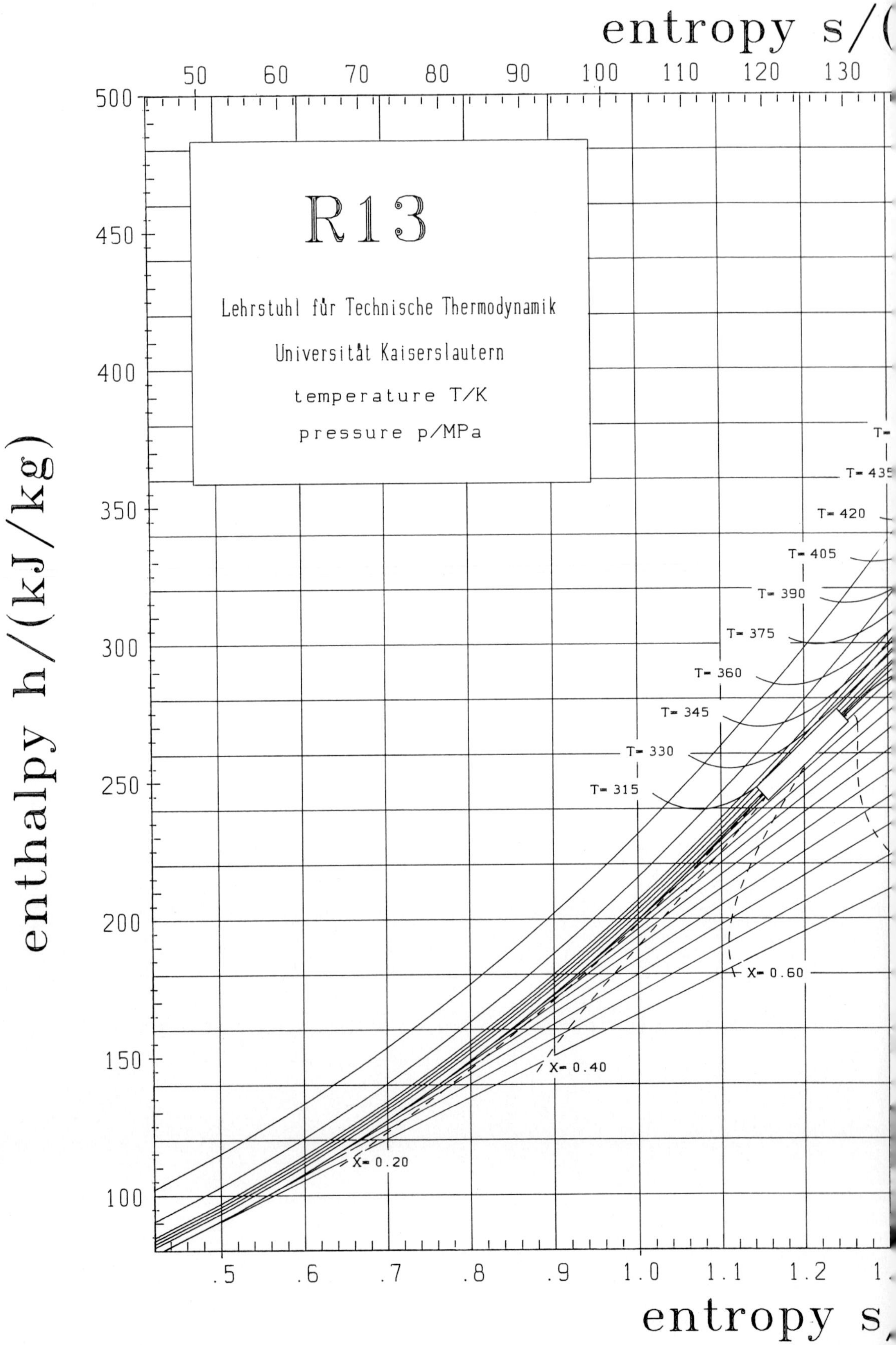

entropy s/(
50
60
70
80
90
100
110
120
130
500
450
400
350
300
250
200
150
100
R13
Lehrstuhl für Technische Thermodynamik
Universität Kaiserslautern
temperature T/K
pressure p/MPa
enthalpy h/(kJ/kg)
T= 435
T= 420
T= 405
T= 390
T= 375
T= 360
T= 345
T= 330
T= 315
X= 0.60
X= 0.40
X= 0.20
.5
.6
.7
.8
.9
1.0
1.1
1.2
entropy s

Refrigerant R14 (CF_4)

Table R14.1: List of references for p, ρ, T data

Author	n	p/MPa min	p/MPa max	ρ/g cm^{-3} min	ρ/g cm^{-3} max	T/K min	T/K max
DOUSLIN, D.R. *[Dous 61]*	306	1.57	39.89	0.066	1.10	273	623
KLEMM, W. *[Klem 32]*	3	0.08	0.09	0.0031	0.0034	564	565
LANGE, H.B. *[Lang 70]*	53	0.69	8.38	0.0277	0.948	203	368
MARTIN, J.J. *[Mart 71]*	27	1.68	10.47	0.0871	1.06	209	295
SVETLICHNY, P.I. *[Svet 80]*	111	0.25	57.46	1.1584	1.866	93	316
SVETLICHNY, P.I. *[Svet 80a]*	125	0.63	50.78	0.041	1.184	182	374

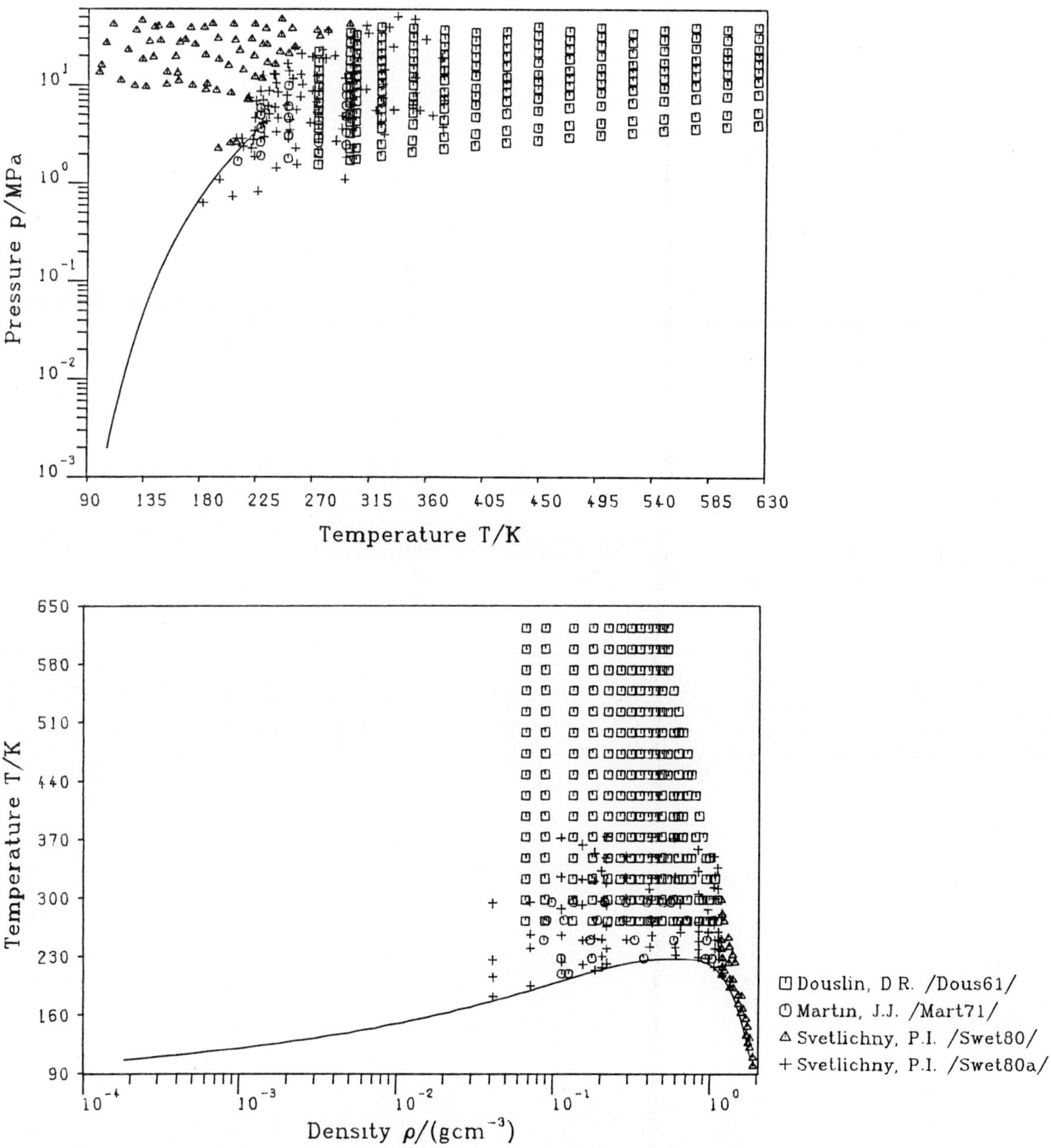

Fig.R14.1: Distribution of p, ρ, T data projected in the p, T and in the T, ρ plane

Table R14.2: Experimental vapor pressure data

Author	Δ / %		
	std	opt	*EOS*
CHARI, N.C.S *[Char 60]*	1.16	1.58	1.30
ENOKIDO, H. *[Enok 69]*	3.00	10.37	3.00
KELLEY, K.K. *[Kell 35]*	3.00	4.51	3.00
KUBIC, W. *[Kubi 80]*	0.10	0.04	0.11
LOBO, L.Q. *[Lobo 79]*	0.61	0.58	1.17
MENZEL, W. *[Menz 33]*	1.52	1.51	1.70
SIMON, M. *[Simo 67]*	0.13	0.12	0.53
SMITH, J.H. *[Smit 69]*	2.33	2.32	2.32

Table R14.3: Experimental saturated liquid density data

Author	Δ / %		
	std	opt	*EOS*
KNOBLER, C.M. *[Knob 65]*	0.57	0.57	0.60
TERRY, M.J. *[Terr 69]*	0.03	0.03	0.08

Table R14.4: Coefficients of the standard vapor pressure equation

Index i	b_i
5	−6.769991 E+00
6	1.116643 E+00
9	−1.879338 E+00
15	−2.466954 E+00

Table R14.5: Coefficients of the quasi-optimal vapor pressure equation

Index i	b_i
1	1.887617 E−04
4	−1.263342 E−01
5	−6.057176 E+00
12	−3.307006 E+00

Table R14.6: Coefficients of the standard saturated liquid density equation

Index i	c_i
8	1.576968 E+00
9	8.385832 E−01
16	2.698069 E−01

Table R14.7: Coefficients of the quasi-optimal saturated liquid density equation

Index i	c_i
3	−4.236862 E−03
8	2.061687 E+00
12	5.963164 E−01

Table R14.8: Coefficients of the *Bender-EOS*

Range of validity: one-phase region

$0.63 \leq p/\text{MPa} \leq 50.78$

$0.04123 \leq \rho/\text{g cm}^{-3} \leq 1.86030$

$98.94 \leq T/\text{K} \leq 623.23$

	Specific units	Molar units
a_{01}	1.570246425 E−01	1.216275210 E+00
a_{02}	8.443193895 E+01	6.539895439 E+02
a_{03}	−6.249138607 E+03	−4.840432848 E+04
a_{04}	1.332634300 E+06	1.032226559 E+07
a_{05}	−6.155901596 E+07	−4.768213696 E+08
a_{06}	−1.416037541 E−01	−9.653189466 E−02
a_{07}	7.881392272 E+01	5.372779367 E+01
a_{08}	−1.065342345 E+03	−7.262485069 E+02
a_{09}	2.621295221 E−01	1.572693239 E−02
a_{10}	−1.029452974 E+02	−6.176388369 E+00
a_{11}	−6.345995921 E−02	−3.350887856 E−04
a_{12}	3.112232893 E+01	1.643358038 E−01
a_{13}	6.846635797 E+00	3.181774174 E−03
a_{14}	−2.980651327 E+06	−2.031922965 E+06
a_{15}	1.683335154 E+09	1.147536891 E+09
a_{16}	−2.500436755 E+11	−1.704558604 E+11
a_{17}	2.758150555 E+06	1.456391292 E+04
a_{18}	−2.361308823 E+09	−1.246846225 E+07
a_{19}	5.274730698 E+11	2.785225717 E+09
a_{20}	2.550000000 E+00	1.975168825 E−02

Table R14.9: Deviations between experimental and predicted p, ρ, T data

Region	n	Δ / %	
		p	ρ
Gas	300	0.38	0.44
Critical	34	1.09	3.59
Liquid	205	2.81	0.30
Total	539	1.81	0.99

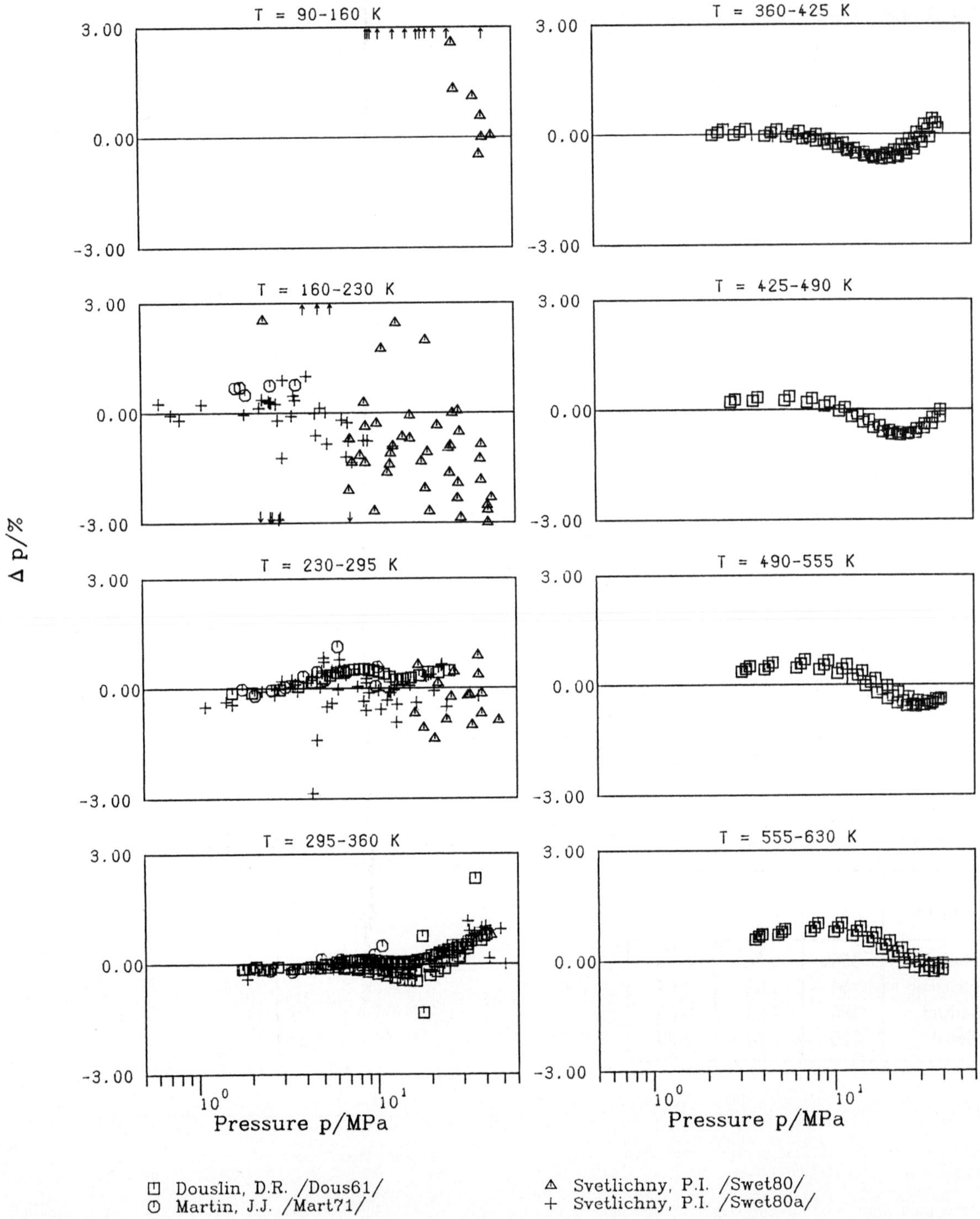

Fig.R14.2: Deviations between experimental and calculated p, ρ, T data

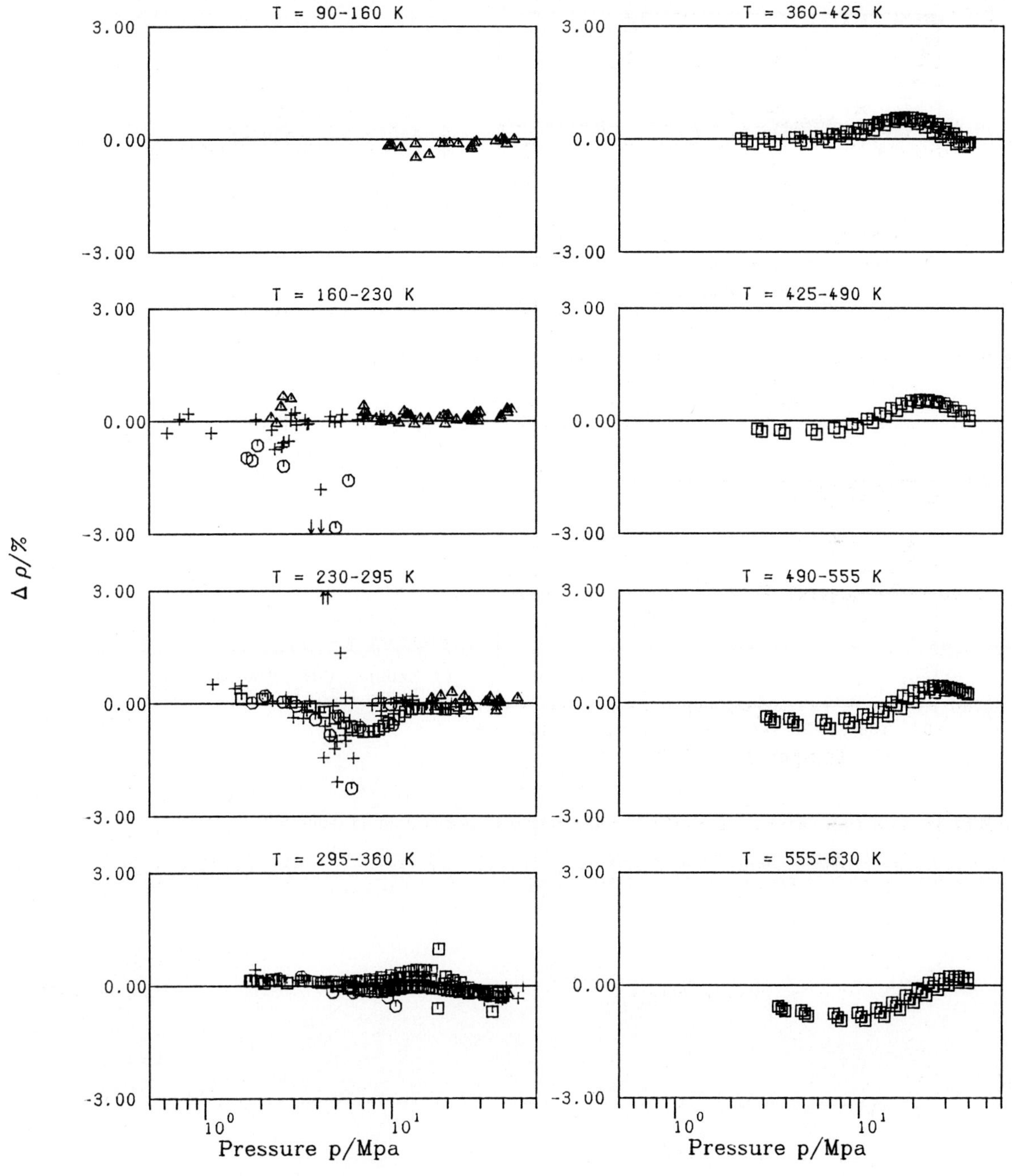

□ Douslin, D.R. /Dous61/
○ Martin, J.J. /Mart71/
△ Svetlichny, P.I. /Swet80/
+ Svetlichny, P.I. /Swet80a/

Fig.R14.2: Deviations between experimental and calculated p, ρ, T data *-continued*

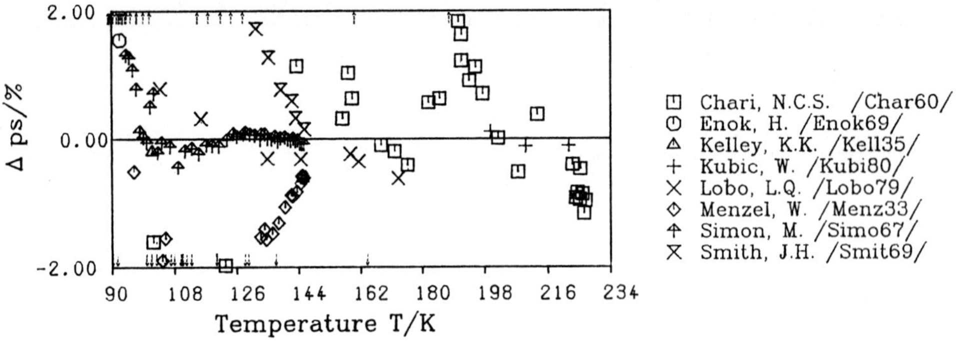

Fig.R14.3: Deviations between experimental and calculated vapor pressures

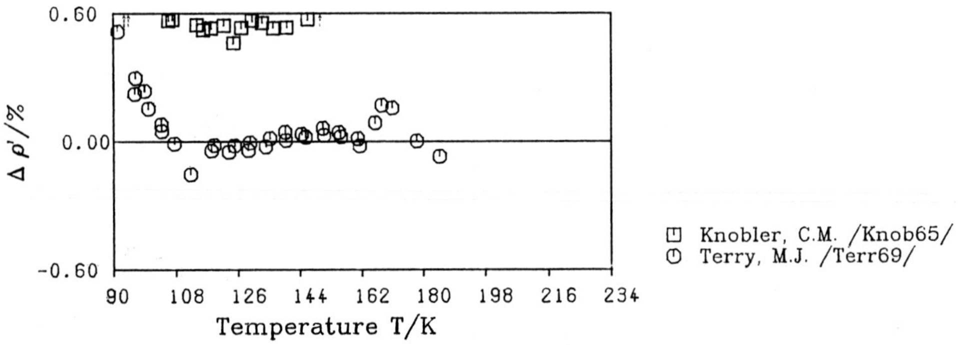

Fig.R14.4: Deviations between experimental and calculated saturated liquid densities

Table R14.10: Information on ideal gas state heat capacity equation

Author	T / K min	T / K max
RODGERS A.S. *[Rodg 74]*	100	600

Coefficients of equation (15)	
Index i	d_i
1	3.266143 E−01
2	8.285805 E−04
3	8.405900 E−04
4	−1.787524 E−06
5	1.149856 E−09

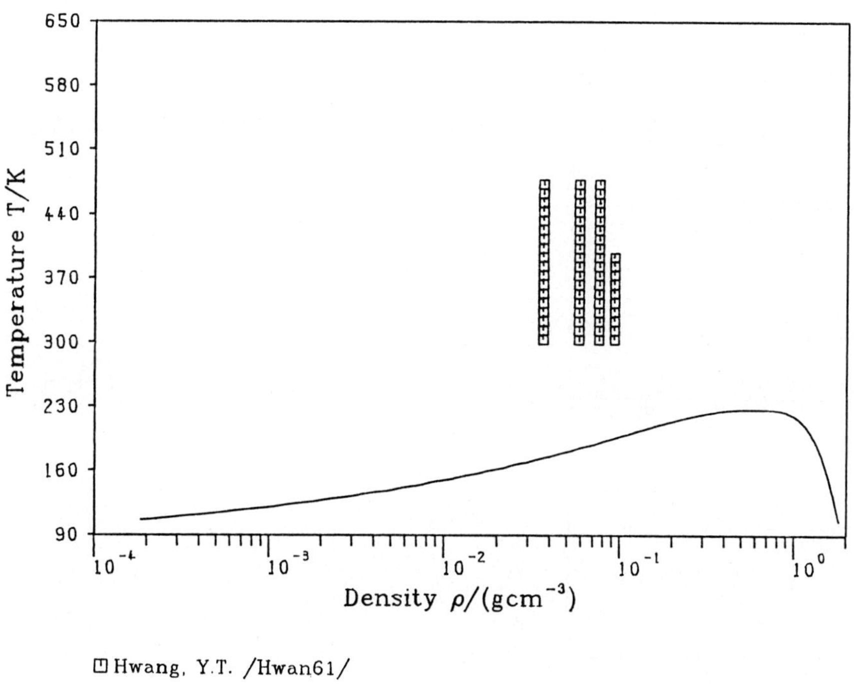

Fig.R14.5: Distribution of $c_v(\rho, T)$ data

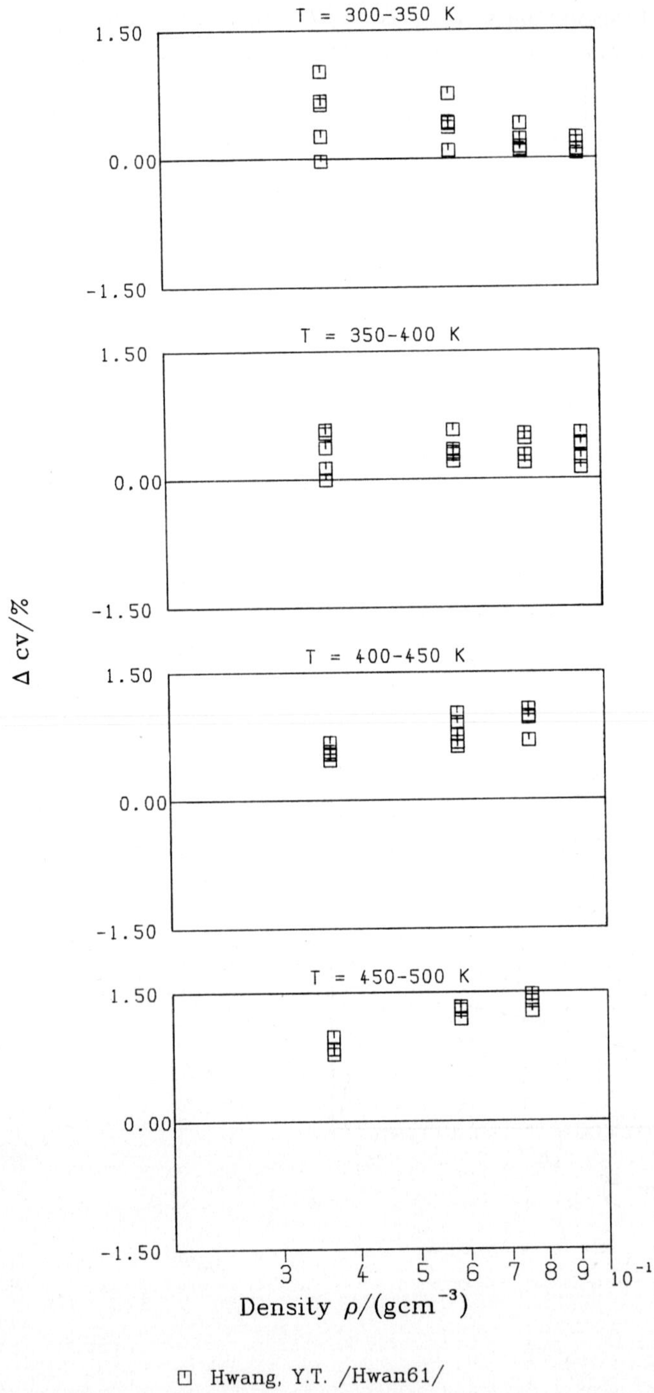

Fig.R14.6: Deviations between experimental and calculated specific isochoric heat capacities c_v

Chart R14.1: Saturation properties

T K	p MPa	v′ dm³/kg	v″ dm³/kg	h′ kJ/kg	h″ kJ/kg	r kJ/kg	s′ kJ/kgK	s″ kJ/kgK
105.0	0.00169	0.5552	5874.6	177.74	331.99	154.25	0.8094	2.2785
106.0	0.00195	0.5565	5120.9	178.75	332.39	153.64	0.8191	2.2684
107.0	0.00225	0.5579	4478.0	179.74	332.79	153.04	0.8283	2.2587
108.0	0.00259	0.5594	3927.8	180.71	333.18	152.47	0.8374	2.2491
109.0	0.00297	0.5608	3455.3	181.66	333.58	151.92	0.8461	2.2399
110.0	0.00340	0.5623	3048.3	182.59	333.98	151.39	0.8546	2.2309
111.0	0.00388	0.5637	2696.6	183.51	334.38	150.87	0.8629	2.2221
112.0	0.00441	0.5652	2391.8	184.42	334.78	150.36	0.8711	2.2136
113.0	0.00500	0.5667	2126.9	185.32	335.18	149.87	0.8790	2.2053
114.0	0.00566	0.5682	1895.9	186.20	335.58	149.38	0.8869	2.1972
115.0	0.00639	0.5698	1694.1	187.08	335.98	148.90	0.8945	2.1893
116.0	0.00719	0.5713	1517.3	187.96	336.39	148.43	0.9021	2.1816
117.0	0.00808	0.5729	1361.9	188.83	336.79	147.96	0.9096	2.1742
118.0	0.00905	0.5745	1225.2	189.69	337.19	147.50	0.9169	2.1669
119.0	0.01012	0.5761	1104.4	190.55	337.59	147.04	0.9242	2.1598
120.0	0.01129	0.5777	997.63	191.41	337.99	146.58	0.9314	2.1529
121.0	0.01257	0.5793	902.93	192.27	338.39	146.12	0.9385	2.1461
122.0	0.01397	0.5809	818.79	193.13	338.80	145.67	0.9455	2.1395
123.0	0.01549	0.5826	743.88	193.98	339.20	145.21	0.9525	2.1331
124.0	0.01715	0.5843	677.04	194.84	339.60	144.76	0.9594	2.1268
125.0	0.01895	0.5860	617.28	195.70	340.00	144.30	0.9663	2.1207
126.0	0.02090	0.5877	563.76	196.55	340.40	143.84	0.9731	2.1147
127.0	0.02301	0.5894	515.73	197.41	340.80	143.38	0.9799	2.1089
128.0	0.02528	0.5911	472.55	198.27	341.20	142.92	0.9866	2.1032
129.0	0.02774	0.5929	433.67	199.14	341.59	142.46	0.9933	2.0977
130.0	0.03039	0.5947	398.59	200.00	341.99	141.99	1.0000	2.0922
131.0	0.03323	0.5965	366.88	200.87	342.39	141.52	1.0066	2.0869
132.0	0.03629	0.5983	338.19	201.74	342.78	141.05	1.0132	2.0818
133.0	0.03957	0.6001	312.17	202.61	343.18	140.57	1.0198	2.0767
134.0	0.04308	0.6019	288.55	203.48	343.57	140.09	1.0263	2.0718
135.0	0.04683	0.6038	267.06	204.36	343.96	139.60	1.0328	2.0669
136.0	0.05084	0.6057	247.50	205.24	344.35	139.11	1.0393	2.0622
137.0	0.05512	0.6076	229.65	206.12	344.74	138.62	1.0458	2.0576
138.0	0.05968	0.6095	213.35	207.01	345.13	138.12	1.0522	2.0531
139.0	0.06454	0.6114	198.44	207.90	345.52	137.62	1.0586	2.0486
140.0	0.06970	0.6134	184.78	208.79	345.90	137.11	1.0650	2.0443
141.0	0.07518	0.6154	172.26	209.69	346.28	136.59	1.0713	2.0401
142.0	0.08099	0.6174	160.76	210.59	346.66	136.08	1.0777	2.0359
143.0	0.08715	0.6194	150.18	211.49	347.04	135.55	1.0840	2.0319
144.0	0.09367	0.6214	140.45	212.40	347.42	135.02	1.0903	2.0279
145.0	0.10057	0.6235	131.48	213.31	347.79	134.48	1.0965	2.0240
146.0	0.10785	0.6256	123.20	214.22	348.16	133.94	1.1028	2.0202
147.0	0.11554	0.6277	115.55	215.14	348.53	133.39	1.1090	2.0164
148.0	0.12365	0.6298	108.48	216.06	348.90	132.84	1.1152	2.0128
149.0	0.13219	0.6320	101.93	216.98	349.26	132.28	1.1214	2.0092
150.0	0.14118	0.6342	95.864	217.91	349.62	131.71	1.1276	2.0057
151.0	0.15064	0.6364	90.232	218.84	349.98	131.14	1.1337	2.0022
152.0	0.16057	0.6386	85.001	219.78	350.34	130.56	1.1399	1.9988
153.0	0.17100	0.6409	80.138	220.72	350.69	129.97	1.1460	1.9955
154.0	0.18195	0.6432	75.612	221.66	351.04	129.38	1.1521	1.9922

Chart R14.1: Saturation properties—*continued*

T K	p MPa	v' dm³/kg	v'' dm³/kg	h' kJ/kg	h'' kJ/kg	r kJ/kg	s' kJ/kgK	s'' kJ/kgK
155.0	0.19342	0.6455	71.396	222.61	351.38	128.77	1.1582	1.9890
156.0	0.20544	0.6478	67.465	223.56	351.72	128.16	1.1642	1.9858
157.0	0.21802	0.6502	63.796	224.51	352.06	127.55	1.1703	1.9827
158.0	0.23118	0.6526	60.369	225.47	352.40	126.92	1.1763	1.9796
159.0	0.24493	0.6550	57.164	226.44	352.73	126.29	1.1823	1.9766
160.0	0.25929	0.6575	54.165	227.40	353.05	125.65	1.1883	1.9736
161.0	0.27429	0.6600	51.356	228.37	353.38	125.00	1.1943	1.9707
162.0	0.28993	0.6626	48.724	229.35	353.69	124.35	1.2003	1.9678
163.0	0.30623	0.6652	46.254	230.33	354.01	123.68	1.2062	1.9650
164.0	0.32322	0.6678	43.935	231.31	354.32	123.01	1.2122	1.9622
165.0	0.34091	0.6704	41.756	232.30	354.62	122.32	1.2181	1.9595
166.0	0.35932	0.6731	39.707	233.29	354.92	121.63	1.2240	1.9567
167.0	0.37846	0.6759	37.778	234.28	355.21	120.93	1.2299	1.9541
168.0	0.39836	0.6786	35.962	235.28	355.50	120.22	1.2358	1.9514
169.0	0.41903	0.6815	34.251	236.29	355.79	119.50	1.2417	1.9488
170.0	0.44049	0.6843	32.637	237.30	356.06	118.77	1.2476	1.9462
171.0	0.46276	0.6872	31.113	238.31	356.34	118.02	1.2534	1.9436
172.0	0.48587	0.6902	29.674	239.33	356.60	117.27	1.2593	1.9411
173.0	0.50982	0.6932	28.314	240.35	356.86	116.51	1.2651	1.9386
174.0	0.53464	0.6963	27.028	241.38	357.12	115.73	1.2709	1.9361
175.0	0.56034	0.6994	25.811	242.42	357.36	114.95	1.2768	1.9336
176.0	0.58696	0.7026	24.658	243.46	357.60	114.15	1.2826	1.9311
177.0	0.61450	0.7058	23.566	244.50	357.84	113.34	1.2884	1.9287
178.0	0.64298	0.7091	22.531	245.55	358.06	112.51	1.2942	1.9263
179.0	0.67244	0.7125	21.548	246.60	358.28	111.68	1.3000	1.9239
180.0	0.70288	0.7159	20.615	247.67	358.49	110.83	1.3058	1.9215
181.0	0.73433	0.7194	19.729	248.73	358.70	109.96	1.3115	1.9191
182.0	0.76680	0.7230	18.887	249.81	358.89	109.09	1.3173	1.9167
183.0	0.80033	0.7266	18.086	250.89	359.08	108.19	1.3231	1.9143
184.0	0.83493	0.7303	17.324	251.97	359.26	107.29	1.3289	1.9120
185.0	0.87061	0.7341	16.599	253.06	359.43	106.36	1.3347	1.9096
186.0	0.90741	0.7380	15.908	254.16	359.59	105.42	1.3404	1.9072
187.0	0.94535	0.7420	15.249	255.27	359.74	104.47	1.3462	1.9049
188.0	0.98444	0.7461	14.620	256.38	359.88	103.49	1.3520	1.9025
189.0	1.02471	0.7502	14.021	257.50	360.00	102.50	1.3578	1.9001
190.0	1.06618	0.7545	13.448	258.63	360.12	101.49	1.3636	1.8977
191.0	1.10888	0.7589	12.901	259.77	360.23	100.46	1.3694	1.8954
192.0	1.15283	0.7634	12.379	260.91	360.32	99.41	1.3752	1.8930
193.0	1.19804	0.7680	11.879	262.07	360.41	98.34	1.3810	1.8905
194.0	1.24455	0.7727	11.400	263.23	360.48	97.25	1.3868	1.8881
195.0	1.29239	0.7776	10.942	264.40	360.53	96.13	1.3927	1.8856
196.0	1.34156	0.7826	10.504	265.58	360.57	94.99	1.3985	1.8832
197.0	1.39210	0.7878	10.083	266.78	360.60	93.82	1.4044	1.8806
198.0	1.44404	0.7932	9.6799	267.98	360.61	92.63	1.4103	1.8781
199.0	1.49740	0.7987	9.2929	269.19	360.61	91.41	1.4162	1.8755
200.0	1.55221	0.8044	8.9214	270.42	360.58	90.16	1.4221	1.8729
201.0	1.60849	0.8102	8.5645	271.66	360.54	88.88	1.4280	1.8703
202.0	1.66627	0.8163	8.2214	272.91	360.48	87.57	1.4340	1.8676
203.0	1.72559	0.8227	7.8914	274.18	360.40	86.23	1.4400	1.8648
204.0	1.78646	0.8293	7.5737	275.46	360.30	84.84	1.4461	1.8620

Chart R14.1: Saturation properties—*continued*

T K	p MPa	v' dm³/kg	v'' dm³/kg	h' kJ/kg	h'' kJ/kg	r kJ/kg	s' kJ/kgK	s'' kJ/kgK
205.0	1.84893	0.8361	7.2676	276.75	360.18	83.42	1.4522	1.8591
206.0	1.91302	0.8432	6.9726	278.07	360.03	81.96	1.4583	1.8562
207.0	1.97876	0.8507	6.6880	279.40	359.85	80.45	1.4645	1.8531
208.0	2.04619	0.8585	6.4132	280.75	359.65	78.90	1.4707	1.8500
209.0	2.11534	0.8666	6.1477	282.12	359.41	77.29	1.4770	1.8468
210.0	2.18626	0.8752	5.8907	283.51	359.14	75.63	1.4834	1.8435
211.0	2.25896	0.8843	5.6419	284.93	358.84	73.91	1.4898	1.8400
212.0	2.33351	0.8939	5.4006	286.38	358.49	72.12	1.4963	1.8365
213.0	2.40994	0.9040	5.1664	287.85	358.10	70.25	1.5029	1.8327
214.0	2.48829	0.9149	4.9386	289.36	357.67	68.31	1.5096	1.8288
215.0	2.56862	0.9265	4.7166	290.90	357.18	66.27	1.5165	1.8247
216.0	2.65097	0.9389	4.4999	292.49	356.62	64.14	1.5235	1.8204
217.0	2.73541	0.9524	4.2878	294.12	356.00	61.88	1.5307	1.8158
218.0	2.82200	0.9672	4.0795	295.80	355.29	59.49	1.5380	1.8109
219.0	2.91080	0.9834	3.8742	297.55	354.49	56.94	1.5456	1.8056
220.0	3.00190	1.0014	3.6709	299.38	353.58	54.19	1.5536	1.7999
221.0	3.09539	1.0218	3.4683	301.30	352.52	51.21	1.5618	1.7936
222.0	3.19138	1.0453	3.2645	303.35	351.28	47.93	1.5706	1.7865
223.0	3.29002	1.0729	3.0570	305.56	349.80	44.25	1.5801	1.7785
224.0	3.39146	1.1068	2.8416	308.00	347.99	40.00	1.5905	1.7691
225.0	3.49597	1.1510	2.6106	310.80	345.66	34.86	1.6025	1.7574
226.0	3.60392	1.2154	2.3460	314.29	342.36	28.07	1.6174	1.7416
227.0	3.71923	1.8670	2.2459	333.93	341.54	7.62	1.7033	1.7368

Chart R14.2: Saturation properties

p MPa	T K	v' dm³/kg	v'' dm³/kg	h' kJ/kg	h'' kJ/kg	r kJ/kg	s' kJ/kgK	s'' kJ/kgK
0.0021	106.50	0.5573	4783.7	179.26	332.59	153.33	0.8238	2.2635
0.0022	106.83	0.5577	4579.9	179.58	332.72	153.14	0.8268	2.2603
0.0023	107.14	0.5582	4393.4	179.88	332.84	152.96	0.8297	2.2573
0.0024	107.45	0.5586	4221.9	180.18	332.96	152.79	0.8324	2.2544
0.0025	107.74	0.5590	4063.8	180.46	333.08	152.62	0.8350	2.2516
0.0026	108.02	0.5594	3917.5	180.73	333.19	152.46	0.8375	2.2489
0.0027	108.29	0.5598	3781.7	180.99	333.30	152.31	0.8399	2.2464
0.0028	108.56	0.5602	3655.3	181.24	333.41	152.16	0.8423	2.2439
0.0029	108.82	0.5605	3537.4	181.49	333.51	152.02	0.8445	2.2416
0.0030	109.06	0.5609	3427.2	181.72	333.61	151.89	0.8467	2.2393
0.0031	109.31	0.5612	3323.8	181.95	333.70	151.76	0.8487	2.2371
0.0032	109.54	0.5616	3226.7	182.17	333.80	151.63	0.8508	2.2350
0.0033	109.77	0.5619	3135.3	182.38	333.89	151.51	0.8527	2.2329
0.0034	110.00	0.5623	3049.1	182.59	333.98	151.39	0.8546	2.2309
0.0035	110.22	0.5626	2967.7	182.79	334.07	151.27	0.8564	2.2290
0.0036	110.43	0.5629	2890.7	182.99	334.15	151.16	0.8582	2.2271
0.0037	110.64	0.5632	2817.8	183.18	334.24	151.05	0.8600	2.2252
0.0038	110.84	0.5635	2748.5	183.37	334.32	150.95	0.8617	2.2235
0.0039	111.04	0.5638	2682.7	183.55	334.40	150.85	0.8633	2.2217
0.0040	111.24	0.5641	2620.1	183.73	334.48	150.75	0.8649	2.2201
0.0041	111.43	0.5644	2560.5	183.90	334.55	150.65	0.8665	2.2184
0.0042	111.62	0.5646	2503.6	184.07	334.63	150.55	0.8680	2.2168
0.0043	111.80	0.5649	2449.2	184.24	334.70	150.46	0.8695	2.2153
0.0044	111.98	0.5652	2397.3	184.40	334.77	150.37	0.8709	2.2137
0.0045	112.16	0.5654	2347.6	184.56	334.84	150.28	0.8723	2.2123
0.0046	112.33	0.5657	2300.0	184.72	334.91	150.20	0.8737	2.2108
0.0047	112.50	0.5660	2254.3	184.87	334.98	150.11	0.8751	2.2094
0.0048	112.67	0.5662	2210.5	185.02	335.05	150.03	0.8764	2.2080
0.0049	112.83	0.5665	2168.5	185.17	335.11	149.95	0.8777	2.2066
0.0050	113.00	0.5667	2128.0	185.31	335.18	149.87	0.8790	2.2053
0.0051	113.15	0.5669	2089.1	185.45	335.24	149.79	0.8803	2.2040
0.0052	113.31	0.5672	2051.7	185.59	335.31	149.71	0.8815	2.2027
0.0053	113.46	0.5674	2015.6	185.73	335.37	149.64	0.8827	2.2015
0.0054	113.62	0.5676	1980.8	185.86	335.43	149.56	0.8839	2.2003
0.0055	113.77	0.5679	1947.3	186.00	335.49	149.49	0.8850	2.1991
0.0056	113.91	0.5681	1914.9	186.13	335.55	149.42	0.8862	2.1979
0.0057	114.06	0.5683	1883.6	186.26	335.61	149.35	0.8873	2.1967
0.0058	114.20	0.5685	1853.3	186.38	335.66	149.28	0.8884	2.1956
0.0059	114.34	0.5688	1824.0	186.51	335.72	149.21	0.8895	2.1945
0.0060	114.48	0.5690	1795.7	186.63	335.78	149.15	0.8906	2.1934
0.0061	114.62	0.5692	1768.3	186.75	335.83	149.08	0.8916	2.1923
0.0062	114.75	0.5694	1741.7	186.87	335.88	149.02	0.8926	2.1913
0.0063	114.88	0.5696	1716.0	186.98	335.94	148.95	0.8937	2.1902
0.0064	115.02	0.5698	1691.0	187.10	335.99	148.89	0.8947	2.1892
0.0065	115.15	0.5700	1666.8	187.21	336.04	148.83	0.8957	2.1882
0.0066	115.27	0.5702	1643.3	187.32	336.09	148.77	0.8966	2.1872
0.0067	115.40	0.5704	1620.5	187.44	336.14	148.71	0.8976	2.1862
0.0068	115.53	0.5706	1598.3	187.54	336.19	148.65	0.8985	2.1853
0.0069	115.65	0.5708	1576.7	187.65	336.24	148.59	0.8995	2.1843
0.0070	115.77	0.5710	1555.8	187.76	336.29	148.53	0.9004	2.1834

Chart R14.2: Saturation properties–*continued*

p MPa	T K	v' dm^3/kg	v'' dm^3/kg	h' kJ/kg	h'' kJ/kg	r kJ/kg	s' kJ/kgK	s'' kJ/kgK
0.0071	115.89	0.5711	1535.4	187.86	336.34	148.48	0.9013	2.1825
0.0072	116.01	0.5713	1515.5	187.97	336.39	148.42	0.9022	2.1816
0.0073	116.13	0.5715	1496.2	188.07	336.44	148.37	0.9031	2.1807
0.0074	116.24	0.5717	1477.4	188.17	336.48	148.31	0.9039	2.1798
0.0076	116.47	0.5721	1441.2	188.37	336.58	148.20	0.9056	2.1781
0.0078	116.70	0.5724	1406.8	188.56	336.67	148.10	0.9073	2.1764
0.0080	116.92	0.5728	1374.1	188.76	336.75	148.00	0.9089	2.1748
0.0082	117.13	0.5731	1342.9	188.94	336.84	147.90	0.9105	2.1732
0.0084	117.34	0.5734	1313.2	189.12	336.92	147.80	0.9121	2.1717
0.0086	117.55	0.5737	1284.8	189.30	337.01	147.71	0.9136	2.1701
0.0088	117.75	0.5741	1257.6	189.48	337.09	147.61	0.9151	2.1687
0.0090	117.95	0.5744	1231.6	189.65	337.17	147.52	0.9165	2.1672
0.0092	118.14	0.5747	1206.7	189.82	337.25	147.43	0.9180	2.1658
0.0094	118.34	0.5750	1182.9	189.98	337.32	147.34	0.9194	2.1645
0.0096	118.52	0.5753	1159.9	190.14	337.40	147.25	0.9207	2.1631
0.0098	118.71	0.5756	1137.9	190.30	337.47	147.17	0.9221	2.1618
0.0100	118.89	0.5759	1116.8	190.46	337.55	147.09	0.9234	2.1605
0.0102	119.07	0.5762	1096.4	190.61	337.62	147.00	0.9247	2.1593
0.0104	119.25	0.5765	1076.8	190.77	337.69	146.92	0.9260	2.1581
0.0106	119.42	0.5767	1058.0	190.91	337.76	146.84	0.9272	2.1569
0.0108	119.59	0.5770	1039.8	191.06	337.83	146.77	0.9284	2.1557
0.0110	119.76	0.5773	1022.2	191.21	337.90	146.69	0.9296	2.1545
0.0115	120.17	0.5779	980.88	191.56	338.06	146.50	0.9326	2.1517
0.0120	120.56	0.5786	942.89	191.90	338.22	146.32	0.9354	2.1490
0.0125	120.95	0.5792	907.84	192.22	338.37	146.15	0.9381	2.1465
0.0130	121.31	0.5798	875.40	192.54	338.52	145.98	0.9407	2.1440
0.0135	121.67	0.5804	845.28	192.85	338.66	145.82	0.9432	2.1417
0.0140	122.02	0.5810	817.24	193.14	338.80	145.66	0.9457	2.1394
0.0145	122.36	0.5815	791.07	193.43	338.94	145.51	0.9480	2.1372
0.0150	122.68	0.5821	766.59	193.71	339.07	145.36	0.9503	2.1351
0.0155	123.00	0.5826	743.63	193.99	339.20	145.21	0.9525	2.1331
0.0160	123.31	0.5831	722.06	194.25	339.32	145.07	0.9547	2.1311
0.0165	123.62	0.5836	701.76	194.51	339.44	144.93	0.9568	2.1292
0.0170	123.91	0.5841	682.60	194.76	339.56	144.80	0.9588	2.1274
0.0175	124.20	0.5846	664.51	195.01	339.68	144.67	0.9608	2.1256
0.0180	124.48	0.5851	647.38	195.25	339.79	144.54	0.9628	2.1239
0.0185	124.76	0.5856	631.15	195.49	339.90	144.41	0.9646	2.1222
0.0190	125.03	0.5860	615.75	195.72	340.01	144.29	0.9665	2.1205
0.0195	125.29	0.5865	601.10	195.95	340.11	144.17	0.9683	2.1190
0.0200	125.55	0.5869	587.17	196.17	340.22	144.05	0.9701	2.1174
0.0210	126.05	0.5878	561.22	196.60	340.42	143.82	0.9735	2.1144
0.0220	126.53	0.5886	537.55	197.01	340.61	143.60	0.9767	2.1116
0.0230	127.00	0.5894	515.86	197.41	340.80	143.39	0.9799	2.1089
0.0240	127.45	0.5902	495.93	197.80	340.98	143.18	0.9829	2.1064
0.0250	127.88	0.5909	477.53	198.17	341.15	142.98	0.9858	2.1039
0.0260	128.30	0.5917	460.50	198.53	341.32	142.78	0.9887	2.1016
0.0270	128.71	0.5924	444.68	198.88	341.48	142.60	0.9914	2.0993
0.0280	129.10	0.5931	429.96	199.22	341.63	142.41	0.9940	2.0971
0.0290	129.48	0.5938	416.21	199.55	341.79	142.23	0.9966	2.0950
0.0300	129.86	0.5944	403.35	199.88	341.94	142.06	0.9991	2.0930

Chart R14.2: Saturation properties—*continued*

p MPa	T K	v' dm³/kg	v'' dm³/kg	h' kJ/kg	h'' kJ/kg	r kJ/kg	s' kJ/kgK	s'' kJ/kgK
0.0310	130.22	0.5951	391.29	200.19	342.08	141.89	1.0015	2.0911
0.0320	130.58	0.5957	379.96	200.50	342.22	141.72	1.0038	2.0892
0.0330	130.92	0.5963	369.29	200.80	342.36	141.56	1.0061	2.0874
0.0340	131.26	0.5969	359.22	201.09	342.49	141.40	1.0083	2.0856
0.0350	131.59	0.5975	349.71	201.38	342.62	141.24	1.0105	2.0839
0.0360	131.91	0.5981	340.71	201.66	342.75	141.09	1.0126	2.0822
0.0370	132.22	0.5987	332.17	201.93	342.87	140.94	1.0147	2.0806
0.0380	132.53	0.5992	324.07	202.20	342.99	140.80	1.0167	2.0791
0.0390	132.83	0.5998	316.38	202.46	343.11	140.65	1.0187	2.0776
0.0400	133.13	0.6003	309.05	202.72	343.23	140.51	1.0206	2.0761
0.0410	133.42	0.6009	302.06	202.97	343.34	140.37	1.0225	2.0746
0.0420	133.70	0.6014	295.40	203.22	343.45	140.23	1.0244	2.0732
0.0430	133.98	0.6019	289.03	203.46	343.56	140.10	1.0262	2.0719
0.0440	134.25	0.6024	282.95	203.70	343.67	139.97	1.0280	2.0705
0.0450	134.52	0.6029	277.12	203.94	343.78	139.84	1.0297	2.0692
0.0460	134.78	0.6034	271.54	204.17	343.88	139.71	1.0314	2.0680
0.0470	135.04	0.6039	266.18	204.40	343.98	139.58	1.0331	2.0667
0.0480	135.30	0.6044	261.05	204.62	344.08	139.46	1.0348	2.0655
0.0490	135.55	0.6048	256.11	204.84	344.18	139.34	1.0364	2.0643
0.0500	135.79	0.6053	251.36	205.06	344.27	139.22	1.0380	2.0632
0.0510	136.04	0.6058	246.79	205.27	344.37	139.10	1.0396	2.0620
0.0520	136.28	0.6062	242.40	205.48	344.46	138.98	1.0411	2.0609
0.0530	136.51	0.6067	238.16	205.69	344.55	138.86	1.0426	2.0598
0.0540	136.74	0.6071	234.07	205.90	344.64	138.75	1.0441	2.0588
0.0550	136.97	0.6075	230.13	206.10	344.73	138.63	1.0456	2.0577
0.0560	137.20	0.6080	226.32	206.30	344.82	138.52	1.0470	2.0567
0.0570	137.42	0.6084	222.64	206.49	344.91	138.41	1.0485	2.0557
0.0580	137.64	0.6088	219.08	206.69	344.99	138.30	1.0499	2.0547
0.0590	137.85	0.6092	215.64	206.88	345.07	138.19	1.0513	2.0537
0.0600	138.07	0.6096	212.30	207.07	345.16	138.09	1.0526	2.0528
0.0610	138.28	0.6100	209.08	207.26	345.24	137.98	1.0540	2.0518
0.0620	138.48	0.6104	205.95	207.44	345.32	137.88	1.0553	2.0509
0.0630	138.69	0.6108	202.92	207.62	345.40	137.77	1.0566	2.0500
0.0640	138.89	0.6112	199.98	207.80	345.47	137.67	1.0579	2.0491
0.0650	139.09	0.6116	197.13	207.98	345.55	137.57	1.0592	2.0482
0.0660	139.29	0.6120	194.36	208.16	345.63	137.47	1.0604	2.0474
0.0670	139.48	0.6124	191.67	208.33	345.70	137.37	1.0617	2.0465
0.0680	139.68	0.6128	189.05	208.50	345.78	137.27	1.0629	2.0457
0.0690	139.87	0.6131	186.51	208.67	345.85	137.18	1.0641	2.0449
0.0700	140.06	0.6135	184.04	208.84	345.92	137.08	1.0653	2.0441
0.0710	140.24	0.6139	181.64	209.01	345.99	136.98	1.0665	2.0433
0.0720	140.43	0.6142	179.30	209.17	346.06	136.89	1.0677	2.0425
0.0730	140.61	0.6146	177.02	209.34	346.13	136.80	1.0688	2.0417
0.0740	140.79	0.6150	174.80	209.50	346.20	136.70	1.0700	2.0410
0.0760	141.15	0.6157	170.53	209.82	346.34	136.52	1.0722	2.0395
0.0780	141.49	0.6164	166.47	210.13	346.47	136.34	1.0744	2.0380
0.0800	141.83	0.6170	162.60	210.44	346.60	136.16	1.0766	2.0366
0.0820	142.17	0.6177	158.92	210.74	346.73	135.99	1.0787	2.0352
0.0840	142.50	0.6184	155.41	211.04	346.85	135.82	1.0808	2.0339
0.0860	142.82	0.6190	152.05	211.33	346.97	135.65	1.0828	2.0326

Chart R14.2: Saturation properties–*continued*

p MPa	T K	v' dm³/kg	v'' dm³/kg	h' kJ/kg	h'' kJ/kg	r kJ/kg	s' kJ/kgK	s'' kJ/kgK
0.0880	143.13	0.6197	148.84	211.61	347.09	135.48	1.0848	2.0313
0.0900	143.44	0.6203	145.76	211.89	347.21	135.32	1.0868	2.0301
0.0920	143.75	0.6209	142.82	212.17	347.32	135.15	1.0887	2.0289
0.0940	144.05	0.6215	139.99	212.44	347.44	134.99	1.0906	2.0277
0.0960	144.34	0.6221	137.28	212.71	347.55	134.84	1.0924	2.0266
0.0980	144.63	0.6227	134.68	212.97	347.66	134.68	1.0942	2.0254
0.1000	144.92	0.6233	132.17	213.23	347.76	134.53	1.0960	2.0243
0.1020	145.20	0.6239	129.76	213.49	347.87	134.38	1.0978	2.0232
0.1040	145.48	0.6245	127.44	213.74	347.97	134.23	1.0995	2.0222
0.1060	145.75	0.6251	125.20	213.99	348.07	134.08	1.1012	2.0211
0.1080	146.02	0.6256	123.04	214.24	348.17	133.93	1.1029	2.0201
0.1100	146.29	0.6262	120.96	214.48	348.27	133.79	1.1046	2.0191
0.1150	146.93	0.6275	116.06	215.08	348.51	133.43	1.1086	2.0167
0.1200	147.56	0.6289	111.55	215.65	348.74	133.09	1.1125	2.0144
0.1250	148.16	0.6302	107.39	216.21	348.96	132.75	1.1162	2.0122
0.1300	148.75	0.6314	103.53	216.75	349.17	132.42	1.1199	2.0101
0.1350	149.32	0.6327	99.952	217.28	349.38	132.10	1.1234	2.0081
0.1400	149.87	0.6339	96.617	217.79	349.58	131.78	1.1268	2.0061
0.1450	150.41	0.6351	93.504	218.29	349.77	131.48	1.1301	2.0042
0.1500	150.93	0.6362	90.589	218.78	349.96	131.18	1.1333	2.0024
0.1550	151.45	0.6374	87.855	219.26	350.14	130.88	1.1365	2.0007
0.1600	151.94	0.6385	85.285	219.73	350.32	130.59	1.1395	1.9990
0.1650	152.43	0.6396	82.865	220.18	350.49	130.31	1.1425	1.9973
0.1700	152.91	0.6407	80.581	220.63	350.66	130.03	1.1454	1.9958
0.1750	153.37	0.6417	78.423	221.07	350.82	129.75	1.1482	1.9942
0.1800	153.83	0.6428	76.379	221.50	350.98	129.48	1.1510	1.9927
0.1850	154.27	0.6438	74.442	221.92	351.13	129.21	1.1537	1.9913
0.1900	154.71	0.6448	72.602	222.33	351.28	128.95	1.1564	1.9899
0.1950	155.13	0.6458	70.853	222.74	351.43	128.69	1.1590	1.9885
0.2000	155.55	0.6468	69.188	223.13	351.57	128.44	1.1615	1.9872
0.2100	156.37	0.6487	66.086	223.91	351.85	127.94	1.1664	1.9846
0.2200	157.15	0.6506	63.254	224.66	352.11	127.45	1.1712	1.9822
0.2300	157.91	0.6524	60.659	225.39	352.37	126.98	1.1758	1.9799
0.2400	158.65	0.6542	58.272	226.10	352.61	126.52	1.1802	1.9777
0.2500	159.36	0.6559	56.067	226.78	352.84	126.06	1.1845	1.9755
0.2600	160.05	0.6576	54.026	227.45	353.07	125.62	1.1886	1.9735
0.2700	160.72	0.6593	52.129	228.10	353.29	125.19	1.1926	1.9715
0.2800	161.37	0.6610	50.362	228.73	353.49	124.76	1.1965	1.9697
0.2900	162.00	0.6626	48.712	229.35	353.70	124.34	1.2003	1.9678
0.3000	162.62	0.6642	47.168	229.96	353.89	123.93	1.2040	1.9661
0.3100	163.23	0.6657	45.719	230.55	354.08	123.53	1.2076	1.9644
0.3200	163.81	0.6673	44.357	231.13	354.26	123.13	1.2111	1.9627
0.3300	164.39	0.6688	43.074	231.69	354.43	122.74	1.2145	1.9611
0.3400	164.95	0.6703	41.863	232.25	354.60	122.36	1.2178	1.9596
0.3500	165.50	0.6718	40.718	232.79	354.77	121.98	1.2211	1.9581
0.3600	166.04	0.6732	39.635	233.32	354.93	121.61	1.2242	1.9566
0.3700	166.56	0.6747	38.607	233.85	355.09	121.24	1.2273	1.9552
0.3800	167.08	0.6761	37.631	234.36	355.24	120.87	1.2304	1.9538
0.3900	167.58	0.6775	36.704	234.87	355.38	120.52	1.2334	1.9525
0.4000	168.08	0.6789	35.820	235.36	355.53	120.16	1.2363	1.9512

Chart R14.2: Saturation properties–*continued*

p MPa	T K	v' dm^3/kg	v'' dm^3/kg	h' kJ/kg	h'' kJ/kg	r kJ/kg	s' kJ/kgK	s'' kJ/kgK
0.4100	168.57	0.6802	34.978	235.85	355.66	119.81	1.2392	1.9499
0.4200	169.05	0.6816	34.174	236.33	355.80	119.46	1.2420	1.9487
0.4300	169.52	0.6829	33.406	236.81	355.93	119.12	1.2447	1.9474
0.4400	169.98	0.6843	32.672	237.28	356.06	118.78	1.2474	1.9462
0.4500	170.43	0.6856	31.969	237.73	356.18	118.45	1.2501	1.9451
0.4600	170.88	0.6869	31.294	238.19	356.30	118.12	1.2527	1.9439
0.4700	171.32	0.6882	30.648	238.63	356.42	117.79	1.2553	1.9428
0.4800	171.75	0.6895	30.027	239.07	356.54	117.46	1.2578	1.9417
0.4900	172.18	0.6907	29.430	239.51	356.65	117.14	1.2603	1.9406
0.5000	172.59	0.6920	28.857	239.94	356.76	116.82	1.2627	1.9396
0.5100	173.01	0.6932	28.304	240.36	356.86	116.50	1.2651	1.9385
0.5200	173.41	0.6945	27.772	240.78	356.97	116.19	1.2675	1.9375
0.5300	173.82	0.6957	27.260	241.19	357.07	115.88	1.2699	1.9365
0.5400	174.21	0.6969	26.765	241.60	357.17	115.57	1.2722	1.9355
0.5500	174.60	0.6982	26.288	242.00	357.27	115.26	1.2744	1.9346
0.5600	174.99	0.6994	25.826	242.40	357.36	114.96	1.2767	1.9336
0.5700	175.37	0.7006	25.381	242.80	357.45	114.66	1.2789	1.9327
0.5800	175.74	0.7018	24.950	243.19	357.54	114.36	1.2811	1.9318
0.5900	176.11	0.7029	24.533	243.57	357.63	114.06	1.2832	1.9309
0.6000	176.48	0.7041	24.129	243.95	357.72	113.76	1.2853	1.9300
0.6100	176.84	0.7053	23.738	244.33	357.80	113.47	1.2874	1.9291
0.6200	177.20	0.7065	23.359	244.70	357.88	113.18	1.2895	1.9282
0.6300	177.55	0.7076	22.991	245.07	357.96	112.89	1.2916	1.9274
0.6400	177.90	0.7088	22.635	245.44	358.04	112.60	1.2936	1.9265
0.6500	178.24	0.7099	22.289	245.80	358.12	112.31	1.2956	1.9257
0.6600	178.58	0.7111	21.953	246.16	358.19	112.03	1.2975	1.9249
0.6700	178.92	0.7122	21.626	246.52	358.27	111.75	1.2995	1.9241
0.6800	179.25	0.7133	21.309	246.87	358.34	111.47	1.3014	1.9233
0.6900	179.58	0.7145	21.000	247.22	358.41	111.19	1.3033	1.9225
0.7000	179.91	0.7156	20.700	247.57	358.47	110.91	1.3052	1.9217
0.7100	180.23	0.7167	20.408	247.91	358.54	110.63	1.3071	1.9209
0.7200	180.55	0.7178	20.124	248.25	358.61	110.36	1.3089	1.9202
0.7300	180.86	0.7189	19.847	248.59	358.67	110.08	1.3108	1.9194
0.7400	181.18	0.7200	19.577	248.92	358.73	109.81	1.3126	1.9187
0.7600	181.79	0.7222	19.058	249.58	358.85	109.27	1.3161	1.9172
0.7800	182.40	0.7244	18.564	250.23	358.97	108.73	1.3196	1.9158
0.8000	182.99	0.7266	18.094	250.87	359.08	108.20	1.3231	1.9144
0.8200	183.57	0.7287	17.646	251.51	359.18	107.68	1.3264	1.9130
0.8400	184.14	0.7309	17.218	252.13	359.28	107.16	1.3297	1.9116
0.8600	184.71	0.7330	16.809	252.74	359.38	106.64	1.3330	1.9103
0.8800	185.26	0.7351	16.417	253.35	359.47	106.12	1.3362	1.9090
0.9000	185.80	0.7372	16.042	253.94	359.56	105.61	1.3393	1.9077
0.9200	186.34	0.7393	15.683	254.53	359.64	105.11	1.3424	1.9064
0.9400	186.86	0.7414	15.339	255.11	359.72	104.60	1.3454	1.9052
0.9600	187.38	0.7435	15.007	255.69	359.79	104.10	1.3484	1.9040
0.9800	187.89	0.7456	14.689	256.26	359.86	103.60	1.3514	1.9028
1.0000	188.39	0.7477	14.383	256.82	359.93	103.11	1.3543	1.9016
1.0200	188.88	0.7497	14.089	257.37	359.99	102.62	1.3571	1.9004
1.0400	189.37	0.7518	13.805	257.92	360.05	102.13	1.3599	1.8992
1.0600	189.85	0.7539	13.531	258.46	360.11	101.64	1.3627	1.8981

Chart R14.2: Saturation properties–*continued*

p MPa	T K	v' dm³/kg	v'' dm³/kg	h' kJ/kg	h'' kJ/kg	r kJ/kg	s' kJ/kgK	s'' kJ/kgK
1.0800	190.33	0.7559	13.267	259.00	360.16	101.16	1.3655	1.8970
1.1000	190.79	0.7580	13.012	259.53	360.21	100.68	1.3682	1.8958
1.1500	191.94	0.7631	12.411	260.84	360.32	99.48	1.3748	1.8931
1.2000	193.04	0.7682	11.858	262.12	360.41	98.29	1.3812	1.8904
1.2500	194.12	0.7733	11.346	263.36	360.48	97.12	1.3875	1.8878
1.3000	195.16	0.7784	10.872	264.59	360.54	95.95	1.3936	1.8852
1.3500	196.17	0.7835	10.431	265.78	360.58	94.79	1.3995	1.8827
1.4000	197.15	0.7886	10.020	266.96	360.60	93.64	1.4053	1.8803
1.4500	198.11	0.7938	9.6354	268.12	360.61	92.50	1.4109	1.8778
1.5000	199.05	0.7989	9.2748	269.25	360.61	91.35	1.4165	1.8754
1.5500	199.96	0.8041	8.9359	270.37	360.58	90.21	1.4219	1.8730
1.6000	200.85	0.8094	8.6169	271.47	360.55	89.08	1.4272	1.8707
1.6500	201.72	0.8146	8.3158	272.56	360.50	87.94	1.4324	1.8683
1.7000	202.57	0.8199	8.0311	273.63	360.44	86.81	1.4375	1.8660
1.7500	203.40	0.8253	7.7615	274.69	360.36	85.67	1.4425	1.8637
1.8000	204.22	0.8307	7.5057	275.74	360.28	84.54	1.4474	1.8614
1.8500	205.02	0.8362	7.2625	276.78	360.17	83.40	1.4523	1.8591
1.9000	205.80	0.8418	7.0311	277.80	360.06	82.26	1.4571	1.8568
1.9500	206.57	0.8474	6.8104	278.82	359.93	81.11	1.4618	1.8545
2.0000	207.32	0.8531	6.5997	279.83	359.79	79.96	1.4664	1.8522
2.1000	208.78	0.8648	6.2052	281.82	359.47	77.65	1.4756	1.8475
2.2000	210.19	0.8769	5.8426	283.78	359.09	75.30	1.4846	1.8428
2.3000	211.55	0.8895	5.5075	285.73	358.65	72.92	1.4934	1.8381
2.4000	212.87	0.9027	5.1961	287.66	358.16	70.50	1.5021	1.8332
2.5000	214.15	0.9165	4.9055	289.58	357.60	68.02	1.5106	1.8283
2.6000	215.38	0.9311	4.6328	291.50	356.97	65.47	1.5192	1.8231
2.7000	216.58	0.9467	4.3756	293.43	356.27	62.84	1.5276	1.8178
2.8000	217.75	0.9633	4.1316	295.37	355.48	60.11	1.5361	1.8122
2.9000	218.88	0.9814	3.8988	297.34	354.59	57.25	1.5447	1.8063
3.0000	219.98	1.0011	3.6751	299.34	353.60	54.25	1.5534	1.8000
3.1000	221.05	1.0229	3.4584	301.40	352.46	51.06	1.5623	1.7933
3.2000	222.09	1.0475	3.2463	303.54	351.16	47.62	1.5714	1.7859
3.3000	223.10	1.0760	3.0360	305.79	349.64	43.85	1.5811	1.7776
3.4000	224.08	1.1100	2.8232	308.21	347.82	39.61	1.5914	1.7682
3.5000	225.04	1.1529	2.6013	310.91	345.56	34.64	1.6030	1.7569
3.6000	225.96	1.2125	2.3564	314.14	342.51	28.36	1.6167	1.7423
3.7000	226.81	1.5909	2.2877	327.01	342.12	15.11	1.6729	1.7395

Chart R14.3: Thermodynamic properties

p/MPa	0.60			0.80			1.00		
T K	v dm³/kg	h kJ/kg	s kJ/kgK	v dm³/kg	h kJ/kg	s kJ/kgK	v dm³/kg	h kJ/kg	s kJ/kgK
110.0	0.5618	182.83	0.8538	0.5617	182.91	0.8535	0.5616	182.99	0.8532
120.0	0.5772	191.64	0.9304	0.5770	191.72	0.9301	0.5769	191.79	0.9298
130.0	0.5941	200.21	0.9990	0.5939	200.28	0.9986	0.5937	200.35	0.9983
140.0	0.6127	208.97	1.0639	0.6125	209.04	1.0635	0.6122	209.11	1.0631
150.0	0.6334	218.05	1.1266	0.6331	218.11	1.1261	0.6327	218.17	1.1257
160.0	0.6568	227.49	1.1875	0.6563	227.54	1.1870	0.6559	227.59	1.1865
170.0	0.6838	237.33	1.2471	0.6832	237.36	1.2465	0.6826	237.40	1.2459
180.0	24.865	359.81	1.9417	0.7155	247.67	1.3054	0.7146	247.69	1.3047
190.0	26.872	365.71	1.9736	19.267	363.52	1.9382	14.623	361.03	1.9074
200.0	28.793	371.60	2.0038	20.840	369.74	1.9701	16.022	367.70	1.9416
210.0	30.653	377.55	2.0328	22.334	375.92	2.0002	17.315	374.17	1.9732
220.0	32.469	383.57	2.0609	23.774	382.12	2.0291	18.540	380.59	2.0031
230.0	34.252	389.69	2.0881	25.175	388.38	2.0569	19.718	387.02	2.0317
240.0	36.008	395.93	2.1146	26.546	394.74	2.0840	20.861	393.51	2.0593
250.0	37.744	402.29	2.1406	27.893	401.19	2.1103	21.978	400.07	2.0861
260.0	39.462	408.78	2.1660	29.221	407.77	2.1361	23.074	406.74	2.1122
270.0	41.166	415.40	2.1910	30.534	414.46	2.1613	24.154	413.50	2.1377
280.0	42.858	422.15	2.2156	31.835	421.28	2.1861	25.220	420.39	2.1628
290.0	44.540	429.04	2.2397	33.124	428.22	2.2105	26.274	427.39	2.1873
300.0	46.212	436.07	2.2636	34.404	435.30	2.2345	27.319	434.52	2.2115
310.0	47.877	443.23	2.2870	35.676	442.50	2.2581	28.356	441.77	2.2353
320.0	49.535	450.52	2.3102	36.941	449.83	2.2814	29.385	449.15	2.2587
330.0	51.188	457.94	2.3330	38.200	457.29	2.3044	30.408	456.64	2.2818
340.0	52.835	465.50	2.3556	39.453	464.88	2.3270	31.426	464.26	2.3045
350.0	54.477	473.18	2.3778	40.702	472.59	2.3493	32.438	472.01	2.3270
360.0	56.115	480.98	2.3998	41.946	480.42	2.3714	33.446	479.87	2.3491
370.0	57.749	488.90	2.4215	43.187	488.37	2.3932	34.451	487.84	2.3709
380.0	59.380	496.94	2.4430	44.424	496.43	2.4147	35.451	495.93	2.3925
390.0	61.007	505.09	2.4641	45.658	504.61	2.4359	36.449	504.13	2.4138
400.0	62.632	513.35	2.4850	46.889	512.89	2.4569	37.443	512.43	2.4348
410.0	64.254	521.72	2.5057	48.117	521.28	2.4776	38.435	520.84	2.4556
420.0	65.873	530.19	2.5261	49.343	529.77	2.4981	39.425	529.35	2.4761
430.0	67.491	538.75	2.5463	50.566	538.35	2.5183	40.412	537.95	2.4963
440.0	69.106	547.42	2.5662	51.788	547.03	2.5382	41.398	546.65	2.5163
450.0	70.720	556.17	2.5859	53.008	555.80	2.5579	42.381	555.44	2.5361
460.0	72.331	565.01	2.6053	54.225	564.66	2.5774	43.362	564.31	2.5556
470.0	73.941	573.94	2.6245	55.442	573.60	2.5966	44.342	573.27	2.5748
480.0	75.550	582.95	2.6435	56.657	582.63	2.6156	45.321	582.30	2.5939
490.0	77.157	592.04	2.6622	57.870	591.73	2.6344	46.298	591.42	2.6127
500.0	78.763	601.20	2.6807	59.082	600.91	2.6529	47.274	600.61	2.6312
510.0	80.367	610.44	2.6990	60.293	610.16	2.6712	48.248	609.87	2.6496
520.0	81.971	619.75	2.7171	61.502	619.48	2.6893	49.221	619.21	2.6677
530.0	83.573	629.14	2.7350	62.711	628.87	2.7072	50.194	628.61	2.6856
540.0	85.174	638.59	2.7526	63.918	638.33	2.7249	51.165	638.08	2.7033
550.0	86.775	648.10	2.7701	65.125	647.86	2.7424	52.135	647.62	2.7208

Chart R14.3: Thermodynamic properties–*continued*

p/MPa	0.60			0.80			1.00		
T K	v dm^3/kg	h kJ/kg	s kJ/kgK	v dm^3/kg	h kJ/kg	s kJ/kgK	v dm^3/kg	h kJ/kg	s kJ/kgK
560.0	88.374	657.69	2.7874	66.331	657.46	2.7597	53.104	657.23	2.7381
570.0	89.972	667.34	2.8044	67.535	667.12	2.7768	54.073	666.90	2.7552
580.0	91.570	677.06	2.8214	68.739	676.85	2.7937	55.041	676.64	2.7722
590.0	93.167	686.85	2.8381	69.942	686.65	2.8105	56.007	686.44	2.7889
600.0	94.763	696.71	2.8547	71.145	696.51	2.8271	56.974	696.32	2.8055
610.0	96.359	706.64	2.8711	72.347	706.45	2.8435	57.939	706.26	2.8220
620.0	97.954	716.65	2.8873	73.548	716.46	2.8598	58.904	716.29	2.8383
p/MPa	2.00			4.00			6.00		
110.0	0.5609	183.40	0.8517	0.5596	184.20	0.8489	0.5583	185.01	0.8461
120.0	0.5761	192.18	0.9282	0.5745	192.95	0.9250	0.5729	193.73	0.9219
130.0	0.5927	200.72	0.9965	0.5907	201.45	0.9931	0.5889	202.20	0.9897
140.0	0.6109	209.45	1.0612	0.6085	210.14	1.0574	0.6062	210.84	1.0538
150.0	0.6311	218.48	1.1235	0.6280	219.12	1.1194	0.6251	219.76	1.1153
160.0	0.6537	227.86	1.1840	0.6497	228.41	1.1793	0.6459	228.98	1.1748
170.0	0.6797	237.60	1.2431	0.6743	238.03	1.2376	0.6693	238.50	1.2325
180.0	0.7104	247.78	1.3012	0.7028	248.03	1.2948	0.6960	248.35	1.2888
190.0	0.7484	258.54	1.3594	0.7370	258.49	1.3514	0.7274	258.59	1.3442
200.0	0.7992	270.22	1.4193	0.7801	269.61	1.4084	0.7654	269.33	1.3992
210.0	6.9091	362.49	1.8651	0.8394	281.77	1.4677	0.8140	280.77	1.4550
220.0	7.8768	371.31	1.9062	0.9381	296.11	1.5343	0.8815	293.30	1.5133
230.0	8.6906	379.20	1.9412	1.9382	338.11	1.7192	0.9916	307.90	1.5781
240.0	9.4240	386.70	1.9732	3.3704	366.12	1.8392	1.2470	327.73	1.6624
250.0	10.107	394.02	2.0030	4.0150	378.23	1.8887	1.8322	352.75	1.7646
260.0	10.756	401.27	2.0315	4.5193	388.14	1.9276	2.3790	370.44	1.8341
270.0	11.380	408.52	2.0588	4.9571	397.15	1.9616	2.8005	383.36	1.8829
280.0	11.985	415.80	2.0853	5.3554	405.72	1.9928	3.1521	394.26	1.9225
290.0	12.575	423.14	2.1111	5.7273	414.06	2.0220	3.4624	404.17	1.9573
300.0	13.152	430.56	2.1362	6.0801	422.27	2.0499	3.7459	413.53	1.9890
310.0	13.720	438.07	2.1609	6.4186	430.44	2.0766	4.0107	422.57	2.0187
320.0	14.279	445.67	2.1850	6.7458	438.60	2.1026	4.2615	431.43	2.0468
330.0	14.831	453.37	2.2087	7.0640	446.78	2.1277	4.5016	440.19	2.0738
340.0	15.377	461.18	2.2320	7.3747	455.01	2.1523	4.7333	448.90	2.0998
350.0	15.918	469.09	2.2549	7.6792	463.29	2.1763	4.9580	457.60	2.1250
360.0	16.454	477.10	2.2775	7.9784	471.63	2.1998	5.1770	466.30	2.1495
370.0	16.985	485.21	2.2997	8.2729	480.05	2.2228	5.3912	475.04	2.1735
380.0	17.513	493.43	2.3216	8.5635	488.54	2.2455	5.6012	483.82	2.1969
390.0	18.038	501.75	2.3432	8.8505	497.10	2.2677	5.8076	492.64	2.2198
400.0	18.559	510.16	2.3645	9.1344	505.75	2.2896	6.0108	501.52	2.2423
410.0	19.078	518.68	2.3856	9.4155	514.47	2.3112	6.2113	510.46	2.2643
420.0	19.595	527.28	2.4063	9.6941	523.27	2.3324	6.4093	519.46	2.2860
430.0	20.109	535.98	2.4267	9.9705	532.15	2.3533	6.6050	528.52	2.3073
440.0	20.621	544.76	2.4469	10.245	541.11	2.3739	6.7987	537.64	2.3283
450.0	21.131	553.63	2.4669	10.517	550.14	2.3942	6.9906	546.83	2.3490
460.0	21.640	562.58	2.4865	10.788	559.24	2.4142	7.1809	556.08	2.3693
470.0	22.146	571.61	2.5060	11.057	568.41	2.4339	7.3696	565.39	2.3893
480.0	22.652	580.71	2.5251	11.325	577.65	2.4533	7.5570	574.76	2.4090
490.0	23.156	589.89	2.5441	11.591	586.96	2.4725	7.7430	584.19	2.4285
500.0	23.658	599.15	2.5628	11.856	596.34	2.4915	7.9279	593.69	2.4477

Chart R14.3: Thermodynamic properties–*continued*

p/MPa	2.00			4.00			6.00		
T K	v dm³/kg	h kJ/kg	s kJ/kgK	v dm³/kg	h kJ/kg	s kJ/kgK	v dm³/kg	h kJ/kg	s kJ/kgK
510.0	24.160	608.47	2.5812	12.120	605.78	2.5102	8.1116	603.24	2.4666
520.0	24.660	617.86	2.5994	12.382	615.28	2.5286	8.2943	612.84	2.4852
530.0	25.159	627.32	2.6175	12.644	624.84	2.5468	8.4761	622.51	2.5036
540.0	25.658	636.84	2.6353	12.905	634.47	2.5648	8.6570	632.23	2.5218
550.0	26.155	646.43	2.6529	13.165	644.16	2.5826	8.8370	642.01	2.5397
560.0	26.651	656.09	2.6703	13.424	653.90	2.6002	9.0163	651.84	2.5575
570.0	27.147	665.81	2.6875	13.682	663.71	2.6175	9.1948	661.73	2.5750
580.0	27.642	675.59	2.7045	13.940	673.58	2.6347	9.3726	671.68	2.5923
590.0	28.136	685.44	2.7213	14.197	683.51	2.6517	9.5497	681.69	2.6094
600.0	28.629	695.36	2.7380	14.453	693.51	2.6685	9.7262	691.76	2.6263
610.0	29.122	705.34	2.7545	14.708	703.57	2.6851	9.9021	701.90	2.6431
620.0	29.614	715.40	2.7708	14.963	713.70	2.7016	10.077	712.10	2.6597
p/MPa	8.00			10.00			15.00		
110.0	0.5571	185.83	0.8433	0.5559	186.64	0.8406	0.5530	188.68	0.8340
120.0	0.5715	194.51	0.9189	0.5700	195.29	0.9159	0.5666	197.27	0.9087
130.0	0.5871	202.95	0.9865	0.5853	203.70	0.9832	0.5812	205.61	0.9755
140.0	0.6039	211.55	1.0502	0.6018	212.27	1.0468	0.5968	214.11	1.0384
150.0	0.6223	220.43	1.1114	0.6196	221.10	1.1077	0.6135	222.84	1.0987
160.0	0.6424	229.58	1.1705	0.6391	230.20	1.1664	0.6315	231.81	1.1566
170.0	0.6648	239.01	1.2277	0.6605	239.55	1.2231	0.6511	241.01	1.2123
180.0	0.6900	248.74	1.2833	0.6844	249.17	1.2780	0.6724	250.42	1.2661
190.0	0.7190	258.79	1.3376	0.7116	259.08	1.3316	0.6961	260.04	1.3181
200.0	0.7533	269.25	1.3912	0.7431	269.32	1.3841	0.7226	269.90	1.3687
210.0	0.7953	280.23	1.4448	0.7804	279.96	1.4360	0.7526	280.01	1.4180
220.0	0.8491	291.91	1.4991	0.8261	291.12	1.4879	0.7870	290.41	1.4664
230.0	0.9229	304.62	1.5556	0.8840	302.93	1.5404	0.8270	301.14	1.5141
240.0	1.0344	318.89	1.6163	0.9608	315.57	1.5942	0.8741	312.21	1.5612
250.0	1.2192	335.33	1.6834	1.0665	329.20	1.6498	0.9299	323.64	1.6078
260.0	1.4985	352.89	1.7523	1.2117	343.69	1.7066	0.9960	335.38	1.6539
270.0	1.8145	368.76	1.8122	1.3953	358.31	1.7618	1.0733	347.33	1.6990
280.0	2.1109	382.23	1.8612	1.5993	372.15	1.8122	1.1616	359.35	1.7427
290.0	2.3776	393.98	1.9024	1.8047	384.85	1.8568	1.2590	371.25	1.7845
300.0	2.6195	404.68	1.9387	2.0016	396.52	1.8963	1.3627	382.90	1.8240
310.0	2.8428	414.73	1.9717	2.1879	407.42	1.9320	1.4697	394.22	1.8611
320.0	3.0519	424.37	2.0023	2.3641	417.76	1.9649	1.5778	405.21	1.8960
330.0	3.2501	433.76	2.0312	2.5316	427.73	1.9956	1.6854	415.88	1.9288
340.0	3.4397	442.99	2.0587	2.6918	437.45	2.0246	1.7915	426.30	1.9599
350.0	3.6223	452.12	2.0852	2.8459	446.99	2.0522	1.8957	436.51	1.9895
360.0	3.7992	461.20	2.1108	2.9949	456.43	2.0788	1.9978	446.55	2.0178
370.0	3.9713	470.26	2.1356	3.1395	465.80	2.1045	2.0978	456.48	2.0450
380.0	4.1393	479.33	2.1598	3.2805	475.14	2.1294	2.1959	466.33	2.0713
390.0	4.3038	488.41	2.1834	3.4182	484.46	2.1536	2.2922	476.12	2.0967
400.0	4.4653	497.52	2.2064	3.5531	493.79	2.1772	2.3868	485.87	2.1214
410.0	4.6241	506.67	2.2290	3.6856	503.13	2.2003	2.4798	495.60	2.1454
420.0	4.7805	515.86	2.2512	3.8159	512.50	2.2229	2.5713	505.33	2.1689
430.0	4.9348	525.09	2.2729	3.9442	521.90	2.2450	2.6616	515.06	2.1918
440.0	5.0872	534.38	2.2943	4.0708	531.34	2.2667	2.7506	524.81	2.2142
450.0	5.2378	543.72	2.3152	4.1958	540.81	2.2880	2.8386	534.58	2.2361

Chart R14.3: Thermodynamic properties–*continued*

p/MPa	8.00			10.00			15.00		
T K	v dm³/kg	h kJ/kg	s kJ/kgK	v dm³/kg	h kJ/kg	s kJ/kgK	v dm³/kg	h kJ/kg	s kJ/kgK
460.0	5.3869	553.11	2.3359	4.3193	550.33	2.3089	2.9255	544.37	2.2577
470.0	5.5346	562.55	2.3562	4.4416	559.90	2.3295	3.0114	554.19	2.2788
480.0	5.6809	572.04	2.3762	4.5626	569.51	2.3497	3.0965	564.04	2.2995
490.0	5.8260	581.59	2.3959	4.6825	579.17	2.3696	3.1808	573.92	2.3199
500.0	5.9701	591.19	2.4153	4.8014	588.87	2.3892	3.2643	583.84	2.3399
510.0	6.1130	600.85	2.4344	4.9193	598.62	2.4086	3.3471	593.79	2.3596
520.0	6.2551	610.55	2.4532	5.0364	608.42	2.4276	3.4292	603.78	2.3790
530.0	6.3962	620.31	2.4718	5.1526	618.27	2.4463	3.5107	613.81	2.3981
540.0	6.5365	630.12	2.4902	5.2681	628.16	2.4648	3.5916	623.88	2.4170
550.0	6.6760	639.99	2.5083	5.3828	638.10	2.4831	3.6719	634.00	2.4355
560.0	6.8148	649.90	2.5261	5.4968	648.09	2.5011	3.7518	644.15	2.4538
570.0	6.9529	659.87	2.5438	5.6103	658.14	2.5189	3.8311	654.35	2.4719
580.0	7.0903	669.90	2.5612	5.7231	668.23	2.5364	3.9100	664.59	2.4897
590.0	7.2271	679.98	2.5784	5.8353	678.38	2.5538	3.9884	674.89	2.5073
600.0	7.3633	690.12	2.5955	5.9470	688.59	2.5709	4.0664	685.23	2.5247
610.0	7.4990	700.32	2.6123	6.0583	698.85	2.5879	4.1440	695.63	2.5418
620.0	7.6342	710.59	2.6290	6.1690	709.18	2.6047	4.2212	706.09	2.5589
p/MPa	20.00			40.00			50.00		
110.0	0.5503	190.74	0.8276	0.5407	199.03	0.8038	0.5365	203.22	0.7929
120.0	0.5634	199.26	0.9018	0.5522	207.38	0.8765	0.5474	211.49	0.8649
130.0	0.5774	207.55	0.9681	0.5644	215.49	0.9414	0.5588	219.55	0.9295
140.0	0.5922	215.98	1.0306	0.5770	223.75	1.0026	0.5706	227.75	0.9902
150.0	0.6080	224.64	1.0903	0.5901	232.20	1.0610	0.5828	236.15	1.0481
160.0	0.6248	233.51	1.1476	0.6037	240.84	1.1167	0.5953	244.71	1.1034
170.0	0.6429	242.59	1.2026	0.6178	249.62	1.1699	0.6082	253.41	1.1561
180.0	0.6623	251.84	1.2555	0.6326	258.51	1.2207	0.6216	262.20	1.2064
190.0	0.6835	261.26	1.3064	0.6481	267.50	1.2693	0.6355	271.07	1.2544
200.0	0.7066	270.86	1.3556	0.6644	276.58	1.3159	0.6499	280.01	1.3002
210.0	0.7322	280.64	1.4034	0.6816	285.73	1.3606	0.6649	289.02	1.3442
220.0	0.7606	290.63	1.4498	0.6996	294.98	1.4036	0.6805	298.10	1.3864
230.0	0.7924	300.82	1.4951	0.7187	304.31	1.4450	0.6968	307.24	1.4270
240.0	0.8280	311.22	1.5394	0.7388	313.73	1.4851	0.7139	316.45	1.4662
250.0	0.8681	321.83	1.5827	0.7601	323.24	1.5240	0.7316	325.74	1.5042
260.0	0.9131	332.63	1.6250	0.7824	332.84	1.5616	0.7500	335.11	1.5409
270.0	0.9633	343.56	1.6663	0.8059	342.51	1.5981	0.7692	344.54	1.5765
280.0	1.0186	354.58	1.7064	0.8305	352.26	1.6335	0.7891	354.05	1.6111
290.0	1.0788	365.62	1.7451	0.8561	362.07	1.6680	0.8097	363.63	1.6447
300.0	1.1432	376.61	1.7824	0.8828	371.93	1.7014	0.8309	373.26	1.6773
310.0	1.2109	387.50	1.8181	0.9103	381.83	1.7339	0.8527	382.96	1.7091
320.0	1.2809	398.26	1.8522	0.9386	391.77	1.7654	0.8750	392.70	1.7401
330.0	1.3525	408.88	1.8849	0.9677	401.74	1.7961	0.8978	402.49	1.7702
340.0	1.4249	419.36	1.9162	0.9974	411.72	1.8259	0.9210	412.32	1.7995
350.0	1.4975	429.70	1.9462	1.0275	421.72	1.8549	0.9446	422.19	1.8281
360.0	1.5700	439.94	1.9750	1.0581	431.73	1.8831	0.9684	432.09	1.8560
370.0	1.6421	450.08	2.0028	1.0889	441.76	1.9105	0.9926	442.02	1.8832
380.0	1.7137	460.15	2.0297	1.1201	451.79	1.9373	1.0169	451.98	1.9098
390.0	1.7846	470.16	2.0557	1.1514	461.83	1.9634	1.0413	461.97	1.9357
400.0	1.8548	480.14	2.0809	1.1828	471.89	1.9889	1.0659	471.99	1.9611

Chart R14.3: Thermodynamic properties–*continued*

p/MPa	20.00			40.00			50.00		
T K	v dm³/kg	h kJ/kg	s kJ/kgK	v dm³/kg	h kJ/kg	s kJ/kgK	v dm³/kg	h kJ/kg	s kJ/kgK
410.0	1.9243	490.09	2.1055	1.2143	481.96	2.0137	1.0906	482.03	1.9859
420.0	1.9931	500.03	2.1295	1.2458	492.05	2.0380	1.1153	492.11	2.0102
430.0	2.0611	509.97	2.1528	1.2773	502.16	2.0618	1.1401	502.21	2.0340
440.0	2.1286	519.91	2.1757	1.3089	512.29	2.0851	1.1649	512.34	2.0572
450.0	2.1953	529.86	2.1981	1.3404	522.43	2.1079	1.1896	522.49	2.0801
460.0	2.2615	539.83	2.2200	1.3718	532.61	2.1303	1.2144	532.68	2.1025
470.0	2.3270	549.82	2.2415	1.4032	542.80	2.1522	1.2391	542.90	2.1244
480.0	2.3921	559.84	2.2625	1.4345	553.03	2.1737	1.2638	553.14	2.1460
490.0	2.4565	569.88	2.2832	1.4658	563.28	2.1949	1.2885	563.42	2.1672
500.0	2.5205	579.95	2.3036	1.4970	573.56	2.2156	1.3132	573.73	2.1880
510.0	2.5840	590.04	2.3236	1.5281	583.87	2.2360	1.3378	584.06	2.2085
520.0	2.6471	600.18	2.3433	1.5592	594.21	2.2561	1.3623	594.43	2.2286
530.0	2.7097	610.34	2.3626	1.5901	604.58	2.2759	1.3868	604.84	2.2484
540.0	2.7719	620.54	2.3817	1.6210	614.99	2.2953	1.4113	615.28	2.2679
550.0	2.8338	630.78	2.4005	1.6518	625.43	2.3145	1.4357	625.75	2.2872
560.0	2.8952	641.05	2.4190	1.6825	635.90	2.3334	1.4600	636.26	2.3061
570.0	2.9563	651.37	2.4372	1.7132	646.42	2.3520	1.4843	646.80	2.3248
580.0	3.0171	661.72	2.4553	1.7438	656.97	2.3703	1.5086	657.39	2.3432
590.0	3.0775	672.13	2.4730	1.7743	667.57	2.3884	1.5328	668.02	2.3614
600.0	3.1376	682.58	2.4906	1.8047	678.21	2.4063	1.5570	678.70	2.3793
610.0	3.1974	693.08	2.5080	1.8351	688.90	2.4240	1.5811	689.43	2,3970
620.0	3.2570	703.64	2.5251	1.8654	699.65	2.4415	1.6052	700.20	2.4145

References for R14

[Altu80] ALTUNIN, W.W.: Thermophysical properties of tetrafluoromethane on liquid-vapor equilibrium lines; Tr. Mosk. Energ. In-ta (1980)451 3/7

[Bady63] BADYLKES, I.S.: Thermodynamic properties of Freon-14; Cholodilnaja technika 5(1963) 70/74

[Bhad68] BHADA, R.K.: Pressure-Volume-Temperature Behavior of Carbon Tetrafluoride Using A Variable Volume Cell of Bellows Design; Ph.D. Dissertation, University of Michigan Ann Arbor (1968)

[Blag68] BLAGOI, Y.P.; BUTKO, A.E.; MIKHAILENKO, S.A. AND YAKUBA, V.V.: Velocity of Sound in Liquid Propene, Carbon Tetrafluoride and Ammonia; Russian Journal of Physical Chemistry 42(1968)5 564/565

[Char60] CHARI, N.C.S.: Thermodynamic Properties of Carbontetrafluoride; Ph.D. Thesis University of Michigan (1960)

[Dous61] DOUSLIN, D.R. ; HARRISON, R.H. ; MOORE , R.T. AND MCCULLOUGH, J.P.: Tetrafluoromethane: P-V-T and Intermolecular Potential Energy Relations; J. Chem. Phys. 35(1961)4 1357/1366

[DuPo] - : Preliminary Thermodynamic Properties of Freon-14; E.I.Dupont de Nemours Bulletin T-14, Organic Chemicals Department, Freon Products Division, Wilmington Delaware

[Enok69] ENOKIDO, H. ; SHINODA, T. AND MASHIKO, Y.: Thermodynamic Properties of Carbontetrafluoride from 4K to its Meltingpoint; Bulletin of the Chemical Society of Japan 42(1969)12 3415/3421

[Hwan61] HWANG, Y.T.: The constant heat capacities of gaseous tetrafluoromethane, chlorodifluoromethane, dichlorotetrafluoroethane and chloropentafluoroethane; Ph.D. Dissertation University of Michigan (1961)

[Kell35] KELLEY, K.K.: Contributions to the Data on Theoretical Metallurgy (Part 3); U.S.Department of the Interior, Bureau of Mines Washington (1935), Bulletin 383

[Klem32] KLEMM, W. AND HENKEL, P.: Über einige Physikalische Eigenschaften von SF_6, SeF_6, TeF_6 und CF_4; Z. Anorg. Allgem. Chem. 207(1932)73/86

[Knob65] KNOBLER, C.M. AND PINGS, C.J.: Saturated Liquid Density of Carbon Tetrafluoride from 90K to 150K; J. Chem. Eng. Data 10(1965)2 129/130

[Kubi80] KUBIC, W.L. AND STEIN, F.P.: An Experimental and Correlative Study of the Vapor-Liquid Equilibria of the Tetrafluoromethane-Chlorotrifluoromethane System; Fluid Phase Equilibria 5(1980/81) 289/304

[Lang70] LANGE, H.B. AND STEIN, F.P.: Volumetric behavior of a polar-nonpolar-gas mixture: trifluoromethane-tetrafluoromethane system; J. Chem. Eng. Data 15(1970)1 56/60

[Lobo79] LOBO, L.Q. AND STAVELEY, L.A.K.: The Vapor Pressure of Tetrafluoromethane; Cryogenics 19(1979)6 335/338

[Mart71] MARTIN, J.J. AND BHADA, R.K.: P-V-T behavior of carbon tetrafluoride using a variable volume cell of bellows design; AIChE J. 17(1971)3 683; see also *[Bhad68]*

[Menz33] MENZEL, W. AND MOHRY, F.: Die Dampfdrücke des CF_4 und NF_3 und der Tripelpunkt des CF_4; Z. Anorg. Allgem. Chem. 210(1933) 257/263

[Rass79] RASSKAZOV, D.S.: Experimental Study of the isothermal Joule Thompson coefficients in R14; Thermophysical Properties of Matter and Substances 13(1979) 39/43

[Rodg74] RODGERS, A.S.; CHAO, J.; WILHOIT, R.C. AND ZWOLINSKI, B.J.: Ideal Gas Thermodynamic Properties of Eight Chloro- and Fluoromethanes; J. Phys. Chem. Ref. Data 3(1974)1 117/140

[Ruff31] RUFF, O. AND KEIM, R.: Das Kohlenstoff-4-fluorid (Tetrafluormethan); Z. Anorg. Allgem. Chem. 192(1931) 249/256

[Simo67] SIMON, M. ; KNOBLER, C.M. AND DUNCAN A.G.: The Vapor Pressure of Tetrafluoromethane from 86K to 146K; Cryogenics 7(1967)3 138/140

[Smit69] SMITH, J.H. AND PACE E.L.: The Thermodynamic Properties of Carbon Tetrafluoride from 12K to its Boiling Point; J. Phys. Chem. 73(1969)12 4232/4236

[Svet80] SVETLICHNY, P.I. AND PORICHANSKIJ, E.G.: Issledowanje termitscheskich swoistw Freona-14; I. Vyss. ucebn. Zaved. Energ. 4(1980) 118/121

[Svet80a] SVETLICHNY, P.I.: Study of compressibility of R14 in the vapor phase; I. Vyss. ucebn. Zaved. Energ. 6(1980) 74/77

[Terr69] TERRY, M.J. ; LYNCH, J.T. ; BUNCLARK, M. ; MANSELL, K.R. AND STAVELEY, L.A.K.: The Densities of Liquid Argon, Krypton, Xenon, Nitrogen, Carbon Monoxide, Methane and Carbon Tetrafluoride along the Orthobaric Liquid Curve; J. Chem. Therm. (1969)1 413/424

R 14. ph

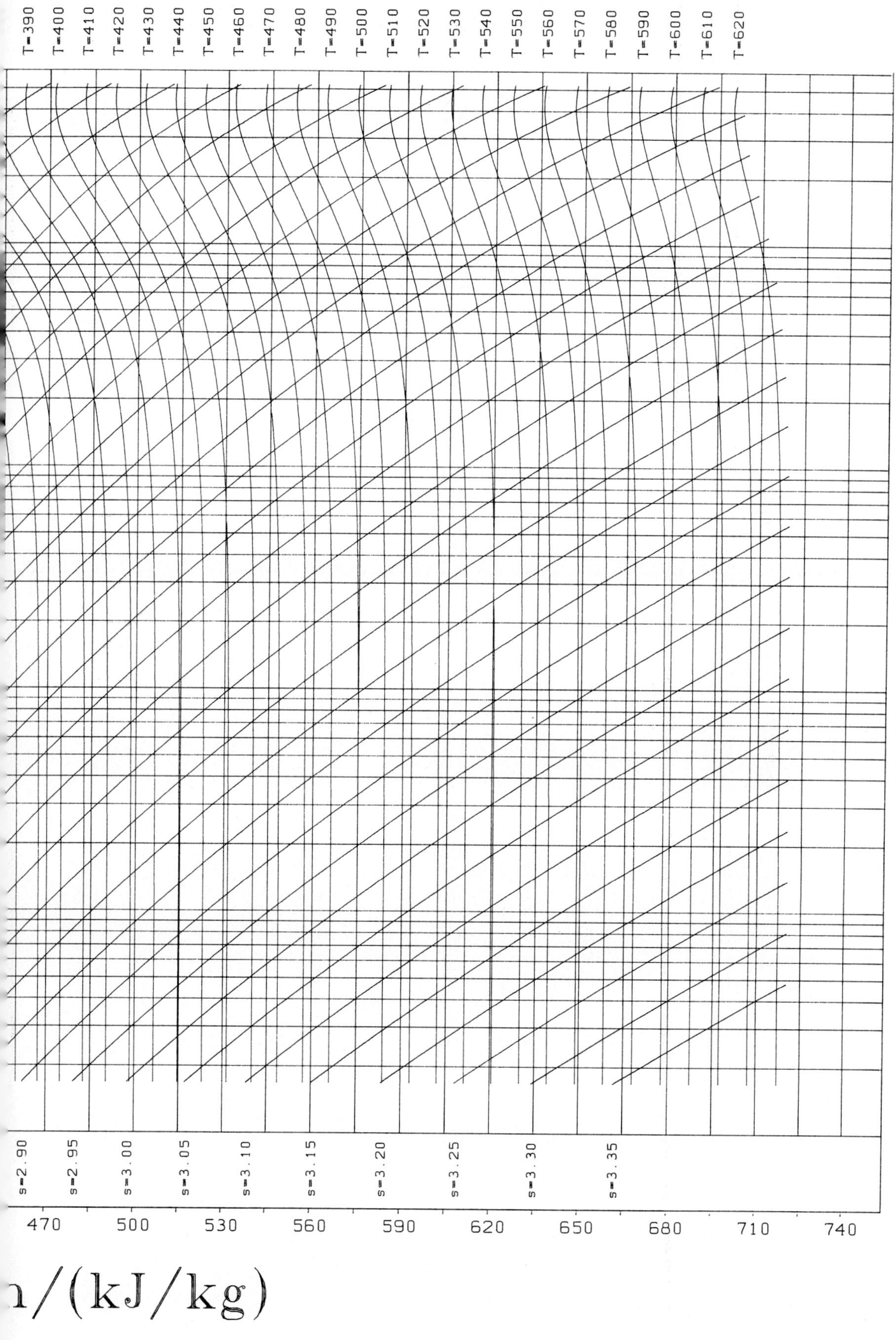

T=390
T=400
T=410
T=420
T=430
T=440
T=450
T=460
T=470
T=480
T=490
T=500
T=510
T=520
T=530
T=540
T=550
T=560
T=570
T=580
T=590
T=600
T=610
T=620
s=2.90
s=2.95
s=3.00
s=3.05
s=3.10
s=3.15
s=3.20
s=3.25
s=3.30
s=3.35
470
500
530
560
590
620
650
680
710
740
h/(kJ/kg)

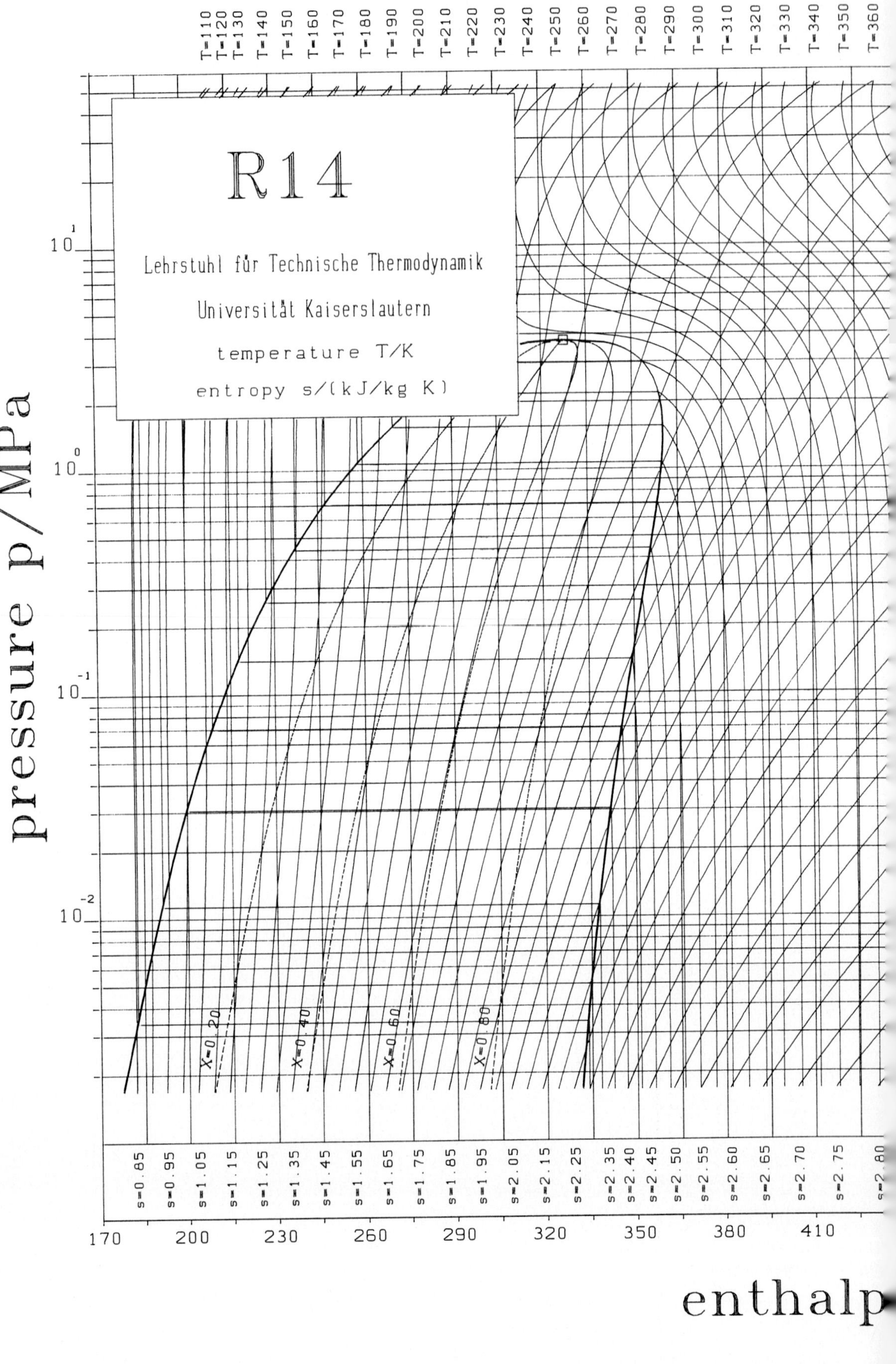
R14
Lehrstuhl für Technische Thermodynamik
Universität Kaiserslautern
temperature T/K
entropy s/(kJ/kg K)
pressure p/MPa
enthalp
T=110 T=120 T=130 T=140 T=150 T=160 T=170 T=180 T=190 T=200 T=210 T=220 T=230 T=240 T=250 T=260 T=270 T=280 T=290 T=300 T=310 T=320 T=330 T=340 T=350 T=360
s=0.85 s=0.95 s=1.05 s=1.15 s=1.25 s=1.35 s=1.45 s=1.55 s=1.65 s=1.75 s=1.85 s=1.95 s=2.05 s=2.15 s=2.25 s=2.35 s=2.40 s=2.45 s=2.50 s=2.55 s=2.60 s=2.65 s=2.70 s=2.75
X=0.20 X=0.40 X=0.60 X=0.80
170 200 230 260 290 320 350 380 410

R 14. Ts

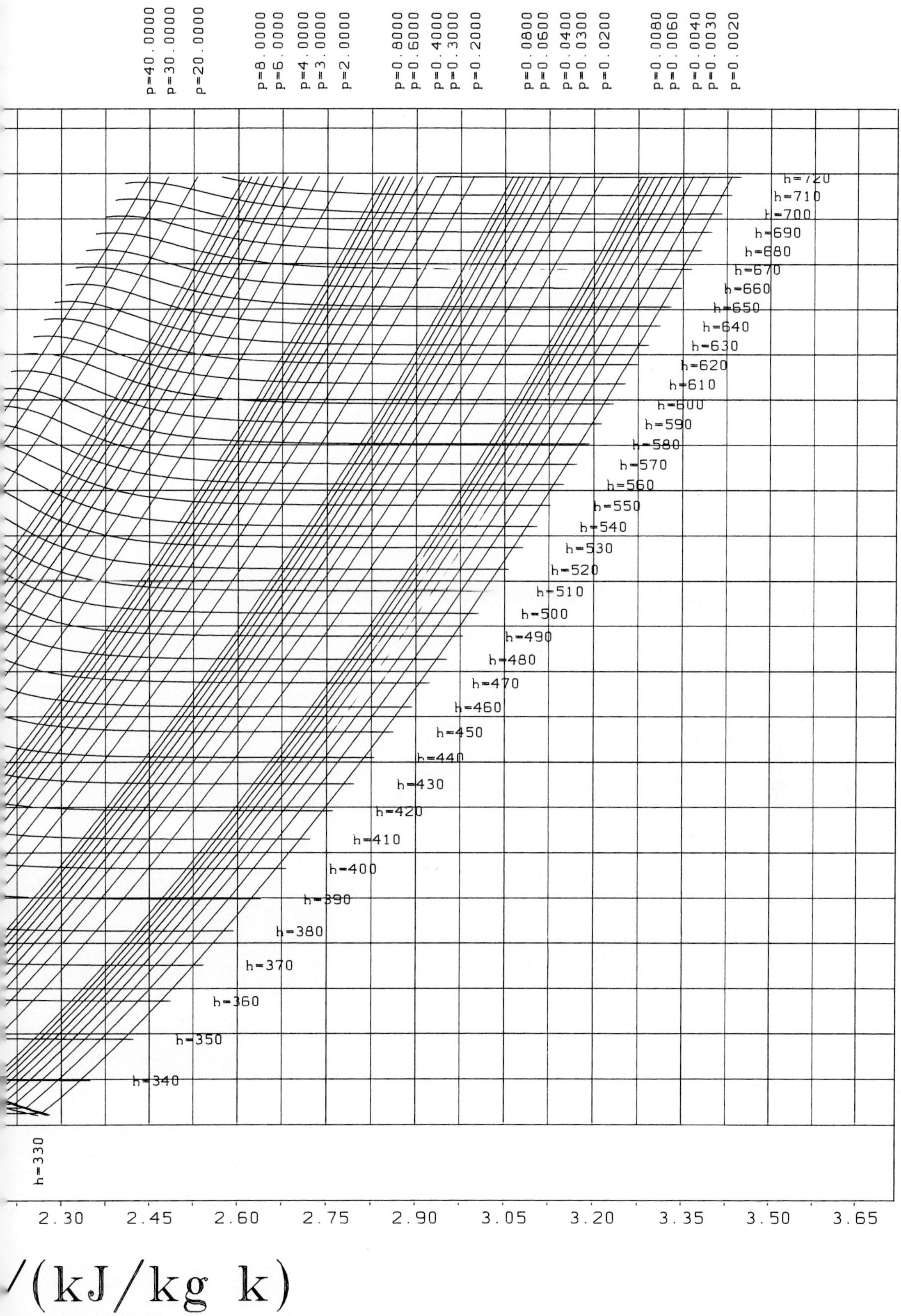
p=40.0000
p=30.0000
p=20.0000
p=8.0000
p=6.0000
p=4.0000
p=3.0000
p=2.0000
p=0.8000
p=0.6000
p=0.4000
p=0.3000
p=0.2000
p=0.0800
p=0.0600
p=0.0400
p=0.0300
p=0.0200
p=0.0080
p=0.0060
p=0.0040
p=0.0030
p=0.0020
h=720
h=710
h=700
h=690
h=680
h=670
h=660
h=650
h=640
h=630
h=620
h=610
h=600
h=590
h=580
h=570
h=560
h=550
h=540
h=530
h=520
h=510
h=500
h=490
h=480
h=470
h=460
h=450
h=440
h=430
h=420
h=410
h=400
h=390
h=380
h=370
h=360
h=350
h=340
h=330
2.30
2.45
2.60
2.75
2.90
3.05
3.20
3.35
3.50
3.65
/(kJ/kg k)

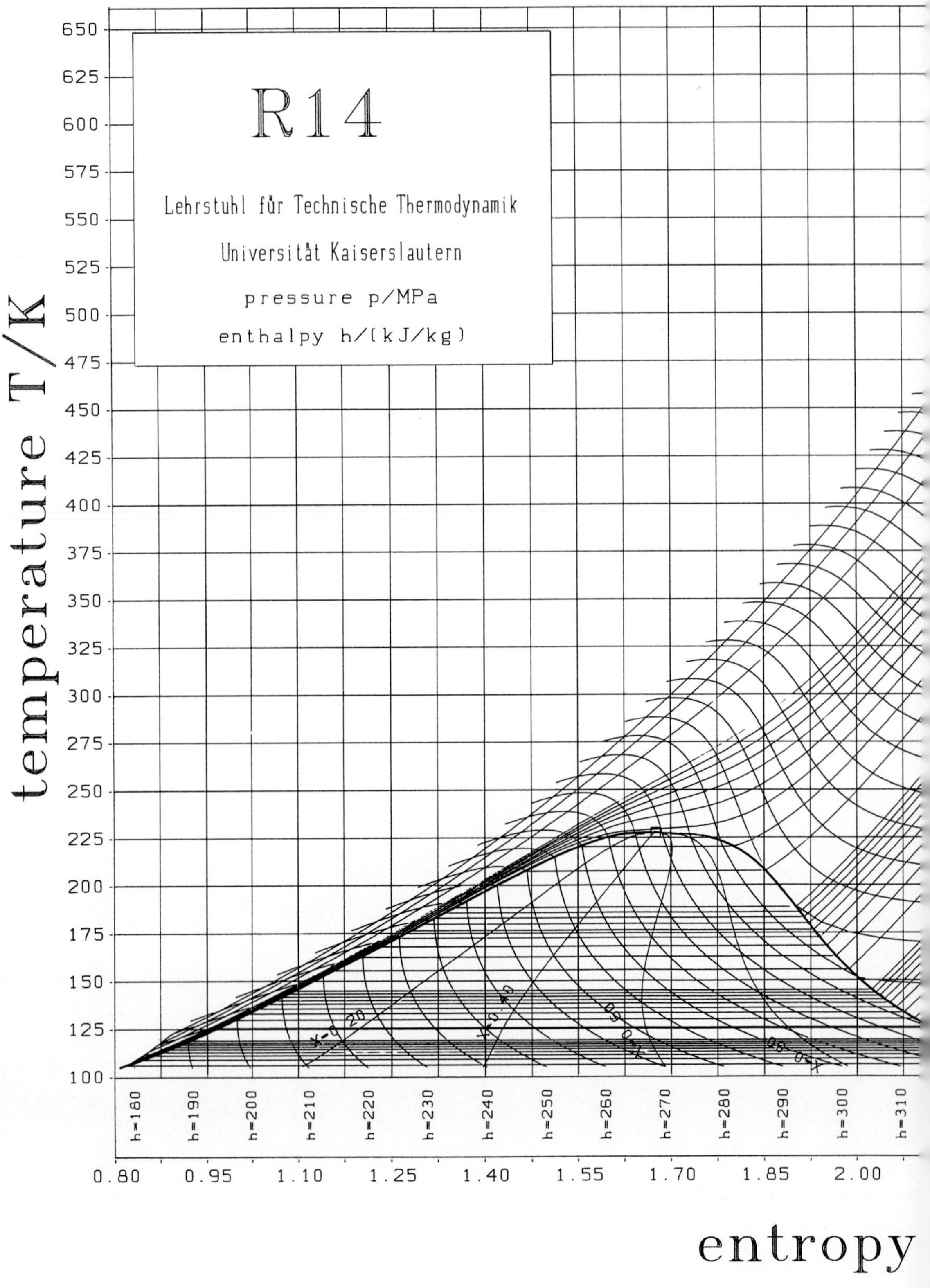
R14
Lehrstuhl für Technische Thermodynamik
Universität Kaiserslautern
pressure p/MPa
enthalpy h/(kJ/kg)
temperature T/K
entropy
650
625
600
575
550
525
500
475
450
425
400
375
350
325
300
275
250
225
200
175
150
125
100
h=180
h=190
h=200
h=210
h=220
h=230
h=240
h=250
h=260
h=270
h=280
h=290
h=300
h=310
0.80
0.95
1.10
1.25
1.40
1.55
1.70
1.85
2.00
x=0.20
x=0.40

R 14. hs

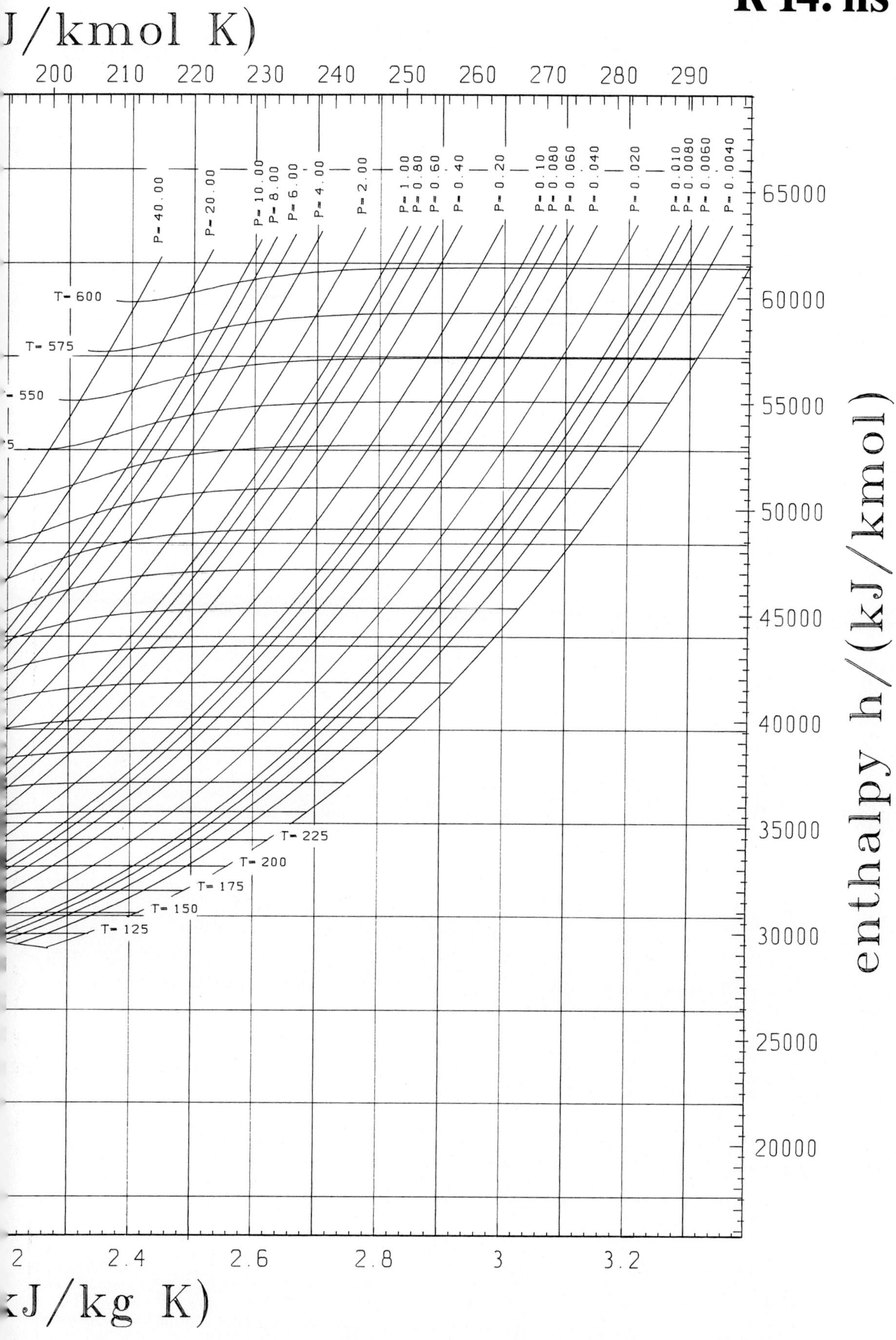
J/kmol K)
200
210
220
230
240
250
260
270
280
290
P= 40.00
P= 20.00
P= 10.00
P= 8.00
P= 6.00
P= 4.00
P= 2.00
P= 1.00
P= 0.80
P= 0.60
P= 0.40
P= 0.20
P= 0.10
P= 0.080
P= 0.060
P= 0.040
P= 0.020
P= 0.010
P= 0.0080
P= 0.0060
P= 0.0040
T= 600
T= 575
= 550
5
T= 225
T= 200
T= 175
T= 150
T= 125
65000
60000
55000
50000
45000
40000
35000
30000
25000
20000
enthalpy h/(kJ/kmol)
2
2.4
2.6
2.8
3
3.2
kJ/kg K)

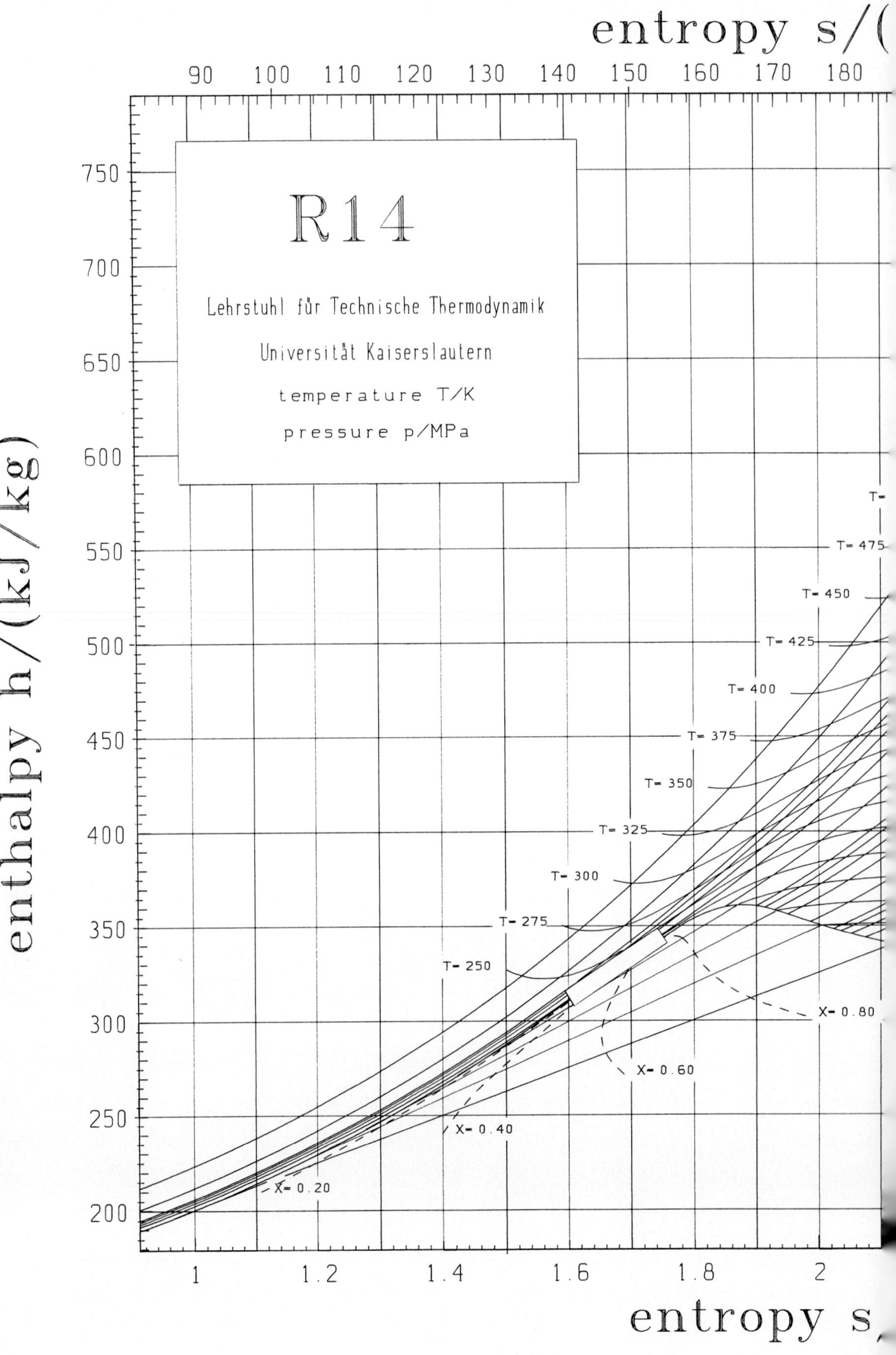

entropy s/(
R14
Lehrstuhl für Technische Thermodynamik
Universität Kaiserslautern
temperature T/K
pressure p/MPa
enthalpy h/(kJ/kg)
T= 475
T= 450
T= 425
T= 400
T= 375
T= 350
T= 325
T= 300
T= 275
T= 250
X= 0.80
X= 0.60
X= 0.40
X= 0.20
entropy s

Refrigerant R21 ($CHCl_2F$)

Table R21.1: List of references for p, ρ, T data

Author	n	p/MPa		ρ/g cm^{-3}		T/K	
		min	max	min	max	min	max
BENNING, A.F. *[Benn 40]*	40	0.03	2.1	0.001	0.07	296	445
DEVANEY, W.E. *[DeVa 76]*	263	0.6	137.9	1.01	1.77	143	414
KALAFATI, D.D. *[Kala 68]*	210	0.33	18.7	0.014	1.31	298	473
SOLOV'EV, A.N. *[Solo 68]*	295	0.24	6.8	0.010	0.54	310	470

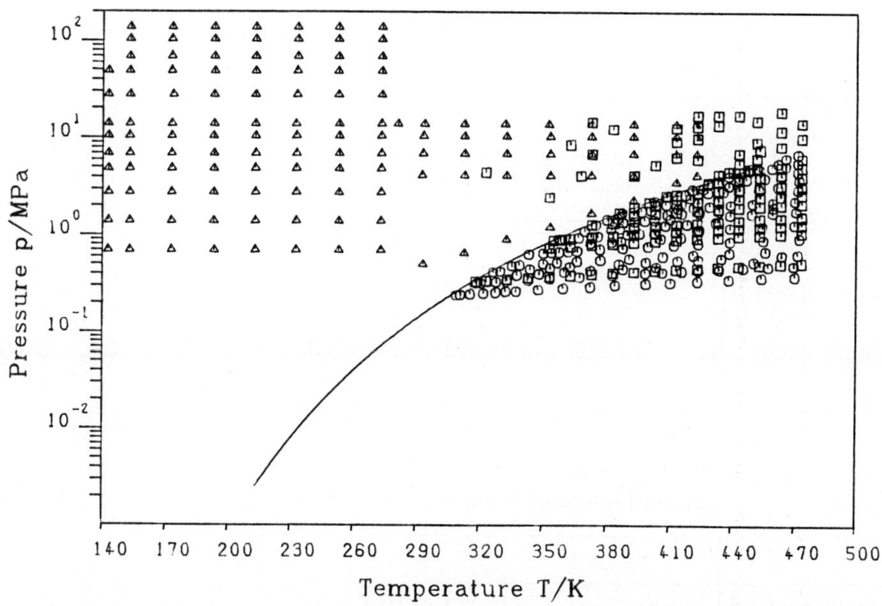

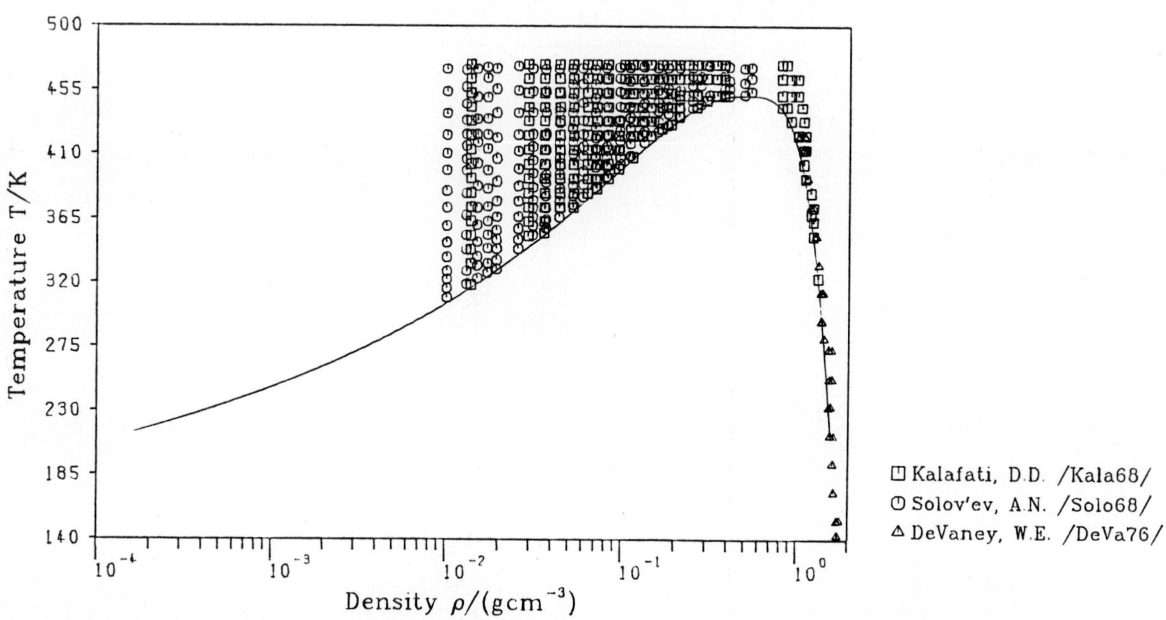

Fig.R21.1: Distribution of p, ρ, T data projected in the p, T and in the T, ρ plane

Table R21.2: Experimental vapor pressure data

Author	Δ / %		
	std	opt	*EOS*
BENNING, A.F. *[Benn 40]*	0.35	0.33	0.28
DEVANEY, W.E. *[DeVa 76]*	1.32	1.30	0.74
KALAFATI, D.D. *[Kala 68]*	0.32	0.91	0.22
SOLOV'EV, A.N. *[Solo 68]*	0.35	0.32	0.41

Table R21.3: Experimental saturated liquid density data

Author	Δ / %		
	std	opt	*EOS*
BENNING, A.F. *[Benn 40b]*	0.05	0.05	0.08

Table R21.4: Coefficients of the standard vapor pressure equation

Index i	b_i
5	−6.898203 E+00
6	1.165844 E+00
9	−2.330523 E+00
15	−3.751456 E+00

Table R21.5: Coefficients of the quasi-optimal vapor pressure equation

Index i	b_i
5	−6.981375 E+00
6	1.425995 E+00
9	−3.070314 E+00
20	−6.461655 E+00

Table R21.6: Coefficients of the standard saturated liquid density equation

Index i	c_i
8	1.671526 E+00
9	8.596728 E−01
16	3.346563 E−01

Table R21.7: Coefficients of the quasi-optimal saturated liquid density equation

Index i	c_i
8	1.683401 E+00
9	8.276885 E−01
14	2.593503 E−01

Table R21.8: Coefficients of the *Bender-EOS*

Range of validity: one-phase region

$0.24 \leq p/\text{MPa} \leq 137.9$

$0.00999 \leq \rho/\text{g cm}^{-3} \leq 1.769$

$142.80 \leq T/\text{K} \leq 473.19$

	Specific units	Molar units
a_{01}	7.751117622 E−02	8.210391804 E−01
a_{02}	5.078747469 E+01	5.379676664 E+02
a_{03}	2.718628821 E+04	2.879714756 E+05
a_{04}	−2.066888946 E+06	−2.189357573 E+07
a_{05}	1.790425448 E+09	1.896512883 E+10
a_{06}	5.049899964 E−01	5.505314168 E−01
a_{07}	−7.138878898 E+02	−7.782683108 E+02
a_{08}	3.933805307 E+05	4.288566952 E+05
a_{09}	−3.511823431 E−01	−3.940322624 E−02
a_{10}	3.379866157 E+02	3.792264432 E+01
a_{11}	5.098568419 E−01	5.887719879 E−03
a_{12}	−5.284928054 E+02	−6.102924077 E+00
a_{13}	1.383389359 E+02	1.644156281 E−01
a_{14}	−1.130989746 E+08	−1.232985587 E+08
a_{15}	2.084803895 E+10	2.272817383 E+10
a_{16}	−1.004536096 E+12	−1.095127991 E+12
a_{17}	−1.757608534 E+07	−2.029649473 E+05
a_{18}	−1.094813860 E+09	−1.264268084 E+07
a_{19}	1.493265890 E+11	1.724392132 E+09
a_{20}	2.700000000 E−01	2.859982128 E−03

Table R21.9: Deviations between experimental and predicted p, ρ, T data

Region	n	Δ / %	
		p	ρ
Gas	441	0.26	0.34
Critical	27	0.25	3.66
Liquid	226	7.70	0.24
Total	694	4.46	0.79

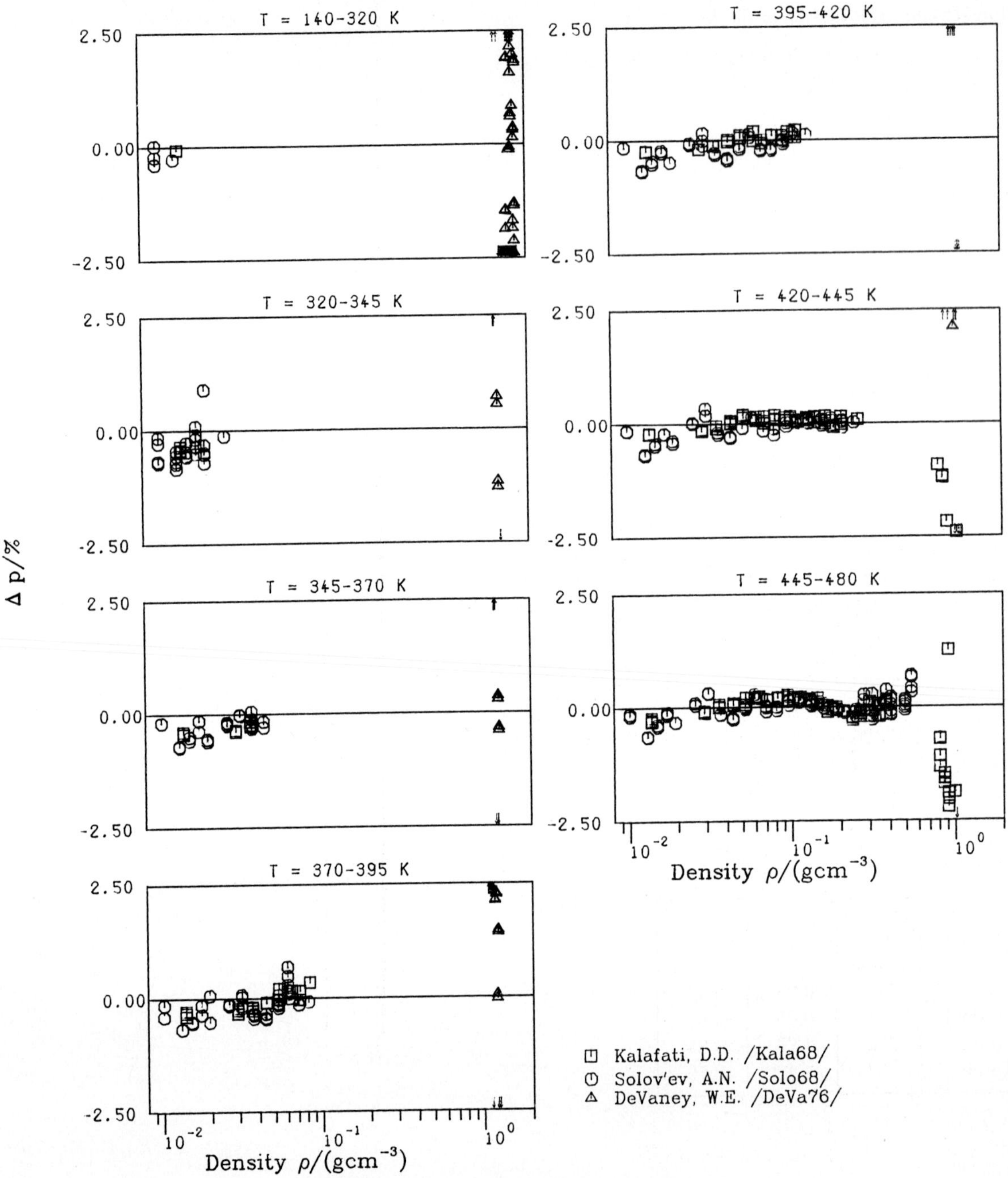

Fig.R21.2: Deviations between experimental and calculated p, ρ, T data

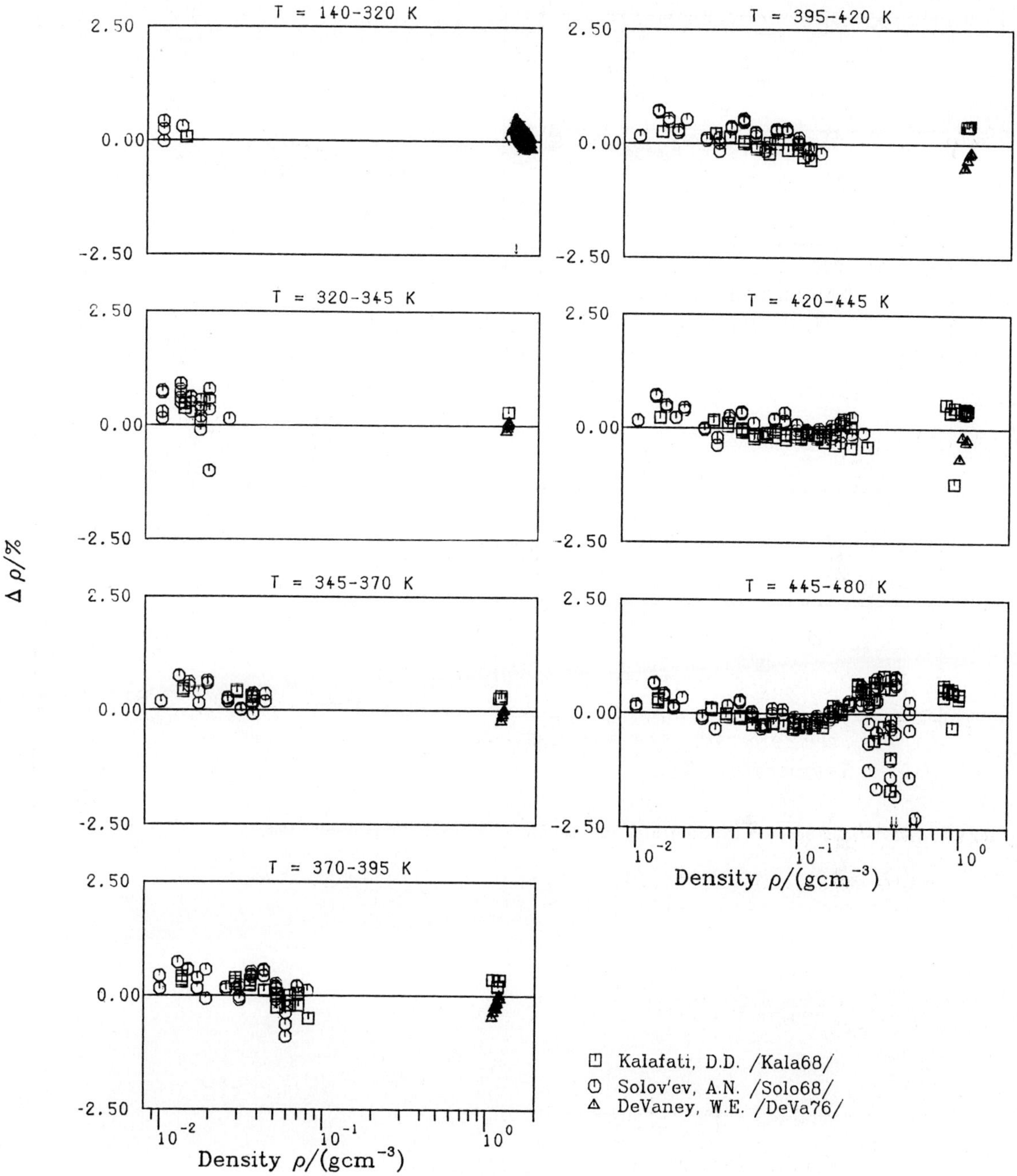

Fig.R21.2: Deviations between experimental and calculated p, ρ, T data *-continued*

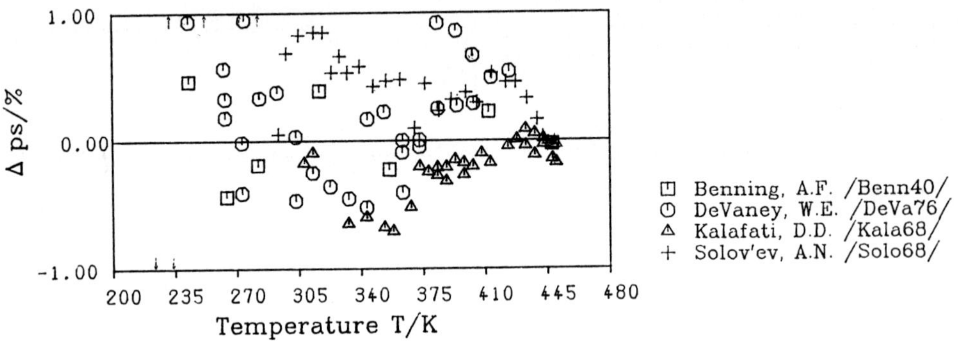

Fig.R21.3: Deviations between experimental and calculated vapor pressures

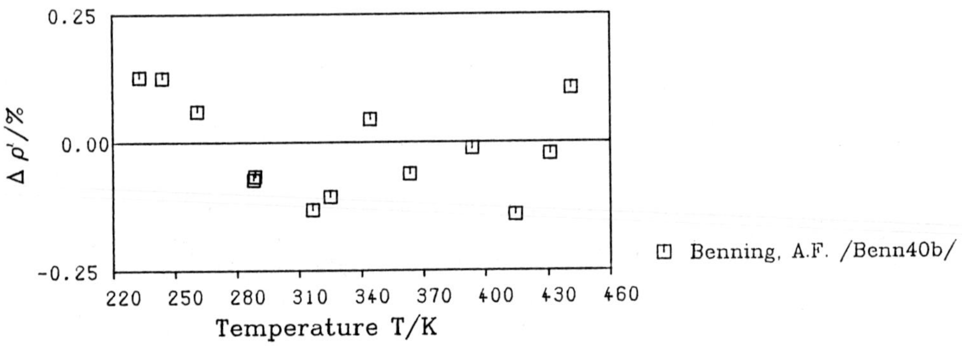

Fig.R21.4: Deviations between experimental and calculated saturated liquid densities

Table R21.10: Information on ideal gas state heat capacity equation

Author	T / K	
	min	max
CHEN, S.S. *[Chen 76]*	100	600
Coefficients of equation (15)		
Index i	d_i	
1	2.376576 E−01	
2	1.271433 E−03	
3	3.241352 E−07	
4	−2.492428 E−09	
5	1.717208 E−12	

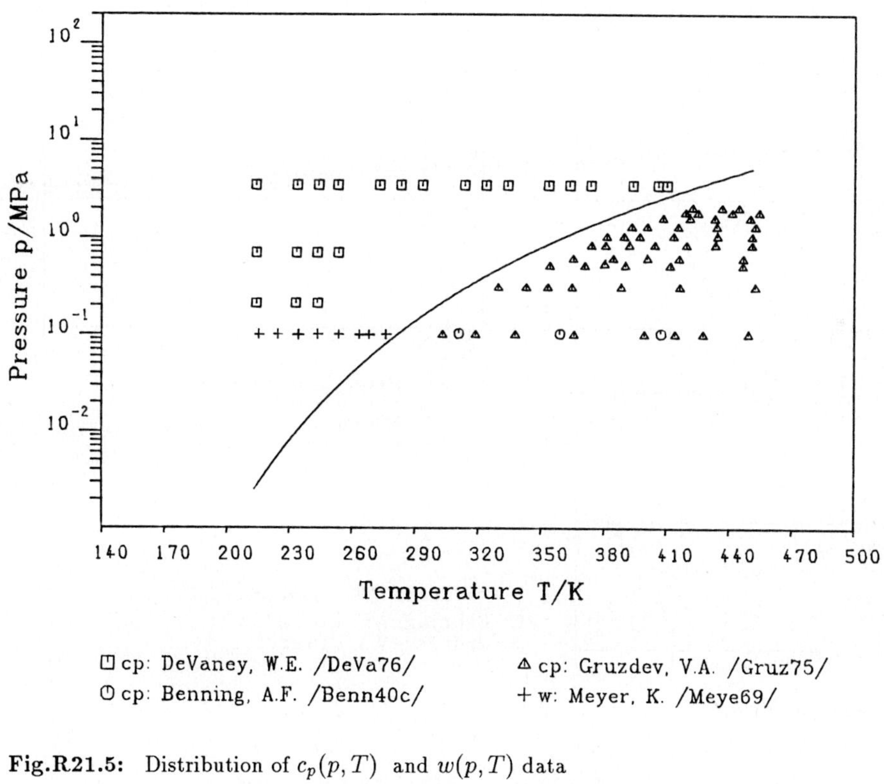

Fig.R21.5: Distribution of $c_p(p,T)$ and $w(p,T)$ data

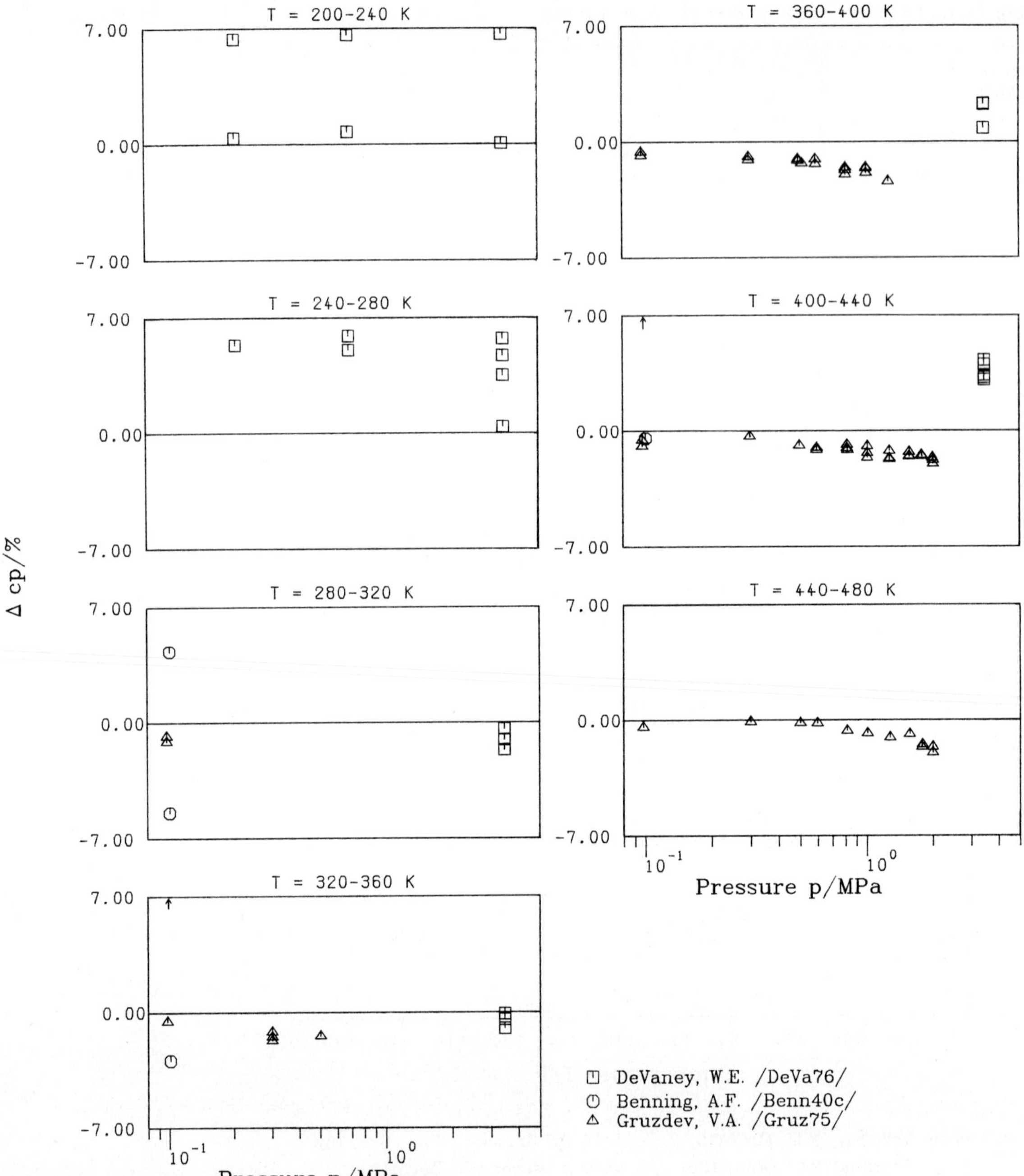

Fig.R21.6: Deviations between experimental and calculated specific isobaric heat capacities c_p

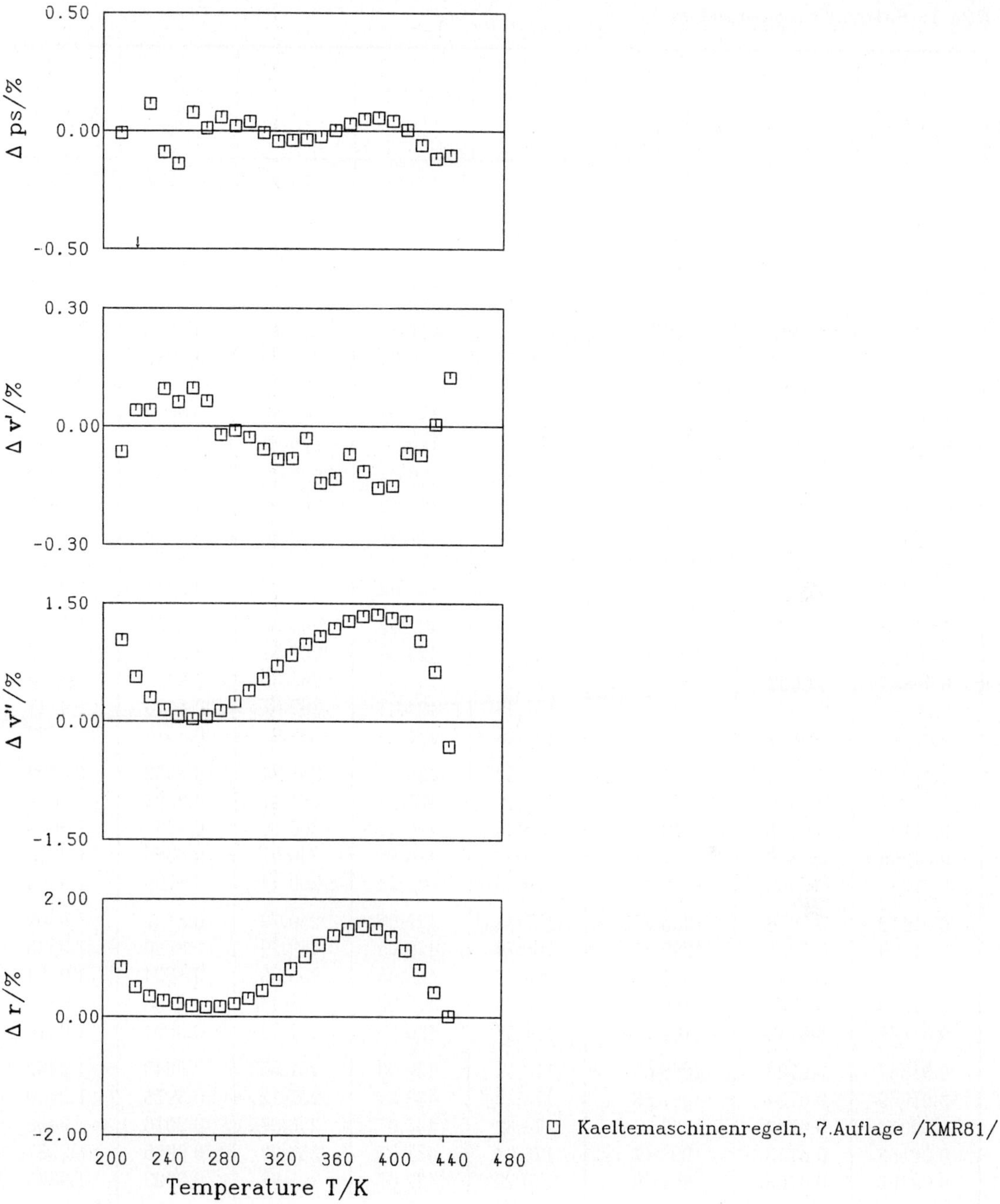

Fig.R21.7: Deviations between saturation properties from the *Kältemaschinenregeln* and those calculated by the *Bender-EOS*

Chart R21.1: Saturation properties

T K	p MPa	v' dm³/kg	v'' dm³/kg	h' kJ/kg	h'' kJ/kg	r kJ/kg	s' kJ/kgK	s'' kJ/kgK
215.0	0.00286	0.6452	6048.9	144.58	415.06	270.48	0.7724	2.0305
216.0	0.00308	0.6460	5652.5	145.54	415.55	270.01	0.7769	2.0269
217.0	0.00331	0.6469	5286.1	146.50	416.05	269.55	0.7813	2.0235
218.0	0.00355	0.6478	4947.1	147.45	416.54	269.09	0.7857	2.0200
219.0	0.00381	0.6486	4633.2	148.40	417.04	268.64	0.7900	2.0167
220.0	0.00408	0.6495	4342.3	149.34	417.54	268.20	0.7943	2.0134
221.0	0.00437	0.6504	4072.5	150.28	418.03	267.76	0.7985	2.0101
222.0	0.00467	0.6513	3822.0	151.21	418.53	267.32	0.8028	2.0069
223.0	0.00500	0.6521	3589.4	152.14	419.03	266.89	0.8069	2.0037
224.0	0.00534	0.6530	3373.2	153.07	419.53	266.46	0.8111	2.0006
225.0	0.00570	0.6539	3172.1	153.99	420.03	266.03	0.8152	1.9976
226.0	0.00609	0.6548	2984.9	154.91	420.52	265.61	0.8193	1.9945
227.0	0.00649	0.6557	2810.5	155.84	421.02	265.19	0.8233	1.9916
228.0	0.00692	0.6566	2647.9	156.76	421.52	264.77	0.8274	1.9887
229.0	0.00737	0.6575	2496.3	157.68	422.02	264.35	0.8314	1.9858
230.0	0.00784	0.6584	2354.8	158.59	422.52	263.93	0.8354	1.9829
231.0	0.00834	0.6593	2222.6	159.51	423.03	263.51	0.8394	1.9802
232.0	0.00887	0.6602	2099.0	160.43	423.53	263.09	0.8434	1.9774
233.0	0.00943	0.6611	1983.5	161.35	424.03	262.67	0.8473	1.9747
234.0	0.01001	0.6621	1875.4	162.28	424.53	262.26	0.8513	1.9720
235.0	0.01062	0.6630	1774.2	163.20	425.03	261.84	0.8552	1.9694
236.0	0.01127	0.6639	1679.3	164.12	425.53	261.41	0.8591	1.9668
237.0	0.01194	0.6648	1590.4	165.05	426.04	260.99	0.8630	1.9643
238.0	0.01265	0.6658	1507.0	165.97	426.54	260.57	0.8669	1.9618
239.0	0.01340	0.6667	1428.8	166.90	427.04	260.14	0.8708	1.9593
240.0	0.01418	0.6676	1355.2	167.83	427.55	259.72	0.8747	1.9569
241.0	0.01499	0.6686	1286.2	168.76	428.05	259.29	0.8786	1.9545
242.0	0.01585	0.6695	1221.3	169.69	428.55	258.86	0.8824	1.9521
243.0	0.01675	0.6705	1160.2	170.63	429.06	258.43	0.8863	1.9498
244.0	0.01769	0.6715	1102.7	171.57	429.56	257.99	0.8901	1.9475
245.0	0.01867	0.6724	1048.6	172.51	430.06	257.56	0.8940	1.9452
246.0	0.01969	0.6734	997.58	173.45	430.56	257.12	0.8978	1.9430
247.0	0.02076	0.6744	949.51	174.39	431.07	256.67	0.9016	1.9408
248.0	0.02188	0.6753	904.17	175.34	431.57	256.23	0.9055	1.9387
249.0	0.02305	0.6763	861.40	176.29	432.07	255.78	0.9093	1.9365
250.0	0.02427	0.6773	821.02	177.24	432.58	255.33	0.9131	1.9344
251.0	0.02554	0.6783	782.88	178.20	433.08	254.88	0.9169	1.9324
252.0	0.02687	0.6793	746.83	179.16	433.58	254.43	0.9207	1.9303
253.0	0.02825	0.6803	712.76	180.11	434.09	253.97	0.9245	1.9283
254.0	0.02969	0.6813	680.53	181.08	434.59	253.51	0.9283	1.9264
255.0	0.03119	0.6823	650.03	182.04	435.09	253.05	0.9321	1.9244
256.0	0.03274	0.6833	621.16	183.01	435.59	252.58	0.9359	1.9225
257.0	0.03437	0.6843	593.81	183.98	436.09	252.11	0.9397	1.9206
258.0	0.03605	0.6853	567.90	184.95	436.60	251.64	0.9434	1.9188
259.0	0.03781	0.6864	543.33	185.93	437.10	251.17	0.9472	1.9170
260.0	0.03963	0.6874	520.03	186.91	437.60	250.69	0.9510	1.9152
261.0	0.04152	0.6884	497.92	187.89	438.10	250.21	0.9547	1.9134
262.0	0.04348	0.6895	476.94	188.87	438.60	249.73	0.9585	1.9116
263.0	0.04552	0.6905	457.01	189.86	439.10	249.24	0.9622	1.9099
264.0	0.04764	0.6916	438.08	190.85	439.60	248.75	0.9660	1.9082

Chart R21.1: Saturation properties—*continued*

T K	p MPa	v' dm³/kg	v'' dm³/kg	h' kJ/kg	h'' kJ/kg	r kJ/kg	s' kJ/kgK	s'' kJ/kgK
265.0	0.04983	0.6926	420.08	191.84	440.10	248.26	0.9697	1.9065
266.0	0.05211	0.6937	402.97	192.83	440.60	247.77	0.9734	1.9049
267.0	0.05446	0.6948	386.70	193.82	441.09	247.27	0.9772	1.9033
268.0	0.05691	0.6958	371.22	194.82	441.59	246.77	0.9809	1.9017
269.0	0.05944	0.6969	356.47	195.82	442.09	246.26	0.9846	1.9001
270.0	0.06205	0.6980	342.44	196.83	442.58	245.76	0.9883	1.8985
271.0	0.06476	0.6991	329.06	197.83	443.08	245.25	0.9920	1.8970
272.0	0.06757	0.7002	316.32	198.84	443.58	244.74	0.9957	1.8955
273.0	0.07047	0.7013	304.17	199.85	444.07	244.22	0.9994	1.8940
274.0	0.07346	0.7024	292.58	200.86	444.56	243.70	1.0031	1.8926
275.0	0.07656	0.7035	281.52	201.87	445.06	243.18	1.0068	1.8911
276.0	0.07976	0.7046	270.97	202.89	445.55	242.66	1.0105	1.8897
277.0	0.08307	0.7058	260.89	203.91	446.04	242.13	1.0142	1.8883
278.0	0.08648	0.7069	251.27	204.93	446.53	241.60	1.0178	1.8869
279.0	0.09001	0.7080	242.07	205.95	447.02	241.07	1.0215	1.8856
280.0	0.09364	0.7092	233.29	206.98	447.51	240.54	1.0252	1.8842
281.0	0.09739	0.7103	224.88	208.00	448.00	240.00	1.0288	1.8829
282.0	0.10126	0.7115	216.85	209.03	448.49	239.46	1.0325	1.8816
283.0	0.10525	0.7127	209.16	210.06	448.98	238.92	1.0361	1.8803
284.0	0.10936	0.7138	201.80	211.09	449.47	238.37	1.0397	1.8791
285.0	0.11360	0.7150	194.75	212.13	449.95	237.82	1.0434	1.8778
286.0	0.11797	0.7162	188.01	213.17	450.44	237.27	1.0470	1.8766
287.0	0.12246	0.7174	181.54	214.20	450.92	236.72	1.0506	1.8754
288.0	0.12709	0.7186	175.35	215.24	451.40	236.16	1.0542	1.8742
289.0	0.13185	0.7198	169.41	216.28	451.89	235.60	1.0578	1.8730
290.0	0.13676	0.7210	163.72	217.33	452.37	235.04	1.0614	1.8719
291.0	0.14180	0.7223	158.26	218.37	452.85	234.47	1.0650	1.8707
292.0	0.14699	0.7235	153.02	219.42	453.33	233.91	1.0685	1.8696
293.0	0.15232	0.7247	147.99	220.47	453.80	233.34	1.0721	1.8685
294.0	0.15780	0.7260	143.16	221.52	454.28	232.76	1.0757	1.8674
295.0	0.16344	0.7272	138.53	222.57	454.76	232.19	1.0792	1.8663
296.0	0.16923	0.7285	134.07	223.62	455.23	231.61	1.0828	1.8652
297.0	0.17517	0.7298	129.79	224.68	455.71	231.03	1.0863	1.8642
298.0	0.18128	0.7311	125.68	225.74	456.18	230.44	1.0899	1.8632
299.0	0.18755	0.7324	121.73	226.79	456.65	229.86	1.0934	1.8621
300.0	0.19399	0.7336	117.92	227.85	457.12	229.27	1.0969	1.8611
301.0	0.20060	0.7350	114.27	228.91	457.59	228.68	1.1004	1.8602
302.0	0.20737	0.7363	110.75	229.98	458.06	228.08	1.1039	1.8592
303.0	0.21433	0.7376	107.36	231.04	458.52	227.48	1.1074	1.8582
304.0	0.22146	0.7389	104.10	232.11	458.99	226.88	1.1109	1.8573
305.0	0.22877	0.7403	100.96	233.17	459.45	226.28	1.1144	1.8563
306.0	0.23627	0.7416	97.930	234.24	459.92	225.67	1.1179	1.8554
307.0	0.24395	0.7430	95.016	235.31	460.38	225.07	1.1214	1.8545
308.0	0.25182	0.7443	92.207	236.38	460.84	224.45	1.1248	1.8536
309.0	0.25988	0.7457	89.499	237.46	461.30	223.84	1.1283	1.8527
310.0	0.26814	0.7471	86.888	238.53	461.75	223.22	1.1317	1.8518
311.0	0.27660	0.7485	84.370	239.61	462.21	222.60	1.1352	1.8510
312.0	0.28526	0.7499	81.941	240.68	462.66	221.98	1.1386	1.8501
313.0	0.29413	0.7513	79.598	241.76	463.12	221.35	1.1421	1.8493
314.0	0.30320	0.7528	77.336	242.84	463.57	220.73	1.1455	1.8484

Chart R21.1: Saturation properties–*continued*

T K	p MPa	v′ dm³/kg	v″ dm³/kg	h′ kJ/kg	h″ kJ/kg	r kJ/kg	s′ kJ/kgK	s″ kJ/kgK
315.0	0.31249	0.7542	75.152	243.92	464.02	220.09	1.1489	1.8476
316.0	0.32199	0.7557	73.044	245.00	464.46	219.46	1.1523	1.8468
317.0	0.33171	0.7571	71.007	246.09	464.91	218.82	1.1557	1.8460
318.0	0.34164	0.7586	69.040	247.17	465.36	218.18	1.1591	1.8452
319.0	0.35180	0.7601	67.139	248.26	465.80	217.54	1.1625	1.8444
320.0	0.36219	0.7616	65.302	249.35	466.24	216.89	1.1659	1.8437
321.0	0.37281	0.7631	63.526	250.44	466.68	216.25	1.1692	1.8429
322.0	0.38366	0.7646	61.809	251.53	467.12	215.59	1.1726	1.8421
323.0	0.39475	0.7661	60.148	252.62	467.56	214.94	1.1760	1.8414
324.0	0.40607	0.7677	58.541	253.71	467.99	214.28	1.1793	1.8407
325.0	0.41764	0.7692	56.987	254.81	468.42	213.62	1.1827	1.8399
326.0	0.42946	0.7708	55.483	255.90	468.86	212.95	1.1860	1.8392
327.0	0.44152	0.7724	54.027	257.00	469.28	212.29	1.1893	1.8385
328.0	0.45384	0.7740	52.617	258.10	469.71	211.61	1.1927	1.8378
329.0	0.46641	0.7756	51.252	259.20	470.14	210.94	1.1960	1.8371
330.0	0.47924	0.7772	49.930	260.30	470.56	210.26	1.1993	1.8364
331.0	0.49234	0.7788	48.650	261.40	470.98	209.58	1.2026	1.8358
332.0	0.50570	0.7804	47.409	262.51	471.40	208.90	1.2059	1.8351
333.0	0.51933	0.7821	46.206	263.61	471.82	208.21	1.2092	1.8344
334.0	0.53323	0.7838	45.040	264.72	472.24	207.52	1.2125	1.8338
335.0	0.54741	0.7855	43.910	265.83	472.65	206.82	1.2158	1.8331
336.0	0.56186	0.7872	42.814	266.94	473.06	206.12	1.2190	1.8325
337.0	0.57660	0.7889	41.751	268.05	473.47	205.42	1.2223	1.8319
338.0	0.59162	0.7906	40.720	269.16	473.88	204.72	1.2256	1.8312
339.0	0.60694	0.7923	39.720	270.28	474.28	204.01	1.2288	1.8306
340.0	0.62254	0.7941	38.749	271.39	474.69	203.29	1.2321	1.8300
341.0	0.63845	0.7959	37.807	272.51	475.09	202.58	1.2353	1.8294
342.0	0.65465	0.7977	36.892	273.63	475.48	201.86	1.2386	1.8288
343.0	0.67115	0.7995	36.004	274.75	475.88	201.13	1.2418	1.8282
344.0	0.68796	0.8013	35.141	275.87	476.27	200.40	1.2450	1.8276
345.0	0.70508	0.8031	34.303	277.00	476.67	199.67	1.2482	1.8270
346.0	0.72252	0.8050	33.489	278.12	477.05	198.93	1.2515	1.8264
347.0	0.74026	0.8069	32.699	279.25	477.44	198.19	1.2547	1.8258
348.0	0.75833	0.8088	31.930	280.38	477.82	197.45	1.2579	1.8252
349.0	0.77673	0.8107	31.183	281.51	478.20	196.70	1.2611	1.8247
350.0	0.79545	0.8126	30.457	282.64	478.58	195.94	1.2643	1.8241
351.0	0.81450	0.8146	29.750	283.77	478.96	195.18	1.2675	1.8235
352.0	0.83389	0.8165	29.064	284.91	479.33	194.42	1.2706	1.8230
353.0	0.85361	0.8185	28.396	286.05	479.70	193.65	1.2738	1.8224
354.0	0.87367	0.8205	27.746	287.19	480.07	192.88	1.2770	1.8219
355.0	0.89408	0.8225	27.114	288.33	480.43	192.11	1.2802	1.8213
356.0	0.91484	0.8246	26.498	289.47	480.80	191.32	1.2833	1.8208
357.0	0.93595	0.8267	25.899	290.62	481.15	190.54	1.2865	1.8202
358.0	0.95741	0.8288	25.316	291.76	481.51	189.75	1.2897	1.8197
359.0	0.97924	0.8309	24.749	292.91	481.86	188.95	1.2928	1.8191
360.0	1.00143	0.8330	24.196	294.06	482.21	188.15	1.2960	1.8186
361.0	1.02398	0.8352	23.658	295.22	482.56	187.34	1.2991	1.8181
362.0	1.04690	0.8373	23.133	296.37	482.90	186.53	1.3023	1.8175
363.0	1.07020	0.8395	22.622	297.53	483.24	185.71	1.3054	1.8170
364.0	1.09388	0.8418	22.125	298.69	483.58	184.89	1.3085	1.8165

Chart R21.1: Saturation properties—*continued*

T K	p MPa	v' dm^3/kg	v'' dm^3/kg	h' kJ/kg	h'' kJ/kg	r kJ/kg	s' kJ/kgK	s'' kJ/kgK
365.0	1.11794	0.8440	21.640	299.85	483.91	184.06	1.3117	1.8159
366.0	1.14238	0.8463	21.167	301.01	484.24	183.23	1.3148	1.8154
367.0	1.16721	0.8486	20.706	302.18	484.57	182.39	1.3179	1.8149
368.0	1.19243	0.8509	20.256	303.35	484.89	181.54	1.3210	1.8144
369.0	1.21805	0.8533	19.818	304.52	485.21	180.69	1.3242	1.8138
370.0	1.24408	0.8557	19.391	305.69	485.52	179.83	1.3273	1.8133
371.0	1.27050	0.8581	18.974	306.87	485.83	178.97	1.3304	1.8128
372.0	1.29733	0.8605	18.567	308.05	486.14	178.09	1.3335	1.8122
373.0	1.32458	0.8630	18.170	309.23	486.45	177.22	1.3366	1.8117
374.0	1.35224	0.8655	17.782	310.41	486.74	176.33	1.3397	1.8112
375.0	1.38032	0.8680	17.404	311.60	487.04	175.44	1.3428	1.8107
376.0	1.40882	0.8706	17.035	312.79	487.33	174.54	1.3459	1.8101
377.0	1.43776	0.8732	16.675	313.98	487.62	173.64	1.3490	1.8096
378.0	1.46712	0.8758	16.323	315.18	487.90	172.72	1.3521	1.8091
379.0	1.49692	0.8785	15.979	316.37	488.18	171.80	1.3552	1.8085
380.0	1.52716	0.8812	15.644	317.57	488.45	170.88	1.3583	1.8080
381.0	1.55784	0.8840	15.316	318.78	488.72	169.94	1.3614	1.8074
382.0	1.58897	0.8867	14.995	319.99	488.98	169.00	1.3645	1.8069
383.0	1.62056	0.8895	14.682	321.20	489.24	168.04	1.3676	1.8063
384.0	1.65260	0.8924	14.376	322.41	489.49	167.08	1.3707	1.8058
385.0	1.68510	0.8953	14.077	323.63	489.74	166.12	1.3738	1.8052
386.0	1.71807	0.8982	13.784	324.85	489.99	165.14	1.3768	1.8047
387.0	1.75150	0.9012	13.498	326.07	490.22	164.15	1.3799	1.8041
388.0	1.78541	0.9042	13.218	327.30	490.46	163.16	1.3830	1.8035
389.0	1.81980	0.9073	12.945	328.53	490.68	162.15	1.3861	1.8030
390.0	1.85467	0.9104	12.677	329.77	490.90	161.13	1.3892	1.8024
391.0	1.89003	0.9136	12.415	331.01	491.12	160.11	1.3923	1.8018
392.0	1.92588	0.9168	12.159	332.25	491.33	159.07	1.3954	1.8012
393.0	1.96222	0.9201	11.908	333.50	491.53	158.03	1.3985	1.8006
394.0	1.99907	0.9234	11.663	334.75	491.72	156.97	1.4016	1.8000
395.0	2.03642	0.9268	11.422	336.01	491.91	155.90	1.4047	1.7994
396.0	2.07428	0.9302	11.187	337.27	492.09	154.82	1.4078	1.7988
397.0	2.11266	0.9337	10.956	338.54	492.27	153.73	1.4109	1.7981
398.0	2.15156	0.9372	10.730	339.81	492.44	152.63	1.4140	1.7975
399.0	2.19098	0.9408	10.509	341.08	492.60	151.51	1.4171	1.7968
400.0	2.23093	0.9445	10.292	342.36	492.75	150.39	1.4202	1.7962
401.0	2.27142	0.9483	10.080	343.65	492.89	149.24	1.4233	1.7955
402.0	2.31245	0.9521	9.8712	344.94	493.03	148.09	1.4264	1.7948
403.0	2.35403	0.9560	9.6670	346.24	493.16	146.92	1.4296	1.7941
404.0	2.39616	0.9599	9.4667	347.54	493.28	145.73	1.4327	1.7934
405.0	2.43884	0.9640	9.2703	348.85	493.39	144.53	1.4358	1.7927
406.0	2.48209	0.9681	9.0776	350.17	493.49	143.32	1.4390	1.7920
407.0	2.52590	0.9723	8.8885	351.49	493.58	142.09	1.4421	1.7912
408.0	2.57029	0.9766	8.7029	352.82	493.66	140.84	1.4453	1.7905
409.0	2.61526	0.9810	8.5208	354.15	493.73	139.57	1.4484	1.7897
410.0	2.66081	0.9855	8.3420	355.50	493.79	138.29	1.4516	1.7889
411.0	2.70696	0.9901	8.1664	356.85	493.83	136.99	1.4548	1.7881
412.0	2.75370	0.9948	7.9939	358.20	493.87	135.67	1.4580	1.7873
413.0	2.80105	0.9996	7.8244	359.57	493.89	134.32	1.4612	1.7864
414.0	2.84902	1.0045	7.6579	360.95	493.91	132.96	1.4644	1.7855

Chart R21.1: Saturation properties—*continued*

T K	p MPa	v' dm³/kg	v'' dm³/kg	h' kJ/kg	h'' kJ/kg	r kJ/kg	s' kJ/kgK	s'' kJ/kgK
415.0	2.89760	1.0096	7.4942	362.33	493.90	131.57	1.4676	1.7847
416.0	2.94680	1.0147	7.3333	363.72	493.89	130.17	1.4708	1.7837
417.0	2.99664	1.0200	7.1750	365.12	493.86	128.73	1.4741	1.7828
418.0	3.04712	1.0255	7.0192	366.54	493.81	127.28	1.4773	1.7818
419.0	3.09825	1.0311	6.8660	367.96	493.75	125.79	1.4806	1.7808
420.0	3.15003	1.0369	6.7151	369.39	493.67	124.28	1.4839	1.7798
421.0	3.20247	1.0428	6.5665	370.84	493.58	122.74	1.4872	1.7788
422.0	3.25559	1.0489	6.4202	372.30	493.47	121.17	1.4905	1.7777
423.0	3.30939	1.0553	6.2760	373.77	493.33	119.56	1.4939	1.7766
424.0	3.36388	1.0618	6.1338	375.25	493.18	117.93	1.4973	1.7754
425.0	3.41907	1.0685	5.9935	376.75	493.01	116.25	1.5007	1.7742
426.0	3.47496	1.0755	5.8552	378.27	492.81	114.54	1.5041	1.7730
427.0	3.53158	1.0827	5.7185	379.80	492.59	112.79	1.5075	1.7717
428.0	3.58892	1.0902	5.5836	381.35	492.34	110.99	1.5110	1.7703
429.0	3.64701	1.0980	5.4502	382.91	492.07	109.15	1.5145	1.7689
430.0	3.70585	1.1061	5.3184	384.50	491.76	107.26	1.5181	1.7675
431.0	3.76545	1.1146	5.1878	386.11	491.43	105.32	1.5216	1.7660
432.0	3.82583	1.1234	5.0586	387.74	491.06	103.32	1.5253	1.7644
433.0	3.88700	1.1326	4.9305	389.40	490.65	101.25	1.5289	1.7628
434.0	3.94898	1.1423	4.8034	391.08	490.21	99.12	1.5327	1.7610
435.0	4.01178	1.1525	4.6772	392.80	489.72	96.92	1.5364	1.7592
436.0	4.07541	1.1633	4.5518	394.55	489.18	94.64	1.5403	1.7573
437.0	4.13990	1.1747	4.4269	396.33	488.60	92.27	1.5442	1.7553
438.0	4.20526	1.1867	4.3024	398.15	487.95	89.80	1.5482	1.7532
439.0	4.27151	1.1996	4.1781	400.02	487.25	87.22	1.5523	1.7510
440.0	4.33868	1.2134	4.0538	401.94	486.47	84.53	1.5564	1.7486
441.0	4.40679	1.2282	3.9291	403.92	485.61	81.69	1.5607	1.7460
442.0	4.47586	1.2443	3.8038	405.96	484.66	78.70	1.5652	1.7432
443.0	4.54592	1.2618	3.6775	408.08	483.60	75.52	1.5698	1.7403
444.0	4.61701	1.2811	3.5496	410.30	482.43	72.13	1.5746	1.7370
445.0	4.68916	1.3025	3.4195	412.62	481.10	68.48	1.5796	1.7335
446.0	4.76241	1.3265	3.2865	415.08	479.60	64.53	1.5849	1.7296
447.0	4.83681	1.3541	3.1493	417.71	477.88	60.18	1.5905	1.7252
448.0	4.91241	1.3863	3.0063	420.56	475.88	55.32	1.5967	1.7202
449.0	4.98928	1.4251	2.8550	423.72	473.50	49.78	1.6035	1.7144
450.0	5.06750	1.4740	2.6910	427.34	470.56	43.22	1.6113	1.7074
451.0	5.14717	1.5407	2.5054	431.74	466.74	34.99	1.6208	1.6984

Chart R21.2: Saturation properties

p MPa	T K	v' dm³/kg	v'' dm³/kg	h' kJ/kg	h'' kJ/kg	r kJ/kg	s' kJ/kgK	s'' kJ/kgK
0.0031	216.10	0.6461	5614.6	145.64	415.60	269.96	0.7773	2.0266
0.0032	216.54	0.6465	5449.9	146.07	415.82	269.76	0.7793	2.0250
0.0033	216.98	0.6469	5294.9	146.48	416.04	269.56	0.7812	2.0235
0.0034	217.40	0.6472	5148.8	146.88	416.25	269.37	0.7830	2.0221
0.0035	217.81	0.6476	5010.8	147.27	416.45	269.18	0.7848	2.0207
0.0036	218.21	0.6479	4880.2	147.65	416.65	269.00	0.7866	2.0193
0.0037	218.60	0.6483	4756.5	148.02	416.84	268.82	0.7883	2.0180
0.0038	218.98	0.6486	4639.1	148.38	417.03	268.65	0.7899	2.0167
0.0039	219.35	0.6489	4527.6	148.73	417.22	268.48	0.7915	2.0155
0.0040	219.72	0.6493	4421.5	149.08	417.40	268.32	0.7931	2.0143
0.0041	220.08	0.6496	4320.4	149.41	417.58	268.16	0.7946	2.0131
0.0042	220.43	0.6499	4223.9	149.74	417.75	268.01	0.7961	2.0120
0.0043	220.77	0.6502	4131.9	150.07	417.92	267.85	0.7976	2.0108
0.0044	221.11	0.6505	4043.9	150.38	418.09	267.71	0.7990	2.0097
0.0045	221.44	0.6508	3959.7	150.69	418.25	267.56	0.8004	2.0087
0.0046	221.77	0.6510	3879.0	150.99	418.41	267.42	0.8018	2.0076
0.0047	222.08	0.6513	3801.7	151.29	418.57	267.28	0.8031	2.0066
0.0048	222.40	0.6516	3727.5	151.58	418.73	267.15	0.8044	2.0056
0.0049	222.71	0.6519	3656.2	151.87	418.88	267.02	0.8057	2.0047
0.0050	223.01	0.6521	3587.8	152.15	419.03	266.89	0.8070	2.0037
0.0051	223.30	0.6524	3521.9	152.42	419.18	266.76	0.8082	2.0028
0.0052	223.60	0.6527	3458.4	152.69	419.33	266.63	0.8094	2.0019
0.0053	223.88	0.6529	3397.3	152.96	419.47	266.51	0.8106	2.0010
0.0054	224.17	0.6532	3338.4	153.22	419.61	266.39	0.8118	2.0001
0.0055	224.45	0.6534	3281.6	153.48	419.75	266.27	0.8129	1.9993
0.0056	224.72	0.6537	3226.8	153.73	419.89	266.15	0.8140	1.9984
0.0057	224.99	0.6539	3173.8	153.98	420.02	266.04	0.8152	1.9976
0.0058	225.26	0.6541	3122.6	154.23	420.15	265.92	0.8162	1.9968
0.0059	225.52	0.6544	3073.0	154.47	420.28	265.81	0.8173	1.9960
0.0060	225.78	0.6546	3025.1	154.71	420.41	265.70	0.8184	1.9952
0.0061	226.03	0.6548	2978.7	154.95	420.54	265.60	0.8194	1.9944
0.0062	226.29	0.6551	2933.7	155.18	420.67	265.49	0.8204	1.9937
0.0063	226.53	0.6553	2890.2	155.41	420.79	265.38	0.8215	1.9930
0.0064	226.78	0.6555	2847.9	155.63	420.91	265.28	0.8225	1.9922
0.0065	227.02	0.6557	2806.9	155.85	421.03	265.18	0.8234	1.9915
0.0066	227.26	0.6559	2767.1	156.07	421.15	265.08	0.8244	1.9908
0.0067	227.50	0.6561	2728.5	156.29	421.27	264.98	0.8254	1.9901
0.0068	227.73	0.6563	2691.0	156.51	421.39	264.88	0.8263	1.9894
0.0069	227.96	0.6566	2654.5	156.72	421.50	264.79	0.8272	1.9888
0.0070	228.19	0.6568	2619.0	156.93	421.62	264.69	0.8281	1.9881
0.0071	228.41	0.6570	2584.5	157.13	421.73	264.60	0.8290	1.9875
0.0072	228.63	0.6572	2551.0	157.34	421.84	264.50	0.8299	1.9868
0.0073	228.85	0.6574	2518.3	157.54	421.95	264.41	0.8308	1.9862
0.0074	229.07	0.6576	2486.5	157.74	422.06	264.32	0.8317	1.9856
0.0076	229.49	0.6579	2425.3	158.13	422.27	264.14	0.8334	1.9844
0.0078	229.91	0.6583	2367.1	158.51	422.48	263.97	0.8351	1.9832
0.0080	230.32	0.6587	2311.8	158.89	422.68	263.80	0.8367	1.9821
0.0082	230.72	0.6590	2259.1	159.25	422.88	263.63	0.8383	1.9809
0.0084	231.11	0.6594	2208.8	159.61	423.08	263.47	0.8398	1.9799
0.0086	231.49	0.6598	2160.8	159.97	423.27	263.31	0.8414	1.9788

Chart R21.2: Saturation properties—*continued*

p MPa	T K	v' dm³/kg	v'' dm³/kg	h' kJ/kg	h'' kJ/kg	r kJ/kg	s' kJ/kgK	s'' kJ/kgK
0.0088	231.87	0.6601	2114.8	160.31	423.46	263.15	0.8429	1.9778
0.0090	232.24	0.6604	2070.9	160.65	423.65	262.99	0.8443	1.9768
0.0092	232.60	0.6608	2028.8	160.99	423.83	262.84	0.8458	1.9758
0.0094	232.96	0.6611	1988.5	161.31	424.01	262.69	0.8472	1.9748
0.0096	233.30	0.6614	1949.8	161.64	424.18	262.55	0.8485	1.9739
0.0098	233.65	0.6617	1912.6	161.95	424.35	262.40	0.8499	1.9730
0.0100	233.99	0.6620	1876.9	162.26	424.52	262.26	0.8512	1.9721
0.0102	234.32	0.6623	1842.5	162.57	424.69	262.12	0.8525	1.9712
0.0104	234.64	0.6626	1809.4	162.87	424.85	261.98	0.8538	1.9703
0.0106	234.97	0.6629	1777.5	163.17	425.02	261.85	0.8551	1.9695
0.0108	235.28	0.6632	1746.8	163.46	425.17	261.72	0.8563	1.9687
0.0110	235.59	0.6635	1717.1	163.75	425.33	261.59	0.8575	1.9679
0.0115	236.35	0.6642	1647.3	164.45	425.71	261.27	0.8605	1.9659
0.0120	237.08	0.6649	1583.2	165.12	426.08	260.96	0.8634	1.9641
0.0125	237.79	0.6656	1524.0	165.78	426.43	260.66	0.8661	1.9623
0.0130	238.47	0.6662	1469.3	166.41	426.78	260.37	0.8688	1.9606
0.0135	239.14	0.6668	1418.4	167.03	427.11	260.09	0.8714	1.9590
0.0140	239.78	0.6674	1371.1	167.62	427.43	259.81	0.8738	1.9574
0.0145	240.40	0.6680	1327.0	168.20	427.75	259.55	0.8763	1.9559
0.0150	241.01	0.6686	1285.7	168.77	428.05	259.29	0.8786	1.9544
0.0155	241.60	0.6692	1247.0	169.32	428.35	259.03	0.8809	1.9530
0.0160	242.17	0.6697	1210.6	169.85	428.64	258.79	0.8831	1.9517
0.0165	242.73	0.6702	1176.4	170.38	428.92	258.54	0.8852	1.9504
0.0170	243.27	0.6708	1144.1	170.89	429.19	258.31	0.8873	1.9491
0.0175	243.81	0.6713	1113.6	171.38	429.46	258.08	0.8894	1.9479
0.0180	244.33	0.6718	1084.7	171.87	429.72	257.85	0.8914	1.9467
0.0185	244.83	0.6723	1057.4	172.35	429.98	257.63	0.8933	1.9456
0.0190	245.33	0.6727	1031.4	172.82	430.23	257.41	0.8952	1.9445
0.0195	245.82	0.6732	1006.7	173.28	430.47	257.20	0.8971	1.9434
0.0200	246.29	0.6737	983.25	173.72	430.71	256.99	0.8989	1.9424
0.0210	247.21	0.6746	939.55	174.60	431.18	256.58	0.9025	1.9403
0.0220	248.10	0.6754	899.69	175.44	431.62	256.19	0.9059	1.9384
0.0230	248.96	0.6763	863.19	176.25	432.05	255.80	0.9091	1.9366
0.0240	249.78	0.6771	829.63	177.03	432.47	255.43	0.9123	1.9349
0.0250	250.58	0.6779	798.67	177.80	432.87	255.07	0.9153	1.9332
0.0260	251.35	0.6786	770.02	178.53	433.26	254.72	0.9182	1.9317
0.0270	252.10	0.6794	743.42	179.25	433.63	254.38	0.9211	1.9302
0.0280	252.82	0.6801	718.66	179.94	434.00	254.05	0.9238	1.9287
0.0290	253.53	0.6808	695.55	180.62	434.35	253.73	0.9265	1.9273
0.0300	254.21	0.6815	673.92	181.28	434.69	253.41	0.9291	1.9260
0.0310	254.88	0.6822	653.65	181.93	435.03	253.10	0.9316	1.9247
0.0320	255.53	0.6828	634.61	182.55	435.36	252.80	0.9341	1.9234
0.0330	256.16	0.6835	616.68	183.17	435.67	252.51	0.9365	1.9222
0.0340	256.78	0.6841	599.77	183.76	435.98	252.22	0.9388	1.9211
0.0350	257.38	0.6847	583.79	184.35	436.29	251.93	0.9411	1.9199
0.0360	257.97	0.6853	568.67	184.92	436.58	251.66	0.9433	1.9188
0.0370	258.54	0.6859	554.35	185.49	436.87	251.38	0.9455	1.9178
0.0380	259.11	0.6865	540.75	186.04	437.15	251.12	0.9476	1.9168
0.0390	259.66	0.6870	527.82	186.57	437.43	250.85	0.9497	1.9158
0.0400	260.20	0.6876	515.52	187.10	437.70	250.59	0.9517	1.9148

Chart R21.2: Saturation properties–*continued*

p MPa	T K	v' dm³/kg	v'' dm³/kg	h' kJ/kg	h'' kJ/kg	r kJ/kg	s' kJ/kgK	s'' kJ/kgK
0.0410	260.73	0.6881	503.80	187.62	437.96	250.34	0.9537	1.9139
0.0420	261.25	0.6887	492.62	188.13	438.22	250.09	0.9557	1.9129
0.0430	261.76	0.6892	481.94	188.63	438.48	249.85	0.9576	1.9121
0.0440	262.26	0.6897	471.73	189.12	438.73	249.60	0.9594	1.9112
0.0450	262.75	0.6902	461.96	189.61	438.97	249.36	0.9613	1.9103
0.0460	263.23	0.6908	452.60	190.08	439.21	249.13	0.9631	1.9095
0.0470	263.70	0.6913	443.62	190.55	439.45	248.90	0.9649	1.9087
0.0480	264.17	0.6917	435.00	191.01	439.68	248.67	0.9666	1.9079
0.0490	264.62	0.6922	426.73	191.46	439.91	248.45	0.9683	1.9072
0.0500	265.07	0.6927	418.77	191.91	440.13	248.22	0.9700	1.9064
0.0510	265.52	0.6932	411.12	192.35	440.35	248.00	0.9716	1.9057
0.0520	265.95	0.6936	403.75	192.78	440.57	247.79	0.9733	1.9050
0.0530	266.38	0.6941	396.64	193.21	440.79	247.58	0.9749	1.9043
0.0540	266.81	0.6945	389.80	193.63	441.00	247.37	0.9765	1.9036
0.0550	267.22	0.6950	383.19	194.05	441.20	247.16	0.9780	1.9029
0.0560	267.63	0.6954	376.81	194.46	441.41	246.95	0.9795	1.9023
0.0570	268.04	0.6959	370.65	194.86	441.61	246.75	0.9810	1.9016
0.0580	268.44	0.6963	364.69	195.26	441.81	246.55	0.9825	1.9010
0.0590	268.83	0.6967	358.92	195.65	442.00	246.35	0.9840	1.9004
0.0600	269.22	0.6971	353.35	196.04	442.20	246.15	0.9854	1.8998
0.0610	269.60	0.6976	347.94	196.43	442.39	245.96	0.9869	1.8992
0.0620	269.98	0.6980	342.71	196.81	442.57	245.77	0.9883	1.8986
0.0630	270.35	0.6984	337.64	197.18	442.76	245.58	0.9896	1.8980
0.0640	270.72	0.6988	332.72	197.55	442.94	245.39	0.9910	1.8974
0.0650	271.09	0.6992	327.94	197.92	443.12	245.21	0.9924	1.8969
0.0660	271.45	0.6996	323.31	198.28	443.30	245.02	0.9937	1.8963
0.0670	271.80	0.7000	318.81	198.64	443.48	244.84	0.9950	1.8958
0.0680	272.15	0.7004	314.44	198.99	443.65	244.66	0.9963	1.8953
0.0690	272.50	0.7007	310.19	199.34	443.82	244.48	0.9976	1.8948
0.0700	272.84	0.7011	306.05	199.69	443.99	244.30	0.9989	1.8943
0.0710	273.18	0.7015	302.03	200.03	444.16	244.13	1.0001	1.8938
0.0720	273.52	0.7019	298.12	200.37	444.33	243.96	1.0013	1.8933
0.0730	273.85	0.7022	294.31	200.71	444.49	243.78	1.0026	1.8928
0.0740	274.18	0.7026	290.60	201.04	444.65	243.61	1.0038	1.8923
0.0760	274.82	0.7033	283.46	201.69	444.97	243.28	1.0062	1.8914
0.0780	275.45	0.7040	276.67	202.34	445.28	242.95	1.0085	1.8905
0.0800	276.07	0.7047	270.21	202.96	445.59	242.62	1.0108	1.8896
0.0820	276.68	0.7054	264.06	203.58	445.89	242.30	1.0130	1.8888
0.0840	277.28	0.7061	258.19	204.19	446.18	241.99	1.0152	1.8879
0.0860	277.86	0.7067	252.58	204.79	446.47	241.68	1.0173	1.8871
0.0880	278.43	0.7074	247.22	205.37	446.75	241.37	1.0194	1.8863
0.0900	279.00	0.7080	242.09	205.95	447.02	241.07	1.0215	1.8856
0.0920	279.55	0.7087	237.17	206.52	447.30	240.78	1.0235	1.8848
0.0940	280.10	0.7093	232.46	207.08	447.56	240.49	1.0255	1.8841
0.0960	280.63	0.7099	227.93	207.63	447.82	240.20	1.0275	1.8834
0.0980	281.16	0.7105	223.58	208.17	448.08	239.92	1.0294	1.8827
0.1000	281.68	0.7111	219.40	208.70	448.33	239.64	1.0313	1.8820
0.1020	282.19	0.7117	215.38	209.22	448.58	239.36	1.0331	1.8814
0.1040	282.69	0.7123	211.51	209.74	448.83	239.09	1.0350	1.8807
0.1060	283.18	0.7129	207.78	210.25	449.07	238.82	1.0368	1.8801

Chart R21.2: Saturation properties—*continued*

p MPa	T K	v' dm³/kg	v'' dm³/kg	h' kJ/kg	h'' kJ/kg	r kJ/kg	s' kJ/kgK	s'' kJ/kgK
0.1080	283.67	0.7135	204.18	210.76	449.31	238.55	1.0385	1.8795
0.1100	284.15	0.7140	200.71	211.25	449.54	238.29	1.0403	1.8789
0.1150	285.32	0.7154	192.54	212.46	450.11	237.64	1.0445	1.8774
0.1200	286.46	0.7168	185.02	213.64	450.66	237.02	1.0486	1.8760
0.1250	287.55	0.7181	178.09	214.78	451.19	236.41	1.0526	1.8747
0.1300	288.61	0.7194	171.67	215.88	451.70	235.82	1.0564	1.8735
0.1350	289.65	0.7206	165.71	216.96	452.20	235.24	1.0601	1.8723
0.1400	290.65	0.7218	160.16	218.00	452.68	234.67	1.0637	1.8711
0.1450	291.62	0.7230	154.98	219.02	453.14	234.12	1.0672	1.8700
0.1500	292.57	0.7242	150.13	220.02	453.60	233.58	1.0706	1.8690
0.1550	293.49	0.7254	145.59	220.99	454.04	233.05	1.0739	1.8679
0.1600	294.39	0.7265	141.32	221.93	454.47	232.54	1.0771	1.8670
0.1650	295.27	0.7276	137.29	222.86	454.89	232.03	1.0802	1.8660
0.1700	296.13	0.7287	133.50	223.76	455.29	231.53	1.0833	1.8651
0.1750	296.97	0.7297	129.91	224.65	455.69	231.04	1.0862	1.8642
0.1800	297.79	0.7308	126.52	225.52	456.08	230.57	1.0891	1.8634
0.1850	298.60	0.7318	123.30	226.37	456.46	230.09	1.0920	1.8626
0.1900	299.38	0.7328	120.25	227.20	456.83	229.63	1.0947	1.8618
0.1950	300.15	0.7338	117.35	228.02	457.19	229.18	1.0975	1.8610
0.2000	300.91	0.7348	114.59	228.82	457.55	228.73	1.1001	1.8602
0.2100	302.38	0.7368	109.44	230.38	458.24	227.85	1.1053	1.8588
0.2200	303.80	0.7387	104.75	231.89	458.90	227.00	1.1102	1.8575
0.2300	305.17	0.7405	100.45	233.35	459.53	226.18	1.1150	1.8562
0.2400	306.49	0.7423	96.491	234.77	460.14	225.38	1.1196	1.8549
0.2500	307.77	0.7440	92.841	236.14	460.73	224.59	1.1240	1.8538
0.2600	309.01	0.7457	89.461	237.47	461.30	223.83	1.1283	1.8527
0.2700	310.22	0.7474	86.323	238.77	461.85	223.09	1.1325	1.8516
0.2800	311.39	0.7491	83.401	240.03	462.39	222.36	1.1365	1.8506
0.2900	312.54	0.7507	80.673	241.26	462.91	221.64	1.1405	1.8496
0.3000	313.65	0.7523	78.120	242.46	463.41	220.95	1.1443	1.8487
0.3100	314.73	0.7538	75.725	243.63	463.90	220.26	1.1480	1.8478
0.3200	315.79	0.7554	73.475	244.78	464.37	219.59	1.1516	1.8470
0.3300	316.83	0.7569	71.356	245.90	464.83	218.93	1.1551	1.8461
0.3400	317.84	0.7584	69.357	247.00	465.28	218.29	1.1585	1.8453
0.3500	318.82	0.7598	67.468	248.07	465.72	217.65	1.1619	1.8446
0.3600	319.79	0.7613	65.681	249.12	466.15	217.03	1.1652	1.8438
0.3700	320.74	0.7627	63.986	250.15	466.57	216.42	1.1684	1.8431
0.3800	321.67	0.7641	62.377	251.16	466.97	215.81	1.1715	1.8424
0.3900	322.57	0.7655	60.847	252.15	467.37	215.22	1.1745	1.8417
0.4000	323.47	0.7668	59.392	253.13	467.76	214.63	1.1775	1.8411
0.4100	324.34	0.7682	58.004	254.09	468.14	214.05	1.1805	1.8404
0.4200	325.20	0.7695	56.680	255.03	468.51	213.48	1.1833	1.8398
0.4300	326.05	0.7709	55.416	255.95	468.87	212.92	1.1862	1.8392
0.4400	326.88	0.7722	54.206	256.86	469.23	212.37	1.1889	1.8386
0.4500	327.69	0.7735	53.049	257.76	469.58	211.82	1.1916	1.8380
0.4600	328.49	0.7747	51.939	258.64	469.92	211.28	1.1943	1.8375
0.4700	329.28	0.7760	50.876	259.51	470.26	210.75	1.1969	1.8369
0.4800	330.06	0.7773	49.854	260.36	470.59	210.22	1.1995	1.8364
0.4900	330.82	0.7785	48.873	261.21	470.91	209.70	1.2020	1.8359
0.5000	331.58	0.7797	47.930	262.04	471.23	209.19	1.2045	1.8354

Chart R21.2: Saturation properties—*continued*

p MPa	T K	v' dm³/kg	v'' dm³/kg	h' kJ/kg	h'' kJ/kg	r kJ/kg	s' kJ/kgK	s'' kJ/kgK
0.5100	332.32	0.7810	47.022	262.86	471.54	208.68	1.2069	1.8349
0.5200	333.05	0.7822	46.148	263.67	471.84	208.17	1.2093	1.8344
0.5300	333.77	0.7834	45.306	264.46	472.14	207.68	1.2117	1.8339
0.5400	334.48	0.7846	44.493	265.25	472.44	207.18	1.2141	1.8335
0.5500	335.18	0.7858	43.709	266.03	472.72	206.70	1.2163	1.8330
0.5600	335.87	0.7869	42.952	266.80	473.01	206.21	1.2186	1.8326
0.5700	336.55	0.7881	42.221	267.55	473.29	205.73	1.2208	1.8321
0.5800	337.23	0.7893	41.514	268.30	473.56	205.26	1.2230	1.8317
0.5900	337.89	0.7904	40.829	269.04	473.83	204.79	1.2252	1.8313
0.6000	338.55	0.7916	40.167	269.77	474.10	204.33	1.2274	1.8309
0.6100	339.20	0.7927	39.526	270.50	474.36	203.87	1.2295	1.8305
0.6200	339.84	0.7938	38.904	271.21	474.62	203.41	1.2315	1.8301
0.6300	340.47	0.7949	38.302	271.92	474.88	202.96	1.2336	1.8297
0.6400	341.10	0.7961	37.717	272.62	475.13	202.51	1.2356	1.8293
0.6500	341.72	0.7972	37.150	273.31	475.37	202.06	1.2376	1.8289
0.6600	342.33	0.7983	36.599	274.00	475.61	201.62	1.2396	1.8286
0.6700	342.93	0.7994	36.064	274.67	475.85	201.18	1.2416	1.8282
0.6800	343.53	0.8004	35.545	275.34	476.09	200.75	1.2435	1.8279
0.6900	344.12	0.8015	35.039	276.01	476.32	200.31	1.2454	1.8275
0.7000	344.71	0.8026	34.548	276.66	476.55	199.89	1.2473	1.8272
0.7100	345.28	0.8037	34.070	277.32	476.78	199.46	1.2492	1.8268
0.7200	345.86	0.8047	33.605	277.96	477.00	199.04	1.2510	1.8265
0.7300	346.42	0.8058	33.151	278.60	477.22	198.62	1.2528	1.8262
0.7400	346.99	0.8069	32.710	279.23	477.43	198.20	1.2546	1.8258
0.7600	348.09	0.8089	31.861	280.48	477.86	197.38	1.2582	1.8252
0.7800	349.18	0.8110	31.054	281.71	478.27	196.56	1.2616	1.8246
0.8000	350.24	0.8131	30.285	282.91	478.67	195.76	1.2650	1.8240
0.8200	351.29	0.8151	29.552	284.10	479.07	194.97	1.2684	1.8234
0.8400	352.31	0.8171	28.853	285.26	479.45	194.18	1.2716	1.8228
0.8600	353.32	0.8192	28.186	286.41	479.82	193.41	1.2748	1.8222
0.8800	354.31	0.8212	27.547	287.54	480.18	192.64	1.2780	1.8217
0.9000	355.29	0.8231	26.935	288.66	480.54	191.88	1.2811	1.8212
0.9200	356.25	0.8251	26.349	289.75	480.88	191.13	1.2841	1.8206
0.9400	357.19	0.8271	25.787	290.83	481.22	190.39	1.2871	1.8201
0.9600	358.12	0.8290	25.248	291.90	481.55	189.65	1.2900	1.8196
0.9800	359.03	0.8309	24.729	292.95	481.87	188.92	1.2929	1.8191
1.0000	359.94	0.8329	24.231	293.99	482.19	188.20	1.2958	1.8186
1.0200	360.82	0.8348	23.751	295.01	482.50	187.48	1.2986	1.8182
1.0400	361.70	0.8367	23.289	296.03	482.80	186.77	1.3013	1.8177
1.0600	362.56	0.8386	22.844	297.02	483.09	186.07	1.3040	1.8172
1.0800	363.42	0.8405	22.414	298.01	483.38	185.37	1.3067	1.8168
1.1000	364.26	0.8423	21.999	298.99	483.66	184.68	1.3093	1.8163
1.1500	366.31	0.8470	21.023	301.37	484.34	182.97	1.3158	1.8152
1.2000	368.30	0.8516	20.125	303.70	484.98	181.29	1.3220	1.8142
1.2500	370.23	0.8562	19.296	305.96	485.59	179.64	1.3280	1.8132
1.3000	372.10	0.8608	18.527	308.16	486.17	178.01	1.3338	1.8122
1.3500	373.92	0.8653	17.813	310.32	486.72	176.40	1.3395	1.8112
1.4000	375.69	0.8698	17.148	312.42	487.24	174.82	1.3450	1.8103
1.4500	377.42	0.8743	16.527	314.48	487.74	173.26	1.3503	1.8094
1.5000	379.10	0.8788	15.945	316.50	488.21	171.71	1.3555	1.8085

Chart R21.2: Saturation properties–*continued*

p MPa	T K	v' dm³/kg	v'' dm³/kg	h' kJ/kg	h'' kJ/kg	r kJ/kg	s' kJ/kgK	s'' kJ/kgK
1.5500	380.75	0.8833	15.398	318.47	488.65	170.18	1.3606	1.8076
1.6000	382.35	0.8877	14.885	320.41	489.07	168.66	1.3656	1.8067
1.6500	383.92	0.8922	14.400	322.31	489.47	167.16	1.3704	1.8058
1.7000	385.45	0.8966	13.943	324.18	489.85	165.67	1.3752	1.8050
1.7500	386.96	0.9011	13.511	326.02	490.21	164.20	1.3798	1.8041
1.8000	388.43	0.9055	13.101	327.83	490.55	162.73	1.3843	1.8033
1.8500	389.87	0.9100	12.712	329.60	490.87	161.27	1.3888	1.8025
1.9000	391.28	0.9145	12.343	331.36	491.18	159.82	1.3932	1.8016
1.9500	392.67	0.9190	11.991	333.08	491.46	158.38	1.3975	1.8008
2.0000	394.03	0.9235	11.656	334.78	491.73	156.94	1.4017	1.8000
2.1000	396.67	0.9325	11.031	338.12	492.21	154.09	1.4099	1.7983
2.2000	399.23	0.9417	10.459	341.37	492.63	151.26	1.4178	1.7967
2.3000	401.70	0.9509	9.9337	344.55	492.99	148.44	1.4255	1.7950
2.4000	404.09	0.9603	9.4487	347.66	493.29	145.63	1.4330	1.7934
2.5000	406.41	0.9698	8.9995	350.71	493.52	142.82	1.4403	1.7917
2.6000	408.66	0.9795	8.5820	353.70	493.70	140.00	1.4474	1.7900
2.7000	410.85	0.9894	8.1925	356.64	493.83	137.18	1.4543	1.7882
2.8000	412.98	0.9995	7.8282	359.54	493.89	134.35	1.4611	1.7864
2.9000	415.05	1.0098	7.4862	362.40	493.90	131.51	1.4678	1.7846
3.0000	417.07	1.0204	7.1645	365.22	493.86	128.64	1.4743	1.7827
3.1000	419.03	1.0313	6.8608	368.01	493.75	125.74	1.4807	1.7808
3.2000	420.95	1.0425	6.5735	370.77	493.58	122.81	1.4871	1.7788
3.3000	422.83	1.0541	6.3009	373.51	493.36	119.85	1.4933	1.7767
3.4000	424.66	1.0662	6.0416	376.24	493.07	116.83	1.4995	1.7746
3.5000	426.44	1.0787	5.7943	378.95	492.71	113.77	1.5056	1.7724
3.6000	428.19	1.0917	5.5579	381.65	492.29	110.64	1.5117	1.7701
3.7000	429.90	1.1053	5.3313	384.34	491.79	107.45	1.5177	1.7676
3.8000	431.57	1.1196	5.1135	387.04	491.22	104.18	1.5237	1.7651
3.9000	433.21	1.1346	4.9036	389.75	490.56	100.81	1.5297	1.7624
4.0000	434.81	1.1506	4.7007	392.48	489.81	97.34	1.5357	1.7596
4.1000	436.38	1.1676	4.5039	395.22	488.97	93.74	1.5418	1.7566
4.2000	437.92	1.1858	4.3123	398.01	488.01	90.00	1.5479	1.7534
4.3000	439.43	1.2054	4.1252	400.83	486.92	86.09	1.5540	1.7499
4.4000	440.90	1.2267	3.9415	403.72	485.70	81.98	1.5603	1.7462
4.5000	442.35	1.2502	3.7602	406.69	484.31	77.62	1.5668	1.7422
4.6000	443.76	1.2763	3.5802	409.76	482.72	72.96	1.5734	1.7378
4.7000	445.15	1.3059	3.3999	412.98	480.89	67.92	1.5804	1.7329
4.8000	446.51	1.3400	3.2175	416.39	478.76	62.38	1.5877	1.7274
4.9000	447.84	1.3806	3.0301	420.08	476.23	56.16	1.5956	1.7210
5.0000	449.14	1.4311	2.8333	424.19	473.13	48.94	1.6045	1.7135

Chart R21.3: Thermodynamic properties

p/MPa	0.25			0.30			0.40		
T K	v dm³/kg	h kJ/kg	s kJ/kgK	v dm³/kg	h kJ/kg	s kJ/kgK	v dm³/kg	h kJ/kg	s kJ/kgK
220.0	0.6493	149.45	0.7941	0.6493	149.48	0.7940	0.6493	149.52	0.7939
230.0	0.6582	158.70	0.8352	0.6582	158.73	0.8352	0.6581	158.77	0.8351
240.0	0.6675	167.93	0.8745	0.6674	167.95	0.8744	0.6674	168.00	0.8743
250.0	0.6771	177.34	0.9129	0.6771	177.36	0.9128	0.6770	177.40	0.9127
260.0	0.6872	187.00	0.9507	0.6871	187.02	0.9507	0.6871	187.06	0.9506
270.0	0.6978	196.90	0.9881	0.6978	196.92	0.9881	0.6976	196.96	0.9880
280.0	0.7090	207.04	1.0250	0.7089	207.06	1.0249	0.7088	207.09	1.0248
290.0	0.7209	217.37	1.0612	0.7208	217.39	1.0612	0.7207	217.42	1.0611
300.0	0.7336	227.87	1.0968	0.7335	227.89	1.0968	0.7333	227.92	1.0966
310.0	93.697	462.21	1.8586	0.7471	238.54	1.1317	0.7469	238.57	1.1316
320.0	97.486	468.84	1.8796	80.186	467.70	1.8623	0.7615	249.36	1.1658
330.0	101.20	475.50	1.9001	83.378	474.46	1.8831	61.048	472.33	1.8550
340.0	104.86	482.18	1.9201	86.508	481.24	1.9033	63.528	479.30	1.8759
350.0	108.47	488.90	1.9395	89.587	488.04	1.9230	65.951	486.27	1.8961
360.0	112.04	495.67	1.9586	92.623	494.88	1.9423	68.327	493.26	1.9157
370.0	115.57	502.49	1.9773	95.621	501.75	1.9611	70.663	500.26	1.9349
380.0	119.07	509.36	1.9956	98.589	508.68	1.9796	72.965	507.30	1.9537
390.0	122.55	516.28	2.0136	101.53	515.65	1.9977	75.240	514.37	1.9721
400.0	126.00	523.26	2.0313	104.45	522.68	2.0155	77.490	521.48	1.9901
410.0	129.44	530.31	2.0487	107.34	529.75	2.0330	79.718	528.64	2.0077
420.0	132.85	537.41	2.0658	110.22	536.89	2.0501	81.929	535.84	2.0251
430.0	136.26	544.56	2.0826	113.09	544.08	2.0671	84.123	543.09	2.0422
440.0	139.64	551.78	2.0992	115.94	551.32	2.0837	86.303	550.40	2.0590
450.0	143.02	559.06	2.1156	118.78	558.62	2.1001	88.470	557.75	2.0755
460.0	146.38	566.39	2.1317	121.60	565.98	2.1163	90.626	565.15	2.0917
470.0	149.74	573.78	2.1476	124.42	573.39	2.1322	92.772	572.60	2.1078

p/MPa	0.50			0.60			0.80		
220.0	0.6492	149.57	0.7939	0.6491	149.61	0.7938	0.6490	149.71	0.7936
230.0	0.6581	158.82	0.8350	0.6580	158.86	0.8349	0.6579	158.95	0.8347
240.0	0.6673	168.04	0.8742	0.6672	168.09	0.8741	0.6671	168.18	0.8740
250.0	0.6769	177.45	0.9126	0.6768	177.49	0.9125	0.6767	177.58	0.9123
260.0	0.6870	187.10	0.9505	0.6869	187.14	0.9504	0.6867	187.23	0.9502
270.0	0.6975	197.00	0.9879	0.6974	197.04	0.9877	0.6972	197.12	0.9875
280.0	0.7087	207.13	1.0247	0.7086	207.17	1.0246	0.7084	207.25	1.0244
290.0	0.7206	217.46	1.0609	0.7204	217.50	1.0608	0.7202	217.57	1.0606
300.0	0.7332	227.96	1.0965	0.7330	227.99	1.0964	0.7327	228.06	1.0961
310.0	0.7467	238.60	1.1314	0.7465	238.63	1.1313	0.7462	238.70	1.1310
320.0	0.7613	249.39	1.1657	0.7611	249.41	1.1655	0.7607	249.47	1.1652
330.0	0.7771	260.30	1.1993	0.7769	260.33	1.1991	0.7764	260.37	1.1988
340.0	49.696	477.29	1.8534	40.433	475.18	1.8341	0.7936	271.43	1.2317
350.0	51.734	484.44	1.8742	42.224	482.54	1.8554	0.8126	282.64	1.2643
360.0	53.721	491.59	1.8943	43.958	489.86	1.8760	31.688	486.21	1.8452
370.0	55.665	498.73	1.9139	45.645	497.15	1.8960	33.069	493.84	1.8661
380.0	57.573	505.88	1.9329	47.294	504.43	1.9154	34.404	501.41	1.8863
390.0	59.450	513.06	1.9516	48.910	511.72	1.9343	35.701	508.94	1.9058
400.0	61.302	520.26	1.9698	50.499	519.02	1.9528	36.967	516.45	1.9249
410.0	63.132	527.50	1.9877	52.064	526.34	1.9709	38.207	523.96	1.9434

Chart R21.3: Thermodynamic properties–*continued*

p/MPa	0.50			0.60			0.80		
T K	v dm³/kg	h kJ/kg	s kJ/kgK	v dm³/kg	h kJ/kg	s kJ/kgK	v dm³/kg	h kJ/kg	s kJ/kgK
420.0	64.942	534.78	2.0052	53.609	533.70	1.9886	39.425	531.48	1.9615
430.0	66.735	542.10	2.0225	55.137	541.08	2.0060	40.623	539.01	1.9792
440.0	68.514	549.46	2.0394	56.649	548.51	2.0231	41.805	546.56	1.9966
450.0	70.280	556.86	2.0560	58.148	555.97	2.0398	42.972	554.14	2.0136
460.0	72.034	564.31	2.0724	59.635	563.47	2.0563	44.126	561.75	2.0304
470.0	73.777	571.81	2.0885	61.111	571.01	2.0726	45.269	569.39	2.0468
p/MPa	**1.00**			**2.00**			**4.00**		
220.0	0.6489	149.80	0.7934	0.6483	150.25	0.7926	0.6471	151.17	0.7908
230.0	0.6577	159.04	0.8345	0.6571	159.49	0.8336	0.6558	160.40	0.8319
240.0	0.6669	168.26	0.8738	0.6662	168.71	0.8728	0.6648	169.59	0.8710
250.0	0.6765	177.66	0.9121	0.6757	178.10	0.9112	0.6741	178.97	0.9092
260.0	0.6865	187.31	0.9500	0.6856	187.73	0.9490	0.6839	188.58	0.9469
270.0	0.6970	197.21	0.9873	0.6960	197.61	0.9862	0.6941	198.43	0.9841
280.0	0.7081	207.33	1.0241	0.7070	207.72	1.0230	0.7048	208.50	1.0208
290.0	0.7199	217.64	1.0603	0.7186	218.01	1.0591	0.7161	218.76	1.0568
300.0	0.7324	228.13	1.0959	0.7310	228.47	1.0946	0.7282	229.18	1.0921
310.0	0.7459	238.76	1.1307	0.7442	239.08	1.1293	0.7410	239.73	1.1267
320.0	0.7603	249.52	1.1649	0.7583	249.81	1.1634	0.7546	250.40	1.1605
330.0	0.7760	260.42	1.1984	0.7737	260.66	1.1968	0.7694	261.18	1.1937
340.0	0.7931	271.46	1.2314	0.7904	271.65	1.2296	0.7853	272.07	1.2262
350.0	0.8119	282.66	1.2639	0.8087	282.79	1.2619	0.8026	283.09	1.2582
360.0	24.239	482.24	1.8188	0.8290	294.10	1.2938	0.8217	294.26	1.2896
370.0	25.457	490.28	1.8408	0.8519	305.64	1.3254	0.8429	305.61	1.3207
380.0	26.618	498.19	1.8619	0.8782	317.47	1.3569	0.8667	317.18	1.3516
390.0	27.735	506.01	1.8822	0.9092	329.71	1.3887	0.8940	329.05	1.3824
400.0	28.815	513.76	1.9018	12.146	497.58	1.8147	0.9261	341.31	1.4134
410.0	29.866	521.48	1.9209	12.905	506.98	1.8379	0.9654	354.15	1.4451
420.0	30.892	529.18	1.9395	13.608	516.04	1.8598	1.0165	367.91	1.4783
430.0	31.896	536.87	1.9576	14.269	524.85	1.8805	1.0924	383.49	1.5149
440.0	32.882	544.56	1.9752	14.898	533.49	1.9004	5.1560	498.95	1.7805
450.0	33.853	552.27	1.9926	15.501	542.01	1.9195	5.8172	513.04	1.8122
460.0	34.810	559.99	2.0095	16.083	550.44	1.9380	6.3498	525.12	1.8387
470.0	35.755	567.73	2.0262	16.647	558.81	1.9560	6.8139	536.17	1.8625
p/MPa	**6.00**			**8.00**			**10.00**		
220.0	0.6460	152.09	0.7892	0.6449	153.02	0.7875	0.6438	153.94	0.7858
230.0	0.6546	161.31	0.8301	0.6533	162.22	0.8284	0.6522	163.13	0.8267
240.0	0.6634	170.49	0.8692	0.6621	171.38	0.8674	0.6608	172.28	0.8656
250.0	0.6726	179.84	0.9074	0.6711	180.72	0.9055	0.6697	181.60	0.9037
260.0	0.6822	189.43	0.9450	0.6805	190.29	0.9430	0.6790	191.15	0.9411
270.0	0.6922	199.26	0.9821	0.6904	200.09	0.9800	0.6886	200.93	0.9780
280.0	0.7027	209.30	1.0186	0.7007	210.11	1.0164	0.6987	210.92	1.0144
290.0	0.7138	219.53	1.0545	0.7115	220.30	1.0522	0.7093	221.08	1.0500
300.0	0.7255	229.90	1.0896	0.7229	230.64	1.0872	0.7205	231.38	1.0849
310.0	0.7379	240.40	1.1241	0.7350	241.09	1.1215	0.7323	241.80	1.1191

Chart R21.3: Thermodynamic properties—*continued*

p/MPa	6.00			8.00			10.00		
T K	v dm³/kg	h kJ/kg	s kJ/kgK	v dm³/kg	h kJ/kg	s kJ/kgK	v dm³/kg	h kJ/kg	s kJ/kgK
320.0	0.7512	251.01	1.1577	0.7479	251.65	1.1550	0.7448	252.30	1.1524
330.0	0.7653	261.72	1.1907	0.7616	262.30	1.1878	0.7580	262.90	1.1850
340.0	0.7806	272.53	1.2230	0.7762	273.04	1.2199	0.7722	273.57	1.2169
350.0	0.7971	283.46	1.2546	0.7920	283.87	1.2513	0.7873	284.32	1.2481
360.0	0.8151	294.50	1.2857	0.8091	294.80	1.2821	0.8037	295.17	1.2786
370.0	0.8349	305.69	1.3164	0.8278	305.86	1.3124	0.8213	306.11	1.3086
380.0	0.8569	317.06	1.3467	0.8482	317.07	1.3423	0.8406	317.18	1.3381
390.0	0.8815	328.66	1.3768	0.8709	328.45	1.3718	0.8616	328.39	1.3672
400.0	0.9098	340.54	1.4069	0.8963	340.06	1.4012	0.8850	339.78	1.3961
410.0	0.9428	352.81	1.4372	0.9253	351.94	1.4305	0.9110	351.38	1.4247
420.0	0.9827	365.61	1.4681	0.9589	364.18	1.4601	0.9405	363.24	1.4533
430.0	1.0335	379.23	1.5001	0.9990	376.92	1.4900	0.9744	375.44	1.4820
440.0	1.1043	394.24	1.5346	1.0488	390.33	1.5209	1.0143	388.06	1.5110
450.0	1.2248	412.40	1.5754	1.1144	404.80	1.5534	1.0626	401.25	1.5406
460.0	1.7077	448.12	1.6537	1.2098	421.06	1.5891	1.1234	415.23	1.5714
470.0	2.9526	494.98	1.7547	1.3728	440.81	1.6315	1.2036	430.31	1.6038
p/MPa	15.00			20.00			35.00		
220.0	0.6412	156.26	0.7818	0.6388	158.60	0.7778	0.6321	165.65	0.7666
230.0	0.6493	165.42	0.8225	0.6467	167.73	0.8185	0.6394	174.73	0.8069
240.0	0.6577	174.54	0.8613	0.6548	176.83	0.8571	0.6469	183.75	0.8453
250.0	0.6663	183.83	0.8992	0.6631	186.07	0.8949	0.6546	192.92	0.8827
260.0	0.6752	193.33	0.9365	0.6717	195.54	0.9320	0.6624	202.29	0.9195
270.0	0.6845	203.06	0.9732	0.6806	205.22	0.9686	0.6705	211.86	0.9556
280.0	0.6941	212.99	1.0093	0.6899	215.10	1.0045	0.6789	221.60	0.9911
290.0	0.7042	223.08	1.0447	0.6996	225.13	1.0397	0.6875	231.49	1.0257
300.0	0.7148	233.30	1.0794	0.7096	235.28	1.0741	0.6965	241.47	1.0596
310.0	0.7259	243.62	1.1132	0.7202	245.52	1.1077	0.7057	251.53	1.0926
320.0	0.7376	254.02	1.1462	0.7312	255.83	1.1404	0.7153	261.63	1.1246
330.0	0.7500	264.49	1.1784	0.7428	266.20	1.1723	0.7253	271.77	1.1558
340.0	0.7630	275.02	1.2099	0.7550	276.61	1.2034	0.7357	281.92	1.1861
350.0	0.7769	285.61	1.2406	0.7679	287.07	1.2337	0.7466	292.09	1.2156
360.0	0.7917	296.26	1.2706	0.7815	297.57	1.2633	0.7579	302.28	1.2443
370.0	0.8075	306.98	1.2999	0.7959	308.11	1.2922	0.7696	312.48	1.2723
380.0	0.8244	317.78	1.3287	0.8112	318.72	1.3204	0.7819	322.69	1.2995
390.0	0.8426	328.67	1.3570	0.8275	329.38	1.3481	0.7948	332.93	1.3261
400.0	0.8623	339.67	1.3849	0.8448	340.12	1.3753	0.8081	343.19	1.3521
410.0	0.8836	350.80	1.4124	0.8633	350.94	1.4020	0.8221	353.48	1.3775
420.0	0.9070	362.07	1.4395	0.8832	361.85	1.4283	0.8367	363.81	1.4024
430.0	0.9326	373.51	1.4664	0.9045	372.87	1.4543	0.8520	374.18	1.4268
440.0	0.9611	385.14	1.4932	0.9275	384.00	1.4799	0.8680	384.60	1.4507
450.0	0.9928	397.01	1.5198	0.9524	395.27	1.5052	0.8846	395.06	1.4742
460.0	1.0285	409.14	1.5465	0.9795	406.67	1.5303	0.9021	405.58	1.4973
470.0	1.0693	421.57	1.5732	1.0089	418.24	1.5551	0.9203	416.15	1.5201

Chart R21.3: Thermodynamic properties—*continued*

p/MPa	50.00			70.00			90.00		
T K	v dm³/kg	h kJ/kg	s kJ/kgK	v dm³/kg	h kJ/kg	s kJ/kgK	v dm³/kg	h kJ/kg	s kJ/kgK
220.0	0.6263	172.77	0.7561	0.6195	182.34	0.7429	0.6135	191.96	0.7306
230.0	0.6332	181.81	0.7962	0.6259	191.33	0.7829	0.6195	200.93	0.7705
240.0	0.6402	190.78	0.8344	0.6323	200.25	0.8208	0.6255	209.81	0.8083
250.0	0.6473	199.88	0.8715	0.6389	209.29	0.8577	0.6317	218.80	0.8450
260.0	0.6546	209.17	0.9080	0.6456	218.51	0.8939	0.6379	227.96	0.8809
270.0	0.6620	218.66	0.9438	0.6524	227.91	0.9294	0.6442	237.29	0.9161
280.0	0.6697	228.31	0.9789	0.6594	237.46	0.9641	0.6507	246.76	0.9506
290.0	0.6776	238.08	1.0132	0.6665	247.12	0.9980	0.6572	256.33	0.9841
300.0	0.6857	247.94	1.0466	0.6739	256.85	1.0310	0.6640	265.97	1.0168
310.0	0.6941	257.86	1.0791	0.6814	266.63	1.0631	0.6709	275.64	1.0485
320.0	0.7027	267.82	1.1107	0.6891	276.43	1.0942	0.6779	285.32	1.0792
330.0	0.7117	277.79	1.1414	0.6970	286.23	1.1243	0.6851	294.99	1.1090
340.0	0.7209	287.76	1.1712	0.7052	296.03	1.1536	0.6925	304.65	1.1379
350.0	0.7304	297.74	1.2001	0.7136	305.82	1.1820	0.7001	314.29	1.1658
360.0	0.7403	307.71	1.2282	0.7222	315.59	1.2095	0.7079	323.91	1.1929
370.0	0.7505	317.68	1.2555	0.7311	325.35	1.2362	0.7159	333.51	1.2192
380.0	0.7611	327.66	1.2821	0.7403	335.10	1.2622	0.7241	343.09	1.2448
390.0	0.7721	337.63	1.3080	0.7496	344.84	1.2875	0.7324	352.66	1.2696
400.0	0.7834	347.62	1.3333	0.7593	354.58	1.3122	0.7410	362.23	1.2938
410.0	0.7951	357.62	1.3580	0.7692	364.32	1.3362	0.7498	371.79	1.3174
420.0	0.8072	367.63	1.3822	0.7794	374.07	1.3597	0.7587	381.35	1.3405
430.0	0.8197	377.67	1.4058	0.7898	383.83	1.3827	0.7679	390.92	1.3630
440.0	0.8326	387.73	1.4289	0.8005	393.61	1.4052	0.7772	400.50	1.3850
450.0	0.8459	397.82	1.4516	0.8114	403.40	1.4272	0.7867	410.10	1.4066
460.0	0.8597	407.94	1.4738	0.8227	413.22	1.4488	0.7965	419.72	1.4277
470.0	0.8739	418.10	1.4957	0.8341	423.06	1.4699	0.8064	429.36	1.4485

References for R21

[Benn40] BENNING, A.F. AND MCHARNESS, R.C.: Thermodynamic Properties of Fluorochloromethanes and -Ethanes; Ind. Eng. Chem. 32(1940)4 497/499

[Benn40a] BENNING, A.F. AND MCHARNESS, R.C.: Thermodynamic Properties of Fluorochloromethanes and -Ethanes; Ind. Eng. Chem. 32(1940)5 698/701

[Benn40b] BENNING, A.F. AND MCHARNESS, R.C.: Thermodynamic Properties of Fluorochloromethanes and -Ethanes; Ind. Eng. Chem. 32(1940)6 814/816

[Benn40c] BENNING, A.F. ; MCHARNESS, R.C. ; MARKWOOD JR. ,W.H. AND SMITH, W.J.: Thermodynamic Properties of Fluorochloromethanes and -Ethanes; Ind. Eng. Chem. 32(1940)7 976/980

[Chen76] CHEN, S.S.; WILHOIT, R.C. AND ZWOLINSKI, B.J.: Ideal Gas Thermodynamic Properties of Six Chlorofluoromethanes; J. Phys. Chem. Ref. Data 5(1976)3 571/580

[DeVa76] DEVANEY, W.E.: Fluid property measurement study, Contract NAS 9-14685; P-V-T Inc. Houston Texas (1976)

[Gruz75] GRUZDEV, V.A.; SHELUDYAKOV, E.P.; KIRIJANENKO, A.A.; KOLOTOV, J.L.; LAVROV, B.A.; SCHILJAKOV, A.A.; SCHESTOVA, A.I. AND SCHUMSKAJA A.I.: Experimental investigation of the isobaric heat capacity of Freons: R-11,R-12,R-13,R-21,R-22,R-23; Thermophysical properties of Matter and Substances 8(1975) 108/120

[Kala68] KALAFATI, D.D. ; PETROV, E.K. ; RASSKAZOV,D.S. AND KAEKIN,V.S.: An experimental study of the p-v-T dependence of Freon-21; Teploenergetika 15(1968) 80/83

[KMR 81] - : Kältemaschinenregeln, 7. Auflage; Verlag C.F. Müller, Karlsruhe (1981)

[Meye69] MEYER, K.J.: Über den Zusammenhang zwischen der Schallgeschwindigkeit und der Wärmeleitfähigkeit bei flüssigen Fluor-Chlor-Derivaten des Methans und Äthans; Kältetechnik-Klimatisierung 21(1969)9 270/275

[Solo68] SOLOV'EV, A.N. ; SHELUDYAKOV,E.P. AND SHILYAKOV, A.A.: Experimental determination of the pvT relationship for vapors of Freon-21; Zhurnal Pridkladnoi Mekhaniki i Teknicheskoi Fiziki 9(1968)6 145/148

R 21. ph

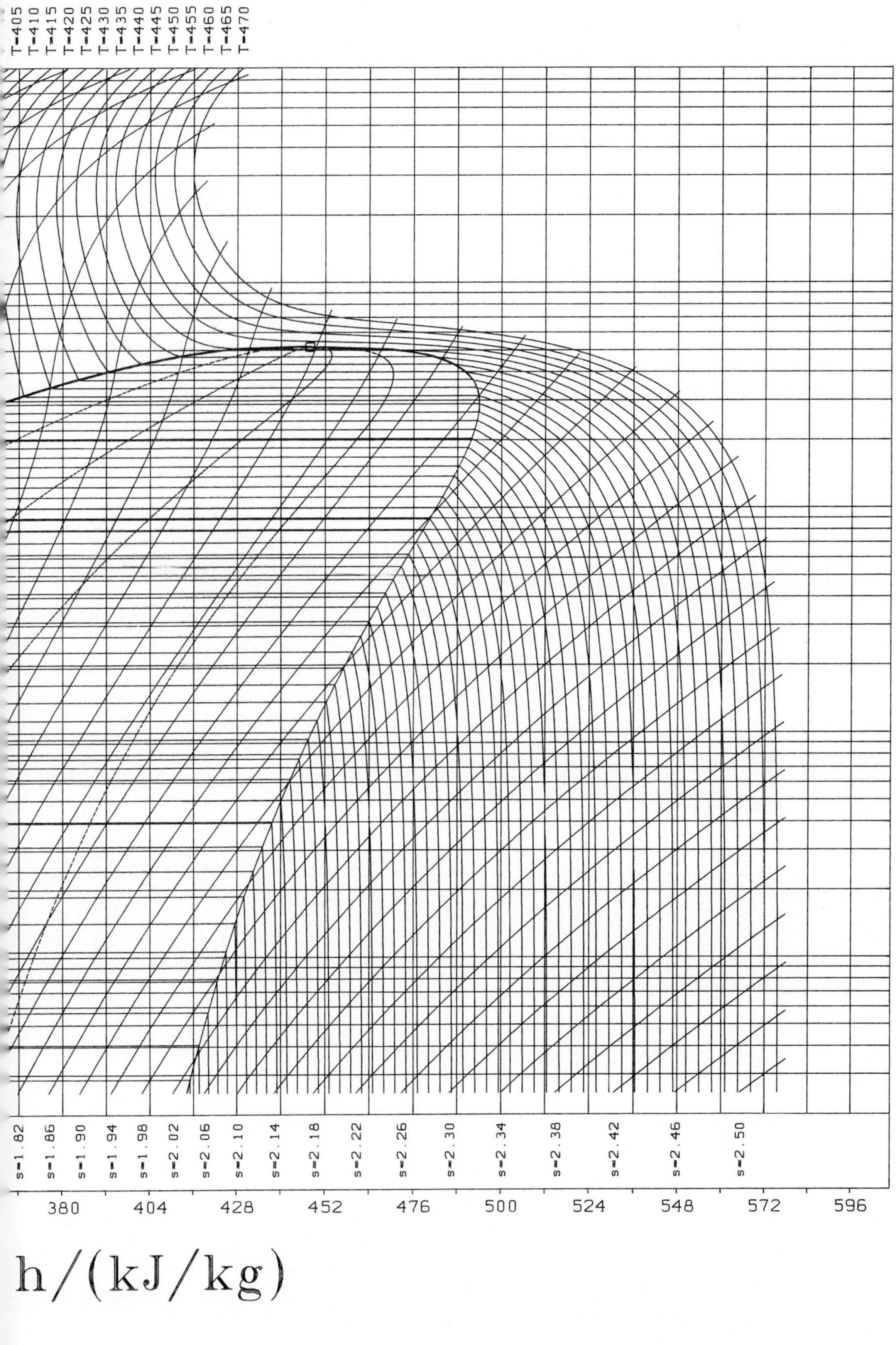
T=405
T=410
T=415
T=420
T=425
T=430
T=435
T=440
T=445
T=450
T=455
T=460
T=465
T=470
s=1.82
s=1.86
s=1.90
s=1.94
s=1.98
s=2.02
s=2.06
s=2.10
s=2.14
s=2.18
s=2.22
s=2.26
s=2.30
s=2.34
s=2.38
s=2.42
s=2.46
s=2.50
380
404
428
452
476
500
524
548
572
596
h/(kJ/kg)

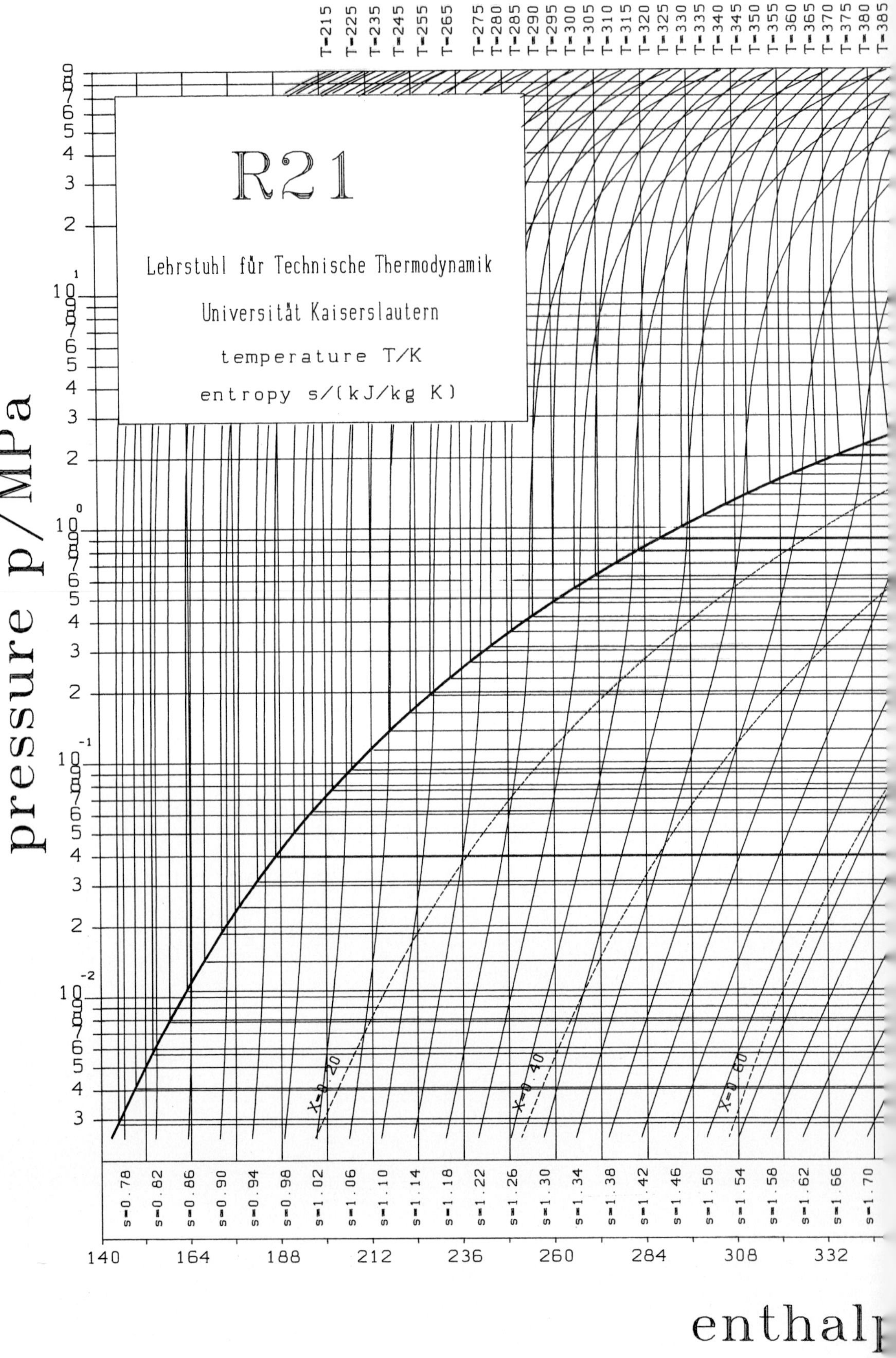
R21
Lehrstuhl für Technische Thermodynamik
Universität Kaiserslautern
temperature T/K
entropy s/(kJ/kg K)
pressure p/MPa

R 21.Ts

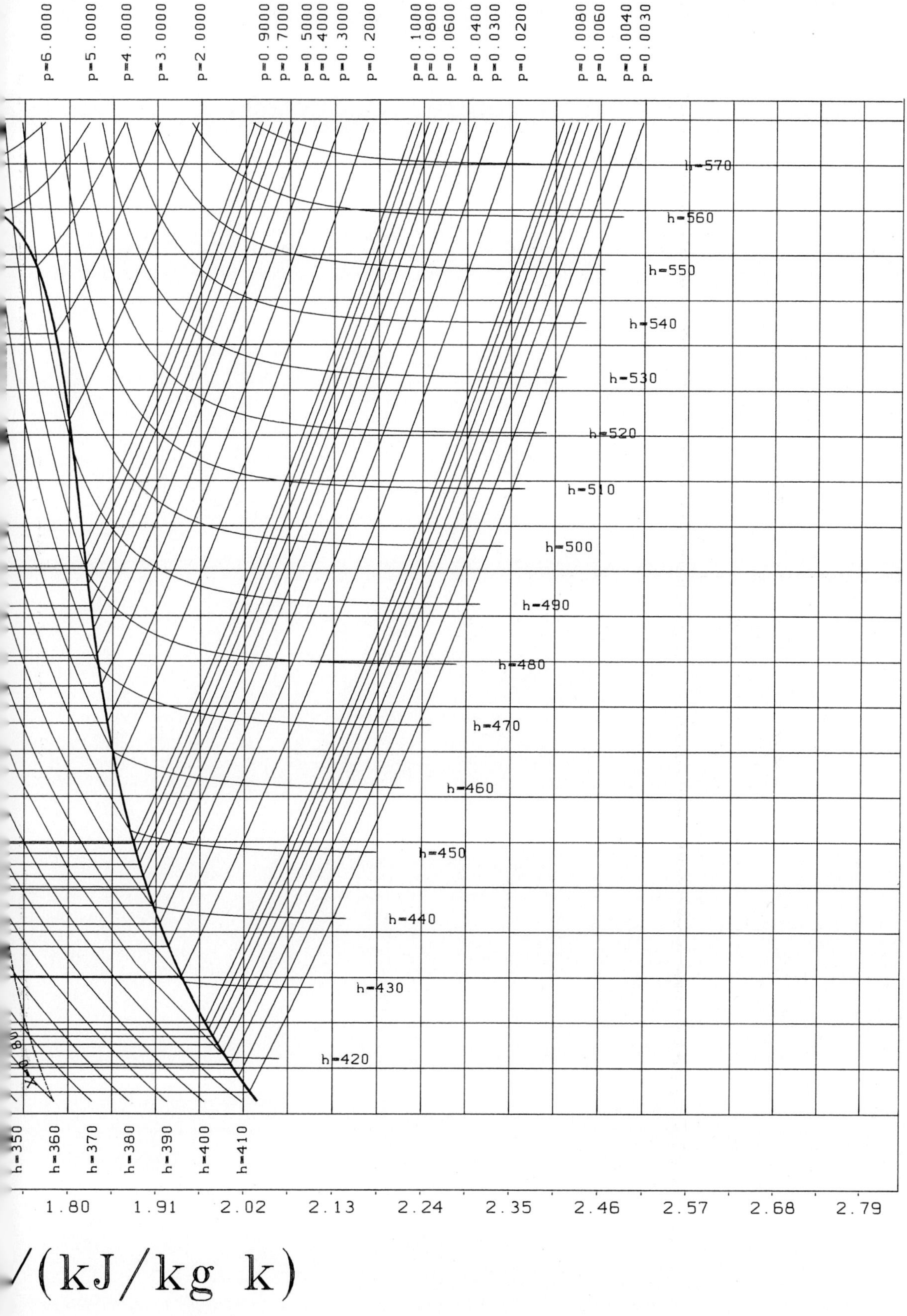

p=6.0000
p=5.0000
p=4.0000
p=3.0000
p=2.0000
p=0.9000
p=0.7000
p=0.5000
p=0.4000
p=0.3000
p=0.2000
p=0.1000
p=0.0800
p=0.0600
p=0.0400
p=0.0300
p=0.0200
p=0.0080
p=0.0060
p=0.0040
p=0.0030
h=570
h=560
h=550
h=540
h=530
h=520
h=510
h=500
h=490
h=480
h=470
h=460
h=450
h=440
h=430
h=420
h=350
h=360
h=370
h=380
h=390
h=400
h=410
1.80
1.91
2.02
2.13
2.24
2.35
2.46
2.57
2.68
2.79
/(kJ/kg k)

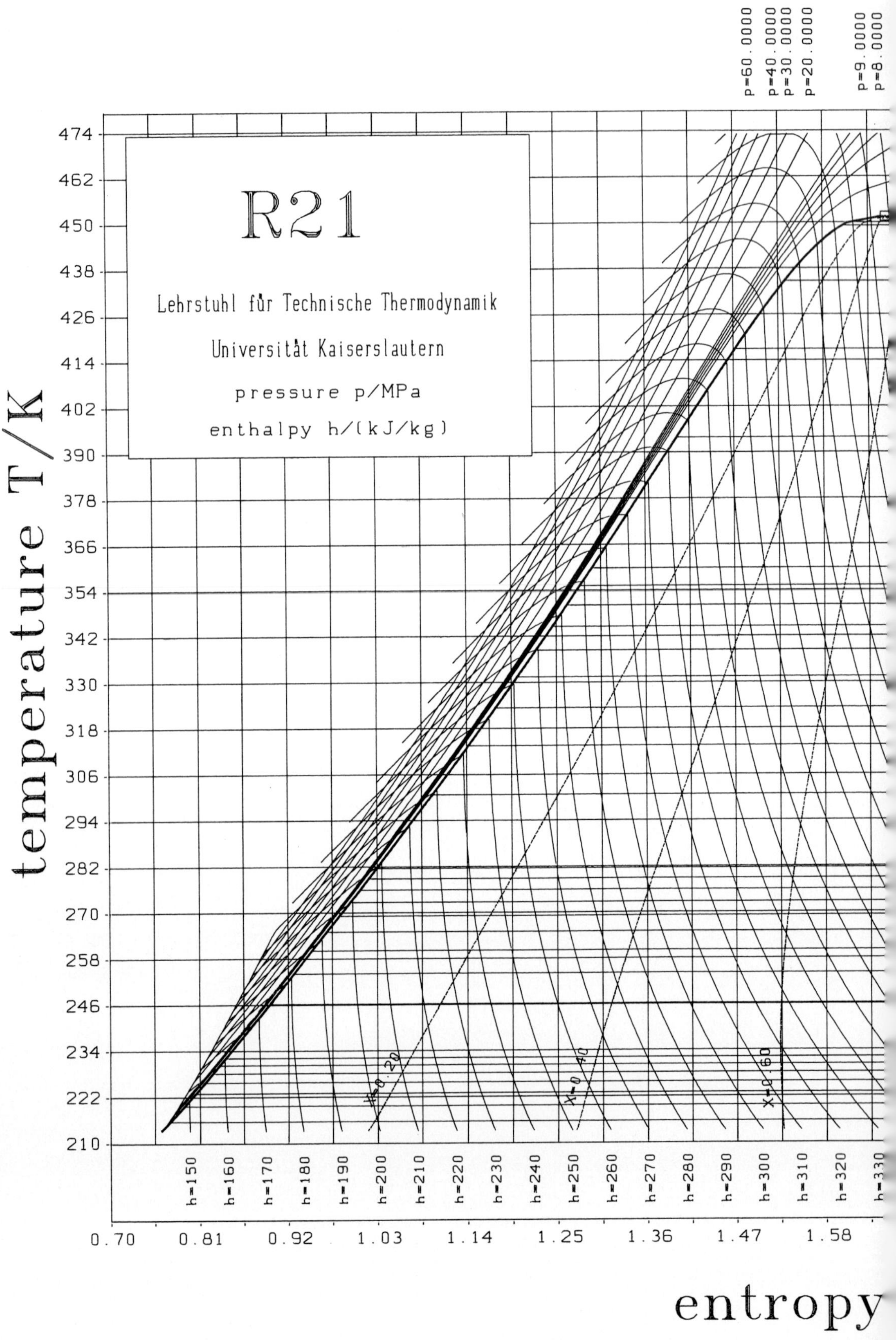

R21
Lehrstuhl für Technische Thermodynamik
Universität Kaiserslautern
pressure p/MPa
enthalpy h/(kJ/kg)
temperature T/K
entropy
p=60.0000
p=40.0000
p=30.0000
p=20.0000
p=9.0000
p=8.0000
474
462
450
438
426
414
402
390
378
366
354
342
330
318
306
294
282
270
258
246
234
222
210
x=0.20
x=0.40
x=0.60
h=150
h=160
h=170
h=180
h=190
h=200
h=210
h=220
h=230
h=240
h=250
h=260
h=270
h=280
h=290
h=300
h=310
h=320
h=330
0.70
0.81
0.92
1.03
1.14
1.25
1.36
1.47
1.58

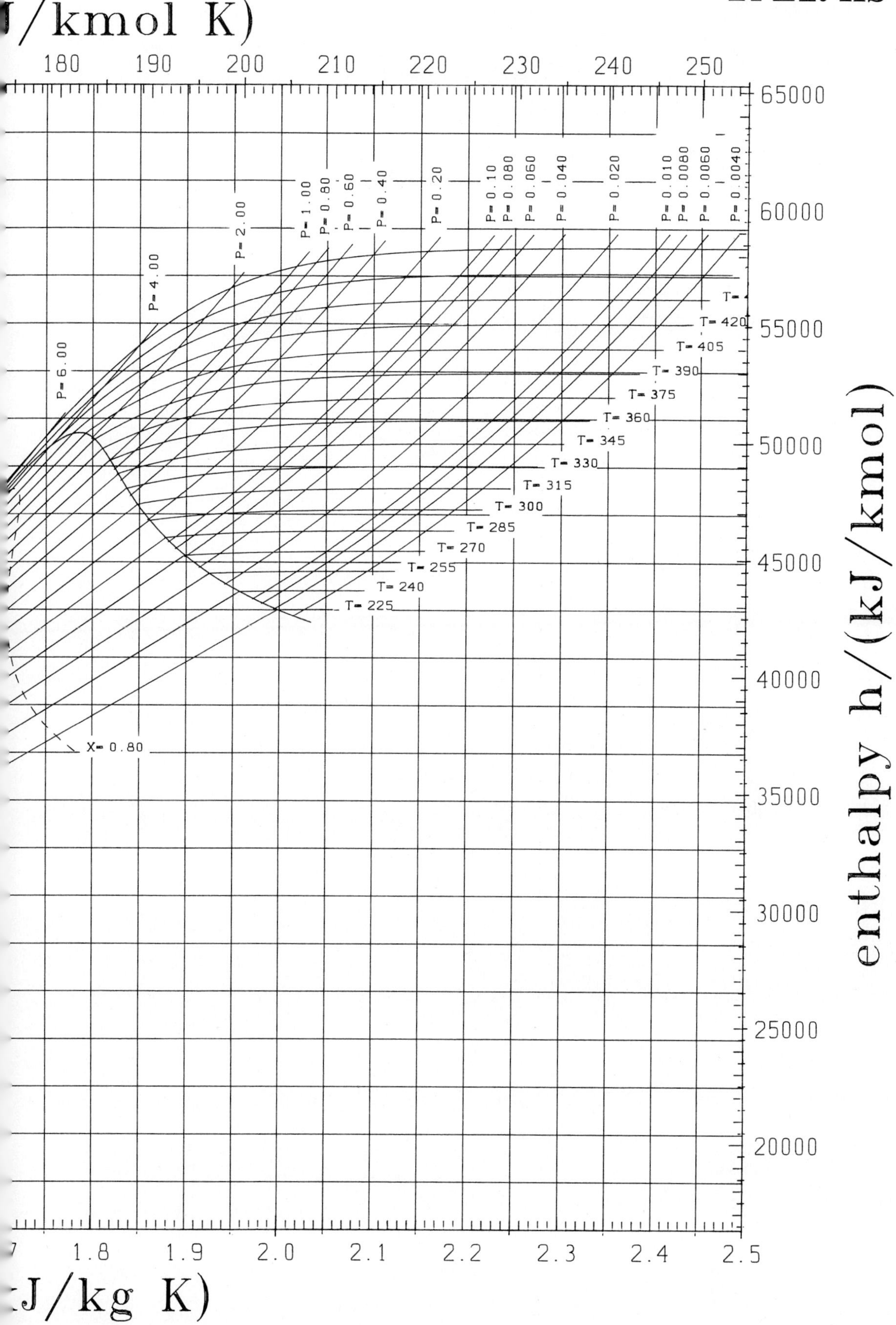
J/kmol K)
180
190
200
210
220
230
240
250
65000
60000
55000
50000
45000
40000
35000
30000
25000
20000
P= 6.00
P= 4.00
P= 2.00
P= 1.00
P= 0.80
P= 0.60
P= 0.40
P= 0.20
P= 0.10
P= 0.080
P= 0.060
P= 0.040
P= 0.020
P= 0.010
P= 0.0080
P= 0.0060
P= 0.0040
T= 420
T= 405
T= 390
T= 375
T= 360
T= 345
T= 330
T= 315
T= 300
T= 285
T= 270
T= 255
T= 240
T= 225
X= 0.80
1.8
1.9
2.0
2.1
2.2
2.3
2.4
2.5
J/kg K)
enthalpy h/(kJ/kmol)

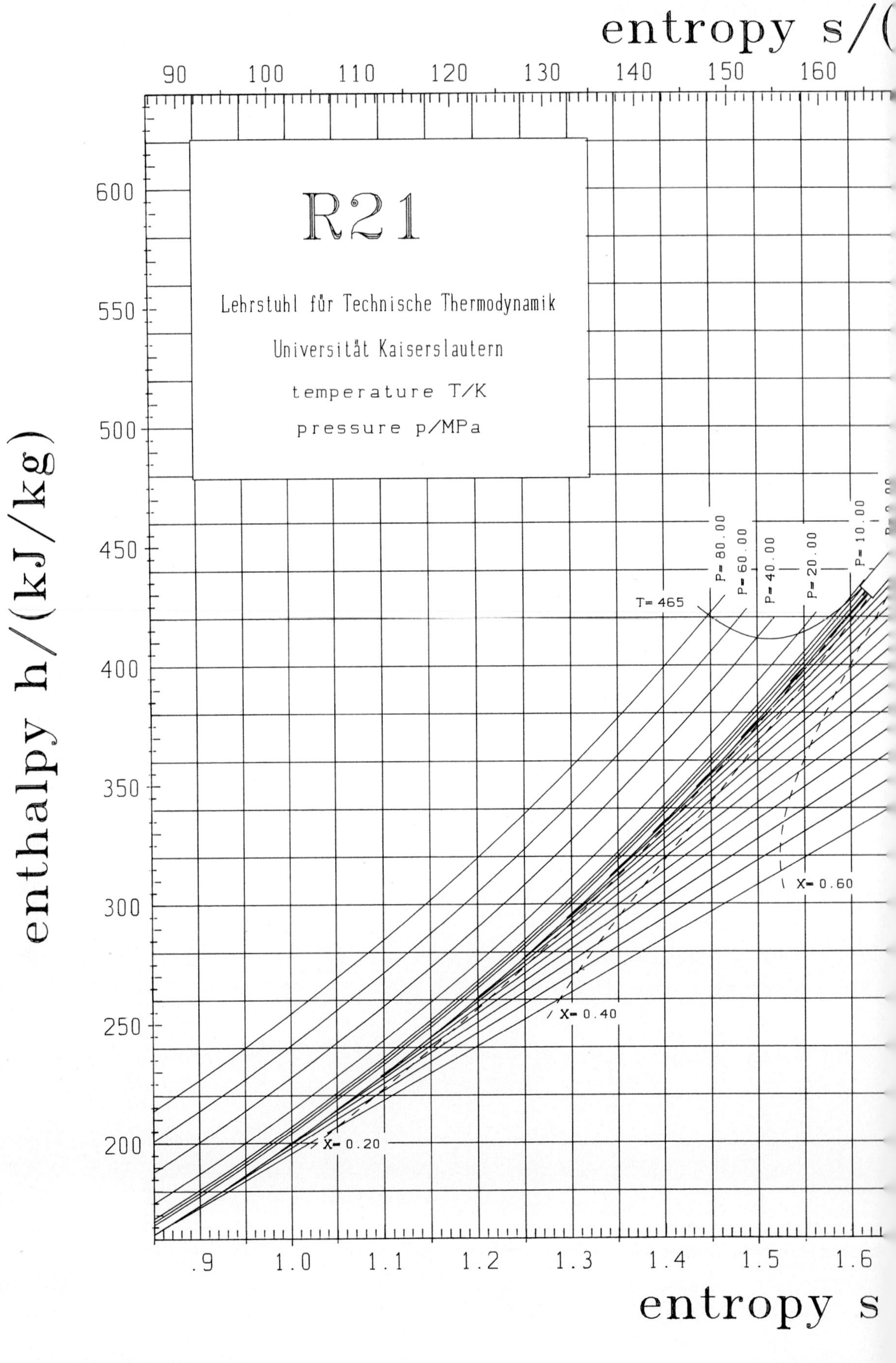

entropy s/(
90 100 110 120 130 140 150 160
R21
Lehrstuhl für Technische Thermodynamik
Universität Kaiserslautern
temperature T/K
pressure p/MPa
enthalpy h/(kJ/kg)
600 550 500 450 400 350 300 250 200
T= 465
P= 80.00
P= 60.00
P= 40.00
P= 20.00
P= 10.00
X= 0.60
X= 0.40
X= 0.20
.9 1.0 1.1 1.2 1.3 1.4 1.5 1.6
entropy s

Refrigerant R22 ($CHClF_2$)

Table R22.1: List of references for p, ρ, T data

Author	n	p/MPa min	p/MPa max	ρ/g cm^{-3} min	ρ/g cm^{-3} max	T/K min	T/K max
BENNING, A.F. *[Benn 40]*	40	0.04	1.85	0.0009	0.08	298	413
KLETSKII, A.V. *[Klet 70]*	63	0.80	5.80	0.028	0.66	294	464
KUMAGAI, A. *[Kuma 78]*	34	9.66	160.86	1.373	1.62	253	293
LAGUTINA, L.M. *[Lagu 66]*	73	0.20	5.85	0.006	0.74	293	373
MICHELS, A. *[Mich 56]*	150	0.65	8.27	0.026	0.36	292	423
OGUCHI, K. *[Oguc 79]*	100	0.23	10.78	0.010	1.21	253	363
TAKAISHI, Y. *[Taka 80]*	27	1.24	9.54	0.725	0.98	334	372
ZANDER, M. *[Zand 68]*	261	0.31	35.02	0.0097	1.25	291	473

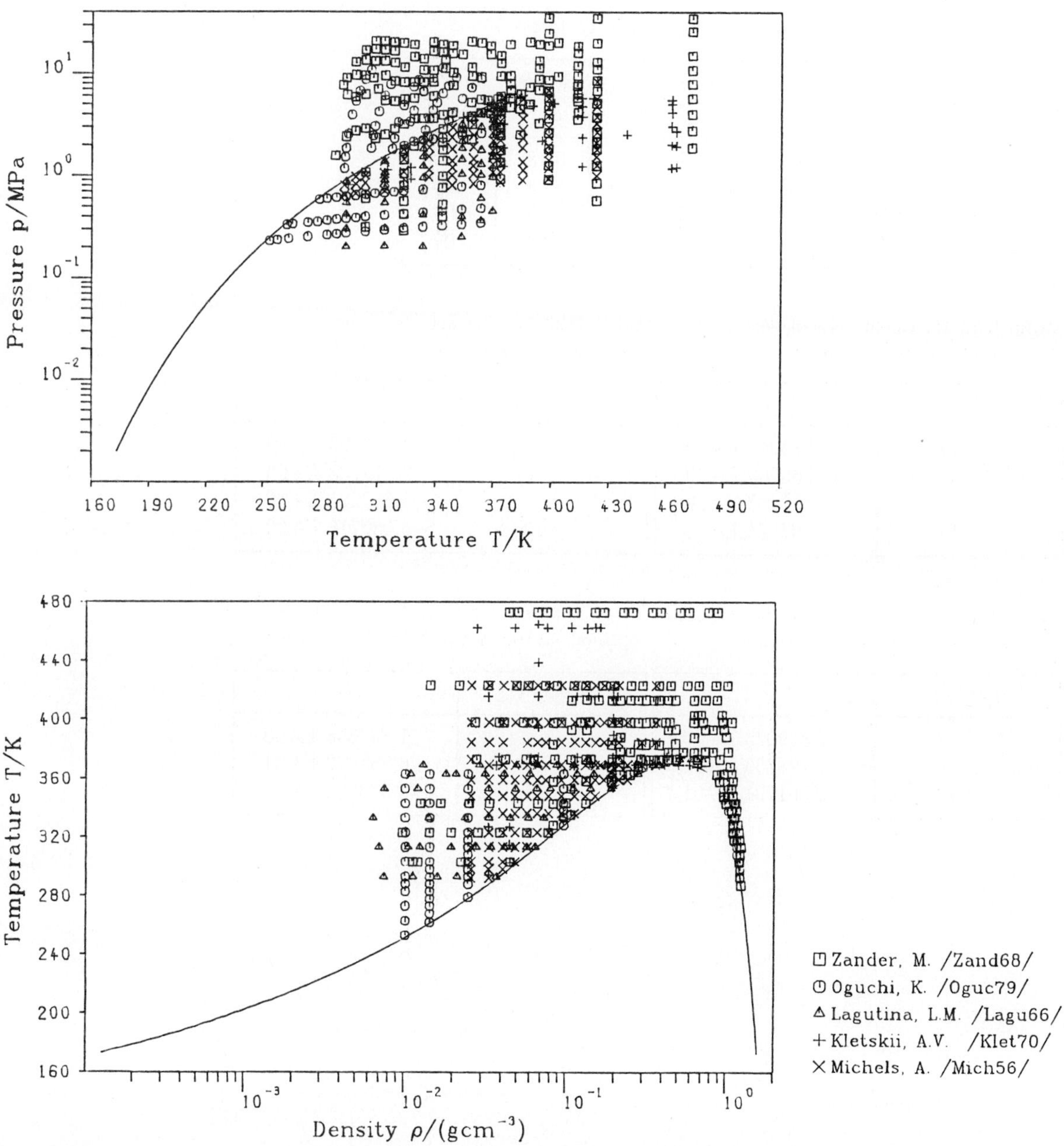

Fig.R22.1: Distribution of p, ρ, T data projected in the p, T and in the T, ρ plane

Table R22.2: Experimental vapor pressure data

Author	Δ / %		
	std	opt	*EOS*
BENNING, A.F. *[Benn 40]*	0.38	0.36	0.36
JÄGER, H.P. *[Jaeg 73]*	0.77	0.78	0.77
KLETSKII, A.V. *[Klet 70]*	0.03	0.02	0.04
LAGUTINA, L.M. *[Lagu 66]*	0.46	0.45	0.45
MICHELS, A. *[Mich 56]*	0.63	0.70	0.35
OGUCHI, K. *[Oguc 79]*	0.05	0.03	0.05
TAKAISHI, Y. *[Taka 80]*	0.04	0.01	0.01
ZANDER, M. *[Zand 68]*	0.14	0.13	0.13

Table R22.3: Experimental saturated liquid density data

Author	Δ / %		
	std	opt	*EOS*
BENNING, A.F. *[Benn 40b]*	0.20	0.18	0.23
KLETSKII, A.V. *[Klet 70]*	0.17	0.03	0.23
OGUCHI, K. *[Oguc 79]*	0.02	0.03	0.09
ZANDER, M. *[Zand 68]*	0.04	0.03	0.07

Table R22.4: Coefficients of the standard vapor pressure equation

Index i	b_i
5	−7.015247 E+00
6	1.260700 E+00
9	−2.487262 E+00
15	−2.707978 E+00

Table R22.5: Coefficients of the quasi-optimal vapor pressure equation

Index i	b_i
5	−7.159523 E+00
6	2.312738 E+00
7	−2.099225 E+00
13	−4.137716 E+00

Table R22.6: Coefficients of the standard saturated liquid density equation

Index i	c_i
8	1.829925 E+00
9	7.798706 E−01
16	5.861791 E−01

Table R22.7: Coefficients of the quasi-optimal saturated liquid density equation

Index i	c_i
8	1.869358 E+00
9	6.555972 E−01
12	3.563298 E−01

Table R22.8: Coefficients of the *Bender-EOS*

Range of validity: one-phase region

$0.20 \leq p/\text{MPa} \leq 35.02$

$0.00637 \leq \rho/\text{g cm}^{-3} \leq 1.249$

$253.12 \leq T/\text{K} \leq 473.19$

	Specific units	Molar units
a_{01}	2.215790361 E−02	1.656261826 E−01
a_{02}	3.157020036 E+01	2.359813393 E+02
a_{03}	7.509214365 E+03	5.612997202 E+04
a_{04}	5.467526809 E+06	4.086873964 E+07
a_{05}	1.598572643 E+08	1.194903133 E+09
a_{06}	5.171656027 E−01	3.342182162 E−01
a_{07}	−2.979490506 E+02	−1.925495425 E+02
a_{08}	7.002529454 E+04	4.525383921 E+04
a_{09}	−6.066207658 E−01	−3.389361718 E−02
a_{10}	2.673547153 E+02	1.493786379 E+01
a_{11}	5.758299319 E−01	2.781602350 E−03
a_{12}	−3.790471928 E+02	−1.831024238 E+00
a_{13}	1.059193104 E+02	4.423602703 E−02
a_{14}	−3.643337733 E+07	−2.354506626 E+07
a_{15}	2.419730094 E+10	1.563750318 E+10
a_{16}	−4.274598365 E+12	−2.762458743 E+12
a_{17}	6.313109058 E+07	3.049608577 E+05
a_{18}	−8.095984950 E+10	−3.910844073 E+08
a_{19}	2.143758161 E+13	1.035563177 E+11
a_{20}	5.000000000 E+00	3.737406424 E−02

Table R22.9: Deviations between experimental and predicted p, ρ, T data

Region	n	Δ / %	
		p	ρ
Gas	433	0.27	0.37
Critical	72	0.33	1.71
Liquid	140	3.11	0.20
Total	645	1.49	0.67

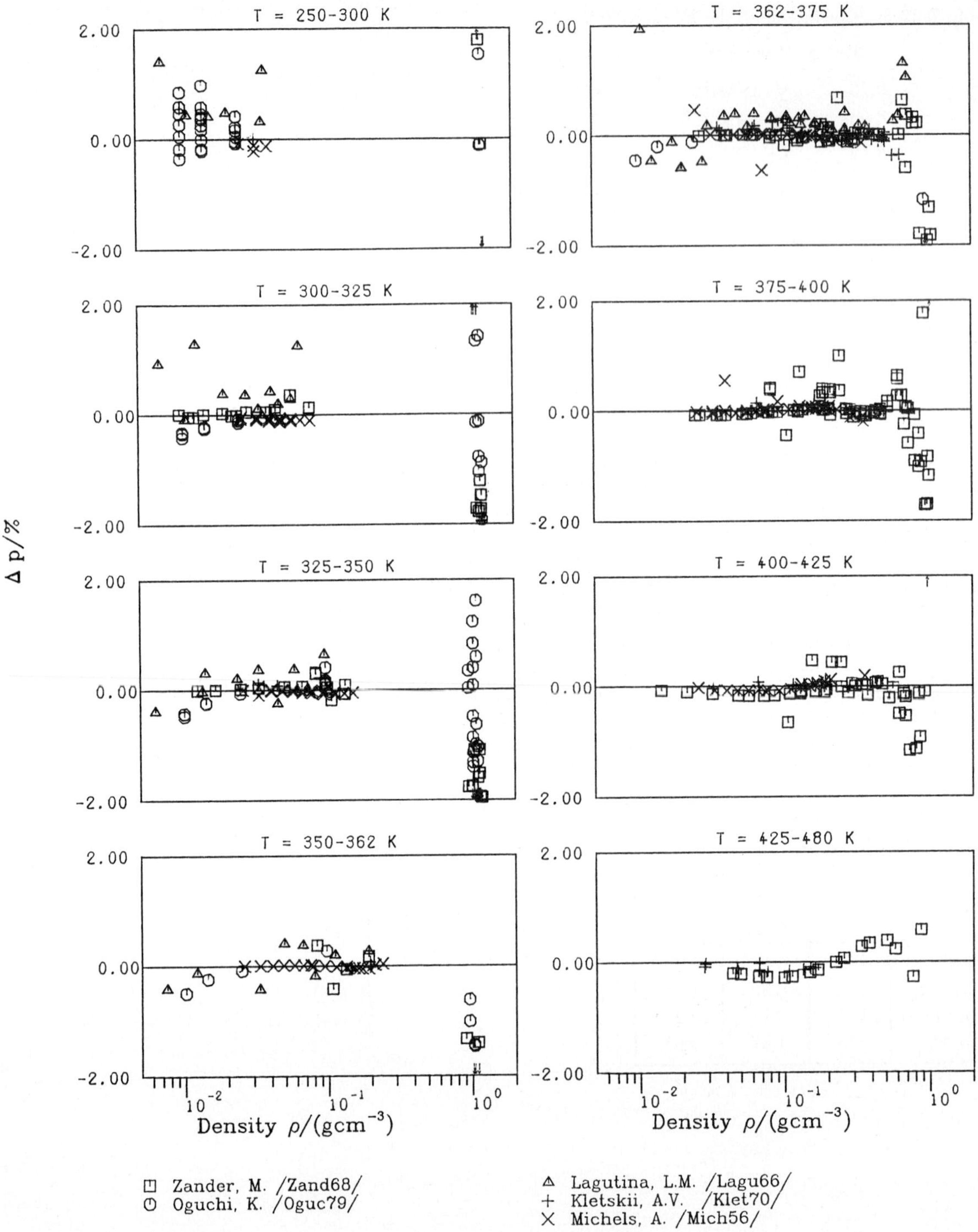

Fig.R22.2: Deviations between experimental and calculated p, ρ, T data

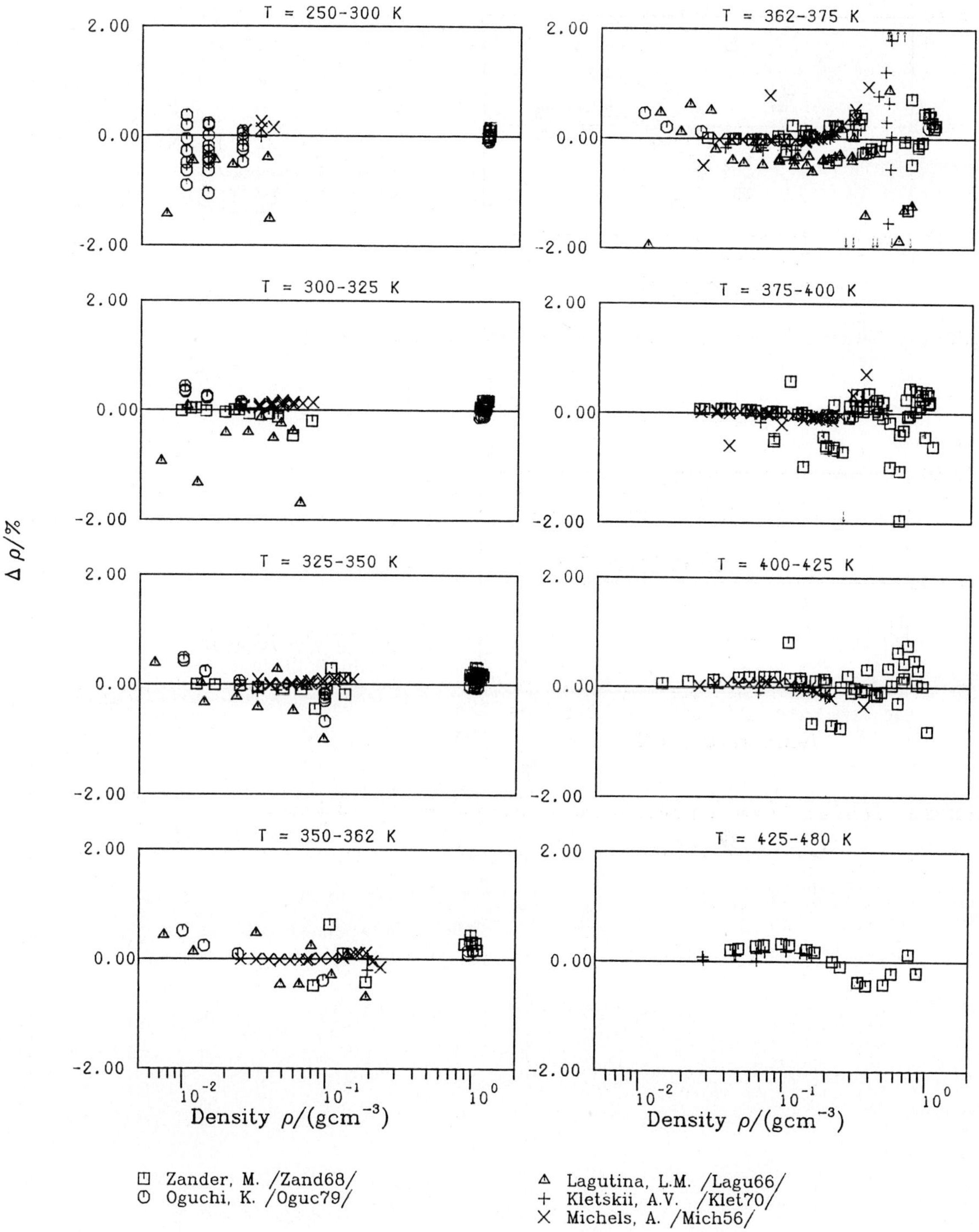

Fig.R22.2: Deviations between experimental and calculated p, ρ, T data *-continued*

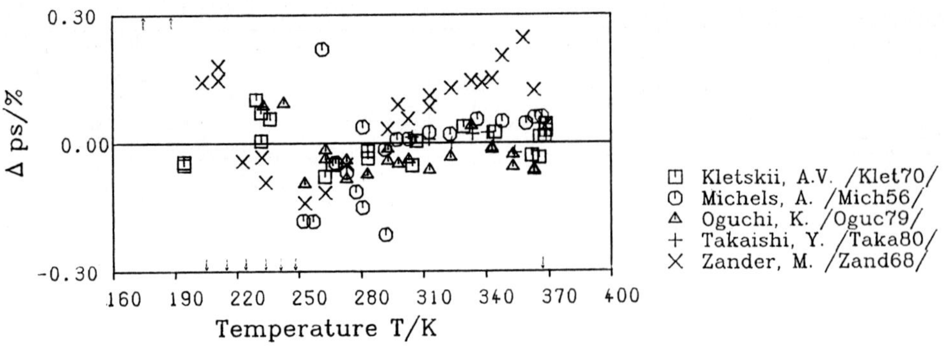

Fig.R22.3: Deviations between experimental and calculated vapor pressures

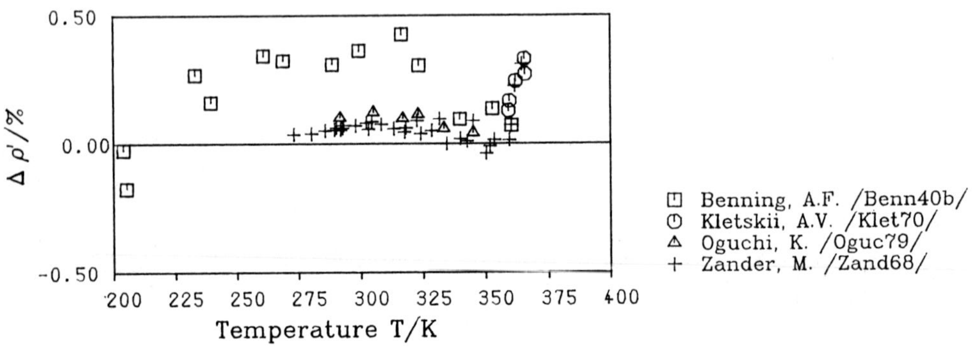

Fig.R22.4: Deviations between experimental and calculated saturated liquid densities

Table R22.10: Information on ideal gas state heat capacity equation

Author	T / K min	T / K max
BARHO, W. *[Barh 66]*	200	550

Coefficients of equation (15)	
Index i	d_i
1	3.115145 E−01
2	7.299535 E−04
3	2.852110 E−06
4	−6.061076 E−09
5	3.505000 E−12

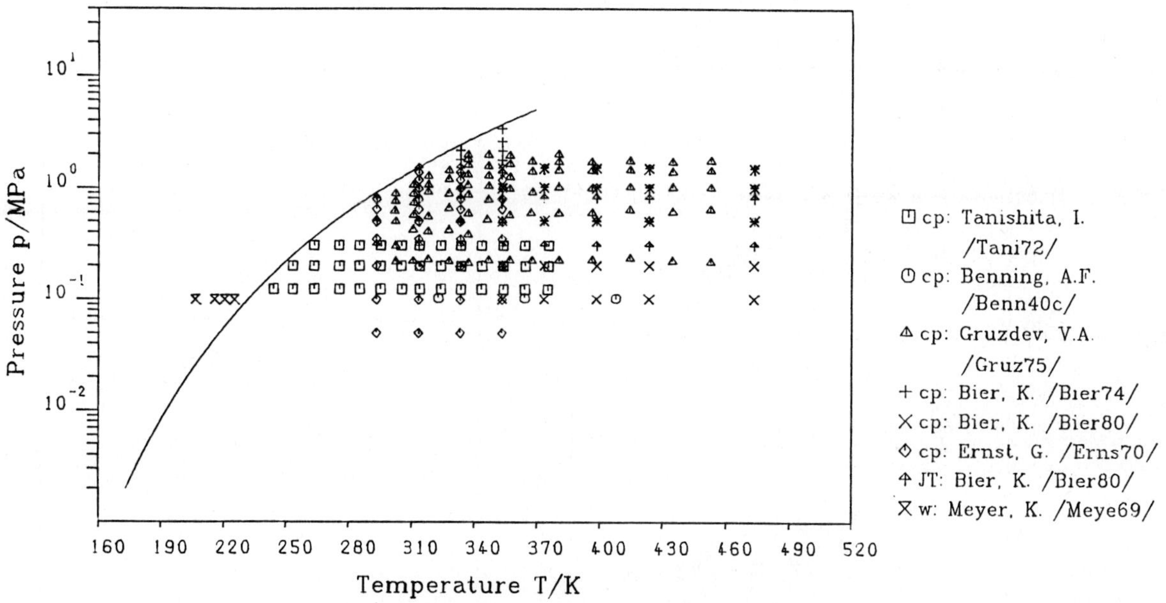

Fig.R22.5: Distribution of $c_p(p,T)$, $\mu(p,T)$ and $w(p,T)$ data

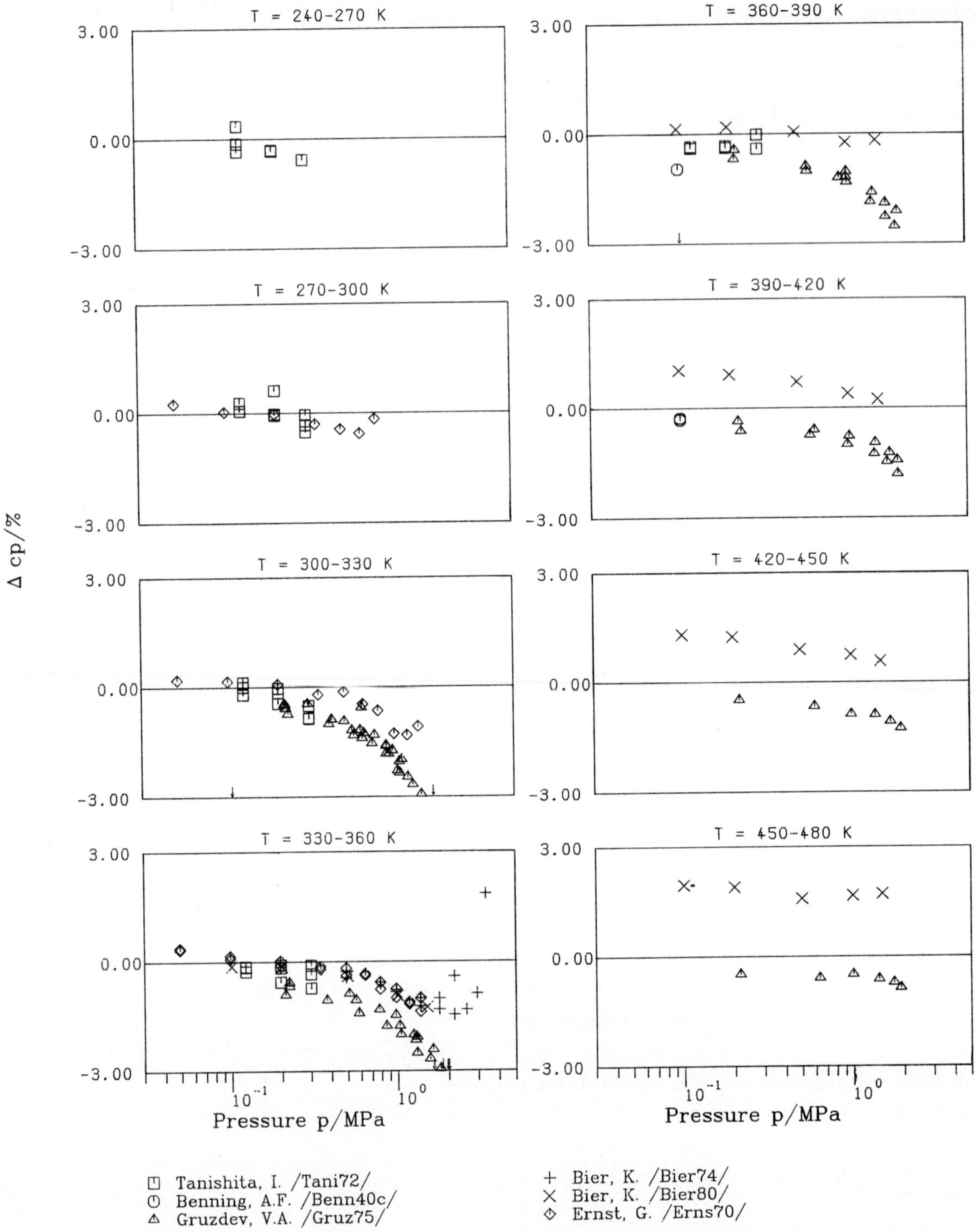

Fig.R22.6: Deviations between experimental and calculated specific isobaric heat capacities c_p

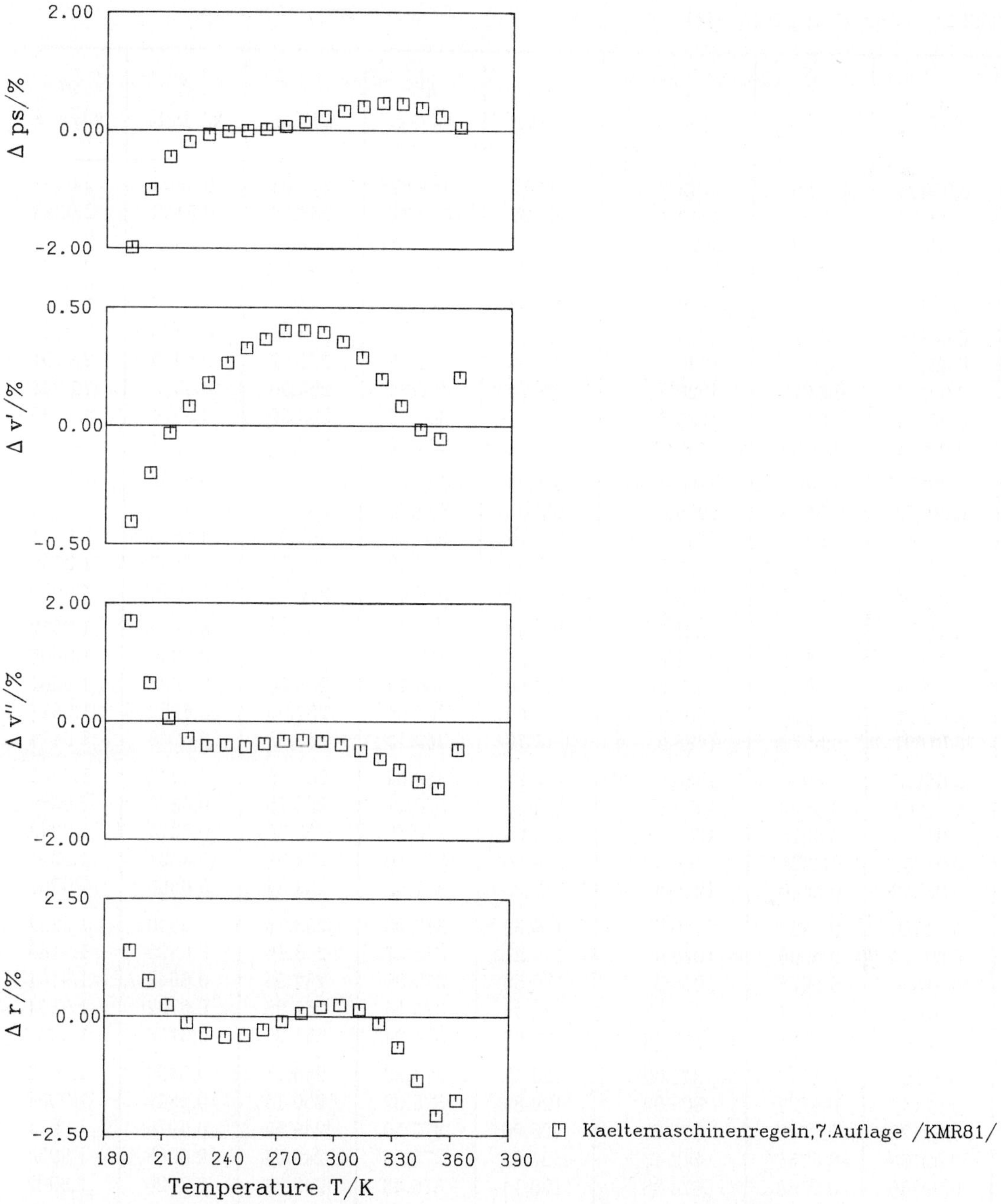

Fig.R22.7: Deviations between saturation properties from the *Kältemaschinenregeln* and those calculated by the *Bender-EOS*

Chart R22.1: Saturation properties

T K	p MPa	v' dm³/kg	v'' dm³/kg	h' kJ/kg	h'' kJ/kg	r kJ/kg	s' kJ/kgK	s'' kJ/kgK
175.0	0.00235	0.6346	7132.9	91.19	359.82	268.63	0.5092	2.0442
176.0	0.00258	0.6358	6549.8	92.27	360.31	268.04	0.5153	2.0383
177.0	0.00282	0.6370	6021.3	93.34	360.79	267.45	0.5214	2.0324
178.0	0.00308	0.6383	5541.8	94.42	361.28	266.86	0.5275	2.0267
179.0	0.00336	0.6395	5106.2	95.50	361.77	266.27	0.5335	2.0211
180.0	0.00366	0.6407	4710.0	96.58	362.26	265.68	0.5395	2.0155
181.0	0.00399	0.6419	4349.3	97.66	362.75	265.09	0.5455	2.0101
182.0	0.00434	0.6431	4020.5	98.74	363.24	264.50	0.5515	2.0048
183.0	0.00471	0.6443	3720.4	99.82	363.73	263.90	0.5574	1.9995
184.0	0.00511	0.6455	3446.2	100.91	364.22	263.31	0.5633	1.9943
185.0	0.00554	0.6467	3195.4	101.99	364.71	262.72	0.5692	1.9893
186.0	0.00600	0.6480	2965.8	103.08	365.20	262.12	0.5750	1.9843
187.0	0.00649	0.6492	2755.3	104.16	365.69	261.53	0.5809	1.9794
188.0	0.00702	0.6504	2562.2	105.25	366.18	260.93	0.5867	1.9746
189.0	0.00758	0.6516	2384.9	106.34	366.67	260.33	0.5924	1.9699
190.0	0.00817	0.6529	2221.9	107.42	367.16	259.74	0.5982	1.9652
191.0	0.00881	0.6541	2071.8	108.51	367.65	259.14	0.6039	1.9606
192.0	0.00949	0.6553	1933.6	109.60	368.15	258.54	0.6096	1.9561
193.0	0.01020	0.6565	1806.2	110.69	368.64	257.95	0.6152	1.9517
194.0	0.01097	0.6578	1688.6	111.78	369.13	257.35	0.6208	1.9474
195.0	0.01178	0.6590	1580.0	112.87	369.62	256.75	0.6264	1.9431
196.0	0.01263	0.6603	1479.5	113.96	370.11	256.15	0.6320	1.9389
197.0	0.01354	0.6615	1386.6	115.05	370.60	255.56	0.6375	1.9348
198.0	0.01451	0.6628	1300.5	116.14	371.10	254.96	0.6431	1.9307
199.0	0.01553	0.6640	1220.8	117.23	371.59	254.36	0.6486	1.9267
200.0	0.01660	0.6653	1146.8	118.32	372.08	253.76	0.6540	1.9228
201.0	0.01774	0.6665	1078.0	119.41	372.57	253.16	0.6595	1.9189
202.0	0.01894	0.6678	1014.2	120.50	373.06	252.56	0.6649	1.9151
203.0	0.02021	0.6691	954.77	121.59	373.55	251.96	0.6702	1.9114
204.0	0.02155	0.6703	899.49	122.68	374.04	251.35	0.6756	1.9077
205.0	0.02295	0.6716	848.00	123.77	374.53	250.75	0.6809	1.9041
206.0	0.02443	0.6729	800.00	124.87	375.02	250.15	0.6862	1.9006
207.0	0.02599	0.6742	755.23	125.96	375.50	249.55	0.6915	1.8971
208.0	0.02763	0.6755	713.43	127.05	375.99	248.94	0.6968	1.8936
209.0	0.02935	0.6768	674.38	128.14	376.48	248.34	0.7020	1.8902
210.0	0.03116	0.6781	637.87	129.23	376.97	247.73	0.7072	1.8869
211.0	0.03305	0.6794	603.71	130.33	377.45	247.13	0.7124	1.8836
212.0	0.03504	0.6807	571.73	131.42	377.94	246.52	0.7176	1.8804
213.0	0.03713	0.6821	541.77	132.51	378.42	245.91	0.7227	1.8772
214.0	0.03931	0.6834	513.69	133.60	378.91	245.30	0.7278	1.8741
215.0	0.04160	0.6848	487.34	134.70	379.39	244.69	0.7329	1.8710
216.0	0.04399	0.6861	462.61	135.79	379.87	244.08	0.7380	1.8680
217.0	0.04649	0.6875	439.38	136.89	380.35	243.47	0.7430	1.8650
218.0	0.04911	0.6888	417.55	137.98	380.83	242.85	0.7480	1.8620
219.0	0.05184	0.6902	397.01	139.08	381.31	242.24	0.7530	1.8591
220.0	0.05469	0.6916	377.69	140.17	381.79	241.62	0.7580	1.8563
221.0	0.05767	0.6930	359.49	141.27	382.27	241.00	0.7630	1.8535
222.0	0.06077	0.6944	342.35	142.36	382.74	240.38	0.7679	1.8507
223.0	0.06401	0.6958	326.19	143.46	383.22	239.76	0.7728	1.8480
224.0	0.06738	0.6972	310.94	144.56	383.69	239.13	0.7777	1.8453

Chart R22.1: Saturation properties—*continued*

T K	p MPa	v' dm³/kg	v'' dm³/kg	h' kJ/kg	h'' kJ/kg	r kJ/kg	s' kJ/kgK	s'' kJ/kgK
225.0	0.07089	0.6986	296.55	145.66	384.16	238.51	0.7826	1.8427
226.0	0.07454	0.7000	282.97	146.76	384.64	237.88	0.7875	1.8401
227.0	0.07835	0.7014	270.13	147.85	385.11	237.25	0.7923	1.8375
228.0	0.08230	0.7029	257.99	148.96	385.58	236.62	0.7972	1.8350
229.0	0.08642	0.7044	246.51	150.06	386.04	235.99	0.8020	1.8325
230.0	0.09069	0.7058	235.65	151.16	386.51	235.35	0.8067	1.8300
231.0	0.09513	0.7073	225.37	152.26	386.97	234.71	0.8115	1.8276
232.0	0.09974	0.7088	215.62	153.36	387.44	234.07	0.8163	1.8252
233.0	0.10452	0.7103	206.39	154.47	387.90	233.43	0.8210	1.8229
234.0	0.10948	0.7118	197.63	155.57	388.36	232.79	0.8257	1.8205
235.0	0.11462	0.7133	189.32	156.68	388.82	232.14	0.8304	1.8182
236.0	0.11996	0.7148	181.44	157.79	389.27	231.49	0.8351	1.8160
237.0	0.12548	0.7163	173.95	158.90	389.73	230.83	0.8398	1.8138
238.0	0.13120	0.7179	166.83	160.00	390.18	230.18	0.8444	1.8116
239.0	0.13713	0.7194	160.07	161.12	390.64	229.52	0.8491	1.8094
240.0	0.14326	0.7210	153.64	162.23	391.09	228.86	0.8537	1.8073
241.0	0.14960	0.7226	147.52	163.34	391.53	228.19	0.8583	1.8052
242.0	0.15616	0.7242	141.70	164.45	391.98	227.53	0.8629	1.8031
243.0	0.16294	0.7258	136.16	165.57	392.42	226.86	0.8675	1.8010
244.0	0.16995	0.7274	130.87	166.69	392.87	226.18	0.8720	1.7990
245.0	0.17719	0.7290	125.84	167.80	393.31	225.50	0.8766	1.7970
246.0	0.18466	0.7307	121.04	168.92	393.75	224.82	0.8811	1.7950
247.0	0.19238	0.7323	116.46	170.04	394.18	224.14	0.8857	1.7931
248.0	0.20035	0.7340	112.10	171.17	394.62	223.45	0.8902	1.7912
249.0	0.20856	0.7356	107.93	172.29	395.05	222.76	0.8947	1.7893
250.0	0.21704	0.7373	103.95	173.42	395.48	222.06	0.8992	1.7874
251.0	0.22578	0.7390	100.14	174.54	395.91	221.36	0.9036	1.7856
252.0	0.23479	0.7408	96.511	175.67	396.33	220.66	0.9081	1.7837
253.0	0.24407	0.7425	93.036	176.80	396.76	219.95	0.9125	1.7819
254.0	0.25363	0.7442	89.713	177.93	397.18	219.24	0.9170	1.7801
255.0	0.26347	0.7460	86.534	179.07	397.59	218.53	0.9214	1.7784
256.0	0.27361	0.7478	83.492	180.20	398.01	217.81	0.9258	1.7766
257.0	0.28404	0.7495	80.579	181.34	398.42	217.08	0.9302	1.7749
258.0	0.29477	0.7514	77.788	182.48	398.83	216.35	0.9346	1.7732
259.0	0.30581	0.7532	75.115	183.62	399.24	215.62	0.9390	1.7715
260.0	0.31716	0.7550	72.553	184.77	399.65	214.88	0.9434	1.7698
261.0	0.32884	0.7569	70.097	185.91	400.05	214.14	0.9477	1.7682
262.0	0.34083	0.7587	67.741	187.06	400.45	213.39	0.9521	1.7666
263.0	0.35316	0.7606	65.481	188.21	400.85	212.64	0.9564	1.7650
264.0	0.36582	0.7625	63.312	189.36	401.24	211.89	0.9608	1.7634
265.0	0.37882	0.7644	61.230	190.51	401.64	211.12	0.9651	1.7618
266.0	0.39217	0.7664	59.230	191.67	402.03	210.36	0.9694	1.7602
267.0	0.40588	0.7683	57.309	192.83	402.41	209.58	0.9737	1.7587
268.0	0.41994	0.7703	55.464	193.99	402.79	208.81	0.9780	1.7571
269.0	0.43437	0.7723	53.689	195.15	403.17	208.02	0.9823	1.7556
270.0	0.44917	0.7743	51.983	196.31	403.55	207.24	0.9866	1.7541
271.0	0.46435	0.7763	50.343	197.48	403.92	206.44	0.9908	1.7526
272.0	0.47991	0.7784	48.764	198.65	404.29	205.64	0.9951	1.7511
273.0	0.49586	0.7804	47.245	199.82	404.66	204.84	0.9994	1.7497
274.0	0.51221	0.7825	45.783	201.00	405.03	204.03	1.0036	1.7482

Chart R22.1: Saturation properties—*continued*

T K	p MPa	v' dm³/kg	v'' dm³/kg	h' kJ/kg	h'' kJ/kg	r kJ/kg	s' kJ/kgK	s'' kJ/kgK
275.0	0.52896	0.7846	44.374	202.18	405.39	203.21	1.0079	1.7468
276.0	0.54612	0.7867	43.018	203.36	405.74	202.39	1.0121	1.7454
277.0	0.56369	0.7889	41.711	204.54	406.10	201.56	1.0163	1.7440
278.0	0.58168	0.7911	40.452	205.73	406.45	200.72	1.0205	1.7426
279.0	0.60010	0.7933	39.238	206.91	406.79	199.88	1.0248	1.7412
280.0	0.61895	0.7955	38.067	208.11	407.13	199.03	1.0290	1.7398
281.0	0.63824	0.7977	36.938	209.30	407.47	198.17	1.0332	1.7384
282.0	0.65798	0.8000	35.849	210.50	407.81	197.31	1.0374	1.7370
283.0	0.67817	0.8023	34.797	211.70	408.14	196.44	1.0416	1.7357
284.0	0.69882	0.8046	33.782	212.90	408.47	195.57	1.0457	1.7344
285.0	0.71994	0.8069	32.803	214.11	408.79	194.68	1.0499	1.7330
286.0	0.74152	0.8093	31.856	215.32	409.11	193.79	1.0541	1.7317
287.0	0.76359	0.8117	30.942	216.53	409.42	192.89	1.0583	1.7304
288.0	0.78614	0.8141	30.059	217.74	409.73	191.99	1.0624	1.7291
289.0	0.80918	0.8166	29.205	218.96	410.04	191.08	1.0666	1.7277
290.0	0.83272	0.8191	28.380	220.19	410.34	190.15	1.0707	1.7264
291.0	0.85676	0.8216	27.582	221.41	410.64	189.22	1.0749	1.7252
292.0	0.88132	0.8241	26.810	222.64	410.93	188.29	1.0790	1.7239
293.0	0.90639	0.8267	26.064	223.87	411.22	187.34	1.0832	1.7226
294.0	0.93199	0.8293	25.341	225.11	411.50	186.39	1.0873	1.7213
295.0	0.95812	0.8319	24.642	226.35	411.78	185.43	1.0915	1.7200
296.0	0.98479	0.8346	23.964	227.60	412.05	184.45	1.0956	1.7188
297.0	1.01201	0.8373	23.309	228.84	412.32	183.47	1.0997	1.7175
298.0	1.03978	0.8400	22.674	230.10	412.58	182.48	1.1039	1.7162
299.0	1.06811	0.8428	22.058	231.35	412.84	181.49	1.1080	1.7150
300.0	1.09700	0.8456	21.462	232.61	413.09	180.48	1.1121	1.7137
301.0	1.12647	0.8485	20.884	233.88	413.33	179.46	1.1162	1.7125
302.0	1.15652	0.8514	20.324	235.14	413.57	178.43	1.1204	1.7112
303.0	1.18715	0.8543	19.781	236.42	413.81	177.39	1.1245	1.7099
304.0	1.21838	0.8573	19.254	237.69	414.04	176.34	1.1286	1.7087
305.0	1.25022	0.8603	18.743	238.98	414.26	175.28	1.1327	1.7074
306.0	1.28266	0.8633	18.247	240.26	414.48	174.21	1.1369	1.7062
307.0	1.31572	0.8665	17.766	241.55	414.69	173.13	1.1410	1.7049
308.0	1.34940	0.8696	17.298	242.85	414.89	172.04	1.1451	1.7037
309.0	1.38372	0.8728	16.845	244.15	415.09	170.93	1.1492	1.7024
310.0	1.41868	0.8761	16.404	245.46	415.27	169.82	1.1533	1.7011
311.0	1.45428	0.8794	15.976	246.77	415.46	168.69	1.1575	1.6999
312.0	1.49054	0.8827	15.560	248.08	415.63	167.55	1.1616	1.6986
313.0	1.52746	0.8861	15.156	249.41	415.80	166.39	1.1657	1.6973
314.0	1.56505	0.8896	14.763	250.73	415.96	165.23	1.1698	1.6960
315.0	1.60331	0.8931	14.381	252.07	416.11	164.04	1.1740	1.6947
316.0	1.64227	0.8967	14.010	253.41	416.26	162.85	1.1781	1.6934
317.0	1.68192	0.9004	13.649	254.75	416.39	161.64	1.1822	1.6921
318.0	1.72227	0.9041	13.297	256.10	416.52	160.42	1.1864	1.6908
319.0	1.76333	0.9079	12.955	257.46	416.64	159.18	1.1905	1.6895
320.0	1.80511	0.9118	12.622	258.83	416.75	157.92	1.1947	1.6882
321.0	1.84761	0.9157	12.298	260.20	416.85	156.65	1.1988	1.6868
322.0	1.89085	0.9197	11.983	261.58	416.94	155.36	1.2030	1.6855
323.0	1.93484	0.9238	11.675	262.96	417.02	154.05	1.2072	1.6841
324.0	1.97958	0.9280	11.376	264.35	417.09	152.73	1.2113	1.6827

Chart R22.1: Saturation properties–*continued*

T K	p MPa	v' dm³/kg	v'' dm³/kg	h' kJ/kg	h'' kJ/kg	r kJ/kg	s' kJ/kgK	s'' kJ/kgK
325.0	2.02508	0.9322	11.084	265.75	417.14	151.39	1.2155	1.6813
326.0	2.07135	0.9366	10.800	267.16	417.19	150.03	1.2197	1.6799
327.0	2.11841	0.9410	10.522	268.58	417.23	148.65	1.2239	1.6785
328.0	2.16625	0.9456	10.252	270.00	417.25	147.25	1.2281	1.6771
329.0	2.21489	0.9503	9.9882	271.44	417.26	145.83	1.2324	1.6756
330.0	2.26433	0.9550	9.7308	272.88	417.26	144.38	1.2366	1.6741
331.0	2.31460	0.9599	9.4797	274.33	417.25	142.91	1.2409	1.6726
332.0	2.36569	0.9649	9.2346	275.80	417.22	141.42	1.2451	1.6711
333.0	2.41763	0.9700	8.9952	277.27	417.18	139.91	1.2494	1.6695
334.0	2.47040	0.9753	8.7614	278.75	417.12	138.37	1.2537	1.6680
335.0	2.52404	0.9807	8.5330	280.25	417.05	136.80	1.2580	1.6664
336.0	2.57855	0.9863	8.3098	281.75	416.96	135.20	1.2623	1.6647
337.0	2.63394	0.9920	8.0915	283.27	416.85	133.58	1.2667	1.6630
338.0	2.69022	0.9979	7.8781	284.80	416.72	131.92	1.2710	1.6613
339.0	2.74740	1.0040	7.6693	286.35	416.58	130.23	1.2754	1.6596
340.0	2.80549	1.0102	7.4650	287.91	416.42	128.51	1.2799	1.6578
341.0	2.86452	1.0167	7.2650	289.48	416.23	126.75	1.2843	1.6560
342.0	2.92448	1.0234	7.0691	291.07	416.02	124.96	1.2888	1.6541
343.0	2.98539	1.0303	6.8772	292.67	415.79	123.12	1.2933	1.6522
344.0	3.04727	1.0375	6.6890	294.30	415.54	121.24	1.2978	1.6503
345.0	3.11012	1.0449	6.5045	295.94	415.26	119.32	1.3024	1.6483
346.0	3.17396	1.0526	6.3235	297.60	414.95	117.35	1.3070	1.6462
347.0	3.23881	1.0607	6.1458	299.29	414.61	115.32	1.3117	1.6440
348.0	3.30468	1.0691	5.9712	300.99	414.24	113.25	1.3164	1.6418
349.0	3.37159	1.0779	5.7996	302.72	413.84	111.11	1.3212	1.6395
350.0	3.43954	1.0871	5.6308	304.48	413.40	108.92	1.3260	1.6372
351.0	3.50857	1.0967	5.4647	306.27	412.92	106.65	1.3309	1.6347
352.0	3.57867	1.1068	5.3010	308.09	412.40	104.31	1.3358	1.6321
353.0	3.64988	1.1176	5.1396	309.94	411.83	101.89	1.3408	1.6295
354.0	3.72220	1.1289	4.9803	311.83	411.21	99.38	1.3460	1.6267
355.0	3.79567	1.1409	4.8228	313.76	410.54	96.77	1.3512	1.6238
356.0	3.87029	1.1538	4.6669	315.74	409.80	94.06	1.3565	1.6207
357.0	3.94610	1.1675	4.5123	317.77	409.00	91.23	1.3619	1.6175
358.0	4.02310	1.1823	4.3588	319.86	408.12	88.26	1.3675	1.6141
359.0	4.10133	1.1984	4.2061	322.02	407.16	85.14	1.3733	1.6105
360.0	4.18080	1.2159	4.0536	324.26	406.10	81.84	1.3793	1.6066
361.0	4.26156	1.2351	3.9009	326.59	404.92	78.33	1.3854	1.6024
362.0	4.34361	1.2565	3.7475	329.03	403.61	74.59	1.3919	1.5979
363.0	4.42700	1.2806	3.5924	331.60	402.14	70.55	1.3987	1.5931
364.0	4.51175	1.3080	3.4348	334.34	400.48	66.14	1.4060	1.5877
365.0	4.59789	1.3401	3.2730	337.30	398.57	61.28	1.4137	1.5816
366.0	4.68547	1.3785	3.1048	340.54	396.34	55.79	1.4223	1.5748
367.0	4.77451	1.4264	2.9266	344.21	393.65	49.44	1.4320	1.5667
368.0	4.86506	1.4900	2.7315	348.55	390.27	41.72	1.4434	1.5568
369.0	4.95714	1.5848	2.5030	354.14	385.60	31.46	1.4582	1.5435

Chart R22.2: Saturation properties

p MPa	T K	v' dm³/kg	v'' dm³/kg	h' kJ/kg	h'' kJ/kg	r kJ/kg	s' kJ/kgK	s'' kJ/kgK
0.0031	178.07	0.6383	5508.3	94.50	361.32	266.82	0.5279	2.0263
0.0032	178.44	0.6388	5346.6	94.89	361.49	266.61	0.5301	2.0242
0.0033	178.79	0.6392	5194.5	95.27	361.67	266.40	0.5322	2.0222
0.0034	179.13	0.6396	5051.1	95.64	361.83	266.19	0.5343	2.0203
0.0035	179.47	0.6400	4915.6	96.00	362.00	266.00	0.5363	2.0185
0.0036	179.80	0.6404	4787.5	96.36	362.16	265.80	0.5383	2.0166
0.0037	180.12	0.6408	4666.1	96.70	362.31	265.61	0.5402	2.0149
0.0038	180.43	0.6412	4550.9	97.04	362.47	265.43	0.5421	2.0132
0.0039	180.74	0.6416	4441.4	97.37	362.62	265.25	0.5439	2.0115
0.0040	181.04	0.6419	4337.2	97.69	362.76	265.07	0.5457	2.0099
0.0041	181.33	0.6423	4238.0	98.01	362.91	264.90	0.5475	2.0083
0.0042	181.62	0.6426	4143.4	98.32	363.05	264.73	0.5492	2.0068
0.0043	181.90	0.6430	4053.0	98.63	363.19	264.56	0.5509	2.0053
0.0044	182.17	0.6433	3966.7	98.93	363.32	264.40	0.5525	2.0038
0.0045	182.44	0.6436	3884.0	99.22	363.45	264.23	0.5541	2.0024
0.0046	182.71	0.6440	3804.9	99.51	363.58	264.08	0.5557	2.0010
0.0047	182.97	0.6443	3729.0	99.79	363.71	263.92	0.5572	1.9997
0.0048	183.23	0.6446	3656.2	100.07	363.84	263.77	0.5587	1.9983
0.0049	183.48	0.6449	3586.3	100.34	363.96	263.62	0.5602	1.9970
0.0050	183.73	0.6452	3519.1	100.61	364.08	263.47	0.5617	1.9958
0.0051	183.97	0.6455	3454.4	100.87	364.20	263.33	0.5631	1.9945
0.0052	184.21	0.6458	3392.2	101.13	364.32	263.19	0.5645	1.9933
0.0053	184.44	0.6461	3332.3	101.39	364.43	263.05	0.5659	1.9921
0.0054	184.68	0.6464	3274.5	101.64	364.55	262.91	0.5673	1.9909
0.0055	184.90	0.6466	3218.7	101.89	364.66	262.77	0.5686	1.9898
0.0056	185.13	0.6469	3164.9	102.13	364.77	262.64	0.5699	1.9886
0.0057	185.35	0.6472	3112.9	102.37	364.88	262.51	0.5712	1.9875
0.0058	185.57	0.6474	3062.6	102.61	364.99	262.38	0.5725	1.9864
0.0059	185.78	0.6477	3014.0	102.84	365.09	262.25	0.5738	1.9854
0.0060	185.99	0.6480	2967.0	103.07	365.19	262.12	0.5750	1.9843
0.0061	186.20	0.6482	2921.5	103.30	365.30	262.00	0.5762	1.9833
0.0062	186.41	0.6485	2877.3	103.52	365.40	261.88	0.5774	1.9823
0.0063	186.61	0.6487	2834.6	103.74	365.50	261.76	0.5786	1.9813
0.0064	186.81	0.6490	2793.1	103.96	365.60	261.64	0.5798	1.9803
0.0065	187.01	0.6492	2752.9	104.18	365.69	261.52	0.5809	1.9793
0.0066	187.21	0.6494	2713.9	104.39	365.79	261.40	0.5821	1.9784
0.0067	187.40	0.6497	2676.0	104.60	365.89	261.29	0.5832	1.9775
0.0068	187.59	0.6499	2639.2	104.80	365.98	261.17	0.5843	1.9765
0.0069	187.78	0.6501	2603.4	105.01	366.07	261.06	0.5854	1.9756
0.0070	187.97	0.6504	2568.6	105.21	366.16	260.95	0.5865	1.9747
0.0071	188.15	0.6506	2534.7	105.41	366.25	260.84	0.5875	1.9739
0.0072	188.33	0.6508	2501.8	105.61	366.34	260.73	0.5886	1.9730
0.0073	188.51	0.6510	2469.8	105.80	366.43	260.63	0.5896	1.9722
0.0074	188.69	0.6512	2438.5	106.00	366.52	260.52	0.5906	1.9713
0.0076	189.04	0.6517	2378.5	106.38	366.69	260.31	0.5926	1.9697
0.0078	189.38	0.6521	2321.4	106.75	366.86	260.11	0.5946	1.9681
0.0080	189.71	0.6525	2267.1	107.11	367.02	259.91	0.5965	1.9665
0.0082	190.04	0.6529	2215.4	107.47	367.18	259.71	0.5984	1.9650
0.0084	190.36	0.6533	2166.1	107.82	367.34	259.52	0.6002	1.9635
0.0086	190.68	0.6537	2119.0	108.16	367.50	259.33	0.6020	1.9621

Chart R22.2: Saturation properties–*continued*

p MPa	T K	v' dm³/kg	v'' dm³/kg	h' kJ/kg	h'' kJ/kg	r kJ/kg	s' kJ/kgK	s'' kJ/kgK
0.0088	190.99	0.6541	2074.0	108.50	367.65	259.15	0.6038	1.9607
0.0090	191.29	0.6544	2030.9	108.83	367.80	258.97	0.6055	1.9593
0.0092	191.59	0.6548	1989.6	109.15	367.94	258.79	0.6072	1.9580
0.0094	191.88	0.6552	1950.0	109.47	368.09	258.62	0.6089	1.9567
0.0096	192.16	0.6555	1912.1	109.78	368.23	258.45	0.6105	1.9554
0.0098	192.45	0.6559	1875.6	110.09	368.37	258.28	0.6121	1.9542
0.0100	192.72	0.6562	1840.5	110.39	368.50	258.11	0.6136	1.9529
0.0102	193.00	0.6565	1806.8	110.69	368.64	257.95	0.6152	1.9517
0.0104	193.26	0.6569	1774.3	110.98	368.77	257.79	0.6167	1.9506
0.0106	193.53	0.6572	1743.1	111.26	368.90	257.63	0.6182	1.9494
0.0108	193.79	0.6575	1712.9	111.55	369.02	257.48	0.6196	1.9483
0.0110	194.04	0.6578	1683.8	111.83	369.15	257.32	0.6211	1.9472
0.0115	194.67	0.6586	1615.4	112.50	369.46	256.95	0.6246	1.9445
0.0120	195.27	0.6593	1552.5	113.16	369.75	256.59	0.6279	1.9420
0.0125	195.85	0.6601	1494.4	113.79	370.04	256.25	0.6312	1.9395
0.0130	196.41	0.6608	1440.7	114.40	370.31	255.91	0.6343	1.9372
0.0135	196.95	0.6614	1390.9	115.00	370.58	255.58	0.6373	1.9350
0.0140	197.48	0.6621	1344.5	115.57	370.84	255.27	0.6402	1.9328
0.0145	197.99	0.6627	1301.2	116.13	371.09	254.96	0.6430	1.9308
0.0150	198.49	0.6634	1260.7	116.67	371.34	254.66	0.6458	1.9288
0.0155	198.97	0.6640	1222.7	117.20	371.57	254.37	0.6484	1.9268
0.0160	199.45	0.6646	1187.0	117.72	371.81	254.09	0.6510	1.9250
0.0165	199.91	0.6651	1153.5	118.22	372.03	253.81	0.6535	1.9232
0.0170	200.35	0.6657	1121.8	118.71	372.25	253.55	0.6559	1.9214
0.0175	200.79	0.6663	1091.9	119.18	372.47	253.28	0.6583	1.9197
0.0180	201.22	0.6668	1063.6	119.65	372.68	253.03	0.6606	1.9181
0.0185	201.64	0.6673	1036.8	120.11	372.88	252.77	0.6629	1.9165
0.0190	202.05	0.6679	1011.3	120.55	373.08	252.53	0.6651	1.9150
0.0195	202.45	0.6684	987.11	120.99	373.28	252.29	0.6673	1.9135
0.0200	202.84	0.6689	964.09	121.42	373.47	252.05	0.6694	1.9120
0.0210	203.60	0.6698	921.23	122.25	373.84	251.60	0.6735	1.9092
0.0220	204.33	0.6708	882.15	123.04	374.20	251.16	0.6774	1.9065
0.0230	205.03	0.6717	846.35	123.81	374.54	250.73	0.6811	1.9040
0.0240	205.71	0.6725	813.45	124.55	374.88	250.32	0.6847	1.9016
0.0250	206.37	0.6734	783.09	125.27	375.20	249.93	0.6882	1.8993
0.0260	207.01	0.6742	755.00	125.96	375.51	249.54	0.6915	1.8970
0.0270	207.62	0.6750	728.91	126.64	375.81	249.17	0.6948	1.8949
0.0280	208.22	0.6758	704.63	127.29	376.10	248.81	0.6979	1.8929
0.0290	208.80	0.6765	681.97	127.92	376.38	248.46	0.7010	1.8909
0.0300	209.36	0.6773	660.77	128.54	376.66	248.12	0.7039	1.8890
0.0310	209.91	0.6780	640.90	129.14	376.92	247.78	0.7068	1.8872
0.0320	210.45	0.6787	622.22	129.73	377.18	247.46	0.7095	1.8854
0.0330	210.97	0.6794	604.64	130.30	377.44	247.14	0.7123	1.8837
0.0340	211.48	0.6801	588.06	130.85	377.69	246.83	0.7149	1.8821
0.0350	211.98	0.6807	572.40	131.40	377.93	246.53	0.7174	1.8805
0.0360	212.46	0.6814	557.57	131.93	378.16	246.24	0.7199	1.8789
0.0370	212.94	0.6820	543.52	132.45	378.39	245.95	0.7224	1.8774
0.0380	213.40	0.6826	530.19	132.95	378.62	245.66	0.7248	1.8759
0.0390	213.86	0.6832	517.52	133.45	378.84	245.39	0.7271	1.8745
0.0400	214.31	0.6838	505.46	133.94	379.05	245.11	0.7294	1.8731

Chart R22.2: Saturation properties–*continued*

p MPa	T K	v' dm^3/kg	v'' dm^3/kg	h' kJ/kg	h'' kJ/kg	r kJ/kg	s' kJ/kgK	s'' kJ/kgK
0.0410	214.74	0.6844	493.96	134.42	379.26	244.85	0.7316	1.8718
0.0420	215.17	0.6850	483.00	134.89	379.47	244.59	0.7338	1.8705
0.0430	215.59	0.6856	472.53	135.35	379.67	244.33	0.7359	1.8692
0.0440	216.00	0.6861	462.52	135.80	379.87	244.08	0.7380	1.8679
0.0450	216.41	0.6867	452.94	136.24	380.07	243.83	0.7400	1.8667
0.0460	216.81	0.6872	443.76	136.67	380.26	243.58	0.7420	1.8655
0.0470	217.20	0.6877	434.95	137.10	380.45	243.34	0.7440	1.8644
0.0480	217.58	0.6883	426.51	137.52	380.63	243.11	0.7459	1.8633
0.0490	217.96	0.6888	418.39	137.94	380.81	242.88	0.7478	1.8621
0.0500	218.33	0.6893	410.59	138.34	380.99	242.65	0.7497	1.8611
0.0510	218.70	0.6898	403.09	138.74	381.17	242.42	0.7515	1.8600
0.0520	219.06	0.6903	395.86	139.14	381.34	242.20	0.7533	1.8590
0.0530	219.41	0.6908	388.90	139.53	381.51	241.98	0.7551	1.8580
0.0540	219.76	0.6912	382.18	139.91	381.68	241.77	0.7568	1.8570
0.0550	220.11	0.6917	375.70	140.29	381.84	241.55	0.7586	1.8560
0.0560	220.45	0.6922	369.45	140.66	382.00	241.34	0.7602	1.8550
0.0570	220.78	0.6927	363.40	141.03	382.16	241.14	0.7619	1.8541
0.0580	221.11	0.6931	357.56	141.39	382.32	240.93	0.7635	1.8532
0.0590	221.44	0.6936	351.91	141.74	382.47	240.73	0.7651	1.8523
0.0600	221.76	0.6940	346.44	142.10	382.63	240.53	0.7667	1.8514
0.0610	222.07	0.6945	341.14	142.44	382.78	240.33	0.7683	1.8505
0.0620	222.39	0.6949	336.01	142.79	382.93	240.14	0.7698	1.8497
0.0630	222.69	0.6953	331.04	143.12	383.07	239.95	0.7713	1.8488
0.0640	223.00	0.6958	326.21	143.46	383.22	239.76	0.7728	1.8480
0.0650	223.30	0.6962	321.53	143.79	383.36	239.57	0.7743	1.8472
0.0660	223.60	0.6966	316.99	144.12	383.50	239.39	0.7758	1.8464
0.0670	223.89	0.6970	312.57	144.44	383.64	239.20	0.7772	1.8456
0.0680	224.18	0.6974	308.29	144.76	383.78	239.02	0.7786	1.8448
0.0690	224.47	0.6978	304.12	145.07	383.91	238.84	0.7800	1.8441
0.0700	224.75	0.6982	300.06	145.38	384.05	238.66	0.7814	1.8433
0.0710	225.03	0.6986	296.12	145.69	384.18	238.49	0.7828	1.8426
0.0720	225.31	0.6990	292.28	146.00	384.31	238.32	0.7841	1.8419
0.0730	225.58	0.6994	288.55	146.30	384.44	238.14	0.7855	1.8411
0.0740	225.85	0.6998	284.91	146.59	384.57	237.97	0.7868	1.8404
0.0760	226.39	0.7006	277.91	147.18	384.82	237.64	0.7894	1.8391
0.0780	226.91	0.7013	271.25	147.76	385.06	237.31	0.7919	1.8377
0.0800	227.42	0.7021	264.92	148.32	385.30	236.99	0.7944	1.8364
0.0820	227.92	0.7028	258.89	148.87	385.54	236.67	0.7968	1.8352
0.0840	228.42	0.7035	253.13	149.41	385.77	236.36	0.7992	1.8339
0.0860	228.90	0.7042	247.63	149.95	386.00	236.05	0.8015	1.8327
0.0880	229.37	0.7049	242.37	150.47	386.22	235.75	0.8038	1.8315
0.0900	229.84	0.7056	237.34	150.98	386.44	235.45	0.8060	1.8304
0.0920	230.30	0.7063	232.52	151.49	386.65	235.16	0.8082	1.8293
0.0940	230.75	0.7069	227.89	151.98	386.86	234.87	0.8103	1.8282
0.0960	231.19	0.7076	223.46	152.47	387.06	234.59	0.8124	1.8271
0.0980	231.63	0.7082	219.19	152.95	387.26	234.31	0.8145	1.8261
0.1000	232.06	0.7089	215.09	153.43	387.46	234.04	0.8165	1.8251
0.1020	232.48	0.7095	211.15	153.89	387.66	233.77	0.8185	1.8241
0.1040	232.89	0.7101	207.35	154.35	387.85	233.50	0.8205	1.8231
0.1060	233.30	0.7107	203.69	154.80	388.04	233.24	0.8224	1.8221

Chart R22.2: Saturation properties—*continued*

p MPa	T K	v' dm³/kg	v'' dm³/kg	h' kJ/kg	h'' kJ/kg	r kJ/kg	s' kJ/kgK	s'' kJ/kgK
0.1080	233.71	0.7113	200.16	155.25	388.22	232.98	0.8243	1.8212
0.1100	234.10	0.7119	196.76	155.69	388.41	232.72	0.8262	1.8203
0.1150	235.07	0.7134	188.75	156.76	388.85	232.09	0.8308	1.8181
0.1200	236.01	0.7148	181.38	157.80	389.28	231.48	0.8351	1.8160
0.1250	236.91	0.7162	174.58	158.80	389.69	230.89	0.8394	1.8140
0.1300	237.79	0.7176	168.28	159.77	390.09	230.32	0.8435	1.8120
0.1350	238.64	0.7189	162.43	160.72	390.48	229.75	0.8474	1.8102
0.1400	239.47	0.7202	156.99	161.64	390.85	229.21	0.8513	1.8084
0.1450	240.28	0.7214	151.91	162.54	391.21	228.67	0.8550	1.8067
0.1500	241.06	0.7227	147.15	163.41	391.56	228.15	0.8586	1.8050
0.1550	241.83	0.7239	142.69	164.26	391.90	227.64	0.8621	1.8034
0.1600	242.57	0.7251	138.50	165.09	392.23	227.14	0.8655	1.8019
0.1650	243.30	0.7263	134.56	165.90	392.56	226.66	0.8688	1.8004
0.1700	244.01	0.7274	130.84	166.69	392.87	226.18	0.8721	1.7990
0.1750	244.70	0.7285	127.32	167.47	393.18	225.71	0.8752	1.7976
0.1800	245.38	0.7296	123.99	168.23	393.47	225.25	0.8783	1.7963
0.1850	246.04	0.7307	120.83	168.97	393.77	224.79	0.8813	1.7950
0.1900	246.69	0.7318	117.84	169.70	394.05	224.35	0.8843	1.7937
0.1950	247.33	0.7329	114.99	170.42	394.33	223.91	0.8872	1.7925
0.2000	247.96	0.7339	112.28	171.12	394.60	223.48	0.8900	1.7913
0.2100	249.17	0.7359	107.23	172.48	395.12	222.64	0.8954	1.7890
0.2200	250.34	0.7379	102.63	173.80	395.63	221.82	0.9007	1.7868
0.2300	251.47	0.7398	98.408	175.08	396.11	221.03	0.9057	1.7847
0.2400	252.57	0.7417	94.527	176.31	396.57	220.26	0.9106	1.7827
0.2500	253.62	0.7436	90.945	177.51	397.02	219.51	0.9153	1.7808
0.2600	254.65	0.7454	87.629	178.67	397.45	218.78	0.9199	1.7790
0.2700	255.65	0.7471	84.549	179.80	397.86	218.06	0.9243	1.7772
0.2800	256.62	0.7489	81.682	180.90	398.27	217.36	0.9285	1.7756
0.2900	257.56	0.7506	79.004	181.98	398.65	216.68	0.9327	1.7739
0.3000	258.48	0.7522	76.499	183.03	399.03	216.00	0.9367	1.7724
0.3100	259.37	0.7539	74.149	184.05	399.39	215.35	0.9406	1.7709
0.3200	260.25	0.7555	71.941	185.05	399.75	214.70	0.9444	1.7694
0.3300	261.10	0.7570	69.861	186.02	400.09	214.07	0.9482	1.7680
0.3400	261.93	0.7586	67.899	186.98	400.42	213.45	0.9518	1.7667
0.3500	262.75	0.7601	66.046	187.92	400.75	212.83	0.9553	1.7654
0.3600	263.54	0.7616	64.291	188.83	401.06	212.23	0.9588	1.7641
0.3700	264.32	0.7631	62.627	189.73	401.37	211.64	0.9622	1.7628
0.3800	265.09	0.7646	61.048	190.62	401.67	211.06	0.9655	1.7616
0.3900	265.84	0.7660	59.547	191.48	401.96	210.48	0.9687	1.7605
0.4000	266.57	0.7675	58.118	192.33	402.25	209.91	0.9719	1.7593
0.4100	267.30	0.7689	56.756	193.17	402.52	209.36	0.9750	1.7582
0.4200	268.00	0.7703	55.456	193.99	402.80	208.80	0.9780	1.7571
0.4300	268.70	0.7717	54.215	194.80	403.06	208.26	0.9810	1.7561
0.4400	269.38	0.7730	53.027	195.60	403.32	207.72	0.9839	1.7550
0.4500	270.06	0.7744	51.891	196.38	403.57	207.19	0.9868	1.7540
0.4600	270.72	0.7757	50.802	197.15	403.82	206.67	0.9896	1.7530
0.4700	271.37	0.7771	49.758	197.91	404.06	206.15	0.9924	1.7521
0.4800	272.01	0.7784	48.755	198.66	404.30	205.64	0.9951	1.7511
0.4900	272.64	0.7797	47.792	199.40	404.53	205.13	0.9978	1.7502
0.5000	273.26	0.7810	46.866	200.12	404.75	204.63	1.0004	1.7493

Chart R22.2: Saturation properties–*continued*

p MPa	T K	v' dm^3/kg	v'' dm^3/kg	h' kJ/kg	h'' kJ/kg	r kJ/kg	s' kJ/kgK	s'' kJ/kgK
0.5100	273.87	0.7822	45.975	200.84	404.98	204.14	1.0030	1.7484
0.5200	274.47	0.7835	45.117	201.55	405.19	203.64	1.0056	1.7476
0.5300	275.06	0.7848	44.290	202.25	405.41	203.16	1.0081	1.7467
0.5400	275.65	0.7860	43.492	202.94	405.62	202.68	1.0106	1.7459
0.5500	276.22	0.7872	42.722	203.62	405.82	202.20	1.0130	1.7451
0.5600	276.79	0.7885	41.979	204.29	406.02	201.73	1.0154	1.7443
0.5700	277.35	0.7897	41.261	204.96	406.22	201.26	1.0178	1.7435
0.5800	277.91	0.7909	40.566	205.62	406.41	200.80	1.0201	1.7427
0.5900	278.45	0.7921	39.894	206.27	406.60	200.34	1.0225	1.7419
0.6000	278.99	0.7933	39.244	206.91	406.79	199.88	1.0247	1.7412
0.6100	279.53	0.7944	38.614	207.54	406.97	199.43	1.0270	1.7404
0.6200	280.05	0.7956	38.004	208.17	407.15	198.98	1.0292	1.7397
0.6300	280.58	0.7968	37.412	208.79	407.33	198.54	1.0314	1.7390
0.6400	281.09	0.7979	36.838	209.41	407.50	198.10	1.0335	1.7383
0.6500	281.60	0.7991	36.281	210.02	407.67	197.66	1.0357	1.7376
0.6600	282.10	0.8002	35.741	210.62	407.84	197.22	1.0378	1.7369
0.6700	282.60	0.8014	35.215	211.21	408.01	196.79	1.0399	1.7362
0.6800	283.09	0.8025	34.705	211.80	408.17	196.36	1.0419	1.7356
0.6900	283.58	0.8036	34.209	212.39	408.33	195.94	1.0440	1.7349
0.7000	284.06	0.8047	33.726	212.97	408.48	195.52	1.0460	1.7343
0.7100	284.53	0.8058	33.257	213.54	408.64	195.10	1.0480	1.7336
0.7200	285.00	0.8070	32.800	214.11	408.79	194.68	1.0499	1.7330
0.7300	285.47	0.8081	32.355	214.67	408.94	194.27	1.0519	1.7324
0.7400	285.93	0.8092	31.921	215.23	409.08	193.85	1.0538	1.7318
0.7600	286.84	0.8113	31.088	216.33	409.37	193.04	1.0576	1.7306
0.7800	287.73	0.8135	30.295	217.41	409.65	192.23	1.0613	1.7294
0.8000	288.60	0.8156	29.540	218.48	409.92	191.44	1.0649	1.7283
0.8200	289.46	0.8177	28.820	219.53	410.18	190.65	1.0685	1.7271
0.8400	290.31	0.8198	28.134	220.56	410.43	189.87	1.0720	1.7261
0.8600	291.13	0.8219	27.478	221.57	410.68	189.10	1.0754	1.7250
0.8800	291.95	0.8240	26.851	222.58	410.91	188.34	1.0788	1.7239
0.9000	292.75	0.8260	26.250	223.56	411.14	187.58	1.0821	1.7229
0.9200	293.53	0.8281	25.675	224.53	411.37	186.83	1.0854	1.7219
0.9400	294.31	0.8301	25.123	225.49	411.58	186.09	1.0886	1.7209
0.9600	295.07	0.8321	24.593	226.44	411.80	185.36	1.0918	1.7199
0.9800	295.82	0.8341	24.084	227.37	412.00	184.63	1.0949	1.7190
1.0000	296.56	0.8361	23.594	228.30	412.20	183.90	1.0979	1.7181
1.0200	297.29	0.8381	23.123	229.21	412.39	183.19	1.1009	1.7171
1.0400	298.01	0.8401	22.669	230.11	412.58	182.48	1.1039	1.7162
1.0600	298.72	0.8420	22.231	230.99	412.76	181.77	1.1068	1.7153
1.0800	299.41	0.8440	21.809	231.87	412.94	181.07	1.1097	1.7144
1.1000	300.10	0.8459	21.402	232.74	413.11	180.37	1.1125	1.7136
1.1500	301.78	0.8507	20.443	234.87	413.52	178.65	1.1195	1.7115
1.2000	303.41	0.8555	19.561	236.94	413.90	176.96	1.1262	1.7094
1.2500	304.99	0.8603	18.746	238.97	414.26	175.29	1.1327	1.7074
1.3000	306.53	0.8650	17.992	240.94	414.59	173.65	1.1390	1.7055
1.3500	308.02	0.8697	17.290	242.87	414.89	172.02	1.1452	1.7036
1.4000	309.47	0.8743	16.637	244.76	415.17	170.41	1.1511	1.7018
1.4500	310.88	0.8790	16.027	246.61	415.44	168.82	1.1570	1.7000
1.5000	312.26	0.8836	15.455	248.43	415.68	167.25	1.1626	1.6983

Chart R22.2: Saturation properties–*continued*

p MPa	T K	v' dm^3/kg	v'' dm^3/kg	h' kJ/kg	h'' kJ/kg	r kJ/kg	s' kJ/kgK	s'' kJ/kgK
1.5500	313.60	0.8882	14.918	250.21	415.90	165.69	1.1682	1.6965
1.6000	314.91	0.8928	14.414	251.95	416.10	164.15	1.1736	1.6949
1.6500	316.20	0.8974	13.938	253.67	416.28	162.61	1.1789	1.6932
1.7000	317.45	0.9020	13.489	255.36	416.45	161.09	1.1841	1.6915
1.7500	318.68	0.9067	13.065	257.02	416.60	159.58	1.1892	1.6899
1.8000	319.88	0.9113	12.662	258.66	416.73	158.07	1.1942	1.6883
1.8500	321.06	0.9159	12.280	260.27	416.85	156.58	1.1991	1.6868
1.9000	322.21	0.9206	11.918	261.86	416.95	155.09	1.2039	1.6852
1.9500	323.34	0.9252	11.572	263.43	417.04	153.61	1.2086	1.6836
2.0000	324.45	0.9299	11.243	264.98	417.11	152.13	1.2132	1.6821
2.1000	326.61	0.9393	10.629	268.03	417.21	149.19	1.2223	1.6791
2.2000	328.70	0.9488	10.068	271.00	417.26	146.26	1.2311	1.6761
2.3000	330.71	0.9585	9.5516	273.91	417.25	143.34	1.2396	1.6731
2.4000	332.66	0.9683	9.0754	276.77	417.19	140.42	1.2479	1.6701
2.5000	334.55	0.9783	8.6343	279.58	417.08	137.50	1.2561	1.6671
2.6000	336.39	0.9885	8.2242	282.34	416.92	134.57	1.2640	1.6641
2.7000	338.17	0.9989	7.8418	285.07	416.70	131.63	1.2718	1.6611
2.8000	339.91	1.0096	7.4840	287.76	416.43	128.67	1.2794	1.6580
2.9000	341.59	1.0206	7.1482	290.42	416.11	125.69	1.2870	1.6549
3.0000	343.24	1.0320	6.8321	293.06	415.74	122.68	1.2944	1.6518
3.1000	344.84	1.0437	6.5338	295.68	415.30	119.63	1.3017	1.6486
3.2000	346.40	1.0558	6.2514	298.28	414.82	116.54	1.3089	1.6453
3.3000	347.93	1.0685	5.9834	300.87	414.27	113.40	1.3161	1.6420
3.4000	349.42	1.0817	5.7284	303.46	413.66	110.20	1.3232	1.6385
3.5000	350.88	1.0955	5.4850	306.05	412.98	106.93	1.3302	1.6350
3.6000	352.30	1.1100	5.2522	308.64	412.23	103.59	1.3373	1.6313
3.7000	353.69	1.1254	5.0287	311.25	411.40	100.16	1.3444	1.6276
3.8000	355.06	1.1417	4.8136	313.88	410.50	96.62	1.3515	1.6236
3.9000	356.39	1.1591	4.6059	316.53	409.50	92.96	1.3586	1.6195
4.0000	357.70	1.1778	4.4045	319.23	408.39	89.16	1.3659	1.6151
4.1000	358.98	1.1981	4.2086	321.98	407.18	85.19	1.3732	1.6105
4.2000	360.24	1.2203	4.0171	324.81	405.83	81.02	1.3807	1.6056
4.3000	361.47	1.2449	3.8289	327.72	404.32	76.60	1.3884	1.6004
4.4000	362.68	1.2725	3.6426	330.75	402.64	71.88	1.3965	1.5947
4.5000	363.86	1.3040	3.4567	333.95	400.72	66.77	1.4049	1.5884
4.6000	365.02	1.3409	3.2690	337.37	398.52	61.15	1.4139	1.5815
4.7000	366.16	1.3856	3.0764	341.11	395.93	54.82	1.4238	1.5735
4.8000	367.28	1.4424	2.8735	345.36	392.78	47.42	1.4350	1.5641
4.9000	368.38	1.5209	2.6499	350.47	388.69	38.22	1.4485	1.5523

Chart R22.3: Thermodynamic properties

p/MPa	0.20			0.30			0.40		
T K	v dm³/kg	h kJ/kg	s kJ/kgK	v dm³/kg	h kJ/kg	s kJ/kgK	v dm³/kg	h kJ/kg	s kJ/kgK
250.0	113.42	395.93	1.7966	0.7372	173.44	0.8990	0.7370	173.47	0.8988
260.0	118.94	402.44	1.8221	77.095	400.07	1.7764	0.7548	184.79	0.9432
270.0	124.34	408.99	1.8468	80.942	406.90	1.8022	59.164	404.69	1.7684
280.0	129.66	415.59	1.8709	84.689	413.73	1.8270	62.146	411.78	1.7942
290.0	134.91	422.26	1.8943	88.357	420.59	1.8511	65.037	418.85	1.8190
300.0	140.10	429.00	1.9171	91.961	427.49	1.8745	67.859	425.93	1.8430
310.0	145.25	435.83	1.9395	95.514	434.45	1.8973	70.623	433.04	1.8663
320.0	150.35	442.74	1.9614	99.024	441.49	1.9196	73.342	440.20	1.8891
330.0	155.42	449.74	1.9830	102.50	448.59	1.9415	76.022	447.42	1.9113
340.0	160.46	456.83	2.0042	105.94	455.77	1.9629	78.670	454.69	1.9330
350.0	165.47	464.02	2.0250	109.36	463.04	1.9840	81.291	462.04	1.9543
360.0	170.47	471.30	2.0455	112.76	470.39	2.0047	83.890	469.47	1.9752
370.0	175.45	478.67	2.0657	116.13	477.83	2.0251	86.468	476.97	1.9958
380.0	180.41	486.14	2.0856	119.49	485.35	2.0452	89.030	484.56	2.0160
390.0	185.35	493.70	2.1053	122.84	492.97	2.0649	91.576	492.22	2.0359
400.0	190.29	501.36	2.1246	126.17	500.67	2.0844	94.110	499.97	2.0555
410.0	195.21	509.10	2.1438	129.49	508.46	2.1037	96.632	507.80	2.0749
420.0	200.13	516.94	2.1626	132.81	516.33	2.1226	99.145	515.72	2.0939
430.0	205.03	524.87	2.1813	136.11	524.29	2.1414	101.65	523.71	2.1128
440.0	209.93	532.88	2.1997	139.41	532.34	2.1599	104.14	531.79	2.1313
450.0	214.82	540.99	2.2179	142.70	540.47	2.1781	106.63	539.95	2.1497
460.0	219.71	549.17	2.2359	145.98	548.68	2.1962	109.11	548.19	2.1678
470.0	224.59	557.44	2.2537	149.26	556.98	2.2140	111.59	556.51	2.1857
p/MPa	0.50			0.60			0.80		
250.0	0.7368	173.50	0.8987	0.7367	173.54	0.8985	0.7363	173.60	0.8982
260.0	0.7546	184.81	0.9430	0.7544	184.84	0.9428	0.7540	184.90	0.9425
270.0	0.7742	196.33	0.9865	0.7739	196.35	0.9863	0.7734	196.39	0.9859
280.0	48.568	409.73	1.7673	39.466	407.56	1.7439	0.7949	208.13	1.0285
290.0	51.008	417.04	1.7929	41.620	415.14	1.7705	29.788	411.05	1.7322
300.0	53.369	424.32	1.8176	43.683	422.64	1.7959	31.508	419.07	1.7594
310.0	55.668	431.59	1.8415	45.678	430.09	1.8204	33.141	426.93	1.7852
320.0	57.916	438.88	1.8646	47.618	437.53	1.8440	34.707	434.71	1.8098
330.0	60.124	446.22	1.8872	49.513	444.99	1.8670	36.222	442.44	1.8336
340.0	62.297	453.60	1.9092	51.373	452.47	1.8893	37.696	450.16	1.8567
350.0	64.443	461.03	1.9308	53.203	460.00	1.9111	39.137	457.89	1.8791
360.0	66.564	468.54	1.9519	55.008	467.59	1.9325	40.550	465.65	1.9009
370.0	68.665	476.11	1.9727	56.791	475.23	1.9534	41.939	473.44	1.9223
380.0	70.748	483.75	1.9930	58.556	482.93	1.9740	43.309	481.27	1.9432
390.0	72.816	491.47	2.0131	60.306	490.71	1.9942	44.662	489.17	1.9637
400.0	74.870	499.27	2.0328	62.041	498.56	2.0140	46.000	497.12	1.9838
410.0	76.913	507.14	2.0523	63.765	506.48	2.0336	47.326	505.13	2.0036
420.0	78.945	515.10	2.0714	65.478	514.47	2.0529	48.640	513.21	2.0231
430.0	80.969	523.13	2.0903	67.182	522.54	2.0718	49.945	521.35	2.0422
440.0	82.984	531.24	2.1090	68.877	530.69	2.0906	51.242	529.57	2.0611
450.0	84.992	539.43	2.1274	70.566	538.91	2.1090	52.531	537.85	2.0797
460.0	86.994	547.70	2.1456	72.248	547.20	2.1273	53.813	546.20	2.0981
470.0	88.990	556.04	2.1635	73.923	555.57	2.1453	55.090	554.62	2.1162

Chart R22.3: Thermodynamic properties—*continued*

p/MPa	1.00			2.00			4.00		
T K	v dm^3/kg	h kJ/kg	s kJ/kgK	v dm^3/kg	h kJ/kg	s kJ/kgK	v dm^3/kg	h kJ/kg	s kJ/kgK
250.0	0.7360	173.66	0.8978	0.7343	173.98	0.8961	0.7310	174.62	0.8929
260.0	0.7536	184.95	0.9421	0.7516	185.23	0.9403	0.7478	185.81	0.9367
270.0	0.7729	196.44	0.9854	0.7705	196.67	0.9834	0.7659	197.16	0.9796
280.0	0.7943	208.17	1.0281	0.7914	208.33	1.0259	0.7858	208.71	1.0216
290.0	0.8184	220.20	1.0703	0.8147	220.28	1.0678	0.8079	220.51	1.0630
300.0	24.114	415.15	1.7279	0.8413	232.58	1.1095	0.8326	232.61	1.1040
310.0	25.556	423.53	1.7554	0.8724	245.34	1.1513	0.8609	245.09	1.1449
320.0	26.916	431.71	1.7814	0.9100	258.74	1.1938	0.8941	258.05	1.1861
330.0	28.214	439.76	1.8062	11.813	423.21	1.7007	0.9345	271.67	1.2280
340.0	29.465	447.75	1.8300	12.734	433.45	1.7313	0.9871	286.31	1.2717
350.0	30.677	455.70	1.8531	13.566	443.10	1.7593	1.0650	302.89	1.3197
360.0	31.859	463.64	1.8754	14.337	452.39	1.7855	4.7011	414.13	1.6311
370.0	33.016	471.59	1.8972	15.062	461.43	1.8102	5.5886	431.95	1.6800
380.0	34.151	479.57	1.9185	15.754	470.32	1.8339	6.2326	445.66	1.7165
390.0	35.268	487.59	1.9393	16.418	479.11	1.8568	6.7724	457.72	1.7479
400.0	36.369	495.65	1.9597	17.061	487.83	1.8788	7.2512	468.88	1.7761
410.0	37.458	503.76	1.9798	17.685	496.51	1.9003	7.6891	479.48	1.8023
420.0	38.534	511.92	1.9994	18.295	505.18	1.9212	8.0976	489.71	1.8270
430.0	39.601	520.15	2.0188	18.892	513.85	1.9416	8.4835	499.68	1.8504
440.0	40.659	528.43	2.0378	19.478	522.54	1.9616	8.8517	509.47	1.8729
450.0	41.709	536.78	2.0566	20.055	531.24	1.9811	9.2057	519.13	1.8946
460.0	42.752	545.19	2.0751	20.623	539.98	2.0003	9.5477	528.69	1.9157
470.0	43.789	553.67	2.0933	21.184	548.75	2.0192	9.8799	538.19	1.9361
p/MPa	6.00			8.00			10.00		
250.0	0.7279	175.29	0.8897	0.7250	175.97	0.8866	0.7221	176.67	0.8836
260.0	0.7441	186.41	0.9333	0.7407	187.03	0.9300	0.7375	187.68	0.9268
270.0	0.7617	197.68	0.9758	0.7576	198.23	0.9723	0.7539	198.81	0.9688
280.0	0.7807	209.13	1.0175	0.7760	209.60	1.0136	0.7715	210.10	1.0098
290.0	0.8017	220.81	1.0585	0.7960	221.16	1.0542	0.7907	221.56	1.0501
300.0	0.8249	232.75	1.0989	0.8179	232.96	1.0942	0.8116	233.24	1.0897
310.0	0.8510	245.00	1.1391	0.8423	245.03	1.1337	0.8346	245.16	1.1287
320.0	0.8810	257.63	1.1792	0.8698	257.42	1.1731	0.8601	257.35	1.1674
330.0	0.9162	270.77	1.2196	0.9014	270.20	1.2124	0.8889	269.86	1.2059
340.0	0.9592	284.57	1.2608	0.9385	283.47	1.2520	0.9219	282.76	1.2444
350.0	1.0149	299.39	1.3038	0.9836	297.40	1.2924	0.9605	296.13	1.2832
360.0	1.0954	315.96	1.3504	1.0413	312.27	1.3343	1.0071	310.12	1.3226
370.0	1.2445	336.77	1.4074	1.1212	328.62	1.3791	1.0656	324.94	1.3632
380.0	1.9649	381.05	1.5252	1.2474	347.62	1.4297	1.1429	340.92	1.4058
390.0	3.1312	421.63	1.6308	1.4861	371.63	1.4920	1.2517	358.56	1.4516
400.0	3.7545	441.63	1.6815	1.9215	400.60	1.5654	1.4122	378.35	1.5017
410.0	4.2233	457.06	1.7196	2.4185	426.25	1.6287	1.6403	399.90	1.5549
420.0	4.6172	470.48	1.7520	2.8404	446.13	1.6767	1.9201	421.37	1.6067
430.0	4.9656	482.77	1.7809	3.1933	462.52	1.7153	2.2107	440.98	1.6528
440.0	5.2831	494.34	1.8075	3.4997	476.93	1.7484	2.4845	458.35	1.6928
450.0	5.5779	505.43	1.8324	3.7741	490.11	1.7780	2.7351	473.89	1.7277
460.0	5.8554	516.17	1.8560	4.0252	502.47	1.8052	2.9645	488.12	1.7590
470.0	6.1192	526.66	1.8786	4.2587	514.26	1.8305	3.1764	501.39	1.7875

Chart R22.3: Thermodynamic properties—*continued*

p/MPa	15.00			20.00			35.00		
T K	v dm³/kg	h kJ/kg	s kJ/kgK	v dm³/kg	h kJ/kg	s kJ/kgK	v dm³/kg	h kJ/kg	s kJ/kgK
250.0	0.7156	178.47	0.8764	0.7096	180.33	0.8696	0.6945	186.22	0.8511
260.0	0.7300	189.35	0.9191	0.7232	191.12	0.9119	0.7063	196.77	0.8925
270.0	0.7452	200.35	0.9606	0.7376	201.99	0.9530	0.7187	207.38	0.9325
280.0	0.7615	211.47	1.0011	0.7528	212.98	0.9929	0.7317	218.08	0.9714
290.0	0.7790	222.73	1.0406	0.7690	224.09	1.0319	0.7453	228.86	1.0092
300.0	0.7978	234.17	1.0794	0.7862	235.34	1.0701	0.7594	239.74	1.0461
310.0	0.8181	245.79	1.1175	0.8045	246.75	1.1075	0.7742	250.73	1.0822
320.0	0.8401	257.61	1.1550	0.8242	258.31	1.1442	0.7898	261.84	1.1174
330.0	0.8642	269.66	1.1921	0.8453	270.05	1.1803	0.8060	273.05	1.1519
340.0	0.8908	281.95	1.2288	0.8682	281.98	1.2159	0.8230	284.38	1.1857
350.0	0.9205	294.52	1.2652	0.8929	294.09	1.2510	0.8409	295.82	1.2189
360.0	0.9538	307.41	1.3015	0.9200	306.42	1.2857	0.8596	307.36	1.2514
370.0	0.9918	320.64	1.3378	0.9496	318.96	1.3201	0.8792	319.02	1.2834
380.0	1.0357	334.29	1.3741	0.9822	331.73	1.3542	0.8998	330.77	1.3147
390.0	1.0871	348.41	1.4108	1.0184	344.75	1.3880	0.9215	342.63	1.3455
400.0	1.1483	363.05	1.4479	1.0588	358.02	1.4216	0.9443	354.59	1.3758
410.0	1.2214	378.24	1.4854	1.1039	371.56	1.4550	0.9682	366.64	1.4055
420.0	1.3086	393.94	1.5232	1.1545	385.35	1.4882	0.9934	378.77	1.4348
430.0	1.4109	410.03	1.5611	1.2110	399.37	1.5212	1.0198	390.99	1.4635
440.0	1.5273	426.25	1.5984	1.2737	413.57	1.5539	1.0476	403.28	1.4918
450.0	1.6546	442.32	1.6345	1.3426	427.89	1.5861	1.0767	415.64	1.5196
460.0	1.7885	457.99	1.6689	1.4172	442.25	1.6176	1.1072	428.05	1.5468
470.0	1.9250	473.11	1.7014	1.4967	456.55	1.6484	1.1391	440.52	1.5736

References for R22

[Barh66] BARHO, W.: Berechnung der Molwärme der Fluor-Chlor-Derivate von Methan und Ethan im Zustand idealer Gase; Dissertation Universität Karlsruhe (1966)

[Benn40] BENNING, A.F. AND MCHARNESS, R.C.: Thermodynamic Properties of Fluorochloromethanes and -Ethanes; Ind. Eng. Chem. 32(1940)4 497/499

[Benn40a] BENNING, A.F. AND MCHARNESS, R.C.: Thermodynamic Properties of Fluorochloromethanes and -Ethanes; Ind. Eng. Chem. 32(1940)5 698/701

[Benn40b] BENNING, A.F. AND MCHARNESS, R.C.: Thermodynamic Properties of Fluorochloromethanes and -Ethanes; Ind. Eng. Chem. 32(1940)6 814/816

[Benn40c] BENNING, A.F. ; MCHARNESS, R.C. ; MARKWOOD JR. ,W.H. AND SMITH, W.J.: Thermodynamic Properties of Fluorochloromethanes and -Ethanes; Ind. Eng. Chem. 32(1940)7 976/980

[Bier74] BIER, K. ; ERNST, G, AND MAURER, G.: Flow apparatus for measuring the heat capacity and the Joule-Thomson coefficient of gases; J. Chem. Therm. 6(1974)11 1027/1037

[Bier80] BIER, K. ; MAURER, G. AND SAND, H.: Bestimmung konsistenter zwischenmolekularer Potentiale realer Gase aus Meßwerten des isothermen Drosseleffektes und des 2. Virialkoeffizienten. Teil I: Meßwerte des isothermen Drosseleffektes von Propan, Schwefelhexafluorid, Difluorchlormethan und Propylen; Ber. Bunsen Ges. Phys. Chem. 84(1980)5 430/437

[Erns70] ERNST, G. AND BÜSSER, J.: Ideal and real gas state heat capacities cp of C_3H_8, $i-C_4H_{10}$, C_2F_5Cl, CH_2ClCF_3, $CF_2ClCFCl_2$ and CHF_2Cl; J. Chem. Therm. (1970)2 787/791

[Gruz75] GRUZDEV, V.A. AND SUMSKAJA, A.I.: Experimental Investigation of Isobaric Heat Capacity of Freons: R-11, R-12, R-13, R-21, R-22, R-23; Thermophysical Properties of Matter and Substances (1975)8 108/120

[Hwan61] HWANG, Y.T.: The constant heat capacities of gaseous tetrafluoromethane, chlorodifluoromethane, dichlorotetrafluoroethane and chloropentafluoroethane; Ph.D. Dissertation University of Michigan (1961)

[Jaeg73] JÄGER, H.P.: Dichte- und Dampfdruckmeßwerte von verschiedenen Kältemitteln; Die Kälte (1973)7 276/278

[Klet70] KLETSKII, A.V.: Thermophysical properties of Freon-22; Comittee of Standards, Measures and Measuring Instruments of the USSR Council on Ministers, Moskau (1970)

[KMR 81] - : Kältemaschinenregeln, 7. Auflage; Verlag C.F. Müller, Karlsruhe (1981)

[Kuma78] KUMAGAI, A. AND IWASAKI, H.: Pressure-volume- temperature relationships of several polar liquids; J. Chem. Eng. Data 23(1978)3 193/195

[Lagu66] LAGUTINA, L.M.: Experimental investigation of p-v-T properties of Freon-22; Kol. Teknika 43(1966)12 25/28

[Meye69] MEYER, K.J.: Über den Zusammenhang zwischen der Schallgeschwindigkeit und der Wärmeleitfähigkeit bei flüssigen Fluor-Chlor-Derivaten des Methans und Äthans; Kältetechnik-Klimatisierung 21(1969)9 270/275

[Mich56] MICHELS, A.: Unpublished p-v-T data from R22; Personal communication from A.Michels to D.E.Kvalnes, E.I. DuPont de Nemours, Wilmington Delaware (1956)

[Neil57] NEILSON, E.F. AND WHITE, D.: The Heat Capacity, Heat of Fusion, Heat of Transition and Heat of Vaporization of Chlorodifluoromethane between 16K and the Boiling Point; J. Am. Chem. Soc. 79(1957) 5618/5621

[Oguc79] OGUCHI, K. ; SAGARA, H. ; MATSUSHITA, Y. ; WATANABE, K. AND TANISHITA, I.: Experimental study of p-v-T properties for fluorocarbon refrigerant R22; Transactions of the Japan Society of Mechanical Engineers, series B 45(1979) 1522/1528

[Taka80] TAKAISHI, Y. ; UEMATSU, M. AND WATANABE, K.: PVT and vapor pressure data from R-22; private communication to R.B. Stewart (1980)

[Tani72] TANISHITA, I. ; WATANABE, K. ; KIJIMA, J. ; OGUCHI, K. ; KONDO, H. AND HAKASHIMA, A.: Status Report on the Thermodynamic Properties Research of Fluorocarbon Refrigerants; Second Report Keio University, Tokyo (1972)

[Zand68] ZANDER, M.: Pressure-volume-temperature behavior of Chlorodifluoromethane (Freon 22) in the gaseous and liquid states; Physikalisch-Techn. Bundesanstalt Braunschweig (1968)

R 22. ph

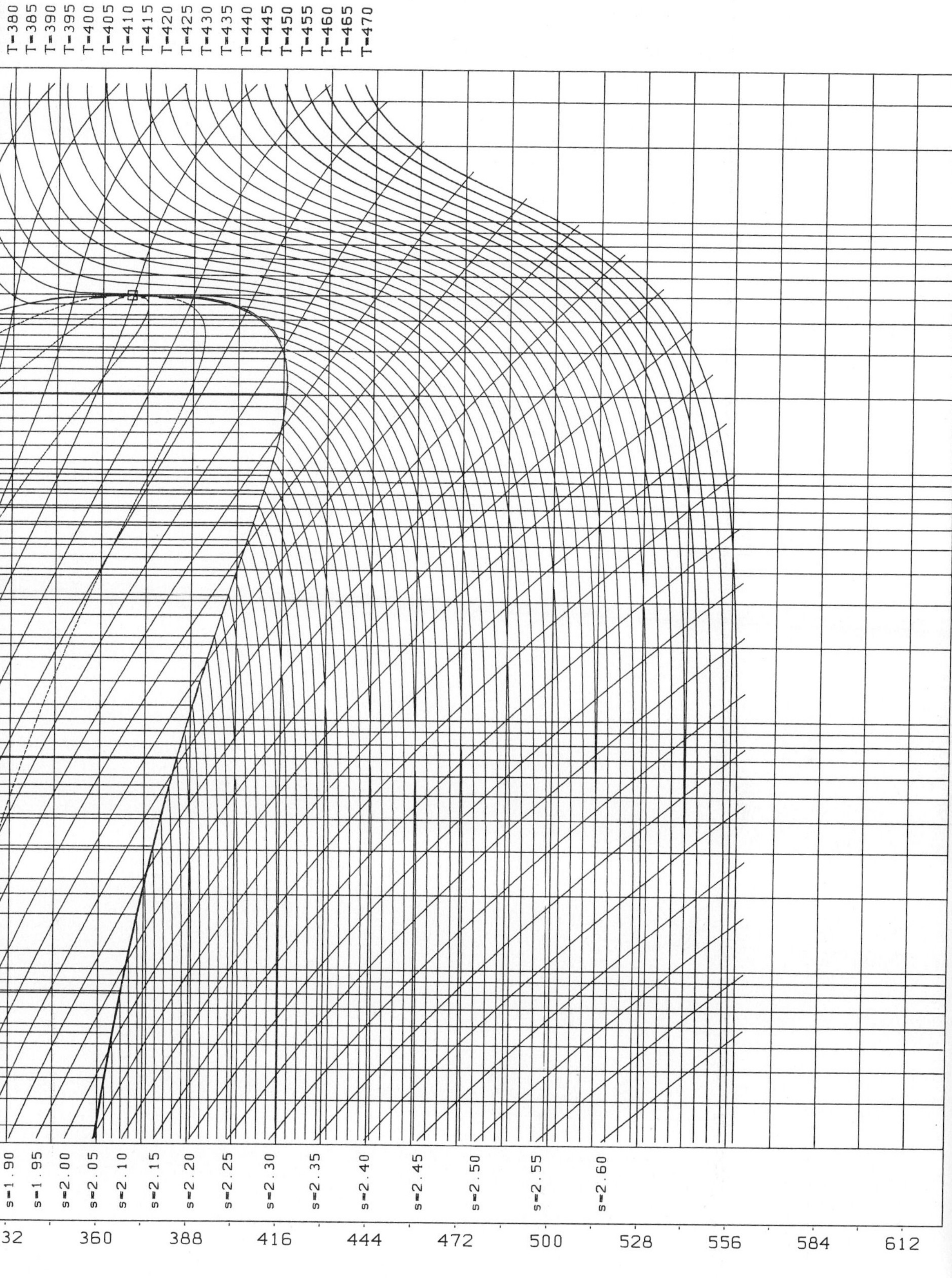
T=380
T=385
T=390
T=395
T=400
T=405
T=410
T=415
T=420
T=425
T=430
T=435
T=440
T=445
T=450
T=455
T=460
T=465
T=470
s=1.90
s=1.95
s=2.00
s=2.05
s=2.10
s=2.15
s=2.20
s=2.25
s=2.30
s=2.35
s=2.40
s=2.45
s=2.50
s=2.55
s=2.60
32
360
388
416
444
472
500
528
556
584
612
y h/(kJ/kg)

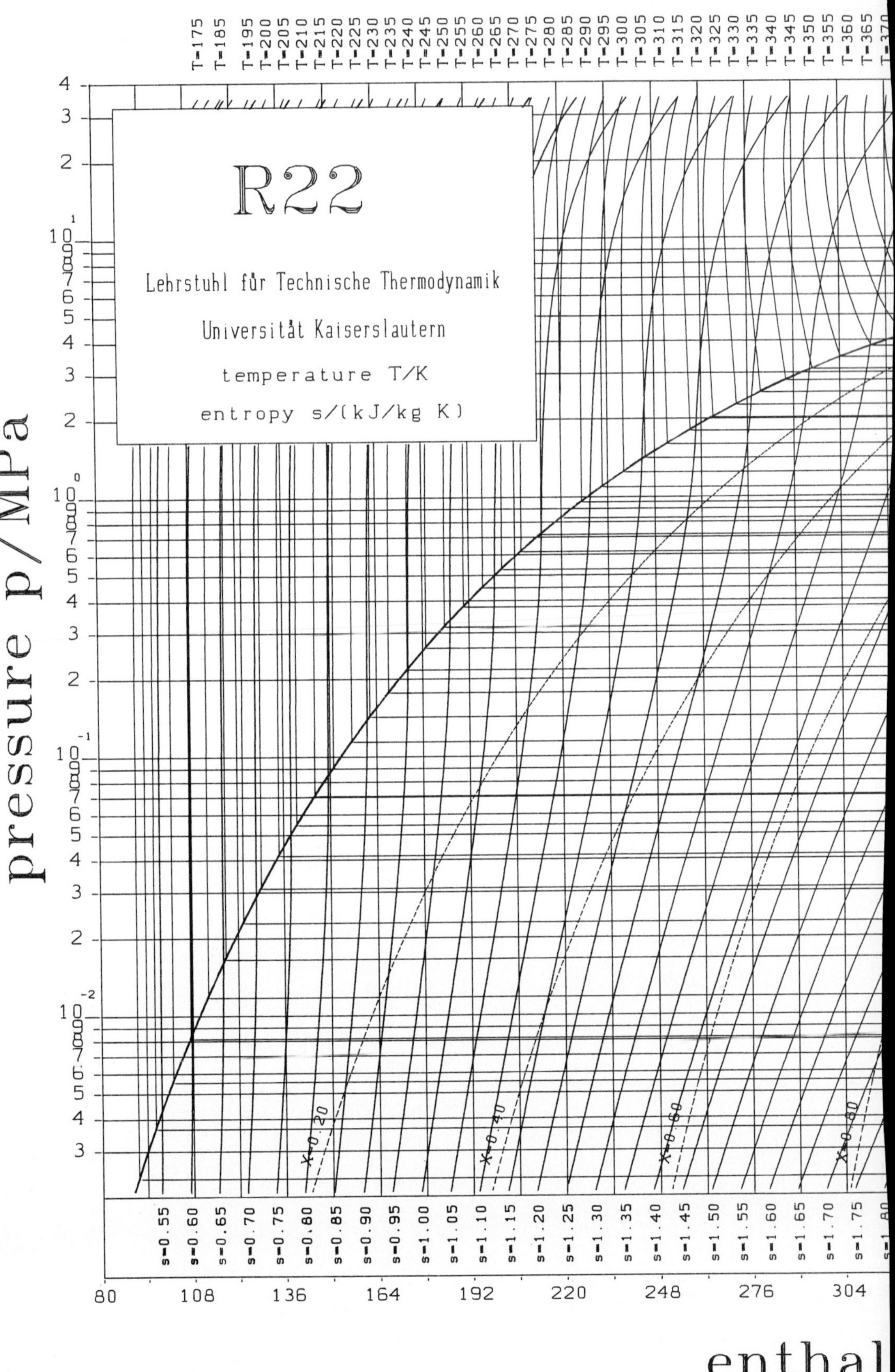
R22
Lehrstuhl für Technische Thermodynamik
Universität Kaiserslautern
temperature T/K
entropy s/(kJ/kg K)
pressure p/MPa
enthal

R 22.Ts

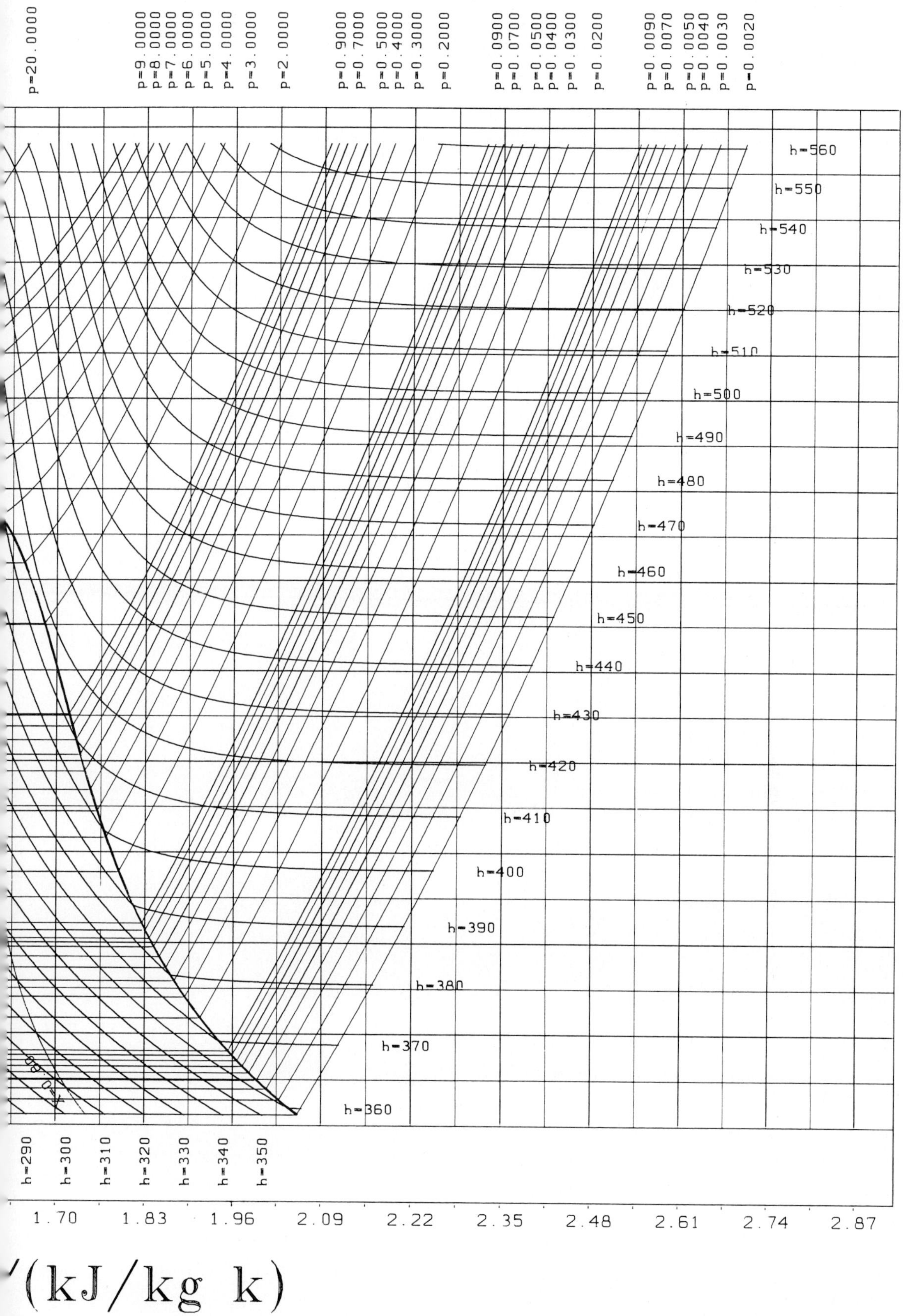

p=20.0000
p=9.0000
p=8.0000
p=7.0000
p=6.0000
p=5.0000
p=4.0000
p=3.0000
p=2.0000
p=0.9000
p=0.7000
p=0.5000
p=0.4000
p=0.3000
p=0.2000
p=0.0900
p=0.0700
p=0.0500
p=0.0400
p=0.0300
p=0.0200
p=0.0090
p=0.0070
p=0.0050
p=0.0040
p=0.0030
p=0.0020
h=560
h=550
h=540
h=530
h=520
h=510
h=500
h=490
h=480
h=470
h=460
h=450
h=440
h=430
h=420
h=410
h=400
h=390
h=380
h=370
h=360
h=290
h=300
h=310
h=320
h=330
h=340
h=350
1.70
1.83
1.96
2.09
2.22
2.35
2.48
2.61
2.74
2.87
'(kJ/kg k)

R 22. Ts

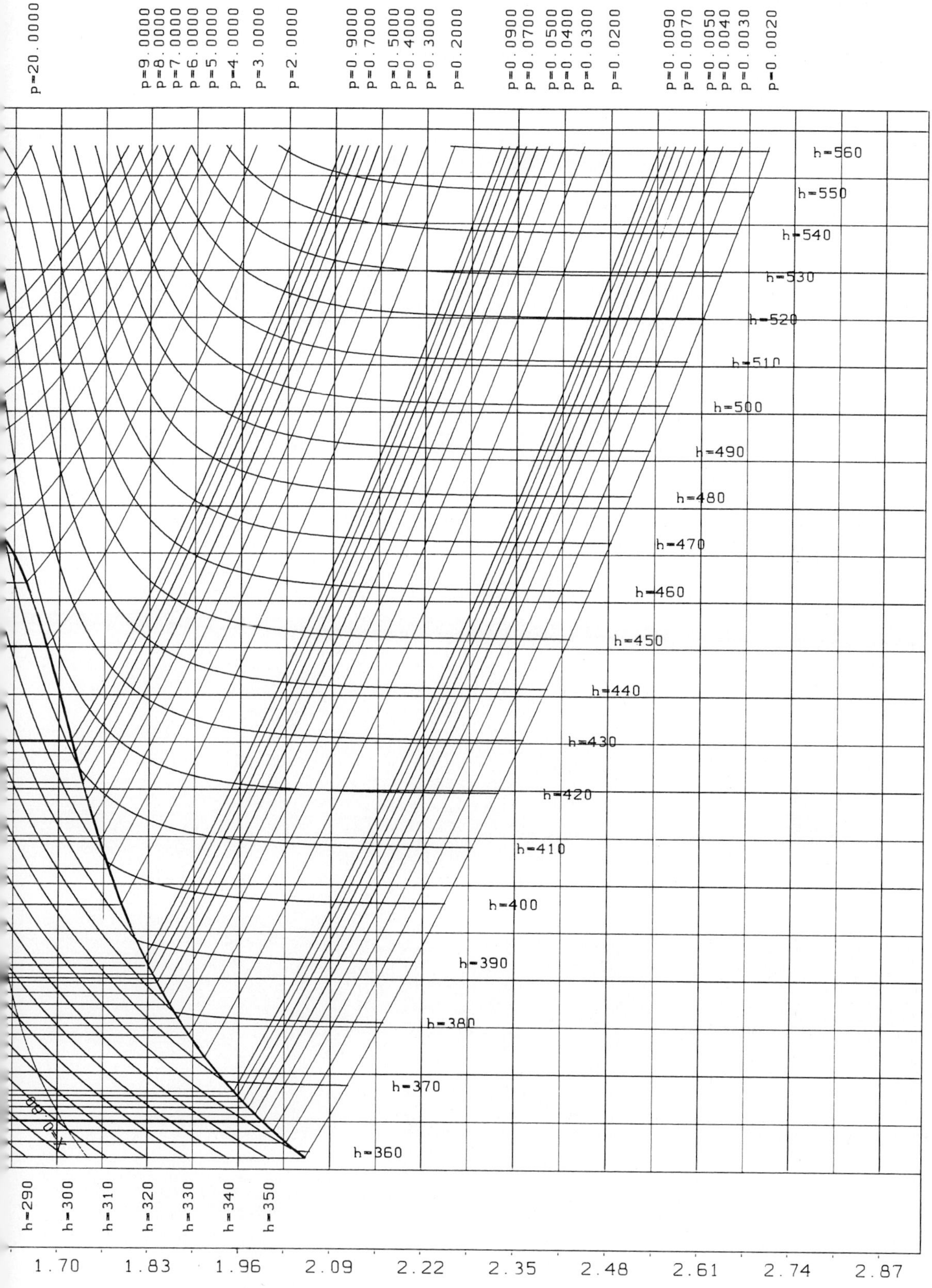

p=20.0000
p=9.0000
p=8.0000
p=7.0000
p=6.0000
p=5.0000
p=4.0000
p=3.0000
p=2.0000
p=0.9000
p=0.7000
p=0.5000
p=0.4000
p=0.3000
p=0.2000
p=0.0900
p=0.0700
p=0.0500
p=0.0400
p=0.0300
p=0.0200
p=0.0090
p=0.0070
p=0.0050
p=0.0040
p=0.0030
p=0.0020
h=560
h=550
h=540
h=530
h=520
h=510
h=500
h=490
h=480
h=470
h=460
h=450
h=440
h=430
h=420
h=410
h=400
h=390
h=380
h=370
h=360
h=290
h=300
h=310
h=320
h=330
h=340
h=350
1.70
1.83
1.96
2.09
2.22
2.35
2.48
2.61
2.74
2.87

/(kJ/kg k)

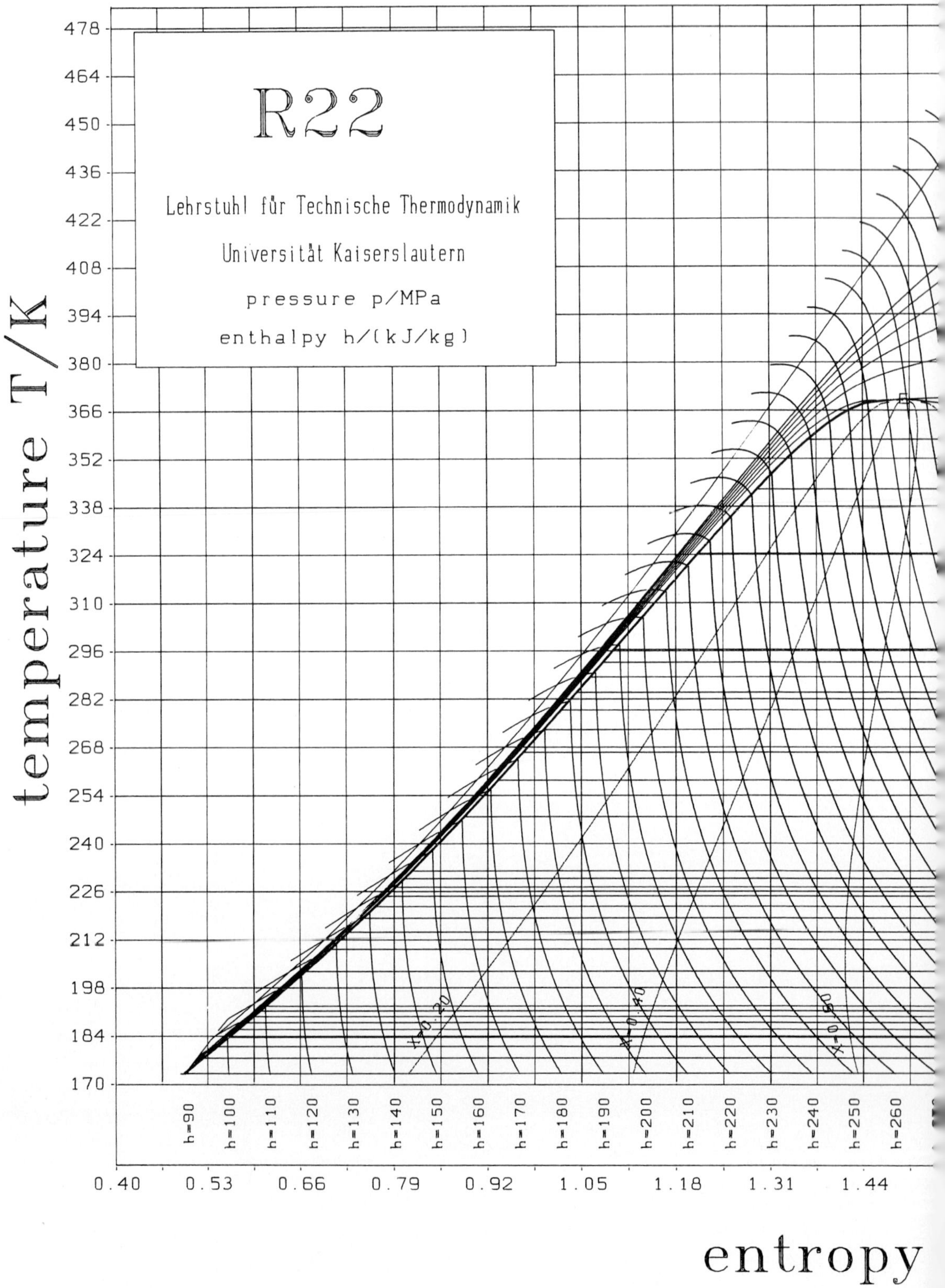
R22
Lehrstuhl für Technische Thermodynamik
Universität Kaiserslautern
pressure p/MPa
enthalpy h/(kJ/kg)
temperature T/K
478
464
450
436
422
408
394
380
366
352
338
324
310
296
282
268
254
240
226
212
198
184
170
X=0.20
X=0.40
X=0.60
h=90
h=100
h=110
h=120
h=130
h=140
h=150
h=160
h=170
h=180
h=190
h=200
h=210
h=220
h=230
h=240
h=250
h=260
0.40
0.53
0.66
0.79
0.92
1.05
1.18
1.31
1.44
entropy

R 22. hs

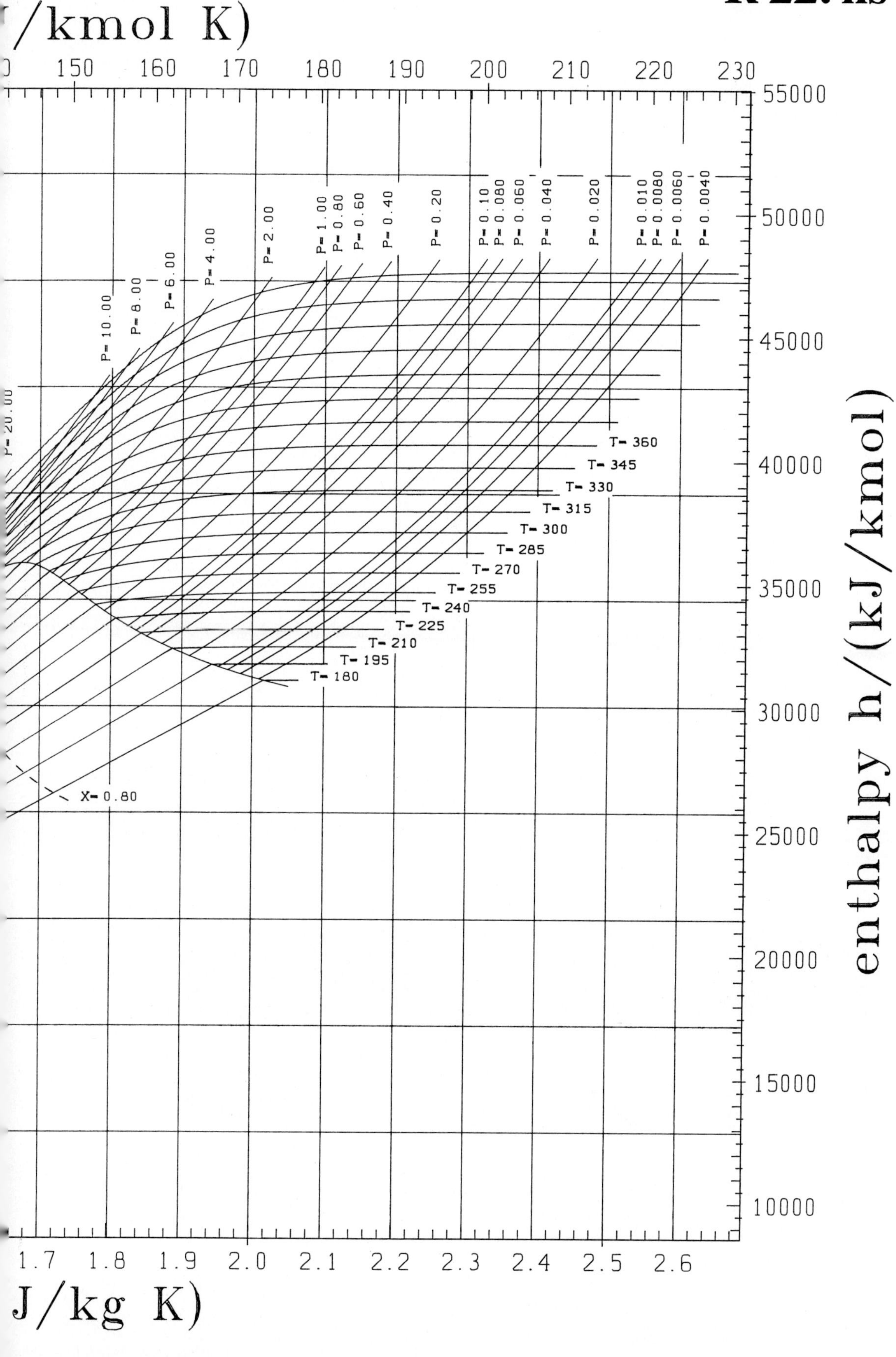

/kmol K)
150 160 170 180 190 200 210 220 230
55000
50000
45000
40000
35000
30000
25000
20000
15000
10000
enthalpy h/(kJ/kmol)
P= 20.00
P= 10.00
P= 8.00
P= 6.00
P= 4.00
P= 2.00
P= 1.00
P= 0.80
P= 0.60
P= 0.40
P= 0.20
P= 0.10
P= 0.080
P= 0.060
P= 0.040
P= 0.020
P= 0.010
P= 0.0080
P= 0.0060
P= 0.0040
T= 360
T= 345
T= 330
T= 315
T= 300
T= 285
T= 270
T= 255
T= 240
T= 225
T= 210
T= 195
T= 180
X= 0.80
1.7 1.8 1.9 2.0 2.1 2.2 2.3 2.4 2.5 2.6
J/kg K)

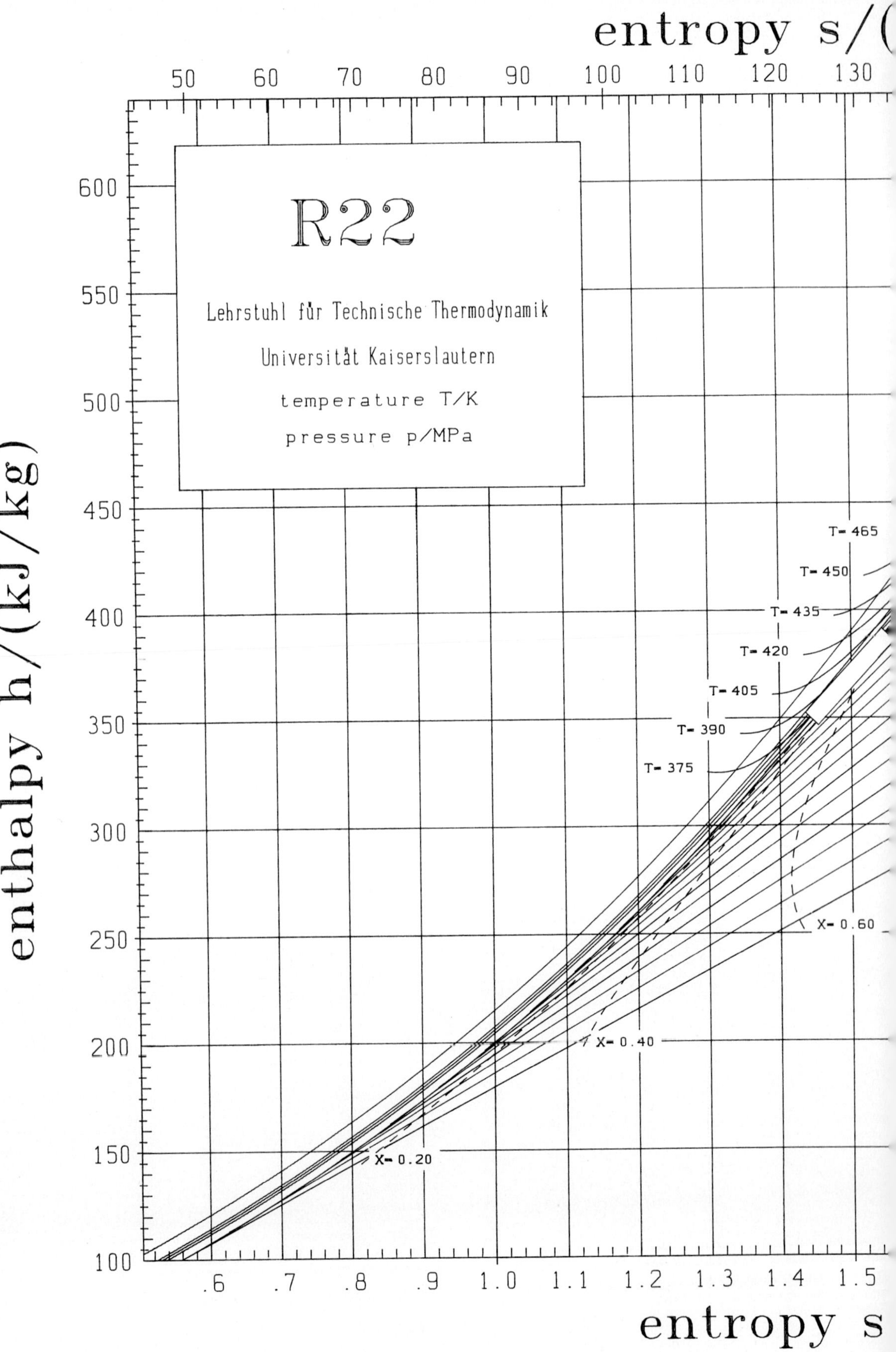

entropy s/(
50
60
70
80
90
100
110
120
130
R22
Lehrstuhl für Technische Thermodynamik
Universität Kaiserslautern
temperature T/K
pressure p/MPa
enthalpy h/(kJ/kg)
600
550
500
450
400
350
300
250
200
150
100
T= 465
T= 450
T= 435
T= 420
T= 405
T= 390
T= 375
X= 0.60
X= 0.40
X= 0.20
.6
.7
.8
.9
1.0
1.1
1.2
1.3
1.4
1.5
entropy s

Refrigerant R23 (CHF_3)

Table **R23.1:** List of references for p, ρ, T data

Author	n	p/MPa min	p/MPa max	ρ/g cm^{-3} min	ρ/g cm^{-3} max	T/K min	T/K max
BELZILLE, J.L. *[Belz 76]*	325	0.11	16.61	0.003	0.79	273	473
BIER, K. *[Bier 85]*	84	0.20	58.64	0.001	0.24	333	423
BÜHNER, K. *[Bueh 82]*	44	0.20	56.02	0.001	0.22	373	-
HOU, Y.C. *[Hou 55]*	79	0.38	13.77	0.016	0.99	222	391
LANGE, H.B. *[Lang 70]*	20	1.38	4.48	0.033	0.19	273	368
RASSKAZOV, D.S. *[Rass 75]*	110	0.57	17.67	0.027	1.04	258	313
TIMOSHENKO, N.I. *[Timo 75]*	148	0.40	4.17	0.013	0.15	243	363
WAGNER, W. *[Wagn 68]*	84	0.17	3.34	0.007	0.09	203	353
WATANABE, K. *[Wata 83]*	71	4.02	11.59	0.239	0.92	286	382

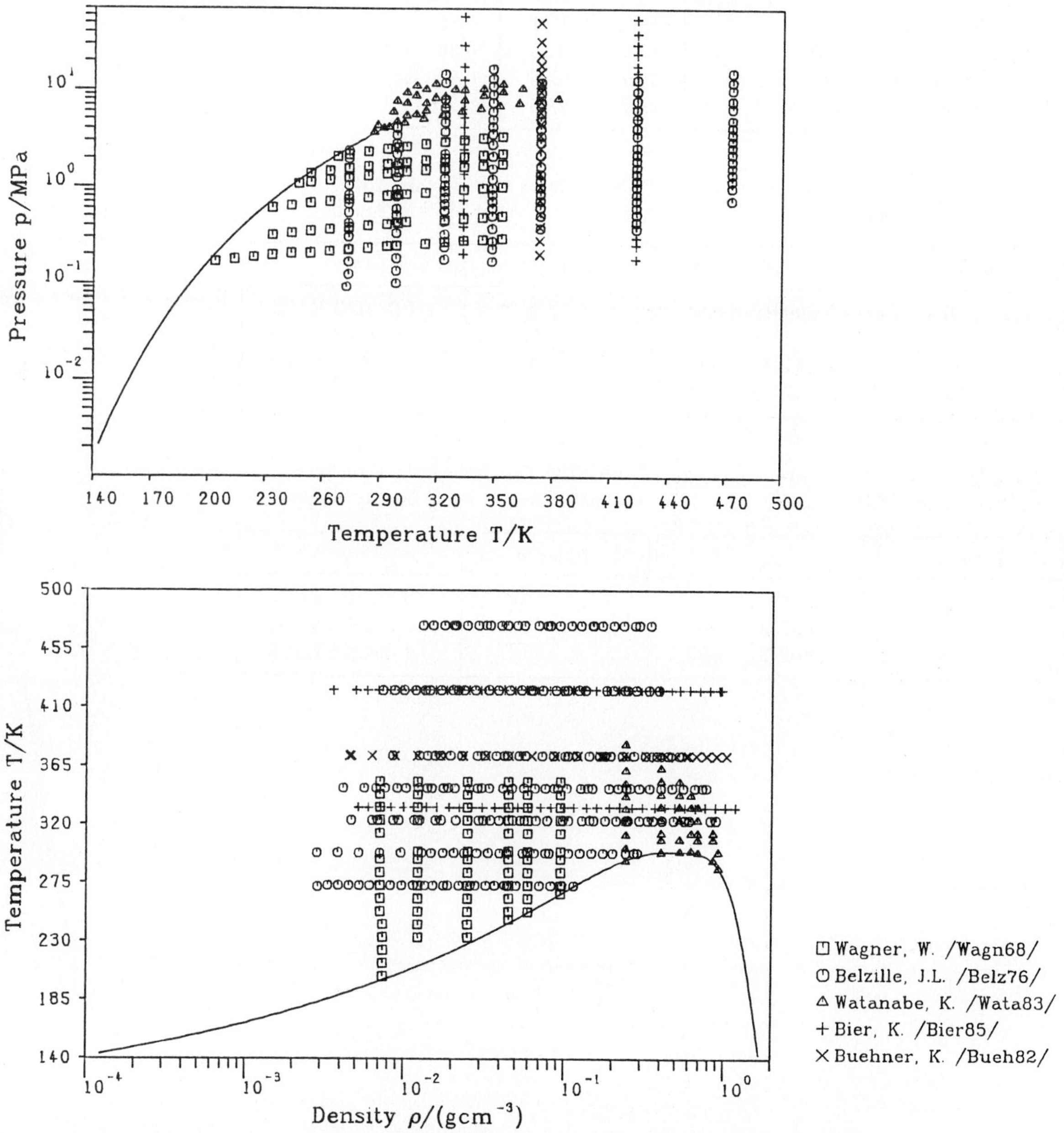

Fig.R23.1: Distribution of p, ρ, T data projected in the p, T and in the T, ρ plane

Table R23.2: Experimental vapor pressure data

Author	Δ / %		
	std	opt	*EOS*
HOU, Y.C. *[Hou 55]*	0.33	0.25	0.34
RASSKAZOV, D.S. *[Rass 75]*	0.49	0.51	0.50
WAGNER, W. *[Wagn 68]*	0.30	0.32	0.31
WATANABE, K. *[Wata 83]*	0.06	0.04	0.06

Table R23.3: Experimental saturated liquid density data

Author	Δ / %		
	std	opt	*EOS*
HOU, Y.C. *[Hou 55]*	0.96	1.31	0.80
LÖFFLER, H.J. *[Loef 68]*	0.19	0.15	0.36
THOMAS, W. *[Doer 68]*	0.79	0.86	0.19
WATANABE, K. *[Wata 83]*	1.59	2.85	4.24

Table R23.4: Coefficients of the standard vapor pressure equation

Index i	b_i
5	−7.300919 E+00
6	1.473088 E+00
9	−3.241386 E+00
15	2.557942 E−01

Table R23.5: Coefficients of the quasi-optimal vapor pressure equation

Index i	b_i
5	−7.390113 E+00
6	1.951468 E+00
8	−2.860027 E+00
21	−9.989847 E+00

Table R23.6: Coefficients of the standard saturated liquid density equation

Index i	c_i
8	1.965963 E+00
9	6.691346 E−01
16	1.402722 E+00

Table R23.7: Coefficients of the quasi-optimal saturated liquid density equation

Index i	c_i
8	2.035536 E+00
9	−1.375653 E−01
10	1.036218 E+00

Table R23.8: Coefficients of the *Bender-EOS*

Range of validity: one-phase region

$$0.09 \leq p/\mathrm{MPa} \leq 58.64$$
$$0.00283 \leq \rho/\mathrm{g\ cm^{-3}} \leq 1.16564$$
$$203.18 \leq T/\mathrm{K} \leq 473.17$$

	Specific units	Molar units
a_{01}	2.013258865 E−01	9.870324432 E−01
a_{02}	1.857623313 E+02	9.107296184 E+02
a_{03}	−5.052515739 E+04	−2.477076864 E+05
a_{04}	1.428594037 E+07	7.003911357 E+07
a_{05}	−5.548501933 E+08	−2.720242049 E+09
a_{06}	2.787489184 E−02	9.568875472 E−03
a_{07}	8.650525750 E+00	2.969547080 E+00
a_{08}	1.616373260 E+04	5.548676037 E+03
a_{09}	3.383573102 E−01	8.132782908 E−03
a_{10}	−1.376472676 E+02	−3.308500544 E+00
a_{11}	−1.592395335 E−01	−2.679972985 E−04
a_{12}	4.807158535 E+01	8.090362191 E−02
a_{13}	2.754540029 E+01	3.245970075 E−03
a_{14}	−8.509755009 E+06	−2.921223388 E+06
a_{15}	5.032460342 E+09	1.727539845 E+09
a_{16}	−4.559906977 E+11	−1.565322021 E+11
a_{17}	−3.505810264 E+07	−5.900216230 E+04
a_{18}	1.391359913 E+10	2.341633952 E+07
a_{19}	−2.580124256 E+11	−4.342303168 E+08
a_{20}	2.600000000 E+00	1.274691694 E−02

Table R23.9: Deviations between experimental and predicted p, ρ, T data

Region	n	Δ / %	
		p	ρ
Gas	486	0.22	0.27
Critical	83	0.24	1.88
Liquid	39	1.67	0.62
Total	608	0.48	0.76

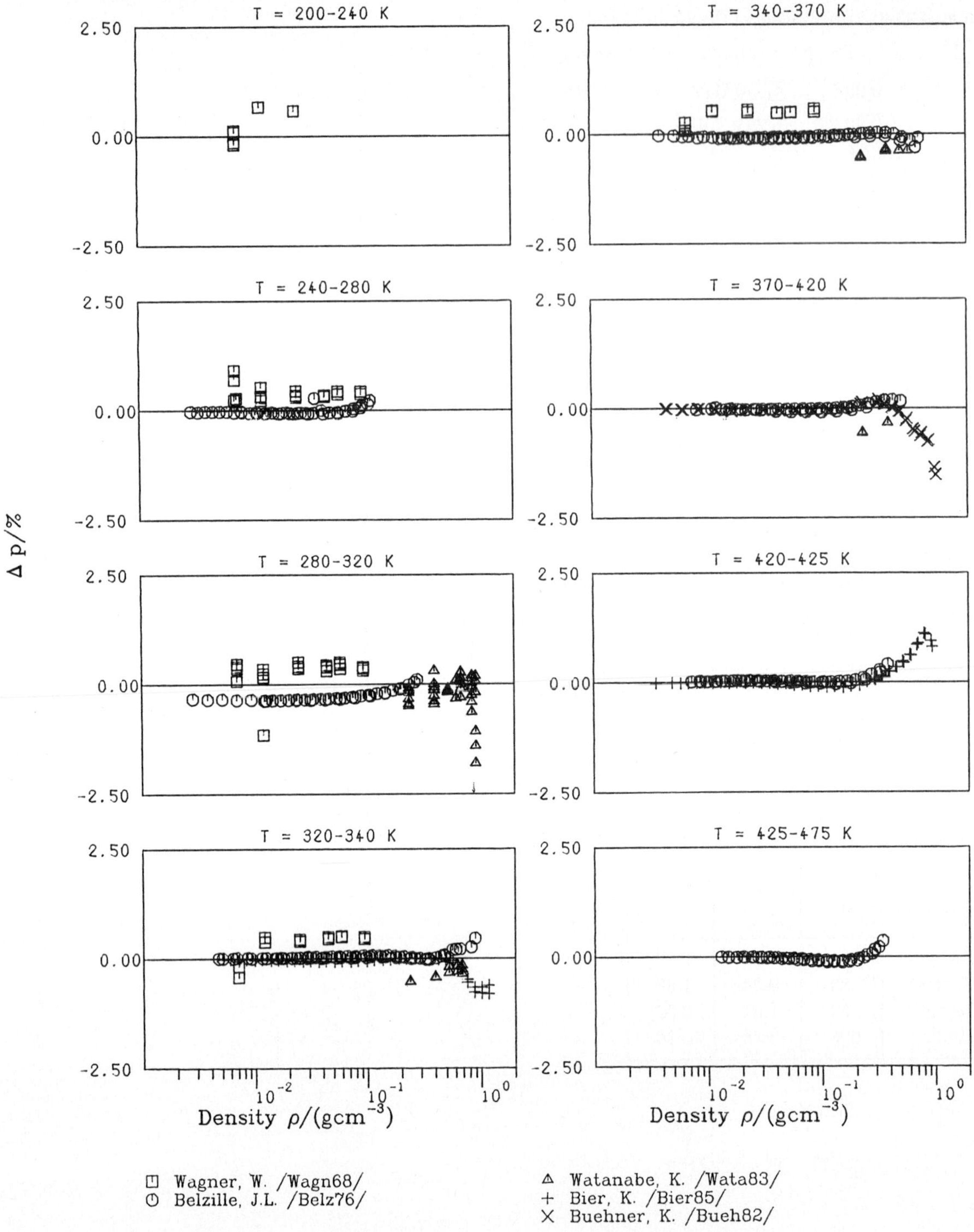

Fig.R23.2: Deviations between experimental and calculated p, ρ, T data

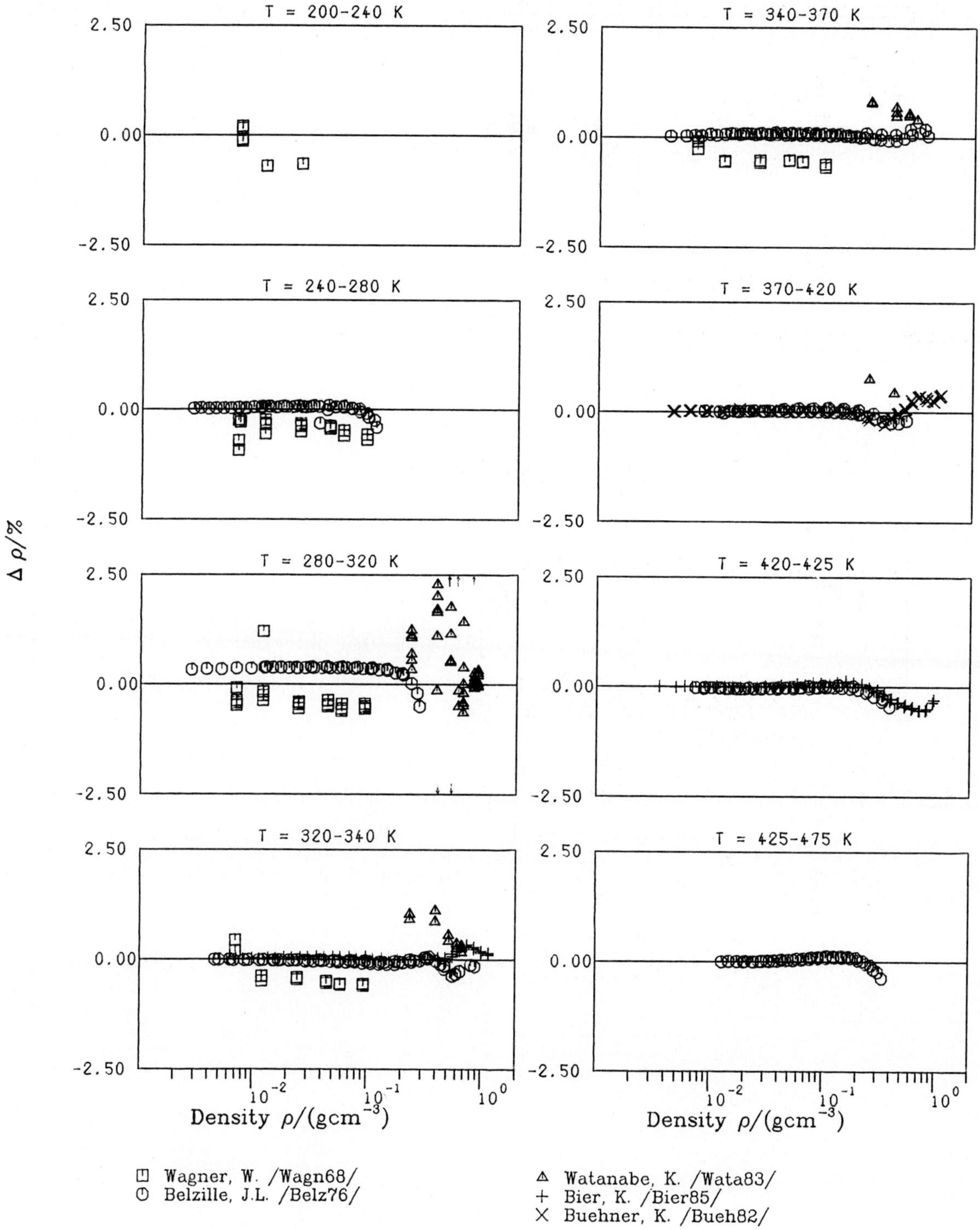

Fig.R23.2: Deviations between experimental and calculated p, ρ, T data *-continued*

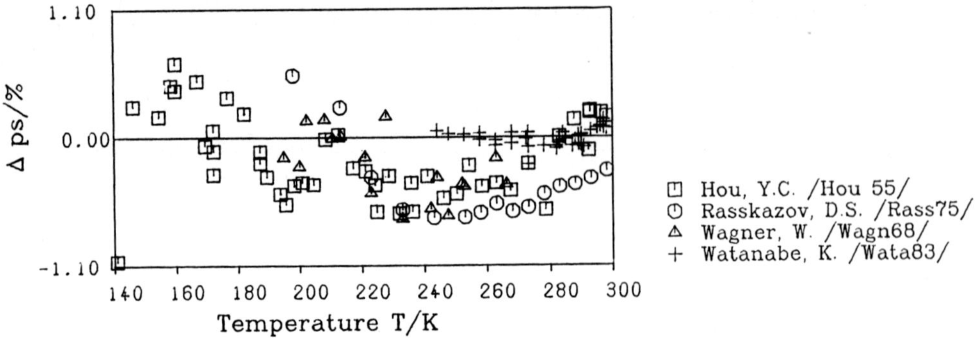

Fig.R23.3: Deviations between experimental and calculated vapor pressures

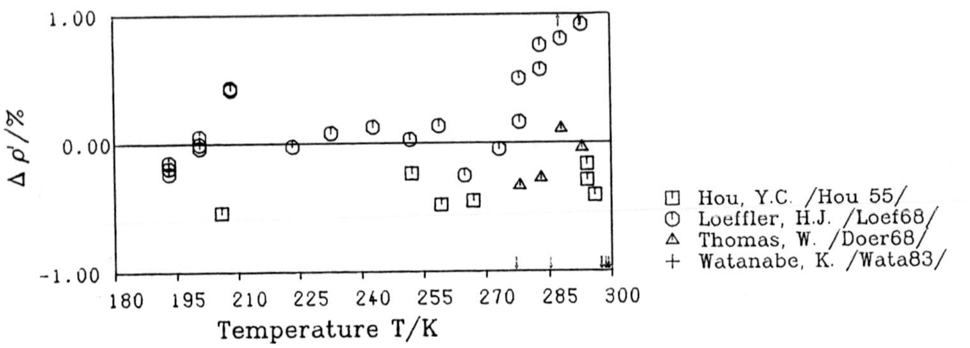

Fig.R23.4: Deviations between experimental and calculated saturated liquid densities

Table R23.10: Information on ideal gas state heat capacity equation

Author	T / K min	T / K max
RODGERS, A.S.*[Rodg 74]*	100	600
Coefficients of equation (15)		
Index i	d_i	
1	4.762244 E−01	
2	−6.564150 E−04	
3	8.788848 E−06	
4	−1.495237 E−08	
5	8.250166 E−12	

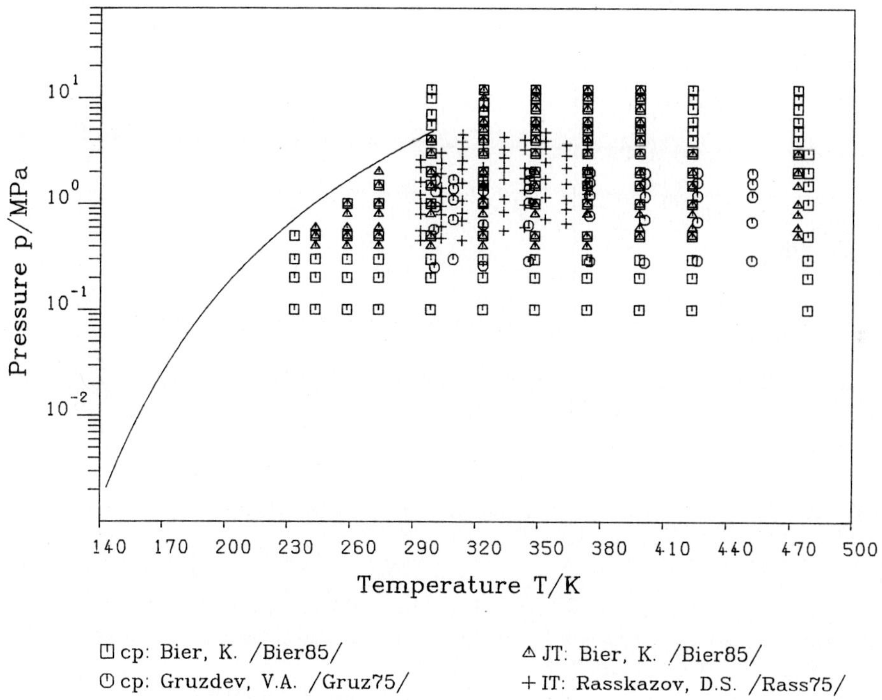

Fig.R23.5: Distribution of $c_p(p,T)$, $IT(p,T)$ and $\mu(p,T)$ data

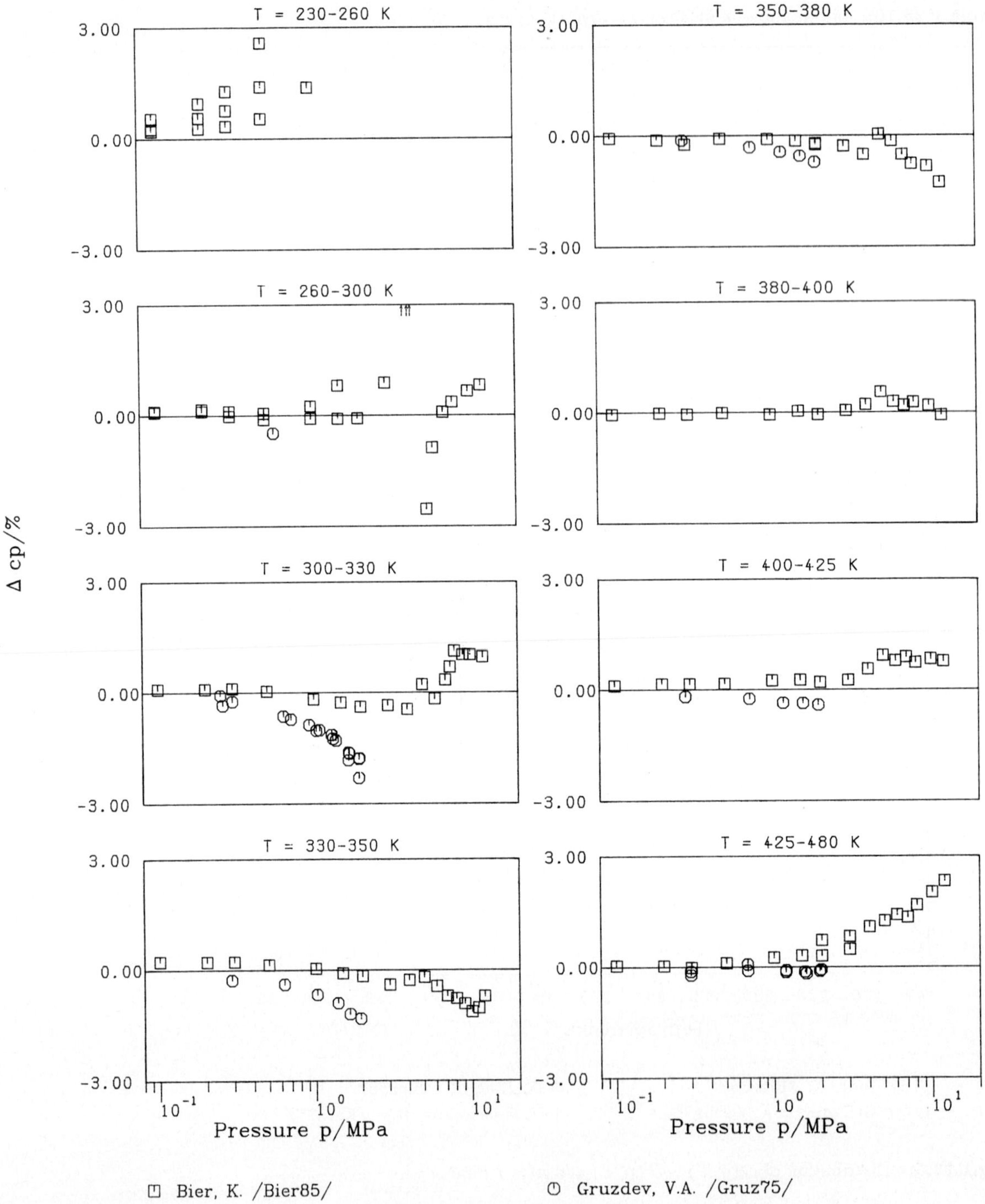

Fig.R23.6: Deviations between experimental and calculated specific isobaric heat capacities c_p

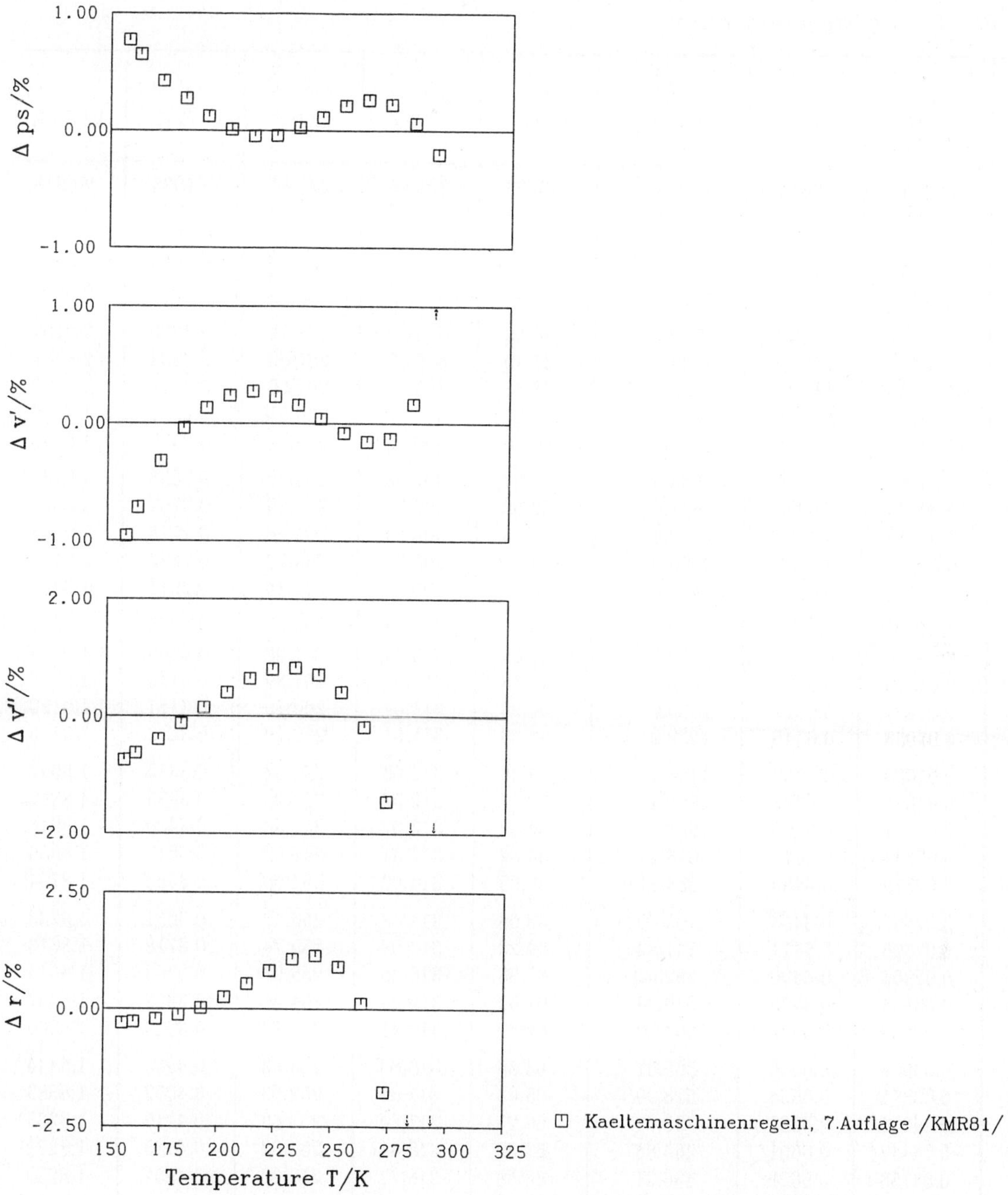

Fig.R23.7: Deviations between saturation properties from the *Kältemaschinenregeln* and those calculated by the *Bender-EOS*

Chart R23.1: Saturation properties

T K	p MPa	v' dm³/kg	v'' dm³/kg	h' kJ/kg	h'' kJ/kg	r kJ/kg	s' kJ/kgK	s'' kJ/kgK
145.0	0.00257	0.6014	6676.4	31.08	301.95	270.88	0.1935	2.0616
146.0	0.00287	0.6031	6033.1	32.14	302.46	270.32	0.2008	2.0524
147.0	0.00319	0.6048	5460.6	33.20	302.97	269.77	0.2080	2.0432
148.0	0.00354	0.6065	4950.0	34.25	303.48	269.23	0.2152	2.0343
149.0	0.00392	0.6082	4494.0	35.30	303.99	268.69	0.2223	2.0255
150.0	0.00434	0.6100	4086.1	36.35	304.49	268.14	0.2293	2.0169
151.0	0.00480	0.6117	3720.7	37.40	305.00	267.60	0.2363	2.0085
152.0	0.00530	0.6134	3392.7	38.45	305.51	267.05	0.2432	2.0002
153.0	0.00584	0.6151	3097.9	39.51	306.02	266.51	0.2501	1.9920
154.0	0.00642	0.6169	2832.6	40.56	306.52	265.96	0.2570	1.9840
155.0	0.00706	0.6186	2593.5	41.62	307.03	265.40	0.2638	1.9761
156.0	0.00774	0.6204	2377.6	42.68	307.53	264.85	0.2707	1.9684
157.0	0.00849	0.6221	2182.5	43.75	308.04	264.28	0.2775	1.9608
158.0	0.00929	0.6239	2005.9	44.82	308.54	263.72	0.2843	1.9534
159.0	0.01015	0.6256	1845.8	45.90	309.04	263.14	0.2911	1.9461
160.0	0.01108	0.6274	1700.5	46.98	309.54	262.57	0.2978	1.9389
161.0	0.01208	0.6291	1568.5	48.06	310.04	261.98	0.3046	1.9318
162.0	0.01316	0.6309	1448.4	49.15	310.54	261.39	0.3113	1.9248
163.0	0.01431	0.6327	1339.0	50.25	311.04	260.79	0.3181	1.9180
164.0	0.01555	0.6345	1239.2	51.35	311.54	260.19	0.3248	1.9113
165.0	0.01688	0.6363	1148.1	52.46	312.03	259.57	0.3315	1.9047
166.0	0.01830	0.6381	1064.8	53.57	312.52	258.95	0.3383	1.8982
167.0	0.01981	0.6399	988.61	54.69	313.02	258.33	0.3450	1.8918
168.0	0.02143	0.6417	918.77	55.82	313.51	257.69	0.3517	1.8856
169.0	0.02315	0.6435	854.71	56.95	314.00	257.05	0.3584	1.8794
170.0	0.02499	0.6453	795.89	58.08	314.48	256.40	0.3651	1.8733
171.0	0.02695	0.6471	741.82	59.23	314.97	255.74	0.3718	1.8673
172.0	0.02903	0.6490	692.07	60.38	315.45	255.07	0.3785	1.8615
173.0	0.03124	0.6508	646.24	61.53	315.93	254.40	0.3852	1.8557
174.0	0.03359	0.6527	604.00	62.69	316.41	253.72	0.3918	1.8500
175.0	0.03608	0.6545	565.01	63.86	316.89	253.03	0.3985	1.8444
176.0	0.03871	0.6564	528.99	65.03	317.36	252.33	0.4052	1.8389
177.0	0.04151	0.6582	495.69	66.21	317.84	251.63	0.4119	1.8335
178.0	0.04446	0.6601	464.86	67.39	318.31	250.92	0.4185	1.8281
179.0	0.04758	0.6620	436.31	68.58	318.77	250.20	0.4252	1.8229
180.0	0.05088	0.6639	409.84	69.77	319.24	249.47	0.4318	1.8177
181.0	0.05436	0.6658	385.27	70.97	319.70	248.73	0.4384	1.8126
182.0	0.05803	0.6677	362.45	72.18	320.16	247.99	0.4450	1.8076
183.0	0.06189	0.6696	341.24	73.38	320.62	247.23	0.4516	1.8027
184.0	0.06596	0.6716	321.50	74.60	321.07	246.47	0.4582	1.7978
185.0	0.07025	0.6735	303.13	75.82	321.52	245.71	0.4648	1.7930
186.0	0.07475	0.6755	286.00	77.04	321.97	244.93	0.4714	1.7882
187.0	0.07948	0.6774	270.03	78.27	322.42	244.15	0.4780	1.7836
188.0	0.08445	0.6794	255.13	79.50	322.86	243.36	0.4845	1.7790
189.0	0.08966	0.6814	241.20	80.74	323.29	242.56	0.4911	1.7745
190.0	0.09512	0.6834	228.19	81.98	323.73	241.75	0.4976	1.7700
191.0	0.10084	0.6854	216.01	83.22	324.16	240.94	0.5041	1.7656
192.0	0.10684	0.6874	204.61	84.47	324.59	240.12	0.5106	1.7612
193.0	0.11311	0.6894	193.94	85.73	325.01	239.29	0.5171	1.7569
194.0	0.11967	0.6915	183.93	86.99	325.43	238.45	0.5236	1.7527

Chart R23.1: Saturation properties—*continued*

T K	p MPa	v' dm³/kg	v'' dm³/kg	h' kJ/kg	h'' kJ/kg	r kJ/kg	s' kJ/kgK	s'' kJ/kgK
195.0	0.12653	0.6935	174.53	88.25	325.85	237.60	0.5301	1.7485
196.0	0.13369	0.6956	165.72	89.51	326.26	236.75	0.5365	1.7444
197.0	0.14117	0.6977	157.43	90.78	326.67	235.89	0.5429	1.7404
198.0	0.14898	0.6998	149.65	92.05	327.08	235.02	0.5494	1.7363
199.0	0.15712	0.7019	142.33	93.33	327.48	234.15	0.5558	1.7324
200.0	0.16560	0.7040	135.43	94.61	327.87	233.26	0.5621	1.7285
201.0	0.17443	0.7061	128.94	95.89	328.27	232.37	0.5685	1.7246
202.0	0.18363	0.7083	122.82	97.18	328.65	231.47	0.5749	1.7208
203.0	0.19320	0.7105	117.05	98.47	329.04	230.57	0.5812	1.7170
204.0	0.20316	0.7127	111.60	99.77	329.42	229.65	0.5875	1.7133
205.0	0.21351	0.7149	106.46	101.06	329.79	228.73	0.5938	1.7096
206.0	0.22427	0.7171	101.61	102.36	330.16	227.80	0.6001	1.7059
207.0	0.23543	0.7193	97.018	103.67	330.53	226.86	0.6064	1.7023
208.0	0.24703	0.7216	92.677	104.97	330.89	225.91	0.6127	1.6988
209.0	0.25906	0.7239	88.569	106.28	331.24	224.96	0.6189	1.6953
210.0	0.27154	0.7262	84.679	107.60	331.59	223.99	0.6251	1.6918
211.0	0.28448	0.7285	80.995	108.91	331.94	223.02	0.6313	1.6883
212.0	0.29788	0.7308	77.503	110.23	332.28	222.04	0.6375	1.6849
213.0	0.31176	0.7332	74.191	111.56	332.61	221.05	0.6437	1.6815
214.0	0.32614	0.7355	71.049	112.88	332.94	220.06	0.6499	1.6782
215.0	0.34102	0.7379	68.066	114.21	333.26	219.05	0.6560	1.6749
216.0	0.35641	0.7404	65.233	115.54	333.58	218.04	0.6621	1.6716
217.0	0.37233	0.7428	62.541	116.88	333.89	217.02	0.6683	1.6683
218.0	0.38878	0.7453	59.982	118.22	334.20	215.98	0.6744	1.6651
219.0	0.40578	0.7478	57.548	119.56	334.50	214.94	0.6804	1.6619
220.0	0.42334	0.7503	55.231	120.90	334.80	213.89	0.6865	1.6587
221.0	0.44147	0.7529	53.026	122.25	335.08	212.83	0.6926	1.6556
222.0	0.46019	0.7554	50.926	123.60	335.37	211.76	0.6986	1.6525
223.0	0.47951	0.7580	48.924	124.96	335.64	210.69	0.7046	1.6494
224.0	0.49943	0.7607	47.016	126.32	335.91	209.60	0.7106	1.6463
225.0	0.51997	0.7633	45.197	127.68	336.18	208.50	0.7166	1.6433
226.0	0.54114	0.7660	43.460	129.04	336.43	207.39	0.7226	1.6403
227.0	0.56296	0.7688	41.803	130.41	336.68	206.27	0.7286	1.6373
228.0	0.58543	0.7715	40.220	131.78	336.93	205.14	0.7345	1.6343
229.0	0.60858	0.7743	38.708	133.16	337.16	204.00	0.7405	1.6313
230.0	0.63241	0.7772	37.263	134.54	337.39	202.85	0.7464	1.6284
231.0	0.65693	0.7800	35.881	135.92	337.61	201.69	0.7523	1.6254
232.0	0.68216	0.7829	34.559	137.31	337.83	200.52	0.7582	1.6225
233.0	0.70811	0.7859	33.295	138.70	338.03	199.33	0.7641	1.6196
234.0	0.73480	0.7889	32.084	140.10	338.23	198.14	0.7700	1.6168
235.0	0.76223	0.7919	30.925	141.50	338.42	196.93	0.7759	1.6139
236.0	0.79042	0.7950	29.814	142.90	338.61	195.71	0.7818	1.6110
237.0	0.81939	0.7981	28.749	144.31	338.78	194.47	0.7876	1.6082
238.0	0.84914	0.8013	27.729	145.72	338.95	193.23	0.7935	1.6053
239.0	0.87969	0.8045	26.750	147.14	339.11	191.97	0.7993	1.6025
240.0	0.91106	0.8077	25.811	148.56	339.26	190.69	0.8052	1.5997
241.0	0.94325	0.8110	24.909	149.99	339.40	189.40	0.8110	1.5969
242.0	0.97629	0.8144	24.044	151.43	339.53	188.10	0.8168	1.5941
243.0	1.01018	0.8178	23.213	152.86	339.65	186.79	0.8226	1.5913
244.0	1.04494	0.8213	22.414	154.31	339.76	185.46	0.8284	1.5885

Chart R23.1: Saturation properties—*continued*

T K	p MPa	v' dm³/kg	v'' dm³/kg	h' kJ/kg	h'' kJ/kg	r kJ/kg	s' kJ/kgK	s'' kJ/kgK
245.0	1.08058	0.8248	21.647	155.76	339.86	184.11	0.8342	1.5857
246.0	1.11712	0.8284	20.908	157.21	339.96	182.75	0.8400	1.5829
247.0	1.15458	0.8321	20.199	158.67	340.04	181.37	0.8458	1.5801
248.0	1.19296	0.8358	19.516	160.14	340.11	179.97	0.8516	1.5773
249.0	1.23228	0.8396	18.858	161.61	340.17	178.56	0.8574	1.5745
250.0	1.27256	0.8435	18.225	163.09	340.22	177.13	0.8632	1.5717
251.0	1.31381	0.8475	17.616	164.58	340.26	175.68	0.8690	1.5689
252.0	1.35604	0.8515	17.028	166.07	340.28	174.21	0.8748	1.5661
253.0	1.39928	0.8556	16.462	167.57	340.30	172.72	0.8806	1.5633
254.0	1.44353	0.8598	15.917	169.08	340.30	171.21	0.8864	1.5605
255.0	1.48882	0.8641	15.390	170.60	340.28	169.68	0.8922	1.5577
256.0	1.53516	0.8685	14.883	172.13	340.26	168.13	0.8980	1.5548
257.0	1.58256	0.8729	14.393	173.66	340.22	166.56	0.9039	1.5520
258.0	1.63104	0.8775	13.919	175.20	340.17	164.96	0.9097	1.5491
259.0	1.68061	0.8822	13.463	176.75	340.10	163.34	0.9155	1.5462
260.0	1.73130	0.8870	13.021	178.32	340.01	161.70	0.9214	1.5433
261.0	1.78312	0.8920	12.595	179.89	339.91	160.03	0.9272	1.5404
262.0	1.83609	0.8970	12.182	181.47	339.80	158.33	0.9331	1.5374
263.0	1.89023	0.9022	11.783	183.06	339.66	156.60	0.9390	1.5344
264.0	1.94554	0.9075	11.398	184.67	339.51	154.84	0.9449	1.5314
265.0	2.00206	0.9130	11.024	186.29	339.34	153.06	0.9508	1.5284
266.0	2.05979	0.9187	10.663	187.92	339.15	151.24	0.9567	1.5253
267.0	2.11876	0.9245	10.313	189.56	338.94	149.38	0.9627	1.5222
268.0	2.17899	0.9305	9.9733	191.22	338.71	147.50	0.9687	1.5190
269.0	2.24049	0.9366	9.6445	192.89	338.46	145.57	0.9747	1.5159
270.0	2.30329	0.9430	9.3256	194.57	338.18	143.61	0.9807	1.5126
271.0	2.36740	0.9496	9.0162	196.28	337.88	141.60	0.9868	1.5093
272.0	2.43284	0.9564	8.7159	198.00	337.55	139.56	0.9929	1.5060
273.0	2.49965	0.9635	8.4243	199.74	337.20	137.46	0.9991	1.5026
274.0	2.56783	0.9708	8.1410	201.50	336.82	135.32	1.0053	1.4991
275.0	2.63741	0.9785	7.8657	203.27	336.40	133.13	1.0115	1.4956
276.0	2.70842	0.9864	7.5978	205.08	335.96	130.88	1.0178	1.4920
277.0	2.78088	0.9947	7.3372	206.90	335.48	128.58	1.0241	1.4883
278.0	2.85480	1.0033	7.0834	208.75	334.96	126.21	1.0305	1.4845
279.0	2.93022	1.0123	6.8361	210.63	334.40	123.78	1.0370	1.4806
280.0	3.00717	1.0218	6.5949	212.53	333.80	121.27	1.0435	1.4766
281.0	3.08566	1.0318	6.3595	214.47	333.16	118.69	1.0501	1.4725
282.0	3.16574	1.0422	6.1295	216.44	332.46	116.02	1.0568	1.4683
283.0	3.24742	1.0533	5.9047	218.45	331.71	113.26	1.0637	1.4639
284.0	3.33074	1.0651	5.6845	220.50	330.91	110.40	1.0706	1.4593
285.0	3.41574	1.0776	5.4688	222.60	330.03	107.43	1.0776	1.4546
286.0	3.50244	1.0909	5.2570	224.75	329.09	104.34	1.0848	1.4496
287.0	3.59089	1.1053	5.0487	226.95	328.06	101.10	1.0922	1.4445
288.0	3.68112	1.1207	4.8436	229.23	326.94	97.71	1.0997	1.4390
289.0	3.77318	1.1375	4.6410	231.57	325.72	94.15	1.1075	1.4333
290.0	3.86711	1.1559	4.4404	234.00	324.38	90.37	1.1155	1.4272
291.0	3.96296	1.1761	4.2410	236.54	322.90	86.36	1.1239	1.4207
292.0	4.06078	1.1988	4.0421	239.19	321.25	82.06	1.1326	1.4136
293.0	4.16065	1.2243	3.8425	242.00	319.41	77.41	1.1418	1.4060
294.0	4.26262	1.2537	3.6406	245.00	317.32	72.32	1.1516	1.3975

Chart R23.1: Saturation properties–*continued*

T K	p MPa	v' dm^3/kg	v'' dm^3/kg	h' kJ/kg	h'' kJ/kg	r kJ/kg	s' kJ/kgK	s'' kJ/kgK
295.0	4.36678	1.2884	3.4345	248.26	314.91	66.65	1.1622	1.3881
296.0	4.47321	1.3307	3.2207	251.86	312.07	60.20	1.1739	1.3773
297.0	4.58203	1.3847	2.9936	256.00	308.59	52.59	1.1874	1.3644
298.0	4.69337	1.4602	2.7417	261.06	304.07	43.02	1.2038	1.3482
299.0	4.81453	2.2085	2.5934	290.91	301.63	10.72	1.3031	1.3389

Chart R23.2: Saturation properties

p MPa	T K	v' dm³/kg	v'' dm³/kg	h' kJ/kg	h'' kJ/kg	r kJ/kg	s' kJ/kgK	s'' kJ/kgK
0.0031	146.74	0.6044	5603.5	32.92	302.84	269.92	0.2062	2.0456
0.0032	147.04	0.6049	5439.1	33.24	302.99	269.75	0.2083	2.0429
0.0033	147.33	0.6054	5284.4	33.55	303.14	269.59	0.2104	2.0403
0.0034	147.62	0.6059	5138.5	33.85	303.28	269.44	0.2125	2.0377
0.0035	147.90	0.6063	5000.7	34.14	303.43	269.29	0.2144	2.0352
0.0036	148.17	0.6068	4870.4	34.42	303.56	269.14	0.2164	2.0328
0.0037	148.43	0.6073	4746.8	34.70	303.70	268.99	0.2182	2.0305
0.0038	148.69	0.6077	4629.7	34.98	303.83	268.85	0.2201	2.0282
0.0039	148.94	0.6081	4518.3	35.24	303.96	268.72	0.2219	2.0260
0.0040	149.19	0.6086	4412.4	35.50	304.08	268.58	0.2236	2.0239
0.0041	149.43	0.6090	4311.4	35.76	304.21	268.45	0.2253	2.0218
0.0042	149.67	0.6094	4215.2	36.01	304.33	268.32	0.2270	2.0197
0.0043	149.90	0.6098	4123.2	36.25	304.45	268.19	0.2286	2.0177
0.0044	150.13	0.6102	4035.4	36.49	304.56	268.07	0.2302	2.0158
0.0045	150.36	0.6106	3951.3	36.73	304.68	267.95	0.2318	2.0139
0.0046	150.58	0.6109	3870.8	36.96	304.79	267.83	0.2333	2.0120
0.0047	150.79	0.6113	3793.6	37.18	304.90	267.71	0.2348	2.0102
0.0048	151.00	0.6117	3719.5	37.41	305.00	267.60	0.2363	2.0084
0.0049	151.21	0.6120	3648.4	37.62	305.11	267.48	0.2377	2.0067
0.0050	151.42	0.6124	3580.0	37.84	305.21	267.37	0.2392	2.0050
0.0051	151.62	0.6127	3514.2	38.05	305.31	267.26	0.2406	2.0033
0.0052	151.81	0.6131	3450.9	38.26	305.42	267.16	0.2419	2.0017
0.0053	152.01	0.6134	3389.9	38.46	305.51	267.05	0.2433	2.0001
0.0054	152.20	0.6138	3331.1	38.67	305.61	266.95	0.2446	1.9985
0.0055	152.39	0.6141	3274.4	38.86	305.71	266.84	0.2459	1.9970
0.0056	152.57	0.6144	3219.6	39.06	305.80	266.74	0.2472	1.9955
0.0057	152.76	0.6147	3166.7	39.25	305.89	266.64	0.2484	1.9940
0.0058	152.94	0.6150	3115.6	39.44	305.98	266.54	0.2497	1.9925
0.0059	153.11	0.6153	3066.1	39.63	306.07	266.44	0.2509	1.9911
0.0060	153.29	0.6156	3018.3	39.81	306.16	266.35	0.2521	1.9897
0.0061	153.46	0.6159	2971.9	40.00	306.25	266.25	0.2533	1.9883
0.0062	153.63	0.6162	2927.1	40.17	306.34	266.16	0.2545	1.9869
0.0063	153.80	0.6165	2883.6	40.35	306.42	266.07	0.2556	1.9856
0.0064	153.97	0.6168	2841.4	40.53	306.50	265.98	0.2568	1.9843
0.0065	154.13	0.6171	2800.5	40.70	306.59	265.89	0.2579	1.9830
0.0066	154.29	0.6174	2760.7	40.87	306.67	265.80	0.2590	1.9817
0.0067	154.45	0.6176	2722.2	41.04	306.75	265.71	0.2601	1.9804
0.0068	154.61	0.6179	2684.7	41.21	306.83	265.62	0.2612	1.9792
0.0069	154.76	0.6182	2648.3	41.37	306.91	265.54	0.2622	1.9780
0.0070	154.91	0.6185	2612.9	41.53	306.98	265.45	0.2633	1.9768
0.0071	155.07	0.6187	2578.4	41.69	307.06	265.37	0.2643	1.9756
0.0072	155.22	0.6190	2544.9	41.85	307.14	265.28	0.2653	1.9744
0.0073	155.36	0.6192	2512.3	42.01	307.21	265.20	0.2663	1.9733
0.0074	155.51	0.6195	2480.5	42.16	307.28	265.12	0.2673	1.9722
0.0076	155.80	0.6200	2419.4	42.47	307.43	264.96	0.2693	1.9700
0.0078	156.08	0.6205	2361.4	42.77	307.57	264.80	0.2712	1.9678
0.0080	156.35	0.6210	2306.1	43.06	307.71	264.65	0.2731	1.9657
0.0082	156.62	0.6214	2253.5	43.35	307.85	264.50	0.2749	1.9637
0.0084	156.89	0.6219	2203.3	43.63	307.98	264.35	0.2767	1.9617
0.0086	157.15	0.6224	2155.3	43.91	308.11	264.20	0.2785	1.9597

Chart R23.2: Saturation properties–*continued*

p MPa	T K	v' dm³/kg	v'' dm³/kg	h' kJ/kg	h'' kJ/kg	r kJ/kg	s' kJ/kgK	s'' kJ/kgK
0.0088	157.40	0.6228	2109.5	44.18	308.24	264.06	0.2802	1.9578
0.0090	157.65	0.6232	2065.7	44.45	308.36	263.92	0.2819	1.9560
0.0092	157.89	0.6237	2023.7	44.71	308.49	263.78	0.2836	1.9542
0.0094	158.13	0.6241	1983.4	44.97	308.61	263.64	0.2852	1.9524
0.0096	158.37	0.6245	1944.7	45.22	308.72	263.51	0.2868	1.9506
0.0098	158.60	0.6249	1907.6	45.47	308.84	263.37	0.2884	1.9489
0.0100	158.83	0.6253	1871.9	45.71	308.96	263.24	0.2899	1.9473
0.0102	159.05	0.6257	1837.6	45.95	309.07	263.11	0.2914	1.9457
0.0104	159.27	0.6261	1804.6	46.19	309.18	262.99	0.2929	1.9441
0.0106	159.49	0.6265	1772.7	46.43	309.29	262.86	0.2944	1.9425
0.0108	159.70	0.6268	1742.0	46.66	309.39	262.74	0.2958	1.9410
0.0110	159.91	0.6272	1712.4	46.88	309.50	262.62	0.2972	1.9395
0.0115	160.43	0.6281	1642.8	47.44	309.76	262.32	0.3007	1.9358
0.0120	160.92	0.6290	1578.7	47.97	310.00	262.03	0.3040	1.9324
0.0125	161.40	0.6298	1519.7	48.49	310.24	261.75	0.3073	1.9290
0.0130	161.86	0.6307	1465.0	49.00	310.47	261.47	0.3104	1.9258
0.0135	162.30	0.6314	1414.3	49.48	310.69	261.21	0.3134	1.9228
0.0140	162.73	0.6322	1367.0	49.96	310.91	260.95	0.3163	1.9198
0.0145	163.15	0.6330	1323.0	50.42	311.12	260.70	0.3191	1.9170
0.0150	163.56	0.6337	1281.8	50.87	311.32	260.45	0.3219	1.9142
0.0155	163.96	0.6344	1243.1	51.31	311.52	260.21	0.3245	1.9116
0.0160	164.35	0.6351	1206.8	51.73	311.71	259.97	0.3271	1.9090
0.0165	164.72	0.6358	1172.6	52.15	311.89	259.74	0.3297	1.9065
0.0170	165.09	0.6364	1140.4	52.56	312.07	259.52	0.3321	1.9041
0.0175	165.45	0.6371	1109.9	52.96	312.25	259.30	0.3345	1.9018
0.0180	165.80	0.6377	1081.1	53.35	312.42	259.08	0.3369	1.8995
0.0185	166.14	0.6383	1053.8	53.73	312.59	258.87	0.3392	1.8973
0.0190	166.47	0.6389	1027.9	54.10	312.76	258.66	0.3414	1.8952
0.0195	166.80	0.6395	1003.3	54.47	312.92	258.45	0.3436	1.8931
0.0200	167.12	0.6401	979.87	54.83	313.08	258.25	0.3458	1.8911
0.0210	167.74	0.6412	936.25	55.52	313.38	257.86	0.3499	1.8872
0.0220	168.34	0.6423	896.47	56.20	313.67	257.47	0.3540	1.8835
0.0230	168.91	0.6433	860.03	56.85	313.95	257.10	0.3578	1.8799
0.0240	169.47	0.6443	826.54	57.48	314.22	256.75	0.3615	1.8765
0.0250	170.00	0.6453	795.64	58.09	314.48	256.40	0.3651	1.8733
0.0260	170.52	0.6463	767.05	58.68	314.74	256.06	0.3686	1.8702
0.0270	171.03	0.6472	740.50	59.26	314.98	255.72	0.3720	1.8672
0.0280	171.51	0.6481	715.79	59.82	315.22	255.40	0.3752	1.8643
0.0290	171.99	0.6489	692.72	60.36	315.44	255.08	0.3784	1.8616
0.0300	172.45	0.6498	671.15	60.89	315.67	254.78	0.3815	1.8589
0.0310	172.89	0.6506	650.92	61.41	315.88	254.47	0.3845	1.8563
0.0320	173.33	0.6514	631.91	61.91	316.09	254.18	0.3874	1.8538
0.0330	173.76	0.6522	614.02	62.41	316.29	253.89	0.3902	1.8514
0.0340	174.17	0.6530	597.15	62.89	316.49	253.60	0.3930	1.8491
0.0350	174.57	0.6537	581.21	63.36	316.69	253.32	0.3957	1.8468
0.0360	174.97	0.6545	566.12	63.82	316.87	253.05	0.3983	1.8446
0.0370	175.36	0.6552	551.82	64.28	317.06	252.78	0.4009	1.8424
0.0380	175.74	0.6559	538.25	64.72	317.24	252.52	0.4034	1.8404
0.0390	176.11	0.6566	525.36	65.15	317.41	252.26	0.4059	1.8383
0.0400	176.47	0.6572	513.08	65.58	317.59	252.00	0.4083	1.8364

Chart R23.2: Saturation properties–*continued*

p MPa	T K	v' dm³/kg	v'' dm³/kg	h' kJ/kg	h'' kJ/kg	r kJ/kg	s' kJ/kgK	s'' kJ/kgK
0.0410	176.82	0.6579	501.39	66.00	317.75	251.75	0.4107	1.8344
0.0420	177.17	0.6586	490.23	66.41	317.92	251.51	0.4130	1.8326
0.0430	177.51	0.6592	479.58	66.81	318.08	251.26	0.4153	1.8307
0.0440	177.85	0.6598	469.39	67.21	318.24	251.02	0.4175	1.8289
0.0450	178.18	0.6605	459.64	67.60	318.39	250.79	0.4197	1.8272
0.0460	178.50	0.6611	450.30	67.98	318.54	250.56	0.4218	1.8255
0.0470	178.82	0.6617	441.35	68.36	318.69	250.33	0.4239	1.8238
0.0480	179.13	0.6623	432.75	68.73	318.83	250.10	0.4260	1.8222
0.0490	179.44	0.6628	424.50	69.10	318.98	249.88	0.4281	1.8206
0.0500	179.74	0.6634	416.56	69.46	319.12	249.66	0.4301	1.8191
0.0510	180.04	0.6640	408.92	69.82	319.26	249.44	0.4320	1.8175
0.0520	180.33	0.6645	401.57	70.17	319.39	249.23	0.4340	1.8160
0.0530	180.62	0.6651	394.49	70.51	319.52	249.01	0.4359	1.8146
0.0540	180.90	0.6656	387.65	70.85	319.66	248.80	0.4378	1.8131
0.0550	181.18	0.6661	381.06	71.19	319.78	248.60	0.4396	1.8117
0.0560	181.45	0.6667	374.70	71.52	319.91	248.39	0.4414	1.8103
0.0570	181.73	0.6672	368.55	71.84	320.04	248.19	0.4432	1.8090
0.0580	181.99	0.6677	362.61	72.17	320.16	247.99	0.4450	1.8076
0.0590	182.26	0.6682	356.86	72.49	320.28	247.79	0.4467	1.8063
0.0600	182.52	0.6687	351.29	72.80	320.40	247.60	0.4485	1.8050
0.0610	182.77	0.6692	345.91	73.11	320.51	247.40	0.4502	1.8038
0.0620	183.03	0.6697	340.69	73.42	320.63	247.21	0.4518	1.8025
0.0630	183.28	0.6702	335.63	73.72	320.74	247.02	0.4535	1.8013
0.0640	183.52	0.6707	330.72	74.02	320.86	246.84	0.4551	1.8001
0.0650	183.77	0.6711	325.96	74.32	320.97	246.65	0.4567	1.7989
0.0660	184.01	0.6716	321.34	74.61	321.08	246.47	0.4583	1.7977
0.0670	184.25	0.6721	316.85	74.90	321.18	246.28	0.4599	1.7966
0.0680	184.48	0.6725	312.49	75.18	321.29	246.10	0.4614	1.7955
0.0690	184.71	0.6730	308.25	75.47	321.39	245.93	0.4630	1.7943
0.0700	184.94	0.6734	304.12	75.75	321.50	245.75	0.4645	1.7932
0.0710	185.17	0.6738	300.11	76.03	321.60	245.57	0.4660	1.7922
0.0720	185.40	0.6743	296.21	76.30	321.70	245.40	0.4674	1.7911
0.0730	185.62	0.6747	292.41	76.57	321.80	245.23	0.4689	1.7900
0.0740	185.84	0.6751	288.71	76.84	321.90	245.06	0.4703	1.7890
0.0760	186.27	0.6760	281.59	77.37	322.09	244.72	0.4732	1.7870
0.0780	186.69	0.6768	274.82	77.89	322.28	244.39	0.4760	1.7850
0.0800	187.11	0.6776	268.38	78.40	322.46	244.06	0.4787	1.7831
0.0820	187.51	0.6784	262.25	78.90	322.64	243.74	0.4814	1.7812
0.0840	187.91	0.6792	256.39	79.39	322.82	243.43	0.4840	1.7794
0.0860	188.30	0.6800	250.80	79.88	322.99	243.11	0.4865	1.7776
0.0880	188.69	0.6808	245.46	80.35	323.16	242.81	0.4890	1.7759
0.0900	189.06	0.6815	240.34	80.82	323.32	242.51	0.4915	1.7742
0.0920	189.43	0.6822	235.44	81.28	323.48	242.21	0.4939	1.7725
0.0940	189.80	0.6830	230.74	81.73	323.64	241.91	0.4963	1.7709
0.0960	190.16	0.6837	226.22	82.17	323.80	241.62	0.4986	1.7693
0.0980	190.51	0.6844	221.89	82.61	323.95	241.34	0.5009	1.7677
0.1000	190.86	0.6851	217.72	83.04	324.10	241.06	0.5032	1.7662
0.1020	191.20	0.6858	213.72	83.47	324.25	240.78	0.5054	1.7647
0.1040	191.53	0.6865	209.85	83.89	324.39	240.50	0.5076	1.7633
0.1060	191.86	0.6871	206.14	84.30	324.53	240.23	0.5097	1.7618

Chart R23.2: Saturation properties–*continued*

p MPa	T K	v' dm^3/kg	v'' dm^3/kg	h' kJ/kg	h'' kJ/kg	r kJ/kg	s' kJ/kgK	s'' kJ/kgK
0.1080	192.19	0.6878	202.55	84.71	324.67	239.96	0.5119	1.7604
0.1100	192.51	0.6884	199.09	85.11	324.81	239.69	0.5139	1.7590
0.1150	193.29	0.6900	190.94	86.09	325.14	239.04	0.5190	1.7557
0.1200	194.05	0.6916	183.46	87.05	325.45	238.41	0.5239	1.7525
0.1250	194.78	0.6931	176.55	87.97	325.76	237.79	0.5286	1.7495
0.1300	195.49	0.6945	170.15	88.87	326.05	237.19	0.5332	1.7465
0.1350	196.18	0.6960	164.21	89.74	326.34	236.60	0.5377	1.7437
0.1400	196.85	0.6974	158.68	90.59	326.61	236.02	0.5419	1.7410
0.1450	197.50	0.6987	153.52	91.41	326.87	235.46	0.5461	1.7384
0.1500	198.13	0.7000	148.69	92.22	327.13	234.91	0.5502	1.7358
0.1550	198.74	0.7013	144.16	93.00	327.38	234.37	0.5541	1.7334
0.1600	199.34	0.7026	139.90	93.77	327.61	233.84	0.5580	1.7310
0.1650	199.93	0.7039	135.89	94.52	327.85	233.32	0.5617	1.7287
0.1700	200.50	0.7051	132.11	95.26	328.07	232.82	0.5654	1.7265
0.1750	201.06	0.7063	128.54	95.98	328.29	232.32	0.5689	1.7244
0.1800	201.61	0.7075	125.16	96.68	328.50	231.82	0.5724	1.7223
0.1850	202.15	0.7086	121.96	97.37	328.71	231.34	0.5758	1.7202
0.1900	202.67	0.7098	118.92	98.05	328.91	230.87	0.5791	1.7182
0.1950	203.18	0.7109	116.03	98.71	329.11	230.40	0.5824	1.7163
0.2000	203.69	0.7120	113.28	99.36	329.30	229.94	0.5856	1.7144
0.2100	204.67	0.7141	108.15	100.63	329.67	229.04	0.5917	1.7108
0.2200	205.61	0.7162	103.48	101.85	330.02	228.16	0.5977	1.7074
0.2300	206.52	0.7182	99.198	103.04	330.35	227.31	0.6034	1.7041
0.2400	207.40	0.7202	95.261	104.19	330.67	226.48	0.6089	1.7009
0.2500	208.25	0.7221	91.627	105.30	330.97	225.67	0.6142	1.6979
0.2600	209.08	0.7240	88.264	106.38	331.27	224.88	0.6194	1.6950
0.2700	209.88	0.7259	85.140	107.44	331.55	224.11	0.6244	1.6922
0.2800	210.66	0.7277	82.232	108.46	331.82	223.36	0.6292	1.6895
0.2900	211.42	0.7294	79.518	109.46	332.08	222.62	0.6339	1.6869
0.3000	212.15	0.7312	76.978	110.44	332.33	221.89	0.6385	1.6844
0.3100	212.87	0.7329	74.596	111.39	332.57	221.18	0.6429	1.6819
0.3200	213.58	0.7345	72.357	112.32	332.80	220.48	0.6473	1.6796
0.3300	214.26	0.7362	70.250	113.23	333.02	219.79	0.6515	1.6773
0.3400	214.93	0.7378	68.262	114.12	333.24	219.12	0.6556	1.6751
0.3500	215.59	0.7394	66.383	114.99	333.45	218.46	0.6596	1.6729
0.3600	216.23	0.7409	64.605	115.85	333.65	217.81	0.6635	1.6708
0.3700	216.86	0.7425	62.920	116.69	333.85	217.16	0.6674	1.6688
0.3800	217.47	0.7440	61.321	117.51	334.04	216.53	0.6711	1.6668
0.3900	218.07	0.7455	59.800	118.31	334.22	215.91	0.6748	1.6649
0.4000	218.66	0.7469	58.353	119.11	334.40	215.29	0.6784	1.6630
0.4100	219.24	0.7484	56.973	119.89	334.57	214.69	0.6819	1.6611
0.4200	219.81	0.7498	55.658	120.65	334.74	214.09	0.6854	1.6593
0.4300	220.37	0.7513	54.401	121.40	334.90	213.50	0.6888	1.6576
0.4400	220.92	0.7527	53.199	122.14	335.06	212.92	0.6921	1.6559
0.4500	221.46	0.7540	52.049	122.87	335.22	212.34	0.6953	1.6542
0.4600	221.99	0.7554	50.946	123.59	335.36	211.77	0.6985	1.6525
0.4700	222.51	0.7568	49.890	124.30	335.51	211.21	0.7017	1.6509
0.4800	223.03	0.7581	48.875	124.99	335.65	210.66	0.7048	1.6493
0.4900	223.53	0.7594	47.900	125.68	335.79	210.11	0.7078	1.6478
0.5000	224.03	0.7608	46.964	126.35	335.92	209.57	0.7108	1.6462

Chart R23.2: Saturation properties–*continued*

p MPa	T K	v' dm^3/kg	v'' dm^3/kg	h' kJ/kg	h'' kJ/kg	r kJ/kg	s' kJ/kgK	s'' kJ/kgK
0.5100	224.52	0.7621	46.062	127.02	336.05	209.03	0.7137	1.6447
0.5200	225.00	0.7634	45.194	127.68	336.18	208.50	0.7166	1.6433
0.5300	225.48	0.7646	44.357	128.33	336.30	207.97	0.7195	1.6418
0.5400	225.95	0.7659	43.550	128.97	336.42	207.45	0.7223	1.6404
0.5500	226.41	0.7672	42.772	129.60	336.54	206.93	0.7251	1.6390
0.5600	226.87	0.7684	42.020	130.23	336.65	206.42	0.7278	1.6377
0.5700	227.32	0.7696	41.294	130.85	336.76	205.92	0.7305	1.6363
0.5800	227.76	0.7709	40.592	131.46	336.87	205.41	0.7331	1.6350
0.5900	228.20	0.7721	39.913	132.06	336.97	204.92	0.7357	1.6337
0.6000	228.63	0.7733	39.255	132.65	337.08	204.42	0.7383	1.6324
0.6100	229.06	0.7745	38.619	133.24	337.18	203.93	0.7408	1.6311
0.6200	229.48	0.7757	38.002	133.83	337.27	203.45	0.7433	1.6299
0.6300	229.90	0.7769	37.404	134.40	337.37	202.97	0.7458	1.6287
0.6400	230.31	0.7781	36.824	134.97	337.46	202.49	0.7483	1.6275
0.6500	230.72	0.7792	36.261	135.54	337.55	202.02	0.7507	1.6263
0.6600	231.12	0.7804	35.715	136.09	337.64	201.55	0.7531	1.6251
0.6700	231.52	0.7815	35.184	136.65	337.73	201.08	0.7554	1.6239
0.6800	231.92	0.7827	34.669	137.19	337.81	200.62	0.7577	1.6228
0.6900	232.31	0.7838	34.168	137.74	337.89	200.16	0.7600	1.6216
0.7000	232.69	0.7850	33.680	138.27	337.97	199.70	0.7623	1.6205
0.7100	233.07	0.7861	33.206	138.80	338.05	199.25	0.7646	1.6194
0.7200	233.45	0.7872	32.745	139.33	338.13	198.80	0.7668	1.6183
0.7300	233.82	0.7883	32.295	139.85	338.20	198.35	0.7690	1.6173
0.7400	234.19	0.7895	31.858	140.37	338.27	197.90	0.7711	1.6162
0.7600	234.92	0.7917	31.016	141.39	338.41	197.02	0.7754	1.6141
0.7800	235.63	0.7938	30.215	142.39	338.54	196.15	0.7796	1.6121
0.8000	236.33	0.7960	29.454	143.37	338.67	195.30	0.7837	1.6101
0.8200	237.02	0.7982	28.727	144.34	338.79	194.45	0.7877	1.6081
0.8400	237.70	0.8003	28.035	145.29	338.90	193.61	0.7917	1.6062
0.8600	238.36	0.8024	27.373	146.23	339.01	192.77	0.7956	1.6043
0.8800	239.01	0.8045	26.740	147.16	339.11	191.95	0.7994	1.6025
0.9000	239.65	0.8066	26.135	148.07	339.21	191.14	0.8031	1.6007
0.9200	240.28	0.8086	25.554	148.96	339.30	190.33	0.8068	1.5989
0.9400	240.90	0.8107	24.998	149.85	339.38	189.53	0.8104	1.5972
0.9600	241.51	0.8127	24.463	150.72	339.46	188.74	0.8139	1.5955
0.9800	242.11	0.8148	23.950	151.58	339.54	187.96	0.8174	1.5938
1.0000	242.70	0.8168	23.457	152.43	339.61	187.18	0.8209	1.5921
1.0200	243.29	0.8188	22.982	153.27	339.68	186.41	0.8243	1.5905
1.0400	243.86	0.8208	22.524	154.10	339.75	185.64	0.8276	1.5889
1.0600	244.43	0.8228	22.084	154.92	339.81	184.88	0.8309	1.5873
1.0800	244.98	0.8248	21.659	155.73	339.86	184.13	0.8341	1.5857
1.1000	245.53	0.8267	21.248	156.53	339.91	183.38	0.8373	1.5842
1.1500	246.88	0.8316	20.283	158.49	340.03	181.53	0.8451	1.5805
1.2000	248.18	0.8365	19.395	160.40	340.12	179.72	0.8527	1.5768
1.2500	249.44	0.8413	18.575	162.27	340.19	177.93	0.8600	1.5733
1.3000	250.67	0.8461	17.816	164.08	340.25	176.16	0.8671	1.5699
1.3500	251.86	0.8509	17.110	165.86	340.28	174.42	0.8740	1.5665
1.4000	253.02	0.8557	16.453	167.60	340.30	172.70	0.8807	1.5633
1.4500	254.14	0.8604	15.840	169.30	340.30	170.99	0.8873	1.5601
1.5000	255.24	0.8651	15.265	170.97	340.28	169.31	0.8936	1.5570

Chart R23.2: Saturation properties–*continued*

p MPa	T K	v' dm^3/kg	v'' dm^3/kg	h' kJ/kg	h'' kJ/kg	r kJ/kg	s' kJ/kgK	s'' kJ/kgK
1.5500	256.32	0.8699	14.726	172.61	340.25	167.64	0.8999	1.5539
1.6000	257.36	0.8746	14.219	174.22	340.20	165.98	0.9060	1.5509
1.6500	258.39	0.8793	13.742	175.80	340.14	164.34	0.9119	1.5480
1.7000	259.39	0.8841	13.291	177.35	340.07	162.71	0.9178	1.5451
1.7500	260.36	0.8888	12.865	178.89	339.98	161.09	0.9235	1.5422
1.8000	261.32	0.8936	12.461	180.39	339.88	159.48	0.9291	1.5394
1.8500	262.26	0.8984	12.078	181.88	339.76	157.88	0.9346	1.5366
1.9000	263.18	0.9032	11.714	183.35	339.64	156.29	0.9400	1.5339
1.9500	264.08	0.9080	11.367	184.80	339.50	154.70	0.9454	1.5312
2.0000	264.96	0.9128	11.038	186.23	339.35	153.12	0.9506	1.5285
2.1000	266.68	0.9226	10.422	189.04	339.01	149.97	0.9608	1.5232
2.2000	268.34	0.9326	9.8591	191.79	338.63	146.84	0.9708	1.5180
2.3000	269.95	0.9427	9.3419	194.49	338.20	143.71	0.9804	1.5128
2.4000	271.50	0.9530	8.8647	197.14	337.72	140.58	0.9899	1.5077
2.5000	273.01	0.9635	8.4228	199.75	337.20	137.45	0.9991	1.5026
2.6000	274.46	0.9743	8.0121	202.32	336.63	134.31	1.0081	1.4975
2.7000	275.88	0.9854	7.6289	204.86	336.01	131.15	1.0170	1.4924
2.8000	277.26	0.9969	7.2704	207.38	335.35	127.97	1.0258	1.4873
2.9000	278.60	1.0087	6.9338	209.88	334.63	124.76	1.0344	1.4822
3.0000	279.91	1.0209	6.6169	212.36	333.86	121.51	1.0429	1.4770
3.1000	281.18	1.0336	6.3176	214.82	333.04	118.21	1.0513	1.4718
3.2000	282.42	1.0468	6.0341	217.29	332.15	114.87	1.0597	1.4664
3.3000	283.63	1.0607	5.7647	219.75	331.21	111.46	1.0680	1.4610
3.4000	284.82	1.0752	5.5081	222.21	330.20	107.99	1.0763	1.4555
3.5000	285.97	1.0906	5.2628	224.69	329.11	104.42	1.0846	1.4498
3.6000	287.10	1.1068	5.0277	227.18	327.95	100.77	1.0930	1.4439
3.7000	288.21	1.1241	4.8015	229.70	326.70	96.99	1.1013	1.4379
3.8000	289.29	1.1426	4.5831	232.26	325.35	93.08	1.1098	1.4316
3.9000	290.35	1.1627	4.3714	234.87	323.88	89.02	1.1184	1.4250
4.0000	291.38	1.1845	4.1652	237.53	322.29	84.76	1.1272	1.4180
4.1000	292.40	1.2085	3.9634	240.28	320.55	80.27	1.1362	1.4107
4.2000	293.39	1.2352	3.7644	243.14	318.63	75.49	1.1455	1.4028
4.3000	294.36	1.2656	3.5668	246.15	316.49	70.34	1.1553	1.3943
4.4000	295.31	1.3007	3.3683	249.35	314.07	64.72	1.1657	1.3849
4.5000	296.25	1.3427	3.1659	252.83	311.27	58.44	1.1771	1.3743
4.6000	297.16	1.3952	2.9547	256.75	307.94	51.19	1.1898	1.3621
4.7000	298.06	1.4657	2.7256	261.40	303.76	42.36	1.2049	1.3471
4.8000	298.91	1.9190	2.2963	281.29	293.41	12.11	1.2710	1.3115

Chart R23.3: Thermodynamic properties

p/MPa	0.10			0.15			0.20		
T K	v dm³/kg	h kJ/kg	s kJ/kgK	v dm³/kg	h kJ/kg	s kJ/kgK	v dm³/kg	h kJ/kg	s kJ/kgK
200.0	229.66	329.84	1.7956	150.38	328.35	1.7420	0.7040	94.620	0.5621
210.0	242.51	336.17	1.8265	159.31	334.88	1.7738	117.66	333.56	1.7350
220.0	255.20	342.57	1.8563	168.06	341.45	1.8044	124.46	340.30	1.7664
230.0	267.75	349.07	1.8851	176.67	348.08	1.8338	131.11	347.07	1.7965
240.0	280.19	355.66	1.9132	185.17	354.78	1.8624	137.64	353.89	1.8255
250.0	292.56	362.37	1.9406	193.58	361.59	1.8902	144.08	360.79	1.8537
260.0	304.85	369.20	1.9673	201.92	368.49	1.9172	150.45	367.78	1.8811
270.0	317.09	376.15	1.9936	210.20	375.51	1.9437	156.76	374.87	1.9079
280.0	329.28	383.23	2.0193	218.44	382.65	1.9697	163.02	382.07	1.9340
290.0	341.43	390.44	2.0446	226.64	389.91	1.9952	169.24	389.39	1.9597
300.0	353.55	397.79	2.0695	234.80	397.31	2.0202	175.42	396.82	1.9849
310.0	365.64	405.27	2.0941	242.93	404.83	2.0449	181.58	404.38	2.0097
320.0	377.71	412.89	2.1183	251.05	412.48	2.0692	187.71	412.07	2.0341
330.0	389.76	420.65	2.1421	259.14	420.27	2.0932	193.82	419.89	2.0582
340.0	401.79	428.55	2.1657	267.21	428.20	2.1168	199.92	427.85	2.0819
350.0	413.80	436.58	2.1890	275.27	436.26	2.1402	206.00	435.93	2.1054
360.0	425.81	444.75	2.2120	283.31	444.45	2.1633	212.07	444.14	2.1285
370.0	437.80	453.06	2.2348	291.35	452.78	2.1861	218.12	452.49	2.1514
380.0	449.78	461.50	2.2573	299.37	461.23	2.2086	224.17	460.96	2.1740
390.0	461.75	470.08	2.2796	307.39	469.82	2.2309	230.20	469.57	2.1963
400.0	473.72	478.78	2.3016	315.40	478.54	2.2530	236.23	478.30	2.2184
410.0	485.68	487.61	2.3234	323.40	487.39	2.2749	242.25	487.16	2.2403
420.0	497.63	496.57	2.3450	331.39	496.36	2.2965	248.27	496.14	2.2619
430.0	509.58	505.65	2.3664	339.38	505.45	2.3179	254.28	505.25	2.2834
440.0	521.52	514.86	2.3875	347.37	514.66	2.3391	260.29	514.47	2.3046
450.0	533.46	524.18	2.4085	355.35	523.99	2.3600	266.29	523.81	2.3256
460.0	545.39	533.62	2.4292	363.32	533.44	2.3808	272.29	533.27	2.3463
470.0	557.33	543.17	2.4498	371.29	543.00	2.4013	278.28	542.84	2.3669
p/MPa	0.30			0.40			0.50		
200.0	0.7039	94.648	0.5619	0.7038	94.675	0.5616	0.7037	94.703	0.5614
210.0	0.7261	107.60	0.6251	0.7260	107.63	0.6248	0.7258	107.65	0.6246
220.0	80.785	337.91	1.7102	58.869	335.40	1.6675	0.7502	120.92	0.6863
230.0	85.493	344.99	1.7417	62.631	342.83	1.7006	48.864	340.57	1.6667
240.0	90.072	352.07	1.7718	66.251	350.18	1.7319	51.923	348.23	1.6993
250.0	94.553	359.17	1.8008	69.762	357.51	1.7618	54.863	355.80	1.7302
260.0	98.954	366.34	1.8289	73.188	364.86	1.7906	57.711	363.34	1.7598
270.0	103.29	373.57	1.8562	76.547	372.25	1.8185	60.487	370.90	1.7883
280.0	107.58	380.89	1.8828	79.852	379.70	1.8456	63.206	378.49	1.8159
290.0	111.83	388.32	1.9089	83.114	387.23	1.8720	65.879	386.14	1.8428
300.0	116.04	395.85	1.9344	86.339	394.86	1.8979	68.514	393.86	1.8690
310.0	120.22	403.49	1.9595	89.533	402.59	1.9232	71.118	401.67	1.8946
320.0	124.37	411.25	1.9841	92.702	410.42	1.9481	73.696	409.58	1.9197
330.0	128.51	419.13	2.0084	95.850	418.37	1.9725	76.252	417.59	1.9443
340.0	132.63	427.14	2.0323	98.979	426.43	1.9966	78.789	425.71	1.9686
350.0	136.73	435.27	2.0558	102.09	434.61	2.0203	81.310	433.94	1.9924
360.0	140.82	443.53	2.0791	105.19	442.91	2.0437	83.817	442.29	2.0159
370.0	144.90	451.91	2.1021	108.28	451.34	2.0668	86.312	450.76	2.0391
380.0	148.96	460.42	2.1248	111.36	459.88	2.0896	88.797	459.34	2.0620
390.0	153.02	469.06	2.1472	114.43	468.55	2.1121	91.272	468.04	2.0846

Chart R23.3: Thermodynamic properties–*continued*

p/MPa	0.30			0.40			0.50		
T K	v dm³/kg	h kJ/kg	s kJ/kgK	v dm³/kg	h kJ/kg	s kJ/kgK	v dm³/kg	h kJ/kg	s kJ/kgK
400.0	157.07	477.82	2.1694	117.49	477.34	2.1343	93.740	476.86	2.1070
410.0	161.11	486.71	2.1913	120.54	486.25	2.1563	96.200	485.80	2.1290
420.0	165.15	495.71	2.2130	123.59	495.28	2.1781	98.655	494.86	2.1508
430.0	169.18	504.84	2.2345	126.63	504.43	2.1996	101.10	504.03	2.1724
440.0	173.21	514.08	2.2558	129.67	513.70	2.2209	103.55	513.31	2.1938
450.0	177.23	523.44	2.2768	132.70	523.08	2.2420	105.99	522.71	2.2149
460.0	181.25	532.92	2.2976	135.73	532.57	2.2629	108.42	532.22	2.2358
470.0	185.26	542.50	2.3182	138.76	542.17	2.2835	110.85	541.83	2.2565

p/MPa	0.60			0.80			1.00		
200.0	0.7036	94.731	0.5612	0.7033	94.787	0.5608	0.7031	94.842	0.5604
210.0	0.7257	107.68	0.6244	0.7254	107.72	0.6239	0.7252	107.77	0.6234
220.0	0.7500	120.94	0.6861	0.7496	120.97	0.6855	0.7493	121.01	0.6850
230.0	39.638	338.19	1.6373	0.7768	134.56	0.7459	0.7763	134.58	0.7453
240.0	42.340	346.19	1.6713	30.275	341.86	1.6235	0.8074	148.56	0.8048
250.0	44.908	354.03	1.7033	32.408	350.32	1.6580	24.836	346.31	1.6193
260.0	47.377	361.79	1.7338	34.420	358.56	1.6903	26.599	355.13	1.6539
270.0	49.768	369.52	1.7629	36.343	366.67	1.7209	28.255	363.68	1.6862
280.0	52.100	377.25	1.7910	38.198	374.72	1.7502	29.833	372.08	1.7167
290.0	54.383	385.02	1.8183	39.999	382.74	1.7784	31.351	380.39	1.7459
300.0	56.627	392.85	1.8448	41.757	390.79	1.8057	32.823	388.68	1.7740
310.0	58.838	400.75	1.8708	43.480	398.88	1.8322	34.256	396.96	1.8011
320.0	61.023	408.73	1.8961	45.175	407.02	1.8580	35.660	405.27	1.8275
330.0	63.184	416.81	1.9210	46.846	415.24	1.8833	37.038	413.64	1.8533
340.0	65.327	424.99	1.9454	48.497	423.54	1.9081	38.395	422.06	1.8784
350.0	67.453	433.28	1.9694	50.130	431.93	1.9324	39.734	430.56	1.9031
360.0	69.565	441.67	1.9930	51.749	440.41	1.9563	41.058	439.15	1.9272
370.0	71.665	450.18	2.0163	53.356	449.00	1.9799	42.369	447.82	1.9510
380.0	73.755	458.80	2.0393	54.952	457.70	2.0030	43.669	456.59	1.9744
390.0	75.835	467.53	2.0620	56.538	466.50	2.0259	44.960	465.47	1.9974
400.0	77.907	476.38	2.0844	58.116	475.41	2.0485	46.242	474.44	2.0202
410.0	79.972	485.34	2.1066	59.687	484.43	2.0707	47.516	483.51	2.0426
420.0	82.031	494.42	2.1284	61.251	493.56	2.0927	48.784	492.69	2.0647
430.0	84.084	503.62	2.1501	62.810	502.80	2.1145	50.046	501.98	2.0865
440.0	86.132	512.92	2.1715	64.363	512.15	2.1360	51.303	511.37	2.1081
450.0	88.175	522.34	2.1926	65.912	521.60	2.1572	52.555	520.86	2.1295
460.0	90.215	531.86	2.2135	67.457	531.16	2.1782	53.803	530.46	2.1505
470.0	92.251	541.50	2.2343	68.998	540.83	2.1990	55.048	540.15	2.1714

p/MPa	2.00			4.00			6.00		
200.0	0.7021	95.124	0.5583	0.7001	95.699	0.5541	0.6982	96.290	0.5501
210.0	0.7238	108.02	0.6212	0.7213	108.53	0.6167	0.7188	109.06	0.6124
220.0	0.7475	121.21	0.6825	0.7442	121.62	0.6776	0.7410	122.08	0.6729
230.0	0.7739	134.70	0.7425	0.7694	134.99	0.7370	0.7652	135.32	0.7318
240.0	0.8041	148.57	0.8015	0.7979	148.66	0.7952	0.7922	148.83	0.7893
250.0	0.8399	162.97	0.8603	0.8310	162.75	0.8527	0.8231	162.67	0.8458
260.0	0.8849	178.19	0.9200	0.8711	177.45	0.9104	0.8595	176.98	0.9019
270.0	11.701	345.58	1.5518	0.9227	193.12	0.9695	0.9043	191.98	0.9585
280.0	12.863	356.85	1.5928	0.9971	210.60	1.0330	0.9631	208.08	1.0171
290.0	13.896	367.23	1.6292	1.1435	233.11	1.1119	1.0493	226.16	1.0805

Chart R23.3: Thermodynamic properties—*continued*

p/MPa	2.00			4.00			6.00		
T K	v dm³/kg	h kJ/kg	s kJ/kgK	v dm³/kg	h kJ/kg	s kJ/kgK	v dm³/kg	h kJ/kg	s kJ/kgK
300.0	14.843	377.10	1.6627	5.3125	343.68	1.4905	1.2145	249.11	1.1582
310.0	15.731	386.64	1.6940	6.1592	359.91	1.5438	2.0964	299.83	1.3240
320.0	16.573	395.99	1.7236	6.8384	373.40	1.5866	3.2781	339.05	1.4489
330.0	17.382	405.21	1.7520	7.4299	385.57	1.6240	3.9591	359.66	1.5123
340.0	18.163	414.36	1.7794	7.9659	396.96	1.6581	4.4856	375.72	1.5603
350.0	18.921	423.49	1.8058	8.4631	407.86	1.6897	4.9343	389.71	1.6009
360.0	19.662	432.62	1.8315	8.9314	418.45	1.7195	5.3347	402.54	1.6370
370.0	20.387	441.77	1.8566	9.3776	428.81	1.7479	5.7021	414.63	1.6701
380.0	21.099	450.95	1.8811	9.8060	439.03	1.7751	6.0452	426.23	1.7011
390.0	21.800	460.19	1.9051	10.220	449.16	1.8014	6.3696	437.50	1.7303
400.0	22.492	469.50	1.9286	10.622	459.24	1.8270	6.6793	448.52	1.7583
410.0	23.176	478.87	1.9518	11.014	469.29	1.8518	6.9770	459.39	1.7851
420.0	23.852	488.32	1.9745	11.398	479.34	1.8760	7.2648	470.14	1.8110
430.0	24.522	497.84	1.9970	11.774	489.41	1.8997	7.5443	480.82	1.8361
440.0	25.186	507.45	2.0190	12.144	499.50	1.9229	7.8168	491.46	1.8606
450.0	25.846	517.14	2.0408	12.508	509.63	1.9457	8.0831	502.07	1.8844
460.0	26.501	526.92	2.0623	12.868	519.81	1.9680	8.3443	512.67	1.9077
470.0	27.152	536.79	2.0835	13.223	530.03	1.9900	8.6008	523.29	1.9306
p/MPa	8.00			10.00			15.00		
200.0	0.6964	96.895	0.5461	0.6945	97.514	0.5423	0.6902	99.114	0.5330
210.0	0.7165	109.61	0.6082	0.7142	110.18	0.6041	0.7088	111.69	0.5943
220.0	0.7379	122.56	0.6684	0.7350	123.06	0.6640	0.7283	124.43	0.6536
230.0	0.7613	135.70	0.7268	0.7575	136.12	0.7220	0.7490	137.31	0.7108
240.0	0.7870	149.07	0.7837	0.7822	149.36	0.7784	0.7713	150.31	0.7662
250.0	0.8160	162.71	0.8394	0.8096	162.83	0.8334	0.7956	163.45	0.8198
260.0	0.8495	176.71	0.8943	0.8407	176.59	0.8873	0.8224	176.76	0.8720
270.0	0.8895	191.23	0.9491	0.8770	190.75	0.9408	0.8524	190.30	0.9231
280.0	0.9391	206.51	1.0047	0.9205	205.46	0.9943	0.8865	204.14	0.9734
290.0	1.0047	222.94	1.0623	0.9748	220.95	1.0486	0.9260	218.36	1.0233
300.0	1.0998	241.29	1.1245	1.0457	237.55	1.1049	0.9726	233.05	1.0731
310.0	1.2624	263.31	1.1966	1.1449	255.80	1.1647	1.0284	248.33	1.1232
320.0	1.6090	292.59	1.2895	1.2945	276.45	1.2302	1.0964	264.30	1.1739
330.0	2.1882	324.91	1.3890	1.5286	300.02	1.3027	1.1801	281.01	1.2253
340.0	2.7281	349.63	1.4629	1.8526	324.75	1.3766	1.2831	298.46	1.2774
350.0	3.1706	368.73	1.5183	2.2049	347.28	1.4419	1.4078	316.48	1.3296
360.0	3.5475	384.88	1.5638	2.5357	366.67	1.4965	1.5528	334.68	1.3809
370.0	3.8814	399.32	1.6033	2.8351	383.63	1.5430	1.7126	352.58	1.4300
380.0	4.1849	412.69	1.6390	3.1074	398.93	1.5838	1.8796	369.80	1.4759
390.0	4.4662	425.35	1.6719	3.3581	413.09	1.6206	2.0472	386.15	1.5184
400.0	4.7305	437.50	1.7027	3.5917	426.45	1.6545	2.2113	401.65	1.5576
410.0	4.9812	449.30	1.7318	3.8117	439.24	1.6860	2.3701	416.39	1.5940
420.0	5.2210	460.84	1.7596	4.0206	451.62	1.7159	2.5231	430.50	1.6280
430.0	5.4518	472.19	1.7863	4.2204	463.68	1.7442	2.6704	444.10	1.6600
440.0	5.6750	483.42	1.8121	4.4127	475.51	1.7714	2.8124	457.28	1.6903
450.0	5.8918	494.54	1.8371	4.5984	487.17	1.7976	2.9496	470.14	1.7192
460.0	6.1032	505.61	1.8614	4.7787	498.70	1.8230	3.0826	482.74	1.7469
470.0	6.3097	516.63	1.8852	4.9542	510.14	1.8476	3.2117	495.14	1.7736

Chart R23.3: Thermodynamic properties–*continued*

p/MPa	20.00			35.00			58.00		
T K	v dm³/kg	h kJ/kg	s kJ/kgK	v dm³/kg	h kJ/kg	s kJ/kgK	v dm³/kg	h kJ/kg	s kJ/kgK
200.0	0.6862	100.78	0.5241	0.6755	106.12	0.4997	0.6618	114.99	0.4672
210.0	0.7038	113.28	0.5851	0.6909	118.48	0.5601	0.6749	127.31	0.5273
220.0	0.7222	125.92	0.6439	0.7066	130.94	0.6180	0.6879	139.69	0.5849
230.0	0.7415	138.66	0.7005	0.7227	143.44	0.6736	0.7011	152.06	0.6399
240.0	0.7619	151.49	0.7551	0.7394	155.95	0.7268	0.7143	164.39	0.6924
250.0	0.7838	164.39	0.8078	0.7567	168.46	0.7779	0.7277	176.65	0.7424
260.0	0.8075	177.40	0.8588	0.7748	180.96	0.8269	0.7415	188.84	0.7902
270.0	0.8334	190.55	0.9084	0.7938	193.47	0.8741	0.7555	200.96	0.8360
280.0	0.8621	203.87	0.9569	0.8140	206.01	0.9197	0.7700	213.03	0.8799
290.0	0.8941	217.41	1.0044	0.8354	218.60	0.9639	0.7849	225.06	0.9221
300.0	0.9301	231.23	1.0512	0.8583	231.26	1.0068	0.8004	237.07	0.9628
310.0	0.9710	245.36	1.0976	0.8828	244.02	1.0487	0.8166	249.09	1.0022
320.0	1.0178	259.86	1.1436	0.9091	256.90	1.0896	0.8334	261.13	1.0405
330.0	1.0715	274.73	1.1893	0.9374	269.91	1.1296	0.8510	273.22	1.0777
340.0	1.1331	289.97	1.2348	0.9678	283.06	1.1688	0.8693	285.35	1.1139
350.0	1.2034	305.56	1.2800	1.0005	296.35	1.2074	0.8885	297.56	1.1493
360.0	1.2828	321.40	1.3247	1.0354	309.77	1.2452	0.9085	309.84	1.1839
370.0	1.3711	337.38	1.3684	1.0727	323.32	1.2823	0.9294	322.21	1.2178
380.0	1.4672	353.34	1.4110	1.1122	336.96	1.3187	0.9511	334.66	1.2510
390.0	1.5694	369.13	1.4520	1.1540	350.69	1.3543	0.9737	347.20	1.2835
400.0	1.6755	384.62	1.4912	1.1979	364.47	1.3892	0.9972	359.82	1.3155
410.0	1.7835	399.73	1.5285	1.2437	378.28	1.4233	1.0215	372.52	1.3468
420.0	1.8918	414.42	1.5640	1.2912	392.08	1.4566	1.0466	385.28	1.3776
430.0	1.9993	428.72	1.5976	1.3401	405.84	1.4890	1.0723	398.11	1.4078
440.0	2.1053	442.65	1.6296	1.3902	419.56	1.5205	1.0988	410.99	1.4374
450.0	2.2093	456.24	1.6602	1.4412	433.20	1.5511	1.1259	423.91	1.4664
460.0	2.3112	469.55	1.6894	1.4928	446.75	1.5809	1.1534	436.86	1.4949
470.0	2.4109	482.62	1.7175	1.5448	460.21	1.6099	1.1815	449.84	1.5228

References for R23

[Belz76] Belzille, J.L. ; Kaliaguine, S, and Ramalho, R.S.: PVT study of Trifluoromethane by the Burnett method; The Canadian Journal of Chemical Engineering 54(1976) 446/450

[Bier85] Bier, K.: private communications (1985) ; see also *[Frey86]* and *[Bott86]* ;

[Bott86] Bott, T.: Untersuchung des realen Mischverhaltens einer gasförmigen Mischung aus Schwefelhexafluorid und Trifluormethan anhand von kalorischen Messungen; Dissertation, Universität Karlsruhe (1986)

[Bueh82] Bühner, K.: Messung der thermischen Eigenschaften von binären Gasgemischen aus Helium, Schwefelhexafluorid und Trifluormethan und ihre Darstellung durch Zustandsgleichungen ; Dissertation, Universität Karlsruhe (1982)

[Frey86] Freyhof, R.: Untersuchungen zur Temperaturabhängigkeit des thermischen Verhaltens binärer Gasmischungen in einem großen Dichtebereich ; Dissertation, Universität Karlsruhe (1986)

[Doer68] Döring, R. and Löffler, H.J.: Thermodynamische Eigenschaften von Trifluormethan (R23); Kältetechnik-Klimatisierung 20(1968)11 342/348

[Gell79] Geller, V.Z. ; Poritschanskij, E.G. ; Svetliunij, P.I. and Elkin, JU.G.: Density and EOS of liquid Freons 13, 22, 23 and 152A; Cholodil'naja technika i technologija 29(1979) 43/48

[Gruz75] Gruzdev, V.A. and Sumskaja, A.I.: Experimental Investigation of Isobaric Heat Capacity of Freons: R-11, R-12, R-13, R-21, R-22, R-23; Thermophysical Properties of Matter and Substances (1975)8 108/120

[Hou55] Hou, Y.C.: Physical and thermodynamic properties of trifluoromethane; Dissertation University of Michigan (1955)

[KMR 81] - : Kältemaschinenregeln, 7. Auflage; Verlag C.F. Müller, Karlsruhe (1981)

[Lang70] Lange Jr., H.B. and Stein F.P.: Volumetric behavior of a polar-nonpolar gas mixture: trifluoromethane-tetrafluoromethane system; J. Chem. Eng. Data 15(1970)1 56/60

[Loef68] Döring, R. and Löffler, H.J.: see *[Doer68]* ;

[Rass75] Rassakov, D.S. ; Petrov,E.K. ; Spiridonov, G.A. and Uschmaikin, S.R.: Investigation of the p-v-T dependence of Freon-23; Thermophysical properties of matter and substances 8(1975) 4/16

[Rodg74] Rodgers, A.S.; Chao, J.; Wilhoit, R.C. and Zwolinski, B.J.: Ideal Gas Thermodynamic Properties of Eight Chloro- and Fluoromethanes; J. Phys. Chem. Ref. Data 3(1974)1 117/140

[Solo78] Solov'ev, A.N. ; Suchinin, G.I. ; Stoljarov, N.N. and Tschaschkin, Ju.R.: Experimental Determination of the Heat of Vaporization and Specific Heat along the Saturation Line of Freon 23; Cholodil'naja technika i technologija (1978)6 30/33

[Timo75] Timoshenko, N.I.: Refractive index, polarisibility and density of Freon-23; Thermophysical Properties of Matter and Substances 8(1975) 17/26

[Wagn68] Wagner, W.: Thermodynamische Eigenschaften von Trifluormethan (R23); Kältetechnik-Klimatisierung 20(1968)8 238/240

[Wata83] Watanabe, K.: An experimental study of thermodynamic properties of trifluoromethane ; 8th Symposium on Thermophysical Properties (1982) 380/386

R 23. ph

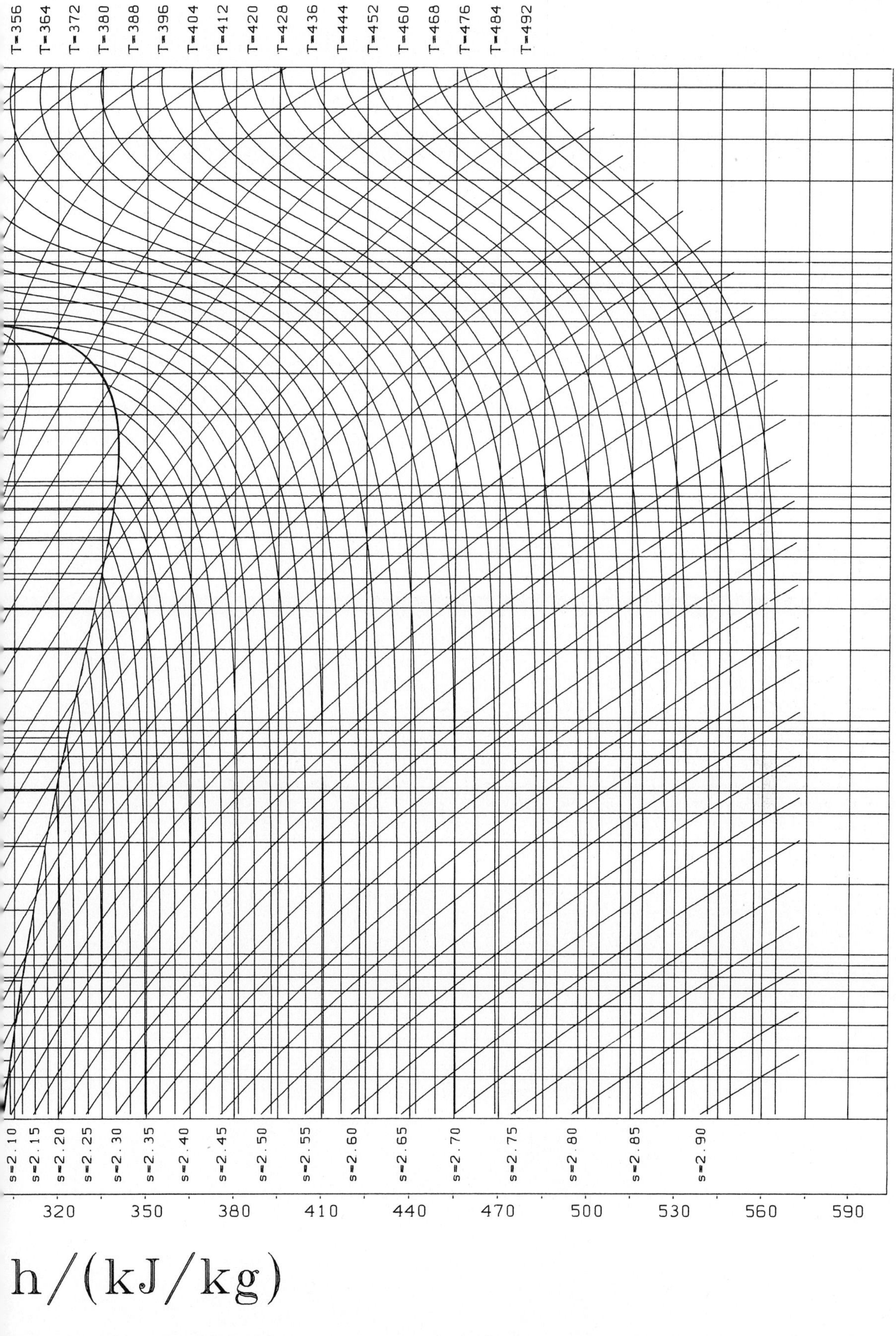
T=356
T=364
T=372
T=380
T=388
T=396
T=404
T=412
T=420
T=428
T=436
T=444
T=452
T=460
T=468
T=476
T=484
T=492
s=2.10
s=2.15
s=2.20
s=2.25
s=2.30
s=2.35
s=2.40
s=2.45
s=2.50
s=2.55
s=2.60
s=2.65
s=2.70
s=2.75
s=2.80
s=2.85
s=2.90
320
350
380
410
440
470
500
530
560
590
h/(kJ/kg)

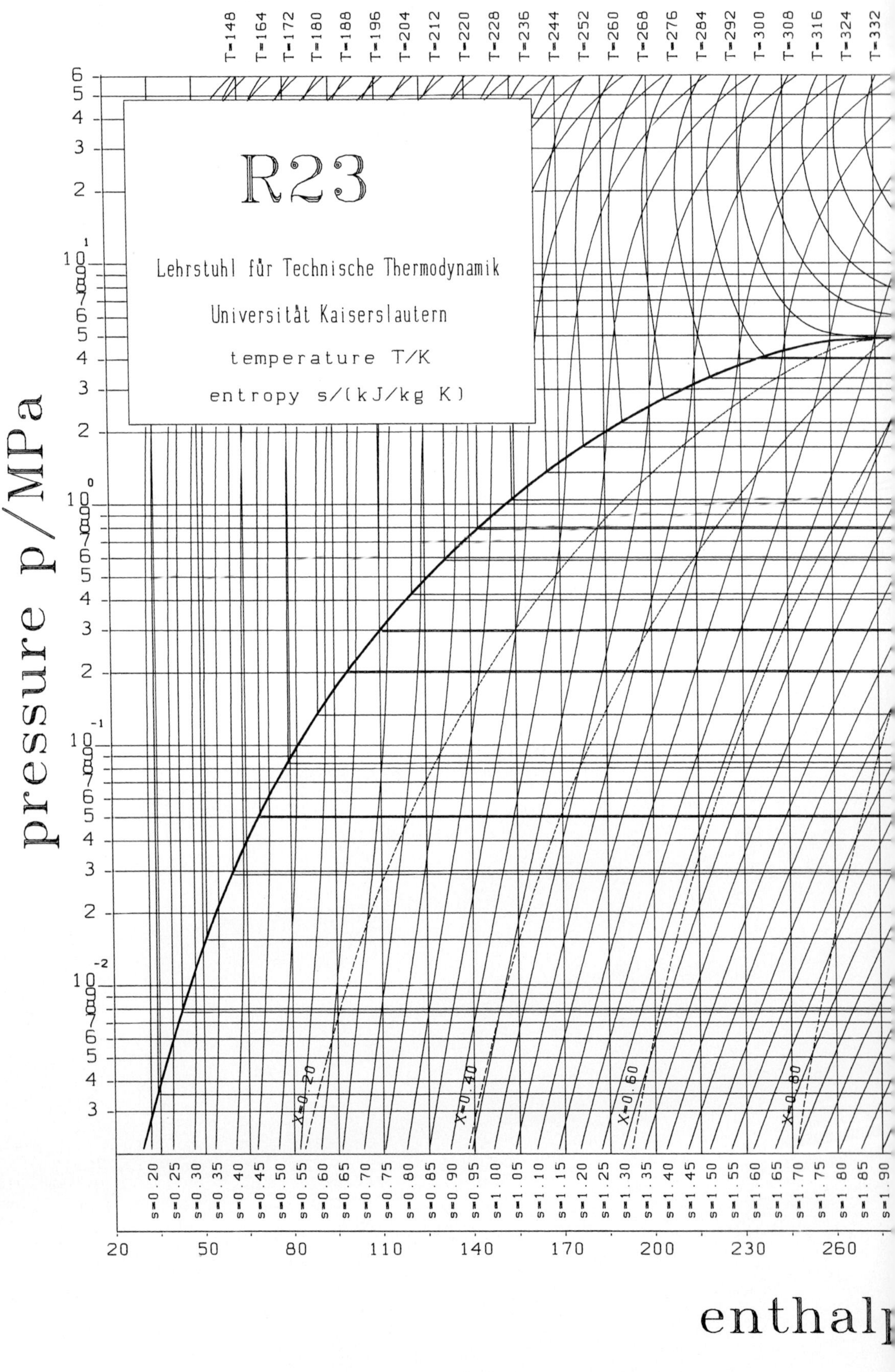
R23
Lehrstuhl für Technische Thermodynamik
Universität Kaiserslautern
temperature T/K
entropy s/(kJ/kg K)
pressure p/MPa
enthal
T=148
T=164
T=172
T=180
T=188
T=196
T=204
T=212
T=220
T=228
T=236
T=244
T=252
T=260
T=268
T=276
T=284
T=292
T=300
T=308
T=316
T=324
T=332
X=0.20
X=0.40
X=0.60
X=0.80
s=0.20
s=0.25
s=0.30
s=0.35
s=0.40
s=0.45
s=0.50
s=0.55
s=0.60
s=0.65
s=0.70
s=0.75
s=0.80
s=0.85
s=0.90
s=0.95
s=1.00
s=1.05
s=1.10
s=1.15
s=1.20
s=1.25
s=1.30
s=1.35
s=1.40
s=1.45
s=1.50
s=1.55
s=1.60
s=1.65
s=1.70
s=1.75
s=1.80
s=1.85
s=1.90
20
50
80
110
140
170
200
230
260

R 23.Ts

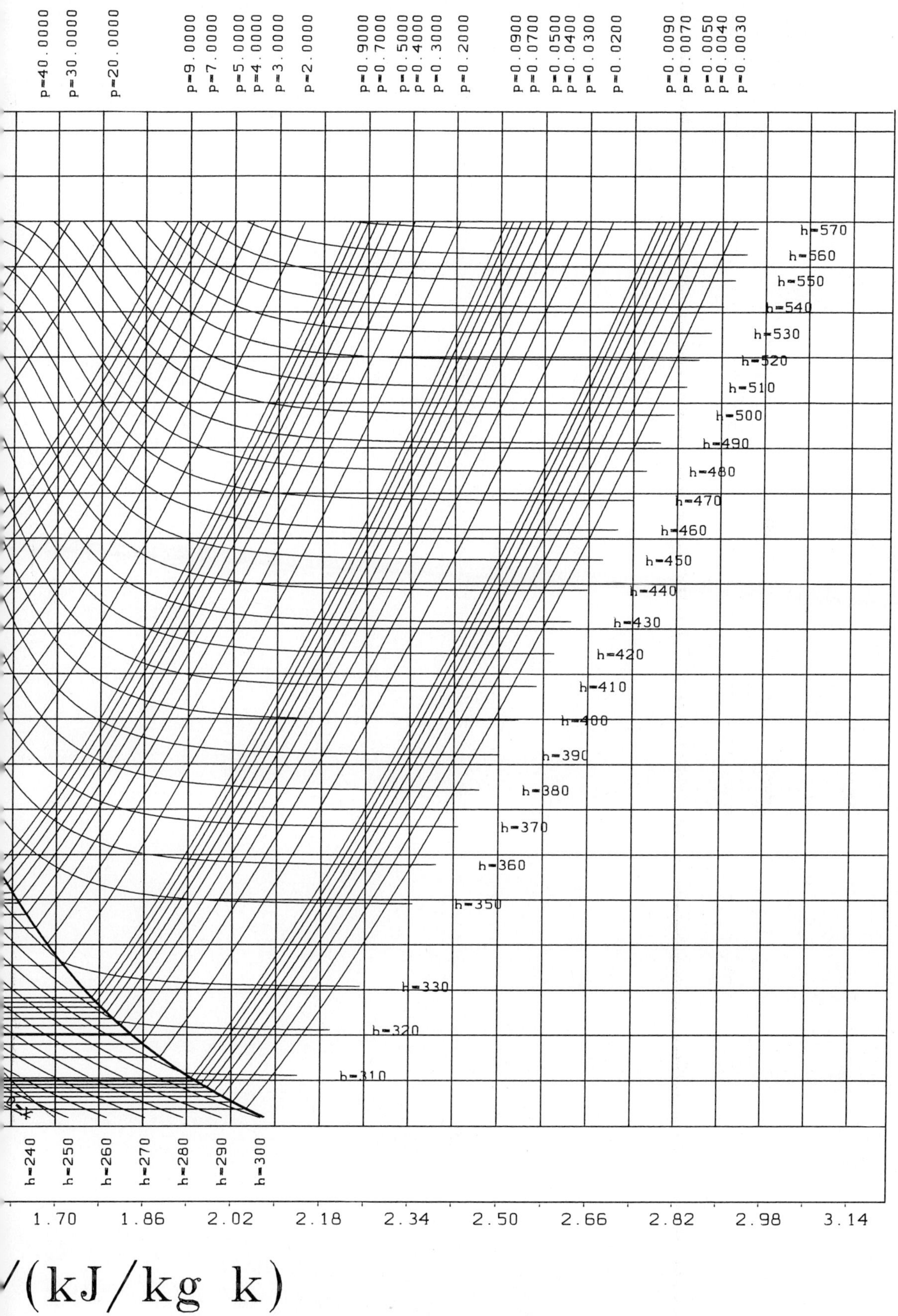

p=40.0000
p=30.0000
p=20.0000
p=9.0000
p=7.0000
p=5.0000
p=4.0000
p=3.0000
p=2.0000
p=0.9000
p=0.7000
p=0.5000
p=0.4000
p=0.3000
p=0.2000
p=0.0900
p=0.0700
p=0.0500
p=0.0400
p=0.0300
p=0.0200
p=0.0090
p=0.0070
p=0.0050
p=0.0040
p=0.0030
h=570
h=560
h=550
h=540
h=530
h=520
h=510
h=500
h=490
h=480
h=470
h=460
h=450
h=440
h=430
h=420
h=410
h=400
h=390
h=380
h=370
h=360
h=350
h=330
h=320
h=310
h=240
h=250
h=260
h=270
h=280
h=290
h=300
1.70
1.86
2.02
2.18
2.34
2.50
2.66
2.82
2.98
3.14
/(kJ/kg k)

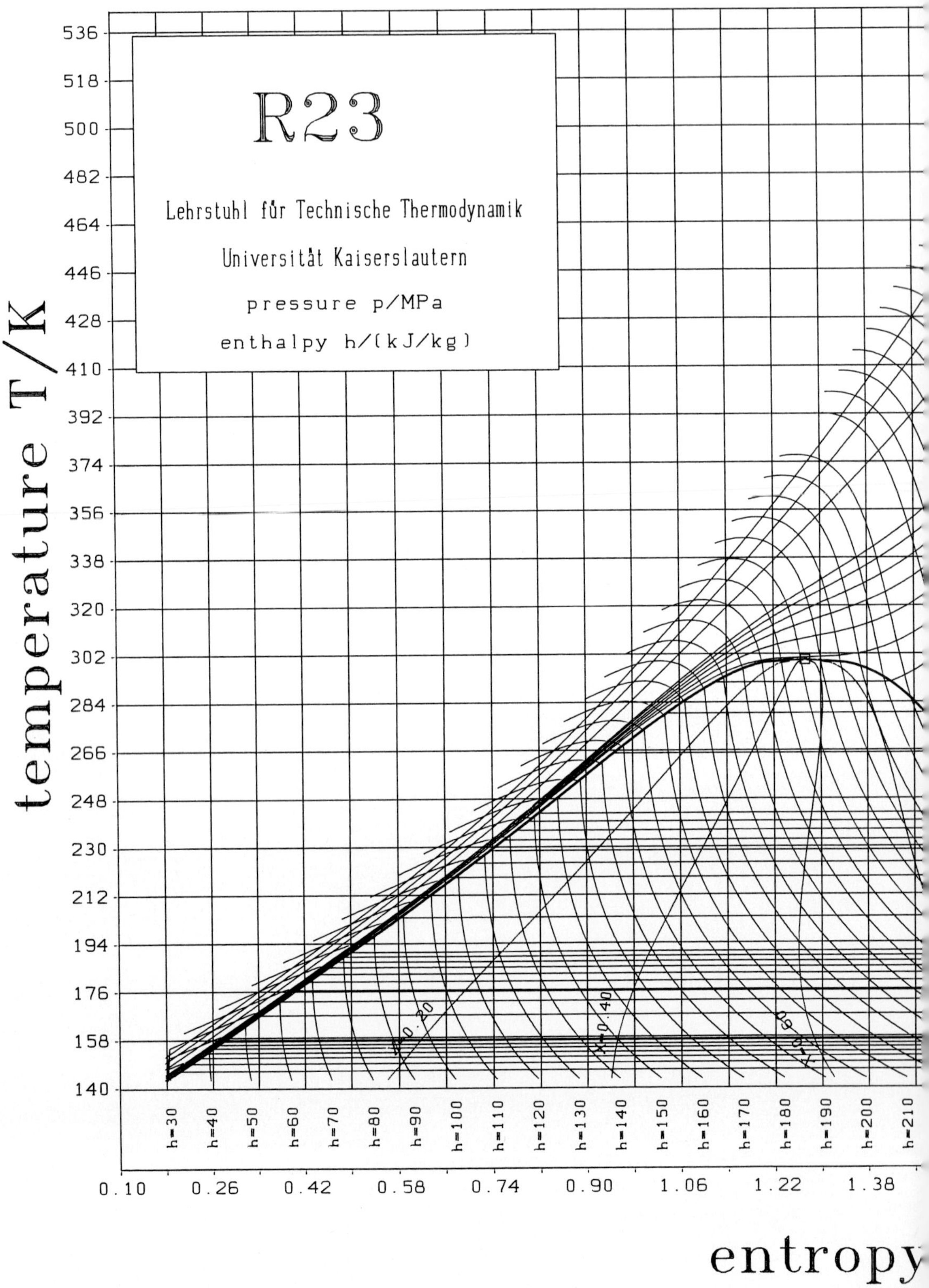

R23
Lehrstuhl für Technische Thermodynamik
Universität Kaiserslautern
pressure p/MPa
enthalpy h/(kJ/kg)
temperature T/K
entropy
536
518
500
482
464
446
428
410
392
374
356
338
320
302
284
266
248
230
212
194
176
158
140
0.10
0.26
0.42
0.58
0.74
0.90
1.06
1.22
1.38
h=30
h=40
h=50
h=60
h=70
h=80
h=90
h=100
h=110
h=120
h=130
h=140
h=150
h=160
h=170
h=180
h=190
h=200
h=210
X=0.20
X=0.40
X=0.60

R 23. hs

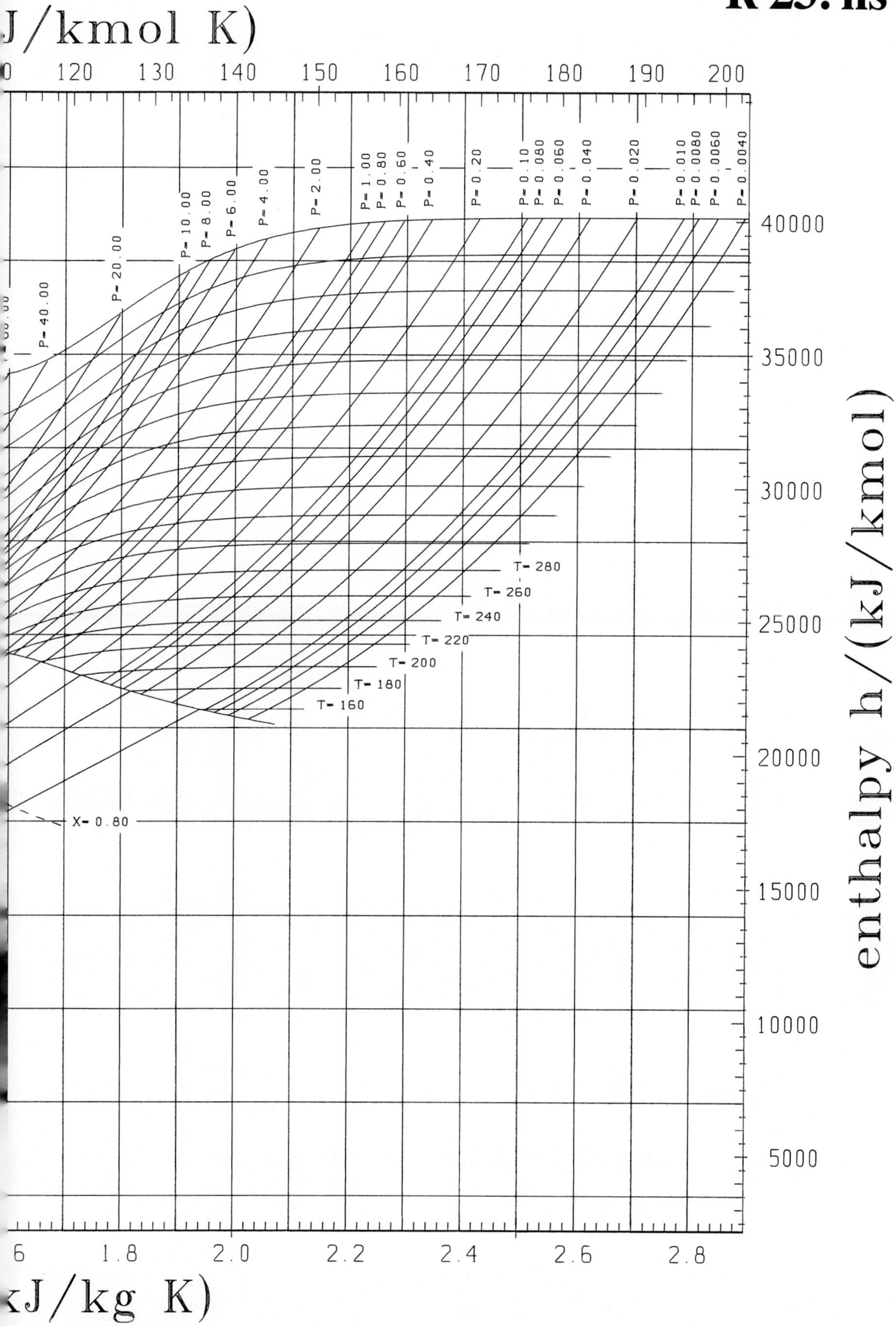

J/kmol K)
0
120
130
140
150
160
170
180
190
200
P= 40.00
P= 20.00
P= 10.00
P= 8.00
P= 6.00
P= 4.00
P= 2.00
P= 1.00
P= 0.80
P= 0.60
P= 0.40
P= 0.20
P= 0.10
P= 0.080
P= 0.060
P= 0.040
P= 0.020
P= 0.010
P= 0.0080
P= 0.0060
P= 0.0040
T= 280
T= 260
T= 240
T= 220
T= 200
T= 180
T= 160
X= 0.80
40000
35000
30000
25000
20000
15000
10000
5000
enthalpy h/(kJ/kmol)
6
1.8
2.0
2.2
2.4
2.6
2.8
kJ/kg K)

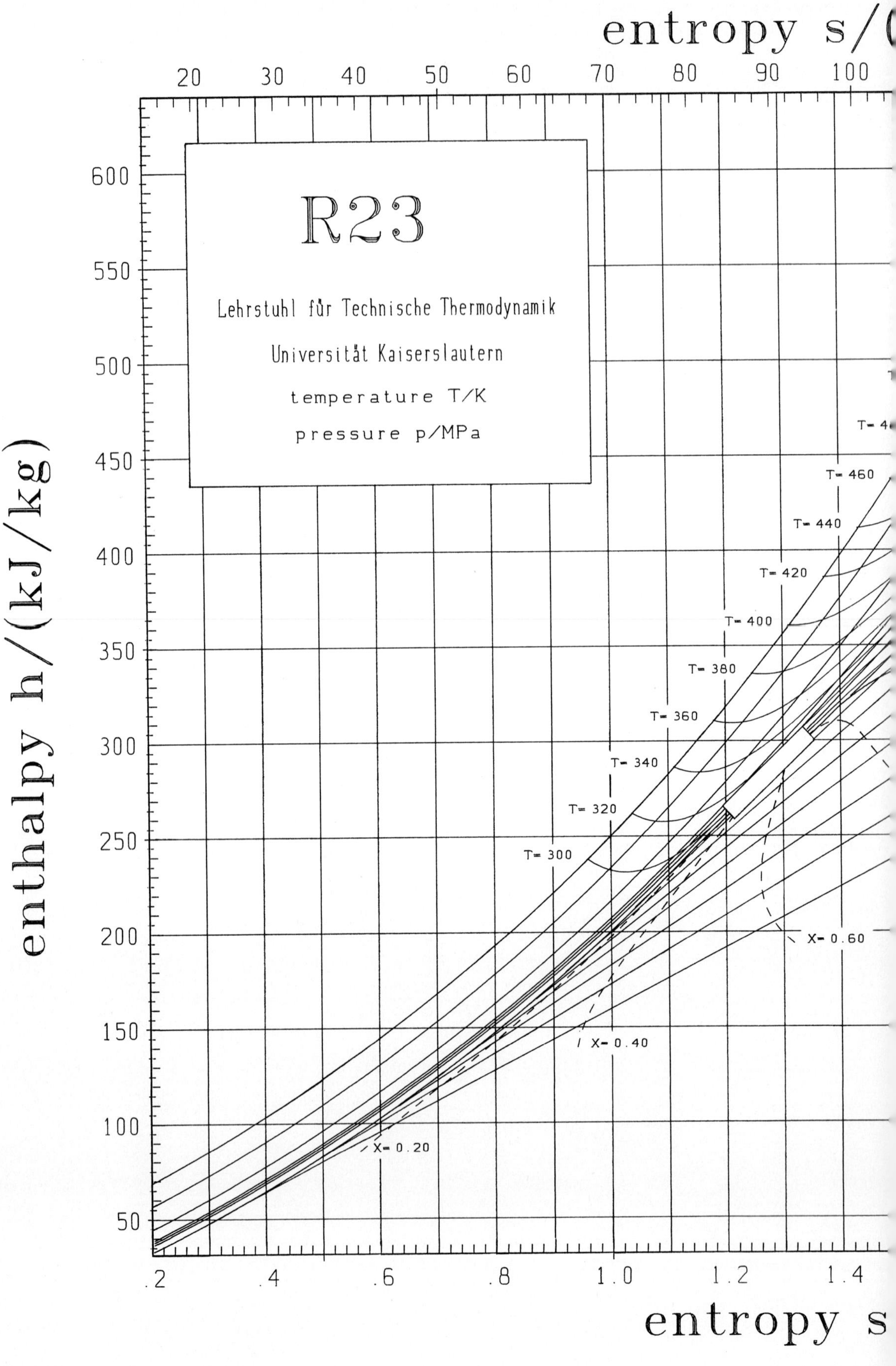

entropy s/(
20
30
40
50
60
70
80
90
100
R23
Lehrstuhl für Technische Thermodynamik
Universität Kaiserslautern
temperature T/K
pressure p/MPa
enthalpy h/(kJ/kg)
600
550
500
450
400
350
300
250
200
150
100
50
T= 460
T= 440
T= 420
T= 400
T= 380
T= 360
T= 340
T= 320
T= 300
X= 0.60
X= 0.40
X= 0.20
.2
.4
.6
.8
1.0
1.2
1.4
entropy s

Refrigerant R12B1 ($CBrClF_2$)

Table R12B1.1: List of references for p, ρ, T data

Author	n	p/MPa min	p/MPa max	ρ/g cm^{-3} min	ρ/g cm^{-3} max	T/K min	T/K max
DÖRING, R. *[Doer 75]*	78	0.14	1.05	0.0104	0.07	284	368
PERELSTEIN, I.I. *[Pere 73]*	48	0.94	6.20	0.0609	0.61	361	462

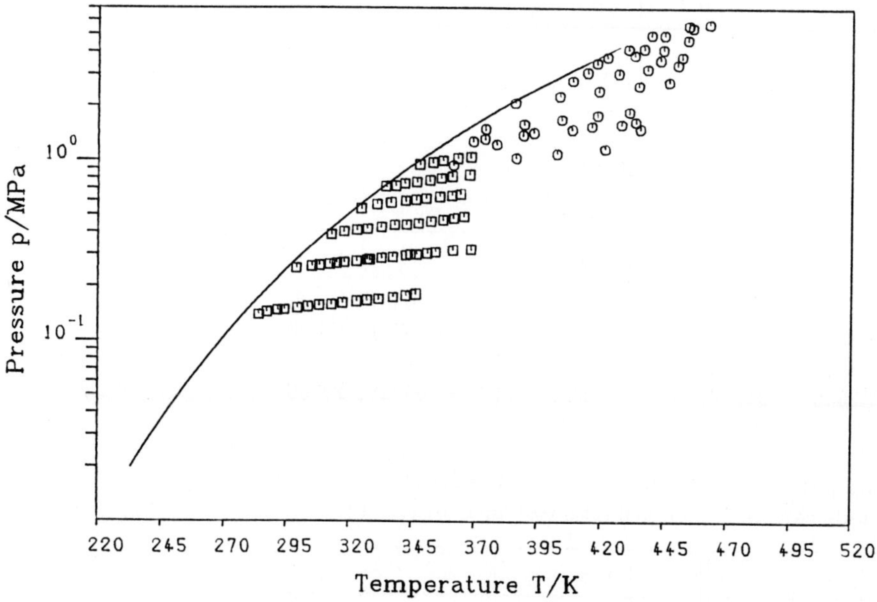

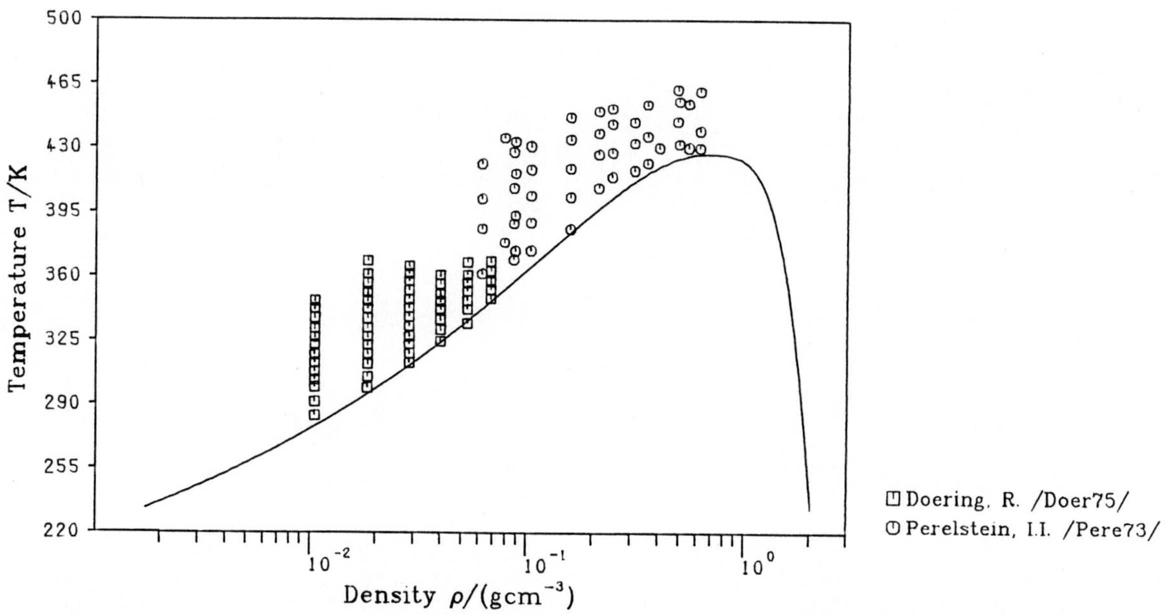

Fig.R12B1.1: Distribution of p, ρ, T data projected in the p, T and in the T, ρ plane

Table R12B1.2: Experimental vapor pressure data

Author	Δ / % std	opt	*EOS*
Döring, R. *[Doer 75]*	2.40	2.40	2.40
Perelstein, J.J. *[Pere 73]*	0.05	0.05	0.06

Table R12B1.3: Experimental saturated liquid density data

Author	Δ / % std	opt	*EOS*
Döring, R. *[Doer 75]*	0.05	0.06	0.15
Perelstein, J.J. *[Pere 73]*	0.24	0.18	0.05

Table R12B1.4: Coefficients of the standard vapor pressure equation

Index i	b_i
5	−6.715733 E+00
6	1.002208 E+00
9	−1.994656 E+00
15	−7.742518 E−01

Table R12B1.5: Coefficients of the quasi-optimal vapor pressure equation

Index i	b_i
5	−6.743818 E+00
6	1.167810 E+00
8	−1.347062 E+00
12	−1.547285 E+00

Table R12B1.6: Coefficients of the standard saturated liquid density equation

Index i	c_i
8	2.021474 E+00
9	6.728795 E−01
16	4.656096 E−01

Table R12B1.7: Coefficients of the quasi-optimal saturated liquid density equation

Index i	c_i
8	2.193948 E+00
10	6.478416 E−01
15	1.146254 E−01

Table R12B1.8: Coefficients of the *Bender-EOS*

Range of validity: one-phase region

$$0.94 \leq p/\mathrm{MPa} \leq 6.20$$
$$0.06097 \leq \rho/\mathrm{g\ cm^{-3}} \leq 0.612$$
$$361.03 \leq T/\mathrm{K} \leq 462.41$$

	Specific units	Molar units
a_{01}	9.304996919 E−02	2.544659551 E+00
a_{02}	5.370447241 E+01	1.468668930 E+03
a_{03}	1.853478820 E+04	5.068752437 E+05
a_{04}	−4.995831746 E+06	−1.366221943 E+08
a_{05}	6.310751196 E+08	1.725816080 E+10
a_{06}	−1.191483710 E−01	−5.388380900 E−01
a_{07}	5.393792079 E+01	2.439295308 E+02
a_{08}	1.634919553 E+04	7.393780731 E+04
a_{09}	2.986084863 E−02	2.233207366 E−02
a_{10}	−1.480514431 E+00	−1.107234350 E+00
a_{11}	9.151154779 E−02	1.131773531 E−02
a_{12}	−9.481726991 E+01	−1.172657211 E+01
a_{13}	2.721693016 E+01	5.566465188 E−01
a_{14}	−7.125002870 E+06	−3.222220251 E+07
a_{15}	4.372082961 E+09	1.977236292 E+10
a_{16}	−1.156062057 E+12	−5.228189572 E+12
a_{17}	3.540280722 E+07	4.378459453 E+06
a_{18}	−3.086865892 E+10	−3.817696450 E+09
a_{19}	5.824394999 E+12	7.203348927 E+11
a_{20}	2.700000000 E+00	7.383753963 E−02

Table R12B1.9: Deviations between experimental and predicted p, ρ, T data

Region	n	Δ / %	
		p	ρ
Gas	39	0.09	0.19
Critical	9	0.12	0.54
Liquid	0	–	–
Total	48	0.12	0.35

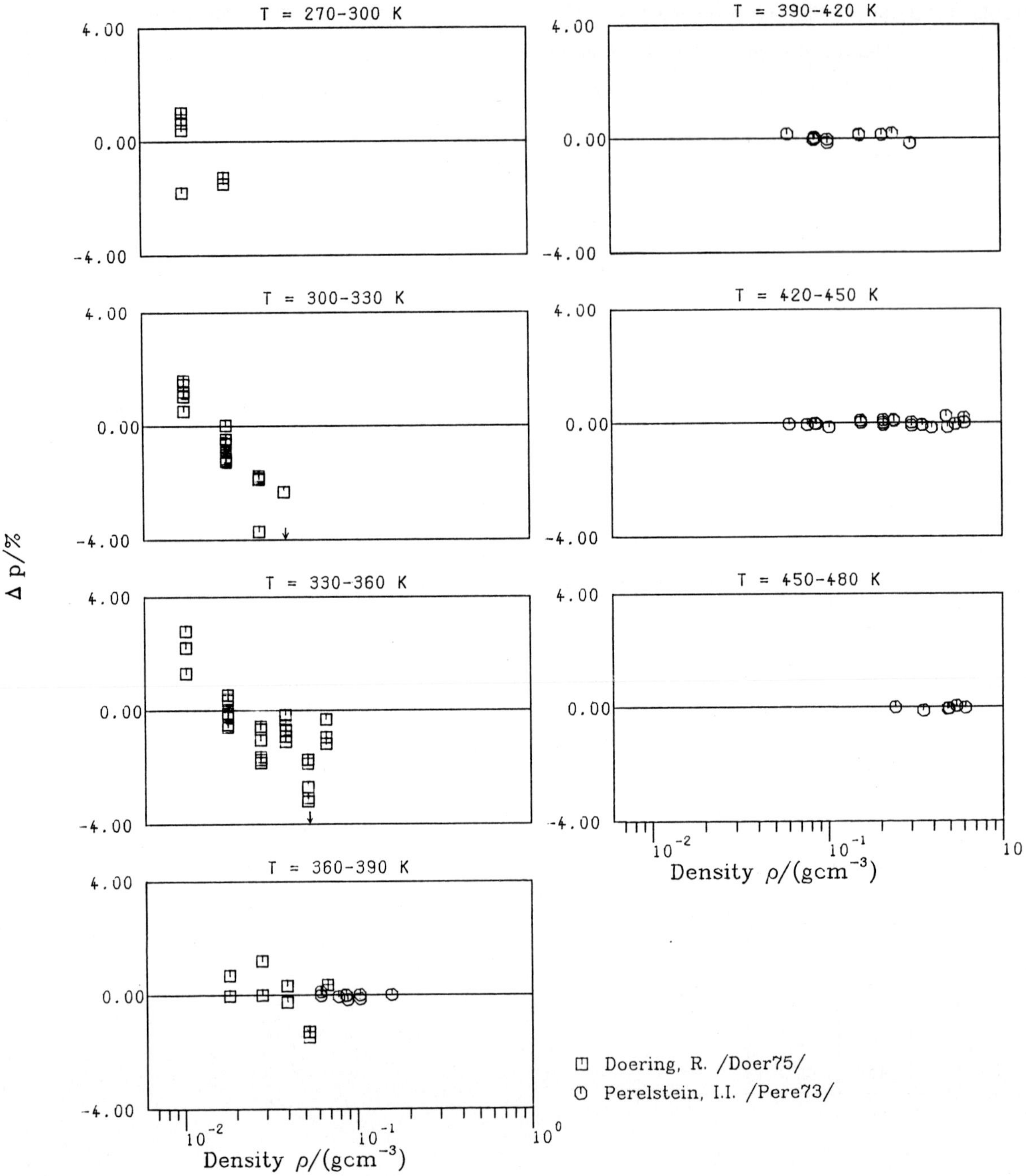

Fig.R12B1.2: Deviations between experimental and calculated p, ρ, T data

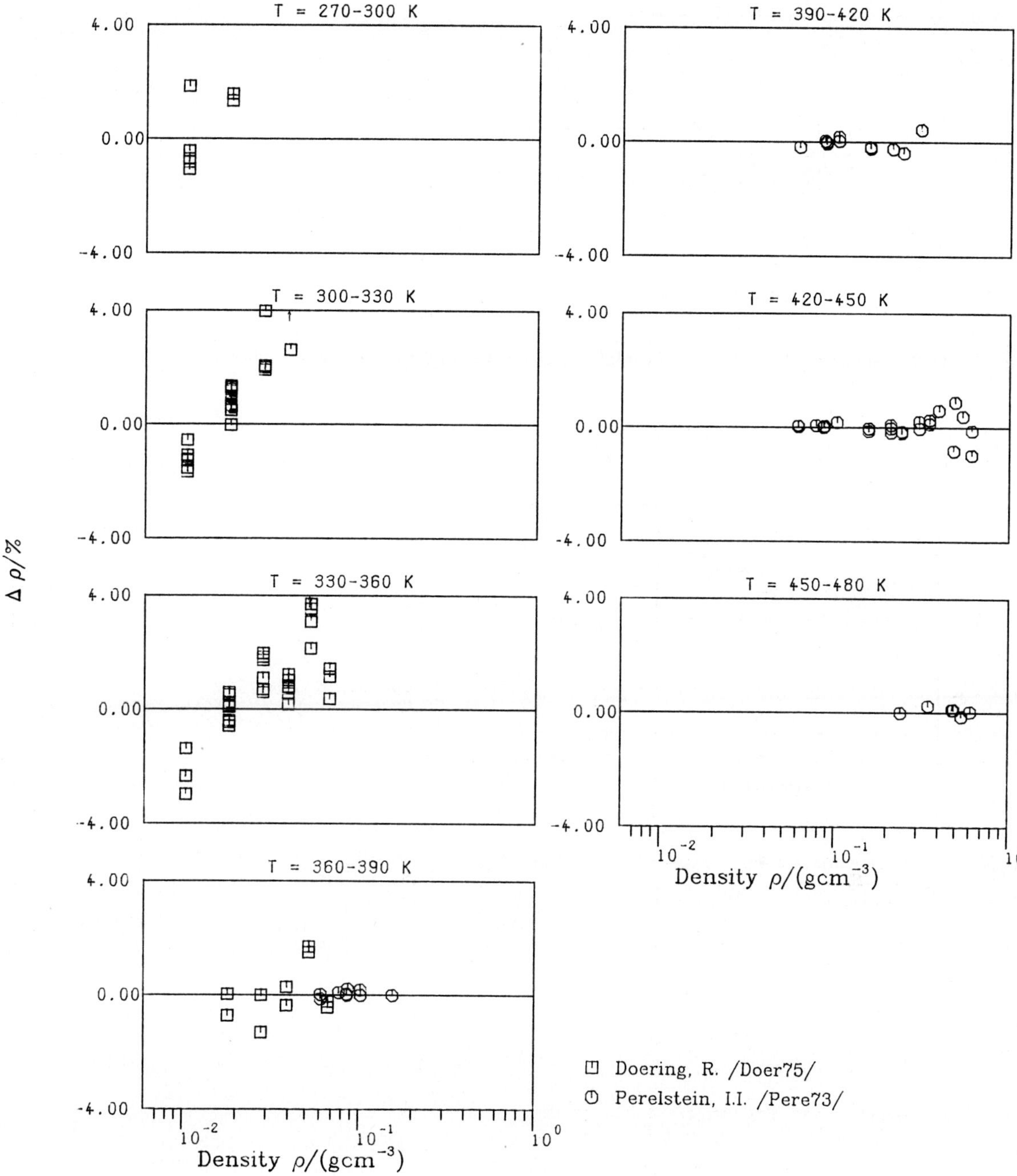

Fig.R12B1.2: Deviations between experimental and calculated p, ρ, T data *-continued*

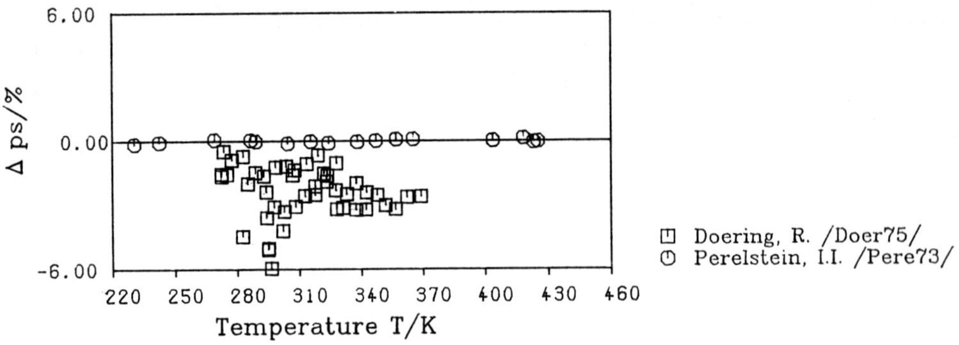

Fig.R12B1.3: Deviations between experimental and calculated vapor pressures

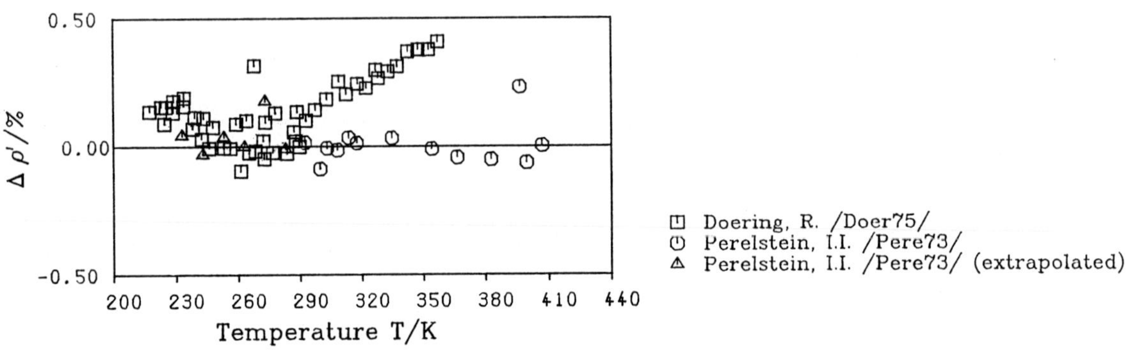

Fig.R12B1.4: Deviations between experimental and calculated saturated liquid densities

Table R12B1.10: Information on ideal gas state heat capacity equation

Author	T / K min	T / K max
Gelles, E. *[Gell 53]*	200	500

Coefficients of equation (15)	
Index i	d_i
1	4.591353 E−02
2	2.887161 E−03
3	−9.858437 E−06
4	2.147228 E−08
5	−1.861243 E−11

Chart R12B1.1: Saturation properties

T K	p MPa	v' dm³/kg	v'' dm³/kg	h' kJ/kg	h'' kJ/kg	r kJ/kg	s' kJ/kgK	s'' kJ/kgK
235.0	0.02176	0.4981	537.14	176.51	319.39	142.87	0.9078	1.5158
236.0	0.02291	0.4989	511.95	177.07	319.77	142.71	0.9101	1.5148
237.0	0.02412	0.4996	488.16	177.62	320.16	142.54	0.9125	1.5139
238.0	0.02538	0.5004	465.67	178.18	320.54	142.36	0.9148	1.5130
239.0	0.02669	0.5012	444.41	178.75	320.93	142.18	0.9172	1.5121
240.0	0.02806	0.5019	424.30	179.32	321.32	142.00	0.9196	1.5112
241.0	0.02949	0.5027	405.28	179.89	321.70	141.81	0.9220	1.5104
242.0	0.03097	0.5035	387.26	180.47	322.09	141.62	0.9244	1.5096
243.0	0.03251	0.5043	370.20	181.05	322.48	141.43	0.9268	1.5088
244.0	0.03412	0.5051	354.03	181.64	322.87	141.23	0.9292	1.5080
245.0	0.03579	0.5058	338.70	182.23	323.25	141.03	0.9316	1.5072
246.0	0.03752	0.5066	324.17	182.82	323.64	140.82	0.9340	1.5064
247.0	0.03933	0.5074	310.37	183.41	324.03	140.61	0.9364	1.5057
248.0	0.04120	0.5083	297.28	184.01	324.42	140.40	0.9388	1.5049
249.0	0.04314	0.5091	284.85	184.62	324.80	140.19	0.9412	1.5042
250.0	0.04516	0.5099	273.04	185.22	325.19	139.97	0.9436	1.5035
251.0	0.04725	0.5107	261.82	185.83	325.58	139.75	0.9461	1.5028
252.0	0.04942	0.5115	251.14	186.44	325.97	139.53	0.9485	1.5022
253.0	0.05167	0.5124	240.99	187.06	326.36	139.30	0.9509	1.5015
254.0	0.05400	0.5132	231.34	187.68	326.74	139.07	0.9534	1.5009
255.0	0.05642	0.5140	222.14	188.30	327.13	138.84	0.9558	1.5003
256.0	0.05892	0.5149	213.39	188.92	327.52	138.60	0.9582	1.4996
257.0	0.06150	0.5157	205.05	189.55	327.91	138.36	0.9607	1.4990
258.0	0.06418	0.5166	197.10	190.17	328.30	138.12	0.9631	1.4985
259.0	0.06695	0.5175	189.52	190.81	328.68	137.88	0.9655	1.4979
260.0	0.06982	0.5183	182.29	191.44	329.07	137.63	0.9680	1.4973
261.0	0.07278	0.5192	175.40	192.08	329.46	137.38	0.9704	1.4968
262.0	0.07584	0.5201	168.82	192.71	329.85	137.13	0.9728	1.4963
263.0	0.07900	0.5210	162.53	193.36	330.23	136.88	0.9753	1.4957
264.0	0.08226	0.5219	156.53	194.00	330.62	136.62	0.9777	1.4952
265.0	0.08563	0.5228	150.80	194.65	331.01	136.36	0.9802	1.4947
266.0	0.08911	0.5237	145.32	195.29	331.39	136.10	0.9826	1.4943
267.0	0.09270	0.5246	140.08	195.95	331.78	135.84	0.9850	1.4938
268.0	0.09640	0.5255	135.07	196.60	332.17	135.57	0.9875	1.4933
269.0	0.10022	0.5264	130.27	197.25	332.55	135.30	0.9899	1.4929
270.0	0.10416	0.5273	125.68	197.91	332.94	135.03	0.9923	1.4924
271.0	0.10821	0.5283	121.29	198.57	333.32	134.75	0.9948	1.4920
272.0	0.11239	0.5292	117.08	199.24	333.71	134.47	0.9972	1.4916
273.0	0.11670	0.5302	113.05	199.90	334.09	134.19	0.9996	1.4912
274.0	0.12113	0.5311	109.19	200.57	334.48	133.91	1.0021	1.4908
275.0	0.12569	0.5321	105.48	201.24	334.86	133.62	1.0045	1.4904
276.0	0.13039	0.5331	101.93	201.91	335.24	133.34	1.0069	1.4900
277.0	0.13522	0.5340	98.527	202.58	335.63	133.05	1.0093	1.4897
278.0	0.14019	0.5350	95.259	203.26	336.01	132.75	1.0118	1.4893
279.0	0.14531	0.5360	92.122	203.93	336.39	132.46	1.0142	1.4889
280.0	0.15056	0.5370	89.111	204.61	336.77	132.16	1.0166	1.4886
281.0	0.15597	0.5380	86.219	205.29	337.15	131.86	1.0190	1.4883
282.0	0.16152	0.5390	83.440	205.98	337.53	131.55	1.0214	1.4880
283.0	0.16722	0.5400	80.770	206.66	337.91	131.25	1.0239	1.4876
284.0	0.17308	0.5411	78.204	207.35	338.29	130.94	1.0263	1.4873

Chart R12B1.1: Saturation properties–*continued*

T K	p MPa	v' dm³/kg	v'' dm³/kg	h' kJ/kg	h'' kJ/kg	r kJ/kg	s' kJ/kgK	s'' kJ/kgK
285.0	0.17909	0.5421	75.737	208.04	338.67	130.63	1.0287	1.4870
286.0	0.18527	0.5431	73.364	208.73	339.05	130.31	1.0311	1.4867
287.0	0.19161	0.5442	71.081	209.43	339.42	130.00	1.0335	1.4865
288.0	0.19811	0.5453	68.884	210.12	339.80	129.68	1.0359	1.4862
289.0	0.20479	0.5463	66.770	210.82	340.18	129.35	1.0383	1.4859
290.0	0.21163	0.5474	64.734	211.52	340.55	129.03	1.0407	1.4857
291.0	0.21865	0.5485	62.773	212.23	340.93	128.70	1.0431	1.4854
292.0	0.22585	0.5496	60.885	212.93	341.30	128.37	1.0455	1.4852
293.0	0.23323	0.5507	59.065	213.64	341.67	128.04	1.0479	1.4849
294.0	0.24079	0.5518	57.311	214.34	342.04	127.70	1.0503	1.4847
295.0	0.24853	0.5529	55.621	215.05	342.41	127.36	1.0527	1.4845
296.0	0.25647	0.5540	53.990	215.77	342.78	127.02	1.0551	1.4843
297.0	0.26459	0.5552	52.418	216.48	343.15	126.67	1.0575	1.4840
298.0	0.27292	0.5563	50.901	217.20	343.52	126.32	1.0599	1.4838
299.0	0.28144	0.5574	49.437	217.91	343.89	125.97	1.0623	1.4836
300.0	0.29016	0.5586	48.024	218.63	344.25	125.62	1.0647	1.4834
301.0	0.29908	0.5598	46.659	219.36	344.62	125.26	1.0671	1.4832
302.0	0.30821	0.5610	45.342	220.08	344.98	124.90	1.0695	1.4831
303.0	0.31755	0.5621	44.070	220.81	345.35	124.54	1.0719	1.4829
304.0	0.32711	0.5633	42.840	221.53	345.71	124.17	1.0742	1.4827
305.0	0.33688	0.5646	41.652	222.26	346.07	123.81	1.0766	1.4825
306.0	0.34687	0.5658	40.504	223.00	346.43	123.43	1.0790	1.4824
307.0	0.35708	0.5670	39.394	223.73	346.79	123.06	1.0814	1.4822
308.0	0.36752	0.5682	38.320	224.46	347.15	122.68	1.0837	1.4820
309.0	0.37818	0.5695	37.282	225.20	347.50	122.30	1.0861	1.4819
310.0	0.38908	0.5707	36.278	225.94	347.86	121.91	1.0885	1.4817
311.0	0.40021	0.5720	35.306	226.68	348.21	121.53	1.0908	1.4816
312.0	0.41158	0.5733	34.365	227.43	348.56	121.14	1.0932	1.4815
313.0	0.42319	0.5746	33.454	228.17	348.91	120.74	1.0956	1.4813
314.0	0.43505	0.5759	32.573	228.92	349.26	120.34	1.0979	1.4812
315.0	0.44715	0.5772	31.719	229.67	349.61	119.94	1.1003	1.4811
316.0	0.45950	0.5785	30.892	230.42	349.96	119.54	1.1027	1.4809
317.0	0.47211	0.5799	30.091	231.17	350.30	119.13	1.1050	1.4808
318.0	0.48497	0.5812	29.314	231.93	350.65	118.72	1.1074	1.4807
319.0	0.49810	0.5826	28.562	232.68	350.99	118.31	1.1097	1.4806
320.0	0.51148	0.5840	27.832	233.44	351.33	117.89	1.1121	1.4805
321.0	0.52514	0.5853	27.125	234.20	351.67	117.47	1.1144	1.4804
322.0	0.53906	0.5867	26.439	234.97	352.01	117.04	1.1168	1.4802
323.0	0.55326	0.5882	25.773	235.73	352.34	116.61	1.1191	1.4801
324.0	0.56773	0.5896	25.128	236.50	352.68	116.18	1.1214	1.4800
325.0	0.58249	0.5910	24.502	237.27	353.01	115.74	1.1238	1.4799
326.0	0.59752	0.5925	23.894	238.04	353.34	115.31	1.1261	1.4798
327.0	0.61285	0.5939	23.304	238.81	353.67	114.86	1.1285	1.4797
328.0	0.62846	0.5954	22.731	239.58	354.00	114.42	1.1308	1.4796
329.0	0.64436	0.5969	22.174	240.36	354.32	113.96	1.1331	1.4795
330.0	0.66057	0.5984	21.634	241.14	354.65	113.51	1.1355	1.4794
331.0	0.67707	0.5999	21.109	241.92	354.97	113.05	1.1378	1.4793
332.0	0.69387	0.6015	20.599	242.70	355.29	112.59	1.1401	1.4793
333.0	0.71099	0.6030	20.104	243.48	355.61	112.12	1.1425	1.4792
334.0	0.72841	0.6046	19.622	244.27	355.92	111.65	1.1448	1.4791

Chart R12B1.1: Saturation properties–*continued*

T K	p MPa	v' dm^3/kg	v'' dm^3/kg	h' kJ/kg	h'' kJ/kg	r kJ/kg	s' kJ/kgK	s'' kJ/kgK
335.0	0.74614	0.6062	19.154	245.06	356.23	111.18	1.1471	1.4790
336.0	0.76419	0.6078	18.699	245.85	356.55	110.70	1.1494	1.4789
337.0	0.78257	0.6094	18.257	246.64	356.85	110.22	1.1517	1.4788
338.0	0.80126	0.6110	17.826	247.43	357.16	109.73	1.1541	1.4787
339.0	0.82028	0.6127	17.407	248.23	357.47	109.24	1.1564	1.4786
340.0	0.83964	0.6144	17.000	249.02	357.77	108.74	1.1587	1.4785
341.0	0.85932	0.6161	16.604	249.82	358.07	108.24	1.1610	1.4784
342.0	0.87935	0.6178	16.218	250.63	358.37	107.74	1.1633	1.4783
343.0	0.89971	0.6195	15.843	251.43	358.66	107.23	1.1656	1.4782
344.0	0.92042	0.6212	15.477	252.23	358.95	106.72	1.1679	1.4782
345.0	0.94148	0.6230	15.121	253.04	359.24	106.20	1.1702	1.4781
346.0	0.96289	0.6248	14.774	253.85	359.53	105.68	1.1725	1.4780
347.0	0.98465	0.6266	14.437	254.66	359.81	105.15	1.1748	1.4779
348.0	1.00677	0.6284	14.108	255.48	360.10	104.62	1.1771	1.4778
349.0	1.02926	0.6302	13.788	256.29	360.37	104.08	1.1794	1.4777
350.0	1.05211	0.6321	13.475	257.11	360.65	103.54	1.1817	1.4776
351.0	1.07533	0.6340	13.171	257.93	360.92	103.00	1.1840	1.4775
352.0	1.09892	0.6359	12.874	258.75	361.19	102.44	1.1863	1.4774
353.0	1.12289	0.6378	12.585	259.57	361.46	101.89	1.1886	1.4773
354.0	1.14724	0.6398	12.303	260.40	361.73	101.33	1.1909	1.4771
355.0	1.17197	0.6418	12.028	261.23	361.99	100.76	1.1932	1.4770
356.0	1.19709	0.6438	11.760	262.06	362.25	100.19	1.1955	1.4769
357.0	1.22261	0.6458	11.498	262.89	362.50	99.61	1.1978	1.4768
358.0	1.24852	0.6479	11.243	263.72	362.75	99.03	1.2001	1.4767
359.0	1.27482	0.6500	10.994	264.56	363.00	98.44	1.2024	1.4766
360.0	1.30153	0.6521	10.751	265.40	363.24	97.84	1.2046	1.4764
361.0	1.32865	0.6542	10.514	266.24	363.48	97.24	1.2069	1.4763
362.0	1.35618	0.6564	10.282	267.08	363.72	96.64	1.2092	1.4762
363.0	1.38412	0.6586	10.056	267.93	363.95	96.02	1.2115	1.4760
364.0	1.41248	0.6609	9.8350	268.78	364.18	95.41	1.2138	1.4759
365.0	1.44126	0.6631	9.6192	269.63	364.41	94.78	1.2161	1.4757
366.0	1.47047	0.6654	9.4085	270.48	364.63	94.15	1.2183	1.4756
367.0	1.50011	0.6677	9.2026	271.33	364.85	93.51	1.2206	1.4754
368.0	1.53018	0.6701	9.0014	272.19	365.06	92.87	1.2229	1.4752
369.0	1.56069	0.6725	8.8048	273.05	365.27	92.22	1.2252	1.4751
370.0	1.59164	0.6750	8.6126	273.91	365.47	91.56	1.2274	1.4749
371.0	1.62304	0.6774	8.4247	274.78	365.67	90.89	1.2297	1.4747
372.0	1.65489	0.6800	8.2410	275.65	365.87	90.22	1.2320	1.4745
373.0	1.68720	0.6825	8.0613	276.52	366.06	89.54	1.2343	1.4743
374.0	1.71996	0.6851	7.8856	277.39	366.24	88.85	1.2366	1.4741
375.0	1.75319	0.6878	7.7137	278.27	366.43	88.16	1.2388	1.4739
376.0	1.78688	0.6905	7.5455	279.14	366.60	87.46	1.2411	1.4737
377.0	1.82105	0.6932	7.3810	280.03	366.77	86.74	1.2434	1.4735
378.0	1.85569	0.6960	7.2199	280.91	366.93	86.02	1.2457	1.4732
379.0	1.89082	0.6988	7.0622	281.80	367.09	85.30	1.2479	1.4730
380.0	1.92643	0.7017	6.9079	282.69	367.25	84.56	1.2502	1.4727
381.0	1.96253	0.7047	6.7567	283.58	367.39	83.81	1.2525	1.4725
382.0	1.99913	0.7077	6.6087	284.48	367.53	83.06	1.2548	1.4722
383.0	2.03622	0.7108	6.4636	285.38	367.67	82.29	1.2571	1.4719
384.0	2.07382	0.7139	6.3215	286.28	367.80	81.51	1.2594	1.4716

Chart R12B1.1: Saturation properties–*continued*

T K	p MPa	v' dm³/kg	v'' dm³/kg	h' kJ/kg	h'' kJ/kg	r kJ/kg	s' kJ/kgK	s'' kJ/kgK
385.0	2.11193	0.7171	6.1823	287.19	367.92	80.73	1.2617	1.4713
386.0	2.15056	0.7204	6.0458	288.10	368.03	79.93	1.2639	1.4710
387.0	2.18970	0.7237	5.9120	289.02	368.14	79.12	1.2662	1.4707
388.0	2.22937	0.7271	5.7807	289.94	368.24	78.30	1.2685	1.4703
389.0	2.26957	0.7306	5.6520	290.86	368.33	77.47	1.2708	1.4700
390.0	2.31030	0.7342	5.5258	291.79	368.41	76.63	1.2731	1.4696
391.0	2.35158	0.7378	5.4019	292.72	368.49	75.77	1.2754	1.4692
392.0	2.39340	0.7416	5.2803	293.65	368.55	74.90	1.2778	1.4688
393.0	2.43578	0.7454	5.1609	294.60	368.61	74.01	1.2801	1.4684
394.0	2.47871	0.7494	5.0437	295.54	368.66	73.12	1.2824	1.4680
395.0	2.52221	0.7534	4.9285	296.50	368.70	72.20	1.2847	1.4675
396.0	2.56627	0.7576	4.8153	297.45	368.72	71.27	1.2871	1.4671
397.0	2.61092	0.7619	4.7041	298.42	368.74	70.32	1.2894	1.4666
398.0	2.65614	0.7664	4.5947	299.39	368.75	69.36	1.2918	1.4660
399.0	2.70196	0.7709	4.4871	300.36	368.74	68.37	1.2941	1.4655
400.0	2.74837	0.7756	4.3813	301.35	368.72	67.37	1.2965	1.4649
401.0	2.79539	0.7805	4.2770	302.34	368.69	66.35	1.2989	1.4644
402.0	2.84301	0.7856	4.1744	303.34	368.64	65.30	1.3013	1.4637
403.0	2.89126	0.7908	4.0732	304.35	368.58	64.23	1.3037	1.4631
404.0	2.94013	0.7963	3.9735	305.37	368.51	63.14	1.3061	1.4624
405.0	2.98964	0.8020	3.8751	306.40	368.41	62.02	1.3086	1.4617
406.0	3.03979	0.8079	3.7780	307.44	368.30	60.87	1.3110	1.4610
407.0	3.09059	0.8141	3.6821	308.49	368.17	59.68	1.3135	1.4602
408.0	3.14205	0.8206	3.5873	309.55	368.02	58.47	1.3160	1.4593
409.0	3.19418	0.8274	3.4934	310.63	367.85	57.21	1.3186	1.4585
410.0	3.24700	0.8346	3.4005	311.73	367.65	55.92	1.3211	1.4575
411.0	3.30050	0.8422	3.3083	312.84	367.43	54.58	1.3237	1.4566
412.0	3.35471	0.8503	3.2168	313.98	367.17	53.20	1.3264	1.4555
413.0	3.40963	0.8589	3.1258	315.13	366.89	51.76	1.3291	1.4544
414.0	3.46527	0.8682	3.0352	316.32	366.57	50.25	1.3318	1.4532
415.0	3.52165	0.8781	2.9447	317.53	366.21	48.68	1.3346	1.4519
416.0	3.57879	0.8890	2.8542	318.78	365.80	47.02	1.3375	1.4506
417.0	3.63669	0.9009	2.7634	320.08	365.35	45.27	1.3405	1.4491
418.0	3.69536	0.9141	2.6719	321.42	364.83	43.40	1.3436	1.4475
419.0	3.75483	0.9289	2.5793	322.84	364.24	41.40	1.3469	1.4457
420.0	3.81511	0.9459	2.4850	324.33	363.56	39.23	1.3503	1.4437
421.0	3.87620	0.9658	2.3884	325.94	362.78	36.83	1.3540	1.4415
422.0	3.93811	0.9897	2.2881	327.71	361.85	34.15	1.3580	1.4389
423.0	4.00083	1.0199	2.1827	329.70	360.75	31.04	1.3626	1.4360
424.0	4.06432	1.0605	2.0690	332.07	359.37	27.30	1.3680	1.4324
425.0	4.12848	1.1203	1.9413	335.09	357.57	22.48	1.3750	1.4279
426.0	4.19308	1.2216	1.7844	339.39	354.91	15.51	1.3849	1.4213

Chart R12B1.2: Saturation properties

p MPa	T K	v' dm³/kg	v'' dm³/kg	h' kJ/kg	h'' kJ/kg	r kJ/kg	s' kJ/kgK	s'' kJ/kgK
0.0260	238.48	0.5008	455.40	178.45	320.73	142.28	0.9160	1.5126
0.0270	239.23	0.5013	439.75	178.88	321.02	142.14	0.9177	1.5119
0.0280	239.96	0.5019	425.18	179.29	321.30	142.01	0.9195	1.5113
0.0290	240.66	0.5024	411.58	179.70	321.57	141.87	0.9212	1.5107
0.0300	241.35	0.5030	398.85	180.09	321.84	141.74	0.9228	1.5101
0.0310	242.02	0.5035	386.92	180.48	322.10	141.62	0.9244	1.5095
0.0320	242.67	0.5040	375.70	180.86	322.35	141.49	0.9260	1.5090
0.0330	243.31	0.5045	365.14	181.23	322.60	141.37	0.9275	1.5085
0.0340	243.93	0.5050	355.18	181.59	322.84	141.24	0.9290	1.5080
0.0350	244.53	0.5055	345.77	181.95	323.07	141.12	0.9304	1.5075
0.0360	245.12	0.5059	336.86	182.30	323.30	141.00	0.9319	1.5071
0.0370	245.70	0.5064	328.42	182.64	323.53	140.88	0.9332	1.5066
0.0380	246.27	0.5069	320.40	182.98	323.74	140.77	0.9346	1.5062
0.0390	246.82	0.5073	312.78	183.31	323.96	140.65	0.9360	1.5058
0.0400	247.36	0.5077	305.53	183.63	324.17	140.54	0.9373	1.5054
0.0410	247.90	0.5082	298.62	183.95	324.38	140.42	0.9385	1.5050
0.0420	248.42	0.5086	292.03	184.26	324.58	140.31	0.9398	1.5046
0.0430	248.93	0.5090	285.73	184.57	324.78	140.20	0.9410	1.5043
0.0440	249.43	0.5094	279.71	184.88	324.97	140.09	0.9423	1.5039
0.0450	249.92	0.5098	273.94	185.17	325.16	139.99	0.9435	1.5036
0.0460	250.41	0.5102	268.42	185.47	325.35	139.88	0.9446	1.5032
0.0470	250.88	0.5106	263.12	185.76	325.53	139.78	0.9458	1.5029
0.0480	251.35	0.5110	258.03	186.04	325.72	139.67	0.9469	1.5026
0.0490	251.81	0.5114	253.15	186.33	325.89	139.57	0.9480	1.5023
0.0500	252.26	0.5117	248.45	186.60	326.07	139.47	0.9491	1.5020
0.0510	252.71	0.5121	243.93	186.88	326.24	139.37	0.9502	1.5017
0.0520	253.14	0.5125	239.58	187.15	326.41	139.27	0.9513	1.5014
0.0530	253.57	0.5128	235.39	187.41	326.58	139.17	0.9523	1.5011
0.0540	254.00	0.5132	231.34	187.67	326.74	139.07	0.9534	1.5009
0.0550	254.42	0.5135	227.44	187.93	326.91	138.97	0.9544	1.5006
0.0560	254.83	0.5139	223.67	188.19	327.07	138.88	0.9554	1.5004
0.0570	255.24	0.5142	220.03	188.44	327.22	138.78	0.9564	1.5001
0.0580	255.64	0.5146	216.51	188.69	327.38	138.69	0.9573	1.4999
0.0590	256.03	0.5149	213.10	188.94	327.53	138.59	0.9583	1.4996
0.0600	256.42	0.5152	209.81	189.18	327.69	138.50	0.9593	1.4994
0.0610	256.81	0.5156	206.62	189.42	327.83	138.41	0.9602	1.4992
0.0620	257.19	0.5159	203.52	189.66	327.98	138.32	0.9611	1.4989
0.0630	257.56	0.5162	200.52	189.90	328.13	138.23	0.9620	1.4987
0.0640	257.93	0.5165	197.62	190.13	328.27	138.14	0.9629	1.4985
0.0650	258.30	0.5169	194.79	190.36	328.41	138.05	0.9638	1.4983
0.0660	258.66	0.5172	192.05	190.59	328.55	137.96	0.9647	1.4981
0.0670	259.02	0.5175	189.39	190.82	328.69	137.88	0.9656	1.4979
0.0680	259.37	0.5178	186.80	191.04	328.83	137.79	0.9664	1.4977
0.0690	259.72	0.5181	184.29	191.26	328.96	137.70	0.9673	1.4975
0.0700	260.06	0.5184	181.85	191.48	329.10	137.62	0.9681	1.4973
0.0710	260.40	0.5187	179.47	191.70	329.23	137.53	0.9690	1.4971
0.0720	260.74	0.5190	177.15	191.91	329.36	137.45	0.9698	1.4969
0.0730	261.07	0.5193	174.90	192.12	329.49	137.37	0.9706	1.4967
0.0740	261.40	0.5196	172.70	192.33	329.62	137.28	0.9714	1.4966
0.0760	262.05	0.5201	168.48	192.75	329.87	137.12	0.9730	1.4962

Chart R12B1.2: Saturation properties—*continued*

p MPa	T K	v' dm³/kg	v'' dm³/kg	h' kJ/kg	h'' kJ/kg	r kJ/kg	s' kJ/kgK	s'' kJ/kgK
0.0780	262.69	0.5207	164.46	193.16	330.11	136.96	0.9745	1.4959
0.0800	263.31	0.5212	160.64	193.56	330.35	136.80	0.9760	1.4956
0.0820	263.92	0.5218	157.00	193.95	330.59	136.64	0.9775	1.4953
0.0840	264.52	0.5223	153.52	194.34	330.82	136.49	0.9790	1.4950
0.0860	265.11	0.5229	150.20	194.72	331.05	136.33	0.9804	1.4947
0.0880	265.68	0.5234	147.02	195.09	331.27	136.18	0.9818	1.4944
0.0900	266.25	0.5239	143.98	195.46	331.49	136.03	0.9832	1.4941
0.0920	266.81	0.5244	141.07	195.82	331.71	135.89	0.9846	1.4939
0.0940	267.35	0.5249	138.27	196.18	331.92	135.74	0.9859	1.4936
0.0960	267.89	0.5254	135.59	196.53	332.13	135.60	0.9872	1.4934
0.0980	268.42	0.5259	133.02	196.88	332.33	135.45	0.9885	1.4931
0.1000	268.94	0.5264	130.54	197.22	332.53	135.31	0.9898	1.4929
0.1020	269.46	0.5268	128.15	197.55	332.73	135.17	0.9910	1.4927
0.1040	269.96	0.5273	125.86	197.89	332.92	135.04	0.9922	1.4924
0.1060	270.46	0.5278	123.65	198.22	333.12	134.90	0.9935	1.4922
0.1080	270.95	0.5282	121.51	198.54	333.30	134.76	0.9946	1.4920
0.1100	271.43	0.5287	119.45	198.86	333.49	134.63	0.9958	1.4918
0.1150	272.61	0.5298	114.61	199.64	333.94	134.30	0.9987	1.4913
0.1200	273.75	0.5309	110.15	200.40	334.38	133.98	1.0014	1.4909
0.1250	274.85	0.5319	106.03	201.14	334.80	133.67	1.0041	1.4905
0.1300	275.92	0.5330	102.22	201.85	335.21	133.36	1.0067	1.4901
0.1350	276.95	0.5340	98.680	202.55	335.61	133.06	1.0092	1.4897
0.1400	277.96	0.5350	95.383	203.23	335.99	132.76	1.0117	1.4893
0.1450	278.94	0.5360	92.305	203.89	336.37	132.47	1.0140	1.4890
0.1500	279.89	0.5369	89.424	204.54	336.73	132.19	1.0164	1.4886
0.1550	280.82	0.5378	86.721	205.17	337.09	131.91	1.0186	1.4883
0.1600	281.73	0.5388	84.181	205.79	337.43	131.64	1.0208	1.4880
0.1650	282.61	0.5396	81.789	206.40	337.77	131.37	1.0229	1.4878
0.1700	283.48	0.5405	79.532	206.99	338.09	131.10	1.0250	1.4875
0.1750	284.32	0.5414	77.398	207.57	338.41	130.84	1.0271	1.4872
0.1800	285.15	0.5423	75.379	208.15	338.73	130.58	1.0291	1.4870
0.1850	285.96	0.5431	73.464	208.70	339.03	130.33	1.0310	1.4868
0.1900	286.75	0.5439	71.647	209.25	339.33	130.08	1.0329	1.4865
0.1950	287.52	0.5448	69.918	209.79	339.62	129.83	1.0348	1.4863
0.2000	288.29	0.5456	68.273	210.32	339.91	129.58	1.0366	1.4861
0.2100	289.76	0.5471	65.208	211.36	340.46	129.11	1.0402	1.4857
0.2200	291.19	0.5487	62.411	212.36	341.00	128.64	1.0436	1.4854
0.2300	292.57	0.5502	59.847	213.33	341.51	128.18	1.0469	1.4850
0.2400	293.90	0.5517	57.489	214.27	342.00	127.73	1.0501	1.4847
0.2500	295.19	0.5531	55.311	215.19	342.48	127.30	1.0532	1.4844
0.2600	296.44	0.5545	53.295	216.08	342.95	126.87	1.0562	1.4842
0.2700	297.65	0.5559	51.422	216.95	343.39	126.45	1.0591	1.4839
0.2800	298.83	0.5573	49.677	217.79	343.83	126.03	1.0619	1.4837
0.2900	299.98	0.5586	48.048	218.62	344.25	125.63	1.0647	1.4834
0.3000	301.10	0.5599	46.523	219.43	344.66	125.23	1.0673	1.4832
0.3100	302.19	0.5612	45.093	220.22	345.05	124.83	1.0699	1.4830
0.3200	303.26	0.5625	43.748	220.99	345.44	124.45	1.0725	1.4828
0.3300	304.30	0.5637	42.481	221.75	345.82	124.07	1.0749	1.4827
0.3400	305.31	0.5649	41.286	222.49	346.18	123.69	1.0774	1.4825
0.3500	306.31	0.5661	40.157	223.22	346.54	123.32	1.0797	1.4823

Chart R12B1.2: Saturation properties–*continued*

p MPa	T K	v' dm³/kg	v'' dm³/kg	h' kJ/kg	h'' kJ/kg	r kJ/kg	s' kJ/kgK	s'' kJ/kgK
0.3600	307.28	0.5673	39.087	223.94	346.89	122.95	1.0820	1.4822
0.3700	308.23	0.5685	38.073	224.64	347.23	122.59	1.0843	1.4820
0.3800	309.17	0.5697	37.111	225.33	347.56	122.23	1.0865	1.4819
0.3900	310.08	0.5709	36.195	226.00	347.89	121.88	1.0887	1.4817
0.4000	310.98	0.5720	35.323	226.67	348.20	121.53	1.0908	1.4816
0.4100	311.86	0.5731	34.493	227.32	348.51	121.19	1.0929	1.4815
0.4200	312.73	0.5742	33.700	227.97	348.82	120.85	1.0949	1.4814
0.4300	313.58	0.5753	32.942	228.60	349.12	120.51	1.0969	1.4813
0.4400	314.41	0.5764	32.218	229.23	349.41	120.18	1.0989	1.4811
0.4500	315.23	0.5775	31.524	229.84	349.69	119.85	1.1008	1.4810
0.4600	316.04	0.5786	30.859	230.45	349.97	119.52	1.1027	1.4809
0.4700	316.83	0.5797	30.222	231.05	350.25	119.20	1.1046	1.4808
0.4800	317.62	0.5807	29.609	231.64	350.52	118.88	1.1065	1.4807
0.4900	318.39	0.5817	29.021	232.22	350.78	118.56	1.1083	1.4807
0.5000	319.14	0.5828	28.455	232.79	351.04	118.25	1.1101	1.4806
0.5100	319.89	0.5838	27.911	233.36	351.29	117.93	1.1118	1.4805
0.5200	320.63	0.5848	27.387	233.92	351.54	117.62	1.1135	1.4804
0.5300	321.35	0.5858	26.881	234.47	351.79	117.32	1.1152	1.4803
0.5400	322.07	0.5868	26.394	235.02	352.03	117.01	1.1169	1.4802
0.5500	322.77	0.5878	25.923	235.56	352.27	116.71	1.1186	1.4802
0.5600	323.47	0.5888	25.469	236.09	352.50	116.41	1.1202	1.4801
0.5700	324.15	0.5898	25.030	236.62	352.73	116.11	1.1218	1.4800
0.5800	324.83	0.5908	24.605	237.14	352.96	115.82	1.1234	1.4799
0.5900	325.50	0.5918	24.194	237.65	353.18	115.52	1.1250	1.4799
0.6000	326.16	0.5927	23.796	238.16	353.40	115.23	1.1265	1.4798
0.6100	326.82	0.5937	23.411	238.67	353.61	114.94	1.1280	1.4797
0.6200	327.46	0.5946	23.038	239.16	353.82	114.66	1.1295	1.4797
0.6300	328.10	0.5956	22.676	239.66	354.03	114.37	1.1310	1.4796
0.6400	328.73	0.5965	22.324	240.15	354.23	114.09	1.1325	1.4796
0.6500	329.35	0.5974	21.984	240.63	354.44	113.81	1.1340	1.4795
0.6600	329.97	0.5984	21.653	241.11	354.64	113.53	1.1354	1.4794
0.6700	330.57	0.5993	21.331	241.58	354.83	113.25	1.1368	1.4794
0.6800	331.18	0.6002	21.019	242.05	355.02	112.97	1.1382	1.4793
0.6900	331.77	0.6011	20.715	242.52	355.21	112.70	1.1396	1.4793
0.7000	332.36	0.6020	20.419	242.98	355.40	112.42	1.1410	1.4792
0.7100	332.94	0.6030	20.132	243.44	355.59	112.15	1.1423	1.4792
0.7200	333.52	0.6039	19.852	243.89	355.77	111.88	1.1437	1.4791
0.7300	334.09	0.6048	19.579	244.34	355.95	111.61	1.1450	1.4791
0.7400	334.66	0.6056	19.314	244.78	356.13	111.34	1.1463	1.4790
0.7600	335.77	0.6074	18.803	245.66	356.47	110.81	1.1489	1.4789
0.7800	336.86	0.6092	18.317	246.53	356.81	110.28	1.1514	1.4788
0.8000	337.93	0.6109	17.855	247.38	357.14	109.76	1.1539	1.4787
0.8200	338.99	0.6127	17.414	248.21	357.46	109.25	1.1563	1.4786
0.8400	340.02	0.6144	16.993	249.04	357.77	108.73	1.1587	1.4785
0.8600	341.03	0.6161	16.590	249.85	358.08	108.23	1.1611	1.4784
0.8800	342.03	0.6178	16.206	250.65	358.37	107.72	1.1634	1.4783
0.9000	343.01	0.6195	15.837	251.44	358.66	107.22	1.1657	1.4782
0.9200	343.98	0.6212	15.484	252.22	358.95	106.73	1.1679	1.4782
0.9400	344.93	0.6229	15.146	252.99	359.22	106.24	1.1701	1.4781
0.9600	345.87	0.6245	14.820	253.74	359.49	105.75	1.1722	1.4780

Chart R12B1.2: Saturation properties–*continued*

p MPa	T K	v' dm³/kg	v'' dm³/kg	h' kJ/kg	h'' kJ/kg	r kJ/kg	s' kJ/kgK	s'' kJ/kgK
0.9800	346.79	0.6262	14.508	254.49	359.75	105.26	1.1744	1.4779
1.0000	347.70	0.6278	14.207	255.23	360.01	104.78	1.1764	1.4778
1.0200	348.59	0.6295	13.918	255.96	360.26	104.30	1.1785	1.4777
1.0400	349.47	0.6311	13.639	256.68	360.51	103.83	1.1805	1.4776
1.0600	350.34	0.6328	13.370	257.39	360.74	103.36	1.1825	1.4775
1.0800	351.20	0.6344	13.111	258.09	360.98	102.89	1.1845	1.4775
1.1000	352.05	0.6360	12.861	258.79	361.21	102.42	1.1864	1.4774
1.1500	354.11	0.6400	12.272	260.49	361.76	101.26	1.1912	1.4771
1.2000	356.11	0.6440	11.730	262.15	362.27	100.12	1.1958	1.4769
1.2500	358.06	0.6480	11.229	263.77	362.77	98.99	1.2002	1.4767
1.3000	359.94	0.6520	10.765	265.35	363.23	97.88	1.2045	1.4764
1.3500	361.78	0.6559	10.333	266.90	363.67	96.77	1.2087	1.4762
1.4000	363.56	0.6599	9.9311	268.41	364.08	95.68	1.2128	1.4759
1.4500	365.30	0.6638	9.5553	269.88	364.48	94.59	1.2167	1.4757
1.5000	367.00	0.6677	9.2033	271.33	364.85	93.52	1.2206	1.4754
1.5500	368.65	0.6717	8.8728	272.75	365.20	92.45	1.2244	1.4751
1.6000	370.27	0.6756	8.5619	274.15	365.53	91.38	1.2281	1.4749
1.6500	371.85	0.6796	8.2687	275.51	365.84	90.33	1.2317	1.4746
1.7000	373.39	0.6835	7.9919	276.86	366.13	89.27	1.2352	1.4743
1.7500	374.90	0.6875	7.7299	278.18	366.41	88.23	1.2386	1.4739
1.8000	376.39	0.6915	7.4817	279.48	366.67	87.18	1.2420	1.4736
1.8500	377.84	0.6955	7.2460	280.77	366.91	86.14	1.2453	1.4733
1.9000	379.26	0.6996	7.0219	282.03	367.13	85.11	1.2485	1.4729
1.9500	380.65	0.7037	6.8086	283.27	367.34	84.07	1.2517	1.4726
2.0000	382.02	0.7078	6.6052	284.50	367.54	83.04	1.2548	1.4722
2.1000	384.69	0.7161	6.2254	286.91	367.88	80.97	1.2609	1.4714
2.2000	387.26	0.7246	5.8775	289.26	368.16	78.91	1.2668	1.4706
2.3000	389.75	0.7333	5.5573	291.55	368.39	76.84	1.2726	1.4697
2.4000	392.16	0.7422	5.2615	293.80	368.56	74.76	1.2781	1.4688
2.5000	394.49	0.7514	4.9868	296.01	368.68	72.67	1.2836	1.4678
2.6000	396.76	0.7609	4.7310	298.18	368.74	70.56	1.2888	1.4667
2.7000	398.96	0.7707	4.4917	300.32	368.74	68.42	1.2940	1.4655
2.8000	401.10	0.7810	4.2670	302.44	368.68	66.25	1.2991	1.4643
2.9000	403.18	0.7918	4.0552	304.53	368.57	64.04	1.3041	1.4630
3.0000	405.21	0.8032	3.8549	306.61	368.39	61.78	1.3091	1.4616
3.1000	407.18	0.8153	3.6646	308.68	368.14	59.46	1.3140	1.4600
3.2000	409.11	0.8282	3.4831	310.75	367.83	57.07	1.3189	1.4584
3.3000	410.99	0.8421	3.3092	312.83	367.43	54.60	1.3237	1.4566
3.4000	412.83	0.8574	3.1417	314.93	366.94	52.01	1.3286	1.4546
3.5000	414.62	0.8742	2.9793	317.06	366.35	49.29	1.3336	1.4524
3.6000	416.37	0.8932	2.8208	319.25	365.64	46.39	1.3386	1.4500
3.7000	418.08	0.9152	2.6646	321.53	364.78	43.25	1.3439	1.4473
3.8000	419.75	0.9414	2.5087	323.95	363.74	39.79	1.3494	1.4442
3.9000	421.39	0.9744	2.3502	326.60	362.44	35.84	1.3555	1.4405
4.0000	422.99	1.0195	2.1841	329.68	360.76	31.09	1.3625	1.4360
4.1000	424.56	1.0905	2.0002	333.64	358.44	24.80	1.3716	1.4300
4.2000	426.11	1.2369	1.7642	339.99	354.52	14.54	1.3863	1.4204

Chart R12B1.3: Thermodynamic properties

p/MPa	0.10			0.15			0.20		
T K	v dm³/kg	h kJ/kg	s kJ/kgK	v dm³/kg	h kJ/kg	s kJ/kgK	v dm³/kg	h kJ/kg	s kJ/kgK
360.0	178.39	375.36	1.6296	118.04	374.95	1.6085	87.860	374.54	1.5932
370.0	183.56	380.39	1.6434	121.54	380.01	1.6223	90.525	379.61	1.6071
380.0	188.73	385.49	1.6570	125.03	385.12	1.6360	93.179	384.75	1.6208
390.0	193.88	390.65	1.6704	128.51	390.29	1.6494	95.825	389.93	1.6343
400.0	199.03	395.86	1.6836	131.99	395.52	1.6626	98.462	395.18	1.6476
410.0	204.17	401.12	1.6966	135.45	400.80	1.6756	101.09	400.47	1.6606
420.0	209.31	406.43	1.7094	138.91	406.12	1.6885	103.72	405.81	1.6735
430.0	214.44	411.79	1.7220	142.37	411.49	1.7011	106.33	411.19	1.6861
440.0	219.56	417.18	1.7344	145.82	416.89	1.7135	108.94	416.60	1.6986
450.0	224.68	422.61	1.7466	149.26	422.33	1.7258	111.55	422.05	1.7108
460.0	229.80	428.07	1.7586	152.70	427.80	1.7378	114.15	427.53	1.7229
p/MPa	0.30			0.40			0.50		
360.0	57.661	373.70	1.5712	42.543	372.84	1.5551	33.457	371.94	1.5422
370.0	59.493	378.82	1.5853	43.963	378.00	1.5693	34.632	377.15	1.5565
380.0	61.314	383.99	1.5990	45.370	383.21	1.5832	35.792	382.41	1.5705
390.0	63.126	389.21	1.6126	46.766	388.47	1.5968	36.941	387.71	1.5842
400.0	64.928	394.48	1.6260	48.153	393.77	1.6103	38.080	393.05	1.5978
410.0	66.723	399.80	1.6391	49.531	399.13	1.6235	39.210	398.44	1.6111
420.0	68.510	405.17	1.6520	50.901	404.52	1.6365	40.331	403.86	1.6241
430.0	70.291	410.57	1.6647	52.265	409.95	1.6493	41.446	409.32	1.6370
440.0	72.066	416.01	1.6773	53.623	415.42	1.6618	42.554	414.82	1.6496
450.0	73.836	421.49	1.6895	54.975	420.91	1.6742	43.656	420.33	1.6620
460.0	75.601	426.98	1.7016	56.322	426.43	1.6863	44.753	425.87	1.6742
p/MPa	0.60			0.80			1.00		
360.0	27.385	371.02	1.5312	19.758	369.06	1.5129	15.133	366.92	1.4973
370.0	28.399	376.28	1.5456	20.579	374.46	1.5277	15.850	372.50	1.5126
380.0	29.398	381.59	1.5598	21.382	379.88	1.5421	16.543	378.07	1.5275
390.0	30.384	386.94	1.5737	22.168	385.33	1.5563	17.216	383.63	1.5419
400.0	31.359	392.32	1.5873	22.942	390.80	1.5701	17.873	389.21	1.5561
410.0	32.324	397.74	1.6007	23.704	396.30	1.5837	18.517	394.80	1.5699
420.0	33.280	403.20	1.6138	24.456	401.83	1.5970	19.150	400.41	1.5834
430.0	34.229	408.69	1.6268	25.200	407.38	1.6101	19.773	406.04	1.5966
440.0	35.171	414.21	1.6394	25.936	412.96	1.6229	20.387	411.69	1.6096
450.0	36.107	419.75	1.6519	26.665	418.56	1.6355	20.994	417.34	1.6223
460.0	37.037	425.31	1.6641	27.389	424.17	1.6478	21.594	423.01	1.6348

Chart R12B1.3: Thermodynamic properties–*continued*

p/MPa	2.00			4.00			6.20		
T K	v dm^3/kg	h kJ/kg	s kJ/kgK	v dm^3/kg	h kJ/kg	s kJ/kgK	v dm^3/kg	h kJ/kg	s kJ/kgK
360.0	0.6487	265.30	1.2031	0.6404	265.12	1.1990	0.6326	265.07	1.1950
370.0	0.6723	273.79	1.2264	0.6609	273.37	1.2216	0.6510	273.13	1.2171
380.0	0.7010	282.65	1.2500	0.6848	281.84	1.2442	0.6716	281.35	1.2390
390.0	7.0444	373.08	1.4866	0.7134	290.61	1.2670	0.6951	289.75	1.2608
400.0	7.5406	379.71	1.5034	0.7497	299.81	1.2903	0.7226	298.35	1.2826
410.0	7.9962	386.12	1.5192	0.8015	309.81	1.3150	0.7557	307.23	1.3045
420.0	8.4234	392.40	1.5343	0.9081	322.46	1.3454	0.7979	316.53	1.3269
430.0	8.8292	398.58	1.5489	2.8317	374.02	1.4672	0.8572	326.56	1.3505
440.0	9.2185	404.70	1.5629	3.3106	384.68	1.4917	0.9581	338.23	1.3773
450.0	9.5944	410.76	1.5766	3.6730	393.38	1.5112	1.1910	354.09	1.4130
460.0	9.9593	416.79	1.5898	3.9822	401.24	1.5285	1.5867	372.02	1.4524

References for R12B1

[Bady66] BADYLKES, I.S.: Thermodynamic properties of Freon 12B1 (CF_2ClBr); Kol. Teknika (1966)2 18/21

[Bady69] BADYLKES, I.S.: Thermodynamic tables and i-log p-diagram for R12B1; Proceedings of the 12th International Congress of Refrigeration, Madrid (1967)2.59 365/370

[Doer75] DÖRING, R.: Experimental and analytical investigations of thermodynamic properties of the refrigerant R12B1; 14th International Congress of Refrigeration, Moskau (1975) B1.2 62/67

[Doer76] DÖRING,R.: Kaltron 12B1; Kali-Chemie Hannover

[Gell53] GELLES, E. AND PITZER, K.S.: Thermodynamic functions of the halogenated methanes; J. Am. Chem. Soc. , 75(1953) 5259/5267

[Pere73] PERELSTEIN, I.I. AND ALESCHIN, JU.P.: Thermodynamische Eigenschaften von Freon 12B1; Kol. Teknika (1973)1 30/33

[Scho72] SCHOLTEN, W.: Vorläufige thermodynamische Eigenschaften von R12B1 (Bromchlordifluormethan); Kältetechnik-Klimatisierung 24(1972)2 45/48

R 12 B1. ph

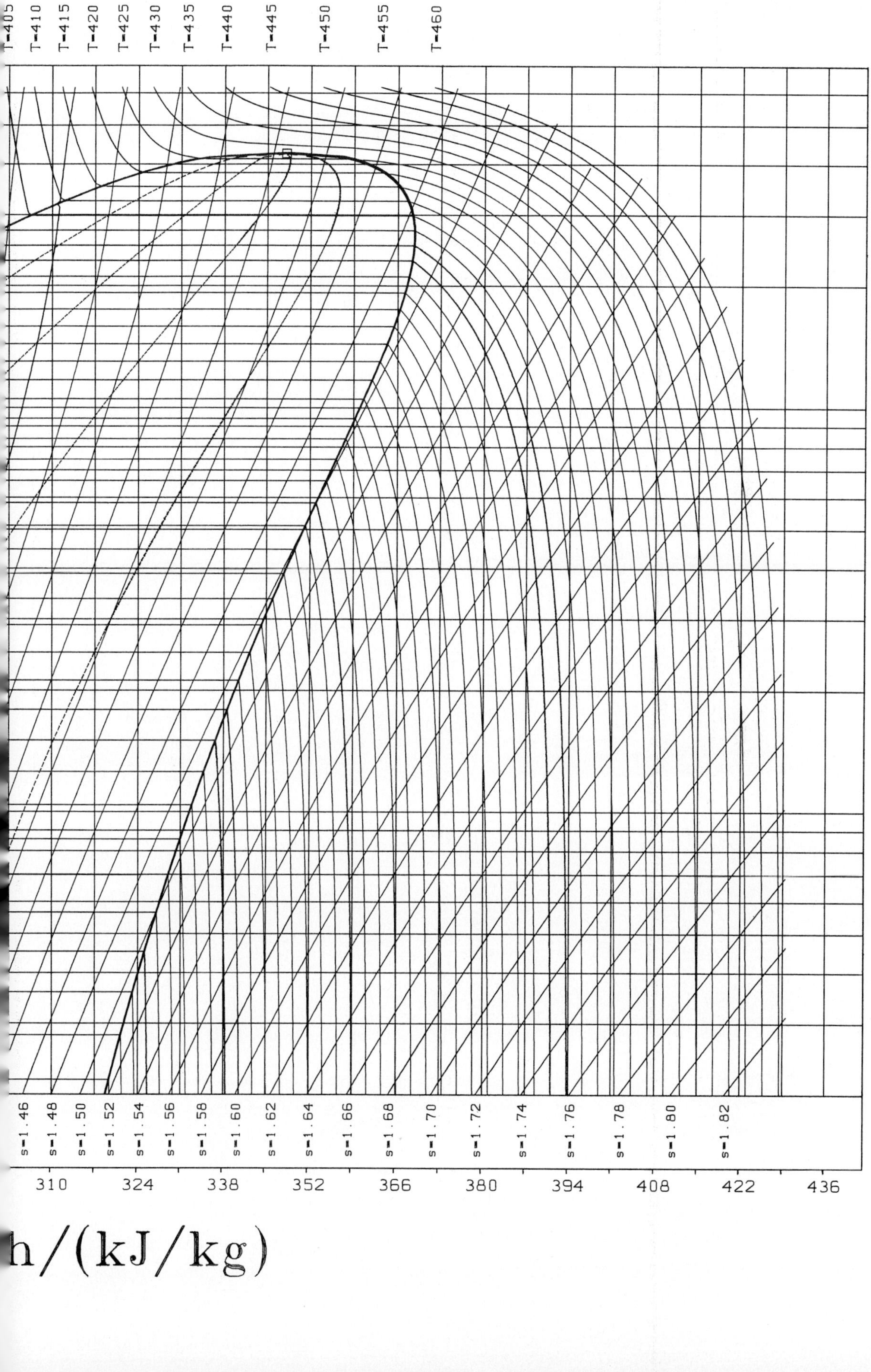
T=405
T=410
T=415
T=420
T=425
T=430
T=435
T=440
T=445
T=450
T=455
T=460
s=1.46
s=1.48
s=1.50
s=1.52
s=1.54
s=1.56
s=1.58
s=1.60
s=1.62
s=1.64
s=1.66
s=1.68
s=1.70
s=1.72
s=1.74
s=1.76
s=1.78
s=1.80
s=1.82
310
324
338
352
366
380
394
408
422
436
h/(kJ/kg)

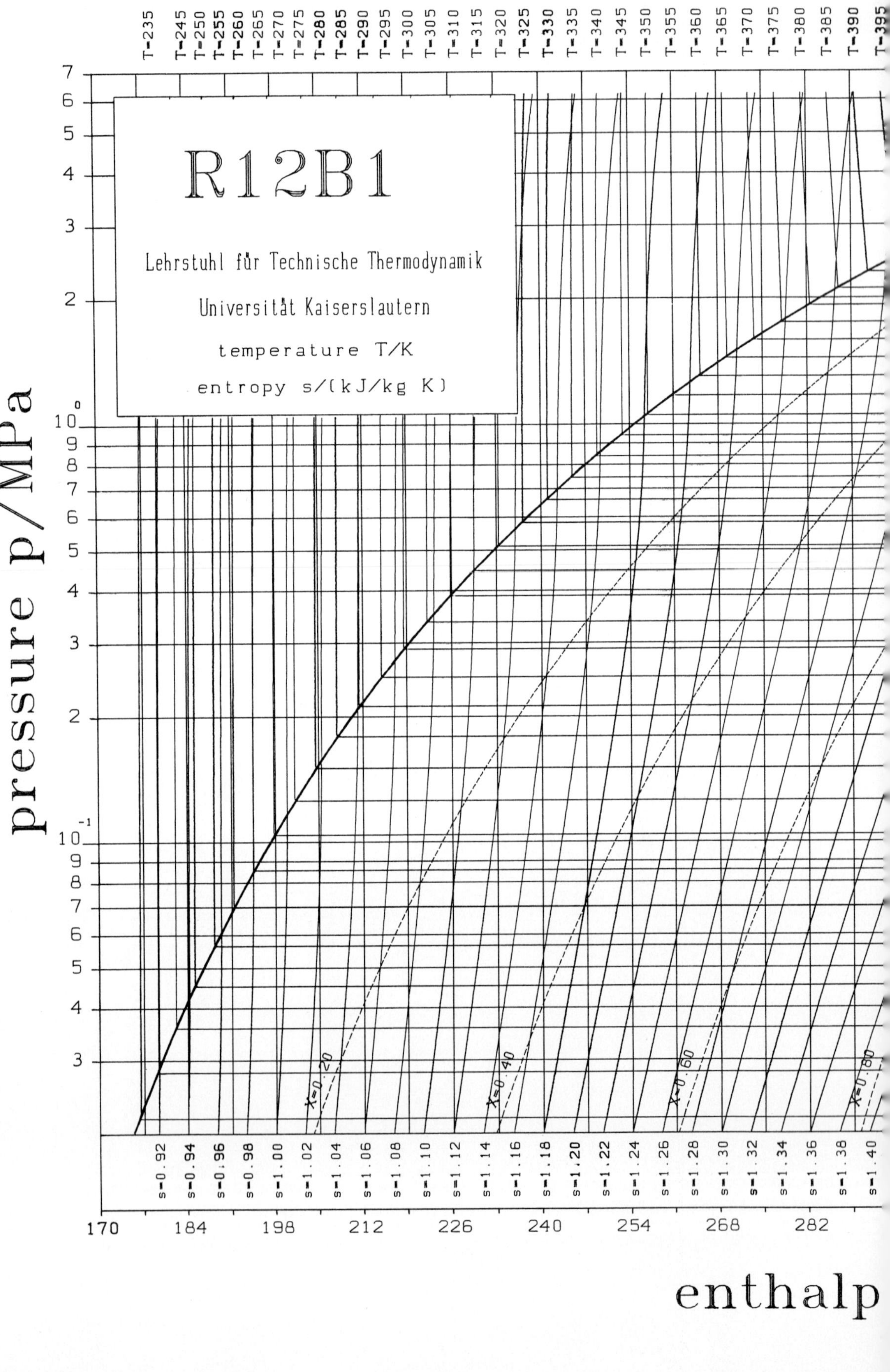
R12B1
Lehrstuhl für Technische Thermodynamik
Universität Kaiserslautern
temperature T/K
entropy s/(kJ/kg K)
pressure p/MPa
enthalp

R 12 B1. Ts

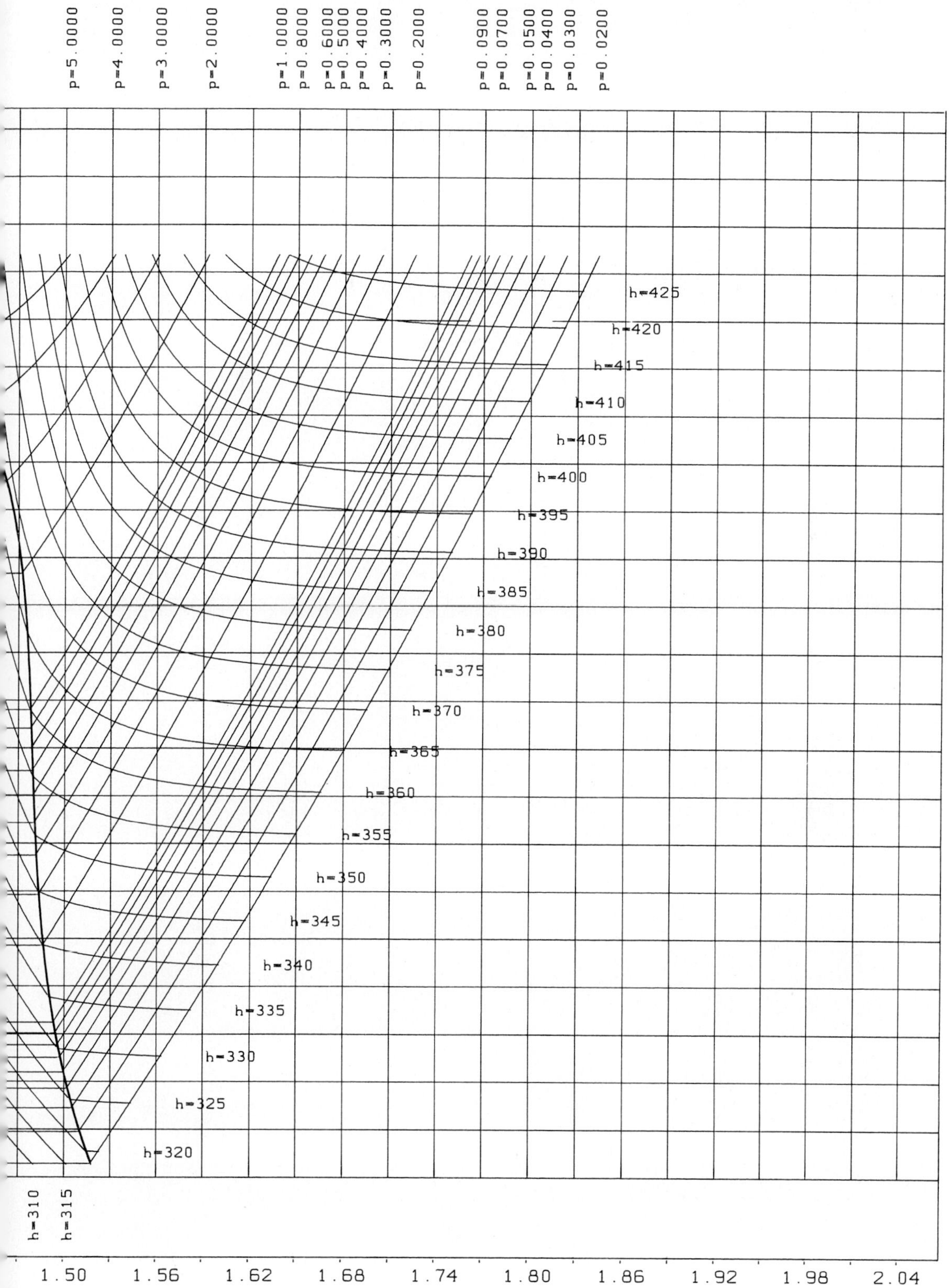

/(kJ/kg k)

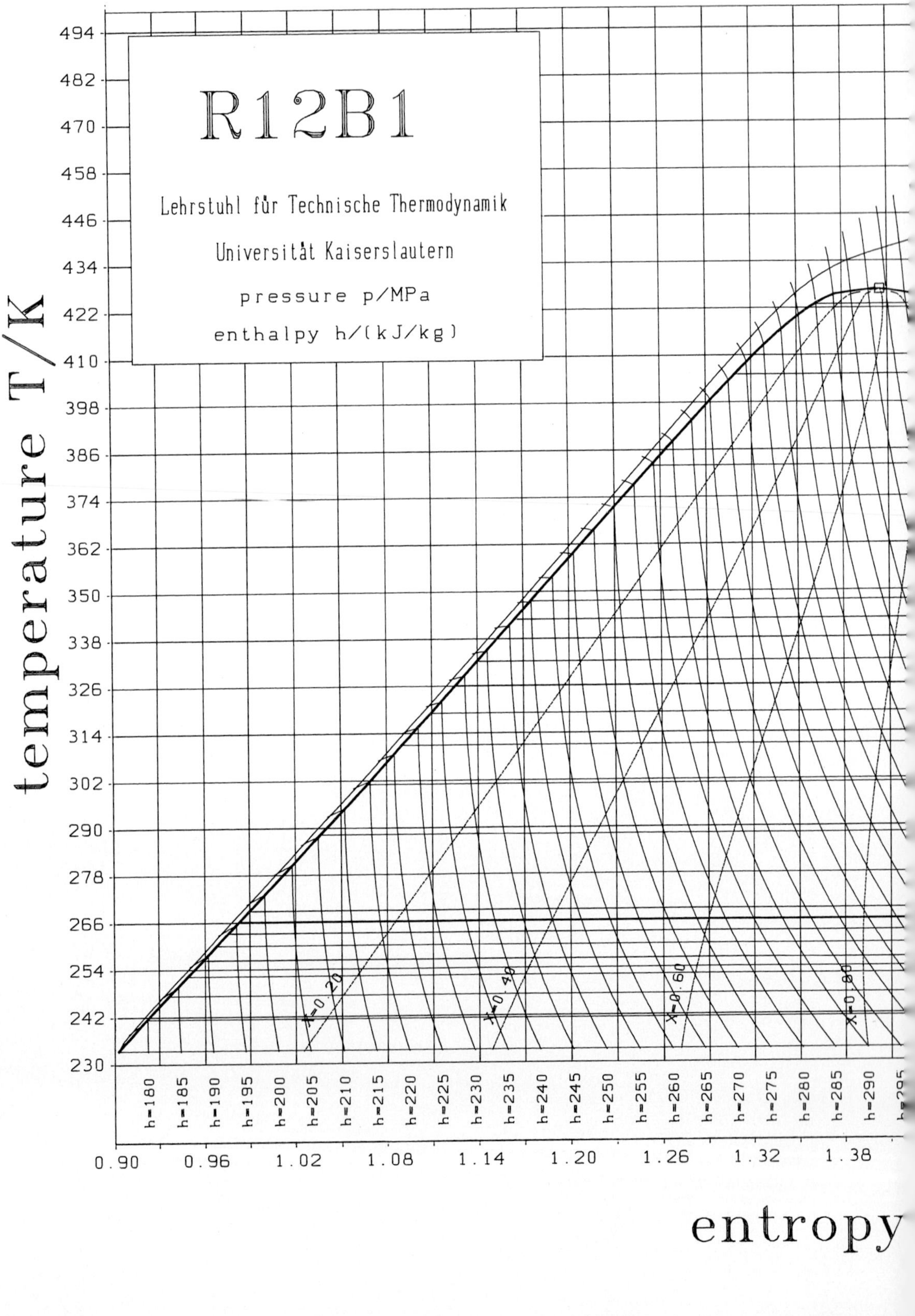

R12B1
Lehrstuhl für Technische Thermodynamik
Universität Kaiserslautern
pressure p/MPa
enthalpy h/(kJ/kg)
temperature T/K
entropy

R 12 B1. hs

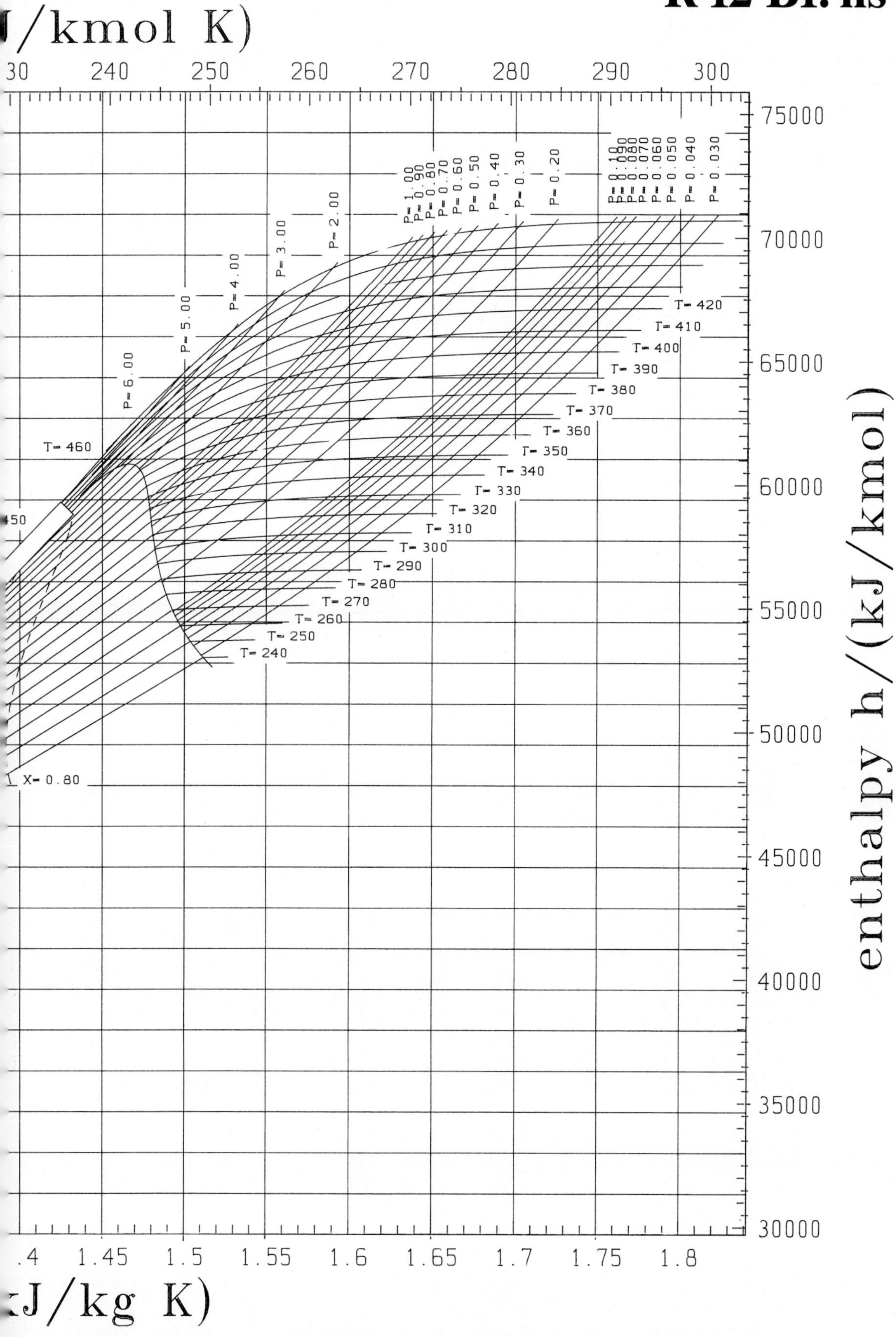

J/kmol K)
30 240 250 260 270 280 290 300
75000
70000
65000
60000
55000
50000
45000
40000
35000
30000
enthalpy h/(kJ/kmol)
.4 1.45 1.5 1.55 1.6 1.65 1.7 1.75 1.8
kJ/kg K)
P= 1.00
P= 0.90
P= 0.80
P= 0.70
P= 0.60
P= 0.50
P= 0.40
P= 0.30
P= 0.20
P= 0.10
P= 0.090
P= 0.080
P= 0.070
P= 0.060
P= 0.050
P= 0.040
P= 0.030
P= 2.00
P= 3.00
P= 4.00
P= 5.00
P= 6.00
T= 460
450
T= 420
T= 410
T= 400
T= 390
T= 380
T= 370
T= 360
T= 350
T= 340
T= 330
T= 320
T= 310
T= 300
T= 290
T= 280
T= 270
T= 260
T= 250
T= 240
X= 0.80

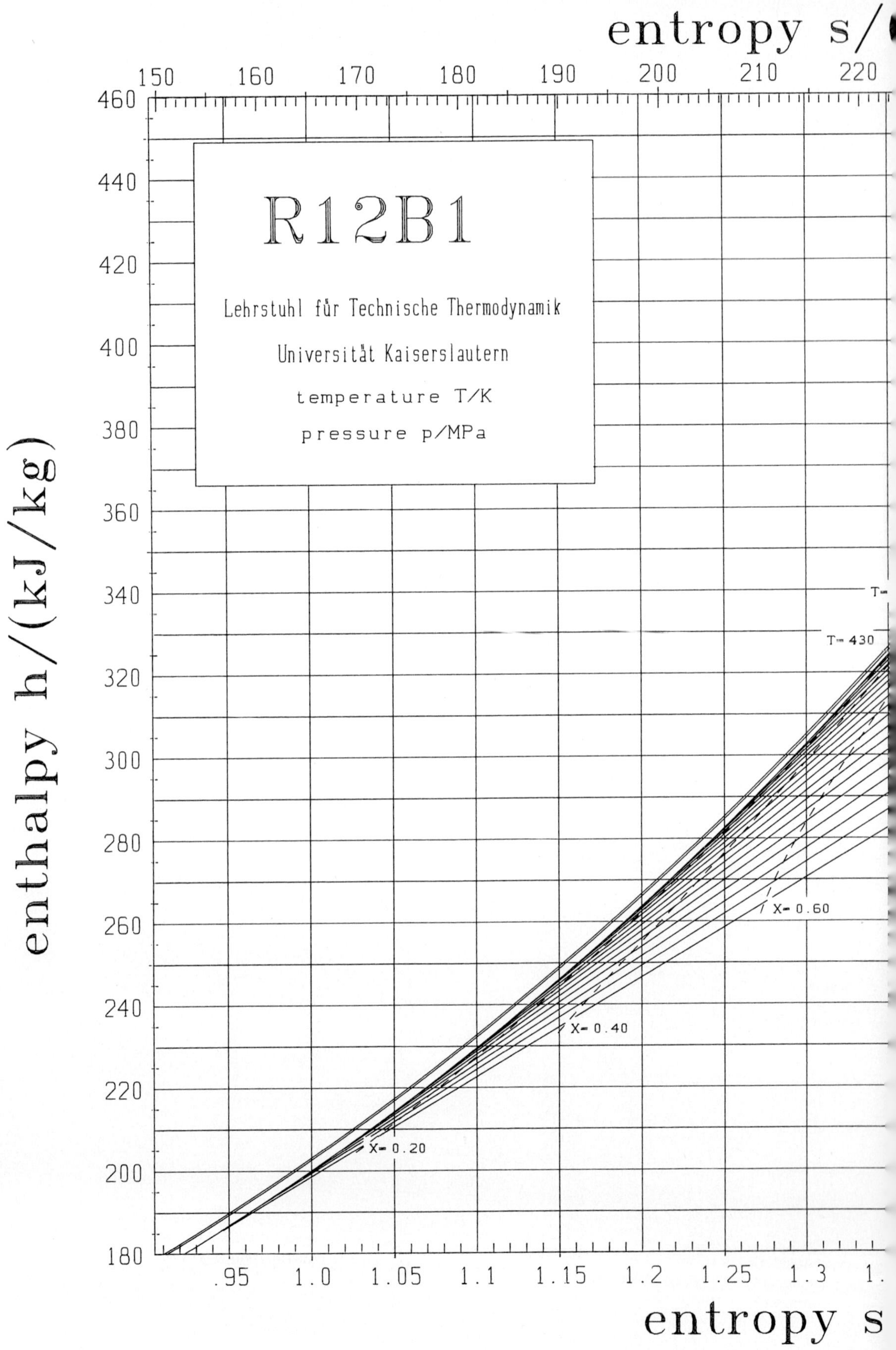

entropy s/
150
160
170
180
190
200
210
220
460
440
420
400
380
360
340
320
300
280
260
240
220
200
180
R12B1
Lehrstuhl für Technische Thermodynamik
Universität Kaiserslautern
temperature T/K
pressure p/MPa
enthalpy h/(kJ/kg)
T= 430
X= 0.60
X= 0.40
X= 0.20
.95
1.0
1.05
1.1
1.15
1.2
1.25
1.3
entropy s

Refrigerant R13B1 ($CBrF_3$)

Table R13B1.1: List of references for p, ρ, T data

Author	n	p/MPa		ρ/g cm^{-3}		T/K	
		min	max	min	max	min	max
MARTIN, J.J. *[Mart 53]*	72	0.14	13.15	0.0116	1.60	231	444
PERELSTEIN, I.I. *[Pere 71]*	109	0.82	10.41	0.0608	0.98	283	424

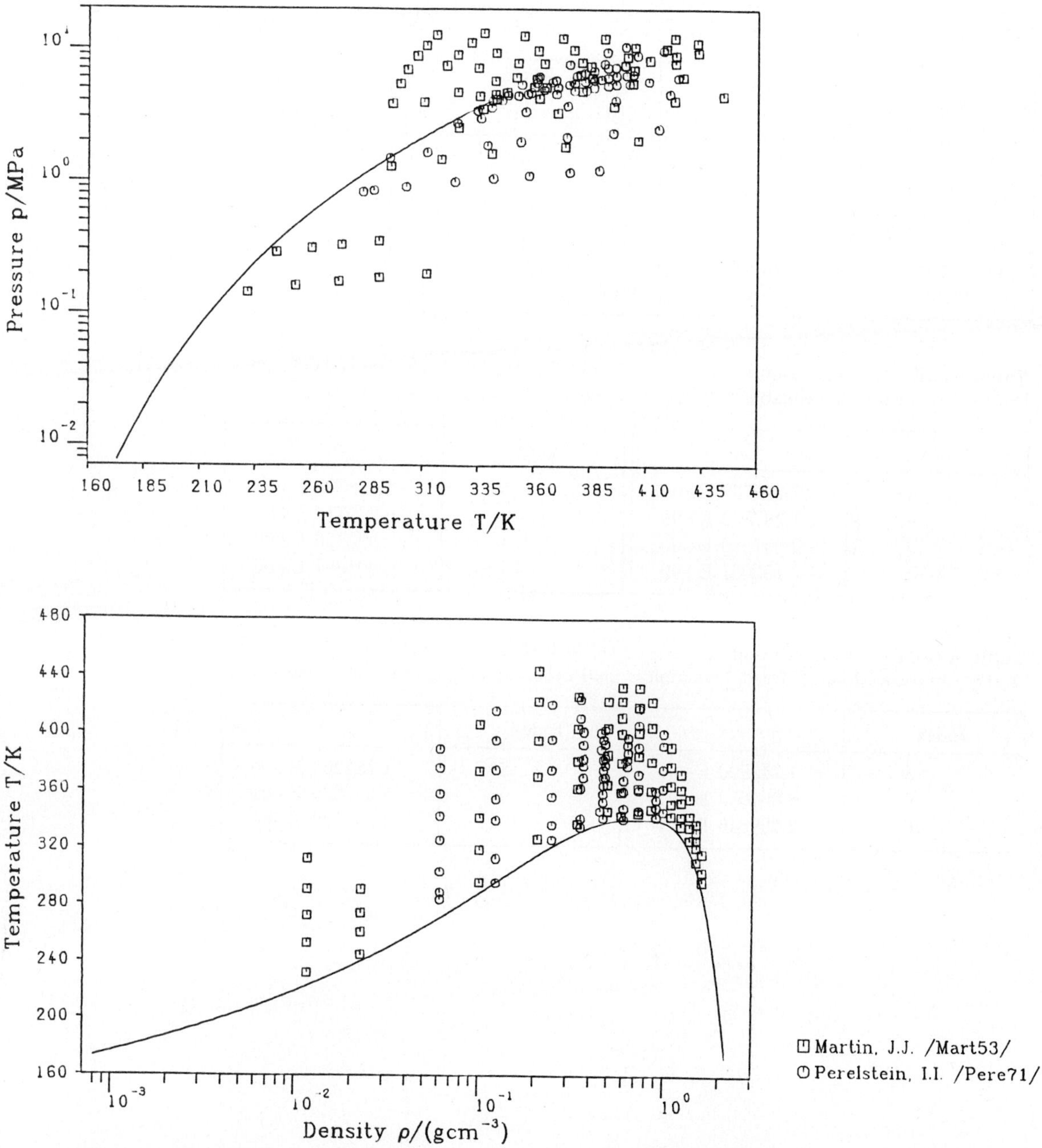

Fig.R13B1.1: Distribution of p, ρ, T data projected in the p, T and in the T, ρ plane

Table R13B1.2: Experimental vapor pressure data

Author	Δ / % std	opt	*EOS*
DuPont *[Doer 80]*	0.30	0.34	0.30
Henrici, H. *[Henr 70]*	0.65	0.65	0.62
Jäger, H.P. *[Jaeg 73]*	0.27	0.28	0.26
Martin, J.J. *[Mart 53]*	0.48	0.53	0.29
McDonald, R.A. *[McDo 59]*	0.37	0.32	0.37
Otto, J. *[Otto 66]*	0.47	0.49	0.43
Perelstein, I.I. *[Pere 71]*	0.09	0.09	0.07

Table R13B1.3: Experimental saturated liquid density data

Author	Δ / % std	opt	*EOS*
DuPont *[Doer 80]*	0.06	0.01	0.11
Henrici, H. *[Henr 70]*	0.26	0.31	0.27
Jäger, H.P. *[Jaeg 73]*	0.22	0.27	0.26
Martin, J.J. *[Mart 53]*	0.15	0.13	0.40
Nikulschin, R.K. *[Niku 67]*	1.00	1.00	1.00
Otto, J. *[Otto 66]*	0.62	0.58	0.62

Table R13B1.4: Coefficients of the standard vapor pressure equation

Index i	b_i
5	−6.794732 E+00
6	1.267832 E+00
9	−2.057210 E+00
15	−3.152061 E+00

Table R13B1.5: Coefficients of the quasi-optimal vapor pressure equation

Index i	b_i
4	−5.658738 E+00
5	−6.279714 E+00
8	2.020196 E+00
10	−4.901367 E+00

Table R13B1.6: Coefficients of the standard saturated liquid density equation

Index i	c_i
8	1.743990 E+00
9	8.004652 E−01
16	2.899846 E−01

Table R13B1.7: Coefficients of the quasi-optimal saturated liquid density equation

Index i	c_i
8	1.782264 E+00
9	6.587279 E−01
11	2.336178 E−01

Table R13B1.8: Coefficients of the *Bender-EOS*

Range of validity: one-phase region

$$0.14 \leq p/\text{MPa} \leq 13.15$$
$$0.01161 \leq \rho/\text{g cm}^{-3} \leq 1.5984$$
$$231.49 \leq T/\text{K} \leq 444.03$$

	Specific units	Molar units
a_{01}	5.681454354 E−02	1.260154808 E+00
a_{02}	5.262864937 E+01	1.167311069 E+03
a_{03}	−6.677623602 E+03	−1.481106591 E+05
a_{04}	2.450831014 E+06	5.435978702 E+07
a_{05}	−8.332292504 E+07	−1.848114551 E+09
a_{06}	−1.381085187 E−02	−4.562123479 E−02
a_{07}	1.999713248 E+01	6.605630736 E+01
a_{08}	6.912686813 E+03	2.283460217 E+04
a_{09}	5.120463557 E−02	2.519057249 E−02
a_{10}	−4.206470319 E+01	−2.069410207 E+01
a_{11}	1.233614864 E−02	9.038378853 E−04
a_{12}	−2.896317805 E+00	−2.122057569 E−01
a_{13}	4.694286640 E+00	5.122273243 E−02
a_{14}	1.062591384 E+06	3.510046411 E+06
a_{15}	−9.631972645 E+08	−3.181718909 E+09
a_{16}	9.113128830 E+10	3.010329804 E+11
a_{17}	−7.501434289 E+06	−5.496107983 E+05
a_{18}	3.847291392 E+09	2.818811459 E+08
a_{19}	−2.721249504 E+11	−1.993789527 E+10
a_{20}	1.600000000 E+00	3.548823184 E−02

Table R13B1.9: Deviations between experimental and predicted p, ρ, T data

Region	n	Δ / %	
		p	ρ
Gas	99	0.75	1.17
Critical	62	2.32	3.68
Liquid	20	2.96	0.31
Total	181	1.85	2.43

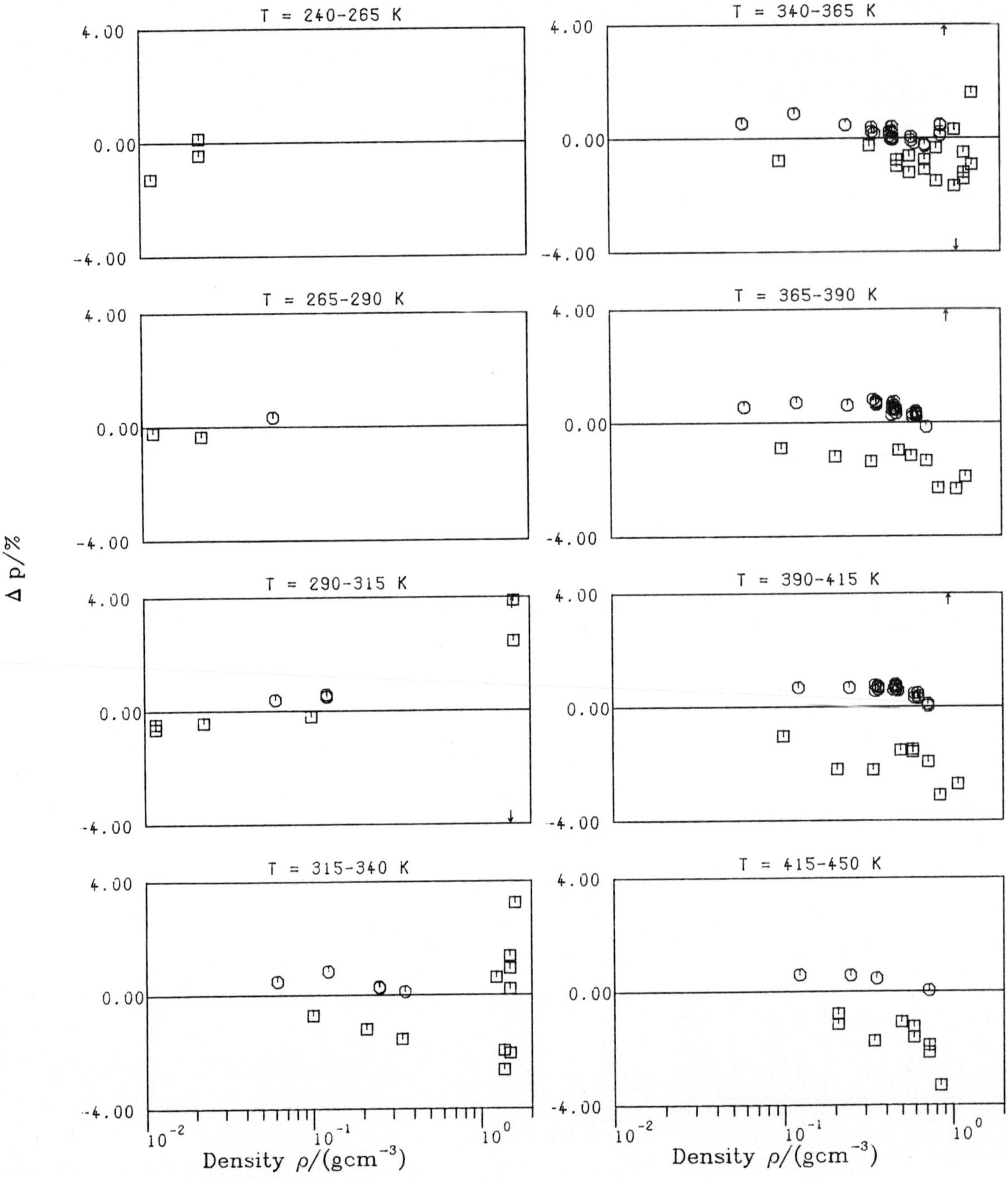

□ Martin, J.J. /Mart53/ ⦶ Perelstein, I.I. /Pere71/

Fig.R13B1.2: Deviations between experimental and calculated p, ρ, T data

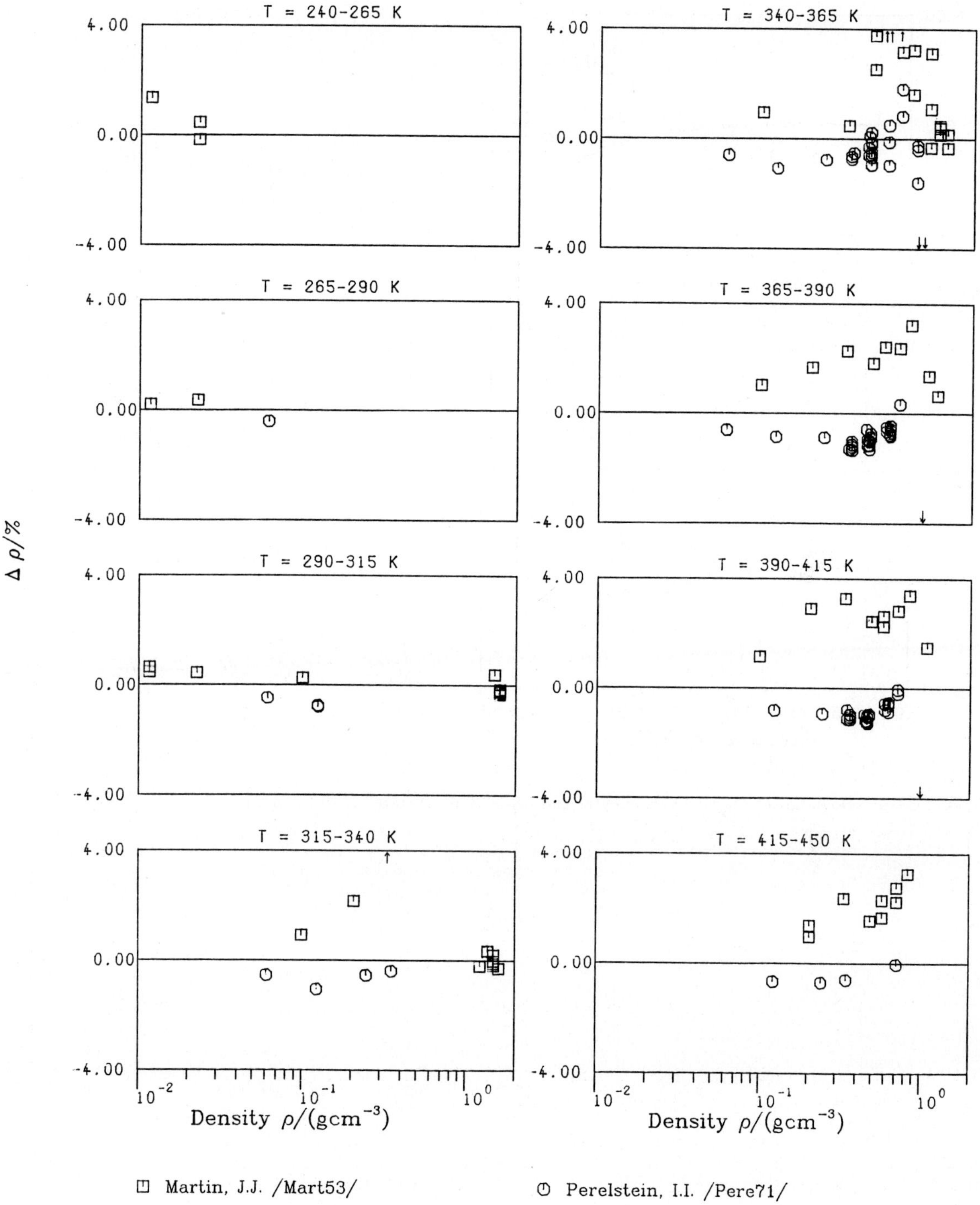

Fig.R13B1.2: Deviations between experimental and calculated p, ρ, T data *-continued*

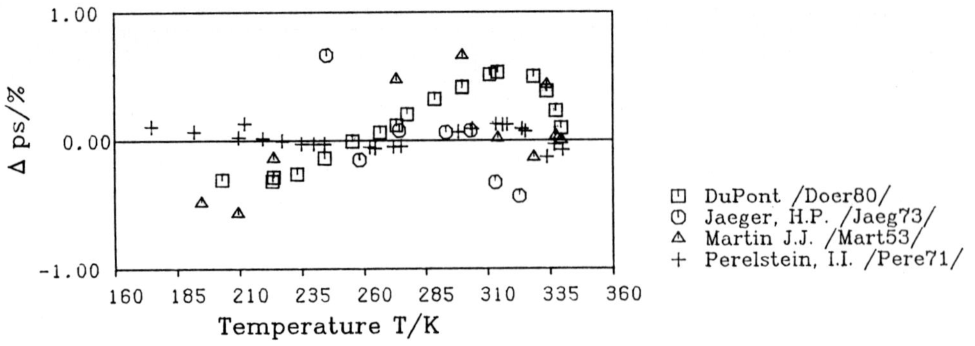

Fig.R13B1.3: Deviations between experimental and calculated vapor pressures

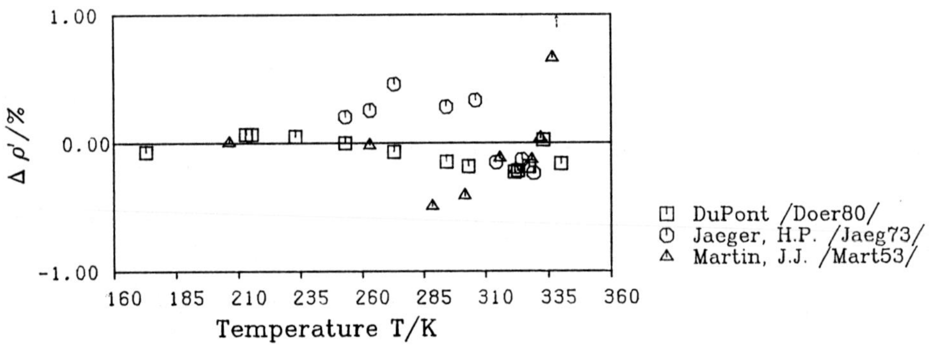

Fig.R13B1.4: Deviations between experimental and calculated saturated liquid densities

Table R13B1.10: Information on ideal gas state heat capacity equation

Author	T / K	
	min	max
GELLES, E. *[Gell 53]*	200	600

Coefficients of equation (15)	
Index i	d_i
1	1.118574 E+00
2	1.680007 E−02
3	−1.938103 E−05
4	9.655307 E−09
5	−1.333723 E−12

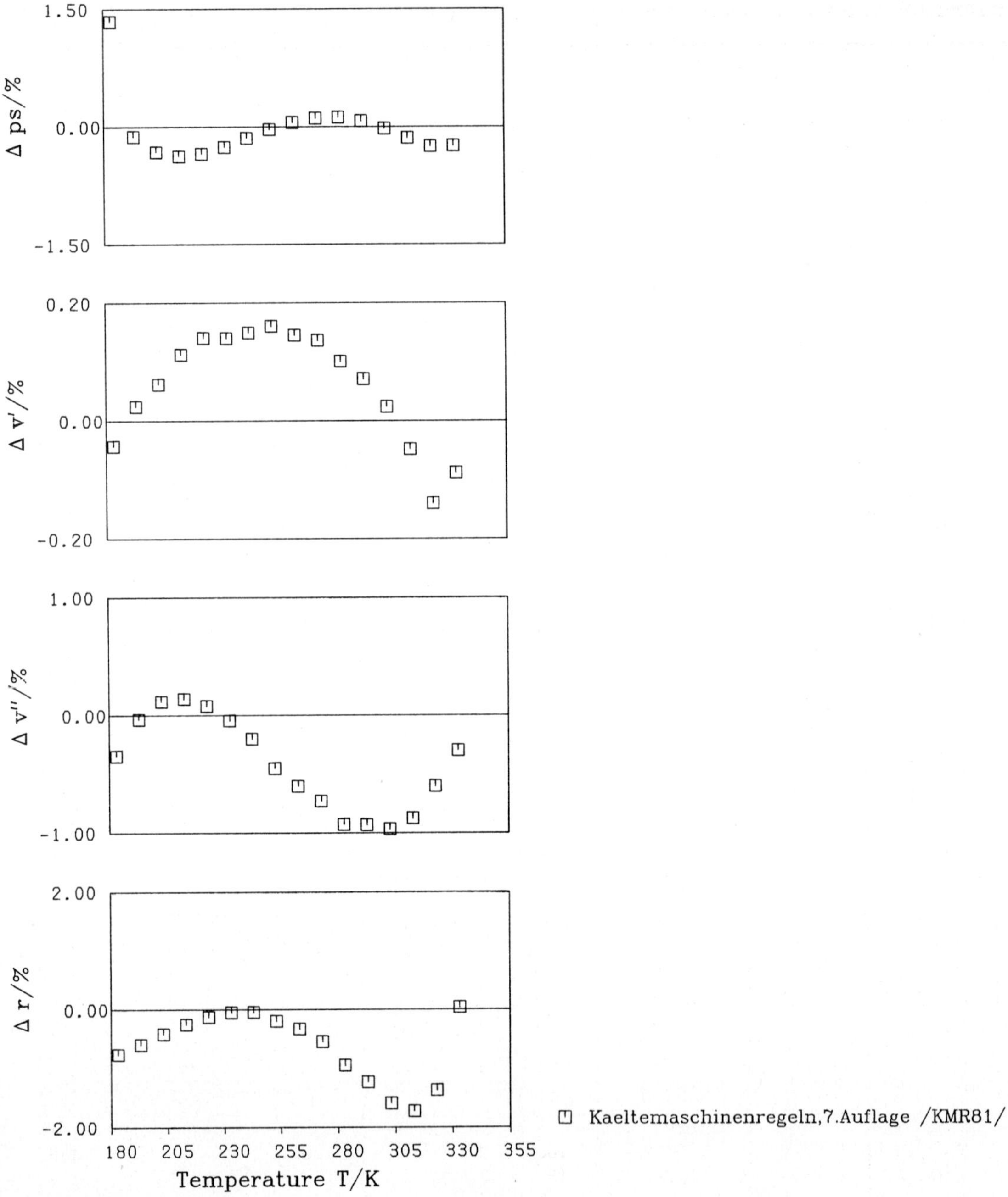

Fig.R13B1.5: Deviations between saturation properties from the *Kältemaschinenregeln* and those calculated by the *Bender-EOS*

Chart R13B1.1: Saturation properties

T K	p MPa	v' dm^3/kg	v'' dm^3/kg	h' kJ/kg	h'' kJ/kg	r kJ/kg	s' kJ/kgK	s'' kJ/kgK
175.0	0.00874	0.4632	1110.7	130.97	262.49	131.52	0.6903	1.4419
176.0	0.00944	0.4641	1034.0	131.65	262.82	131.17	0.6942	1.4395
177.0	0.01018	0.4650	963.44	132.33	263.16	130.83	0.6981	1.4373
178.0	0.01097	0.4658	898.61	133.01	263.50	130.49	0.7019	1.4350
179.0	0.01182	0.4667	838.93	133.69	263.84	130.15	0.7057	1.4328
180.0	0.01271	0.4676	783.94	134.37	264.18	129.82	0.7095	1.4307
181.0	0.01366	0.4685	733.22	135.04	264.52	129.48	0.7132	1.4286
182.0	0.01466	0.4694	686.39	135.72	264.87	129.15	0.7170	1.4265
183.0	0.01573	0.4703	643.11	136.39	265.21	128.81	0.7206	1.4245
184.0	0.01685	0.4712	603.07	137.07	265.55	128.48	0.7243	1.4226
185.0	0.01805	0.4721	565.99	137.74	265.89	128.15	0.7280	1.4207
186.0	0.01931	0.4730	531.63	138.41	266.23	127.82	0.7316	1.4188
187.0	0.02063	0.4739	499.75	139.08	266.57	127.49	0.7352	1.4169
188.0	0.02204	0.4748	470.15	139.75	266.92	127.16	0.7387	1.4151
189.0	0.02352	0.4757	442.65	140.43	267.26	126.83	0.7423	1.4134
190.0	0.02507	0.4767	417.07	141.10	267.60	126.51	0.7458	1.4117
191.0	0.02671	0.4776	393.25	141.77	267.94	126.18	0.7494	1.4100
192.0	0.02844	0.4785	371.07	142.44	268.29	125.85	0.7529	1.4083
193.0	0.03025	0.4795	350.38	143.11	268.63	125.52	0.7563	1.4067
194.0	0.03216	0.4804	331.08	143.78	268.97	125.19	0.7598	1.4051
195.0	0.03416	0.4814	313.05	144.45	269.31	124.87	0.7632	1.4036
196.0	0.03626	0.4823	296.21	145.12	269.66	124.54	0.7667	1.4021
197.0	0.03846	0.4833	280.45	145.79	270.00	124.21	0.7701	1.4006
198.0	0.04076	0.4843	265.70	146.46	270.34	123.88	0.7735	1.3991
199.0	0.04317	0.4853	251.89	147.13	270.68	123.55	0.7768	1.3977
200.0	0.04570	0.4863	238.94	147.80	271.03	123.22	0.7802	1.3963
201.0	0.04834	0.4872	226.79	148.48	271.37	122.89	0.7835	1.3949
202.0	0.05111	0.4882	215.39	149.15	271.71	122.56	0.7869	1.3936
203.0	0.05399	0.4893	204.68	149.82	272.05	122.23	0.7902	1.3923
204.0	0.05700	0.4903	194.62	150.50	272.39	121.90	0.7935	1.3910
205.0	0.06015	0.4913	185.15	151.17	272.73	121.56	0.7968	1.3898
206.0	0.06343	0.4923	176.25	151.84	273.07	121.23	0.8001	1.3885
207.0	0.06685	0.4933	167.86	152.52	273.41	120.89	0.8033	1.3873
208.0	0.07041	0.4944	159.95	153.20	273.75	120.56	0.8066	1.3862
209.0	0.07412	0.4954	152.50	153.87	274.09	120.22	0.8098	1.3850
210.0	0.07798	0.4965	145.46	154.55	274.43	119.88	0.8130	1.3839
211.0	0.08200	0.4976	138.82	155.23	274.77	119.54	0.8162	1.3828
212.0	0.08617	0.4986	132.55	155.91	275.11	119.20	0.8194	1.3817
213.0	0.09051	0.4997	126.63	156.59	275.45	118.86	0.8226	1.3807
214.0	0.09502	0.5008	121.02	157.27	275.78	118.52	0.8258	1.3796
215.0	0.09970	0.5019	115.72	157.95	276.12	118.17	0.8290	1.3786
216.0	0.10456	0.5030	110.70	158.63	276.45	117.82	0.8321	1.3776
217.0	0.10961	0.5041	105.94	159.32	276.79	117.48	0.8353	1.3766
218.0	0.11484	0.5052	101.43	160.00	277.12	117.13	0.8384	1.3757
219.0	0.12025	0.5063	97.160	160.69	277.46	116.77	0.8415	1.3747
220.0	0.12587	0.5075	93.106	161.37	277.79	116.42	0.8447	1.3738
221.0	0.13169	0.5086	89.257	162.06	278.12	116.06	0.8478	1.3729
222.0	0.13771	0.5098	85.603	162.75	278.46	115.71	0.8509	1.3721
223.0	0.14394	0.5109	82.131	163.44	278.79	115.35	0.8539	1.3712
224.0	0.15038	0.5121	78.830	164.13	279.12	114.99	0.8570	1.3704

Chart R13B1.1: Saturation properties–*continued*

T K	p MPa	v' dm³/kg	v'' dm³/kg	h' kJ/kg	h'' kJ/kg	r kJ/kg	s' kJ/kgK	s'' kJ/kgK
225.0	0.15705	0.5133	75.691	164.82	279.45	114.62	0.8601	1.3695
226.0	0.16394	0.5145	72.704	165.51	279.77	114.26	0.8631	1.3687
227.0	0.17106	0.5157	69.861	166.21	280.10	113.89	0.8662	1.3679
228.0	0.17841	0.5169	67.153	166.90	280.43	113.52	0.8692	1.3671
229.0	0.18600	0.5181	64.574	167.60	280.75	113.15	0.8723	1.3664
230.0	0.19384	0.5194	62.114	168.30	281.08	112.78	0.8753	1.3656
231.0	0.20193	0.5206	59.769	169.00	281.40	112.40	0.8783	1.3649
232.0	0.21027	0.5219	57.532	169.70	281.72	112.02	0.8813	1.3642
233.0	0.21887	0.5231	55.397	170.40	282.04	111.64	0.8843	1.3635
234.0	0.22773	0.5244	53.358	171.11	282.36	111.26	0.8873	1.3628
235.0	0.23687	0.5257	51.411	171.81	282.68	110.87	0.8903	1.3621
236.0	0.24628	0.5270	49.549	172.52	283.00	110.48	0.8933	1.3614
237.0	0.25597	0.5283	47.770	173.22	283.32	110.09	0.8962	1.3608
238.0	0.26595	0.5297	46.069	173.93	283.63	109.70	0.8992	1.3601
239.0	0.27621	0.5310	44.441	174.64	283.95	109.30	0.9022	1.3595
240.0	0.28678	0.5324	42.883	175.36	284.26	108.90	0.9051	1.3589
241.0	0.29764	0.5337	41.391	176.07	284.57	108.50	0.9081	1.3583
242.0	0.30882	0.5351	39.962	176.79	284.88	108.09	0.9110	1.3577
243.0	0.32030	0.5365	38.593	177.50	285.19	107.68	0.9139	1.3571
244.0	0.33211	0.5379	37.280	178.22	285.49	107.27	0.9169	1.3565
245.0	0.34424	0.5393	36.022	178.94	285.80	106.86	0.9198	1.3559
246.0	0.35669	0.5408	34.815	179.66	286.10	106.44	0.9227	1.3554
247.0	0.36949	0.5422	33.657	180.39	286.40	106.02	0.9256	1.3548
248.0	0.38262	0.5437	32.545	181.11	286.71	105.59	0.9285	1.3543
249.0	0.39610	0.5452	31.477	181.84	287.00	105.16	0.9314	1.3537
250.0	0.40993	0.5467	30.452	182.57	287.30	104.73	0.9343	1.3532
251.0	0.42412	0.5482	29.467	183.30	287.60	104.30	0.9372	1.3527
252.0	0.43867	0.5497	28.520	184.03	287.89	103.86	0.9400	1.3522
253.0	0.45359	0.5513	27.609	184.77	288.18	103.42	0.9429	1.3517
254.0	0.46888	0.5528	26.734	185.50	288.47	102.97	0.9458	1.3512
255.0	0.48456	0.5544	25.891	186.24	288.76	102.52	0.9487	1.3507
256.0	0.50062	0.5560	25.081	186.98	289.04	102.07	0.9515	1.3502
257.0	0.51708	0.5576	24.300	187.72	289.33	101.61	0.9544	1.3497
258.0	0.53393	0.5593	23.549	188.46	289.61	101.15	0.9572	1.3493
259.0	0.55118	0.5609	22.825	189.21	289.89	100.68	0.9601	1.3488
260.0	0.56885	0.5626	22.127	189.96	290.17	100.21	0.9629	1.3483
261.0	0.58693	0.5643	21.455	190.71	290.44	99.73	0.9658	1.3479
262.0	0.60543	0.5660	20.807	191.46	290.71	99.25	0.9686	1.3474
263.0	0.62437	0.5678	20.182	192.21	290.98	98.77	0.9714	1.3470
264.0	0.64373	0.5695	19.579	192.97	291.25	98.28	0.9743	1.3465
265.0	0.66354	0.5713	18.997	193.73	291.52	97.79	0.9771	1.3461
266.0	0.68379	0.5731	18.435	194.49	291.78	97.29	0.9799	1.3457
267.0	0.70449	0.5750	17.893	195.25	292.04	96.79	0.9827	1.3452
268.0	0.72565	0.5768	17.369	196.02	292.30	96.28	0.9855	1.3448
269.0	0.74728	0.5787	16.863	196.79	292.55	95.77	0.9884	1.3444
270.0	0.76938	0.5806	16.375	197.56	292.81	95.25	0.9912	1.3439
271.0	0.79196	0.5825	15.902	198.33	293.06	94.73	0.9940	1.3435
272.0	0.81501	0.5845	15.446	199.11	293.30	94.20	0.9968	1.3431
273.0	0.83856	0.5865	15.004	199.88	293.55	93.66	0.9996	1.3427
274.0	0.86261	0.5885	14.577	200.66	293.79	93.12	1.0024	1.3422

Chart R13B1.1: Saturation properties–*continued*

T K	p MPa	v' dm³/kg	v'' dm³/kg	h' kJ/kg	h'' kJ/kg	r kJ/kg	s' kJ/kgK	s'' kJ/kgK
275.0	0.88716	0.5906	14.164	201.45	294.02	92.58	1.0052	1.3418
276.0	0.91222	0.5927	13.764	202.23	294.26	92.02	1.0080	1.3414
277.0	0.93779	0.5948	13.376	203.02	294.49	91.47	1.0108	1.3410
278.0	0.96389	0.5969	13.002	203.81	294.71	90.90	1.0136	1.3406
279.0	0.99051	0.5991	12.639	204.61	294.94	90.33	1.0164	1.3401
280.0	1.01767	0.6013	12.287	205.40	295.16	89.76	1.0192	1.3397
281.0	1.04538	0.6036	11.946	206.20	295.37	89.17	1.0219	1.3393
282.0	1.07363	0.6058	11.616	207.00	295.59	88.58	1.0247	1.3389
283.0	1.10244	0.6082	11.296	207.81	295.79	87.98	1.0275	1.3384
284.0	1.13181	0.6105	10.985	208.62	296.00	87.38	1.0303	1.3380
285.0	1.16176	0.6129	10.684	209.43	296.20	86.77	1.0331	1.3376
286.0	1.19228	0.6154	10.392	210.25	296.39	86.15	1.0359	1.3371
287.0	1.22338	0.6179	10.108	211.07	296.59	85.52	1.0387	1.3367
288.0	1.25507	0.6204	9.8330	211.89	296.77	84.88	1.0415	1.3362
289.0	1.28737	0.6230	9.5658	212.71	296.95	84.24	1.0443	1.3358
290.0	1.32027	0.6256	9.3064	213.54	297.13	83.59	1.0471	1.3353
291.0	1.35378	0.6283	9.0543	214.37	297.30	82.93	1.0499	1.3348
292.0	1.38791	0.6310	8.8094	215.21	297.47	82.26	1.0527	1.3344
293.0	1.42267	0.6338	8.5714	216.05	297.63	81.58	1.0555	1.3339
294.0	1.45807	0.6367	8.3401	216.90	297.79	80.89	1.0583	1.3334
295.0	1.49411	0.6396	8.1152	217.74	297.94	80.19	1.0611	1.3329
296.0	1.53080	0.6426	7.8964	218.60	298.08	79.48	1.0639	1.3324
297.0	1.56815	0.6456	7.6836	219.45	298.22	78.77	1.0667	1.3319
298.0	1.60617	0.6487	7.4765	220.32	298.35	78.04	1.0695	1.3314
299.0	1.64486	0.6519	7.2749	221.18	298.48	77.30	1.0723	1.3308
300.0	1.68423	0.6551	7.0786	222.05	298.60	76.54	1.0751	1.3303
301.0	1.72430	0.6584	6.8875	222.93	298.71	75.78	1.0780	1.3297
302.0	1.76507	0.6618	6.7013	223.81	298.81	75.00	1.0808	1.3291
303.0	1.80654	0.6653	6.5199	224.70	298.91	74.21	1.0836	1.3285
304.0	1.84874	0.6689	6.3431	225.59	298.99	73.41	1.0865	1.3279
305.0	1.89166	0.6726	6.1707	226.48	299.07	72.59	1.0893	1.3273
306.0	1.93532	0.6764	6.0025	227.39	299.15	71.76	1.0922	1.3267
307.0	1.97972	0.6802	5.8385	228.30	299.21	70.91	1.0950	1.3260
308.0	2.02488	0.6842	5.6784	229.21	299.26	70.05	1.0979	1.3253
309.0	2.07080	0.6884	5.5221	230.14	299.30	69.17	1.1008	1.3247
310.0	2.11749	0.6926	5.3695	231.06	299.33	68.27	1.1037	1.3239
311.0	2.16497	0.6970	5.2203	232.00	299.35	67.35	1.1066	1.3232
312.0	2.21325	0.7015	5.0746	232.95	299.36	66.42	1.1095	1.3224
313.0	2.26233	0.7061	4.9320	233.90	299.36	65.46	1.1125	1.3216
314.0	2.31223	0.7110	4.7926	234.86	299.34	64.48	1.1154	1.3208
315.0	2.36296	0.7160	4.6562	235.83	299.31	63.48	1.1184	1.3199
316.0	2.41452	0.7212	4.5226	236.81	299.26	62.45	1.1214	1.3190
317.0	2.46695	0.7266	4.3916	237.80	299.20	61.40	1.1244	1.3181
318.0	2.52023	0.7322	4.2633	238.80	299.12	60.32	1.1274	1.3171
319.0	2.57439	0.7381	4.1374	239.81	299.03	59.22	1.1305	1.3161
320.0	2.62945	0.7442	4.0138	240.84	298.91	58.08	1.1336	1.3150
321.0	2.68541	0.7506	3.8924	241.87	298.78	56.90	1.1367	1.3139
322.0	2.74229	0.7573	3.7731	242.93	298.62	55.69	1.1398	1.3128
323.0	2.80011	0.7644	3.6556	244.00	298.43	54.44	1.1430	1.3115
324.0	2.85889	0.7719	3.5398	245.08	298.22	53.14	1.1462	1.3102

Chart R13B1.1: Saturation properties–*continued*

T K	p MPa	v' dm³/kg	v'' dm³/kg	h' kJ/kg	h'' kJ/kg	r kJ/kg	s' kJ/kgK	s'' kJ/kgK
325.0	2.91863	0.7798	3.4256	246.19	297.98	51.79	1.1495	1.3088
326.0	2.97936	0.7883	3.3128	247.32	297.71	50.40	1.1528	1.3074
327.0	3.04110	0.7972	3.2012	248.47	297.41	48.94	1.1562	1.3058
328.0	3.10388	0.8069	3.0905	249.65	297.06	47.41	1.1596	1.3042
329.0	3.16770	0.8173	2.9805	250.86	296.66	45.80	1.1631	1.3024
330.0	3.23261	0.8286	2.8709	252.10	296.22	44.11	1.1668	1.3004
331.0	3.29861	0.8410	2.7613	253.39	295.71	42.32	1.1705	1.2983
332.0	3.36576	0.8546	2.6513	254.73	295.13	40.40	1.1744	1.2960
333.0	3.43407	0.8700	2.5403	256.14	294.46	38.33	1.1784	1.2935
334.0	3.50360	0.8874	2.4274	257.62	293.69	36.08	1.1827	1.2907
335.0	3.57437	0.9077	2.3115	259.20	292.79	33.59	1.1872	1.2875
336.0	3.64645	0.9319	2.1909	260.92	291.70	30.77	1.1922	1.2837
337.0	3.71989	0.9622	2.0625	262.86	290.35	27.50	1.1977	1.2793
338.0	3.79478	1.0028	1.9209	265.14	288.61	23.47	1.2042	1.2737
339.0	3.87120	1.0657	1.7519	268.12	286.08	17.96	1.2128	1.2658
340.0	3.94925	1.2213	1.4818	273.84	280.82	6.98	1.2294	1.2499

Chart R13B1.2: Saturation properties

p MPa	T K	v' dm^3/kg	v'' dm^3/kg	h' kJ/kg	h'' kJ/kg	r kJ/kg	s' kJ/kgK	s'' kJ/kgK
0.0102	177.02	0.4650	961.95	132.35	263.17	130.82	0.6982	1.4372
0.0104	177.28	0.4652	944.72	132.52	263.26	130.74	0.6992	1.4366
0.0106	177.53	0.4654	928.13	132.70	263.34	130.65	0.7001	1.4361
0.0108	177.78	0.4656	912.13	132.87	263.43	130.56	0.7011	1.4355
0.0110	178.03	0.4659	896.69	133.03	263.51	130.48	0.7020	1.4350
0.0115	178.63	0.4664	860.37	133.44	263.72	130.28	0.7043	1.4336
0.0120	179.21	0.4669	826.98	133.83	263.91	130.08	0.7065	1.4324
0.0125	179.77	0.4674	796.17	134.21	264.11	129.89	0.7086	1.4312
0.0130	180.31	0.4679	767.66	134.58	264.29	129.71	0.7107	1.4300
0.0135	180.84	0.4683	741.20	134.93	264.47	129.54	0.7126	1.4289
0.0140	181.35	0.4688	716.56	135.28	264.64	129.36	0.7145	1.4279
0.0145	181.84	0.4692	693.57	135.61	264.81	129.20	0.7164	1.4269
0.0150	182.32	0.4697	672.06	135.94	264.98	129.04	0.7181	1.4259
0.0155	182.79	0.4701	651.89	136.25	265.14	128.88	0.7199	1.4250
0.0160	183.25	0.4705	632.94	136.56	265.29	128.73	0.7215	1.4241
0.0165	183.69	0.4709	615.10	136.86	265.44	128.58	0.7232	1.4232
0.0170	184.12	0.4713	598.28	137.15	265.59	128.44	0.7248	1.4223
0.0175	184.55	0.4717	582.39	137.44	265.74	128.30	0.7263	1.4215
0.0180	184.96	0.4720	567.35	137.71	265.88	128.16	0.7278	1.4207
0.0185	185.37	0.4724	553.09	137.99	266.02	128.03	0.7293	1.4200
0.0190	185.76	0.4728	539.56	138.25	266.15	127.90	0.7307	1.4192
0.0195	186.15	0.4731	526.70	138.51	266.28	127.77	0.7321	1.4185
0.0200	186.53	0.4735	514.47	138.77	266.41	127.65	0.7335	1.4178
0.0210	187.27	0.4741	491.68	139.26	266.66	127.40	0.7361	1.4165
0.0220	187.97	0.4748	470.90	139.74	266.91	127.17	0.7387	1.4152
0.0230	188.66	0.4754	451.86	140.20	267.14	126.95	0.7411	1.4140
0.0240	189.32	0.4760	434.36	140.64	267.37	126.73	0.7434	1.4128
0.0250	189.95	0.4766	418.21	141.07	267.59	126.52	0.7457	1.4117
0.0260	190.57	0.4772	403.26	141.48	267.80	126.32	0.7479	1.4107
0.0270	191.17	0.4778	389.38	141.88	268.00	126.12	0.7500	1.4097
0.0280	191.75	0.4783	376.46	142.27	268.20	125.93	0.7520	1.4087
0.0290	192.31	0.4788	364.40	142.65	268.39	125.75	0.7540	1.4078
0.0300	192.86	0.4793	353.11	143.02	268.58	125.57	0.7559	1.4069
0.0310	193.40	0.4799	342.53	143.37	268.77	125.39	0.7577	1.4061
0.0320	193.92	0.4804	332.59	143.72	268.94	125.22	0.7595	1.4052
0.0330	194.43	0.4808	323.23	144.06	269.12	125.05	0.7613	1.4045
0.0340	194.92	0.4813	314.40	144.40	269.29	124.89	0.7630	1.4037
0.0350	195.41	0.4818	306.05	144.72	269.45	124.73	0.7646	1.4030
0.0360	195.88	0.4822	298.16	145.04	269.62	124.58	0.7663	1.4022
0.0370	196.34	0.4827	290.67	145.35	269.77	124.43	0.7678	1.4015
0.0380	196.80	0.4831	283.57	145.65	269.93	124.28	0.7694	1.4009
0.0390	197.24	0.4835	276.82	145.95	270.08	124.13	0.7709	1.4002
0.0400	197.68	0.4840	270.39	146.24	270.23	123.99	0.7724	1.3996
0.0410	198.10	0.4844	264.26	146.53	270.38	123.85	0.7738	1.3990
0.0420	198.52	0.4848	258.42	146.81	270.52	123.71	0.7752	1.3984
0.0430	198.93	0.4852	252.84	147.08	270.66	123.58	0.7766	1.3978
0.0440	199.33	0.4856	247.50	147.35	270.80	123.44	0.7780	1.3972
0.0450	199.73	0.4860	242.39	147.62	270.93	123.31	0.7793	1.3967
0.0460	200.12	0.4864	237.50	147.88	271.07	123.18	0.7806	1.3961
0.0470	200.50	0.4867	232.81	148.14	271.20	123.06	0.7819	1.3956

Chart R13B1.2: Saturation properties—*continued*

p MPa	T K	v' dm³/kg	v'' dm³/kg	h' kJ/kg	h'' kJ/kg	r kJ/kg	s' kJ/kgK	s'' kJ/kgK
0.0480	200.87	0.4871	228.30	148.39	271.32	122.93	0.7831	1.3951
0.0490	201.24	0.4875	223.97	148.64	271.45	122.81	0.7843	1.3946
0.0500	201.60	0.4879	219.81	148.88	271.57	122.69	0.7856	1.3941
0.0510	201.96	0.4882	215.81	149.12	271.70	122.57	0.7867	1.3937
0.0520	202.31	0.4886	211.95	149.36	271.82	122.46	0.7879	1.3932
0.0530	202.66	0.4889	208.24	149.59	271.94	122.34	0.7891	1.3927
0.0540	203.00	0.4893	204.66	149.82	272.05	122.23	0.7902	1.3923
0.0550	203.34	0.4896	201.20	150.05	272.17	122.12	0.7913	1.3919
0.0560	203.67	0.4899	197.86	150.27	272.28	122.01	0.7924	1.3914
0.0570	204.00	0.4903	194.63	150.49	272.39	121.90	0.7935	1.3910
0.0580	204.32	0.4906	191.52	150.71	272.50	121.79	0.7946	1.3906
0.0590	204.64	0.4909	188.50	150.93	272.61	121.68	0.7956	1.3902
0.0600	204.95	0.4912	185.58	151.14	272.72	121.58	0.7966	1.3898
0.0610	205.26	0.4916	182.75	151.35	272.82	121.48	0.7976	1.3894
0.0620	205.57	0.4919	180.01	151.55	272.93	121.37	0.7986	1.3891
0.0630	205.87	0.4922	177.36	151.76	273.03	121.27	0.7996	1.3887
0.0640	206.17	0.4925	174.78	151.96	273.13	121.17	0.8006	1.3883
0.0650	206.46	0.4928	172.28	152.16	273.23	121.07	0.8016	1.3880
0.0660	206.76	0.4931	169.86	152.35	273.33	120.98	0.8025	1.3876
0.0670	207.04	0.4934	167.50	152.55	273.43	120.88	0.8035	1.3873
0.0680	207.33	0.4937	165.21	152.74	273.53	120.78	0.8044	1.3870
0.0690	207.61	0.4940	162.98	152.93	273.62	120.69	0.8053	1.3866
0.0700	207.89	0.4943	160.82	153.12	273.72	120.60	0.8062	1.3863
0.0710	208.16	0.4946	158.71	153.31	273.81	120.50	0.8071	1.3860
0.0720	208.43	0.4948	156.66	153.49	273.90	120.41	0.8080	1.3857
0.0730	208.70	0.4951	154.67	153.67	273.99	120.32	0.8088	1.3854
0.0740	208.97	0.4954	152.72	153.85	274.08	120.23	0.8097	1.3851
0.0760	209.49	0.4960	148.98	154.21	274.26	120.05	0.8114	1.3845
0.0780	210.01	0.4965	145.43	154.55	274.43	119.88	0.8130	1.3839
0.0800	210.51	0.4970	142.04	154.89	274.60	119.71	0.8147	1.3833
0.0820	211.00	0.4976	138.82	155.23	274.77	119.54	0.8162	1.3828
0.0840	211.48	0.4981	135.74	155.56	274.93	119.38	0.8178	1.3823
0.0860	211.96	0.4986	132.80	155.88	275.09	119.22	0.8193	1.3818
0.0880	212.43	0.4991	129.99	156.20	275.25	119.06	0.8208	1.3813
0.0900	212.88	0.4996	127.30	156.51	275.41	118.90	0.8223	1.3808
0.0920	213.33	0.5001	124.72	156.81	275.56	118.74	0.8237	1.3803
0.0940	213.78	0.5005	122.25	157.12	275.71	118.59	0.8251	1.3799
0.0960	214.21	0.5010	119.87	157.41	275.85	118.44	0.8265	1.3794
0.0980	214.64	0.5015	117.59	157.70	276.00	118.29	0.8278	1.3790
0.1000	215.06	0.5019	115.40	157.99	276.14	118.15	0.8292	1.3785
0.1020	215.48	0.5024	113.29	158.27	276.28	118.00	0.8305	1.3781
0.1040	215.89	0.5029	111.26	158.55	276.42	117.86	0.8318	1.3777
0.1060	216.29	0.5033	109.30	158.83	276.55	117.72	0.8330	1.3773
0.1080	216.69	0.5037	107.41	159.10	276.68	117.58	0.8343	1.3769
0.1100	217.08	0.5042	105.59	159.37	276.82	117.45	0.8355	1.3766
0.1150	218.03	0.5052	101.30	160.02	277.14	117.11	0.8385	1.3757
0.1200	218.95	0.5063	97.353	160.65	277.44	116.79	0.8414	1.3748
0.1250	219.85	0.5073	93.711	161.27	277.74	116.47	0.8442	1.3740
0.1300	220.71	0.5083	90.339	161.86	278.03	116.17	0.8469	1.3732
0.1350	221.55	0.5093	87.207	162.44	278.31	115.87	0.8495	1.3724

Chart R13B1.2: Saturation properties–*continued*

p MPa	T K	v' dm³/kg	v'' dm³/kg	h' kJ/kg	h'' kJ/kg	r kJ/kg	s' kJ/kgK	s'' kJ/kgK
0.1400	222.37	0.5102	84.290	163.00	278.58	115.57	0.8520	1.3717
0.1450	223.17	0.5111	81.567	163.55	278.84	115.29	0.8545	1.3711
0.1500	223.94	0.5120	79.019	164.09	279.10	115.01	0.8568	1.3704
0.1550	224.70	0.5129	76.628	164.61	279.35	114.74	0.8592	1.3698
0.1600	225.43	0.5138	74.381	165.12	279.59	114.47	0.8614	1.3692
0.1650	226.15	0.5147	72.265	165.62	279.82	114.20	0.8636	1.3686
0.1700	226.85	0.5155	70.269	166.11	280.05	113.95	0.8657	1.3680
0.1750	227.54	0.5164	68.382	166.58	280.28	113.69	0.8678	1.3675
0.1800	228.21	0.5172	66.596	167.05	280.50	113.44	0.8699	1.3670
0.1850	228.87	0.5180	64.903	167.51	280.71	113.20	0.8719	1.3665
0.1900	229.51	0.5188	63.295	167.96	280.92	112.96	0.8738	1.3660
0.1950	230.15	0.5196	61.766	168.40	281.12	112.72	0.8757	1.3655
0.2000	230.76	0.5203	60.311	168.83	281.32	112.49	0.8776	1.3651
0.2100	231.97	0.5218	57.601	169.68	281.71	112.04	0.8812	1.3642
0.2200	233.13	0.5233	55.128	170.49	282.09	111.59	0.8847	1.3634
0.2300	234.25	0.5247	52.861	171.28	282.44	111.16	0.8881	1.3626
0.2400	235.34	0.5262	50.775	172.05	282.79	110.74	0.8913	1.3619
0.2500	236.39	0.5275	48.850	172.79	283.12	110.33	0.8944	1.3612
0.2600	237.41	0.5289	47.068	173.51	283.45	109.93	0.8975	1.3605
0.2700	238.40	0.5302	45.412	174.22	283.76	109.54	0.9004	1.3599
0.2800	239.36	0.5315	43.869	174.90	284.06	109.16	0.9032	1.3593
0.2900	240.30	0.5328	42.429	175.57	284.35	108.78	0.9060	1.3587
0.3000	241.21	0.5340	41.081	176.22	284.64	108.41	0.9087	1.3581
0.3100	242.10	0.5352	39.816	176.86	284.91	108.05	0.9113	1.3576
0.3200	242.97	0.5365	38.628	177.48	285.18	107.70	0.9139	1.3571
0.3300	243.82	0.5377	37.508	178.09	285.44	107.35	0.9163	1.3566
0.3400	244.65	0.5388	36.452	178.69	285.69	107.00	0.9188	1.3561
0.3500	245.47	0.5400	35.453	179.28	285.94	106.66	0.9211	1.3557
0.3600	246.26	0.5411	34.508	179.85	286.18	106.33	0.9234	1.3552
0.3700	247.04	0.5423	33.612	180.42	286.42	106.00	0.9257	1.3548
0.3800	247.80	0.5434	32.761	180.97	286.65	105.68	0.9279	1.3544
0.3900	248.55	0.5445	31.952	181.51	286.87	105.36	0.9301	1.3540
0.4000	249.28	0.5456	31.181	182.05	287.09	105.04	0.9322	1.3536
0.4100	250.01	0.5467	30.447	182.57	287.30	104.73	0.9343	1.3532
0.4200	250.71	0.5477	29.746	183.09	287.51	104.42	0.9363	1.3528
0.4300	251.41	0.5488	29.077	183.60	287.72	104.12	0.9383	1.3525
0.4400	252.09	0.5498	28.436	184.10	287.92	103.82	0.9403	1.3521
0.4500	252.76	0.5509	27.823	184.59	288.11	103.52	0.9422	1.3518
0.4600	253.42	0.5519	27.236	185.08	288.30	103.23	0.9441	1.3515
0.4700	254.07	0.5529	26.672	185.55	288.49	102.94	0.9460	1.3512
0.4800	254.71	0.5539	26.131	186.03	288.68	102.65	0.9478	1.3508
0.4900	255.34	0.5549	25.611	186.49	288.86	102.37	0.9496	1.3505
0.5000	255.96	0.5559	25.111	186.95	289.03	102.08	0.9514	1.3502
0.5100	256.57	0.5569	24.630	187.40	289.21	101.80	0.9532	1.3499
0.5200	257.18	0.5579	24.166	187.85	289.38	101.53	0.9549	1.3497
0.5300	257.77	0.5589	23.720	188.29	289.55	101.25	0.9566	1.3494
0.5400	258.35	0.5598	23.289	188.73	289.71	100.98	0.9582	1.3491
0.5500	258.93	0.5608	22.873	189.16	289.87	100.71	0.9599	1.3488
0.5600	259.50	0.5618	22.471	189.59	290.03	100.44	0.9615	1.3486
0.5700	260.06	0.5627	22.083	190.01	290.18	100.18	0.9631	1.3483

Chart R13B1.2: Saturation properties—*continued*

p MPa	T K	v' dm^3/kg	v'' dm^3/kg	h' kJ/kg	h'' kJ/kg	r kJ/kg	s' kJ/kgK	s'' kJ/kgK
0.5800	260.62	0.5637	21.708	190.42	290.34	99.91	0.9647	1.3481
0.5900	261.17	0.5646	21.345	190.83	290.49	99.65	0.9662	1.3478
0.6000	261.71	0.5655	20.993	191.24	290.64	99.39	0.9678	1.3476
0.6100	262.24	0.5664	20.652	191.64	290.78	99.14	0.9693	1.3473
0.6200	262.77	0.5674	20.322	192.04	290.92	98.88	0.9708	1.3471
0.6300	263.29	0.5683	20.002	192.44	291.06	98.63	0.9723	1.3468
0.6400	263.81	0.5692	19.692	192.83	291.20	98.38	0.9737	1.3466
0.6500	264.32	0.5701	19.391	193.21	291.34	98.12	0.9752	1.3464
0.6600	264.82	0.5710	19.098	193.60	291.47	97.88	0.9766	1.3462
0.6700	265.32	0.5719	18.814	193.97	291.60	97.63	0.9780	1.3460
0.6800	265.81	0.5728	18.538	194.35	291.73	97.38	0.9794	1.3457
0.6900	266.30	0.5737	18.269	194.72	291.86	97.14	0.9808	1.3455
0.7000	266.78	0.5746	18.008	195.09	291.99	96.90	0.9821	1.3453
0.7100	267.26	0.5754	17.754	195.45	292.11	96.66	0.9835	1.3451
0.7200	267.73	0.5763	17.506	195.82	292.23	96.42	0.9848	1.3449
0.7300	268.20	0.5772	17.265	196.17	292.35	96.18	0.9861	1.3447
0.7400	268.67	0.5781	17.031	196.53	292.47	95.94	0.9874	1.3445
0.7600	269.58	0.5798	16.579	197.23	292.70	95.47	0.9900	1.3441
0.7800	270.47	0.5815	16.149	197.92	292.92	95.00	0.9925	1.3437
0.8000	271.35	0.5832	15.740	198.60	293.14	94.54	0.9950	1.3434
0.8200	272.21	0.5849	15.350	199.27	293.35	94.08	0.9974	1.3430
0.8400	273.06	0.5866	14.978	199.93	293.56	93.63	0.9997	1.3426
0.8600	273.89	0.5883	14.622	200.58	293.76	93.18	1.0021	1.3423
0.8800	274.71	0.5900	14.282	201.22	293.95	92.74	1.0044	1.3419
0.9000	275.52	0.5916	13.956	201.85	294.14	92.29	1.0066	1.3416
0.9200	276.31	0.5933	13.644	202.47	294.33	91.85	1.0088	1.3413
0.9400	277.09	0.5949	13.344	203.09	294.51	91.42	1.0110	1.3409
0.9600	277.85	0.5966	13.056	203.69	294.68	90.99	1.0132	1.3406
0.9800	278.61	0.5982	12.780	204.29	294.85	90.56	1.0153	1.3403
1.0000	279.35	0.5999	12.514	204.88	295.02	90.13	1.0173	1.3400
1.0200	280.08	0.6015	12.258	205.47	295.18	89.71	1.0194	1.3397
1.0400	280.81	0.6031	12.011	206.05	295.33	89.28	1.0214	1.3394
1.0600	281.52	0.6047	11.773	206.62	295.48	88.87	1.0234	1.3391
1.0800	282.22	0.6064	11.544	207.18	295.63	88.45	1.0254	1.3388
1.1000	282.92	0.6080	11.322	207.74	295.78	88.03	1.0273	1.3385
1.1500	284.61	0.6120	10.800	209.11	296.12	87.01	1.0320	1.3377
1.2000	286.25	0.6160	10.320	210.45	296.44	85.99	1.0366	1.3370
1.2500	287.84	0.6200	9.8762	211.76	296.74	84.99	1.0410	1.3363
1.3000	289.39	0.6240	9.4647	213.03	297.02	83.99	1.0454	1.3356
1.3500	290.89	0.6280	9.0821	214.28	297.28	83.00	1.0495	1.3349
1.4000	292.35	0.6320	8.7254	215.51	297.53	82.02	1.0536	1.3342
1.4500	293.77	0.6360	8.3919	216.70	297.75	81.05	1.0576	1.3335
1.5000	295.16	0.6401	8.0794	217.88	297.96	80.08	1.0615	1.3328
1.5500	296.52	0.6441	7.7858	219.04	298.15	79.11	1.0653	1.3321
1.6000	297.84	0.6482	7.5094	220.18	298.33	78.15	1.0690	1.3314
1.6500	299.13	0.6523	7.2488	221.30	298.49	77.20	1.0727	1.3307
1.7000	300.40	0.6564	7.0024	222.40	298.64	76.24	1.0762	1.3300
1.7500	301.63	0.6606	6.7692	223.49	298.77	75.29	1.0797	1.3293
1.8000	302.84	0.6648	6.5480	224.56	298.89	74.34	1.0832	1.3286
1.8500	304.03	0.6690	6.3379	225.61	299.00	73.38	1.0866	1.3279

Chart R13B1.2: Saturation properties–*continued*

p MPa	T K	v' dm³/kg	v'' dm³/kg	h' kJ/kg	h'' kJ/kg	r kJ/kg	s' kJ/kgK	s'' kJ/kgK
1.9000	305.19	0.6733	6.1380	226.66	299.09	72.43	1.0899	1.3272
1.9500	306.33	0.6776	5.9475	227.69	299.17	71.48	1.0931	1.3265
2.0000	307.45	0.6820	5.7658	228.71	299.23	70.52	1.0963	1.3257
2.1000	309.63	0.6910	5.4259	230.72	299.32	68.60	1.1026	1.3242
2.2000	311.73	0.7002	5.1140	232.69	299.36	66.67	1.1087	1.3226
2.3000	313.76	0.7098	4.8263	234.62	299.34	64.72	1.1147	1.3210
2.4000	315.72	0.7197	4.5597	236.53	299.28	62.74	1.1206	1.3193
2.5000	317.62	0.7301	4.3115	238.42	299.15	60.74	1.1263	1.3175
2.6000	319.47	0.7409	4.0794	240.29	298.98	58.69	1.1319	1.3156
2.7000	321.26	0.7523	3.8614	242.14	298.74	56.59	1.1375	1.3136
2.8000	323.00	0.7644	3.6558	243.99	298.43	54.44	1.1430	1.3115
2.9000	324.69	0.7773	3.4609	245.84	298.06	52.22	1.1485	1.3093
3.0000	326.34	0.7912	3.2752	247.70	297.61	49.91	1.1539	1.3069
3.1000	327.94	0.8063	3.0972	249.57	297.08	47.50	1.1594	1.3043
3.2000	329.50	0.8228	2.9257	251.48	296.45	44.97	1.1649	1.3014
3.3000	331.02	0.8412	2.7590	253.42	295.70	42.28	1.1706	1.2983
3.4000	332.50	0.8621	2.5956	255.43	294.81	39.38	1.1764	1.2948
3.5000	333.95	0.8864	2.4332	257.54	293.74	36.20	1.1824	1.2908
3.6000	335.36	0.9158	2.2690	259.80	292.42	32.62	1.1889	1.2862
3.7000	336.73	0.9533	2.0980	262.31	290.75	28.44	1.1961	1.2806
3.8000	338.07	1.0062	1.9104	265.31	288.46	23.15	1.2047	1.2732
3.9000	339.37	1.1019	1.6742	269.62	284.74	15.12	1.2171	1.2617

Chart R13B1.3: Thermodynamic properties

p/MPa	0.15			0.20			0.30		
T K	v dm³/kg	h kJ/kg	s kJ/kgK	v dm³/kg	h kJ/kg	s kJ/kgK	v dm³/kg	h kJ/kg	s kJ/kgK
230.0	81.532	281.67	1.3818	0.5194	168.30	0.8753	0.5192	168.32	0.8752
240.0	85.629	285.98	1.4001	63.249	285.37	1.3822	0.5323	175.36	0.9051
250.0	89.674	290.34	1.4179	66.374	289.79	1.4003	43.030	288.65	1.3745
260.0	93.675	294.77	1.4353	69.454	294.28	1.4179	45.199	293.24	1.3925
270.0	97.642	299.27	1.4523	72.497	298.82	1.4350	47.326	297.88	1.4100
280.0	101.58	303.84	1.4689	75.509	303.42	1.4518	49.418	302.56	1.4270
290.0	105.49	308.48	1.4851	78.495	308.09	1.4682	51.482	307.30	1.4437
300.0	109.38	313.18	1.5011	81.460	312.83	1.4842	53.523	312.10	1.4599
310.0	113.26	317.96	1.5167	84.407	317.62	1.4999	55.545	316.95	1.4758
320.0	117.12	322.79	1.5321	87.338	322.48	1.5154	57.550	321.85	1.4914
330.0	120.97	327.70	1.5472	90.257	327.41	1.5305	59.541	326.82	1.5067
340.0	124.80	332.67	1.5620	93.163	332.39	1.5454	61.519	331.84	1.5217
350.0	128.63	337.70	1.5766	96.060	337.44	1.5600	63.488	336.92	1.5364
360.0	132.45	342.79	1.5909	98.948	342.55	1.5744	65.447	342.06	1.5508
370.0	136.26	347.94	1.6051	101.83	347.71	1.5886	67.398	347.25	1.5651
380.0	140.06	353.15	1.6190	104.70	352.93	1.6025	69.342	352.49	1.5791
390.0	143.86	358.42	1.6326	107.57	358.21	1.6162	71.280	357.79	1.5928
400.0	147.65	363.74	1.6461	110.43	363.54	1.6297	73.213	363.15	1.6064
410.0	151.44	369.12	1.6594	113.29	368.93	1.6430	75.140	368.55	1.6197
420.0	155.22	374.55	1.6725	116.14	374.37	1.6561	77.064	374.00	1.6329
430.0	159.00	380.02	1.6854	118.99	379.85	1.6690	78.983	379.51	1.6458
440.0	162.78	385.55	1.6981	121.84	385.39	1.6817	80.898	385.06	1.6586
p/MPa	0.40			0.50			0.60		
230.0	0.5191	168.35	0.8750	0.5190	168.37	0.8749	0.5188	168.39	0.8748
240.0	0.5322	175.38	0.9050	0.5320	175.40	0.9048	0.5318	175.42	0.9047
250.0	31.307	287.43	1.3549	0.5465	182.58	0.9341	0.5463	182.60	0.9340
260.0	33.034	292.15	1.3735	25.699	291.00	1.3579	0.5625	189.96	0.9629
270.0	34.712	296.90	1.3914	27.118	295.87	1.3762	22.030	294.79	1.3632
280.0	36.351	301.67	1.4088	28.492	300.75	1.3940	23.234	299.78	1.3813
290.0	37.959	306.49	1.4256	29.831	305.64	1.4112	24.399	304.77	1.3989
300.0	39.542	311.35	1.4421	31.142	310.57	1.4279	25.532	309.78	1.4158
310.0	41.104	316.25	1.4582	32.430	315.54	1.4442	26.641	314.81	1.4323
320.0	42.648	321.21	1.4739	33.700	320.55	1.4601	27.729	319.88	1.4484
330.0	44.177	326.22	1.4894	34.953	325.61	1.4756	28.800	324.98	1.4641
340.0	45.693	331.28	1.5045	36.193	330.71	1.4908	29.856	330.13	1.4795
350.0	47.198	336.39	1.5193	37.421	335.86	1.5058	30.901	335.32	1.4945
360.0	48.694	341.56	1.5338	38.639	341.06	1.5204	31.935	340.55	1.5093
370.0	50.181	346.78	1.5481	39.849	346.31	1.5348	32.959	345.83	1.5237
380.0	51.661	352.05	1.5622	41.051	351.60	1.5489	33.977	351.15	1.5379
390.0	53.135	357.37	1.5760	42.246	356.95	1.5628	34.987	356.52	1.5519
400.0	54.603	362.75	1.5896	43.436	362.34	1.5765	35.991	361.94	1.5656
410.0	56.066	368.17	1.6030	44.620	367.79	1.5899	36.990	367.40	1.5791
420.0	57.524	373.64	1.6162	45.800	373.28	1.6031	37.984	372.91	1.5923
430.0	58.978	379.16	1.6292	46.976	378.81	1.6162	38.974	378.46	1.6054
440.0	60.429	384.72	1.6420	48.147	384.39	1.6290	39.960	384.06	1.6183

Chart R13B1.3: Thermodynamic properties—*continued*

p/MPa	0.80			1.00			2.00		
T K	v dm³/kg	h kJ/kg	s kJ/kgK	v dm³/kg	h kJ/kg	s kJ/kgK	v dm³/kg	h kJ/kg	s kJ/kgK
230.0	0.5186	168.44	0.8745	0.5183	168.49	0.8743	0.5170	168.72	0.8731
240.0	0.5315	175.46	0.9044	0.5312	175.50	0.9041	0.5296	175.71	0.9028
250.0	0.5459	182.63	0.9337	0.5455	182.67	0.9334	0.5435	182.84	0.9319
260.0	0.5620	189.98	0.9625	0.5615	190.01	0.9622	0.5591	190.13	0.9605
270.0	0.5805	197.56	0.9911	0.5798	197.57	0.9907	0.5767	197.63	0.9888
280.0	16.612	297.70	1.3599	12.572	295.38	1.3413	0.5971	205.36	1.0169
290.0	17.573	302.92	1.3782	13.431	300.89	1.3606	0.6215	213.42	1.0452
300.0	18.494	308.11	1.3958	14.238	306.31	1.3790	0.6522	221.92	1.0740
310.0	19.384	313.29	1.4128	15.006	311.68	1.3966	5.9372	301.12	1.3318
320.0	20.250	318.49	1.4293	15.746	317.02	1.4136	6.5346	308.03	1.3538
330.0	21.097	323.70	1.4453	16.462	322.36	1.4300	7.0560	314.47	1.3736
340.0	21.927	328.94	1.4610	17.160	327.70	1.4459	7.5306	320.66	1.3921
350.0	22.744	334.21	1.4762	17.843	333.06	1.4615	7.9731	326.68	1.4095
360.0	23.549	339.51	1.4912	18.512	338.45	1.4767	8.3920	332.62	1.4263
370.0	24.344	344.85	1.5058	19.171	343.86	1.4915	8.7928	338.49	1.4423
380.0	25.131	350.24	1.5202	19.821	349.30	1.5060	9.1793	344.32	1.4579
390.0	25.911	355.66	1.5343	20.463	354.78	1.5202	9.5541	350.13	1.4730
400.0	26.684	361.12	1.5481	21.099	360.29	1.5342	9.9194	355.94	1.4877
410.0	27.451	366.62	1.5617	21.728	365.84	1.5479	10.277	361.75	1.5020
420.0	28.214	372.17	1.5750	22.352	371.42	1.5613	10.627	367.57	1.5161
430.0	28.972	377.76	1.5882	22.971	377.05	1.5746	10.972	373.40	1.5298
440.0	29.726	383.38	1.6011	23.586	382.71	1.5876	11.311	379.25	1.5432
p/MPa	4.00			8.00			13.15		
230.0	0.5145	169.21	0.8707	0.5100	170.23	0.8662	0.5047	171.60	0.8608
240.0	0.5267	176.15	0.9002	0.5213	177.08	0.8954	0.5151	178.36	0.8896
250.0	0.5399	183.21	0.9290	0.5334	184.03	0.9237	0.5262	185.21	0.9175
260.0	0.5545	190.42	0.9573	0.5466	191.10	0.9515	0.5381	192.15	0.9448
270.0	0.5709	197.80	0.9852	0.5611	198.31	0.9787	0.5509	199.20	0.9714
280.0	0.5895	205.37	1.0127	0.5771	205.66	1.0054	0.5647	206.36	0.9974
290.0	0.6110	213.20	1.0402	0.5950	213.18	1.0318	0.5798	213.63	1.0229
300.0	0.6367	221.32	1.0677	0.6152	220.88	1.0579	0.5963	221.03	1.0480
310.0	0.6687	229.87	1.0957	0.6384	228.79	1.0838	0.6144	228.56	1.0727
320.0	0.7118	239.06	1.1249	0.6656	236.95	1.1097	0.6344	236.22	1.0970
330.0	0.7796	249.53	1.1571	0.6982	245.42	1.1358	0.6567	244.03	1.1210
340.0	1.0312	267.52	1.2106	0.7387	254.27	1.1622	0.6817	251.98	1.1448
350.0	2.6486	306.04	1.3231	0.7912	263.63	1.1893	0.7100	260.08	1.1683
360.0	3.1176	316.25	1.3519	0.8633	273.70	1.2177	0.7420	268.34	1.1915
370.0	3.4756	324.59	1.3748	0.9684	284.76	1.2480	0.7786	276.76	1.2146
380.0	3.7811	332.14	1.3949	1.1222	296.85	1.2802	0.8205	285.34	1.2375
390.0	4.0548	339.23	1.4133	1.3180	309.13	1.3121	0.8686	294.07	1.2602
400.0	4.3069	346.04	1.4306	1.5224	320.44	1.3408	0.9237	302.94	1.2826
410.0	4.5433	352.67	1.4469	1.7133	330.53	1.3657	0.9860	311.90	1.3047
420.0	4.7676	359.18	1.4626	1.8868	339.66	1.3877	1.0554	320.90	1.3264
430.0	4.9825	365.59	1.4777	2.0454	348.12	1.4076	1.1307	329.86	1.3475
440.0	5.1897	371.95	1.4923	2.1921	356.10	1.4260	1.2101	338.69	1.3678

References for R13B1

[Doer80] DÖRING, R. AND BUCHWALD, H.: Kaltron 13B1 Thermodynamische Eigenschaften; Kali-Chemie AG Hannover (1980)

[Gell53] GELLES, E. AND PITZER, K.S.: Thermodynamic functions of the halogenated methanes; J. Am. Chem. Soc. , 75(1953) 5259/5267

[Henr70] HENRICI, H. AND MALL, K.: Thermodynamische Eigenschaften von Gemischen aus Trifluormonobrommethan (R13B1) und Ölen; Kältetechnik-Klimatisierung 22(1970)8 259/263

[Jaeg73] JÄGER, H.P.: Dichte- und Dampfdruckmeßwerte von verschiedenen Kältemitteln; Die Kälte (1973)7 276/278

[KMR 81] - : Kältemaschinenregeln, 7. Auflage; Verlag C.F. Müller, Karlsruhe (1981)

[Mart53] MARTIN, J.J. ; WELSHANS, L.M. ; CHOU, C.H. AND CRUKA, G.E.: Data and equations for some thermodynamic properties of Freon-13B1 ($CBrF_3$); Engineering Research Institute, report on project M777, University of Michigan, Willington, Delaware (1953)

[McDo59] MCDONALD, R.A. ; SHRADER, S.A. AND STULL, D.R.: Vapor Pressure and Freezing Points of 30 Organics; J. Chem. Eng. Data 4(1959)4 312/313

[Niku67] NIKULSCHIN, R.K. AND SATWORNICKIJ, JU.G.: The density of Trifluoromonobrommethane (R13B1) at low temperatures; Kol. Teknika (1967)5 83/86

[Otto66] OTTO, J. ; THOMAS, W. AND ZANDER, M.: Die Füllmenge verflüssigter Gase in Behältern; Teil 4: Trifluormonochlormethan (R13) und Trifluormonobrommethan (R13B1); PTB Mitteilungen (1966) 6 526/528

[Pere71] PERELSTEIN, I.I. AND ALESHIN, Y.P.: Experimental investigation of the thermodynamic properties of Freon 13B1; Thermophysical Properties of Matter and Substances 4(1971) 50/64

R 13 B1. ph

T=340
T=345
T=350
T=355
T=360
T=365
T=370
T=375
T=380
T=385
T=390
T=395
T=400
T=405
T=410
T=415
T=420
T=425
T=430
T=435
T=440
s=1.41
s=1.44
s=1.47
s=1.50
s=1.53
s=1.56
s=1.59
s=1.62
s=1.65
s=1.68
s=1.71
s=1.74
s=1.77
s=1.80
s=1.83
s=1.86
260
274
288
302
316
330
344
358
372
386
h/(kJ/kg)

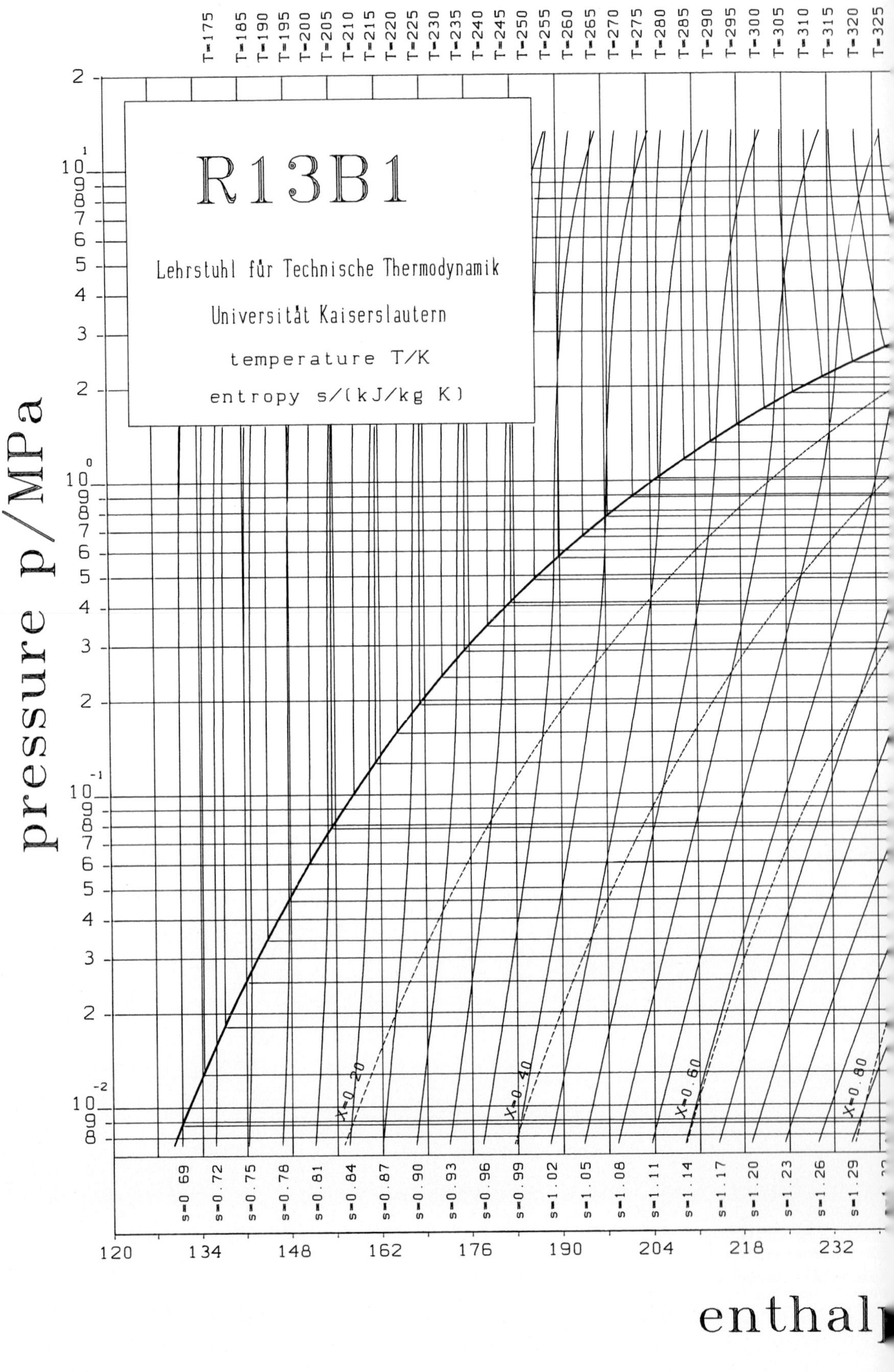

R13B1
Lehrstuhl für Technische Thermodynamik
Universität Kaiserslautern
temperature T/K
entropy s/(kJ/kg K)
pressure p/MPa
enthal

R 13 B1. Ts

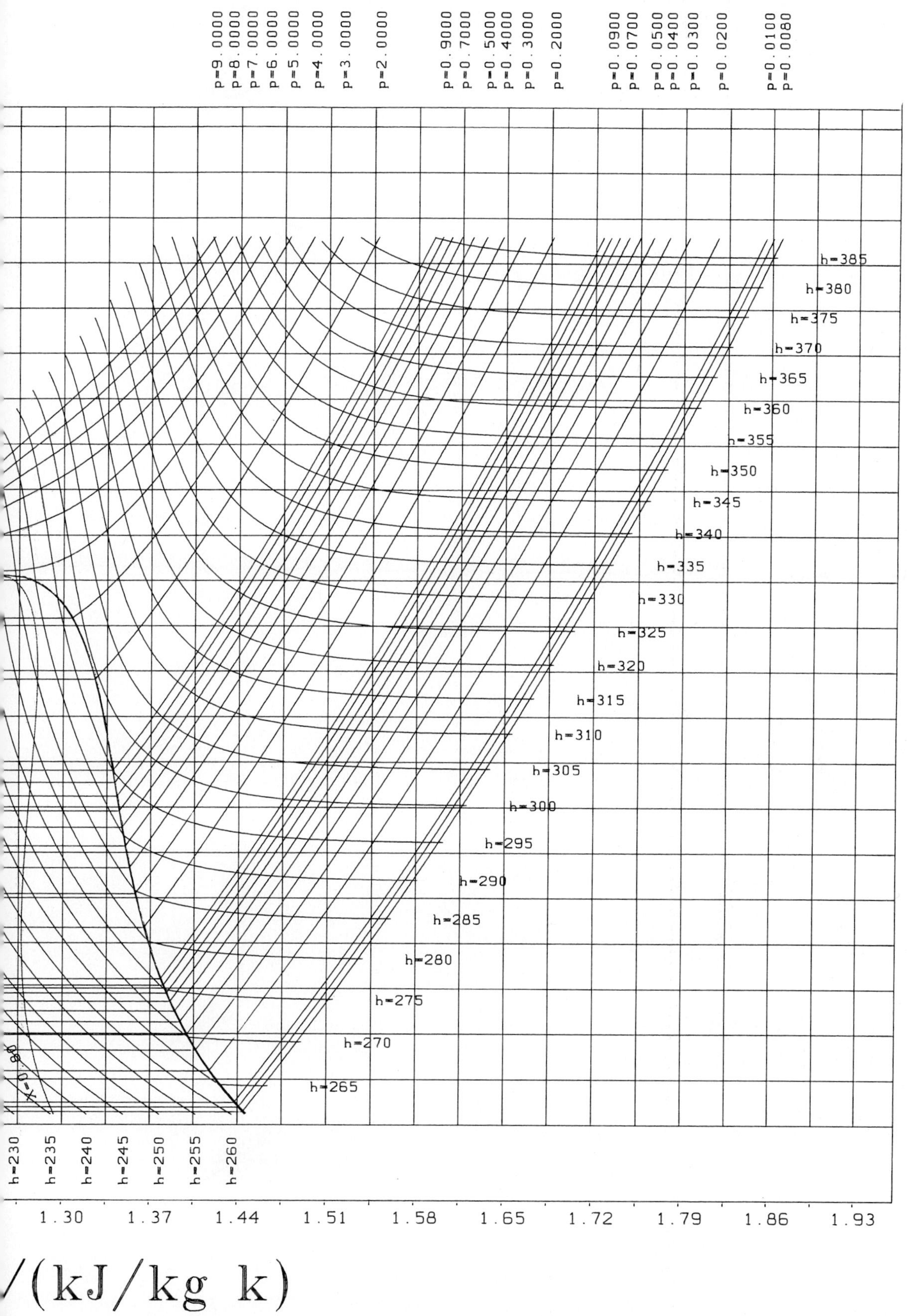

p=9.0000
p=8.0000
p=7.0000
p=6.0000
p=5.0000
p=4.0000
p=3.0000
p=2.0000
p=0.9000
p=0.7000
p=0.5000
p=0.4000
p=0.3000
p=0.2000
p=0.0900
p=0.0700
p=0.0500
p=0.0400
p=0.0300
p=0.0200
p=0.0100
p=0.0080
h=385
h=380
h=375
h=370
h=365
h=360
h=355
h=350
h=345
h=340
h=335
h=330
h=325
h=320
h=315
h=310
h=305
h=300
h=295
h=290
h=285
h=280
h=275
h=270
h=265
x=0.80
h=230
h=235
h=240
h=245
h=250
h=255
h=260
1.30
1.37
1.44
1.51
1.58
1.65
1.72
1.79
1.86
1.93
/(kJ/kg k)

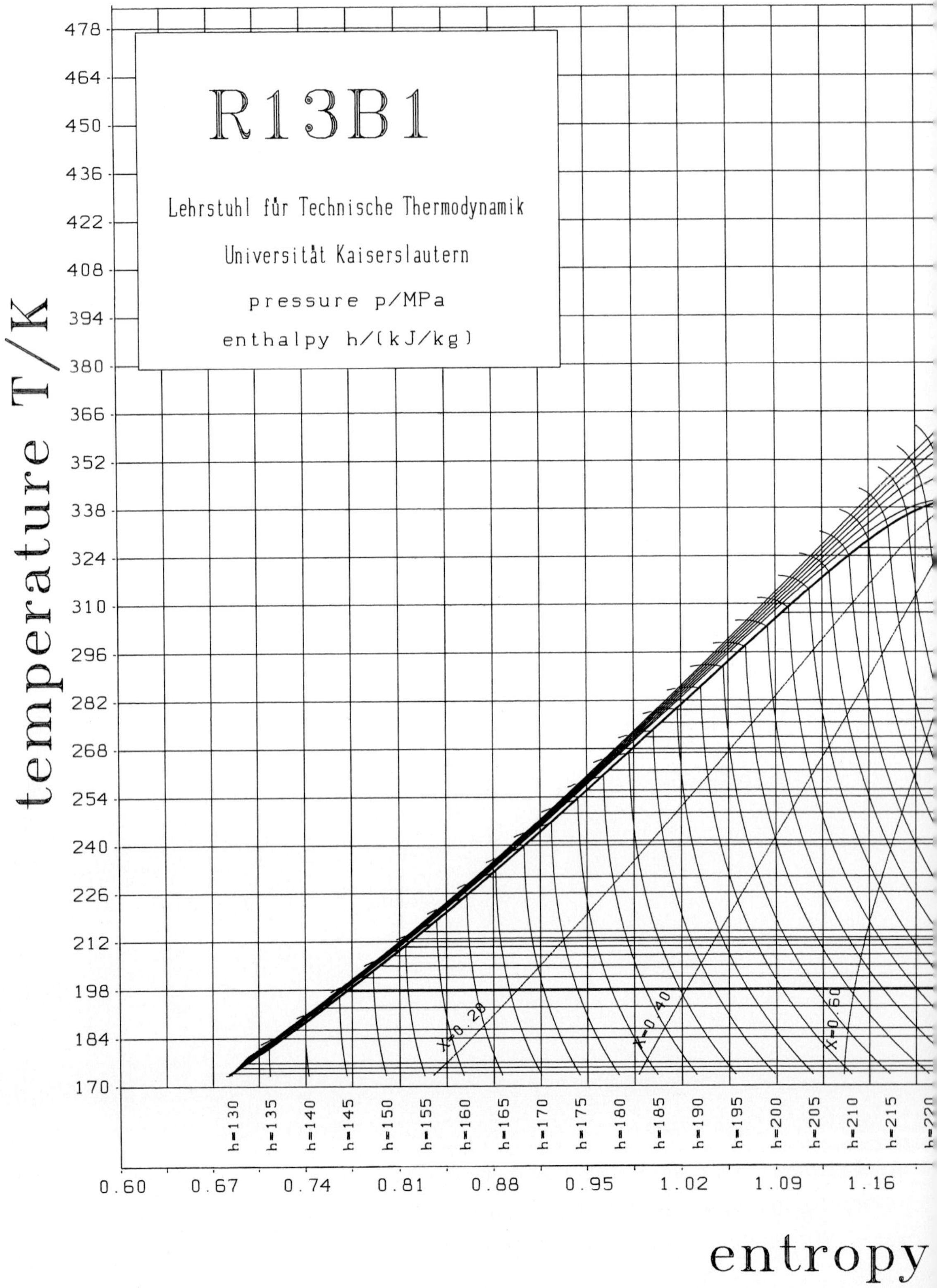

R13B1
Lehrstuhl für Technische Thermodynamik
Universität Kaiserslautern
pressure p/MPa
enthalpy h/(kJ/kg)
temperature T/K
478
464
450
436
422
408
394
380
366
352
338
324
310
296
282
268
254
240
226
212
198
184
170
X=0.20
X=0.40
X=0.60
h=130
h=135
h=140
h=145
h=150
h=155
h=160
h=165
h=170
h=175
h=180
h=185
h=190
h=195
h=200
h=205
h=210
h=215
0.60
0.67
0.74
0.81
0.88
0.95
1.02
1.09
1.16
entropy

R 13 B1. hs

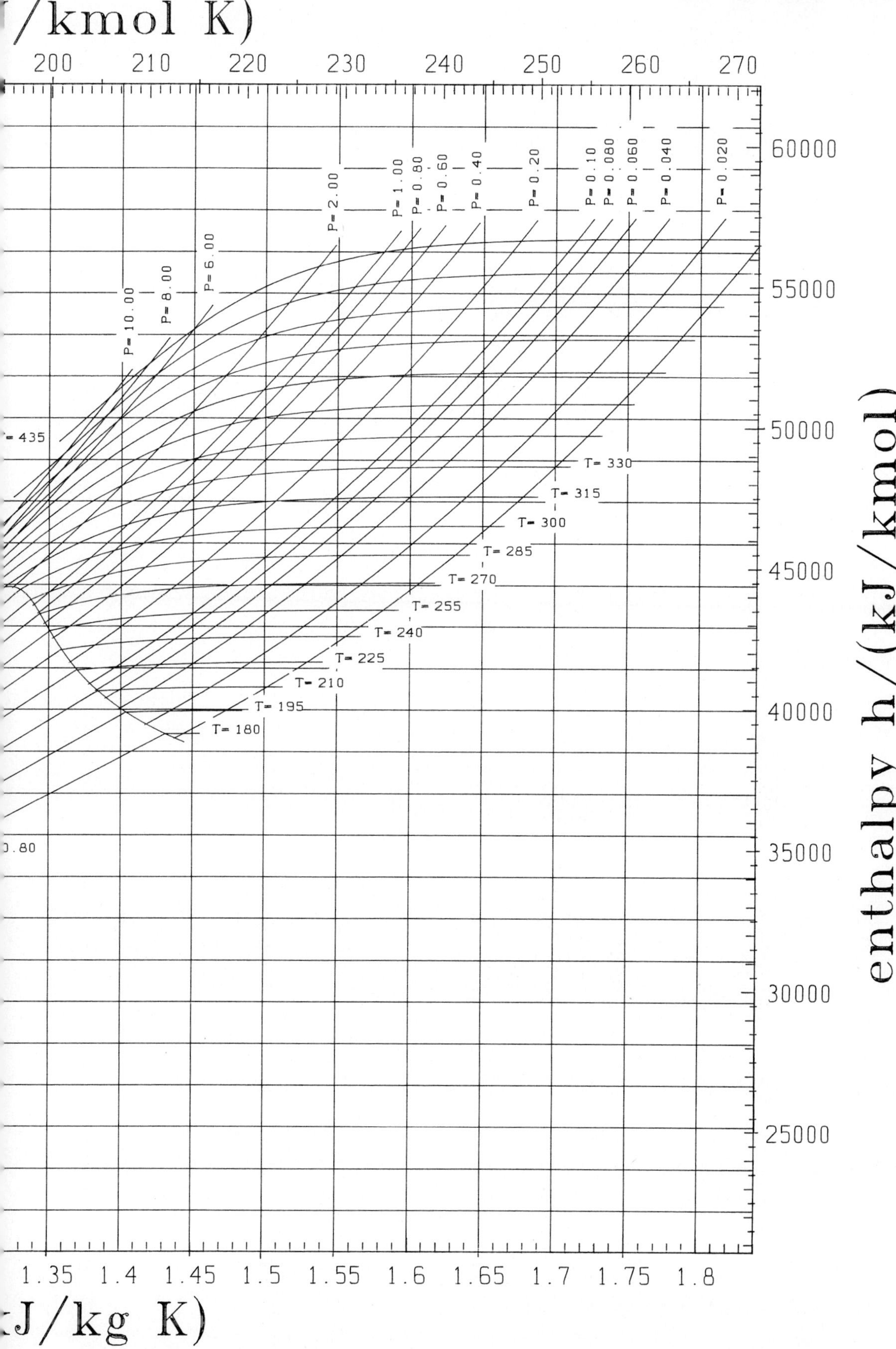

/kmol K)
200 210 220 230 240 250 260 270
P= 10.00
P= 8.00
P= 6.00
P= 2.00
P= 1.00
P= 0.80
P= 0.60
P= 0.40
P= 0.20
P= 0.10
P= 0.080
P= 0.060
P= 0.040
P= 0.020
= 435
T= 330
T= 315
T= 300
T= 285
T= 270
T= 255
T= 240
T= 225
T= 210
T= 195
T= 180
.80
60000
55000
50000
45000
40000
35000
30000
25000
enthalpy h/(kJ/kmol)
1.35 1.4 1.45 1.5 1.55 1.6 1.65 1.7 1.75 1.8
J/kg K)

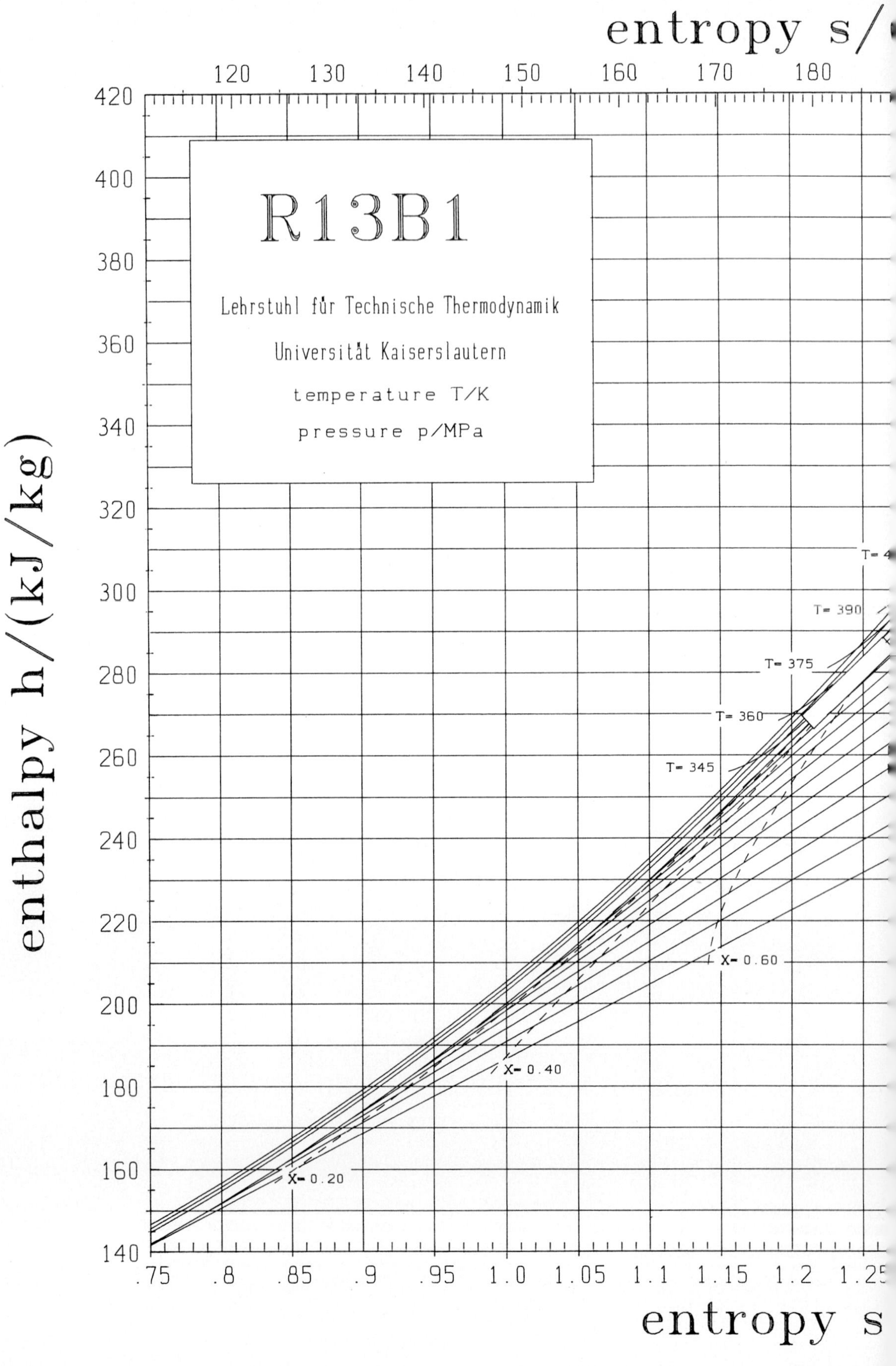

entropy s/
R13B1
Lehrstuhl für Technische Thermodynamik
Universität Kaiserslautern
temperature T/K
pressure p/MPa
enthalpy h/(kJ/kg)
T= 345
T= 360
T= 375
T= 390
X= 0.20
X= 0.40
X= 0.60
entropy s

Refrigerant R113 (CCl_2FCClF_2)

Table R113.1: List of references for p, ρ, T data

Author	n	p/MPa min	p/MPa max	ρ/g cm^{-3} min	ρ/g cm^{-3} max	T/K min	T/K max
BENNING, A.F. *[Benn 40]*	27	0.03	2.12	0.0027	0.15	298	488
GELLER, V.Z. *[Gell 73]*	162	0.52	24.32	0.981	1.67	253	508
MASTROIANNI, M.J. *[Mast 78]*	42	0.82	6.21	0.0558	0.69	405	538
RIEDEL, L. *[Ried 38]*	17	0.04	0.19	0.0028	0.01	294	359
SCHEINDLIN, A. *[Sche 54]*	59	0.46	3.43	0.916	1.60	274	477

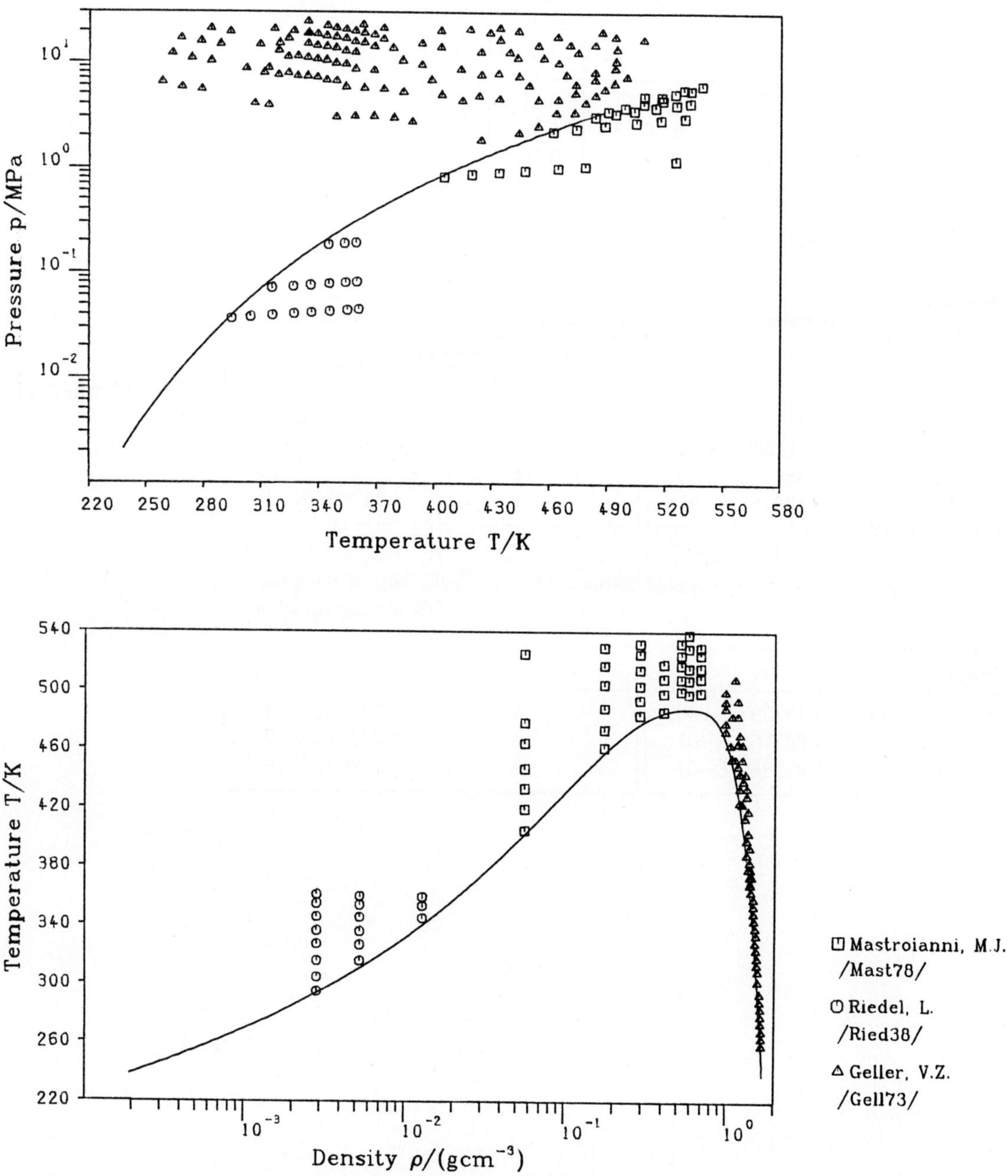

Fig.R113.1: Distribution of p, ρ, T data projected in the p, T and in the T, ρ plane

Table R113.2: Experimental vapor pressure data

Author	Δ / % std	opt	*EOS*
BENNING, A.F. *[Benn 40]*	0.96	1.28	0.83
GELLER, V.Z. *[Gell 73a]*	0.68	0.69	0.69
HIRAOKA, H. *[Hira 63]*	8.04	7.99	8.03
MASTROIANNI, M.J. *[Mast 78]*	0.19	0.18	0.21
RIEDEL, L. *[Ried 38]*	0.14	0.14	0.15

Table R113.3: Experimental saturated liquid density data

Author	Δ / % std	opt	*EOS*
BENNING, A.F. *[Benn 40b]*	0.12	0.11	0.11
GELLER, V.Z. *[Gell 73a]*	0.11	0.10	0.12
MASTROIANNI, M.J. *[Mast 78]*	0.12	0.16	0.12
RIEDEL, L. *[Ried 38]*	0.06	0.05	0.06
SHIMANSKAYA, E.T. *[Shim 81]*	1.35	1.50	2.40

Table R113.4: Coefficients of the standard vapor pressure equation

Index i	b_i
5	−7.311712 E+00
6	1.621082 E+00
9	−3.488387 E+00
15	−6.436411 E−01

Table R113.5: Coefficients of the quasi-optimal vapor pressure equation

Index i	b_i
3	−1.986967 E−02
5	−6.620194 E+00
14	−1.291625 E+01
21	4.457367 E+01

Table R113.6: Coefficients of the standard saturated liquid density equation

Index i	c_i
8	1.757368 E+00
9	7.663568 E−01
16	4.019289 E−01

Table R113.7: Coefficients of the quasi-optimal saturated liquid density equation

Index i	c_i
8	1.807474 E+00
9	6.206942 E−01
12	3.178801 E−01

Table R113.8: Coefficients of the *Bender-EOS*

Range of validity: one-phase region

$$0.04 \le p/\mathrm{MPa} \le 24.32$$
$$0.00284 \le \rho/\mathrm{g\ cm^{-3}} \le 1.6668$$
$$257.97 \le T/\mathrm{K} \le 538.19$$

	Specific units	Molar units
a_{01}	6.397589689 E−02	2.246514393 E+00
a_{02}	9.418013958 E+01	3.307136741 E+03
a_{03}	−3.105761025 E+04	−1.090588360 E+06
a_{04}	1.338483727 E+07	4.700087228 E+08
a_{05}	−9.328570498 E+08	−3.275728659 E+10
a_{06}	4.121477022 E−02	2.712015183 E−01
a_{07}	−9.720340748 E−01	−6.396180677 E+00
a_{08}	1.711679504 E+04	1.126319710 E+05
a_{09}	3.424587979 E−02	4.222736408 E−02
a_{10}	−5.000728239 E+01	−6.166218339 E+01
a_{11}	3.485811787 E−02	8.054451858 E−03
a_{12}	−3.344872015 E+01	−7.728790958 E+00
a_{13}	2.574535238 E+01	1.114749555 E+00
a_{14}	−3.865301300 E+06	−2.543446380 E+07
a_{15}	4.663581500 E+09	3.068730886 E+10
a_{16}	−2.052662294 E+12	−1.350693277 E+13
a_{17}	−6.029488258 E+07	−1.393196933 E+07
a_{18}	3.079021189 E+10	7.114505732 E+09
a_{19}	1.752293554 E+12	4.048917422 E+11
a_{20}	2.900000000 E+00	1.018335351 E−01

Table R113.9: Deviations between experimental and predicted p, ρ, T data

Region	n	Δ / %	
		p	ρ
Gas	36	0.16	0.32
Critical	20	0.20	0.89
Liquid	134	3.89	0.15
Total	190	3.43	0.36

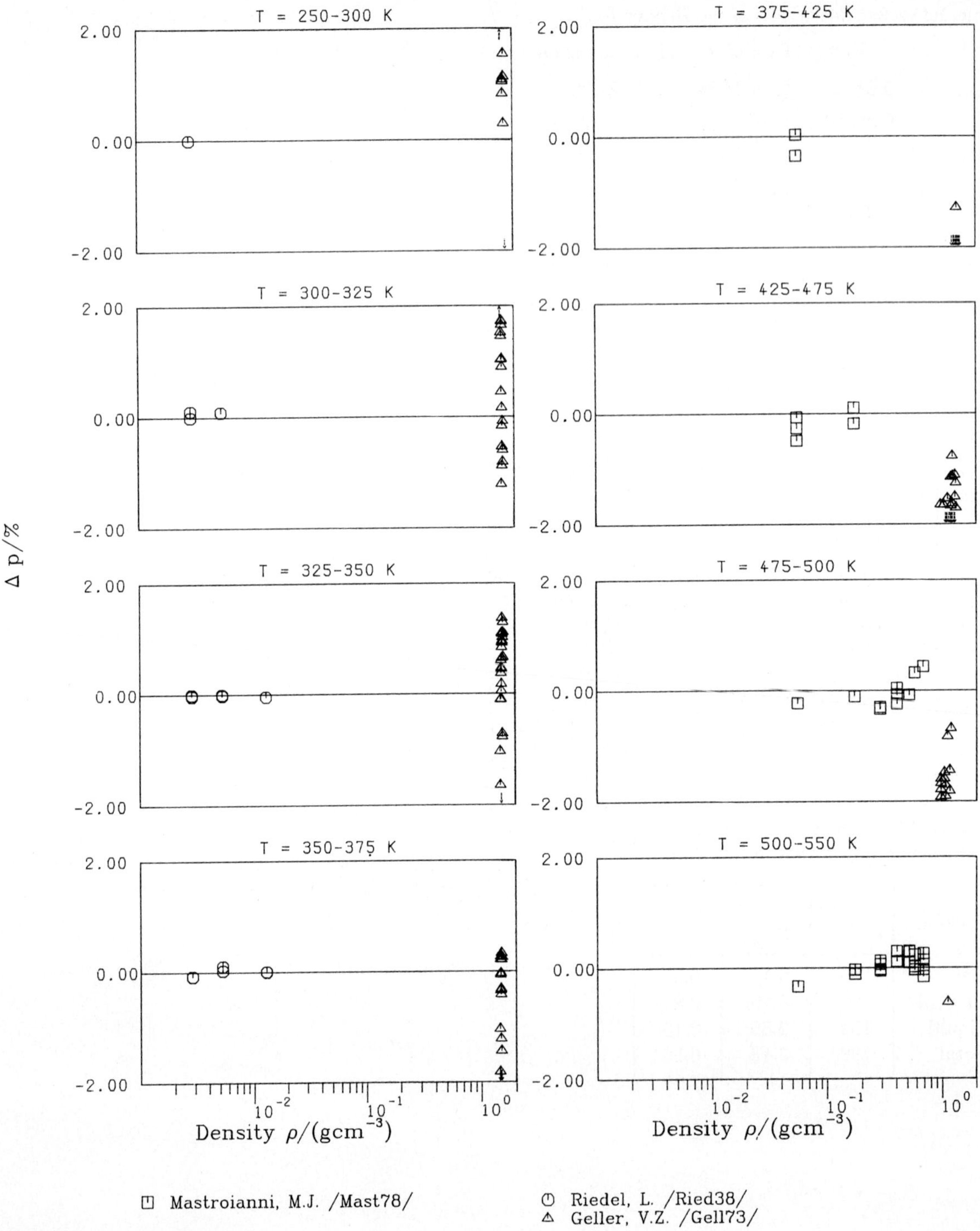

Fig.R113.2: Deviations between experimental and calculated p, ρ, T data

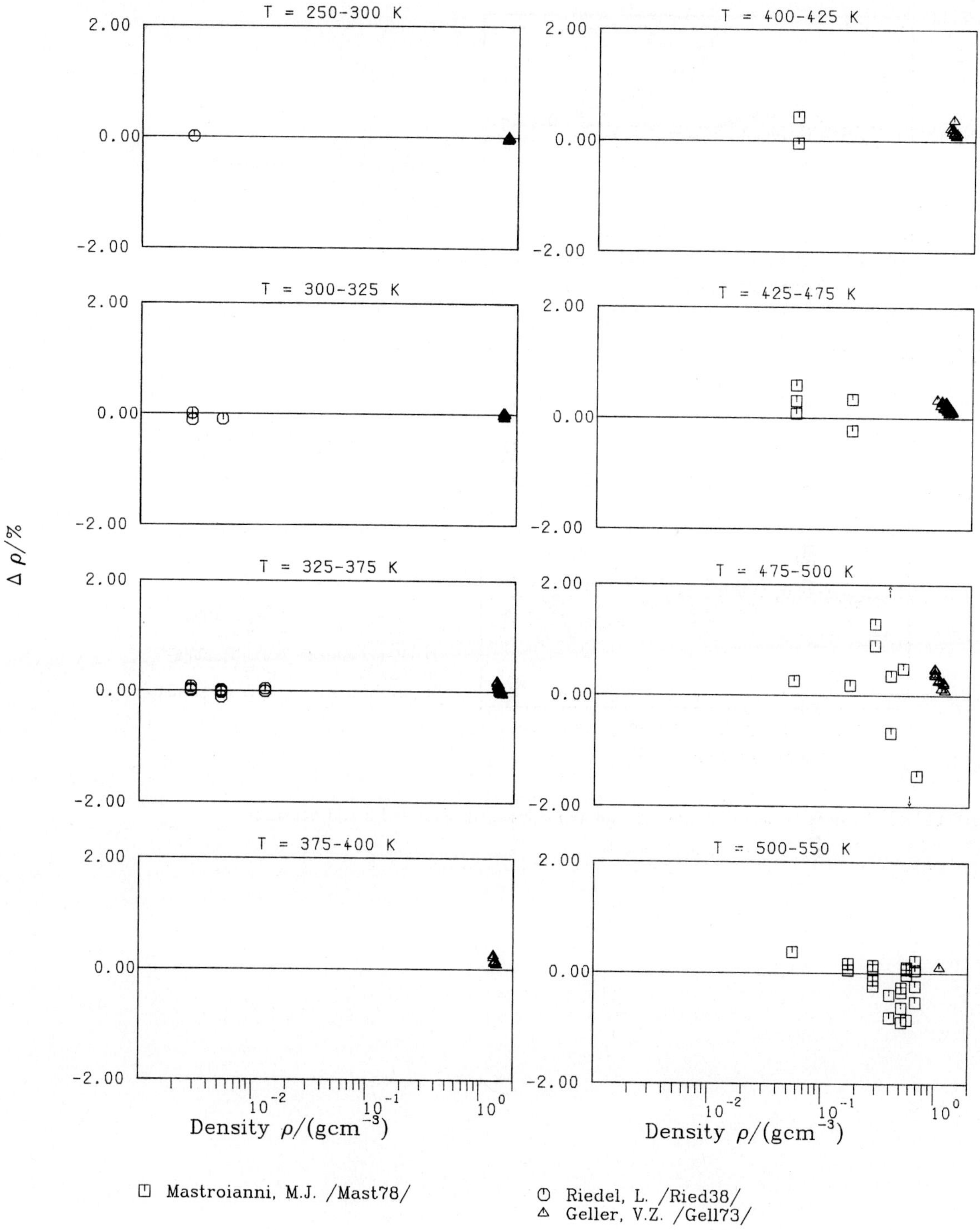

Fig.R113.2: Deviations between experimental and calculated p, ρ, T data *-continued*

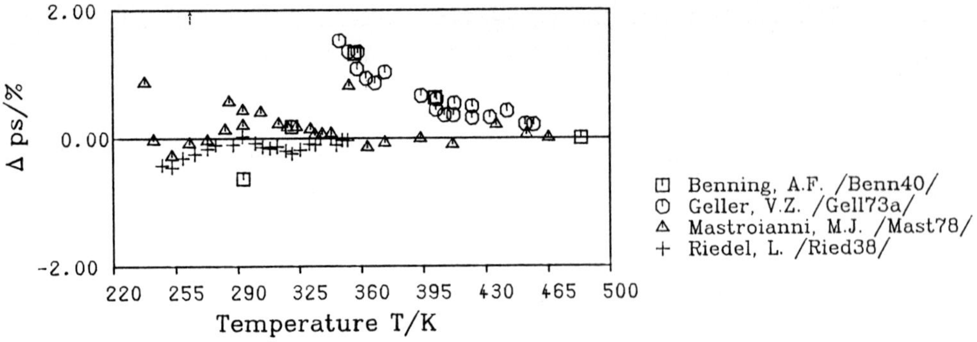

Fig.R113.3: Deviations between experimental and calculated vapor pressures

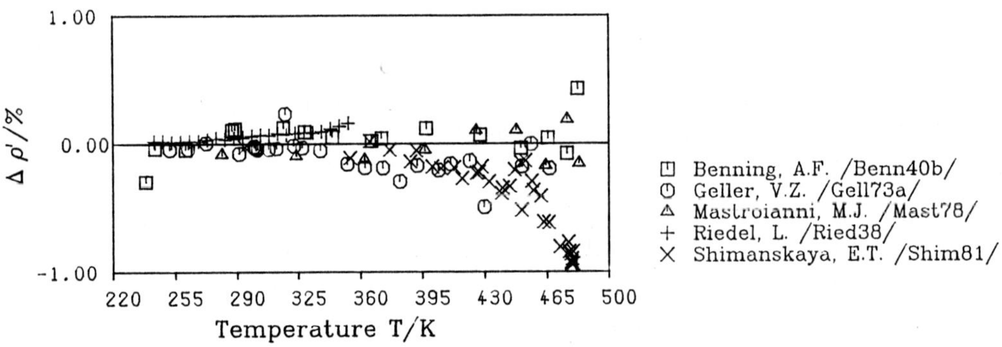

Fig.R113.4: Deviations between experimental and calculated saturated liquid densities

Table R113.10: Information on ideal gas state heat capacity equation

Author	T / K min	T / K max
BARHO, W. *[Barh 66]* ERNST, G. *[Erns 70]*	255	650

Coefficients of equation (15)	
Index i	d_i
1	7.533772 E−01
2	−3.440014 E−03
3	1.740469 E−05
4	−2.835975 E−08
5	1.565591 E−11

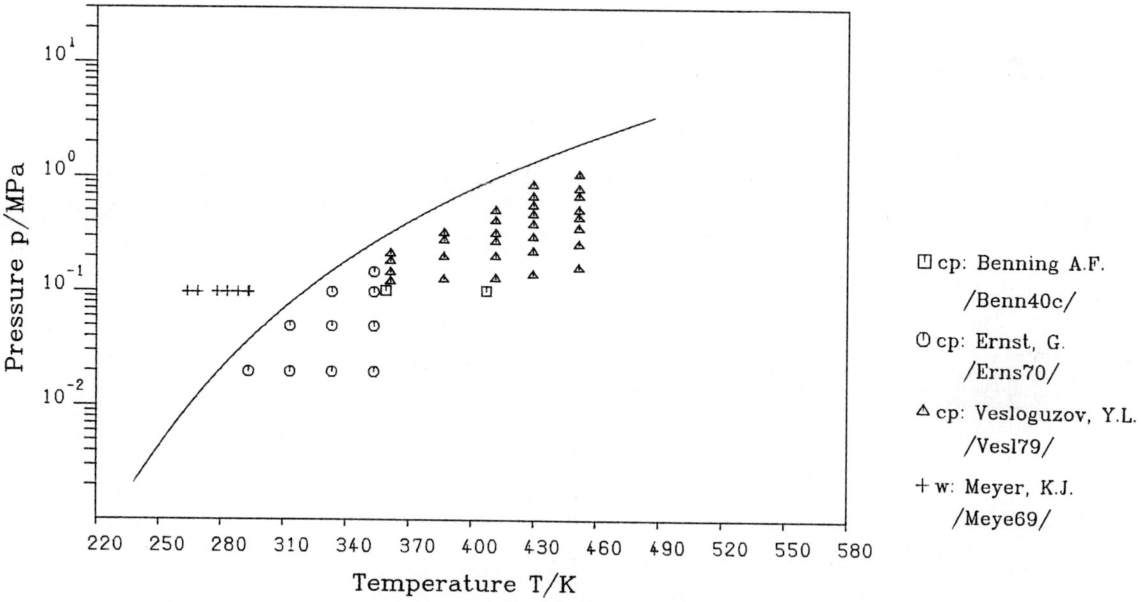

Fig.R113.5: Distribution of $c_p(p,T)$ and $w(p,T)$ data

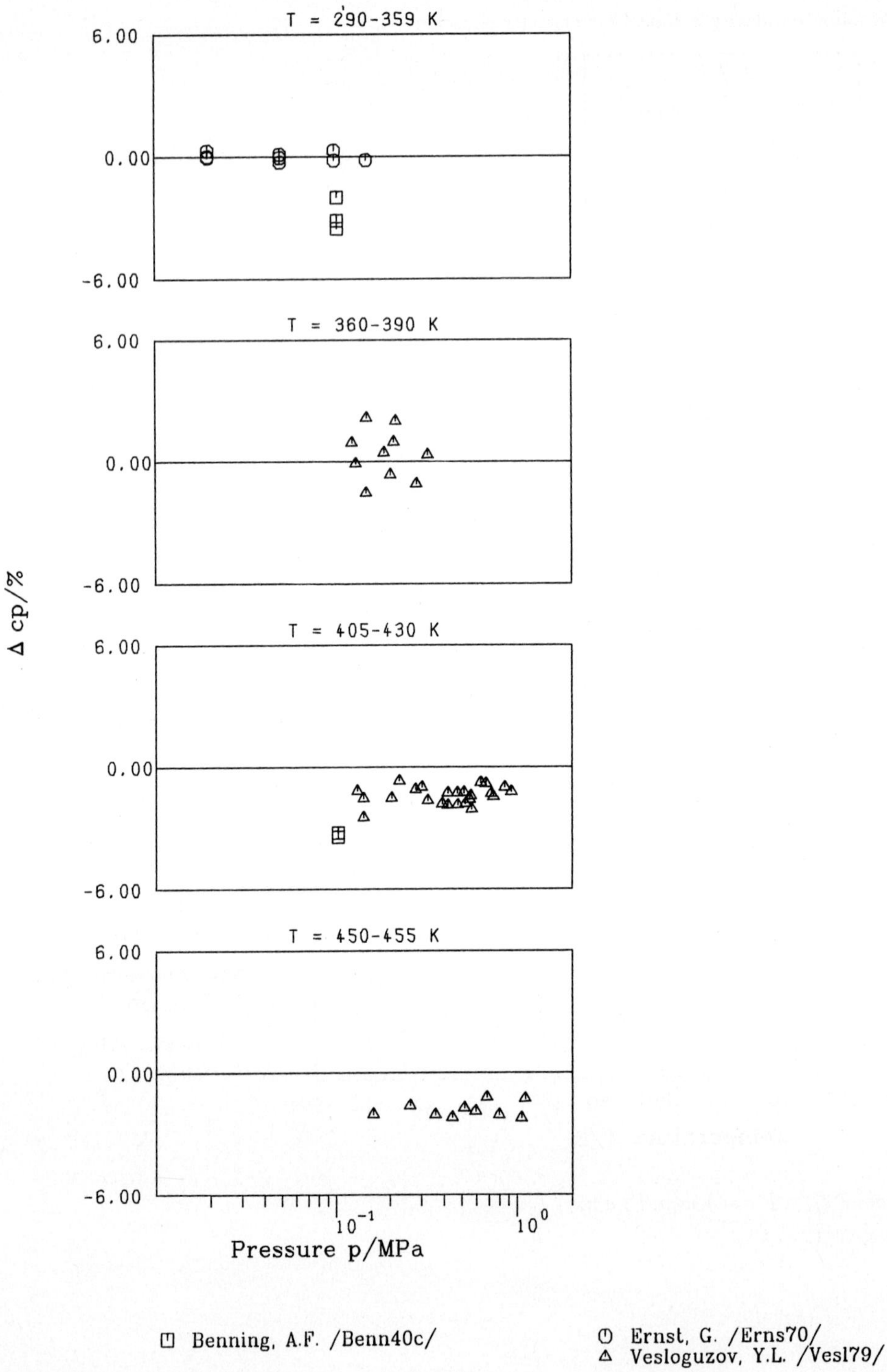

Fig.R113.6: Deviations between experimental and calculated specific isobaric heat capacities c_p

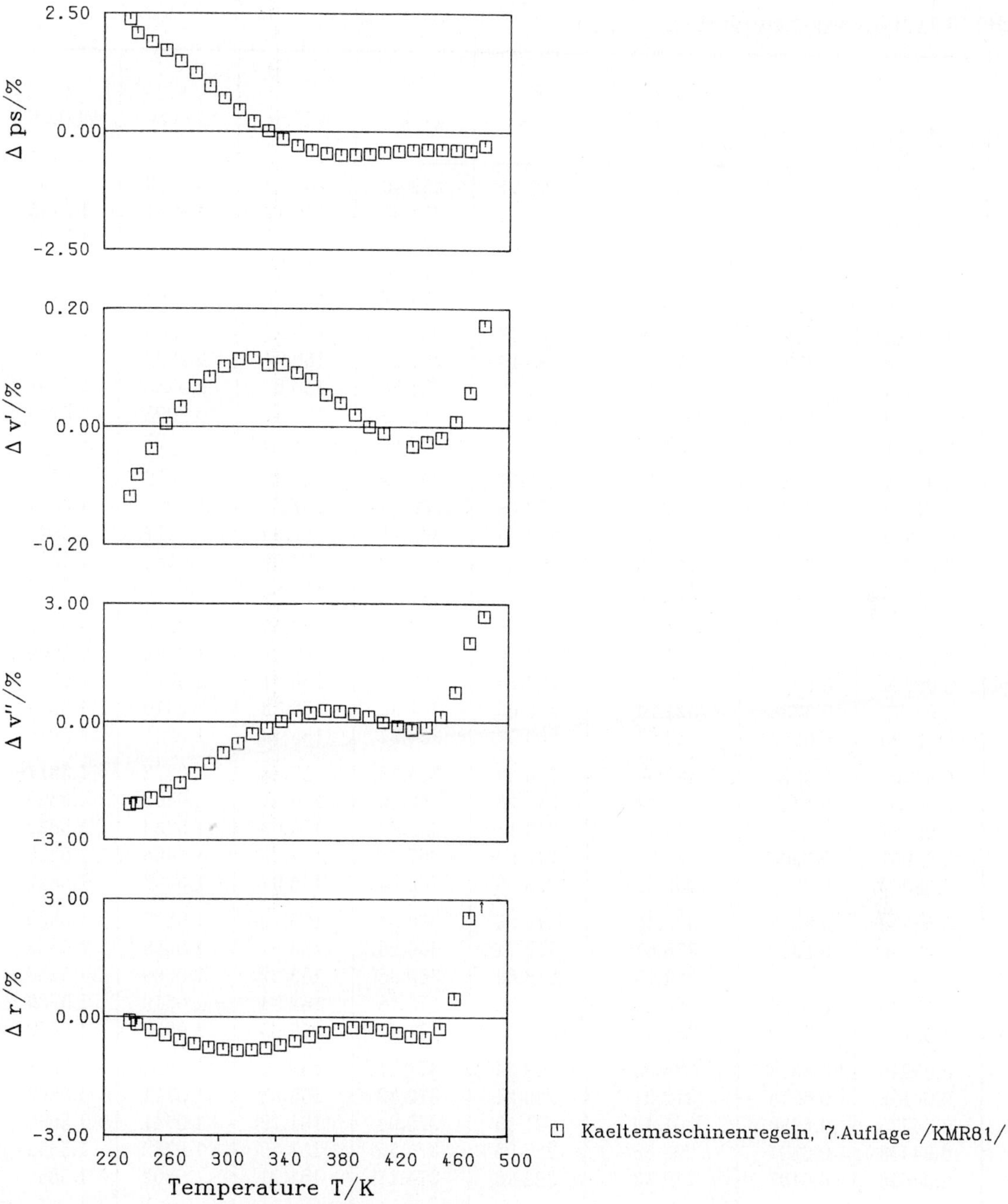

Fig.R113.7: Deviations between saturation properties from the *Kältemaschinenregeln* and those calculated by the *Bender-EOS*

Chart R113.1: Saturation properties

T K	p MPa	v' dm³/kg	v'' dm³/kg	h' kJ/kg	h'' kJ/kg	r kJ/kg	s' kJ/kgK	s'' kJ/kgK
265.00	0.00999	0.6102	1168.8	192.97	353.46	160.48	0.9739	1.5795
266.00	0.01052	0.6110	1113.7	193.83	354.06	160.23	0.9771	1.5795
267.00	0.01107	0.6119	1061.8	194.69	354.66	159.97	0.9803	1.5795
268.00	0.01165	0.6127	1012.7	195.55	355.26	159.71	0.9836	1.5795
269.00	0.01225	0.6135	966.22	196.41	355.87	159.46	0.9868	1.5795
270.00	0.01287	0.6144	922.28	197.27	356.47	159.20	0.9900	1.5796
271.00	0.01353	0.6153	880.71	198.14	357.07	158.94	0.9932	1.5796
272.00	0.01421	0.6161	841.35	199.00	357.68	158.67	0.9963	1.5797
273.00	0.01491	0.6170	804.06	199.87	358.28	158.41	0.9995	1.5798
274.00	0.01565	0.6178	768.73	200.74	358.89	158.15	1.0027	1.5799
275.00	0.01642	0.6187	735.24	201.61	359.49	157.89	1.0059	1.5800
276.00	0.01721	0.6196	703.47	202.48	360.10	157.62	1.0090	1.5801
277.00	0.01804	0.6205	673.33	203.35	360.71	157.35	1.0122	1.5803
278.00	0.01890	0.6213	644.72	204.23	361.32	157.09	1.0153	1.5804
279.00	0.01980	0.6222	617.56	205.11	361.92	156.82	1.0185	1.5806
280.00	0.02073	0.6231	591.75	205.98	362.53	156.55	1.0216	1.5807
281.00	0.02169	0.6240	567.23	206.86	363.14	156.28	1.0248	1.5809
282.00	0.02269	0.6249	543.91	207.74	363.75	156.01	1.0279	1.5811
283.00	0.02373	0.6258	521.74	208.63	364.36	155.73	1.0310	1.5813
284.00	0.02480	0.6267	500.64	209.51	364.97	155.46	1.0341	1.5815
285.00	0.02592	0.6276	480.56	210.40	365.58	155.18	1.0372	1.5817
286.00	0.02707	0.6285	461.44	211.28	366.19	154.91	1.0403	1.5820
287.00	0.02827	0.6294	443.22	212.17	366.80	154.63	1.0434	1.5822
288.00	0.02951	0.6303	425.87	213.06	367.41	154.35	1.0465	1.5825
289.00	0.03079	0.6313	409.33	213.96	368.03	154.07	1.0496	1.5827
290.00	0.03212	0.6322	393.55	214.85	368.64	153.79	1.0527	1.5830
291.00	0.03349	0.6331	378.51	215.74	369.25	153.51	1.0558	1.5833
292.00	0.03491	0.6341	364.15	216.64	369.86	153.22	1.0589	1.5836
293.00	0.03638	0.6350	350.44	217.54	370.48	152.94	1.0619	1.5839
294.00	0.03789	0.6359	337.36	218.44	371.09	152.65	1.0650	1.5842
295.00	0.03946	0.6369	324.85	219.34	371.71	152.37	1.0680	1.5845
296.00	0.04108	0.6378	312.91	220.24	372.32	152.08	1.0711	1.5849
297.00	0.04275	0.6388	301.49	221.15	372.94	151.79	1.0741	1.5852
298.00	0.04448	0.6397	290.58	222.05	373.55	151.50	1.0772	1.5856
299.00	0.04626	0.6407	280.13	222.96	374.17	151.21	1.0802	1.5859
300.00	0.04810	0.6417	270.14	223.87	374.78	150.91	1.0833	1.5863
301.00	0.04999	0.6427	260.58	224.78	375.40	150.62	1.0863	1.5867
302.00	0.05195	0.6436	251.43	225.69	376.01	150.32	1.0893	1.5871
303.00	0.05396	0.6446	242.66	226.60	376.63	150.03	1.0923	1.5874
304.00	0.05604	0.6456	234.26	227.52	377.25	149.73	1.0953	1.5878
305.00	0.05818	0.6466	226.21	228.44	377.86	149.43	1.0983	1.5883
306.00	0.06038	0.6476	218.50	229.35	378.48	149.13	1.1013	1.5887
307.00	0.06265	0.6486	211.10	230.27	379.10	148.82	1.1043	1.5891
308.00	0.06499	0.6496	204.01	231.19	379.72	148.52	1.1073	1.5895
309.00	0.06740	0.6506	197.20	232.12	380.33	148.22	1.1103	1.5900
310.00	0.06987	0.6516	190.67	233.04	380.95	147.91	1.1133	1.5904
311.00	0.07242	0.6526	184.40	233.96	381.57	147.60	1.1162	1.5909
312.00	0.07504	0.6537	178.38	234.89	382.19	147.29	1.1192	1.5913
313.00	0.07773	0.6547	172.60	235.82	382.80	146.98	1.1222	1.5918
314.00	0.08050	0.6557	167.04	236.75	383.42	146.67	1.1251	1.5923

Chart R113.1: Saturation properties–*continued*

T K	p MPa	v' dm³/kg	v'' dm³/kg	h' kJ/kg	h'' kJ/kg	r kJ/kg	s' kJ/kgK	s'' kJ/kgK
315.00	0.08335	0.6568	161.70	237.68	384.04	146.36	1.1281	1.5927
316.00	0.08627	0.6578	156.57	238.61	384.66	146.05	1.1310	1.5932
317.00	0.08928	0.6589	151.63	239.55	385.28	145.73	1.1340	1.5937
318.00	0.09236	0.6599	146.88	240.48	385.90	145.41	1.1369	1.5942
319.00	0.09553	0.6610	142.32	241.42	386.52	145.10	1.1399	1.5947
320.00	0.09879	0.6621	137.92	242.36	387.13	144.78	1.1428	1.5952
321.00	0.10212	0.6631	133.69	243.30	387.75	144.46	1.1457	1.5957
322.00	0.10555	0.6642	129.62	244.24	388.37	144.13	1.1486	1.5963
323.00	0.10907	0.6653	125.70	245.18	388.99	143.81	1.1516	1.5968
324.00	0.11267	0.6664	121.92	246.12	389.61	143.48	1.1545	1.5973
325.00	0.11637	0.6675	118.28	247.07	390.23	143.16	1.1574	1.5979
326.00	0.12016	0.6686	114.77	248.02	390.85	142.83	1.1603	1.5984
327.00	0.12405	0.6697	111.39	248.97	391.47	142.50	1.1632	1.5990
328.00	0.12804	0.6708	108.13	249.92	392.08	142.17	1.1661	1.5995
329.00	0.13212	0.6720	104.98	250.87	392.70	141.84	1.1689	1.6001
330.00	0.13630	0.6731	101.95	251.82	393.32	141.50	1.1718	1.6006
331.00	0.14059	0.6742	99.019	252.77	393.94	141.17	1.1747	1.6012
332.00	0.14498	0.6754	96.194	253.73	394.56	140.83	1.1776	1.6018
333.00	0.14947	0.6765	93.466	254.69	395.18	140.49	1.1805	1.6023
334.00	0.15407	0.6777	90.833	255.64	395.79	140.15	1.1833	1.6029
335.00	0.15878	0.6789	88.289	256.60	396.41	139.81	1.1862	1.6035
336.00	0.16360	0.6800	85.832	257.56	397.03	139.47	1.1890	1.6041
337.00	0.16853	0.6812	83.458	258.53	397.65	139.12	1.1919	1.6047
338.00	0.17357	0.6824	81.164	259.49	398.27	138.77	1.1947	1.6053
339.00	0.17873	0.6836	78.946	260.46	398.88	138.43	1.1976	1.6059
340.00	0.18401	0.6848	76.802	261.42	399.50	138.08	1.2004	1.6065
341.00	0.18941	0.6860	74.728	262.39	400.12	137.72	1.2032	1.6071
342.00	0.19492	0.6872	72.723	263.36	400.73	137.37	1.2061	1.6077
343.00	0.20056	0.6884	70.782	264.33	401.35	137.02	1.2089	1.6084
344.00	0.20632	0.6897	68.905	265.31	401.97	136.66	1.2117	1.6090
345.00	0.21221	0.6909	67.088	266.28	402.58	136.30	1.2145	1.6096
346.00	0.21822	0.6922	65.329	267.26	403.20	135.94	1.2173	1.6102
347.00	0.22436	0.6934	63.626	268.23	403.81	135.58	1.2201	1.6109
348.00	0.23064	0.6947	61.976	269.21	404.43	135.21	1.2230	1.6115
349.00	0.23704	0.6960	60.379	270.19	405.04	134.85	1.2258	1.6121
350.00	0.24359	0.6972	58.831	271.17	405.66	134.48	1.2285	1.6128
351.00	0.25026	0.6985	57.331	272.16	406.27	134.11	1.2313	1.6134
352.00	0.25708	0.6998	55.878	273.14	406.88	133.74	1.2341	1.6141
353.00	0.26404	0.7011	54.469	274.13	407.49	133.37	1.2369	1.6147
354.00	0.27113	0.7025	53.103	275.11	408.11	132.99	1.2397	1.6154
355.00	0.27838	0.7038	51.779	276.10	408.72	132.62	1.2425	1.6160
356.00	0.28576	0.7051	50.494	277.09	409.33	132.24	1.2452	1.6167
357.00	0.29330	0.7065	49.248	278.08	409.94	131.86	1.2480	1.6173
358.00	0.30098	0.7078	48.039	279.08	410.55	131.47	1.2508	1.6180
359.00	0.30882	0.7092	46.865	280.07	411.16	131.09	1.2535	1.6187
360.00	0.31681	0.7106	45.726	281.07	411.77	130.70	1.2563	1.6193
361.00	0.32495	0.7120	44.621	282.07	412.38	130.31	1.2590	1.6200
362.00	0.33325	0.7134	43.547	283.07	412.99	129.92	1.2618	1.6207
363.00	0.34171	0.7148	42.505	284.07	413.60	129.53	1.2645	1.6214
364.00	0.35033	0.7162	41.493	285.07	414.20	129.13	1.2673	1.6220

Chart R113.1: Saturation properties—*continued*

T K	p MPa	v' dm³/kg	v'' dm³/kg	h' kJ/kg	h'' kJ/kg	r kJ/kg	s' kJ/kgK	s'' kJ/kgK
365.00	0.35911	0.7176	40.509	286.07	414.81	128.74	1.2700	1.6227
366.00	0.36805	0.7191	39.554	287.08	415.42	128.34	1.2727	1.6234
367.00	0.37717	0.7205	38.625	288.09	416.02	127.94	1.2755	1.6241
368.00	0.38645	0.7220	37.723	289.09	416.63	127.53	1.2782	1.6247
369.00	0.39590	0.7234	36.845	290.10	417.23	127.12	1.2809	1.6254
370.00	0.40552	0.7249	35.992	291.12	417.83	126.72	1.2836	1.6261
371.00	0.41532	0.7264	35.163	292.13	418.43	126.30	1.2863	1.6268
372.00	0.42530	0.7279	34.357	293.14	419.03	125.89	1.2891	1.6275
373.00	0.43545	0.7295	33.572	294.16	419.64	125.47	1.2918	1.6282
374.00	0.44578	0.7310	32.809	295.18	420.24	125.06	1.2945	1.6288
375.00	0.45630	0.7326	32.066	296.20	420.83	124.64	1.2972	1.6295
376.00	0.46700	0.7341	31.343	297.22	421.43	124.21	1.2999	1.6302
377.00	0.47788	0.7357	30.640	298.24	422.03	123.79	1.3026	1.6309
378.00	0.48896	0.7373	29.955	299.27	422.62	123.36	1.3053	1.6316
379.00	0.50022	0.7389	29.288	300.29	423.22	122.93	1.3080	1.6323
380.00	0.51168	0.7405	28.639	301.32	423.81	122.49	1.3106	1.6330
381.00	0.52333	0.7421	28.007	302.35	424.41	122.05	1.3133	1.6337
382.00	0.53518	0.7438	27.391	303.38	425.00	121.62	1.3160	1.6344
383.00	0.54723	0.7455	26.791	304.42	425.59	121.17	1.3187	1.6351
384.00	0.55948	0.7471	26.207	305.45	426.18	120.73	1.3214	1.6358
385.00	0.57193	0.7488	25.637	306.49	426.77	120.28	1.3240	1.6364
386.00	0.58458	0.7505	25.082	307.53	427.36	119.83	1.3267	1.6371
387.00	0.59744	0.7523	24.541	308.57	427.94	119.37	1.3294	1.6378
388.00	0.61052	0.7540	24.013	309.61	428.53	118.92	1.3320	1.6385
389.00	0.62380	0.7558	23.499	310.65	429.11	118.46	1.3347	1.6392
390.00	0.63730	0.7575	22.997	311.70	429.69	117.99	1.3373	1.6399
391.00	0.65101	0.7593	22.508	312.75	430.27	117.53	1.3400	1.6406
392.00	0.66494	0.7611	22.031	313.80	430.85	117.06	1.3427	1.6413
393.00	0.67909	0.7630	21.565	314.85	431.43	116.59	1.3453	1.6420
394.00	0.69346	0.7648	21.111	315.90	432.01	116.11	1.3480	1.6427
395.00	0.70806	0.7667	20.668	316.95	432.58	115.63	1.3506	1.6433
396.00	0.72288	0.7686	20.235	318.01	433.16	115.15	1.3532	1.6440
397.00	0.73793	0.7705	19.813	319.07	433.73	114.66	1.3559	1.6447
398.00	0.75322	0.7724	19.400	320.13	434.30	114.17	1.3585	1.6454
399.00	0.76873	0.7744	18.998	321.19	434.87	113.68	1.3612	1.6461
400.00	0.78448	0.7763	18.605	322.26	435.44	113.18	1.3638	1.6467
401.00	0.80047	0.7783	18.221	323.33	436.00	112.68	1.3664	1.6474
402.00	0.81670	0.7804	17.846	324.39	436.57	112.17	1.3691	1.6481
403.00	0.83318	0.7824	17.480	325.47	437.13	111.66	1.3717	1.6488
404.00	0.84989	0.7845	17.122	326.54	437.69	111.15	1.3743	1.6494
405.00	0.86686	0.7865	16.772	327.61	438.25	110.63	1.3769	1.6501
406.00	0.88407	0.7887	16.430	328.69	438.80	110.11	1.3796	1.6508
407.00	0.90154	0.7908	16.096	329.77	439.36	109.59	1.3822	1.6514
408.00	0.91926	0.7929	15.770	330.85	439.91	109.06	1.3848	1.6521
409.00	0.93723	0.7951	15.450	331.94	440.46	108.52	1.3874	1.6528
410.00	0.95547	0.7974	15.138	333.02	441.01	107.99	1.3900	1.6534
411.00	0.97397	0.7996	14.833	334.11	441.56	107.44	1.3927	1.6541
412.00	0.99273	0.8019	14.534	335.20	442.10	106.90	1.3953	1.6547
413.00	1.01175	0.8042	14.242	336.30	442.64	106.34	1.3979	1.6554
414.00	1.03105	0.8065	13.957	337.39	443.18	105.79	1.4005	1.6560

Chart R113.1: Saturation properties—*continued*

T K	p MPa	v' dm^3/kg	v'' dm^3/kg	h' kJ/kg	h'' kJ/kg	r kJ/kg	s' kJ/kgK	s'' kJ/kgK
415.00	1.05062	0.8089	13.677	338.49	443.71	105.22	1.4031	1.6567
416.00	1.07046	0.8112	13.403	339.59	444.25	104.66	1.4057	1.6573
417.00	1.09058	0.8137	13.136	340.69	444.78	104.08	1.4083	1.6579
418.00	1.11097	0.8161	12.874	341.80	445.31	103.51	1.4109	1.6586
419.00	1.13165	0.8186	12.617	342.91	445.83	102.92	1.4136	1.6592
420.00	1.15261	0.8211	12.366	344.02	446.36	102.34	1.4162	1.6598
421.00	1.17386	0.8237	12.120	345.13	446.87	101.74	1.4188	1.6604
422.00	1.19540	0.8263	11.879	346.25	447.39	101.14	1.4214	1.6610
423.00	1.21723	0.8290	11.643	347.37	447.90	100.54	1.4240	1.6617
424.00	1.23936	0.8316	11.412	348.49	448.42	99.92	1.4266	1.6623
425.00	1.26178	0.8344	11.186	349.62	448.92	99.31	1.4292	1.6629
426.00	1.28451	0.8371	10.964	350.74	449.43	98.68	1.4318	1.6634
427.00	1.30753	0.8399	10.746	351.87	449.93	98.05	1.4344	1.6640
428.00	1.33087	0.8428	10.533	353.01	450.42	97.41	1.4370	1.6646
429.00	1.35451	0.8457	10.324	354.15	450.91	96.77	1.4396	1.6652
430.00	1.37846	0.8486	10.119	355.29	451.40	96.12	1.4422	1.6658
431.00	1.40273	0.8516	9.9186	356.43	451.89	95.46	1.4448	1.6663
432.00	1.42732	0.8547	9.7216	357.58	452.37	94.79	1.4474	1.6669
433.00	1.45223	0.8577	9.5284	358.73	452.85	94.12	1.4500	1.6674
434.00	1.47746	0.8609	9.3389	359.88	453.32	93.44	1.4527	1.6680
435.00	1.50302	0.8641	9.1530	361.04	453.79	92.75	1.4553	1.6685
436.00	1.52891	0.8674	8.9705	362.20	454.25	92.05	1.4579	1.6690
437.00	1.55513	0.8707	8.7915	363.36	454.71	91.34	1.4605	1.6695
438.00	1.58169	0.8741	8.6157	364.53	455.16	90.63	1.4631	1.6700
439.00	1.60859	0.8776	8.4432	365.70	455.61	89.91	1.4657	1.6705
440.00	1.63583	0.8811	8.2737	366.88	456.05	89.17	1.4684	1.6710
441.00	1.66341	0.8847	8.1073	368.06	456.49	88.43	1.4710	1.6715
442.00	1.69135	0.8884	7.9439	369.24	456.92	87.68	1.4736	1.6720
443.00	1.71964	0.8921	7.7833	370.43	457.34	86.91	1.4762	1.6724
444.00	1.74829	0.8959	7.6255	371.63	457.76	86.14	1.4789	1.6729
445.00	1.77730	0.8998	7.4704	372.82	458.18	85.35	1.4815	1.6733
446.00	1.80668	0.9038	7.3180	374.03	458.58	84.55	1.4842	1.6737
447.00	1.83642	0.9079	7.1681	375.23	458.98	83.75	1.4868	1.6741
448.00	1.86653	0.9121	7.0207	376.45	459.37	82.92	1.4894	1.6745
449.00	1.89702	0.9164	6.8758	377.67	459.76	82.09	1.4921	1.6749
450.00	1.92789	0.9208	6.7331	378.89	460.13	81.24	1.4948	1.6753
451.00	1.95915	0.9253	6.5928	380.12	460.50	80.38	1.4974	1.6757
452.00	1.99079	0.9300	6.4546	381.36	460.86	79.50	1.5001	1.6760
453.00	2.02282	0.9347	6.3186	382.60	461.21	78.61	1.5028	1.6763
454.00	2.05526	0.9396	6.1846	383.85	461.55	77.71	1.5055	1.6766
455.00	2.08809	0.9446	6.0527	385.10	461.88	76.78	1.5082	1.6769
456.00	2.12133	0.9498	5.9226	386.36	462.20	75.84	1.5109	1.6772
457.00	2.15498	0.9551	5.7944	387.63	462.51	74.88	1.5136	1.6774
458.00	2.18905	0.9606	5.6680	388.91	462.81	73.90	1.5163	1.6776
459.00	2.22354	0.9662	5.5433	390.19	463.10	72.90	1.5190	1.6778
460.00	2.25845	0.9721	5.4202	391.49	463.37	71.88	1.5218	1.6780
461.00	2.29380	0.9781	5.2987	392.79	463.63	70.84	1.5245	1.6782
462.00	2.32958	0.9844	5.1787	394.10	463.87	69.77	1.5273	1.6783
463.00	2.36581	0.9909	5.0601	395.42	464.10	68.68	1.5301	1.6784
464.00	2.40249	0.9976	4.9429	396.76	464.31	67.56	1.5329	1.6785

Chart R113.1: Saturation properties—*continued*

T K	p MPa	v' dm³/kg	v'' dm³/kg	h' kJ/kg	h'' kJ/kg	r kJ/kg	s' kJ/kgK	s'' kJ/kgK
465.00	2.43962	1.0046	4.8269	398.10	464.51	66.41	1.5357	1.6785
466.00	2.47722	1.0119	4.7120	399.46	464.69	65.23	1.5385	1.6785
467.00	2.51528	1.0196	4.5983	400.83	464.84	64.01	1.5414	1.6784
468.00	2.55382	1.0275	4.4855	402.21	464.98	62.76	1.5442	1.6783
469.00	2.59285	1.0359	4.3736	403.61	465.09	61.47	1.5471	1.6782
470.00	2.63236	1.0447	4.2624	405.03	465.17	60.14	1.5501	1.6780
471.00	2.67238	1.0539	4.1519	406.46	465.23	58.76	1.5530	1.6778
472.00	2.71291	1.0637	4.0419	407.92	465.25	57.33	1.5560	1.6775
473.00	2.75396	1.0740	3.9321	409.40	465.25	55.85	1.5591	1.6771
474.00	2.79554	1.0850	3.8226	410.90	465.20	54.30	1.5621	1.6767
475.00	2.83766	1.0968	3.7129	412.43	465.11	52.68	1.5653	1.6762
476.00	2.88033	1.1095	3.6030	413.99	464.97	50.98	1.5684	1.6756
477.00	2.92357	1.1231	3.4924	415.59	464.78	49.19	1.5717	1.6748
478.00	2.96739	1.1380	3.3809	417.23	464.53	47.30	1.5750	1.6740
479.00	3.01181	1.1544	3.2678	418.92	464.20	45.28	1.5785	1.6730
480.00	3.05685	1.1725	3.1527	420.68	463.78	43.11	1.5820	1.6718
481.00	3.10252	1.1930	3.0347	422.51	463.26	40.75	1.5857	1.6704
482.00	3.14886	1.2164	2.9126	424.43	462.60	38.16	1.5896	1.6688
483.00	3.19589	1.2438	2.7847	426.49	461.76	35.27	1.5937	1.6668
484.00	3.24365	1.2771	2.6483	428.73	460.67	31.95	1.5982	1.6642
485.00	3.29218	1.3197	2.4981	431.25	459.22	27.97	1.6033	1.6610
486.00	3.33984	1.7360	1.7365	444.54	444.55	0.01	1.6305	1.6306
487.00	3.39766	1.4342	1.5708	437.03	441.12	4.09	1.6149	1.6233

Chart R113.2: Saturation properties

p MPa	T K	v' dm³/kg	v'' dm³/kg	h' kJ/kg	h'' kJ/kg	r kJ/kg	s' kJ/kgK	s'' kJ/kgK
0.0102	265.41	0.6105	1145.9	193.32	353.70	160.38	0.9752	1.5795
0.0104	265.78	0.6108	1125.3	193.65	353.93	160.28	0.9764	1.5795
0.0106	266.15	0.6111	1105.5	193.96	354.15	160.19	0.9776	1.5795
0.0108	266.52	0.6114	1086.4	194.28	354.37	160.09	0.9788	1.5795
0.0110	266.88	0.6118	1068.0	194.58	354.59	160.00	0.9800	1.5795
0.0115	267.75	0.6125	1024.6	195.34	355.11	159.78	0.9828	1.5795
0.0120	268.59	0.6132	984.74	196.06	355.62	159.56	0.9855	1.5795
0.0125	269.41	0.6139	947.96	196.76	356.11	159.35	0.9881	1.5796
0.0130	270.20	0.6146	913.92	197.44	356.59	159.14	0.9906	1.5796
0.0135	270.96	0.6152	882.33	198.10	357.05	158.95	0.9930	1.5796
0.0140	271.70	0.6159	852.92	198.74	357.50	158.75	0.9954	1.5797
0.0145	272.42	0.6165	825.47	199.37	357.93	158.57	0.9977	1.5797
0.0150	273.12	0.6171	799.80	199.97	358.35	158.38	0.9999	1.5798
0.0155	273.80	0.6177	775.73	200.56	358.77	158.20	1.0021	1.5799
0.0160	274.46	0.6182	753.11	201.14	359.17	158.03	1.0042	1.5799
0.0165	275.10	0.6188	731.82	201.70	359.56	157.86	1.0062	1.5800
0.0170	275.73	0.6194	711.75	202.25	359.94	157.69	1.0082	1.5801
0.0175	276.35	0.6199	692.78	202.79	360.31	157.53	1.0101	1.5802
0.0180	276.95	0.6204	674.84	203.31	360.68	157.37	1.0120	1.5802
0.0185	277.54	0.6209	657.83	203.82	361.03	157.21	1.0139	1.5803
0.0190	278.11	0.6214	641.69	204.32	361.38	157.06	1.0157	1.5804
0.0195	278.67	0.6219	626.35	204.82	361.72	156.91	1.0175	1.5805
0.0200	279.22	0.6224	611.75	205.30	362.06	156.76	1.0192	1.5806
0.0210	280.29	0.6234	584.57	206.24	362.71	156.47	1.0225	1.5808
0.0220	281.31	0.6243	559.78	207.14	363.33	156.19	1.0257	1.5810
0.0230	282.30	0.6252	537.08	208.01	363.93	155.92	1.0288	1.5812
0.0240	283.26	0.6260	516.20	208.85	364.52	155.66	1.0318	1.5814
0.0250	284.18	0.6269	496.95	209.67	365.08	155.41	1.0347	1.5816
0.0260	285.07	0.6277	479.12	210.46	365.63	155.16	1.0375	1.5818
0.0270	285.94	0.6285	462.58	211.23	366.15	154.92	1.0402	1.5820
0.0280	286.78	0.6292	447.17	211.98	366.67	154.69	1.0428	1.5822
0.0290	287.60	0.6300	432.79	212.70	367.17	154.46	1.0453	1.5824
0.0300	288.39	0.6307	419.34	213.41	367.65	154.24	1.0477	1.5826
0.0310	289.16	0.6314	406.73	214.10	368.13	154.03	1.0501	1.5828
0.0320	289.91	0.6321	394.88	214.77	368.59	153.81	1.0524	1.5830
0.0330	290.65	0.6328	383.72	215.43	369.04	153.61	1.0547	1.5832
0.0340	291.36	0.6335	373.20	216.07	369.47	153.40	1.0569	1.5834
0.0350	292.06	0.6341	363.26	216.70	369.90	153.21	1.0591	1.5836
0.0360	292.75	0.6348	353.85	217.31	370.32	153.01	1.0612	1.5838
0.0370	293.42	0.6354	344.93	217.91	370.73	152.82	1.0632	1.5840
0.0380	294.07	0.6360	336.47	218.50	371.13	152.63	1.0652	1.5842
0.0390	294.71	0.6366	328.43	219.08	371.53	152.45	1.0672	1.5844
0.0400	295.34	0.6372	320.77	219.64	371.91	152.27	1.0691	1.5847
0.0410	295.95	0.6378	313.47	220.20	372.29	152.09	1.0710	1.5849
0.0420	296.55	0.6384	306.51	220.74	372.66	151.92	1.0728	1.5851
0.0430	297.15	0.6389	299.87	221.28	373.03	151.75	1.0746	1.5853
0.0440	297.73	0.6395	293.51	221.81	373.38	151.58	1.0764	1.5855
0.0450	298.30	0.6400	287.43	222.32	373.73	151.41	1.0781	1.5857
0.0460	298.86	0.6406	281.60	222.83	374.08	151.25	1.0798	1.5859
0.0470	299.41	0.6411	276.01	223.33	374.42	151.09	1.0815	1.5861

Chart R113.2: Saturation properties—*continued*

p MPa	T K	v' dm³/kg	v'' dm³/kg	h' kJ/kg	h'' kJ/kg	r kJ/kg	s' kJ/kgK	s'' kJ/kgK
0.0480	299.95	0.6416	270.65	223.82	374.75	150.93	1.0831	1.5863
0.0490	300.48	0.6421	265.50	224.31	375.08	150.77	1.0847	1.5865
0.0500	301.00	0.6427	260.54	224.78	375.40	150.62	1.0863	1.5867
0.0510	301.52	0.6432	255.78	225.25	375.72	150.46	1.0878	1.5869
0.0520	302.03	0.6437	251.19	225.72	376.03	150.31	1.0894	1.5871
0.0530	302.53	0.6441	246.77	226.17	376.34	150.17	1.0909	1.5873
0.0540	303.02	0.6446	242.50	226.62	376.64	150.02	1.0924	1.5875
0.0550	303.50	0.6451	238.39	227.07	376.94	149.88	1.0938	1.5876
0.0560	303.98	0.6456	234.42	227.50	377.24	149.73	1.0953	1.5878
0.0570	304.45	0.6460	230.58	227.93	377.53	149.59	1.0967	1.5880
0.0580	304.92	0.6465	226.87	228.36	377.81	149.45	1.0981	1.5882
0.0590	305.38	0.6470	223.28	228.78	378.10	149.31	1.0995	1.5884
0.0600	305.83	0.6474	219.80	229.20	378.37	149.18	1.1008	1.5886
0.0610	306.27	0.6479	216.44	229.61	378.65	149.04	1.1021	1.5888
0.0620	306.72	0.6483	213.18	230.01	378.92	148.91	1.1035	1.5890
0.0630	307.15	0.6487	210.02	230.41	379.19	148.78	1.1048	1.5892
0.0640	307.58	0.6492	206.96	230.81	379.46	148.65	1.1061	1.5893
0.0650	308.00	0.6496	203.98	231.20	379.72	148.52	1.1073	1.5895
0.0660	308.42	0.6500	201.10	231.58	379.98	148.39	1.1086	1.5897
0.0670	308.84	0.6504	198.29	231.97	380.23	148.27	1.1098	1.5899
0.0680	309.25	0.6509	195.57	232.34	380.48	148.14	1.1110	1.5901
0.0690	309.65	0.6513	192.92	232.72	380.73	148.02	1.1122	1.5902
0.0700	310.05	0.6517	190.35	233.09	380.98	147.90	1.1134	1.5904
0.0710	310.45	0.6521	187.84	233.45	381.23	147.77	1.1146	1.5906
0.0720	310.84	0.6525	185.40	233.81	381.47	147.65	1.1158	1.5908
0.0730	311.22	0.6529	183.03	234.17	381.71	147.53	1.1169	1.5910
0.0740	311.61	0.6533	180.72	234.53	381.94	147.42	1.1181	1.5911
0.0760	312.36	0.6540	176.27	235.23	382.41	147.18	1.1203	1.5915
0.0780	313.10	0.6548	172.04	235.91	382.86	146.95	1.1225	1.5918
0.0800	313.82	0.6555	168.02	236.58	383.31	146.73	1.1246	1.5922
0.0820	314.53	0.6563	164.18	237.24	383.75	146.51	1.1267	1.5925
0.0840	315.23	0.6570	160.52	237.89	384.18	146.29	1.1288	1.5928
0.0860	315.91	0.6577	157.03	238.53	384.60	146.08	1.1308	1.5932
0.0880	316.58	0.6584	153.69	239.15	385.02	145.86	1.1327	1.5935
0.0900	317.24	0.6591	150.49	239.77	385.42	145.66	1.1347	1.5938
0.0920	317.88	0.6598	147.43	240.37	385.82	145.45	1.1366	1.5941
0.0940	318.52	0.6605	144.49	240.97	386.22	145.25	1.1385	1.5945
0.0960	319.15	0.6612	141.67	241.56	386.61	145.05	1.1403	1.5948
0.0980	319.76	0.6618	138.96	242.13	386.99	144.85	1.1421	1.5951
0.1000	320.37	0.6625	136.35	242.70	387.36	144.66	1.1439	1.5954
0.1020	320.96	0.6631	133.85	243.26	387.73	144.47	1.1456	1.5957
0.1040	321.55	0.6637	131.43	243.81	388.09	144.28	1.1473	1.5960
0.1060	322.13	0.6644	129.10	244.36	388.45	144.09	1.1490	1.5963
0.1080	322.70	0.6650	126.86	244.90	388.80	143.91	1.1507	1.5966
0.1100	323.26	0.6656	124.70	245.43	389.15	143.73	1.1523	1.5969
0.1150	324.63	0.6671	119.60	246.72	390.00	143.28	1.1563	1.5977
0.1200	325.96	0.6686	114.92	247.98	390.82	142.84	1.1601	1.5984
0.1250	327.24	0.6700	110.59	249.19	391.61	142.42	1.1639	1.5991
0.1300	328.48	0.6714	106.59	250.38	392.38	142.01	1.1675	1.5998
0.1350	329.69	0.6727	102.87	251.52	393.13	141.61	1.1709	1.6005

Chart R113.2: Saturation properties—*continued*

p MPa	T K	v' dm^3/kg	v'' dm^3/kg	h' kJ/kg	h'' kJ/kg	r kJ/kg	s' kJ/kgK	s'' kJ/kgK
0.1400	330.86	0.6741	99.412	252.64	393.86	141.21	1.1743	1.6011
0.1450	332.00	0.6754	96.180	253.73	394.56	140.83	1.1776	1.6018
0.1500	333.12	0.6767	93.156	254.80	395.25	140.45	1.1808	1.6024
0.1550	334.20	0.6779	90.319	255.83	395.92	140.08	1.1839	1.6030
0.1600	335.26	0.6792	87.654	256.85	396.57	139.72	1.1869	1.6037
0.1650	336.29	0.6804	85.144	257.84	397.21	139.37	1.1898	1.6043
0.1700	337.29	0.6816	82.776	258.81	397.83	139.02	1.1927	1.6049
0.1750	338.28	0.6827	80.538	259.76	398.44	138.68	1.1955	1.6055
0.1800	339.24	0.6839	78.420	260.69	399.03	138.34	1.1983	1.6061
0.1850	340.19	0.6850	76.413	261.60	399.61	138.01	1.2009	1.6066
0.1900	341.11	0.6861	74.506	262.50	400.18	137.69	1.2035	1.6072
0.1950	342.01	0.6872	72.694	263.38	400.74	137.37	1.2061	1.6077
0.2000	342.90	0.6883	70.970	264.24	401.29	137.05	1.2086	1.6083
0.2100	344.63	0.6904	67.757	265.92	402.35	136.43	1.2135	1.6094
0.2200	346.29	0.6925	64.826	267.54	403.38	135.84	1.2182	1.6104
0.2300	347.90	0.6946	62.140	269.11	404.36	135.25	1.2227	1.6114
0.2400	349.45	0.6965	59.670	270.64	405.32	134.68	1.2270	1.6124
0.2500	350.96	0.6985	57.389	272.12	406.24	134.13	1.2312	1.6134
0.2600	352.42	0.7004	55.278	273.56	407.14	133.58	1.2353	1.6143
0.2700	353.84	0.7022	53.317	274.96	408.01	133.05	1.2392	1.6153
0.2800	355.22	0.7041	51.491	276.32	408.85	132.53	1.2431	1.6162
0.2900	356.56	0.7059	49.786	277.65	409.68	132.02	1.2468	1.6171
0.3000	357.87	0.7077	48.190	278.95	410.47	131.52	1.2504	1.6179
0.3100	359.15	0.7094	46.693	280.22	411.25	131.03	1.2539	1.6188
0.3200	360.39	0.7111	45.286	281.46	412.01	130.55	1.2574	1.6196
0.3300	361.61	0.7128	43.961	282.68	412.75	130.08	1.2607	1.6204
0.3400	362.80	0.7145	42.711	283.87	413.47	129.61	1.2640	1.6212
0.3500	363.96	0.7161	41.530	285.03	414.18	129.15	1.2672	1.6220
0.3600	365.10	0.7178	40.412	286.17	414.87	128.70	1.2703	1.6228
0.3700	366.22	0.7194	39.352	287.29	415.55	128.25	1.2733	1.6235
0.3800	367.31	0.7210	38.345	288.39	416.21	127.81	1.2763	1.6243
0.3900	368.38	0.7225	37.388	289.48	416.85	127.38	1.2792	1.6250
0.4000	369.43	0.7241	36.477	290.54	417.49	126.95	1.2821	1.6257
0.4100	370.46	0.7256	35.609	291.58	418.11	126.53	1.2849	1.6264
0.4200	371.47	0.7271	34.780	292.61	418.72	126.11	1.2876	1.6271
0.4300	372.47	0.7286	33.989	293.62	419.31	125.70	1.2903	1.6278
0.4400	373.44	0.7301	33.232	294.61	419.90	125.29	1.2930	1.6285
0.4500	374.40	0.7316	32.507	295.59	420.48	124.89	1.2956	1.6291
0.4600	375.35	0.7331	31.812	296.55	421.04	124.49	1.2981	1.6298
0.4700	376.28	0.7346	31.146	297.50	421.60	124.09	1.3006	1.6304
0.4800	377.19	0.7360	30.507	298.44	422.14	123.70	1.3031	1.6310
0.4900	378.09	0.7374	29.892	299.36	422.68	123.32	1.3055	1.6317
0.5000	378.98	0.7389	29.301	300.27	423.21	122.93	1.3079	1.6323
0.5100	379.85	0.7403	28.733	301.17	423.73	122.56	1.3103	1.6329
0.5200	380.72	0.7417	28.185	302.06	424.24	122.18	1.3126	1.6335
0.5300	381.56	0.7431	27.657	302.93	424.74	121.81	1.3148	1.6341
0.5400	382.40	0.7445	27.148	303.80	425.24	121.44	1.3171	1.6346
0.5500	383.23	0.7458	26.657	304.65	425.72	121.07	1.3193	1.6352
0.5600	384.04	0.7472	26.182	305.50	426.20	120.71	1.3215	1.6358
0.5700	384.85	0.7486	25.724	306.33	426.68	120.35	1.3236	1.6363

Chart R113.2: Saturation properties—*continued*

p MPa	T K	v' dm^3/kg	v'' dm^3/kg	h' kJ/kg	h'' kJ/kg	r kJ/kg	s' kJ/kgK	s'' kJ/kgK
0.5800	385.64	0.7499	25.280	307.15	427.14	119.99	1.3257	1.6369
0.5900	386.42	0.7513	24.851	307.97	427.60	119.64	1.3278	1.6374
0.6000	387.20	0.7526	24.436	308.77	428.06	119.29	1.3299	1.6380
0.6100	387.96	0.7539	24.034	309.57	428.50	118.94	1.3319	1.6385
0.6200	388.72	0.7553	23.644	310.36	428.94	118.59	1.3339	1.6390
0.6300	389.46	0.7566	23.266	311.13	429.38	118.24	1.3359	1.6395
0.6400	390.20	0.7579	22.899	311.91	429.81	117.90	1.3379	1.6400
0.6500	390.93	0.7592	22.543	312.67	430.23	117.56	1.3398	1.6405
0.6600	391.65	0.7605	22.198	313.42	430.65	117.22	1.3417	1.6410
0.6700	392.36	0.7618	21.862	314.17	431.06	116.89	1.3436	1.6415
0.6800	393.06	0.7631	21.536	314.91	431.47	116.56	1.3455	1.6420
0.6900	393.76	0.7644	21.219	315.65	431.87	116.22	1.3473	1.6425
0.7000	394.45	0.7657	20.910	316.37	432.27	115.89	1.3491	1.6430
0.7100	395.13	0.7669	20.610	317.09	432.66	115.57	1.3510	1.6434
0.7200	395.81	0.7682	20.318	317.81	433.05	115.24	1.3527	1.6439
0.7300	396.47	0.7695	20.033	318.51	433.43	114.92	1.3545	1.6443
0.7400	397.14	0.7708	19.756	319.21	433.81	114.59	1.3562	1.6448
0.7600	398.44	0.7733	19.222	320.60	434.55	113.95	1.3597	1.6457
0.7800	399.72	0.7758	18.715	321.96	435.28	113.32	1.3631	1.6466
0.8000	400.97	0.7783	18.232	323.29	435.99	112.69	1.3664	1.6474
0.8200	402.20	0.7808	17.772	324.61	436.68	112.07	1.3696	1.6482
0.8400	403.41	0.7832	17.332	325.91	437.36	111.45	1.3728	1.6490
0.8600	404.60	0.7857	16.912	327.18	438.02	110.84	1.3759	1.6498
0.8800	405.76	0.7882	16.510	328.44	438.67	110.24	1.3790	1.6506
0.9000	406.91	0.7906	16.125	329.68	439.31	109.63	1.3820	1.6514
0.9200	408.04	0.7930	15.756	330.90	439.93	109.04	1.3849	1.6521
0.9400	409.15	0.7955	15.402	332.10	440.55	108.44	1.3878	1.6529
0.9600	410.25	0.7979	15.062	333.29	441.14	107.85	1.3907	1.6536
0.9800	411.32	0.8003	14.736	334.46	441.73	107.27	1.3935	1.6543
1.0000	412.38	0.8027	14.421	335.62	442.31	106.68	1.3963	1.6550
1.0200	413.43	0.8052	14.119	336.77	442.87	106.10	1.3990	1.6557
1.0400	414.46	0.8076	13.827	337.90	443.42	105.53	1.4017	1.6563
1.0600	415.47	0.8100	13.546	339.01	443.97	104.96	1.4044	1.6570
1.0800	416.48	0.8124	13.275	340.12	444.50	104.39	1.4070	1.6576
1.1000	417.46	0.8148	13.013	341.21	445.02	103.82	1.4095	1.6582
1.1500	419.88	0.8208	12.397	343.88	446.29	102.41	1.4158	1.6597
1.2000	422.21	0.8269	11.829	346.49	447.50	101.01	1.4219	1.6612
1.2500	424.48	0.8329	11.304	349.03	448.66	99.63	1.4278	1.6625
1.3000	426.67	0.8390	10.817	351.51	449.76	98.26	1.4336	1.6638
1.3500	428.81	0.8451	10.364	353.93	450.82	96.89	1.4391	1.6651
1.4000	430.89	0.8513	9.9409	356.30	451.83	95.53	1.4445	1.6663
1.4500	432.91	0.8575	9.5455	358.62	452.80	94.18	1.4498	1.6674
1.5000	434.88	0.8637	9.1746	360.90	453.73	92.83	1.4550	1.6684
1.5500	436.81	0.8701	8.8261	363.13	454.62	91.48	1.4600	1.6694
1.6000	438.68	0.8765	8.4977	365.33	455.47	90.14	1.4649	1.6704
1.6500	440.52	0.8829	8.1876	367.49	456.28	88.79	1.4697	1.6713
1.7000	442.31	0.8895	7.8943	369.61	457.05	87.44	1.4744	1.6721
1.7500	444.06	0.8962	7.6163	371.70	457.79	86.09	1.4790	1.6729
1.8000	445.77	0.9029	7.3522	373.75	458.49	84.74	1.4836	1.6736
1.8500	447.45	0.9098	7.1011	375.78	459.16	83.38	1.4880	1.6743

Chart R113.2: Saturation properties—*continued*

p MPa	T K	v' dm³/kg	v'' dm³/kg	h' kJ/kg	h'' kJ/kg	r kJ/kg	s' kJ/kgK	s'' kJ/kgK
1.9000	449.10	0.9168	6.8618	377.78	459.79	82.01	1.4924	1.6750
1.9500	450.71	0.9240	6.6334	379.76	460.39	80.63	1.4967	1.6756
2.0000	452.29	0.9313	6.4151	381.71	460.96	79.25	1.5009	1.6761
2.1000	455.36	0.9464	6.0057	385.55	462.00	76.44	1.5091	1.6770
2.2000	458.32	0.9624	5.6280	389.32	462.90	73.58	1.5172	1.6777
2.3000	461.17	0.9792	5.2777	393.02	463.67	70.65	1.5250	1.6782
2.4000	463.93	0.9972	4.9507	396.67	464.30	67.63	1.5327	1.6785
2.5000	466.60	1.0165	4.6437	400.28	464.78	64.50	1.5402	1.6785
2.6000	469.18	1.0375	4.3533	403.87	465.10	61.23	1.5477	1.6782
2.7000	471.68	1.0605	4.0767	407.46	465.25	57.79	1.5551	1.6776
2.8000	474.11	1.0862	3.8109	411.06	465.19	54.13	1.5625	1.6766
2.9000	476.46	1.1156	3.5526	414.72	464.89	50.18	1.5699	1.6752
3.0000	478.74	1.1499	3.2979	418.47	464.29	45.82	1.5775	1.6733
3.1000	480.95	1.1918	3.0413	422.40	463.29	40.89	1.5855	1.6705
3.2000	483.09	1.2464	2.7733	426.67	461.67	35.00	1.5941	1.6666
3.3000	485.16	1.3277	2.4723	431.69	458.94	27.25	1.6042	1.6604
3.4000	487.05	1.4390	1.5736	437.23	441.25	4.02	1.6153	1.6236

Chart R113.3: Thermodynamic properties

p/MPa	0.05			0.08			0.10		
T K	v dm³/kg	h kJ/kg	s kJ/kgK	v dm³/kg	h kJ/kg	s kJ/kgK	v dm³/kg	h kJ/kg	s kJ/kgK
270.00	0.6144	197.29	0.9899	0.6143	197.30	0.9899	0.6143	197.31	0.9899
280.00	0.6231	205.99	1.0216	0.6231	206.01	1.0216	0.6230	206.01	1.0216
290.00	0.6322	214.86	1.0527	0.6321	214.87	1.0527	0.6321	214.87	1.0526
300.00	0.6417	223.87	1.0832	0.6416	223.88	1.0832	0.6416	223.89	1.0832
310.00	268.97	381.36	1.6062	0.6516	233.04	1.1133	0.6516	233.05	1.1132
320.00	278.28	388.08	1.6275	171.72	387.51	1.6054	0.6621	242.36	1.1428
330.00	287.55	394.90	1.6485	177.68	394.37	1.6265	141.03	394.00	1.6158
340.00	296.78	401.81	1.6691	183.59	401.32	1.6473	145.84	400.98	1.6367
350.00	305.97	408.82	1.6894	189.46	408.36	1.6677	150.61	408.05	1.6571
360.00	315.13	415.92	1.7095	195.30	415.49	1.6878	155.34	415.20	1.6773
370.00	324.27	423.12	1.7292	201.11	422.72	1.7076	160.05	422.45	1.6972
380.00	333.38	430.41	1.7486	206.90	430.03	1.7271	164.73	429.78	1.7167
390.00	342.47	437.79	1.7678	212.67	437.44	1.7463	169.39	437.20	1.7360
400.00	351.54	445.26	1.7867	218.42	444.92	1.7653	174.04	444.70	1.7550
410.00	360.60	452.81	1.8053	224.15	452.49	1.7840	178.66	452.28	1.7737
420.00	369.64	460.44	1.8237	229.86	460.15	1.8024	183.27	459.95	1.7922
430.00	378.67	468.16	1.8419	235.57	467.88	1.8206	187.86	467.69	1.8104
440.00	387.68	475.95	1.8598	241.26	475.68	1.8385	192.45	475.50	1.8283
450.00	396.69	483.81	1.8775	246.94	483.55	1.8562	197.02	483.38	1.8461
460.00	405.69	491.74	1.8949	252.61	491.49	1.8737	201.58	491.33	1.8635
470.00	414.68	499.73	1.9121	258.27	499.50	1.8909	206.14	499.35	1.8808
480.00	423.66	507.79	1.9290	263.93	507.57	1.9079	210.68	507.42	1.8978
490.00	432.64	515.90	1.9458	269.58	515.69	1.9246	215.22	515.55	1.9145
500.00	441.60	524.07	1.9623	275.22	523.87	1.9411	219.75	523.73	1.9311
510.00	450.57	532.29	1.9785	280.85	532.10	1.9574	224.28	531.97	1.9474
520.00	459.52	540.56	1.9946	286.48	540.37	1.9735	228.80	540.25	1.9634
530.00	468.48	548.87	2.0104	292.11	548.69	1.9894	233.32	548.57	1.9793

p/MPa	0.15			0.20			0.30		
270.00	0.6143	197.33	0.9898	0.6142	197.34	0.9898	0.6142	197.38	0.9897
280.00	0.6230	206.03	1.0215	0.6229	206.05	1.0215	0.6229	206.09	1.0214
290.00	0.6321	214.89	1.0526	0.6320	214.91	1.0526	0.6319	214.95	1.0525
300.00	0.6416	223.90	1.0832	0.6415	223.92	1.0831	0.6414	223.96	1.0830
310.00	0.6515	233.07	1.1132	0.6514	233.08	1.1131	0.6513	233.12	1.1130
320.00	0.6620	242.37	1.1427	0.6619	242.39	1.1427	0.6618	242.42	1.1426
330.00	0.6731	251.82	1.1718	0.6730	251.84	1.1718	0.6728	251.87	1.1716
340.00	95.458	400.11	1.6169	0.6848	261.43	1.2004	0.6846	261.46	1.2003
350.00	98.767	407.24	1.6375	72.802	406.41	1.6231	0.6971	271.19	1.2285
360.00	102.04	414.46	1.6579	75.349	413.69	1.6436	48.578	412.06	1.6223
370.00	105.28	421.75	1.6779	77.861	421.04	1.6637	50.379	419.54	1.6428
380.00	108.49	429.13	1.6975	80.343	428.47	1.6835	52.143	427.08	1.6629
390.00	111.68	436.59	1.7169	82.800	435.97	1.7030	53.878	434.68	1.6827
400.00	114.85	444.13	1.7360	85.235	443.55	1.7222	55.588	442.35	1.7021
410.00	118.00	451.75	1.7548	87.650	451.20	1.7411	57.275	450.08	1.7212
420.00	121.13	459.44	1.7733	90.049	458.93	1.7597	58.943	457.87	1.7400
430.00	124.25	467.21	1.7916	92.433	466.73	1.7781	60.595	465.73	1.7585
440.00	127.36	475.05	1.8096	94.803	474.59	1.7962	62.232	473.65	1.7767
450.00	130.45	482.95	1.8274	97.162	482.52	1.8140	63.857	481.63	1.7946
460.00	133.54	490.93	1.8449	99.510	490.51	1.8315	65.471	489.67	1.8123

Chart R113.3: Thermodynamic properties–*continued*

p/MPa	0.15			0.20			0.30		
T K	v dm³/kg	h kJ/kg	s kJ/kgK	v dm³/kg	h kJ/kg	s kJ/kgK	v dm³/kg	h kJ/kg	s kJ/kgK
470.00	136.61	498.96	1.8622	101.85	498.57	1.8489	67.074	497.77	1.8297
480.00	139.68	507.05	1.8792	104.18	506.68	1.8659	68.668	505.92	1.8468
490.00	142.74	515.20	1.8960	106.50	514.84	1.8828	70.255	514.12	1.8638
500.00	145.80	523.40	1.9126	108.82	523.06	1.8994	71.834	522.37	1.8804
510.00	148.85	531.64	1.9289	111.13	531.32	1.9157	73.407	530.66	1.8968
520.00	151.89	539.94	1.9450	113.43	539.63	1.9319	74.974	538.99	1.9130
530.00	154.93	548.28	1.9609	115.73	547.98	1.9478	76.535	547.37	1.9290
p/MPa	0.40			0.50			0.60		
270.00	0.6141	197.42	0.9896	0.6140	197.46	0.9895	0.6139	197.50	0.9895
280.00	0.6228	206.13	1.0213	0.6227	206.16	1.0212	0.6226	206.20	1.0211
290.00	0.6318	214.98	1.0524	0.6317	215.02	1.0523	0.6316	215.06	1.0522
300.00	0.6413	223.99	1.0829	0.6412	224.03	1.0828	0.6410	224.06	1.0827
310.00	0.6512	233.15	1.1129	0.6511	233.18	1.1128	0.6509	233.22	1.1127
320.00	0.6616	242.45	1.1425	0.6615	242.48	1.1424	0.6613	242.52	1.1423
330.00	0.6727	251.90	1.1715	0.6725	251.93	1.1714	0.6723	251.96	1.1713
340.00	0.6844	261.48	1.2001	0.6842	261.51	1.2000	0.6840	261.54	1.1999
350.00	0.6969	271.21	1.2283	0.6967	271.24	1.2282	0.6964	271.26	1.2281
360.00	0.7103	281.09	1.2562	0.7101	281.11	1.2560	0.7098	281.13	1.2559
370.00	36.560	417.92	1.6269	0.7246	291.13	1.2835	0.7243	291.15	1.2833
380.00	37.982	425.60	1.6474	29.425	424.01	1.6344	0.7402	301.33	1.3105
390.00	39.367	433.32	1.6674	30.614	431.86	1.6548	24.732	430.31	1.6438
400.00	40.723	441.08	1.6871	31.768	439.75	1.6748	25.761	438.33	1.6641
410.00	42.054	448.90	1.7064	32.891	447.67	1.6943	26.754	446.37	1.6839
420.00	43.363	456.77	1.7254	33.991	455.63	1.7135	27.720	454.43	1.7033
430.00	44.654	464.70	1.7440	35.069	463.63	1.7323	28.661	462.52	1.7224
440.00	45.929	472.69	1.7624	36.130	471.68	1.7508	29.582	470.65	1.7411
450.00	47.189	480.72	1.7804	37.175	479.78	1.7690	30.487	478.81	1.7594
460.00	48.438	488.81	1.7982	38.207	487.93	1.7869	31.377	487.02	1.7774
470.00	49.675	496.95	1.8157	39.227	496.12	1.8046	32.253	495.26	1.7952
480.00	50.904	505.14	1.8330	40.237	504.35	1.8219	33.119	503.54	1.8126
490.00	52.123	513.38	1.8499	41.238	512.63	1.8390	33.975	511.86	1.8298
500.00	53.335	521.66	1.8667	42.230	520.95	1.8558	34.823	520.22	1.8466
510.00	54.540	529.99	1.8832	43.216	529.31	1.8723	35.663	528.61	1.8633
520.00	55.739	538.35	1.8994	44.195	537.70	1.8886	36.496	537.04	1.8796
530.00	56.932	546.76	1.9154	45.168	546.14	1.9047	37.322	545.51	1.8958
p/MPa	0.80			1.00			2.00		
270.00	0.6138	197.58	0.9893	0.6136	197.65	0.9891	0.6128	198.04	0.9883
280.00	0.6224	206.28	1.0209	0.6222	206.35	1.0208	0.6213	206.73	1.0199
290.00	0.6314	215.13	1.0520	0.6312	215.20	1.0518	0.6302	215.57	1.0509
300.00	0.6408	224.13	1.0825	0.6406	224.20	1.0823	0.6395	224.56	1.0814
310.00	0.6507	233.29	1.1125	0.6504	233.35	1.1123	0.6491	233.69	1.1113
320.00	0.6610	242.58	1.1420	0.6607	242.64	1.1418	0.6593	242.97	1.1408
330.00	0.6720	252.02	1.1711	0.6716	252.08	1.1709	0.6700	252.38	1.1697
340.00	0.6836	261.59	1.1997	0.6832	261.65	1.1994	0.6813	261.93	1.1982
350.00	0.6960	271.31	1.2278	0.6955	271.36	1.2276	0.6934	271.61	1.2263
360.00	0.7093	281.17	1.2556	0.7088	281.21	1.2553	0.7063	281.43	1.2540

Chart R113.3: Thermodynamic properties–*continued*

p/MPa	0.80			1.00			2.00		
T K	v dm³/kg	h kJ/kg	s kJ/kgK	v dm³/kg	h kJ/kg	s kJ/kgK	v dm³/kg	h kJ/kg	s kJ/kgK
370.00	0.7237	291.18	1.2830	0.7231	291.21	1.2827	0.7201	291.39	1.2813
380.00	0.7394	301.35	1.3102	0.7387	301.38	1.3098	0.7352	301.50	1.3082
390.00	0.7568	311.70	1.3370	0.7559	311.71	1.3367	0.7516	311.78	1.3349
400.00	0.7763	322.26	1.3638	0.7751	322.25	1.3634	0.7699	322.23	1.3614
410.00	19.003	443.52	1.6660	0.7970	333.02	1.3899	0.7903	332.88	1.3877
420.00	19.819	451.84	1.6860	14.993	448.91	1.6708	0.8138	343.78	1.4139
430.00	20.602	460.14	1.7056	15.702	457.50	1.6911	0.8413	354.97	1.4403
440.00	21.360	468.45	1.7247	16.376	466.05	1.7107	0.8750	366.57	1.4669
450.00	22.095	476.77	1.7434	17.022	474.56	1.7298	0.9189	378.79	1.4944
460.00	22.813	485.11	1.7617	17.645	483.07	1.7485	6.8915	469.30	1.6944
470.00	23.516	493.47	1.7797	18.249	491.58	1.7668	7.4184	479.47	1.7162
480.00	24.206	501.86	1.7973	18.838	500.09	1.7848	7.8836	489.23	1.7368
490.00	24.884	510.28	1.8147	19.414	508.62	1.8023	8.3080	498.74	1.7564
500.00	25.553	518.72	1.8318	19.978	517.16	1.8196	8.7031	508.09	1.7753
510.00	26.212	527.19	1.8485	20.532	525.72	1.8366	9.0762	517.32	1.7936
520.00	26.865	535.69	1.8650	21.078	534.30	1.8532	9.4318	526.48	1.8114
530.00	27.510	544.22	1.8813	21.616	542.89	1.8696	9.7734	535.57	1.8287
p/MPa	4.00			6.00			8.00		
270.00	0.6112	198.82	0.9866	0.6097	199.60	0.9850	0.6083	200.39	0.9834
280.00	0.6196	207.49	1.0182	0.6179	208.25	1.0165	0.6163	209.03	1.0148
290.00	0.6283	216.31	1.0491	0.6264	217.06	1.0474	0.6246	217.81	1.0457
300.00	0.6373	225.27	1.0795	0.6352	226.00	1.0777	0.6332	226.74	1.0759
310.00	0.6467	234.38	1.1094	0.6444	235.09	1.1075	0.6421	235.80	1.1056
320.00	0.6565	243.63	1.1387	0.6539	244.30	1.1367	0.6514	244.99	1.1348
330.00	0.6669	253.01	1.1676	0.6639	253.65	1.1655	0.6611	254.31	1.1635
340.00	0.6778	262.51	1.1960	0.6744	263.12	1.1938	0.6712	263.74	1.1917
350.00	0.6893	272.14	1.2239	0.6855	272.71	1.2216	0.6819	273.29	1.2193
360.00	0.7015	281.90	1.2514	0.6972	282.42	1.2489	0.6931	282.96	1.2466
370.00	0.7146	291.79	1.2785	0.7096	292.24	1.2758	0.7049	292.73	1.2733
380.00	0.7287	301.82	1.3052	0.7228	302.19	1.3024	0.7175	302.62	1.2997
390.00	0.7439	311.98	1.3316	0.7370	312.27	1.3285	0.7308	312.63	1.3257
400.00	0.7605	322.30	1.3577	0.7524	322.48	1.3544	0.7451	322.74	1.3513
410.00	0.7788	332.77	1.3836	0.7690	332.82	1.3799	0.7605	332.98	1.3766
420.00	0.7992	343.43	1.4093	0.7873	343.31	1.4052	0.7772	343.35	1.4016
430.00	0.8223	354.30	1.4348	0.8074	353.96	1.4303	0.7953	353.84	1.4263
440.00	0.8488	365.42	1.4604	0.8300	364.80	1.4552	0.8151	364.48	1.4507
450.00	0.8803	376.84	1.4861	0.8555	375.83	1.4800	0.8370	375.27	1.4750
460.00	0.9189	388.68	1.5121	0.8849	387.10	1.5047	0.8615	386.22	1.4990
470.00	0.9692	401.11	1.5388	0.9195	398.65	1.5296	0.8890	397.35	1.5230
480.00	1.0421	414.58	1.5672	0.9612	410.55	1.5546	0.9203	408.68	1.5468
490.00	1.1800	430.60	1.6002	1.0135	422.91	1.5801	0.9565	420.23	1.5706
500.00	1.9117	462.32	1.6642	1.0824	435.91	1.6064	0.9989	432.05	1.5945
510.00	2.7903	487.35	1.7138	1.1794	449.86	1.6340	1.0494	444.16	1.6185
520.00	3.2716	502.34	1.7429	1.3260	465.16	1.6637	1.1104	456.61	1.6427
530.00	3.6342	514.90	1.7669	1.5469	481.91	1.6956	1.1852	469.45	1.6671

Chart R113.3: Thermodynamic properties–*continued*

p/MPa	10.00			20.00			24.30		
T K	v dm³/kg	h kJ/kg	s kJ/kgK	v dm³/kg	h kJ/kg	s kJ/kgK	v dm³/kg	h kJ/kg	s kJ/kgK
270.00	0.6068	201.18	0.9818	0.6002	205.20	0.9744	0.5975	206.96	0.9714
280.00	0.6147	209.80	1.0132	0.6074	213.77	1.0056	0.6045	215.51	1.0025
290.00	0.6228	218.57	1.0440	0.6148	222.48	1.0361	0.6117	224.20	1.0329
300.00	0.6312	227.48	1.0742	0.6224	231.32	1.0661	0.6190	233.01	1.0628
310.00	0.6399	236.53	1.1038	0.6302	240.28	1.0955	0.6265	241.95	1.0921
320.00	0.6490	245.70	1.1330	0.6383	249.36	1.1243	0.6342	251.00	1.1209
330.00	0.6584	254.98	1.1615	0.6465	258.55	1.1526	0.6421	260.15	1.1490
340.00	0.6682	264.39	1.1896	0.6551	267.84	1.1803	0.6502	269.41	1.1767
350.00	0.6785	273.91	1.2172	0.6639	277.22	1.2075	0.6586	278.75	1.2037
360.00	0.6893	283.53	1.2443	0.6731	286.70	1.2342	0.6672	288.19	1.2303
370.00	0.7006	293.26	1.2710	0.6825	296.27	1.2604	0.6761	297.70	1.2564
380.00	0.7125	303.09	1.2972	0.6924	305.92	1.2861	0.6853	307.30	1.2820
390.00	0.7252	313.03	1.3230	0.7026	315.64	1.3114	0.6949	316.96	1.3071
400.00	0.7386	323.08	1.3484	0.7132	325.45	1.3362	0.7047	326.71	1.3318
410.00	0.7530	333.23	1.3735	0.7243	335.34	1.3606	0.7149	336.52	1.3560
420.00	0.7684	343.50	1.3982	0.7359	345.29	1.3846	0.7255	346.39	1.3798
430.00	0.7849	353.87	1.4227	0.7480	355.32	1.4082	0.7365	356.34	1.4032
440.00	0.8028	364.37	1.4468	0.7607	365.42	1.4315	0.7480	366.34	1.4262
450.00	0.8222	374.98	1.4706	0.7740	375.59	1.4543	0.7599	376.40	1.4488
460.00	0.8435	385.72	1.4942	0.7879	385.82	1.4768	0.7722	386.52	1.4710
470.00	0.8668	396.59	1.5176	0.8025	396.11	1.4989	0.7851	396.69	1.4929
480.00	0.8927	407.60	1.5408	0.8178	406.46	1.5207	0.7985	406.92	1.5144
490.00	0.9214	418.76	1.5638	0.8339	416.87	1.5422	0.8125	417.19	1.5356
500.00	0.9536	430.06	1.5866	0.8508	427.33	1.5633	0.8270	427.50	1.5564
510.00	0.9898	441.53	1.6094	0.8685	437.84	1.5841	0.8421	437.85	1.5769
520.00	1.0309	453.17	1.6320	0.8872	448.39	1.6046	0.8578	448.24	1.5971
530.00	1.0777	464.98	1.6545	0.9068	458.99	1.6248	0.8742	458.66	1.6170

References for R113

[Barh66] BARHO, W.: Berechnung der Molwärme der Fluor-Chlor-Derivate von Methan und Ethan im Zustand idealer Gase; Dissertation Universität Karlsruhe (1966)

[Benn40] BENNING, A.F. AND MCHARNESS, R.C.: Thermodynamic Properties of Fluorochloromethanes and -Ethanes; Ind. Eng. Chem. 32(1940)4 497/499

[Benn40a] BENNING, A.F. AND MCHARNESS, R.C.: Thermodynamic Properties of Fluorochloromethanes and -Ethanes; Ind. Eng. Chem. 32(1940)5 698/701

[Benn40b] BENNING, A.F. AND MCHARNESS, R.C.: Thermodynamic Properties of Fluorochloromethanes and -Ethanes; Ind. Eng. Chem. 32(1940)6 814/816

[Benn40c] BENNING, A.F. ; MCHARNESS, R.C. ; MARKWOOD, W.H.,JR. AND SMITH, W.J.: Thermodynamic Properties of Fluorochloromethanes and -Ethanes; Ind. Eng. Chem. 32(1940)7 976/980

[Erns70] ERNST, G. AND BÜSSER, J.: Ideal and real gas state heat capacities ; J. Chem. Therm. (1970)2 787/791

[Gell73] GELLER, V.Z.: Investigation of Thermodynamic Properties of Freon 113; Thermophysical Properties of Matter and Substances 7(1973) 135/154

[Gell73a] GELLER, V.Z. ; PORICHANSKIJ, E.G. AND PARAMONOV, I.A.: Thermal Properties of Freon-113; Voprosy Techniceskoj Teplofiziki (1973)4 63/67

[Hira63] HIRAOKA, H. AND HILDEBRAND, J.H.: Solubility Relations of the Isomeric Trichlorotrifluoroethanes; J. Phys. Chem. 67(1963) 916/918

[KMR 81] - : Kältemaschinenregeln, 7. Auflage; Verlag C.F. Müller, Karlsruhe (1981)

[Koke71] KOKERNAL, R.P. AND FELDMAN, C.L.: The velocity of sound in Liquid R12; ASHRAE Journal (1951) 59/72

[Kole81] KOLESOV, V.P. ;KOSARUKINA, E.A. ; ZHOGIN, D.JU. ; POLOZNIKOVA, M.E. AND PENTIN, JU.A.: Heat capacities, phase transitions and thermodynamic functions of 1,1,2,2-tetrafluoro 1,2-dichloro-ethane and 1,1,2-trifluoro 1,2,2-trichloroethane; J. Chem. Therm. 13(1981) 115/129

[Mast78] MASTROIANNI, M.J. ; STAHL, R.F. AND SHELDON, P.N.: Physical and thermodynamic properties of 1,1,2-Trifluorotrichloroethane; J. Chem. Eng. Data 23(1978)2 113/118

[Meye69] MEYER, K.J.: Über den Zusammenhang zwischen der Schallgeschwindigkeit und der Wärmeleitfähigkeit bei flüssigen Fluor-Chlor-Derivaten des Methans und Äthans; Kältetechnik - Klimatisierung 21(1969)9 270/275

[Ried38] RIEDEL, L.: Bestimmung der thermischen Eigenschaften von Trifluor-Trichloräthan; Zeitschrift für die gesamte Kälteindustrie 45(1938)12 221/227

[Sche54] SCHEINDLIN, A. AND BUBUSCHJAN, M.: Experimentelle Bestimmung der Dichte und der Wärmekapazität cp des flüssigen Freons-113; Cholodilnaja Technika 31(1954)4 53/55

[Shim81] SHIMANSKAYA, E.T. ; BEZRUCHKO, I.V. ; BASOK, V.I. AND SHIMANSKIJ, JU.I.: Experimental Determination of the Critical Exponent and of the Asymetric and Nonasymetric Corrections to the Equation of the Coexistence Curve of Freon-113; Sov. Phys. JETP 51(1981)1 139/148

[Vesl79] VESLOGUZOV, Y.A.: Experimental Study of Isobaric Heat Capacity of Freon 113 Vapors; Teplofiziceskie svojstva vescestv i materialov 13(1979) 42/46

R 113. ph

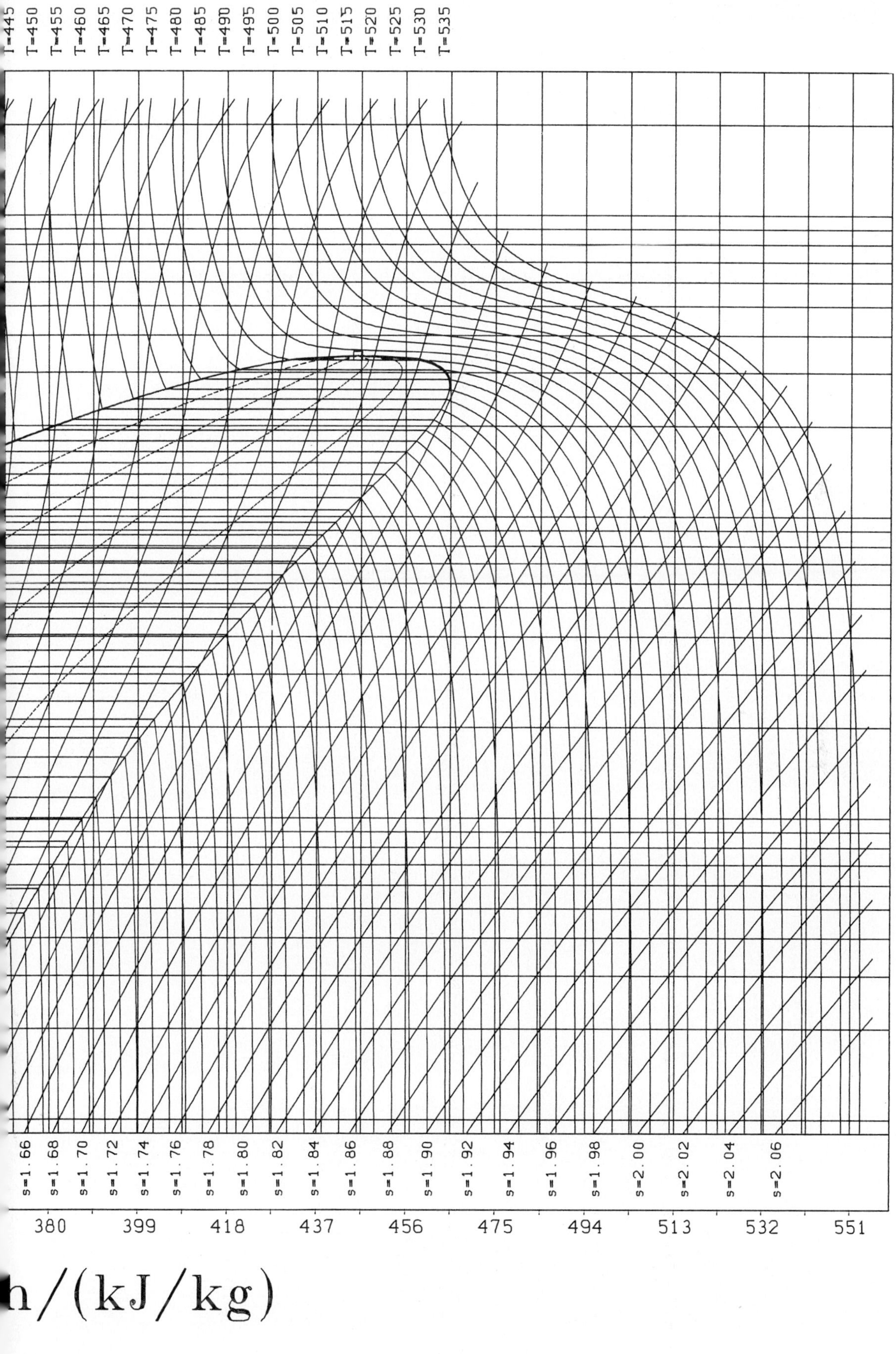

T=445
T=450
T=455
T=460
T=465
T=470
T=475
T=480
T=485
T=490
T=495
T=500
T=505
T=510
T=515
T=520
T=525
T=530
T=535
s=1.66
s=1.68
s=1.70
s=1.72
s=1.74
s=1.76
s=1.78
s=1.80
s=1.82
s=1.84
s=1.86
s=1.88
s=1.90
s=1.92
s=1.94
s=1.96
s=1.98
s=2.00
s=2.02
s=2.04
s=2.06
380
399
418
437
456
475
494
513
532
551
h/(kJ/kg)

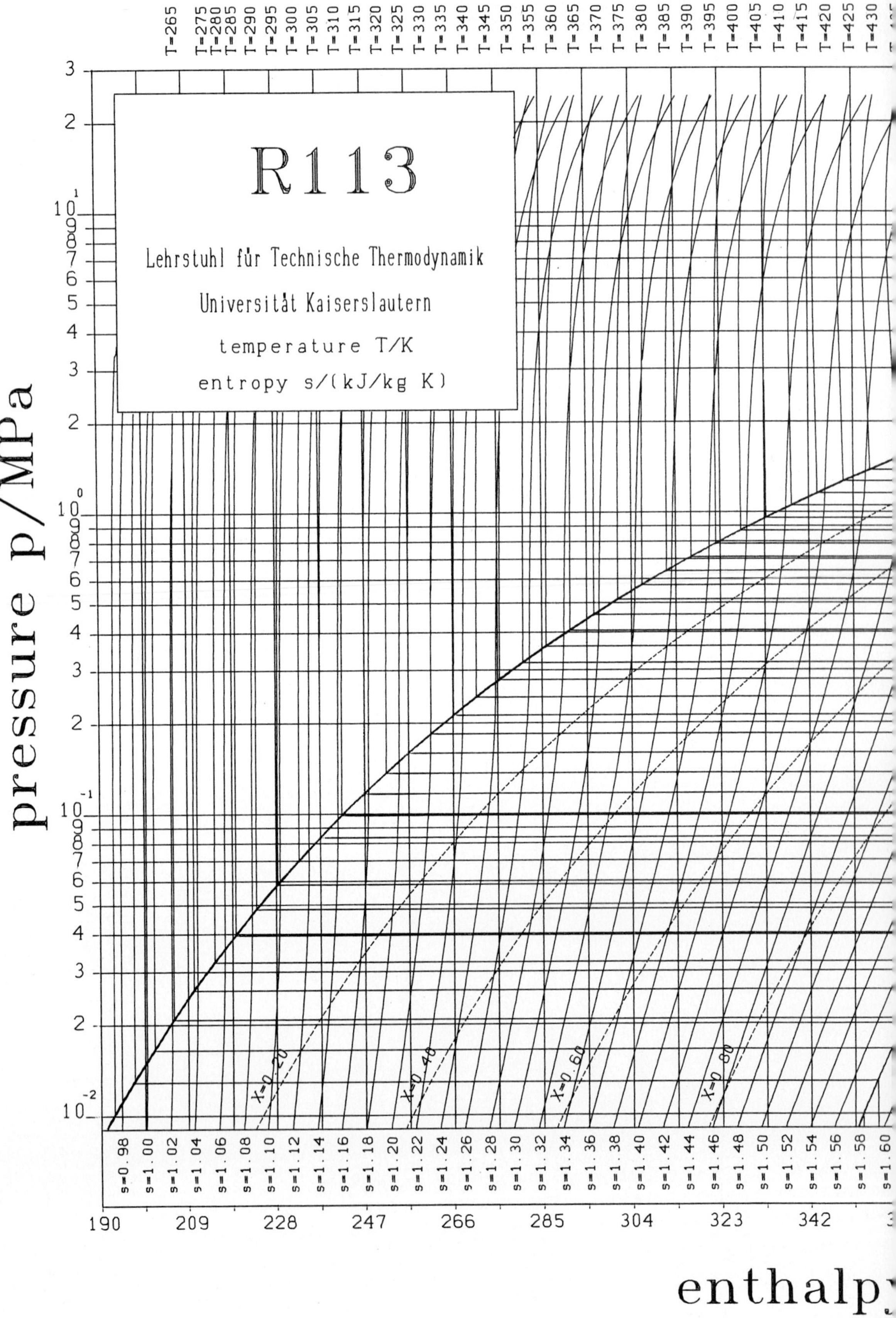
R113
Lehrstuhl für Technische Thermodynamik
Universität Kaiserslautern
temperature T/K
entropy s/(kJ/kg K)
pressure p/MPa
enthalp
T=265 T=275 T=280 T=285 T=290 T=295 T=300 T=305 T=310 T=315 T=320 T=325 T=330 T=335 T=340 T=345 T=350 T=355 T=360 T=365 T=370 T=375 T=380 T=385 T=390 T=395 T=400 T=405 T=410 T=415 T=420 T=425 T=430
s=0.98 s=1.00 s=1.02 s=1.04 s=1.06 s=1.08 s=1.10 s=1.12 s=1.14 s=1.16 s=1.18 s=1.20 s=1.22 s=1.24 s=1.26 s=1.28 s=1.30 s=1.32 s=1.34 s=1.36 s=1.38 s=1.40 s=1.42 s=1.44 s=1.46 s=1.48 s=1.50 s=1.52 s=1.54 s=1.56 s=1.58 s=1.60
X=0.20 X=0.40 X=0.60 X=0.80
190 209 228 247 266 285 304 323 342

R 113. Ts

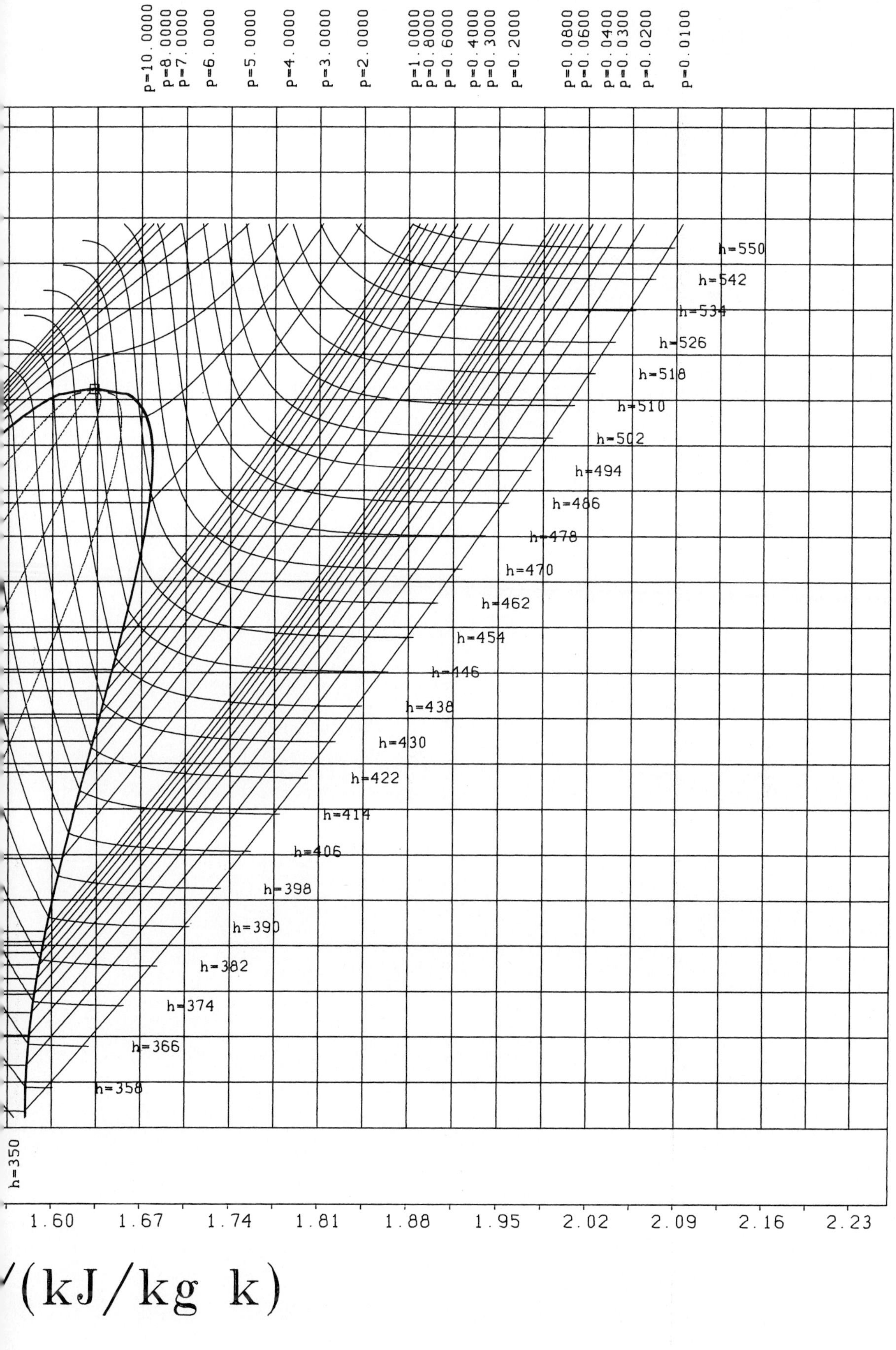
p=10.0000
p=8.0000
p=7.0000
p=6.0000
p=5.0000
p=4.0000
p=3.0000
p=2.0000
p=1.0000
p=0.8000
p=0.6000
p=0.4000
p=0.3000
p=0.2000
p=0.0800
p=0.0600
p=0.0400
p=0.0300
p=0.0200
p=0.0100
h=550
h=542
h=534
h=526
h=518
h=510
h=502
h=494
h=486
h=478
h=470
h=462
h=454
h=446
h=438
h=430
h=422
h=414
h=406
h=398
h=390
h=382
h=374
h=366
h=358
h=350
1.60
1.67
1.74
1.81
1.88
1.95
2.02
2.09
2.16
2.23
/(kJ/kg k)

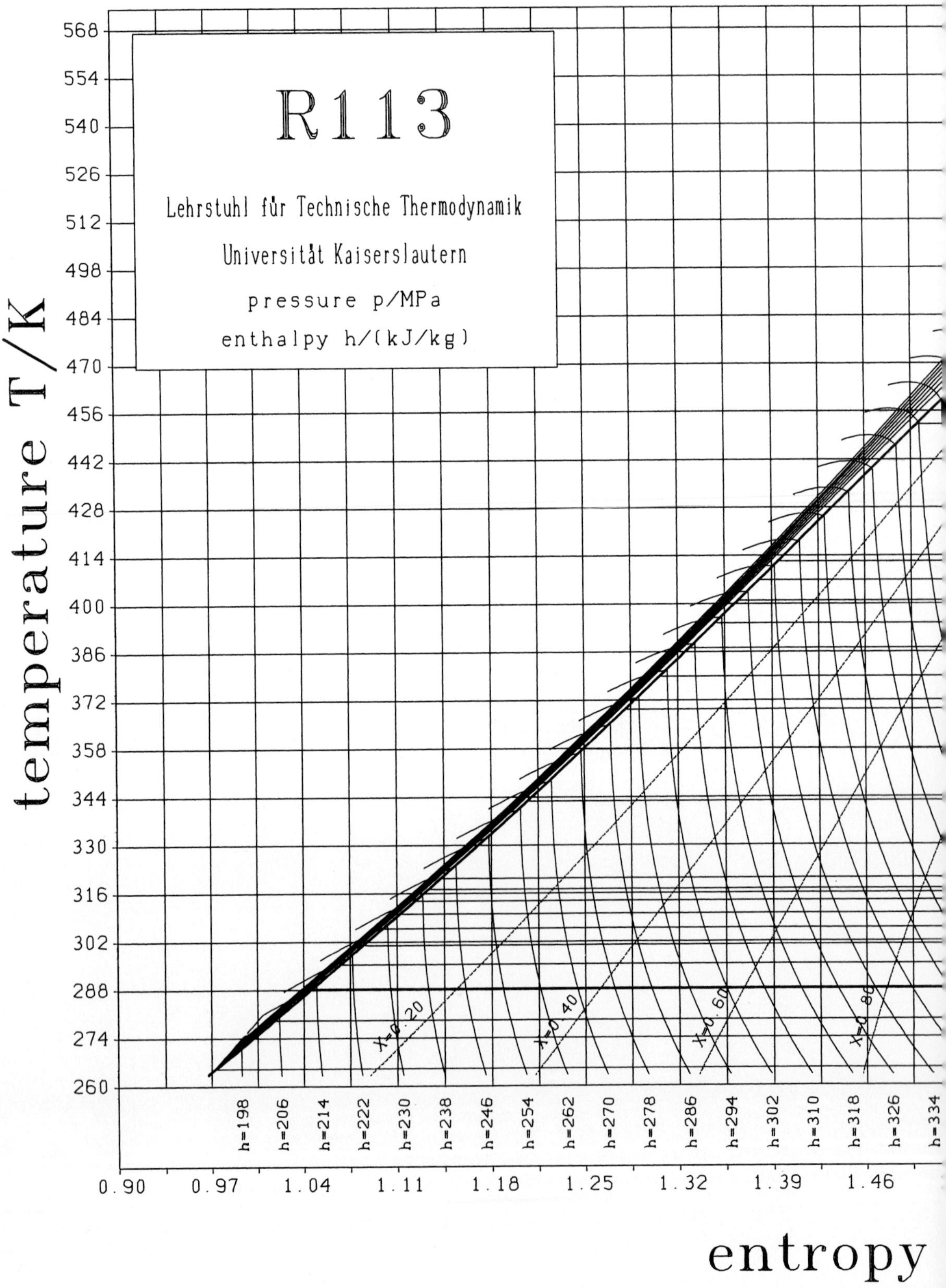

R113
Lehrstuhl für Technische Thermodynamik
Universität Kaiserslautern
pressure p/MPa
enthalpy h/(kJ/kg)
temperature T/K
568
554
540
526
512
498
484
470
456
442
428
414
400
386
372
358
344
330
316
302
288
274
260
X=0.20
X=0.40
X=0.60
X=0.80
h=198
h=206
h=214
h=222
h=230
h=238
h=246
h=254
h=262
h=270
h=278
h=286
h=294
h=302
h=310
h=318
h=326
h=334
0.90
0.97
1.04
1.11
1.18
1.25
1.32
1.39
1.46
entropy

R 113. hs

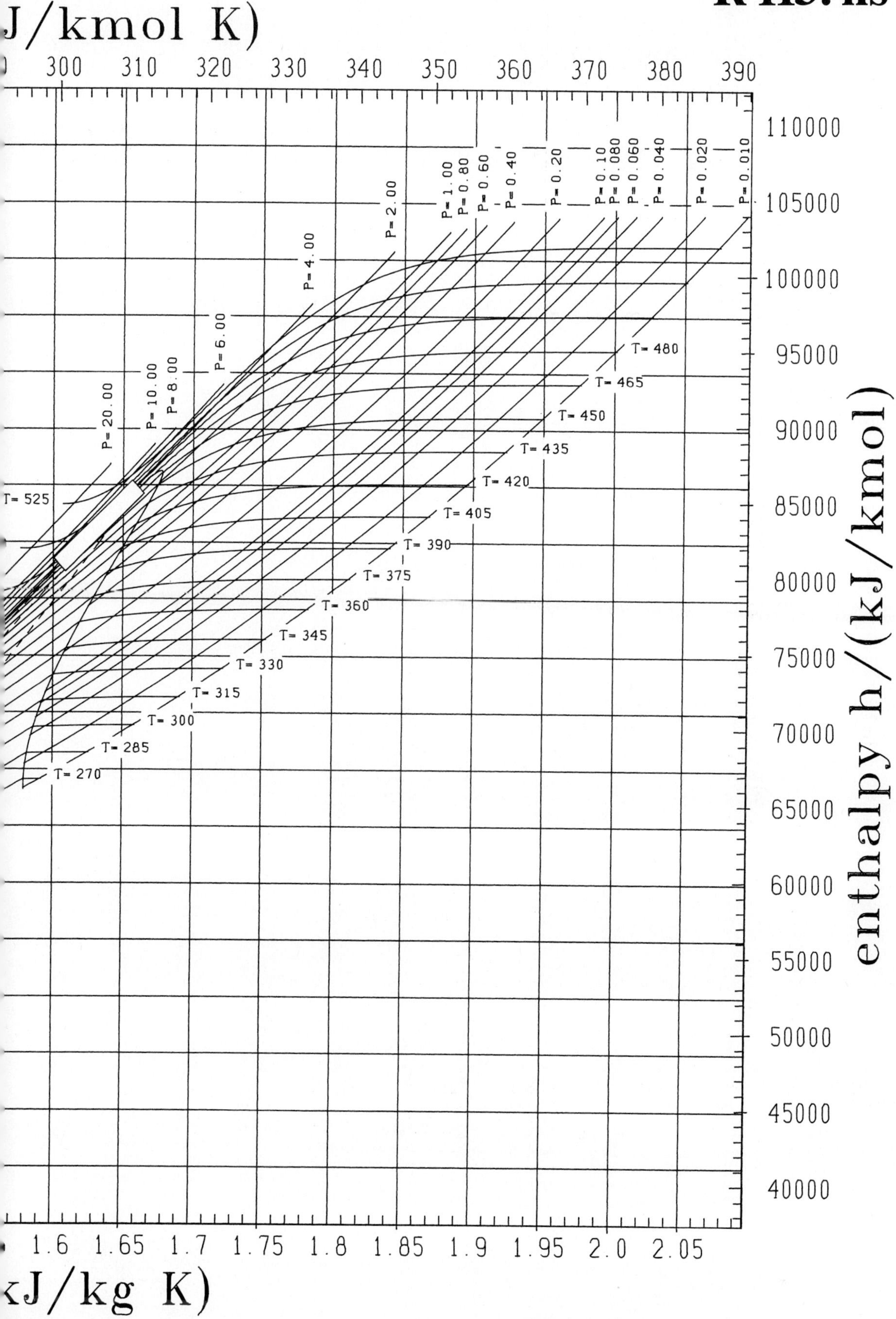
J/kmol K)
300
310
320
330
340
350
360
370
380
390
110000
105000
100000
95000
90000
85000
80000
75000
70000
65000
60000
55000
50000
45000
40000
enthalpy h/(kJ/kmol)
P= 20.00
P= 10.00
P= 8.00
P= 6.00
P= 4.00
P= 2.00
P= 1.00
P= 0.80
P= 0.60
P= 0.40
P= 0.20
P= 0.10
P= 0.080
P= 0.060
P= 0.040
P= 0.020
P= 0.010
T= 525
T= 480
T= 465
T= 450
T= 435
T= 420
T= 405
T= 390
T= 375
T= 360
T= 345
T= 330
T= 315
T= 300
T= 285
T= 270
1.6
1.65
1.7
1.75
1.8
1.85
1.9
1.95
2.0
2.05
kJ/kg K)

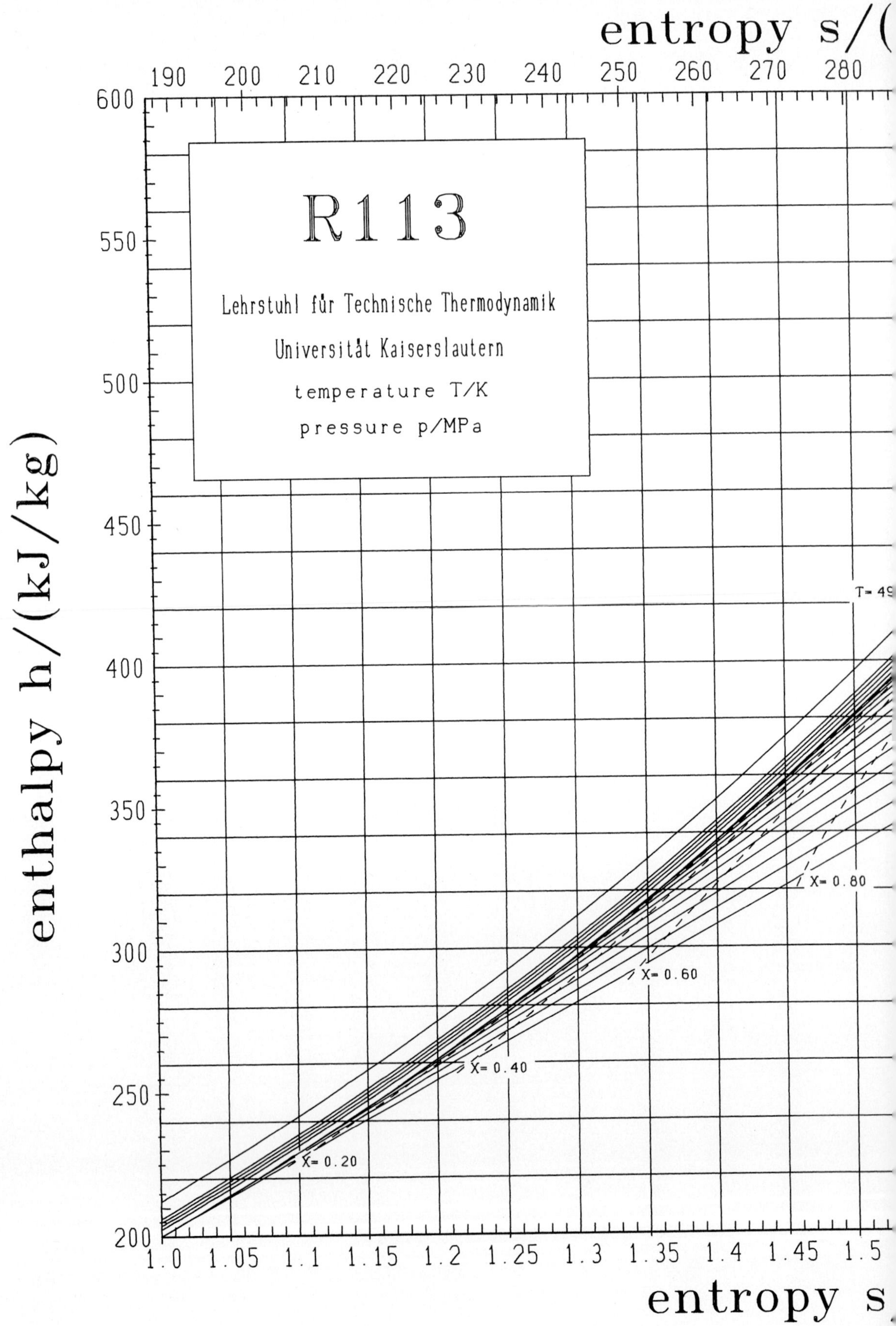

entropy s/(
190 200 210 220 230 240 250 260 270 280
R113
Lehrstuhl für Technische Thermodynamik
Universität Kaiserslautern
temperature T/K
pressure p/MPa
enthalpy h/(kJ/kg)
600
550
500
450
400
350
300
250
200
X= 0.20
X= 0.40
X= 0.60
X= 0.80
1.0 1.05 1.1 1.15 1.2 1.25 1.3 1.35 1.4 1.45 1.5
entropy s

Refrigerant R114 ($CClF_2CClF_2$)

Table R114.1: List of references for p, ρ, T data

Author	n	p/MPa min	p/MPa max	ρ/g cm^{-3} min	ρ/g cm^{-3} max	T/K min	T/K max
GELLER, Z.I. *[Gell 75]*	24	0.93	20.62	1.1341	1.38	323	433
MARTIN, J.J. *[Mart 60]*	47	1.69	10.91	0.1310	0.94	390	486
MCHARNESS, R.C. *[McHa 51]*	17	0.10	2.14	0.0073	0.17	298	444
MEARS, W.H. *[Mear 55]*	26	1.06	3.94	0.0821	0.32	361	444
WILSON D.P. *[Wils 81]*	99	0.67	10.45	0.0632	1.42	310	507

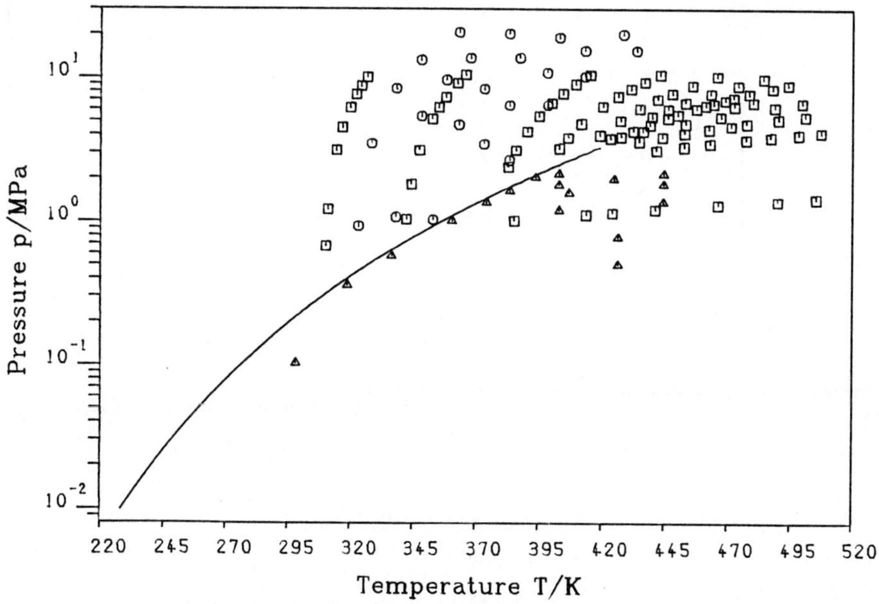

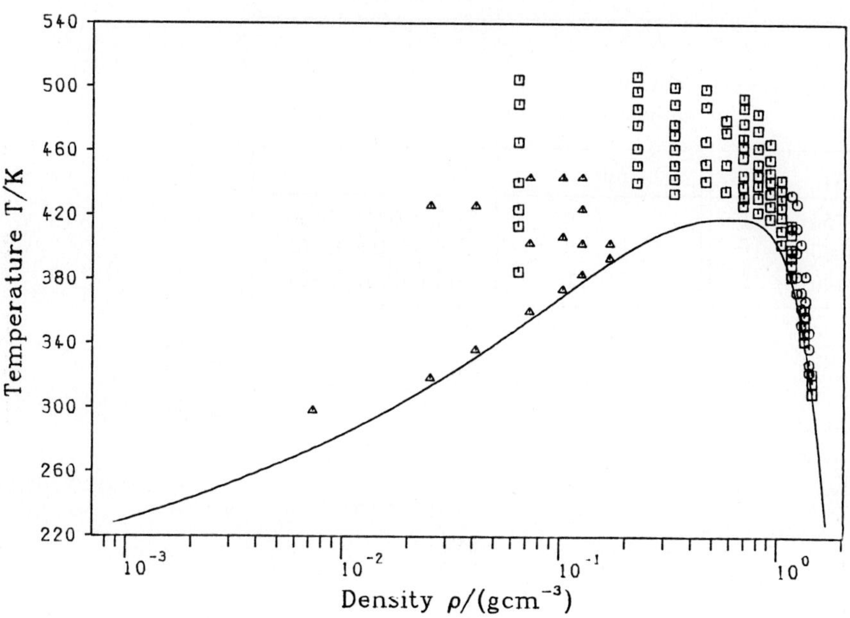

□ Wilson, D.P. /Wils81/ ○ Geller, Z.I, Porichanskii, E.G./Gell75/ △ McHarness, R.C. /McHa51/

Fig.R114.1: Distribution of p, ρ, T data projected in the p, T and in the T, ρ plane

Table R114.2: Experimental vapor pressure data

Author	Δ / % std	opt	*EOS*
Benning, A.F. *[Benn 44]*	0.77	0.76	0.77
Hasegawa, N. *[Hase 85]*	0.19	0.19	0.18
Martin, J.J. *[Mart 60]*	0.99	1.20	0.59
McHarness, R.C. *[McHa 51]*	2.10	2.70	1.20
Thornton, N.V. *[Thor 33]*	4.20	4.40	3.50
Wilson, D.P. *[Wils 81]*	0.33	0.33	0.33

Table R114.3: Experimental saturated liquid density data

Author	Δ / % std	opt	*EOS*
Benning, A.F. *[Benn 44]*	0.12	0.04	0.11
Locke, E.G. *[Lock 34]*	0.11	0.12	0.10
Martin, J.J. *[Mart 60]*	0.83	0.57	3.10
McHarness, R.C. *[McHa 51]*	0.13	0.08	0.18
Wilson, D.P. *[Wils 81]*	0.26	0.20	0.51

Table R114.4: Coefficients of the standard vapor pressure equation

Index i	b_i
5	−7.160325 E+00
6	1.232331 E+00
9	−2.584490 E+00
15	−4.239129 E+00

Table R114.5: Coefficients of the quasi-optimal vapor pressure equation

Index i	b_i
5	−7.167204 E+00
6	1.260366 E+00
9	−2.748553 E+00
18	−1.015734 E+01

Table R114.6: Coefficients of the standard saturated liquid density equation

Index i	c_i
8	1.801754 E+00
9	7.044965 E−01
16	4.643545 E−01

Table R114.7: Coefficients of the quasi-optimal saturated liquid density equation

Index i	c_i
8	1.940245 E+00
10	9.341101 E−01
11	−2.055137 E−01

Table R114.8: Coefficients of the *Bender-EOS*

Range of validity: one-phase region

$0.10 \leq p/\text{MPa} \leq 20.62$

$0.00727 \leq \rho/\text{g cm}^{-3} \leq 1.4816$

$298.26 \leq T/\text{K} \leq 507.15$

	Specific units	Molar units
a_{01}	9.052499776 E−02	2.644874734 E+00
a_{02}	6.998626571 E+01	2.044793267 E+03
a_{03}	2.282318990 E+03	6.668266205 E+04
a_{04}	7.484416422 E+05	2.186726803 E+07
a_{05}	4.259192099 E+07	1.244410920 E+09
a_{06}	−5.405537631 E−02	−2.699564901 E−01
a_{07}	3.731526704 E+01	1.863551640 E+02
a_{08}	5.877697853 E+03	2.935365158 E+04
a_{09}	2.066944940 E−02	1.764420432 E−02
a_{10}	3.384478525 E+01	2.889115693 E+01
a_{11}	7.920822641 E−02	1.155744278 E−02
a_{12}	−1.258926192 E+02	−1.8369263010 E+01
a_{13}	4.928029861 E+01	1.229088305 E+00
a_{14}	−4.423913771 E+07	−2.209334789 E+08
a_{15}	3.985084931 E+10	1.990180467 E+11
a_{16}	−9.201888477 E+12	−4.595490190 E+13
a_{17}	1.276936730 E+08	1.8632058630 E+07
a_{18}	−1.309313891 E+11	−1.910448076 E+10
a_{19}	3.243627152 E+13	4.732846184 E+12
a_{20}	3.600000000 E+00	1.051814336 E−01

Table R114.9: Deviations between experimental and predicted p, ρ, T data

Region	n	Δ / %	
		p	ρ
Gas	40	0.29	0.39
Critical	37	0.14	0.29
Liquid	58	3.83	0.11
Total	135	2.70	0.29

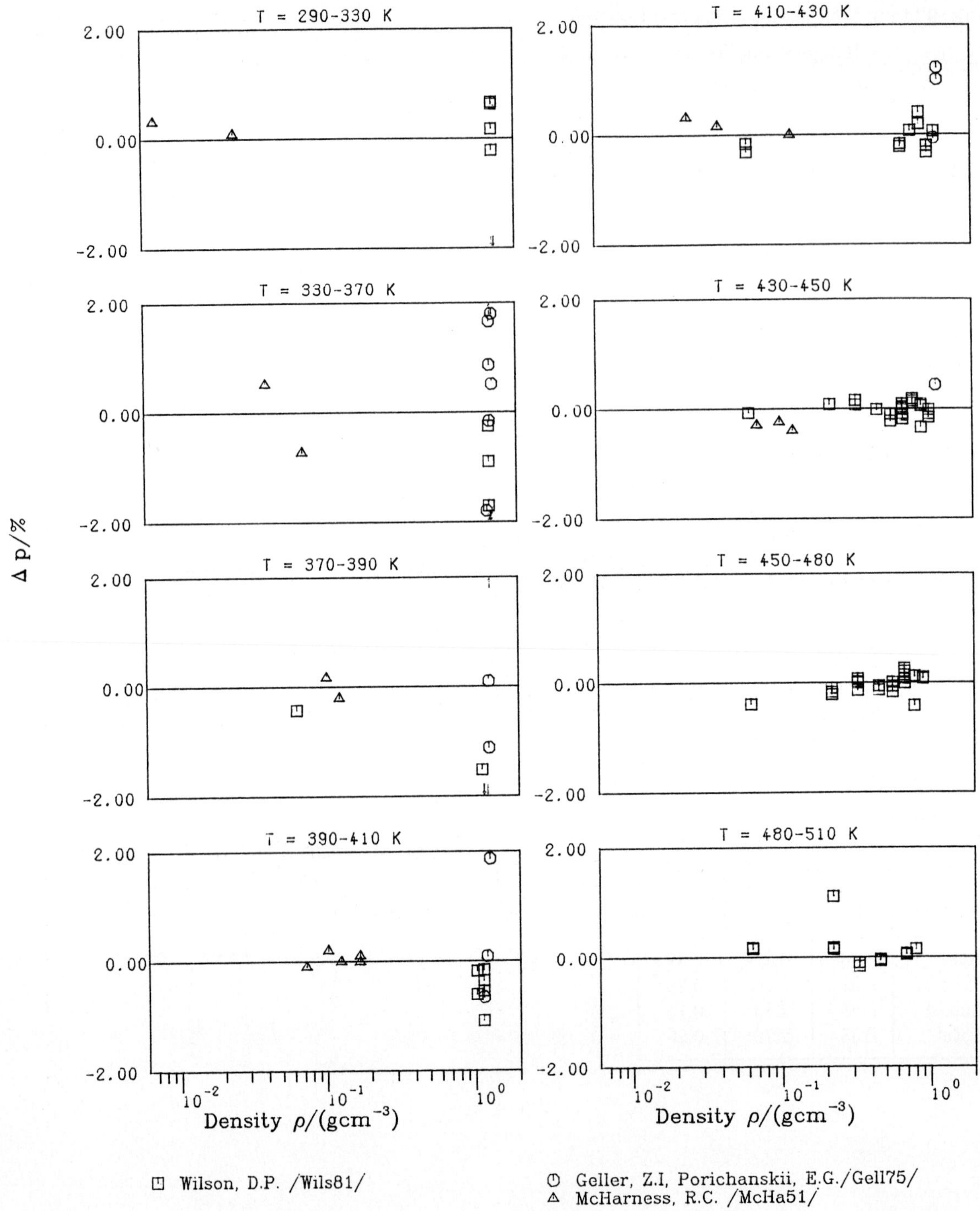

Fig.R114.2: Deviations between experimental and calculated p, ρ, T data

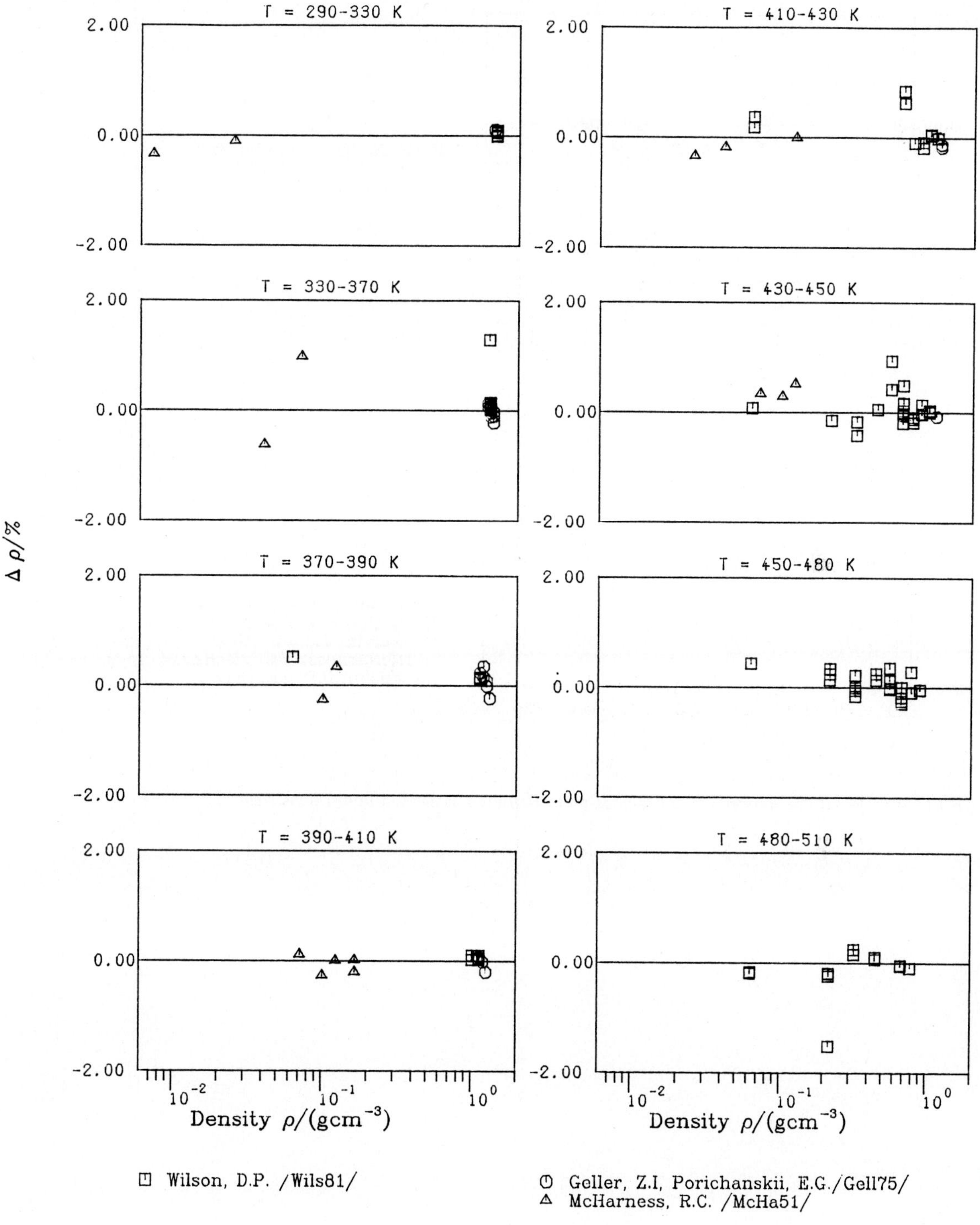

Fig.R114.2: Deviations between experimental and calculated p, ρ, T data *-continued*

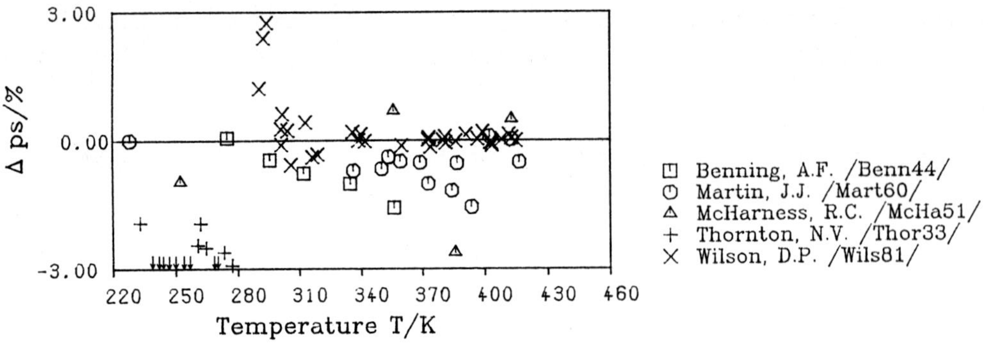

Fig.R114.3: Deviations between experimental and calculated vapor pressures

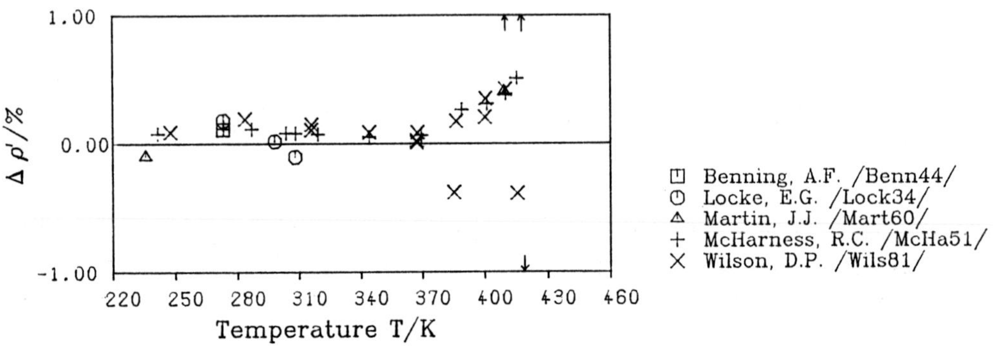

Fig.R114.4: Deviations between experimental and calculated saturated liquid densities

Table R114.10: Information on ideal gas state heat capacity equation

Author	T / K min	T / K max
BARHO, W. *[Barh 66]*	200	550

Coefficients of equation (15)	
Index i	d_i
1	9.765138 E−02
2	3.240861 E−03
3	−5.895364 E−06
4	6.737929 E−09
5	−3.546364 E−12

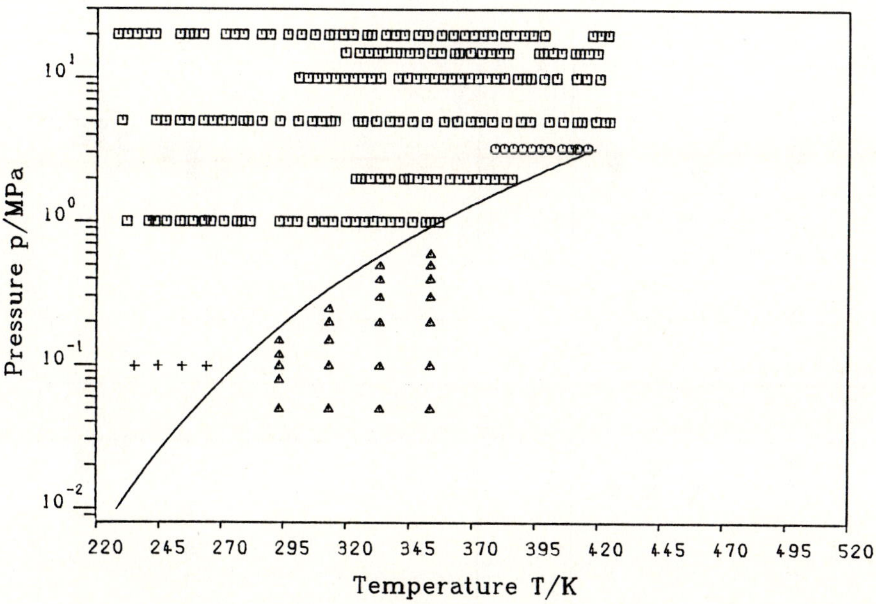

□ cp: Ponomareva, O.P. /Pono77/ △ cp: Ernst, G. /Erns69/
⊙ cp: Ponomareva, O.P. crit. isob/Pono77/ + w: Meyer, K.J. /Meye69/

Fig.R114.5: Distribution of $c_p(p, T)$ and $w(p, T)$ data

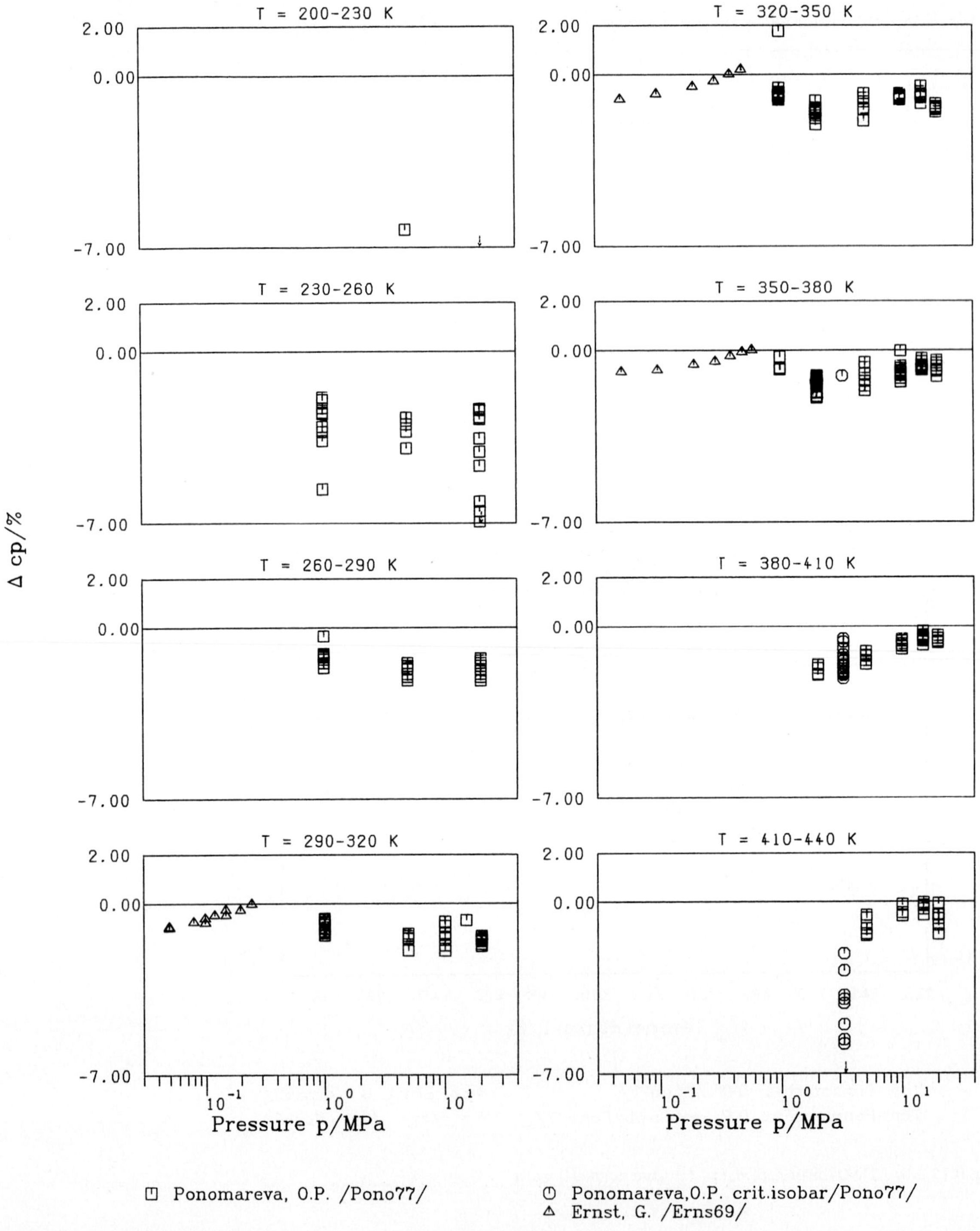

Fig.R114.6: Deviations between experimental and calculated specific isobaric heat capacities c_p

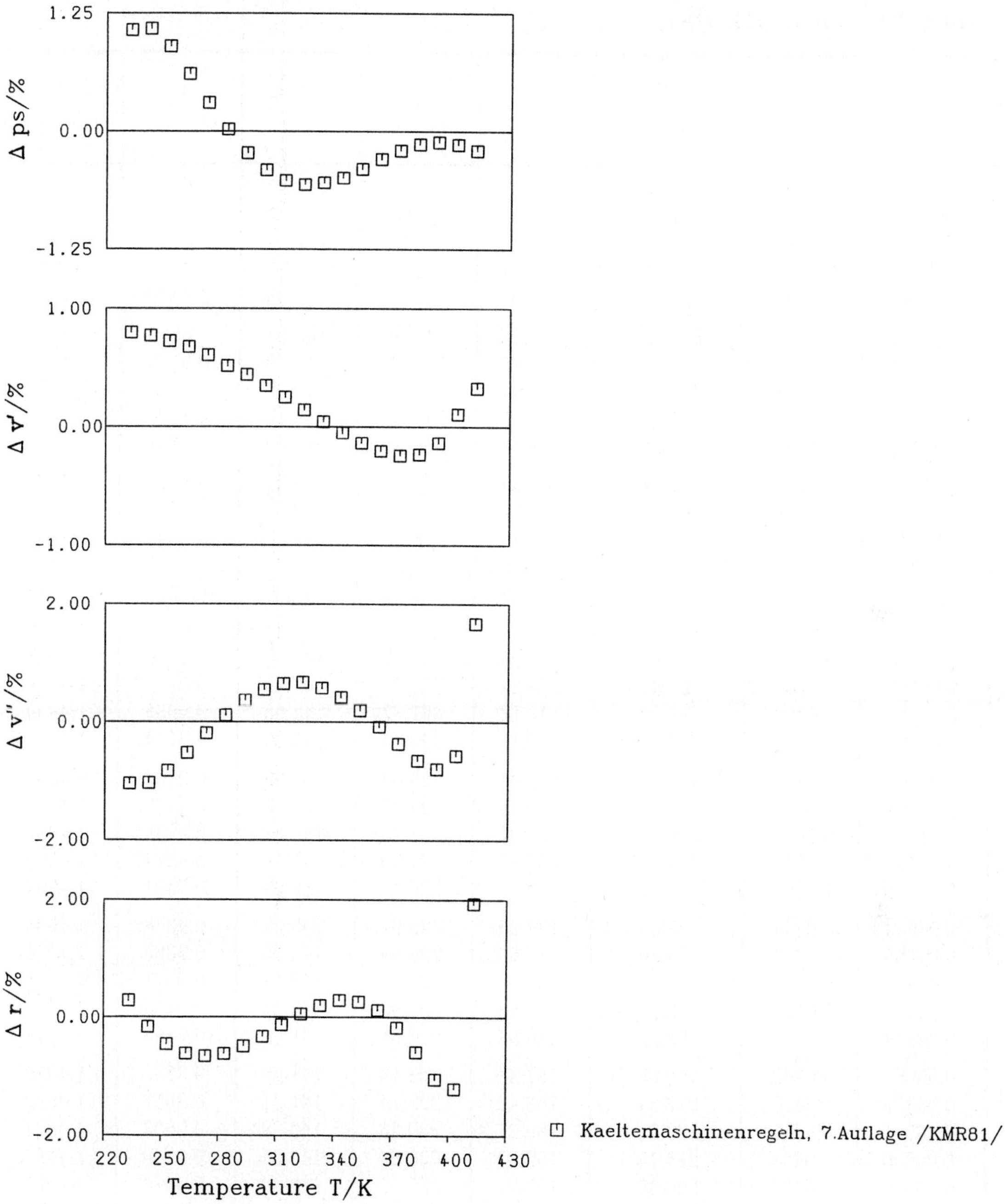

Fig.R114.7: Deviations between saturation properties from the *Kältemaschinenregeln* and those calculated by the *Bender-EOS*

Chart R114.1: Saturation properties

T K	p MPa	v' dm³/kg	v'' dm³/kg	h' kJ/kg	h'' kJ/kg	r kJ/kg	s' kJ/kgK	s'' kJ/kgK
230.0	0.01095	0.6075	1014.9	159.23	311.04	151.80	0.8378	1.4978
231.0	0.01161	0.6084	960.49	160.18	311.63	151.45	0.8419	1.4975
232.0	0.01231	0.6094	909.53	161.12	312.22	151.10	0.8459	1.4972
233.0	0.01305	0.6104	861.77	162.06	312.81	150.75	0.8500	1.4970
234.0	0.01382	0.6114	816.98	163.00	313.41	150.41	0.8540	1.4968
235.0	0.01462	0.6124	774.96	163.95	314.01	150.06	0.8580	1.4966
236.0	0.01547	0.6134	735.51	164.89	314.60	149.72	0.8620	1.4964
237.0	0.01635	0.6144	698.45	165.83	315.20	149.37	0.8660	1.4963
238.0	0.01727	0.6154	663.61	166.77	315.80	149.03	0.8700	1.4962
239.0	0.01824	0.6164	630.84	167.71	316.40	148.69	0.8739	1.4960
240.0	0.01925	0.6174	600.00	168.65	317.00	148.35	0.8778	1.4960
241.0	0.02030	0.6185	570.96	169.59	317.60	148.01	0.8817	1.4959
242.0	0.02140	0.6195	543.59	170.52	318.20	147.67	0.8856	1.4958
243.0	0.02255	0.6205	517.80	171.46	318.80	147.34	0.8895	1.4958
244.0	0.02375	0.6216	493.46	172.40	319.40	147.00	0.8933	1.4958
245.0	0.02500	0.6226	470.50	173.34	320.01	146.66	0.8972	1.4958
246.0	0.02630	0.6236	448.81	174.28	320.61	146.33	0.9010	1.4958
247.0	0.02765	0.6247	428.33	175.22	321.22	145.99	0.9048	1.4959
248.0	0.02906	0.6258	408.96	176.16	321.82	145.66	0.9086	1.4959
249.0	0.03053	0.6268	390.64	177.10	322.43	145.33	0.9124	1.4960
250.0	0.03205	0.6279	373.31	178.04	323.03	144.99	0.9162	1.4961
251.0	0.03364	0.6290	356.90	178.98	323.64	144.66	0.9199	1.4962
252.0	0.03529	0.6300	341.36	179.92	324.25	144.32	0.9236	1.4964
253.0	0.03700	0.6311	326.63	180.87	324.86	143.99	0.9274	1.4965
254.0	0.03878	0.6322	312.67	181.81	325.46	143.66	0.9311	1.4967
255.0	0.04063	0.6333	299.43	182.75	326.07	143.32	0.9348	1.4968
256.0	0.04255	0.6344	286.86	183.69	326.68	142.99	0.9385	1.4970
257.0	0.04453	0.6355	274.93	184.64	327.29	142.65	0.9421	1.4972
258.0	0.04659	0.6366	263.59	185.58	327.90	142.32	0.9458	1.4974
259.0	0.04873	0.6377	252.82	186.53	328.51	141.99	0.9495	1.4977
260.0	0.05094	0.6389	242.58	187.48	329.13	141.65	0.9531	1.4979
261.0	0.05323	0.6400	232.84	188.42	329.74	141.31	0.9567	1.4982
262.0	0.05561	0.6411	223.58	189.37	330.35	140.98	0.9603	1.4984
263.0	0.05806	0.6423	214.75	190.32	330.96	140.64	0.9640	1.4987
264.0	0.06060	0.6434	206.35	191.27	331.57	140.30	0.9676	1.4990
265.0	0.06323	0.6446	198.35	192.22	332.18	139.97	0.9711	1.4993
266.0	0.06595	0.6457	190.72	193.17	332.80	139.63	0.9747	1.4996
267.0	0.06875	0.6469	183.45	194.12	333.41	139.29	0.9783	1.5000
268.0	0.07165	0.6481	176.52	195.07	334.02	138.95	0.9818	1.5003
269.0	0.07465	0.6493	169.90	196.03	334.64	138.61	0.9854	1.5007
270.0	0.07774	0.6505	163.58	196.98	335.25	138.26	0.9889	1.5010
271.0	0.08094	0.6517	157.54	197.94	335.86	137.92	0.9924	1.5014
272.0	0.08423	0.6529	151.78	198.90	336.48	137.58	0.9960	1.5018
273.0	0.08763	0.6541	146.27	199.86	337.09	137.23	0.9995	1.5022
274.0	0.09114	0.6553	141.01	200.82	337.70	136.89	1.0030	1.5026
275.0	0.09475	0.6565	135.97	201.78	338.32	136.54	1.0065	1.5030
276.0	0.09848	0.6578	131.15	202.74	338.93	136.19	1.0100	1.5034
277.0	0.10231	0.6590	126.54	203.70	339.54	135.84	1.0134	1.5038
278.0	0.10627	0.6603	122.13	204.67	340.15	135.49	1.0169	1.5043
279.0	0.11034	0.6615	117.90	205.63	340.77	135.14	1.0203	1.5047

Chart R114.1: Saturation properties—*continued*

T K	p MPa	v' dm³/kg	v'' dm³/kg	h' kJ/kg	h'' kJ/kg	r kJ/kg	s' kJ/kgK	s'' kJ/kgK
280.0	0.11453	0.6628	113.85	206.60	341.38	134.78	1.0238	1.5052
281.0	0.11885	0.6641	109.97	207.57	341.99	134.43	1.0272	1.5056
282.0	0.12329	0.6654	106.25	208.53	342.61	134.07	1.0307	1.5061
283.0	0.12785	0.6666	102.68	209.51	343.22	133.71	1.0341	1.5066
284.0	0.13255	0.6680	99.258	210.48	343.83	133.35	1.0375	1.5071
285.0	0.13738	0.6693	95.974	211.45	344.44	132.99	1.0409	1.5076
286.0	0.14234	0.6706	92.822	212.43	345.05	132.63	1.0443	1.5081
287.0	0.14744	0.6719	89.795	213.40	345.66	132.26	1.0477	1.5086
288.0	0.15269	0.6733	86.888	214.38	346.27	131.90	1.0511	1.5091
289.0	0.15807	0.6746	84.096	215.36	346.88	131.53	1.0545	1.5096
290.0	0.16360	0.6760	81.411	216.34	347.49	131.16	1.0579	1.5101
291.0	0.16927	0.6773	78.831	217.32	348.10	130.78	1.0612	1.5107
292.0	0.17510	0.6787	76.350	218.30	348.71	130.41	1.0646	1.5112
293.0	0.18108	0.6801	73.964	219.29	349.32	130.03	1.0679	1.5117
294.0	0.18721	0.6815	71.667	220.27	349.93	129.66	1.0713	1.5123
295.0	0.19350	0.6829	69.457	221.26	350.54	129.28	1.0746	1.5128
296.0	0.19996	0.6843	67.330	222.25	351.14	128.89	1.0780	1.5134
297.0	0.20657	0.6858	65.281	223.24	351.75	128.51	1.0813	1.5140
298.0	0.21335	0.6872	63.308	224.23	352.35	128.12	1.0846	1.5145
299.0	0.22031	0.6887	61.406	225.23	352.96	127.73	1.0879	1.5151
300.0	0.22743	0.6901	59.574	226.22	353.56	127.34	1.0912	1.5157
301.0	0.23472	0.6916	57.808	227.22	354.17	126.95	1.0945	1.5163
302.0	0.24220	0.6931	56.105	228.22	354.77	126.55	1.0978	1.5169
303.0	0.24985	0.6946	54.462	229.22	355.37	126.15	1.1011	1.5174
304.0	0.25768	0.6961	52.877	230.22	355.97	125.75	1.1044	1.5180
305.0	0.26570	0.6976	51.348	231.22	356.57	125.35	1.1077	1.5186
306.0	0.27391	0.6992	49.872	232.23	357.17	124.94	1.1109	1.5192
307.0	0.28231	0.7007	48.447	233.24	357.77	124.53	1.1142	1.5198
308.0	0.29091	0.7023	47.071	234.24	358.37	124.12	1.1175	1.5205
309.0	0.29970	0.7039	45.742	235.26	358.96	123.71	1.1207	1.5211
310.0	0.30868	0.7055	44.457	236.27	359.56	123.29	1.1240	1.5217
311.0	0.31788	0.7071	43.216	237.28	360.15	122.87	1.1272	1.5223
312.0	0.32727	0.7087	42.017	238.30	360.74	122.45	1.1305	1.5229
313.0	0.33688	0.7103	40.857	239.31	361.34	122.02	1.1337	1.5235
314.0	0.34670	0.7120	39.736	240.33	361.93	121.59	1.1369	1.5242
315.0	0.35673	0.7137	38.651	241.36	362.52	121.16	1.1401	1.5248
316.0	0.36697	0.7153	37.601	242.38	363.10	120.73	1.1434	1.5254
317.0	0.37744	0.7170	36.585	243.40	363.69	120.29	1.1466	1.5260
318.0	0.38813	0.7188	35.602	244.43	364.28	119.85	1.1498	1.5267
319.0	0.39905	0.7205	34.651	245.46	364.86	119.40	1.1530	1.5273
320.0	0.41020	0.7222	33.729	246.49	365.44	118.95	1.1562	1.5279
321.0	0.42158	0.7240	32.837	247.52	366.02	118.50	1.1594	1.5286
322.0	0.43319	0.7258	31.972	248.56	366.60	118.05	1.1626	1.5292
323.0	0.44504	0.7276	31.134	249.59	367.18	117.59	1.1658	1.5298
324.0	0.45714	0.7294	30.322	250.63	367.76	117.13	1.1690	1.5305
325.0	0.46948	0.7312	29.535	251.67	368.33	116.66	1.1721	1.5311
326.0	0.48206	0.7331	28.772	252.72	368.91	116.19	1.1753	1.5317
327.0	0.49490	0.7350	28.032	253.76	369.48	115.72	1.1785	1.5324
328.0	0.50799	0.7369	27.315	254.81	370.05	115.24	1.1816	1.5330
329.0	0.52134	0.7388	26.619	255.86	370.62	114.76	1.1848	1.5336

Chart R114.1: Saturation properties—*continued*

T K	p MPa	v' dm³/kg	v'' dm³/kg	h' kJ/kg	h'' kJ/kg	r kJ/kg	s' kJ/kgK	s'' kJ/kgK
330.0	0.53495	0.7407	25.943	256.91	371.18	114.28	1.1880	1.5343
331.0	0.54882	0.7427	25.287	257.96	371.75	113.79	1.1911	1.5349
332.0	0.56296	0.7446	24.651	259.02	372.31	113.29	1.1943	1.5355
333.0	0.57737	0.7466	24.033	260.07	372.87	112.80	1.1974	1.5362
334.0	0.59205	0.7487	23.433	261.13	373.43	112.30	1.2006	1.5368
335.0	0.60701	0.7507	22.851	262.20	373.99	111.79	1.2037	1.5374
336.0	0.62225	0.7528	22.285	263.26	374.54	111.28	1.2069	1.5380
337.0	0.63778	0.7549	21.735	264.33	375.09	110.76	1.2100	1.5387
338.0	0.65359	0.7570	21.200	265.40	375.64	110.24	1.2131	1.5393
339.0	0.66969	0.7591	20.681	266.47	376.19	109.72	1.2162	1.5399
340.0	0.68608	0.7613	20.176	267.54	376.73	109.19	1.2194	1.5405
341.0	0.70277	0.7635	19.685	268.62	377.28	108.66	1.2225	1.5411
342.0	0.71976	0.7657	19.208	269.70	377.82	108.12	1.2256	1.5418
343.0	0.73706	0.7680	18.743	270.78	378.36	107.58	1.2287	1.5424
344.0	0.75466	0.7703	18.292	271.86	378.89	107.03	1.2319	1.5430
345.0	0.77257	0.7726	17.852	272.95	379.42	106.47	1.2350	1.5436
346.0	0.79080	0.7749	17.424	274.04	379.95	105.91	1.2381	1.5442
347.0	0.80935	0.7773	17.008	275.13	380.48	105.35	1.2412	1.5448
348.0	0.82821	0.7797	16.603	276.23	381.00	104.78	1.2443	1.5454
349.0	0.84740	0.7821	16.208	277.32	381.52	104.20	1.2474	1.5460
350.0	0.86692	0.7846	15.824	278.42	382.04	103.62	1.2505	1.5466
351.0	0.88677	0.7871	15.450	279.53	382.56	103.03	1.2536	1.5471
352.0	0.90696	0.7896	15.085	280.63	383.07	102.43	1.2567	1.5477
353.0	0.92749	0.7922	14.730	281.74	383.58	101.83	1.2598	1.5483
354.0	0.94836	0.7948	14.384	282.85	384.08	101.23	1.2629	1.5489
355.0	0.96958	0.7975	14.046	283.97	384.58	100.61	1.2660	1.5494
356.0	0.99114	0.8002	13.717	285.09	385.08	99.99	1.2691	1.5500
357.0	1.01307	0.8029	13.397	286.21	385.57	99.37	1.2722	1.5505
358.0	1.03535	0.8057	13.084	287.33	386.06	98.73	1.2753	1.5511
359.0	1.05799	0.8086	12.779	288.46	386.55	98.09	1.2784	1.5516
360.0	1.08100	0.8114	12.481	289.59	387.03	97.44	1.2815	1.5522
361.0	1.10438	0.8144	12.191	290.73	387.51	96.78	1.2846	1.5527
362.0	1.12813	0.8173	11.908	291.86	387.98	96.12	1.2877	1.5532
363.0	1.15227	0.8204	11.631	293.01	388.45	95.45	1.2908	1.5537
364.0	1.17678	0.8234	11.361	294.15	388.92	94.77	1.2939	1.5542
365.0	1.20168	0.8266	11.098	295.30	389.38	94.08	1.2970	1.5547
366.0	1.22697	0.8298	10.840	296.45	389.84	93.38	1.3001	1.5552
367.0	1.25266	0.8330	10.589	297.61	390.29	92.68	1.3032	1.5557
368.0	1.27874	0.8363	10.343	298.77	390.73	91.96	1.3063	1.5562
369.0	1.30523	0.8397	10.104	299.93	391.17	91.24	1.3094	1.5566
370.0	1.33213	0.8431	9.8691	301.10	391.61	90.51	1.3125	1.5571
371.0	1.35944	0.8466	9.6400	302.28	392.04	89.76	1.3156	1.5575
372.0	1.38716	0.8502	9.4160	303.45	392.46	89.01	1.3187	1.5579
373.0	1.41531	0.8539	9.1970	304.64	392.88	88.25	1.3218	1.5584
374.0	1.44388	0.8576	8.9828	305.82	393.29	87.47	1.3249	1.5588
375.0	1.47288	0.8614	8.7733	307.02	393.70	86.68	1.3280	1.5592
376.0	1.50231	0.8653	8.5683	308.21	394.10	85.89	1.3311	1.5596
377.0	1.53218	0.8693	8.3677	309.41	394.49	85.08	1.3343	1.5599
378.0	1.56250	0.8734	8.1714	310.62	394.88	84.26	1.3374	1.5603
379.0	1.59327	0.8775	7.9793	311.83	395.26	83.42	1.3405	1.5606

Chart R114.1: Saturation properties–*continued*

T K	p MPa	v' dm³/kg	v'' dm³/kg	h' kJ/kg	h'' kJ/kg	r kJ/kg	s' kJ/kgK	s'' kJ/kgK
380.0	1.62449	0.8818	7.7911	313.05	395.63	82.57	1.3437	1.5610
381.0	1.65617	0.8862	7.6068	314.28	395.99	81.71	1.3468	1.5613
382.0	1.68832	0.8907	7.4263	315.51	396.34	80.84	1.3499	1.5616
383.0	1.72093	0.8953	7.2494	316.74	396.69	79.95	1.3531	1.5618
384.0	1.75402	0.9000	7.0760	317.99	397.03	79.04	1.3563	1.5621
385.0	1.78759	0.9049	6.9061	319.24	397.36	78.12	1.3594	1.5623
386.0	1.82164	0.9099	6.7394	320.50	397.67	77.18	1.3626	1.5626
387.0	1.85619	0.9151	6.5759	321.76	397.98	76.22	1.3658	1.5628
388.0	1.89123	0.9204	6.4155	323.03	398.28	75.25	1.3690	1.5630
389.0	1.92678	0.9259	6.2581	324.31	398.57	74.25	1.3722	1.5631
390.0	1.96283	0.9315	6.1035	325.60	398.84	73.24	1.3755	1.5632
391.0	1.99940	0.9373	5.9518	326.90	399.11	72.21	1.3787	1.5634
392.0	2.03649	0.9434	5.8026	328.21	399.36	71.15	1.3819	1.5634
393.0	2.07411	0.9496	5.6561	329.53	399.59	70.07	1.3852	1.5635
394.0	2.11226	0.9561	5.5120	330.85	399.82	68.96	1.3885	1.5635
395.0	2.15095	0.9629	5.3702	332.19	400.02	67.83	1.3918	1.5635
396.0	2.19018	0.9699	5.2308	333.55	400.22	66.67	1.3951	1.5635
397.0	2.22998	0.9772	5.0934	334.91	400.39	65.48	1.3985	1.5634
398.0	2.27033	0.9848	4.9582	336.29	400.55	64.26	1.4018	1.5633
399.0	2.31125	0.9927	4.8248	337.68	400.69	63.01	1.4052	1.5631
400.0	2.35275	1.0010	4.6933	339.08	400.80	61.72	1.4086	1.5629
401.0	2.39483	1.0098	4.5635	340.51	400.90	60.39	1.4121	1.5627
402.0	2.43750	1.0189	4.4353	341.95	400.97	59.02	1.4156	1.5624
403.0	2.48078	1.0286	4.3085	343.41	401.01	57.60	1.4191	1.5620
404.0	2.52466	1.0389	4.1831	344.89	401.03	56.14	1.4226	1.5616
405.0	2.56917	1.0497	4.0589	346.40	401.02	54.62	1.4262	1.5611
406.0	2.61430	1.0613	3.9356	347.93	400.97	53.04	1.4299	1.5605
407.0	2.66007	1.0736	3.8132	349.49	400.89	51.39	1.4336	1.5599
408.0	2.70648	1.0869	3.6915	351.09	400.76	49.67	1.4374	1.5592
409.0	2.75355	1.1013	3.5701	352.72	400.59	47.87	1.4413	1.5583
410.0	2.80129	1.1168	3.4488	354.39	400.36	45.97	1.4452	1.5574
411.0	2.84972	1.1339	3.3273	356.12	400.08	43.96	1.4493	1.5563
412.0	2.89883	1.1527	3.2052	357.90	399.72	41.83	1.4535	1.5550
413.0	2.94865	1.1736	3.0819	359.74	399.29	39.54	1.4578	1.5536
414.0	2.99918	1.1973	2.9567	361.68	398.75	37.07	1.4624	1.5519
415.0	3.05045	1.2245	2.8288	363.72	398.09	34.38	1.4671	1.5500
416.0	3.10246	1.2564	2.6966	365.89	397.28	31.39	1.4722	1.5477
417.0	3.15523	1.2950	2.5580	368.26	396.26	28.00	1.4777	1.5449
418.0	3.20878	1.3438	2.4091	370.91	394.93	24.02	1.4839	1.5414

Chart R114.2: Saturation properties

p MPa	T K	v' dm^3/kg	v'' dm^3/kg	h' kJ/kg	h'' kJ/kg	r kJ/kg	s' kJ/kgK	s'' kJ/kgK
0.0102	228.82	0.6063	1084.1	158.11	310.34	152.22	0.8329	1.4982
0.0104	229.14	0.6066	1064.7	158.42	310.53	152.11	0.8342	1.4981
0.0106	229.46	0.6069	1045.9	158.72	310.71	152.00	0.8356	1.4980
0.0108	229.77	0.6072	1027.9	159.02	310.90	151.88	0.8368	1.4979
0.0110	230.08	0.6075	1010.4	159.31	311.08	151.78	0.8381	1.4978
0.0115	230.83	0.6083	969.41	160.02	311.53	151.51	0.8412	1.4976
0.0120	231.56	0.6090	931.71	160.70	311.96	151.26	0.8441	1.4974
0.0125	232.26	0.6097	896.93	161.36	312.37	151.01	0.8470	1.4972
0.0130	232.94	0.6104	864.74	162.00	312.78	150.78	0.8497	1.4970
0.0135	233.59	0.6110	834.86	162.62	313.17	150.55	0.8524	1.4969
0.0140	234.23	0.6116	807.05	163.22	313.55	150.33	0.8550	1.4967
0.0145	234.85	0.6123	781.10	163.80	313.92	150.11	0.8574	1.4966
0.0150	235.45	0.6129	756.82	164.37	314.27	149.90	0.8599	1.4965
0.0155	236.04	0.6134	734.06	164.92	314.62	149.70	0.8622	1.4964
0.0160	236.61	0.6140	712.67	165.46	314.97	149.51	0.8645	1.4963
0.0165	237.17	0.6146	692.54	165.98	315.30	149.32	0.8667	1.4963
0.0170	237.71	0.6151	673.56	166.49	315.62	149.13	0.8688	1.4962
0.0175	238.24	0.6157	655.62	166.99	315.94	148.95	0.8709	1.4961
0.0180	238.76	0.6162	638.65	167.48	316.25	148.77	0.8729	1.4961
0.0185	239.26	0.6167	622.57	167.95	316.55	148.60	0.8749	1.4960
0.0190	239.76	0.6172	607.31	168.42	316.85	148.43	0.8769	1.4960
0.0195	240.24	0.6177	592.80	168.87	317.14	148.27	0.8788	1.4959
0.0200	240.72	0.6182	578.99	169.32	317.43	148.11	0.8806	1.4959
0.0210	241.64	0.6191	553.29	170.19	317.98	147.80	0.8842	1.4959
0.0220	242.53	0.6200	529.85	171.02	318.51	147.50	0.8877	1.4958
0.0230	243.38	0.6209	508.38	171.82	319.03	147.21	0.8910	1.4958
0.0240	244.21	0.6218	488.64	172.60	319.53	146.93	0.8941	1.4958
0.0250	245.00	0.6226	470.43	173.35	320.01	146.66	0.8972	1.4958
0.0260	245.78	0.6234	453.58	174.07	320.48	146.40	0.9001	1.4958
0.0270	246.52	0.6242	437.93	174.77	320.93	146.15	0.9030	1.4959
0.0280	247.25	0.6250	423.36	175.46	321.37	145.91	0.9058	1.4959
0.0290	247.96	0.6257	409.76	176.12	321.80	145.67	0.9084	1.4959
0.0300	248.64	0.6264	397.04	176.77	322.21	145.44	0.9110	1.4960
0.0310	249.31	0.6272	385.11	177.40	322.62	145.22	0.9136	1.4961
0.0320	249.96	0.6278	373.90	178.01	323.01	145.00	0.9160	1.4961
0.0330	250.60	0.6285	363.35	178.61	323.40	144.79	0.9184	1.4962
0.0340	251.22	0.6292	353.40	179.19	323.77	144.58	0.9207	1.4963
0.0350	251.83	0.6298	344.00	179.76	324.14	144.38	0.9230	1.4963
0.0360	252.42	0.6305	335.10	180.32	324.50	144.18	0.9252	1.4964
0.0370	253.00	0.6311	326.66	180.86	324.86	143.99	0.9274	1.4965
0.0380	253.56	0.6317	318.66	181.40	325.20	143.80	0.9295	1.4966
0.0390	254.12	0.6323	311.05	181.92	325.54	143.62	0.9315	1.4967
0.0400	254.66	0.6329	303.81	182.43	325.87	143.44	0.9335	1.4968
0.0410	255.20	0.6335	296.91	182.94	326.19	143.26	0.9355	1.4969
0.0420	255.72	0.6341	290.32	183.43	326.51	143.08	0.9374	1.4970
0.0430	256.23	0.6347	284.03	183.91	326.82	142.91	0.9393	1.4971
0.0440	256.74	0.6352	278.02	184.39	327.13	142.74	0.9412	1.4972
0.0450	257.23	0.6358	272.27	184.86	327.43	142.58	0.9430	1.4973
0.0460	257.72	0.6363	266.75	185.31	327.73	142.42	0.9448	1.4974
0.0470	258.19	0.6368	261.47	185.77	328.02	142.26	0.9465	1.4975

Chart R114.2: Saturation properties—*continued*

p MPa	T K	v' dm³/kg	v'' dm³/kg	h' kJ/kg	h'' kJ/kg	r kJ/kg	s' kJ/kgK	s'' kJ/kgK
0.0480	258.66	0.6374	256.39	186.21	328.31	142.10	0.9482	1.4976
0.0490	259.12	0.6379	251.52	186.65	328.59	141.94	0.9499	1.4977
0.0500	259.58	0.6384	246.83	187.08	328.87	141.79	0.9516	1.4978
0.0510	260.03	0.6389	242.32	187.50	329.14	141.64	0.9532	1.4979
0.0520	260.47	0.6394	237.98	187.92	329.41	141.49	0.9548	1.4980
0.0530	260.90	0.6399	233.80	188.33	329.68	141.35	0.9564	1.4981
0.0540	261.33	0.6404	229.76	188.73	329.94	141.20	0.9579	1.4983
0.0550	261.75	0.6408	225.87	189.13	330.19	141.06	0.9594	1.4984
0.0560	262.16	0.6413	222.11	189.52	330.45	140.92	0.9609	1.4985
0.0570	262.57	0.6418	218.48	189.91	330.70	140.79	0.9624	1.4986
0.0580	262.98	0.6422	214.97	190.30	330.94	140.65	0.9639	1.4987
0.0590	263.37	0.6427	211.57	190.67	331.19	140.52	0.9653	1.4988
0.0600	263.77	0.6432	208.28	191.05	331.43	140.38	0.9667	1.4989
0.0610	264.15	0.6436	205.10	191.41	331.67	140.25	0.9681	1.4991
0.0620	264.54	0.6440	202.01	191.78	331.90	140.12	0.9695	1.4992
0.0630	264.91	0.6445	199.02	192.14	332.13	140.00	0.9708	1.4993
0.0640	265.29	0.6449	196.12	192.49	332.36	139.87	0.9722	1.4994
0.0650	265.66	0.6453	193.31	192.84	332.59	139.74	0.9735	1.4995
0.0660	266.02	0.6458	190.58	193.19	332.81	139.62	0.9748	1.4996
0.0670	266.38	0.6462	187.92	193.53	333.03	139.50	0.9761	1.4998
0.0680	266.74	0.6466	185.35	193.87	333.25	139.38	0.9773	1.4999
0.0690	267.09	0.6470	182.84	194.20	333.46	139.26	0.9786	1.5000
0.0700	267.43	0.6474	180.40	194.54	333.68	139.14	0.9798	1.5001
0.0710	267.78	0.6478	178.03	194.86	333.89	139.02	0.9810	1.5002
0.0720	268.12	0.6482	175.72	195.19	334.09	138.91	0.9823	1.5003
0.0730	268.45	0.6486	173.48	195.51	334.30	138.79	0.9834	1.5005
0.0740	268.79	0.6490	171.29	195.82	334.50	138.68	0.9846	1.5006
0.0760	269.44	0.6498	167.08	196.45	334.91	138.46	0.9869	1.5008
0.0780	270.08	0.6506	163.08	197.06	335.30	138.24	0.9892	1.5010
0.0800	270.71	0.6513	159.27	197.66	335.68	138.02	0.9914	1.5013
0.0820	271.33	0.6521	155.64	198.25	336.06	137.81	0.9936	1.5015
0.0840	271.93	0.6528	152.17	198.83	336.43	137.60	0.9957	1.5017
0.0860	272.52	0.6535	148.86	199.40	336.80	137.40	0.9978	1.5020
0.0880	273.11	0.6542	145.70	199.96	337.15	137.20	0.9999	1.5022
0.0900	273.68	0.6549	142.67	200.51	337.51	137.00	1.0019	1.5024
0.0920	274.24	0.6556	139.77	201.05	337.85	136.80	1.0038	1.5027
0.0940	274.79	0.6563	136.99	201.58	338.19	136.61	1.0058	1.5029
0.0960	275.34	0.6569	134.31	202.10	338.52	136.42	1.0076	1.5031
0.0980	275.87	0.6576	131.75	202.62	338.85	136.23	1.0095	1.5033
0.1000	276.40	0.6583	129.28	203.12	339.17	136.05	1.0113	1.5036
0.1020	276.92	0.6589	126.91	203.62	339.49	135.87	1.0131	1.5038
0.1040	277.43	0.6595	124.62	204.12	339.81	135.69	1.0149	1.5040
0.1060	277.93	0.6602	122.42	204.60	340.11	135.51	1.0167	1.5042
0.1080	278.43	0.6608	120.29	205.08	340.42	135.34	1.0184	1.5045
0.1100	278.92	0.6614	118.24	205.55	340.72	135.17	1.0201	1.5047
0.1150	280.11	0.6629	113.42	206.70	341.45	134.74	1.0242	1.5052
0.1200	281.26	0.6644	108.98	207.82	342.15	134.33	1.0281	1.5057
0.1250	282.38	0.6658	104.88	208.90	342.84	133.94	1.0320	1.5063
0.1300	283.46	0.6672	101.09	209.95	343.50	133.55	1.0357	1.5068
0.1350	284.51	0.6686	97.564	210.97	344.14	133.17	1.0393	1.5073

Chart R114.2: Saturation properties–*continued*

p MPa	T K	v' dm³/kg	v'' dm³/kg	h' kJ/kg	h'' kJ/kg	r kJ/kg	s' kJ/kgK	s'' kJ/kgK
0.1400	285.53	0.6700	94.283	211.97	344.77	132.80	1.0427	1.5078
0.1450	286.52	0.6713	91.220	212.94	345.37	132.44	1.0461	1.5083
0.1500	287.49	0.6726	88.353	213.88	345.96	132.08	1.0494	1.5088
0.1550	288.43	0.6738	85.665	214.80	346.54	131.74	1.0526	1.5093
0.1600	289.35	0.6751	83.138	215.70	347.10	131.40	1.0557	1.5098
0.1650	290.25	0.6763	80.758	216.58	347.65	131.06	1.0587	1.5103
0.1700	291.13	0.6775	78.513	217.44	348.18	130.74	1.0616	1.5107
0.1750	291.98	0.6787	76.391	218.29	348.70	130.42	1.0645	1.5112
0.1800	292.82	0.6799	74.383	219.11	349.21	130.10	1.0673	1.5116
0.1850	293.64	0.6810	72.479	219.92	349.71	129.79	1.0701	1.5121
0.1900	294.45	0.6821	70.671	220.71	350.20	129.49	1.0728	1.5125
0.1950	295.23	0.6832	68.952	221.49	350.68	129.19	1.0754	1.5130
0.2000	296.01	0.6843	67.316	222.26	351.15	128.89	1.0780	1.5134
0.2100	297.51	0.6865	64.269	223.74	352.06	128.31	1.0830	1.5143
0.2200	298.96	0.6886	61.488	225.18	352.93	127.75	1.0878	1.5151
0.2300	300.36	0.6907	58.939	226.58	353.78	127.20	1.0924	1.5159
0.2400	301.71	0.6927	56.595	227.93	354.59	126.67	1.0969	1.5167
0.2500	303.02	0.6946	54.431	229.24	355.38	126.14	1.1012	1.5175
0.2600	304.29	0.6966	52.426	230.51	356.15	125.63	1.1053	1.5182
0.2700	305.53	0.6985	50.565	231.75	356.89	125.13	1.1094	1.5190
0.2800	306.73	0.7003	48.831	232.96	357.61	124.64	1.1133	1.5197
0.2900	307.90	0.7021	47.212	234.14	358.30	124.16	1.1171	1.5204
0.3000	309.03	0.7039	45.697	235.29	358.98	123.69	1.1208	1.5211
0.3100	310.14	0.7057	44.276	236.41	359.64	123.23	1.1244	1.5218
0.3200	311.23	0.7075	42.939	237.51	360.29	122.77	1.1280	1.5224
0.3300	312.29	0.7092	41.681	238.59	360.91	122.33	1.1314	1.5231
0.3400	313.32	0.7109	40.494	239.64	361.53	121.88	1.1347	1.5237
0.3500	314.33	0.7125	39.372	240.67	362.12	121.45	1.1380	1.5244
0.3600	315.32	0.7142	38.309	241.68	362.71	121.02	1.1412	1.5250
0.3700	316.29	0.7158	37.302	242.68	363.27	120.60	1.1443	1.5256
0.3800	317.24	0.7175	36.345	243.65	363.83	120.18	1.1474	1.5262
0.3900	318.17	0.7191	35.436	244.61	364.38	119.77	1.1503	1.5268
0.4000	319.09	0.7206	34.570	245.55	364.91	119.36	1.1533	1.5273
0.4100	319.98	0.7222	33.745	246.47	365.43	118.96	1.1561	1.5279
0.4200	320.86	0.7238	32.957	247.38	365.94	118.56	1.1590	1.5285
0.4300	321.73	0.7253	32.205	248.27	366.45	118.17	1.1617	1.5290
0.4400	322.58	0.7268	31.485	249.16	366.94	117.78	1.1644	1.5296
0.4500	323.41	0.7283	30.796	250.02	367.42	117.40	1.1671	1.5301
0.4600	324.23	0.7298	30.136	250.88	367.89	117.02	1.1697	1.5306
0.4700	325.04	0.7313	29.503	251.72	368.36	116.64	1.1723	1.5311
0.4800	325.84	0.7328	28.895	252.55	368.81	116.27	1.1748	1.5316
0.4900	326.62	0.7342	28.310	253.36	369.26	115.90	1.1773	1.5321
0.5000	327.39	0.7357	27.748	254.17	369.70	115.53	1.1797	1.5326
0.5100	328.15	0.7371	27.208	254.97	370.14	115.17	1.1821	1.5331
0.5200	328.90	0.7386	26.687	255.75	370.56	114.81	1.1845	1.5336
0.5300	329.64	0.7400	26.185	256.53	370.98	114.45	1.1868	1.5340
0.5400	330.37	0.7414	25.701	257.29	371.39	114.10	1.1891	1.5345
0.5500	331.08	0.7428	25.233	258.05	371.80	113.75	1.1914	1.5349
0.5600	331.79	0.7442	24.782	258.80	372.19	113.40	1.1936	1.5354
0.5700	332.49	0.7456	24.346	259.54	372.59	113.05	1.1958	1.5358

Chart R114.2: Saturation properties–*continued*

p MPa	T K	v' dm³/kg	v'' dm³/kg	h' kJ/kg	h'' kJ/kg	r kJ/kg	s' kJ/kgK	s'' kJ/kgK
0.5800	333.18	0.7470	23.924	260.27	372.97	112.71	1.1980	1.5363
0.5900	333.86	0.7484	23.516	260.99	373.35	112.37	1.2001	1.5367
0.6000	334.53	0.7497	23.120	261.70	373.73	112.03	1.2022	1.5371
0.6100	335.20	0.7511	22.738	262.41	374.10	111.69	1.2043	1.5375
0.6200	335.85	0.7525	22.367	263.10	374.46	111.35	1.2064	1.5379
0.6300	336.50	0.7538	22.007	263.80	374.82	111.02	1.2084	1.5384
0.6400	337.14	0.7552	21.658	264.48	375.17	110.69	1.2104	1.5388
0.6500	337.77	0.7565	21.319	265.16	375.52	110.36	1.2124	1.5391
0.6600	338.40	0.7578	20.991	265.83	375.86	110.04	1.2144	1.5395
0.6700	339.02	0.7592	20.671	266.49	376.20	109.71	1.2163	1.5399
0.6800	339.63	0.7605	20.361	267.15	376.53	109.39	1.2182	1.5403
0.6900	340.24	0.7618	20.059	267.80	376.86	109.07	1.2201	1.5407
0.7000	340.84	0.7631	19.765	268.44	377.19	108.75	1.2220	1.5410
0.7100	341.43	0.7644	19.479	269.08	377.51	108.43	1.2238	1.5414
0.7200	342.01	0.7657	19.201	269.71	377.82	108.11	1.2257	1.5418
0.7300	342.59	0.7671	18.930	270.34	378.14	107.80	1.2275	1.5421
0.7400	343.17	0.7684	18.666	270.96	378.45	107.48	1.2293	1.5425
0.7600	344.30	0.7709	18.159	272.19	379.05	106.86	1.2328	1.5432
0.7800	345.41	0.7735	17.676	273.40	379.64	106.24	1.2362	1.5438
0.8000	346.50	0.7761	17.216	274.58	380.21	105.63	1.2396	1.5445
0.8200	347.57	0.7786	16.777	275.75	380.78	105.02	1.2430	1.5451
0.8400	348.62	0.7812	16.358	276.90	381.32	104.42	1.2462	1.5457
0.8600	349.65	0.7837	15.958	278.04	381.86	103.82	1.2494	1.5464
0.8800	350.66	0.7862	15.576	279.15	382.38	103.23	1.2526	1.5469
0.9000	351.66	0.7888	15.209	280.25	382.89	102.64	1.2557	1.5475
0.9200	352.64	0.7913	14.858	281.34	383.39	102.05	1.2587	1.5481
0.9400	353.60	0.7938	14.520	282.41	383.88	101.47	1.2617	1.5486
0.9600	354.55	0.7963	14.197	283.47	384.36	100.89	1.2646	1.5492
0.9800	355.49	0.7988	13.885	284.51	384.82	100.31	1.2675	1.5497
1.0000	356.41	0.8013	13.586	285.54	385.28	99.74	1.2704	1.5502
1.0200	357.31	0.8038	13.298	286.56	385.73	99.17	1.2732	1.5507
1.0400	358.21	0.8063	13.020	287.57	386.16	98.60	1.2759	1.5512
1.0600	359.09	0.8088	12.752	288.56	386.59	98.03	1.2787	1.5517
1.0800	359.96	0.8113	12.494	289.54	387.01	97.47	1.2814	1.5521
1.1000	360.81	0.8138	12.244	290.51	387.42	96.91	1.2840	1.5526
1.1500	362.91	0.8201	11.657	292.90	388.41	95.51	1.2905	1.5537
1.2000	364.93	0.8264	11.115	295.22	389.35	94.13	1.2968	1.5547
1.2500	366.90	0.8327	10.615	297.49	390.24	92.75	1.3028	1.5556
1.3000	368.80	0.8390	10.150	299.70	391.09	91.38	1.3087	1.5565
1.3500	370.66	0.8454	9.7182	301.87	391.89	90.02	1.3145	1.5574
1.4000	372.46	0.8519	9.3150	304.00	392.66	88.66	1.3201	1.5581
1.4500	374.21	0.8584	8.9379	306.08	393.38	87.30	1.3256	1.5589
1.5000	375.92	0.8650	8.5841	308.12	394.07	85.95	1.3309	1.5595
1.5500	377.59	0.8717	8.2515	310.12	394.72	84.60	1.3361	1.5601
1.6000	379.22	0.8784	7.9382	312.10	395.34	83.24	1.3412	1.5607
1.6500	380.81	0.8853	7.6422	314.04	395.92	81.88	1.3462	1.5612
1.7000	382.36	0.8923	7.3622	315.95	396.47	80.52	1.3511	1.5617
1.7500	383.88	0.8995	7.0968	317.84	396.99	79.15	1.3559	1.5621
1.8000	385.37	0.9067	6.8447	319.70	397.47	77.78	1.3606	1.5624
1.8500	386.82	0.9141	6.6048	321.53	397.93	76.39	1.3652	1.5627

Chart R114.2: Saturation properties—*continued*

p MPa	T K	v' dm^3/kg	v'' dm^3/kg	h' kJ/kg	h'' kJ/kg	r kJ/kg	s' kJ/kgK	s'' kJ/kgK
1.9000	388.25	0.9217	6.3762	323.35	398.35	75.00	1.3698	1.5630
1.9500	389.65	0.9295	6.1580	325.15	398.75	73.60	1.3743	1.5632
2.0000	391.02	0.9374	5.9493	326.92	399.11	72.19	1.3787	1.5634
2.1000	393.68	0.9540	5.5578	330.43	399.75	69.32	1.3874	1.5635
2.2000	396.25	0.9716	5.1965	333.88	400.26	66.38	1.3959	1.5635
2.3000	398.73	0.9905	4.8611	337.30	400.65	63.35	1.4043	1.5632
2.4000	401.12	1.0109	4.5478	340.68	400.91	60.22	1.4125	1.5626
2.5000	403.44	1.0331	4.2532	344.06	401.02	56.96	1.4206	1.5618
2.6000	405.68	1.0576	3.9744	347.45	400.99	53.54	1.4287	1.5607
2.7000	407.86	1.0850	3.7083	350.87	400.78	49.91	1.4369	1.5593
2.8000	409.97	1.1164	3.4521	354.35	400.37	46.02	1.4451	1.5574
2.9000	412.02	1.1531	3.2023	357.94	399.71	41.78	1.4536	1.5550
3.0000	414.02	1.1977	2.9547	361.71	398.74	37.03	1.4624	1.5519
3.1000	415.95	1.2548	2.7029	365.79	397.32	31.54	1.4720	1.5478
3.2000	417.84	1.3349	2.4343	370.45	395.17	24.72	1.4829	1.5420

Chart R114.3: Thermodynamic properties

p/MPa	0.10			0.15			0.20		
T K	v dm³/kg	h kJ/kg	s kJ/kgK	v dm³/kg	h kJ/kg	s kJ/kgK	v dm³/kg	h kJ/kg	s kJ/kgK
300.0	141.61	355.44	1.5600	92.867	354.74	1.5387	68.437	353.99	1.5230
310.0	146.77	362.51	1.5832	96.424	361.86	1.5620	71.209	361.18	1.5465
320.0	151.89	369.68	1.6060	99.946	369.08	1.5849	73.938	368.45	1.5696
330.0	156.99	376.94	1.6283	103.44	376.38	1.6074	76.634	375.80	1.5922
340.0	162.07	384.30	1.6503	106.91	383.78	1.6295	79.302	383.24	1.6144
350.0	167.14	391.76	1.6719	110.35	391.27	1.6512	81.948	390.76	1.6362
360.0	172.18	399.31	1.6932	113.79	398.85	1.6726	84.573	398.37	1.6577
370.0	177.21	406.95	1.7141	117.20	406.51	1.6936	87.183	406.06	1.6787
380.0	182.23	414.69	1.7347	120.60	414.27	1.7143	89.778	413.84	1.6995
390.0	187.24	422.50	1.7550	123.99	422.11	1.7346	92.360	421.70	1.7199
400.0	192.24	430.41	1.7751	127.37	430.03	1.7547	94.932	429.64	1.7400
410.0	197.23	438.39	1.7948	130.74	438.03	1.7744	97.494	437.66	1.7598
420.0	202.21	446.46	1.8142	134.11	446.11	1.7939	100.05	445.76	1.7793
430.0	207.19	454.60	1.8334	137.46	454.27	1.8131	102.59	453.94	1.7986
440.0	212.16	462.82	1.8523	140.81	462.51	1.8320	105.13	462.19	1.8175
450.0	217.12	471.12	1.8709	144.15	470.81	1.8507	107.67	470.51	1.8362
460.0	222.08	479.49	1.8893	147.49	479.19	1.8691	110.19	478.90	1.8547
470.0	227.03	487.92	1.9074	150.82	487.64	1.8873	112.71	487.36	1.8729
480.0	231.98	496.43	1.9254	154.15	496.16	1.9052	115.23	495.88	1.8908
490.0	236.92	505.00	1.9430	157.47	504.74	1.9229	117.74	504.47	1.9085
500.0	241.86	513.63	1.9605	160.79	513.38	1.9404	120.25	513.13	1.9260
p/MPa	0.30			0.40			0.50		
300.0	0.6900	226.24	1.0911	0.6897	226.26	1.0910	0.6895	226.29	1.0908
310.0	45.889	359.69	1.5234	0.7052	236.29	1.1238	0.7050	236.31	1.1237
320.0	47.852	367.10	1.5469	34.715	365.61	1.5295	0.7219	246.50	1.1560
330.0	49.769	374.57	1.5699	36.266	373.22	1.5530	28.096	371.74	1.5388
340.0	51.650	382.10	1.5924	37.771	380.88	1.5758	29.392	379.55	1.5621
350.0	53.503	389.70	1.6144	39.239	388.58	1.5981	30.643	387.38	1.5848
360.0	55.332	397.38	1.6360	40.679	396.34	1.6200	31.857	395.24	1.6070
370.0	57.141	405.14	1.6573	42.095	404.17	1.6414	33.044	403.15	1.6286
380.0	58.934	412.97	1.6782	43.492	412.06	1.6625	34.208	411.11	1.6499
390.0	60.713	420.88	1.6987	44.872	420.02	1.6832	35.353	419.14	1.6707
400.0	62.479	428.86	1.7189	46.239	428.05	1.7035	36.483	427.22	1.6912
410.0	64.235	436.92	1.7388	47.594	436.15	1.7235	37.600	435.37	1.7113
420.0	65.981	445.05	1.7584	48.939	444.33	1.7432	38.706	443.58	1.7311
430.0	67.719	453.26	1.7777	50.274	452.57	1.7626	39.801	451.86	1.7506
440.0	69.450	461.54	1.7968	51.602	460.88	1.7817	40.888	460.21	1.7697
450.0	71.174	469.88	1.8155	52.923	469.25	1.8005	41.968	468.61	1.7886
460.0	72.892	478.30	1.8340	54.237	477.70	1.8191	43.041	477.08	1.8073
470.0	74.604	486.78	1.8523	55.546	486.20	1.8374	44.107	485.62	1.8256
480.0	76.312	495.33	1.8703	56.849	494.77	1.8554	45.169	494.21	1.8437
490.0	78.015	503.94	1.8880	58.148	503.41	1.8732	46.225	502.87	1.8616
500.0	79.714	512.62	1.9055	59.442	512.10	1.8908	47.277	511.58	1.8792

Chart R114.3: Thermodynamic properties–*continued*

p/MPa	0.60			0.80			1.00		
T K	v dm³/kg	h kJ/kg	s kJ/kgK	v dm³/kg	h kJ/kg	s kJ/kgK	v dm³/kg	h kJ/kg	s kJ/kgK
300.0	0.6893	226.31	1.0907	0.6888	226.36	1.0904	0.6884	226.41	1.0901
310.0	0.7047	236.33	1.1235	0.7042	236.37	1.1232	0.7036	236.41	1.1229
320.0	0.7216	246.52	1.1559	0.7209	246.55	1.1555	0.7203	246.58	1.1551
330.0	0.7404	256.91	1.1878	0.7396	256.93	1.1874	0.7388	256.95	1.1870
340.0	23.756	378.10	1.5501	0.7607	267.54	1.2191	0.7597	267.54	1.2187
350.0	24.874	386.09	1.5732	17.557	383.16	1.5529	0.7837	278.41	1.2502
360.0	25.948	394.07	1.5957	18.485	391.48	1.5764	13.897	388.44	1.5590
370.0	26.989	402.08	1.6177	19.363	399.76	1.5991	14.710	397.12	1.5828
380.0	28.002	410.13	1.6391	20.202	408.01	1.6211	15.466	405.67	1.6056
390.0	28.995	418.22	1.6602	21.013	416.28	1.6425	16.182	414.16	1.6277
400.0	29.969	426.37	1.6808	21.801	424.57	1.6635	16.868	422.63	1.6491
410.0	30.929	434.57	1.7010	22.570	432.89	1.6841	17.530	431.10	1.6700
420.0	31.877	442.82	1.7209	23.324	441.25	1.7042	18.173	439.59	1.6905
430.0	32.813	451.14	1.7405	24.066	449.66	1.7240	18.802	448.11	1.7105
440.0	33.741	459.52	1.7598	24.796	458.12	1.7435	19.418	456.66	1.7302
450.0	34.661	467.96	1.7787	25.518	466.63	1.7626	20.023	465.26	1.7495
460.0	35.573	476.46	1.7974	26.232	475.20	1.7814	20.619	473.89	1.7685
470.0	36.479	485.02	1.8158	26.938	483.82	1.8000	21.207	482.58	1.7872
480.0	37.380	493.64	1.8340	27.638	492.49	1.8182	21.788	491.31	1.8056
490.0	38.275	502.32	1.8519	28.333	501.22	1.8362	22.363	500.09	1.8237
500.0	39.166	511.06	1.8695	29.023	510.00	1.8539	22.933	508.92	1.8415

p/MPa	2.00			4.00			6.00		
300.0	0.6863	226.67	1.0886	0.6822	227.21	1.0859	0.6784	227.77	1.0832
310.0	0.7010	236.63	1.1213	0.6962	237.10	1.1183	0.6918	237.60	1.1155
320.0	0.7172	246.75	1.1534	0.7114	247.13	1.1502	0.7062	247.57	1.1471
330.0	0.7349	257.05	1.1851	0.7279	257.33	1.1815	0.7217	257.68	1.1782
340.0	0.7548	267.56	1.2165	0.7461	267.70	1.2125	0.7385	267.94	1.2088
350.0	0.7773	278.31	1.2477	0.7663	278.27	1.2431	0.7570	278.37	1.2391
360.0	0.8035	289.35	1.2788	0.7891	289.05	1.2735	0.7774	288.97	1.2689
370.0	0.8349	300.76	1.3100	0.8153	300.10	1.3038	0.8002	299.78	1.2985
380.0	0.8747	312.70	1.3419	0.8461	311.46	1.3341	0.8262	310.81	1.3280
390.0	0.9303	325.54	1.3752	0.8838	323.24	1.3647	0.8562	322.11	1.3573
400.0	6.5931	409.36	1.5893	0.9325	335.61	1.3960	0.8917	333.72	1.3867
410.0	7.1708	419.70	1.6148	1.0014	348.95	1.4289	0.9351	345.75	1.4164
420.0	7.6719	429.50	1.6384	1.1204	364.33	1.4659	0.9903	358.31	1.4467
430.0	8.1257	439.00	1.6608	1.5121	387.92	1.5214	1.0642	371.62	1.4780
440.0	8.5473	448.33	1.6822	2.5097	417.91	1.5904	1.1700	386.00	1.5110
450.0	8.9452	457.56	1.7030	3.0560	434.04	1.6267	1.3288	401.73	1.5464
460.0	9.3252	466.73	1.7231	3.4512	446.93	1.6550	1.5563	418.47	1.5832
470.0	9.6910	475.86	1.7428	3.7781	458.52	1.6800	1.8255	434.69	1.6181
480.0	10.045	484.98	1.7620	4.0654	469.43	1.7030	2.0891	449.37	1.6490
490.0	10.390	494.11	1.7808	4.3264	479.93	1.7246	2.3288	462.63	1.6763
500.0	10.727	503.24	1.7992	4.5688	490.17	1.7453	2.5455	474.89	1.7011

Chart R114.3: Thermodynamic properties—*continued*

p/MPa	8.00			10.00			20.00		
T K	v dm^3/kg	h kJ/kg	s kJ/kgK	v dm^3/kg	h kJ/kg	s kJ/kgK	v dm^3/kg	h kJ/kg	s kJ/kgK
300.0	0.6749	228.36	1.0807	0.6715	228.98	1.0782	0.6574	232.30	1.0672
310.0	0.6877	238.14	1.1127	0.6839	238.71	1.1102	0.6678	241.87	1.0986
320.0	0.7014	248.05	1.1442	0.6969	248.56	1.1414	0.6787	251.53	1.1292
330.0	0.7160	258.08	1.1751	0.7109	258.54	1.1721	0.6901	261.29	1.1593
340.0	0.7318	268.26	1.2054	0.7257	268.63	1.2023	0.7020	271.14	1.1887
350.0	0.7489	278.57	1.2353	0.7417	278.86	1.2319	0.7145	281.09	1.2175
360.0	0.7675	289.04	1.2648	0.7589	289.22	1.2611	0.7276	291.12	1.2458
370.0	0.7880	299.68	1.2940	0.7776	299.73	1.2899	0.7412	301.25	1.2735
380.0	0.8107	310.49	1.3228	0.7981	310.38	1.3183	0.7556	311.47	1.3008
390.0	0.8362	321.49	1.3514	0.8206	321.18	1.3463	0.7708	321.77	1.3275
400.0	0.8652	332.71	1.3798	0.8455	332.15	1.3741	0.7867	332.15	1.3538
410.0	0.8986	344.18	1.4081	0.8733	343.29	1.4016	0.8036	342.62	1.3797
420.0	0.9378	355.94	1.4364	0.9047	354.62	1.4289	0.8213	353.16	1.4051
430.0	0.9846	368.03	1.4649	0.9404	366.17	1.4561	0.8401	363.78	1.4300
440.0	1.0416	380.51	1.4936	0.9814	377.93	1.4831	0.8599	374.47	1.4546
450.0	1.1122	393.43	1.5226	1.0288	389.93	1.5101	0.8808	385.22	1.4788
460.0	1.2001	406.79	1.5520	1.0838	402.16	1.5370	0.9029	396.04	1.5026
470.0	1.3079	420.51	1.5815	1.1475	414.60	1.5637	0.9263	406.91	1.5259
480.0	1.4346	434.33	1.6106	1.2205	427.20	1.5903	0.9509	417.84	1.5489
490.0	1.5747	447.94	1.6387	1.3024	439.86	1.6164	0.9769	428.81	1.5716
500.0	1.7206	461.11	1.6653	1.3919	452.47	1.6418	1.0042	439.82	1.5938

References for R114

[Barh66] BARHO, W.: Berechnung der Molwärme der Fluor-Chlor-Derivate von Methan und Ethan im Zustand idealer Gase; Dissertation Universität Karlsruhe (1966)

[Benn44] BENNING, A.F. AND MCHARNESS, R.C.: Thermodynamic properties of Freon 114 Refrigerant, Freon Products Division; E.I. DuPont de Nemours & Co. ,Wilmington Delaware (1944)

[Erns69] ERNST, G.: Zur universellen Darstellung des Realanteils der spezifischen Wärme von Gasen; Chem. Ing. Techn. 41(1969) 544/551

[Gell75] GELLER, Z.I. AND PORICHANSKIJ, E.G.: Thermophysical properties of Freon 114; Thermophysical Properties of Matter and Substances 8(1975) 65/70

[Hase85] HASEGAWA, N. ; UEMATSU, M. AND WATANABE, K.: Measurements of PVTx Properties for the R22 + R114 System; J. Chem. Eng. Data 30(1985)1 32/36

[Hwan61] HWANG, Y.T.: The constant heat capacities of gaseous tetrafluoromethane, chlorodifluoromethane, dichlorotetrafluoroethane and chloropentafluoroethane; Ph.D. Dissertation University of Michigan (1961)

[KMR 81] - : Kältemaschinenregeln, 7. Auflage; Verlag C.F. Müller, Karlsruhe (1981)

[Kole81] KOLESOV, V.P. ; KOSARUKINA, E.A. ; ZHOGIN, D.JU. ; POLOZNIKOVA, M.E. AND PENTIN, JU.A.: Heat capacities, phase transitions, and thermodynamic functions of 1,1,2,2-tetrafluoro 1,2-dichloroethane and 1,1,2-trifluoro 1,2,2-trichloroethane; J. Chem. Therm. 13(1981) 115/129

[Lock34] LOCKE, E.G. ; BRODE, W.R. AND HENNE, A.L.: Fluorochloroethanes and fluoroethylenes; J. Am. Chem. Soc. 56(1934) 1726/1728

[Mart60] MARTIN, J.J.: Thermodynamic properties of dichlorotetrafluoroethane; J. Chem. Eng. Data 5(1960)3 334/336

[McHa51] MCHARNESS, R.C.: unpublished (1951) see [Mart60]

[Mear55] MEARS, W.H.: Thermodynamic Properties of Halogenated Ethanes and Ethylenes. Physical Measurements and Thermodynamic Calculations CH_3CHF_2, CH_3CF_2Cl, CH_3CF_3, CF_3CFCl_2, $CH_2{=}CF_2$ and $CHCl{=}CF_2$; Material on Deposit at the American Documentation Institute 4502(1955)

[Meye69] MEYER, K.J.: Über den Zusammenhang zwischen der Schallgeschwindigkeit und der Wärmeleitfähigkeit bei flüssigen Fluor-Chlor-Derivaten des Methans und Äthans; Kältetechnik-Klimatisierung 21(1969)9 270/275

[Pono77] PONOMAREVA, O.P.: Study of the isobaric heat capacity of Freon 114; Teplofiziceskie svojstva vescestv i materialov 11(1977) 20/26

[Thor33] THORNTON, N.V. ; BURG, A.B. AND SCHLESINGER, H.I.: The behaviour of dichlorodifluoromethane and of chlorotrifluoromethane in the electric discharge; J. Am. Chem. Soc. 55(1933) 3177/3182

[Wils81] WILSON, D.P. AND HULES, K.R.: Experimental study of the thermodynamic properties of 1,2 dichlorotetrafluoroethane; 8th Symposium on Thermophysical Properties, NBS–ASME Gaithersburg (1981)

R 114. ph

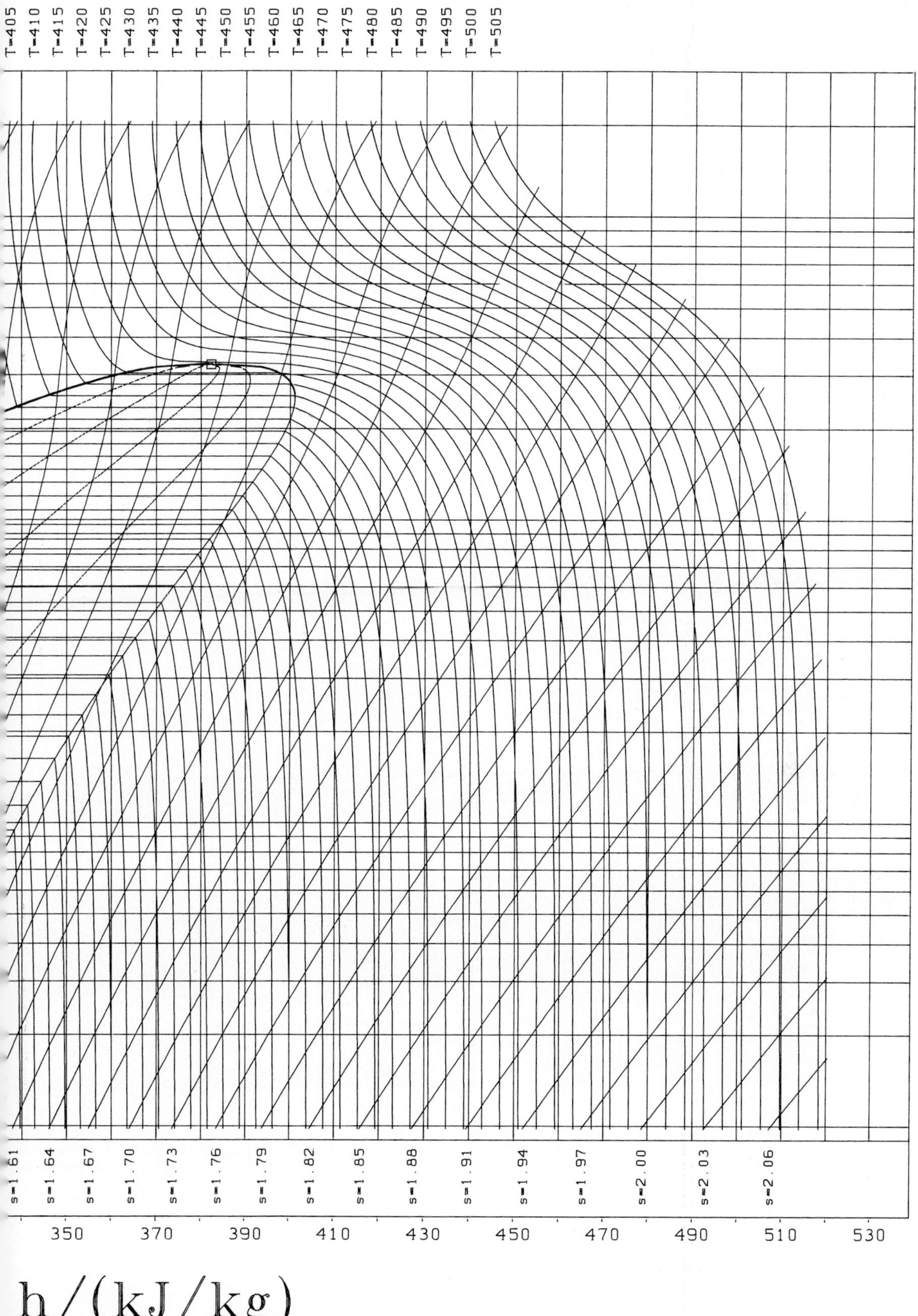
T=405
T=410
T=415
T=420
T=425
T=430
T=435
T=440
T=445
T=450
T=455
T=460
T=465
T=470
T=475
T=480
T=485
T=490
T=495
T=500
T=505
s=1.61
s=1.64
s=1.67
s=1.70
s=1.73
s=1.76
s=1.79
s=1.82
s=1.85
s=1.88
s=1.91
s=1.94
s=1.97
s=2.00
s=2.03
s=2.06
350
370
390
410
430
450
470
490
510
530
h/(kJ/kg)

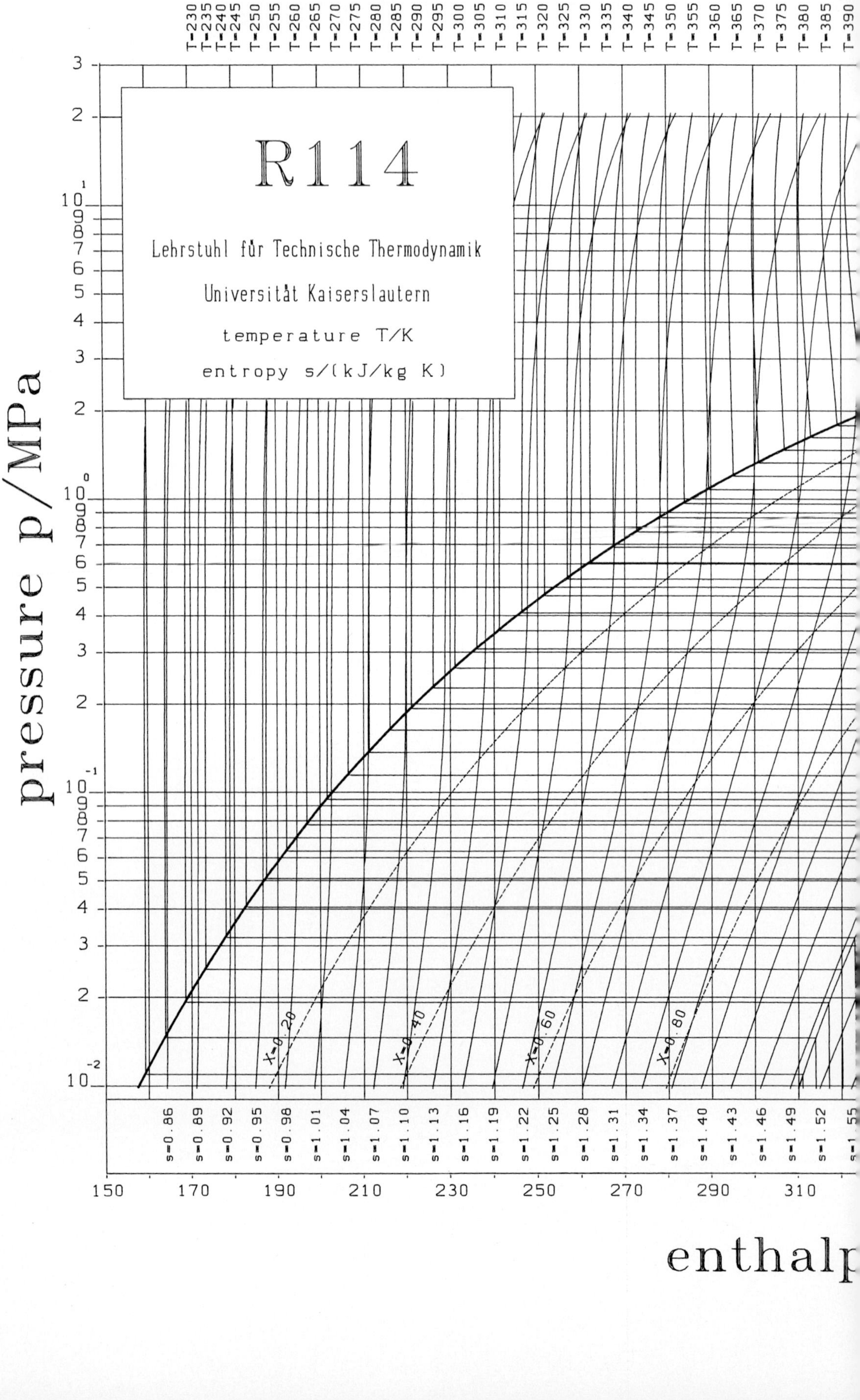
R114
Lehrstuhl für Technische Thermodynamik
Universität Kaiserslautern
temperature T/K
entropy s/(kJ/kg K)
pressure p/MPa

R 114. Ts

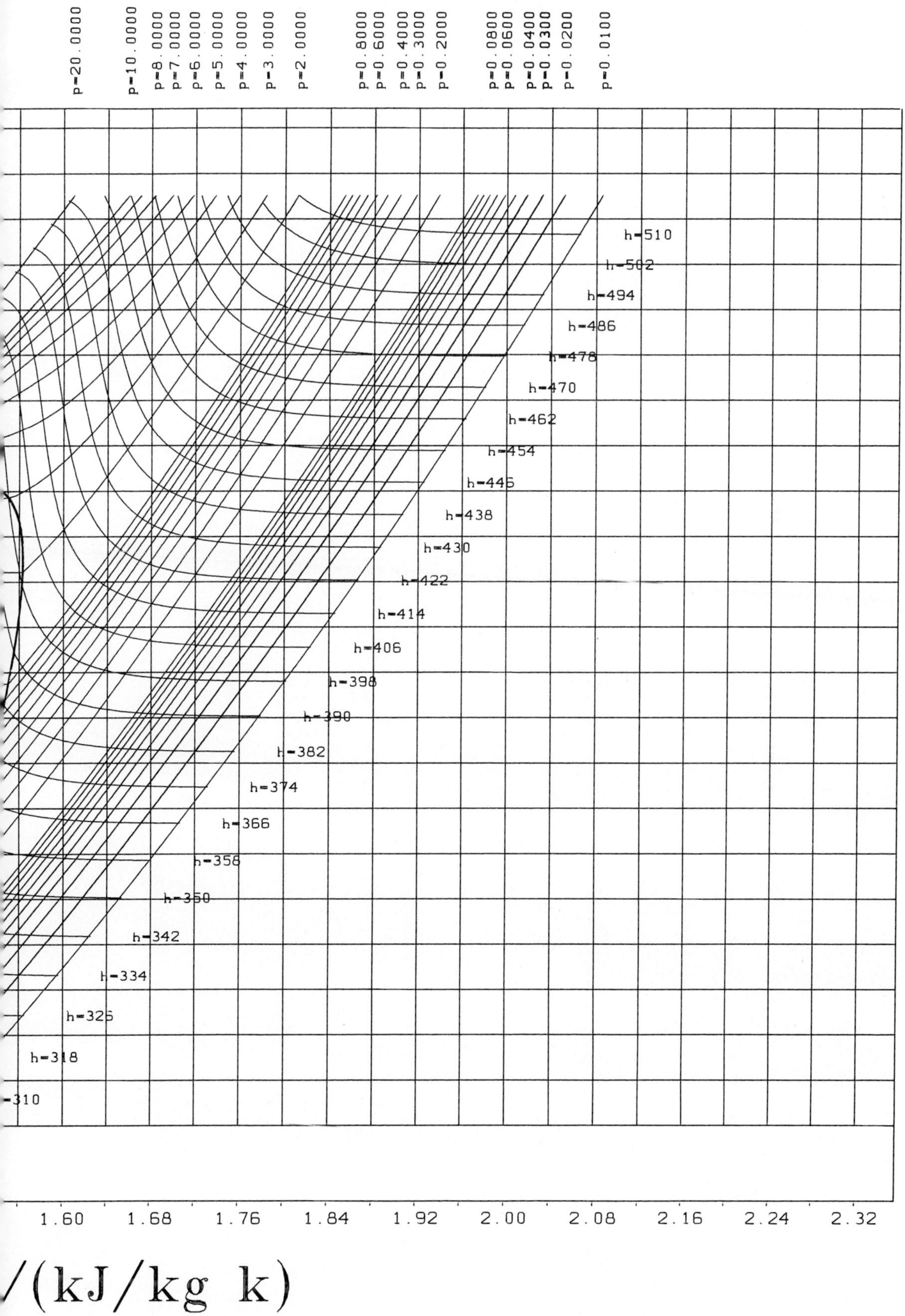
p=20.0000
p=10.0000
p=8.0000
p=7.0000
p=6.0000
p=5.0000
p=4.0000
p=3.0000
p=2.0000
p=0.8000
p=0.6000
p=0.4000
p=0.3000
p=0.2000
p=0.0800
p=0.0600
p=0.0400
p=0.0300
p=0.0200
p=0.0100
h=510
h=502
h=494
h=486
h=478
h=470
h=462
h=454
h=446
h=438
h=430
h=422
h=414
h=406
h=398
h=390
h=382
h=374
h=366
h=358
h=350
h=342
h=334
h=325
h=318
=310
1.60
1.68
1.76
1.84
1.92
2.00
2.08
2.16
2.24
2.32
/(kJ/kg k)

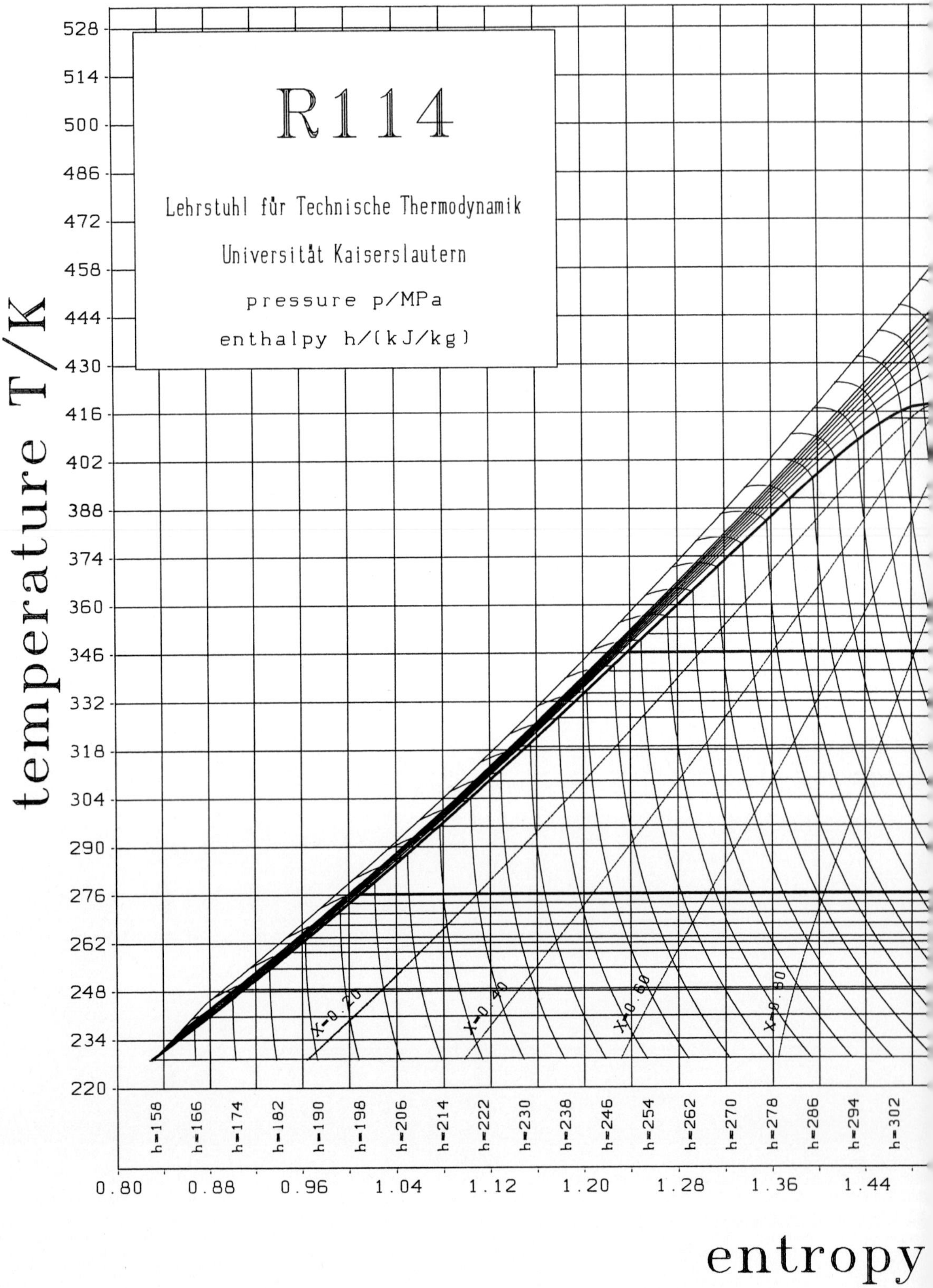

R114
Lehrstuhl für Technische Thermodynamik
Universität Kaiserslautern
pressure p/MPa
enthalpy h/(kJ/kg)
temperature T/K
entropy
528
514
500
486
472
458
444
430
416
402
388
374
360
346
332
318
304
290
276
262
248
234
220
x=0.20
x=0.40
x=0.60
x=0.80
h=158
h=166
h=174
h=182
h=190
h=198
h=206
h=214
h=222
h=230
h=238
h=246
h=254
h=262
h=270
h=278
h=286
h=294
h=302
0.80
0.88
0.96
1.04
1.12
1.20
1.28
1.36
1.44

R 114. hs

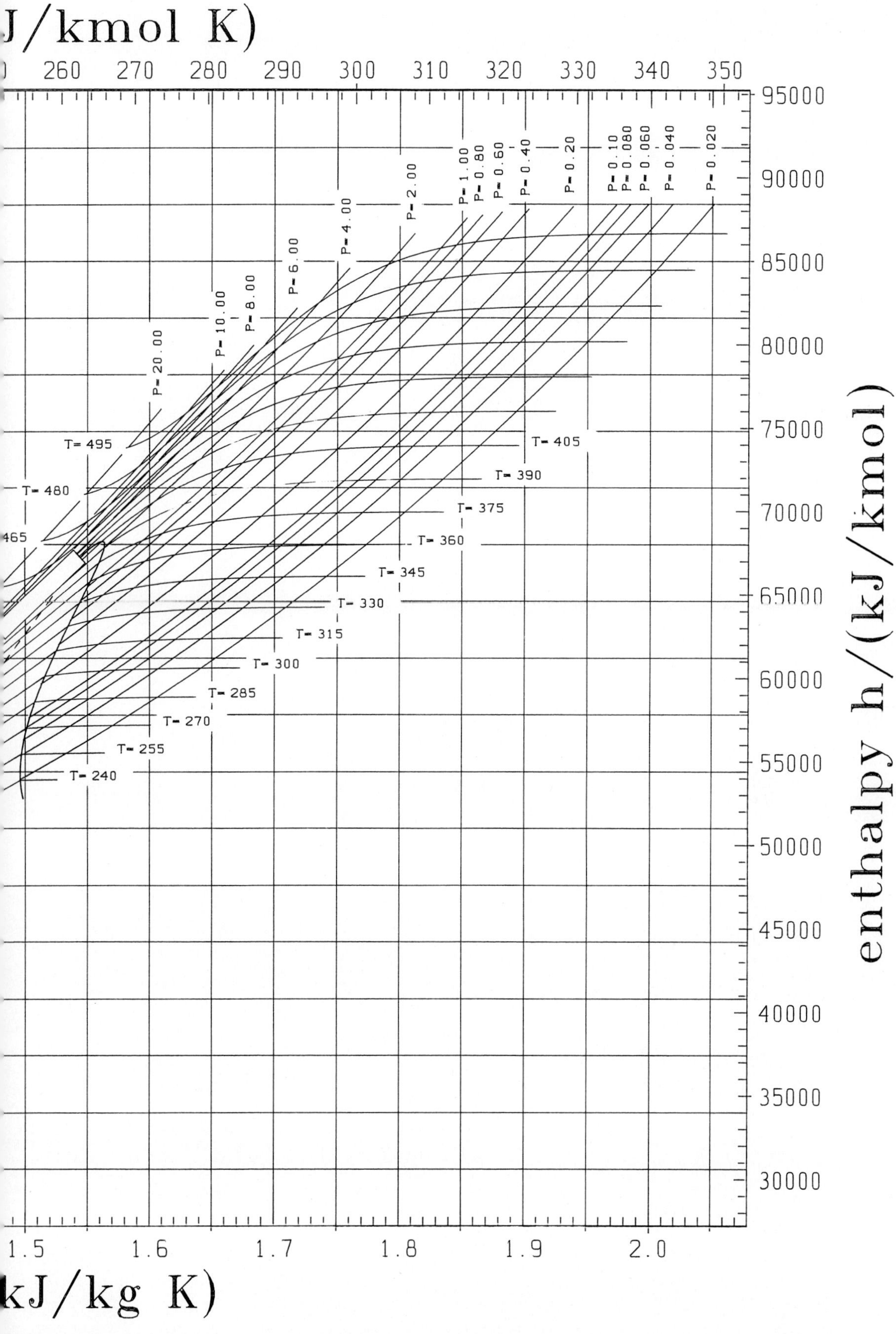

J/kmol K)
260 270 280 290 300 310 320 330 340 350
95000
90000
85000
80000
75000
70000
65000
60000
55000
50000
45000
40000
35000
30000
enthalpy h/(kJ/kmol)
P= 20.00
P= 10.00
P= 8.00
P= 6.00
P= 4.00
P= 2.00
P= 1.00
P= 0.80
P= 0.60
P= 0.40
P= 0.20
P= 0.10
P= 0.080
P= 0.060
P= 0.040
P= 0.020
T= 495
T= 480
465
T= 405
T= 390
T= 375
T= 360
T= 345
T= 330
T= 315
T= 300
T= 285
T= 270
T= 255
T= 240
1.5 1.6 1.7 1.8 1.9 2.0
kJ/kg K)

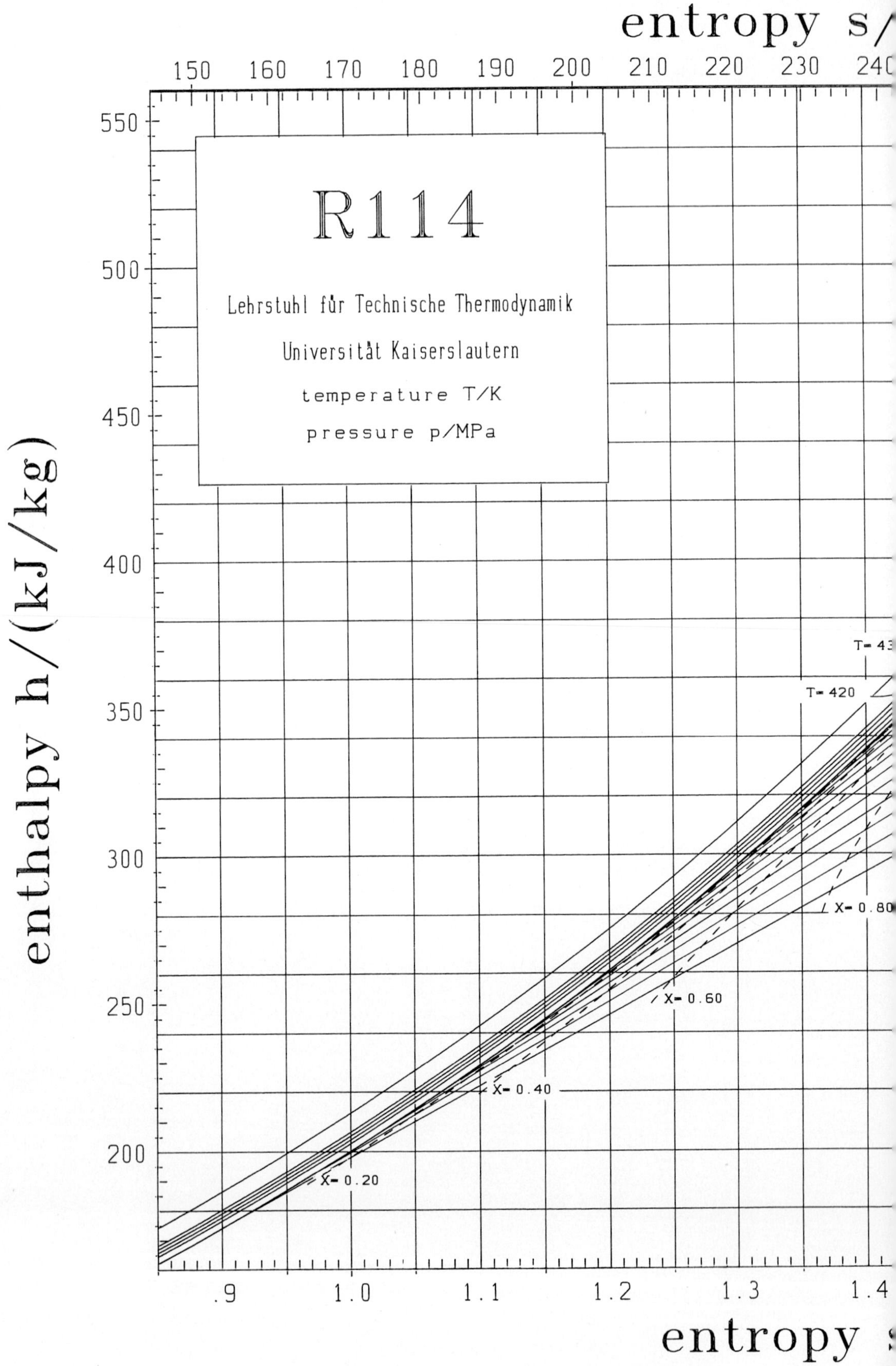
entropy
150 160 170 180 190 200 210 220 230
R114
Lehrstuhl für Technische Thermodynamik
Universität Kaiserslautern
temperature T/K
pressure p/MPa
enthalpy h/(kJ/kg)
550
500
450
400
350
300
250
200
T= 420
X= 0.20
X= 0.40
X= 0.60
X= 0.80
.9 1.0 1.1 1.2 1.3 1.4
entropy

Refrigerant R115 ($CClF_2CF_3$)

Table R115.1: List of references for p, ρ, T data

Author	n	p/MPa min	p/MPa max	ρ/g cm^{-3} min	ρ/g cm^{-3} max	T/K min	T/K max
MATTHIAS, H. *[Matt 66]*	112	0.03	4.40	0.0031	0.59	213	423
MEARS, W.H. *[Mear 66]*	90	1.25	6.99	0.0990	0.88	313	449

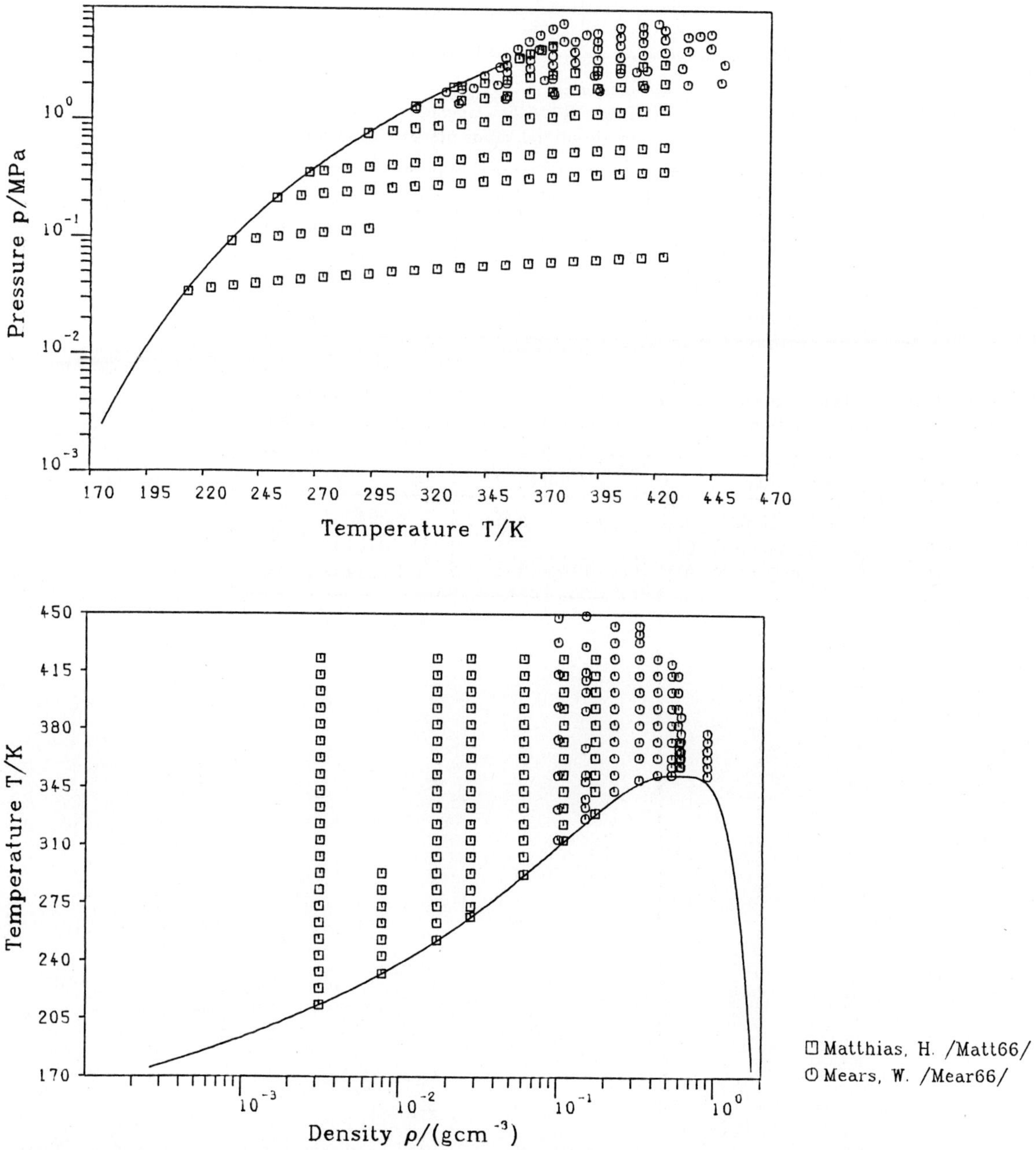

Fig.R115.1: Distribution of p, ρ, T data projected in the p, T and in the T, ρ plane

Table R115.2: Experimental vapor pressure data

Author	Δ / %		
	std	opt	*EOS*
ASTON, J.G. *[Asto 55]*	0.24	0.09	0.06
MATTHIAS, H. *[Matt 66]*	1.45	1.43	1.11
MEARS, W.H. *[Mear 66]*	0.39	0.38	0.45

Table R115.3: Experimental saturated liquid density data

Author	Δ / %		
	std	opt	*EOS*
MEARS, W.H. *[Mear 66]*	0.28	0.06	0.09

Table R115.4: Coefficients of the standard vapor pressure equation

Index i	b_i
5	−7.023396 E+00
6	1.044096 E+00
9	−2.801354 E+00
15	−1.068209 E+00

Table R115.5: Coefficients of the quasi-optimal vapor pressure equation

Index i	b_i
5	−7.679621 E+00
6	4.195499 E+00
7	−4.478978 E+00
19	−9.933725 E+00

Table R115.6: Coefficients of the standard saturated liquid density equation

Index i	c_i
8	1.711552 E+00
9	6.068560 E−01
16	7.606392 E−01

Table R115.7: Coefficients of the quasi-optimal saturated liquid density equation

Index i	c_i
5	6.334964 E−01
8	1.410193 E+00
9	1.091107 E+00

Table **R115.8:** Coefficients of the *Bender-EOS*

Range of validity: one-phase region

$$0.03 \leq p/\text{MPa} \leq 6.99$$
$$0.00313 \leq \rho/\text{g cm}^{-3} \leq 0.88464$$
$$213.10 \leq T/\text{K} \leq 449.26$$

	Specific units	Molar units
a_{01}	2.338967563 E−02	5.581728658 E−01
a_{02}	3.629287331 E+01	8.660956838 E+02
a_{03}	−8.942170397 E+03	−2.133965839 E+05
a_{04}	3.780110251 E+06	9.020881714 E+07
a_{05}	2.379529673 E+08	5.678526364 E+09
a_{06}	4.106890591 E−01	1.514014085 E+00
a_{07}	−2.680773374 E+02	−9.882728937 E+02
a_{08}	5.891167326 E+04	2.171791557 E+05
a_{09}	− 4.113443173 E−01	−2.342580614 E−01
a_{10}	1.824095626 E+02	1.038811252 E+02
a_{11}	4.510191862 E−01	3.967859822 E−02
a_{12}	−2.867270773 E+02	−2.522493243 E+01
a_{13}	7.213444529 E+01	9.803389570 E−01
a_{14}	−1.156640666 E+07	−4.263980793 E+07
a_{15}	4.366061434 E+09	1.609557976 E+10
a_{16}	−4.830345383 E+11	−1.780717257 E+12
a_{17}	1.765759495 E+07	1.553434171 E+06
a_{18}	−1.932739204 E+10	−1.700335256 E+09
a_{19}	3.888796465 E+12	3.421184669 E+11
a_{20}	4.000000000 E+00	9.545628160 E−02

Table **R115.9:** Deviations between experimental and predicted p, ρ, T data

Region	n	Δ / %	
		p	ρ
Gas	155	0.65	0.76
Critical	44	0.66	3.13
Liquid	0	-	-
Total	199	0.66	1.68

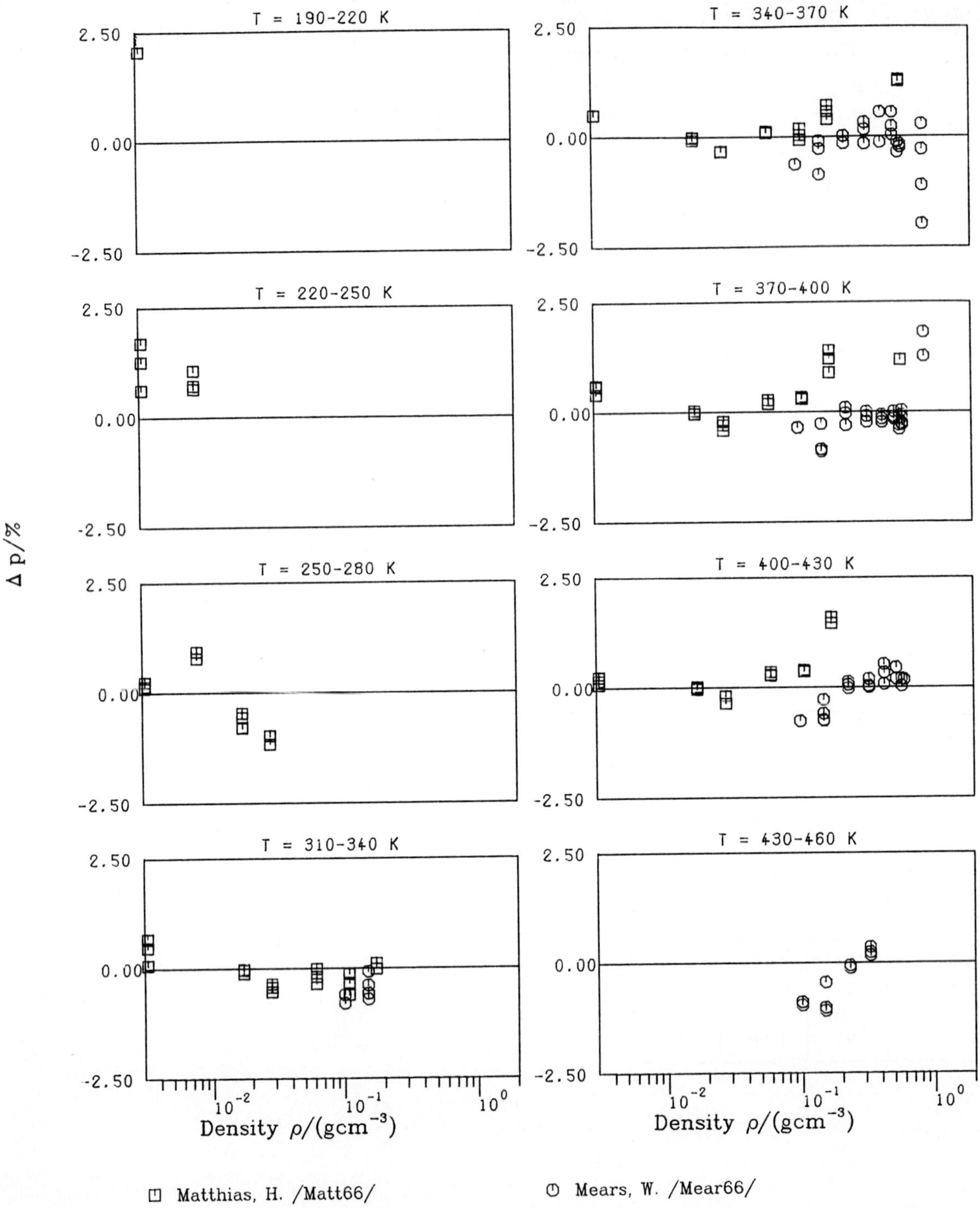

Fig.R115.2: Deviations between experimental and calculated p, ρ, T data

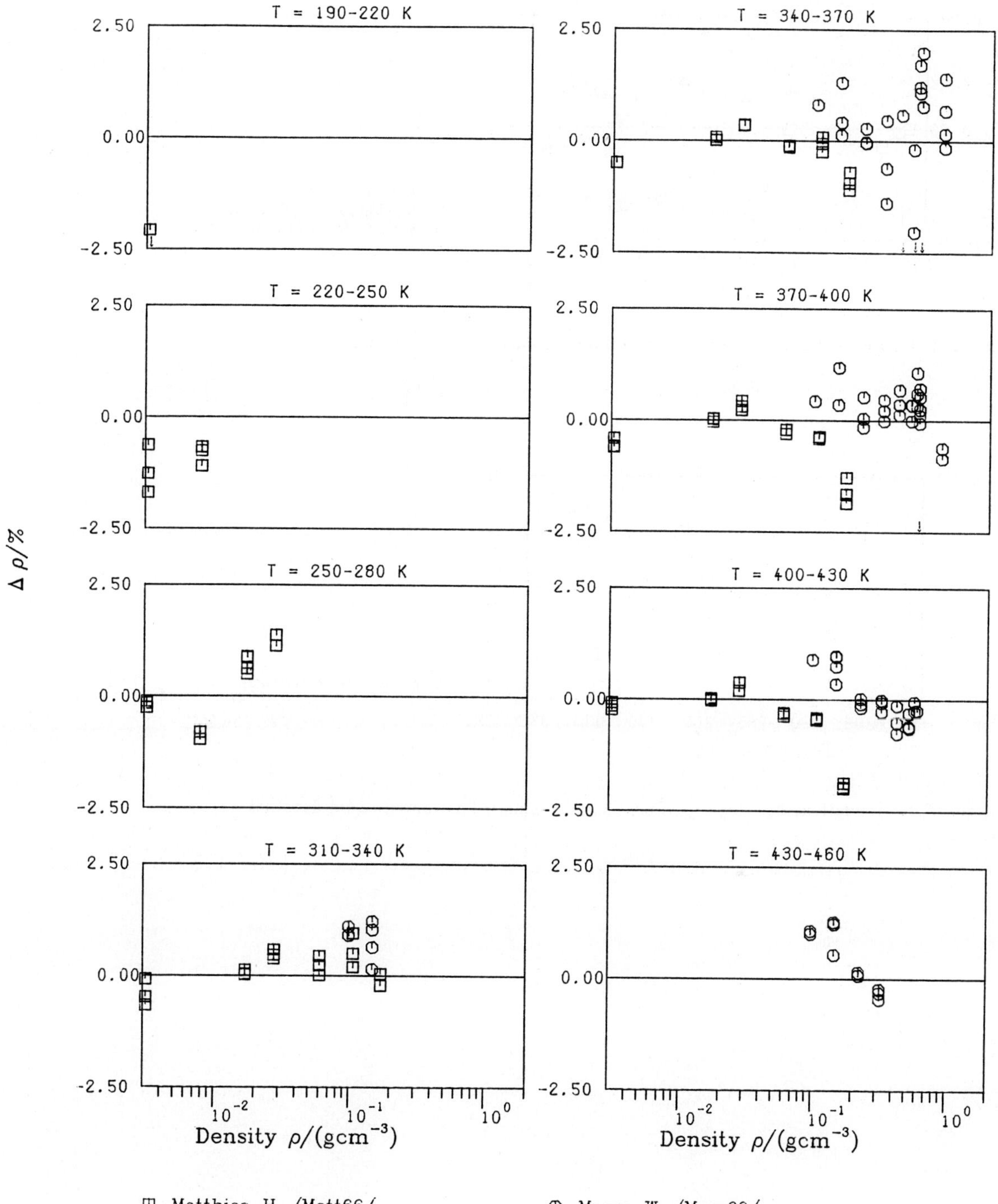

Fig.R115.2: Deviations between experimental and calculated p, ρ, T data *-continued*

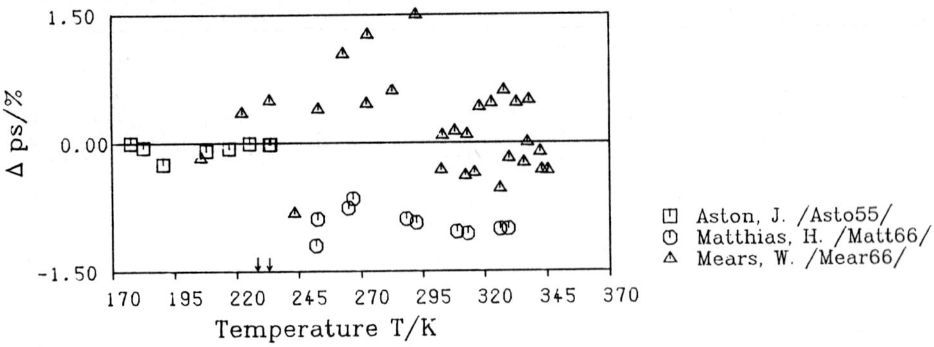

Fig.R115.3: Deviations between experimental and calculated vapor pressures

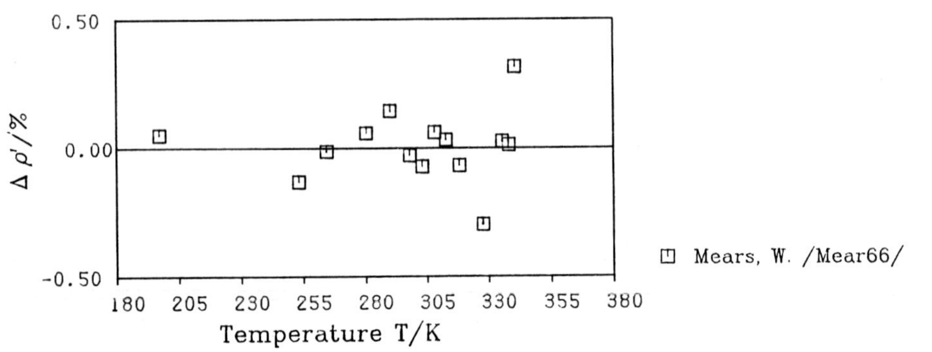

Fig.R115.4: Deviations between experimental and calculated saturated liquid densities

Table R115.10: Information on ideal gas state heat capacity equation

Author	T / K min	T / K max
BARHO, W. *[Barh 66]* ERNST, G. *[Erns 70]*	295	550
Coefficients of equation (15)		
Index i	d_i	
1	1.313895 E−01	
2	2.882158 E−03	
3	−4.406363 E−06	
4	5.603690 E−09	
5	−3.856480 E−12	

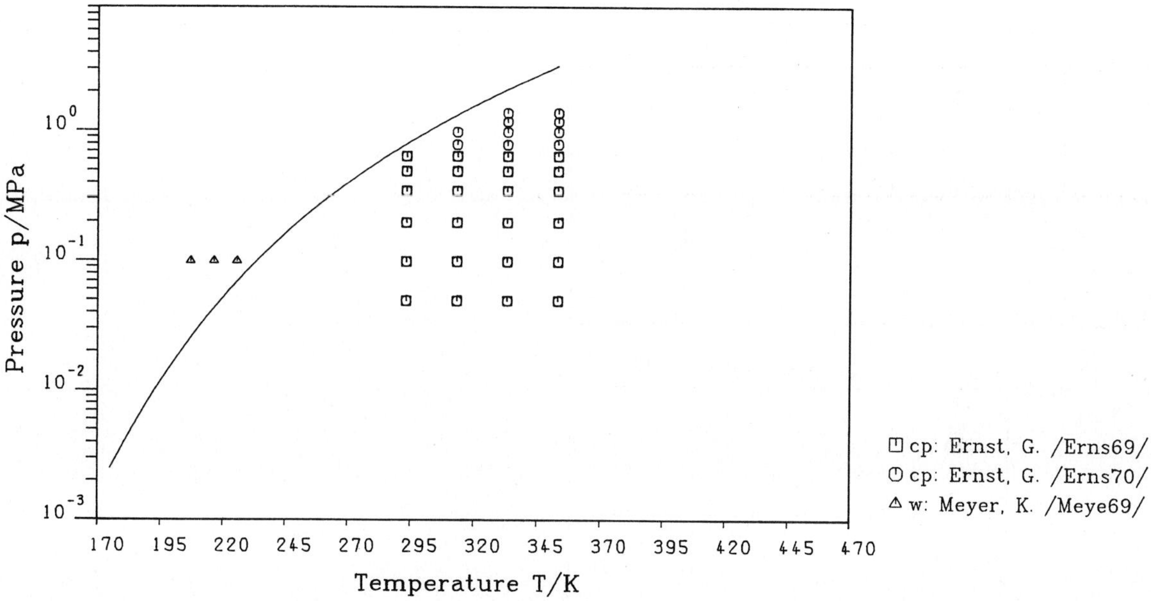

Fig.R115.5: Distribution of $c_p(p,T)$ and $w(p,T)$ data

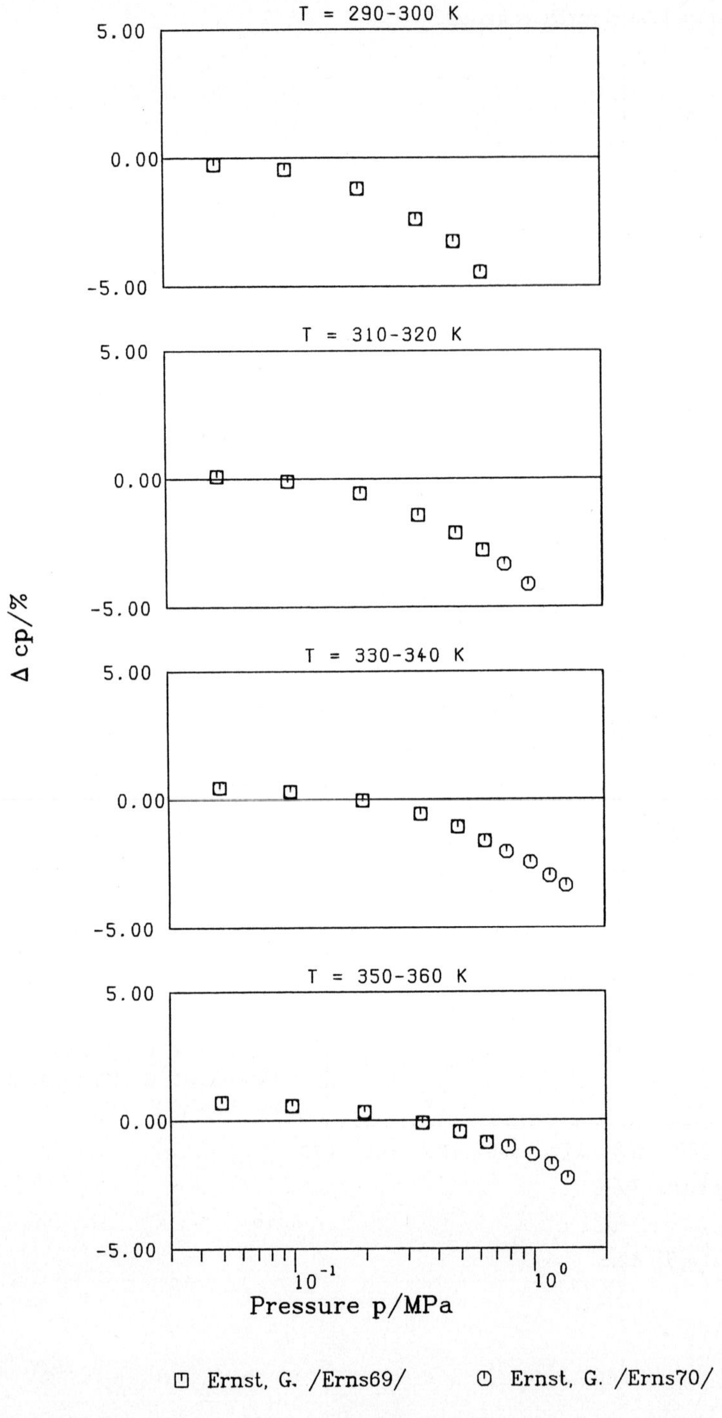

Fig.R115.6: Deviations between experimental and calculated specific isobaric heat capacities c_p

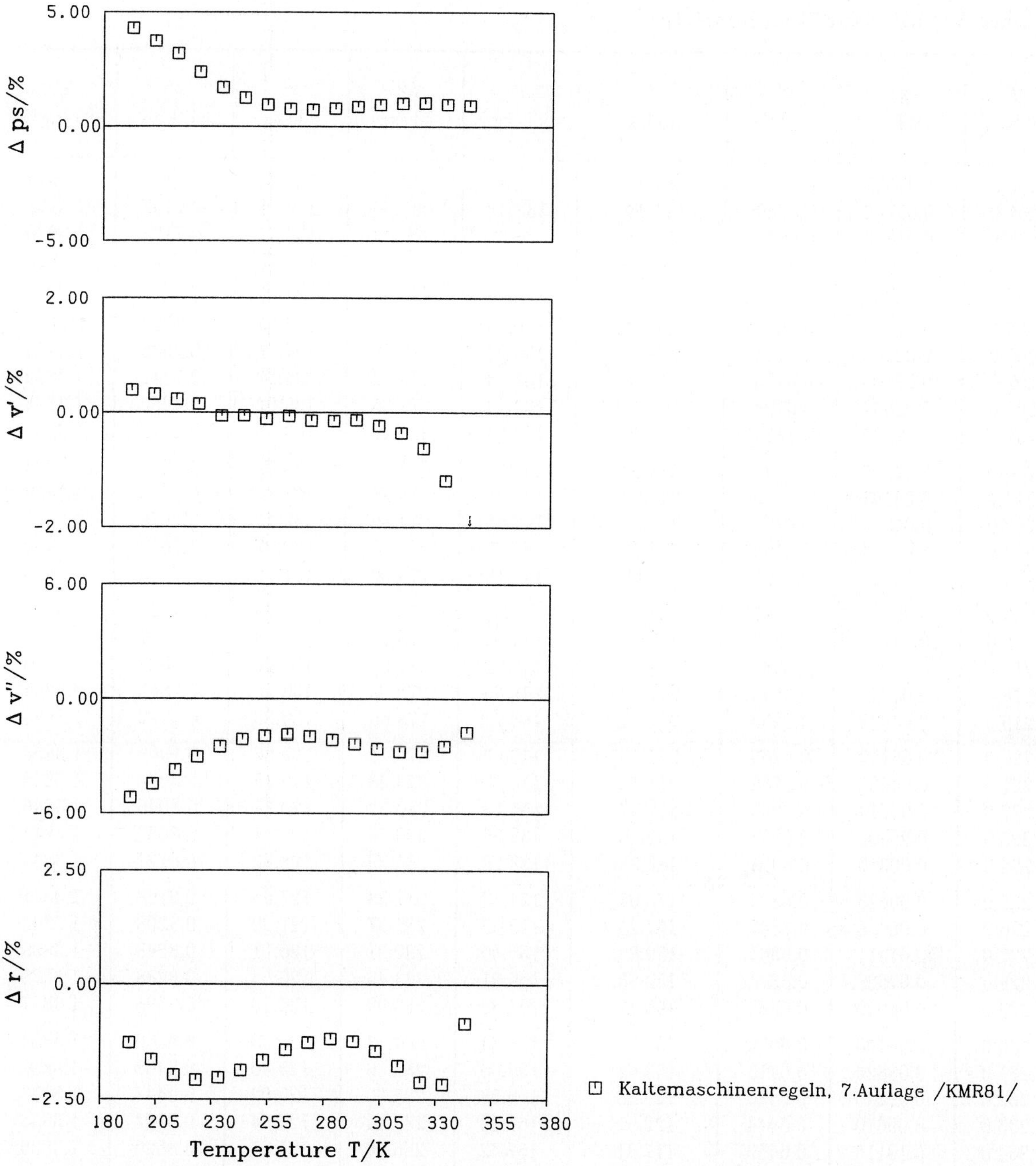

Fig.R115.7: Deviations between saturation properties from the *Kältemaschinenregeln* and those calculated by the *Bender-EOS*

Chart R115.1: Saturation properties

T K	p MPa	v' dm³/kg	v'' dm³/kg	h' kJ/kg	h'' kJ/kg	r kJ/kg	s' kJ/kgK	s'' kJ/kgK
200.0	0.01625	0.6035	652.11	131.52	268.36	136.84	0.7102	1.3944
201.0	0.01732	0.6046	614.14	132.42	268.90	136.48	0.7147	1.3937
202.0	0.01846	0.6057	578.78	133.32	269.43	136.11	0.7191	1.3930
203.0	0.01966	0.6068	545.84	134.23	269.97	135.75	0.7236	1.3923
204.0	0.02091	0.6079	515.12	135.13	270.51	135.38	0.7280	1.3917
205.0	0.02224	0.6090	486.46	136.03	271.05	135.01	0.7324	1.3911
206.0	0.02363	0.6101	459.69	136.94	271.58	134.65	0.7368	1.3905
207.0	0.02508	0.6113	434.68	137.84	272.12	134.28	0.7412	1.3899
208.0	0.02662	0.6124	411.29	138.74	272.66	133.92	0.7456	1.3894
209.0	0.02822	0.6136	389.40	139.65	273.20	133.55	0.7499	1.3889
210.0	0.02990	0.6147	368.90	140.55	273.74	133.19	0.7542	1.3884
211.0	0.03167	0.6159	349.69	141.46	274.28	132.82	0.7585	1.3880
212.0	0.03351	0.6171	331.67	142.36	274.82	132.45	0.7628	1.3876
213.0	0.03544	0.6183	314.76	143.27	275.36	132.09	0.7670	1.3872
214.0	0.03746	0.6195	298.88	144.18	275.90	131.72	0.7713	1.3868
215.0	0.03957	0.6207	283.97	145.08	276.44	131.35	0.7755	1.3865
216.0	0.04177	0.6219	269.94	145.99	276.98	130.99	0.7797	1.3861
217.0	0.04407	0.6231	256.74	146.90	277.52	130.62	0.7839	1.3858
218.0	0.04647	0.6244	244.32	147.80	278.06	130.25	0.7881	1.3855
219.0	0.04898	0.6256	232.62	148.71	278.60	129.88	0.7922	1.3853
220.0	0.05159	0.6269	221.59	149.62	279.14	129.52	0.7963	1.3850
221.0	0.05431	0.6282	211.18	150.53	279.68	129.15	0.8004	1.3848
222.0	0.05714	0.6294	201.37	151.44	280.22	128.78	0.8045	1.3846
223.0	0.06009	0.6307	192.10	152.35	280.75	128.41	0.8086	1.3844
224.0	0.06316	0.6321	183.35	153.26	281.29	128.03	0.8127	1.3843
225.0	0.06635	0.6334	175.07	154.17	281.83	127.66	0.8168	1.3841
226.0	0.06966	0.6347	167.25	155.09	282.37	127.29	0.8208	1.3840
227.0	0.07311	0.6361	159.84	156.00	282.91	126.91	0.8248	1.3839
228.0	0.07669	0.6374	152.83	156.91	283.45	126.54	0.8288	1.3838
229.0	0.08040	0.6388	146.19	157.83	283.99	126.16	0.8328	1.3837
230.0	0.08426	0.6402	139.90	158.74	284.53	125.78	0.8368	1.3837
231.0	0.08826	0.6416	133.94	159.66	285.06	125.40	0.8408	1.3836
232.0	0.09240	0.6430	128.28	160.58	285.60	125.02	0.8447	1.3836
233.0	0.09670	0.6444	122.91	161.50	286.14	124.64	0.8487	1.3836
234.0	0.10116	0.6459	117.81	162.42	286.68	124.25	0.8526	1.3836
235.0	0.10577	0.6473	112.96	163.34	287.21	123.87	0.8565	1.3836
236.0	0.11054	0.6488	108.36	164.27	287.75	123.48	0.8604	1.3837
237.0	0.11549	0.6503	103.98	165.19	288.28	123.09	0.8643	1.3837
238.0	0.12060	0.6518	99.814	166.12	288.82	122.70	0.8682	1.3838
239.0	0.12589	0.6533	95.850	167.05	289.35	122.31	0.8721	1.3838
240.0	0.13135	0.6548	92.076	167.98	289.89	121.91	0.8760	1.3839
241.0	0.13700	0.6563	88.482	168.91	290.42	121.51	0.8798	1.3840
242.0	0.14283	0.6579	85.057	169.84	290.95	121.11	0.8837	1.3841
243.0	0.14886	0.6594	81.792	170.77	291.49	120.71	0.8875	1.3843
244.0	0.15508	0.6610	78.678	171.71	292.02	120.31	0.8913	1.3844
245.0	0.16150	0.6626	75.707	172.65	292.55	119.90	0.8951	1.3845
246.0	0.16813	0.6642	72.871	173.59	293.08	119.49	0.8990	1.3847
247.0	0.17496	0.6659	70.163	174.53	293.61	119.08	0.9028	1.3849
248.0	0.18200	0.6675	67.577	175.47	294.14	118.66	0.9065	1.3850
249.0	0.18926	0.6692	65.106	176.42	294.67	118.25	0.9103	1.3852

Chart R115.1: Saturation properties—*continued*

T K	p MPa	v' dm³/kg	v'' dm³/kg	h' kJ/kg	h'' kJ/kg	r kJ/kg	s' kJ/kgK	s'' kJ/kgK
250.0	0.19674	0.6708	62.743	177.37	295.19	117.83	0.9141	1.3854
251.0	0.20445	0.6725	60.484	178.32	295.72	117.40	0.9179	1.3856
252.0	0.21238	0.6743	58.323	179.27	296.25	116.98	0.9217	1.3858
253.0	0.22055	0.6760	56.254	180.22	296.77	116.55	0.9254	1.3861
254.0	0.22896	0.6777	54.274	181.18	297.30	116.11	0.9292	1.3863
255.0	0.23761	0.6795	52.378	182.14	297.82	115.68	0.9329	1.3865
256.0	0.24651	0.6813	50.561	183.10	298.34	115.24	0.9366	1.3868
257.0	0.25567	0.6831	48.821	184.07	298.86	114.80	0.9404	1.3871
258.0	0.26507	0.6849	47.152	185.03	299.38	114.35	0.9441	1.3873
259.0	0.27474	0.6867	45.551	186.00	299.90	113.90	0.9478	1.3876
260.0	0.28468	0.6886	44.016	186.97	300.42	113.45	0.9515	1.3879
261.0	0.29489	0.6905	42.543	187.95	300.94	112.99	0.9553	1.3882
262.0	0.30537	0.6924	41.129	188.92	301.45	112.53	0.9590	1.3885
263.0	0.31613	0.6943	39.771	189.90	301.97	112.07	0.9627	1.3888
264.0	0.32718	0.6962	38.466	190.88	302.48	111.60	0.9664	1.3891
265.0	0.33852	0.6982	37.213	191.87	303.00	111.13	0.9701	1.3894
266.0	0.35015	0.7002	36.009	192.86	303.51	110.65	0.9737	1.3897
267.0	0.36208	0.7022	34.851	193.85	304.02	110.17	0.9774	1.3901
268.0	0.37432	0.7042	33.737	194.84	304.53	109.69	0.9811	1.3904
269.0	0.38686	0.7062	32.666	195.83	305.03	109.20	0.9848	1.3907
270.0	0.39972	0.7083	31.635	196.83	305.54	108.71	0.9885	1.3911
271.0	0.41290	0.7104	30.643	197.84	306.04	108.21	0.9921	1.3914
272.0	0.42641	0.7125	29.687	198.84	306.55	107.71	0.9958	1.3918
273.0	0.44024	0.7147	28.767	199.85	307.05	107.20	0.9995	1.3921
274.0	0.45440	0.7168	27.881	200.86	307.55	106.69	1.0031	1.3925
275.0	0.46891	0.7190	27.027	201.87	308.05	106.17	1.0068	1.3929
276.0	0.48376	0.7212	26.203	202.89	308.54	105.65	1.0104	1.3932
277.0	0.49895	0.7235	25.410	203.91	309.04	105.13	1.0141	1.3936
278.0	0.51451	0.7258	24.644	204.94	309.53	104.60	1.0177	1.3940
279.0	0.53042	0.7281	23.905	205.96	310.03	104.06	1.0214	1.3944
280.0	0.54669	0.7304	23.193	206.99	310.52	103.52	1.0250	1.3947
281.0	0.56334	0.7328	22.505	208.03	311.00	102.98	1.0287	1.3951
282.0	0.58036	0.7351	21.841	209.06	311.49	102.43	1.0323	1.3955
283.0	0.59776	0.7376	21.199	210.10	311.97	101.87	1.0359	1.3959
284.0	0.61554	0.7400	20.580	211.15	312.46	101.31	1.0396	1.3963
285.0	0.63372	0.7425	19.981	212.20	312.94	100.74	1.0432	1.3967
286.0	0.65229	0.7450	19.403	213.25	313.41	100.17	1.0468	1.3971
287.0	0.67126	0.7475	18.844	214.30	313.89	99.59	1.0505	1.3975
288.0	0.69064	0.7501	18.303	215.36	314.36	99.00	1.0541	1.3979
289.0	0.71043	0.7527	17.780	216.42	314.83	98.41	1.0577	1.3983
290.0	0.73064	0.7554	17.274	217.49	315.30	97.81	1.0614	1.3986
291.0	0.75127	0.7581	16.784	218.56	315.77	97.21	1.0650	1.3990
292.0	0.77233	0.7608	16.310	219.63	316.23	96.60	1.0686	1.3994
293.0	0.79383	0.7636	15.852	220.71	316.69	95.98	1.0722	1.3998
294.0	0.81576	0.7664	15.407	221.79	317.15	95.36	1.0759	1.4002
295.0	0.83813	0.7692	14.977	222.87	317.60	94.73	1.0795	1.4006
296.0	0.86096	0.7721	14.560	223.96	318.06	94.09	1.0831	1.4010
297.0	0.88425	0.7751	14.156	225.06	318.50	93.45	1.0868	1.4014
298.0	0.90799	0.7780	13.764	226.16	318.95	92.79	1.0904	1.4018
299.0	0.93220	0.7811	13.385	227.26	319.39	92.13	1.0940	1.4022

Chart R115.1: Saturation properties—*continued*

T K	p MPa	v' dm³/kg	v'' dm³/kg	h' kJ/kg	h'' kJ/kg	r kJ/kg	s' kJ/kgK	s'' kJ/kgK
300.0	0.95689	0.7841	13.016	228.36	319.83	91.47	1.0976	1.4025
301.0	0.98206	0.7873	12.659	229.47	320.27	90.79	1.1013	1.4029
302.0	1.00771	0.7905	12.312	230.59	320.70	90.11	1.1049	1.4033
303.0	1.03385	0.7937	11.976	231.71	321.13	89.42	1.1085	1.4036
304.0	1.06049	0.7970	11.649	232.83	321.55	88.72	1.1122	1.4040
305.0	1.08764	0.8003	11.332	233.96	321.97	88.01	1.1158	1.4044
306.0	1.11529	0.8037	11.024	235.10	322.39	87.29	1.1194	1.4047
307.0	1.14346	0.8072	10.725	236.23	322.80	86.56	1.1231	1.4050
308.0	1.17215	0.8107	10.434	237.38	323.20	85.82	1.1267	1.4054
309.0	1.20137	0.8143	10.151	238.53	323.60	85.08	1.1304	1.4057
310.0	1.23113	0.8180	9.8767	239.68	324.00	84.32	1.1340	1.4060
311.0	1.26143	0.8217	9.6094	240.84	324.39	83.55	1.1377	1.4063
312.0	1.29228	0.8256	9.3494	242.01	324.78	82.77	1.1413	1.4066
313.0	1.32368	0.8295	9.0964	243.18	325.16	81.98	1.1450	1.4069
314.0	1.35565	0.8334	8.8502	244.35	325.53	81.18	1.1487	1.4072
315.0	1.38818	0.8375	8.6105	245.53	325.90	80.37	1.1523	1.4075
316.0	1.42130	0.8417	8.3770	246.72	326.26	79.54	1.1560	1.4077
317.0	1.45500	0.8459	8.1496	247.92	326.62	78.70	1.1597	1.4080
318.0	1.48929	0.8503	7.9279	249.12	326.97	77.85	1.1634	1.4082
319.0	1.52418	0.8548	7.7119	250.33	327.31	76.98	1.1671	1.4084
320.0	1.55967	0.8593	7.5012	251.54	327.64	76.10	1.1708	1.4086
321.0	1.59579	0.8640	7.2957	252.76	327.97	75.20	1.1745	1.4088
322.0	1.63253	0.8688	7.0951	253.99	328.28	74.29	1.1782	1.4089
323.0	1.66990	0.8738	6.8994	255.23	328.59	73.36	1.1820	1.4091
324.0	1.70791	0.8789	6.7082	256.47	328.88	72.41	1.1857	1.4092
325.0	1.74658	0.8841	6.5214	257.73	329.17	71.44	1.1895	1.4093
326.0	1.78590	0.8895	6.3389	258.99	329.45	70.45	1.1933	1.4094
327.0	1.82590	0.8951	6.1604	260.26	329.71	69.45	1.1970	1.4094
328.0	1.86658	0.9008	5.9858	261.54	329.96	68.42	1.2008	1.4094
329.0	1.90794	0.9068	5.8149	262.84	330.20	67.36	1.2047	1.4094
330.0	1.95001	0.9129	5.6475	264.14	330.42	66.28	1.2085	1.4094
331.0	1.99279	0.9193	5.4835	265.45	330.63	65.18	1.2124	1.4093
332.0	2.03630	0.9259	5.3226	266.78	330.82	64.04	1.2162	1.4091
333.0	2.08055	0.9328	5.1647	268.12	331.00	62.88	1.2201	1.4090
334.0	2.12554	0.9400	5.0097	269.47	331.15	61.68	1.2241	1.4087
335.0	2.17130	0.9476	4.8573	270.84	331.29	60.44	1.2280	1.4085
336.0	2.21784	0.9554	4.7074	272.23	331.40	59.17	1.2320	1.4081
337.0	2.26518	0.9637	4.5596	273.63	331.48	57.85	1.2361	1.4077
338.0	2.31332	0.9724	4.4139	275.06	331.54	56.49	1.2402	1.4073
339.0	2.36230	0.9817	4.2700	276.50	331.57	55.07	1.2443	1.4067
340.0	2.41213	0.9915	4.1276	277.97	331.56	53.59	1.2485	1.4061
341.0	2.46283	1.0019	3.9864	279.47	331.52	52.05	1.2527	1.4053
342.0	2.51442	1.0131	3.8461	281.00	331.42	50.43	1.2570	1.4045
343.0	2.56693	1.0253	3.7063	282.56	331.28	48.72	1.2614	1.4035
344.0	2.62040	1.0384	3.5664	284.16	331.08	46.92	1.2660	1.4023
345.0	2.67484	1.0529	3.4259	285.82	330.81	44.99	1.2706	1.4010
346.0	2.73031	1.0690	3.2840	287.54	330.45	42.91	1.2754	1.3994
347.0	2.78685	1.0872	3.1395	289.33	329.98	40.65	1.2804	1.3975
348.0	2.84450	1.1081	2.9908	291.22	329.37	38.16	1.2856	1.3953
349.0	2.90335	1.1330	2.8356	293.24	328.58	35.34	1.2913	1.3925

Chart R115.1: Saturation properties—*continued*

T K	p MPa	v' dm^3/kg	v'' dm^3/kg	h' kJ/kg	h'' kJ/kg	r kJ/kg	s' kJ/kgK	s'' kJ/kgK
350.0	2.96347	1.1637	2.6695	295.47	327.51	32.04	1.2974	1.3890
351.0	3.02500	1.2048	2.4842	298.02	325.99	27.97	1.3045	1.3842
352.0	3.08814	1.2687	2.2573	301.28	323.57	22.29	1.3136	1.3769
353.0	3.15391	1.4376	1.5284	307.47	309.91	2.44	1.3309	1.3378

Chart R115.2: Saturation properties

p MPa	T K	v' dm³/kg	v'' dm³/kg	h' kJ/kg	h'' kJ/kg	r kJ/kg	s' kJ/kgK	s'' kJ/kgK
0.0210	204.07	0.6080	513.14	135.19	270.55	135.36	0.7283	1.3916
0.0220	204.83	0.6088	491.33	135.87	270.95	135.08	0.7317	1.3912
0.0230	205.56	0.6096	471.35	136.53	271.35	134.81	0.7349	1.3907
0.0240	206.26	0.6104	452.99	137.17	271.73	134.55	0.7380	1.3903
0.0250	206.94	0.6112	436.05	137.79	272.09	134.30	0.7410	1.3900
0.0260	207.60	0.6119	420.37	138.39	272.45	134.06	0.7438	1.3896
0.0270	208.24	0.6127	405.81	138.96	272.79	133.83	0.7466	1.3893
0.0280	208.87	0.6134	392.26	139.53	273.13	133.60	0.7493	1.3890
0.0290	209.47	0.6141	379.62	140.07	273.45	133.38	0.7519	1.3887
0.0300	210.06	0.6148	367.79	140.60	273.77	133.17	0.7545	1.3884
0.0310	210.63	0.6154	356.69	141.12	274.08	132.96	0.7569	1.3882
0.0320	211.18	0.6161	346.27	141.63	274.38	132.75	0.7593	1.3879
0.0330	211.73	0.6167	336.46	142.12	274.67	132.55	0.7616	1.3877
0.0340	212.26	0.6174	327.21	142.60	274.96	132.36	0.7639	1.3875
0.0350	212.78	0.6180	318.47	143.07	275.24	132.17	0.7661	1.3873
0.0360	213.28	0.6186	310.19	143.52	275.51	131.98	0.7682	1.3871
0.0370	213.78	0.6192	302.35	143.97	275.78	131.80	0.7703	1.3869
0.0380	214.26	0.6198	294.91	144.41	276.04	131.63	0.7724	1.3867
0.0390	214.73	0.6203	287.84	144.84	276.29	131.45	0.7744	1.3865
0.0400	215.20	0.6209	281.11	145.26	276.54	131.28	0.7763	1.3864
0.0410	215.65	0.6215	274.69	145.68	276.79	131.11	0.7783	1.3862
0.0420	216.10	0.6220	268.58	146.08	277.03	130.95	0.7801	1.3861
0.0430	216.54	0.6225	262.73	146.48	277.27	130.79	0.7820	1.3860
0.0440	216.97	0.6231	257.14	146.87	277.50	130.63	0.7838	1.3858
0.0450	217.39	0.6236	251.80	147.25	277.73	130.48	0.7855	1.3857
0.0460	217.81	0.6241	246.68	147.63	277.95	130.32	0.7872	1.3856
0.0470	218.21	0.6246	241.76	148.00	278.17	130.17	0.7889	1.3855
0.0480	218.61	0.6251	237.05	148.36	278.39	130.03	0.7906	1.3854
0.0490	219.01	0.6256	232.52	148.72	278.60	129.88	0.7922	1.3853
0.0500	219.40	0.6261	228.17	149.07	278.81	129.74	0.7938	1.3852
0.0510	219.78	0.6266	223.98	149.42	279.02	129.60	0.7954	1.3851
0.0520	220.15	0.6271	219.95	149.76	279.22	129.46	0.7970	1.3850
0.0530	220.52	0.6275	216.06	150.10	279.42	129.32	0.7985	1.3849
0.0540	220.89	0.6280	212.31	150.43	279.62	129.19	0.8000	1.3848
0.0550	221.25	0.6285	208.70	150.76	279.81	129.05	0.8015	1.3848
0.0560	221.60	0.6289	205.21	151.08	280.00	128.92	0.8029	1.3847
0.0570	221.95	0.6294	201.83	151.39	280.19	128.79	0.8043	1.3846
0.0580	222.30	0.6298	198.57	151.71	280.37	128.67	0.8058	1.3846
0.0590	222.64	0.6303	195.42	152.02	280.56	128.54	0.8071	1.3845
0.0600	222.97	0.6307	192.37	152.32	280.74	128.42	0.8085	1.3844
0.0610	223.30	0.6311	189.41	152.62	280.92	128.29	0.8099	1.3844
0.0620	223.63	0.6316	186.55	152.92	281.09	128.17	0.8112	1.3843
0.0630	223.95	0.6320	183.77	153.22	281.27	128.05	0.8125	1.3843
0.0640	224.27	0.6324	181.08	153.51	281.44	127.93	0.8138	1.3842
0.0650	224.58	0.6328	178.47	153.79	281.61	127.82	0.8151	1.3842
0.0660	224.89	0.6332	175.93	154.08	281.78	127.70	0.8163	1.3841
0.0670	225.20	0.6336	173.47	154.36	281.94	127.59	0.8176	1.3841
0.0680	225.50	0.6341	171.08	154.63	282.10	127.47	0.8188	1.3841
0.0690	225.80	0.6345	168.75	154.91	282.27	127.36	0.8200	1.3840
0.0700	226.10	0.6349	166.49	155.18	282.43	127.25	0.8212	1.3840

Chart R115.2: Saturation properties—*continued*

p MPa	T K	v' dm³/kg	v'' dm³/kg	h' kJ/kg	h'' kJ/kg	r kJ/kg	s' kJ/kgK	s'' kJ/kgK
0.0710	226.39	0.6352	164.29	155.44	282.58	127.14	0.8224	1.3840
0.0720	226.68	0.6356	162.15	155.71	282.74	127.03	0.8235	1.3839
0.0730	226.97	0.6360	160.07	155.97	282.89	126.92	0.8247	1.3839
0.0740	227.25	0.6364	158.04	156.23	283.05	126.82	0.8258	1.3839
0.0760	227.81	0.6372	154.13	156.74	283.35	126.61	0.8281	1.3838
0.0780	228.36	0.6379	150.42	157.24	283.64	126.40	0.8303	1.3838
0.0800	228.89	0.6387	146.88	157.73	283.93	126.20	0.8324	1.3837
0.0820	229.42	0.6394	143.52	158.21	284.21	126.00	0.8345	1.3837
0.0840	229.93	0.6401	140.31	158.68	284.49	125.81	0.8365	1.3837
0.0860	230.44	0.6408	137.24	159.15	284.76	125.61	0.8385	1.3837
0.0880	230.94	0.6415	134.30	159.60	285.03	125.43	0.8405	1.3836
0.0900	231.42	0.6422	131.50	160.05	285.29	125.24	0.8424	1.3836
0.0920	231.90	0.6429	128.81	160.49	285.55	125.06	0.8443	1.3836
0.0940	232.38	0.6435	126.23	160.93	285.80	124.88	0.8462	1.3836
0.0960	232.84	0.6442	123.75	161.35	286.05	124.70	0.8480	1.3836
0.0980	233.29	0.6448	121.37	161.77	286.30	124.53	0.8498	1.3836
0.1000	233.74	0.6455	119.09	162.19	286.54	124.35	0.8516	1.3836
0.1020	234.19	0.6461	116.89	162.59	286.78	124.18	0.8533	1.3836
0.1040	234.62	0.6468	114.77	162.99	287.01	124.02	0.8550	1.3836
0.1060	235.05	0.6474	112.73	163.39	287.24	123.85	0.8567	1.3836
0.1080	235.47	0.6480	110.76	163.78	287.46	123.69	0.8584	1.3836
0.1100	235.89	0.6486	108.86	164.16	287.69	123.52	0.8600	1.3836
0.1150	236.90	0.6501	104.39	165.10	288.23	123.13	0.8639	1.3837
0.1200	237.88	0.6516	100.28	166.01	288.76	122.75	0.8678	1.3838
0.1250	238.83	0.6530	96.491	166.89	289.26	122.37	0.8714	1.3838
0.1300	239.76	0.6544	92.981	167.75	289.76	122.01	0.8750	1.3839
0.1350	240.65	0.6558	89.721	168.58	290.23	121.65	0.8785	1.3840
0.1400	241.52	0.6571	86.686	169.39	290.70	121.31	0.8818	1.3841
0.1450	242.36	0.6584	83.853	170.18	291.15	120.97	0.8851	1.3842
0.1500	243.19	0.6597	81.203	170.95	291.58	120.64	0.8882	1.3843
0.1550	243.99	0.6610	78.717	171.70	292.01	120.31	0.8913	1.3844
0.1600	244.77	0.6622	76.381	172.43	292.43	119.99	0.8943	1.3845
0.1650	245.53	0.6635	74.182	173.15	292.83	119.68	0.8972	1.3846
0.1700	246.28	0.6647	72.107	173.85	293.23	119.38	0.9000	1.3847
0.1750	247.01	0.6659	70.147	174.54	293.61	119.08	0.9028	1.3849
0.1800	247.72	0.6670	68.292	175.21	293.99	118.78	0.9055	1.3850
0.1850	248.42	0.6682	66.534	175.87	294.36	118.49	0.9081	1.3851
0.1900	249.10	0.6693	64.864	176.51	294.72	118.20	0.9107	1.3852
0.1950	249.77	0.6705	63.278	177.15	295.07	117.92	0.9132	1.3854
0.2000	250.43	0.6716	61.768	177.77	295.42	117.65	0.9157	1.3855
0.2100	251.70	0.6737	58.955	178.99	296.09	117.10	0.9205	1.3858
0.2200	252.93	0.6759	56.390	180.16	296.74	116.58	0.9252	1.3861
0.2300	254.12	0.6779	54.040	181.30	297.36	116.06	0.9296	1.3863
0.2400	255.27	0.6800	51.878	182.40	297.96	115.56	0.9339	1.3866
0.2500	256.38	0.6820	49.884	183.47	298.54	115.07	0.9381	1.3869
0.2600	257.46	0.6839	48.037	184.51	299.10	114.59	0.9421	1.3872
0.2700	258.51	0.6858	46.323	185.53	299.65	114.12	0.9460	1.3875
0.2800	259.53	0.6877	44.726	186.52	300.18	113.66	0.9498	1.3878
0.2900	260.52	0.6896	43.236	187.48	300.69	113.21	0.9535	1.3880
0.3000	261.49	0.6914	41.841	188.43	301.19	112.77	0.9571	1.3883

Chart R115.2: Saturation properties–*continued*

p MPa	T K	v' dm³/kg	v'' dm³/kg	h' kJ/kg	h'' kJ/kg	r kJ/kg	s' kJ/kgK	s'' kJ/kgK
0.3100	262.43	0.6932	40.533	189.35	301.68	112.33	0.9606	1.3886
0.3200	263.35	0.6950	39.304	190.25	302.15	111.90	0.9640	1.3889
0.3300	264.25	0.6967	38.147	191.13	302.61	111.48	0.9673	1.3892
0.3400	265.13	0.6984	37.055	192.00	303.06	111.07	0.9705	1.3894
0.3500	265.99	0.7001	36.024	192.84	303.50	110.66	0.9737	1.3897
0.3600	266.83	0.7018	35.047	193.67	303.93	110.25	0.9768	1.3900
0.3700	267.65	0.7035	34.122	194.49	304.35	109.86	0.9798	1.3903
0.3800	268.46	0.7051	33.243	195.29	304.76	109.46	0.9828	1.3905
0.3900	269.25	0.7068	32.408	196.08	305.16	109.08	0.9857	1.3908
0.4000	270.02	0.7084	31.613	196.85	305.55	108.70	0.9885	1.3911
0.4100	270.78	0.7100	30.856	197.62	305.93	108.32	0.9913	1.3913
0.4200	271.53	0.7115	30.133	198.37	306.31	107.94	0.9941	1.3916
0.4300	272.26	0.7131	29.443	199.10	306.68	107.57	0.9968	1.3919
0.4400	272.98	0.7146	28.783	199.83	307.04	107.21	0.9994	1.3921
0.4500	273.69	0.7162	28.151	200.55	307.40	106.85	1.0020	1.3924
0.4600	274.39	0.7177	27.545	201.25	307.74	106.49	1.0045	1.3926
0.4700	275.07	0.7192	26.964	201.95	308.08	106.14	1.0070	1.3929
0.4800	275.75	0.7207	26.407	202.64	308.42	105.78	1.0095	1.3931
0.4900	276.41	0.7222	25.871	203.31	308.75	105.44	1.0119	1.3934
0.5000	277.07	0.7236	25.357	203.98	309.07	105.09	1.0143	1.3936
0.5100	277.71	0.7251	24.861	204.64	309.39	104.75	1.0167	1.3939
0.5200	278.35	0.7266	24.384	205.29	309.70	104.41	1.0190	1.3941
0.5300	278.97	0.7280	23.924	205.94	310.01	104.08	1.0213	1.3943
0.5400	279.59	0.7294	23.481	206.57	310.32	103.74	1.0235	1.3946
0.5500	280.20	0.7309	23.053	207.20	310.61	103.41	1.0257	1.3948
0.5600	280.80	0.7323	22.640	207.82	310.91	103.09	1.0279	1.3950
0.5700	281.39	0.7337	22.240	208.43	311.19	102.76	1.0301	1.3953
0.5800	281.98	0.7351	21.854	209.04	311.48	102.44	1.0322	1.3955
0.5900	282.56	0.7365	21.481	209.64	311.76	102.12	1.0343	1.3957
0.6000	283.13	0.7379	21.119	210.24	312.03	101.80	1.0364	1.3959
0.6100	283.69	0.7392	20.769	210.82	312.31	101.48	1.0384	1.3962
0.6200	284.25	0.7406	20.430	211.41	312.57	101.17	1.0405	1.3964
0.6300	284.80	0.7420	20.101	211.98	312.84	100.86	1.0425	1.3966
0.6400	285.34	0.7433	19.782	212.55	313.10	100.55	1.0444	1.3968
0.6500	285.88	0.7447	19.473	213.12	313.35	100.24	1.0464	1.3970
0.6600	286.41	0.7460	19.172	213.68	313.61	99.93	1.0483	1.3972
0.6700	286.93	0.7474	18.880	214.23	313.86	99.63	1.0502	1.3974
0.6800	287.45	0.7487	18.596	214.78	314.10	99.32	1.0521	1.3976
0.6900	287.97	0.7500	18.320	215.32	314.35	99.02	1.0540	1.3978
0.7000	288.48	0.7514	18.052	215.86	314.59	98.72	1.0558	1.3980
0.7100	288.98	0.7527	17.791	216.40	314.82	98.42	1.0577	1.3982
0.7200	289.48	0.7540	17.537	216.93	315.06	98.13	1.0595	1.3984
0.7300	289.97	0.7553	17.290	217.45	315.29	97.83	1.0612	1.3986
0.7400	290.46	0.7566	17.049	217.98	315.51	97.54	1.0630	1.3988
0.7600	291.42	0.7592	16.585	219.00	315.96	96.96	1.0665	1.3992
0.7800	292.36	0.7618	16.144	220.02	316.40	96.38	1.0699	1.3996
0.8000	293.28	0.7644	15.724	221.01	316.82	95.81	1.0733	1.3999
0.8200	294.19	0.7669	15.324	222.00	317.24	95.24	1.0766	1.4003
0.8400	295.08	0.7695	14.942	222.96	317.64	94.68	1.0798	1.4006
0.8600	295.96	0.7720	14.577	223.92	318.04	94.12	1.0830	1.4010

Chart R115.2: Saturation properties–*continued*

p MPa	T K	v' dm³/kg	v'' dm³/kg	h' kJ/kg	h'' kJ/kg	r kJ/kg	s' kJ/kgK	s'' kJ/kgK
0.8800	296.82	0.7745	14.228	224.86	318.42	93.56	1.0861	1.4013
0.9000	297.67	0.7770	13.894	225.79	318.80	93.01	1.0892	1.4016
0.9200	298.50	0.7795	13.574	226.70	319.17	92.47	1.0922	1.4020
0.9400	299.32	0.7820	13.266	227.61	319.53	91.92	1.0952	1.4023
0.9600	300.12	0.7845	12.971	228.50	319.89	91.38	1.0981	1.4026
0.9800	300.92	0.7870	12.688	229.38	320.23	90.85	1.1010	1.4029
1.0000	301.70	0.7895	12.415	230.26	320.57	90.31	1.1038	1.4032
1.0200	302.47	0.7920	12.152	231.12	320.90	89.78	1.1066	1.4034
1.0400	303.23	0.7944	11.899	231.97	321.22	89.25	1.1094	1.4037
1.0600	303.98	0.7969	11.655	232.81	321.54	88.73	1.1121	1.4040
1.0800	304.72	0.7994	11.420	233.65	321.85	88.21	1.1148	1.4043
1.1000	305.45	0.8018	11.193	234.47	322.16	87.69	1.1174	1.4045
1.1500	307.23	0.8080	10.657	236.50	322.89	86.39	1.1239	1.4051
1.2000	308.95	0.8142	10.164	238.47	323.59	85.11	1.1302	1.4057
1.2500	310.62	0.8203	9.7088	240.40	324.25	83.84	1.1363	1.4062
1.3000	312.25	0.8265	9.2861	242.29	324.87	82.58	1.1422	1.4067
1.3500	313.82	0.8327	8.8929	244.15	325.47	81.32	1.1480	1.4072
1.4000	315.36	0.8390	8.5260	245.96	326.03	80.07	1.1537	1.4076
1.4500	316.85	0.8453	8.1827	247.74	326.57	78.83	1.1592	1.4079
1.5000	318.31	0.8517	7.8606	249.49	327.07	77.58	1.1645	1.4083
1.5500	319.73	0.8581	7.5577	251.21	327.55	76.34	1.1698	1.4086
1.6000	321.12	0.8646	7.2723	252.90	328.00	75.10	1.1749	1.4088
1.6500	322.47	0.8711	7.0026	254.57	328.43	73.85	1.1800	1.4090
1.7000	323.79	0.8778	6.7474	256.22	328.82	72.61	1.1849	1.4092
1.7500	325.09	0.8846	6.5053	257.84	329.19	71.36	1.1898	1.4093
1.8000	326.35	0.8915	6.2752	259.44	329.54	70.10	1.1946	1.4094
1.8500	327.59	0.8985	6.0562	261.02	329.86	68.84	1.1993	1.4094
1.9000	328.81	0.9056	5.8472	262.59	330.15	67.56	1.2039	1.4094
1.9500	330.00	0.9129	5.6475	264.14	330.42	66.28	1.2085	1.4094
2.0000	331.17	0.9204	5.4564	265.67	330.66	64.99	1.2130	1.4092
2.1000	333.43	0.9359	5.0971	268.71	331.07	62.36	1.2218	1.4089
2.2000	335.62	0.9524	4.7643	271.70	331.36	59.66	1.2305	1.4083
2.3000	337.72	0.9700	4.4538	274.66	331.53	56.87	1.2390	1.4074
2.4000	339.76	0.9890	4.1619	277.61	331.57	53.95	1.2474	1.4062
2.5000	341.72	1.0099	3.8850	280.57	331.45	50.89	1.2558	1.4047
2.6000	343.62	1.0333	3.6195	283.55	331.16	47.61	1.2642	1.4028
2.7000	345.46	1.0600	3.3615	286.59	330.65	44.06	1.2728	1.4003
2.8000	347.23	1.0917	3.1057	289.75	329.85	40.10	1.2816	1.3971
2.9000	348.94	1.1314	2.8445	293.12	328.63	35.51	1.2909	1.3927
3.0000	350.60	1.1865	2.5621	296.94	326.68	29.74	1.3015	1.3863
3.1000	352.18	1.2860	2.2060	302.05	322.92	20.87	1.3157	1.3750

Chart R115.3: Thermodynamic properties

p/MPa	0.03			0.05			0.08		
T K	v dm³/kg	h kJ/kg	s kJ/kgK	v dm³/kg	h kJ/kg	s kJ/kgK	v dm³/kg	h kJ/kg	s kJ/kgK
210.0	0.6147	140.55	0.7542	0.6147	140.56	0.7542	0.6147	140.57	0.7542
220.0	386.86	279.85	1.4167	228.89	279.19	1.3869	0.6269	149.63	0.7963
230.0	405.82	286.09	1.4444	240.69	285.52	1.4151	158.07	284.80	1.3908
240.0	424.59	292.46	1.4715	252.31	291.97	1.4425	166.13	291.35	1.4187
250.0	443.22	298.96	1.4981	263.78	298.54	1.4693	174.03	298.00	1.4459
260.0	461.74	305.60	1.5241	275.14	305.23	1.4955	181.81	304.75	1.4724
270.0	480.18	312.37	1.5497	286.40	312.04	1.5213	189.49	311.62	1.4983
280.0	498.53	319.28	1.5748	297.58	318.98	1.5465	197.10	318.61	1.5237
290.0	516.83	326.31	1.5995	308.71	326.05	1.5713	204.64	325.72	1.5486
300.0	535.08	333.48	1.6238	319.79	333.24	1.5957	212.13	332.95	1.5731
310.0	553.29	340.78	1.6477	330.82	340.56	1.6197	219.58	340.29	1.5972
320.0	571.47	348.20	1.6712	341.82	348.00	1.6433	227.00	347.76	1.6209
330.0	589.62	355.74	1.6945	352.80	355.57	1.6666	234.38	355.34	1.6443
340.0	607.74	363.41	1.7173	363.74	363.25	1.6895	241.74	363.04	1.6673
350.0	625.84	371.20	1.7399	374.67	371.05	1.7121	249.08	370.86	1.6899
360.0	643.93	379.10	1.7622	385.58	378.96	1.7344	256.41	378.79	1.7123
370.0	661.99	387.12	1.7842	396.47	386.99	1.7564	263.71	386.83	1.7343
380.0	680.05	395.25	1.8058	407.36	395.13	1.7781	271.01	394.99	1.7560
390.0	698.10	403.49	1.8272	418.22	403.38	1.7996	278.29	403.25	1.7775
400.0	716.13	411.84	1.8484	429.08	411.74	1.8207	285.56	411.61	1.7987
410.0	734.16	420.30	1.8693	439.93	420.20	1.8416	292.82	420.08	1.8196
420.0	752.18	428.85	1.8899	450.78	428.76	1.8622	300.08	428.65	1.8402
430.0	770.19	437.51	1.9102	461.61	437.42	1.8826	307.33	437.32	1.8606
440.0	788.20	446.26	1.9304	472.44	446.18	1.9027	314.57	446.08	1.8808
450.0	806.20	455.11	1.9502	483.27	455.03	1.9226	321.80	454.94	1.9007

p/MPa	0.10			0.15			0.20		
210.0	0.6146	140.58	0.7541	0.6146	140.60	0.7541	0.6145	140.62	0.7540
220.0	0.6268	149.64	0.7963	0.6268	149.65	0.7962	0.6267	149.67	0.7961
230.0	0.6402	158.75	0.8368	0.6401	158.77	0.8367	0.6400	158.78	0.8366
240.0	123.00	290.71	1.4012	0.6547	167.98	0.8759	0.6547	168.00	0.8758
250.0	129.13	297.44	1.4287	84.166	296.31	1.4034	0.6708	177.37	0.9141
260.0	135.13	304.27	1.4555	88.400	303.28	1.4308	64.991	302.26	1.4123
270.0	141.02	311.20	1.4816	92.524	310.33	1.4574	68.241	309.44	1.4394
280.0	146.84	318.24	1.5072	96.563	317.47	1.4834	71.398	316.69	1.4658
290.0	152.60	325.39	1.5323	100.53	324.70	1.5087	74.483	324.01	1.4915
300.0	158.30	332.65	1.5569	104.45	332.04	1.5336	77.510	331.42	1.5166
310.0	163.96	340.02	1.5811	108.32	339.47	1.5580	80.489	338.92	1.5412
320.0	169.58	347.51	1.6049	112.15	347.02	1.5819	83.431	346.51	1.5653
330.0	175.17	355.12	1.6283	115.96	354.67	1.6055	86.340	354.21	1.5890
340.0	180.74	362.84	1.6513	119.73	362.43	1.6286	89.223	362.01	1.6123
350.0	186.29	370.67	1.6740	123.49	370.30	1.6514	92.083	369.91	1.6352
360.0	191.82	378.62	1.6964	127.22	378.27	1.6739	94.924	377.92	1.6577
370.0	197.33	386.67	1.7185	130.94	386.35	1.6960	97.749	386.03	1.6799
380.0	202.83	394.84	1.7403	134.65	394.54	1.7179	100.56	394.24	1.7018
390.0	208.32	403.11	1.7617	138.35	402.83	1.7394	103.36	402.55	1.7234
400.0	213.80	411.48	1.7829	142.03	411.23	1.7607	106.15	410.97	1.7447

Chart R115.3: Thermodynamic properties—*continued*

p/MPa	0.10			0.15			0.20		
T K	v dm^3/kg	h kJ/kg	s kJ/kgK	v dm^3/kg	h kJ/kg	s kJ/kgK	v dm^3/kg	h kJ/kg	s kJ/kgK
410.0	219.27	419.96	1.8039	145.71	419.72	1.7816	108.93	419.48	1.7657
420.0	224.73	428.54	1.8245	149.38	428.31	1.8023	111.70	428.08	1.7865
430.0	230.18	437.21	1.8450	153.04	437.00	1.8228	114.46	436.79	1.8070
440.0	235.63	445.98	1.8651	156.69	445.78	1.8430	117.22	445.58	1.8272
450.0	241.07	454.84	1.8850	160.34	454.65	1.8629	119.97	454.47	1.8471
p/MPa	0.30			0.40			0.50		
210.0	0.6144	140.65	0.7539	0.6143	140.69	0.7538	0.6142	140.73	0.7537
220.0	0.6266	149.71	0.7960	0.6264	149.74	0.7959	0.6263	149.78	0.7958
230.0	0.6399	158.81	0.8365	0.6397	158.85	0.8364	0.6396	158.88	0.8362
240.0	0.6545	168.02	0.8757	0.6543	168.05	0.8755	0.6541	168.08	0.8754
250.0	0.6706	177.39	0.9139	0.6704	177.42	0.9138	0.6702	177.44	0.9136
260.0	0.6886	186.98	0.9515	0.6883	186.99	0.9513	0.6881	187.01	0.9511
270.0	43.881	307.56	1.4123	0.7083	196.83	0.9885	0.7080	196.85	0.9882
280.0	46.178	315.05	1.4395	33.503	313.31	1.4193	25.836	311.44	1.4021
290.0	48.391	322.57	1.4659	35.298	321.05	1.4465	27.399	319.45	1.4302
300.0	50.539	330.14	1.4916	37.019	328.80	1.4727	28.875	327.40	1.4572
310.0	52.635	337.77	1.5166	38.681	336.58	1.4983	30.285	335.35	1.4833
320.0	54.689	345.48	1.5411	40.298	344.42	1.5231	31.645	343.32	1.5086
330.0	56.709	353.28	1.5651	41.878	352.32	1.5474	32.965	351.33	1.5332
340.0	58.701	361.16	1.5886	43.427	360.29	1.5712	34.252	359.40	1.5573
350.0	60.669	369.14	1.6117	44.951	368.35	1.5946	35.513	367.54	1.5809
360.0	62.616	377.21	1.6345	46.455	376.48	1.6175	36.751	375.74	1.6040
370.0	64.547	385.37	1.6568	47.940	384.71	1.6400	37.971	384.03	1.6267
380.0	66.464	393.63	1.6789	49.411	393.02	1.6622	39.175	392.39	1.6490
390.0	68.368	401.99	1.7006	50.868	401.42	1.6840	40.365	400.84	1.6709
400.0	70.261	410.44	1.7220	52.315	409.91	1.7055	41.544	409.37	1.6926
410.0	72.145	418.99	1.7431	53.752	418.49	1.7267	42.714	417.99	1.7138
420.0	74.021	427.63	1.7639	55.180	427.16	1.7476	43.874	426.70	1.7348
430.0	75.889	436.36	1.7844	56.601	435.92	1.7682	45.027	435.49	1.7555
440.0	77.751	445.18	1.8047	58.016	444.77	1.7886	46.174	444.36	1.7759
450.0	79.608	454.09	1.8247	59.425	453.70	1.8086	47.314	453.31	1.7960
p/MPa	0.60			0.80			1.00		
210.0	0.6141	140.76	0.7536	0.6138	140.84	0.7533	0.6136	140.91	0.7531
220.0	0.6262	149.81	0.7956	0.6259	149.88	0.7954	0.6257	149.95	0.7951
230.0	0.6394	158.91	0.8361	0.6391	158.97	0.8358	0.6388	159.04	0.8355
240.0	0.6540	168.11	0.8752	0.6536	168.17	0.8749	0.6533	168.23	0.8746
250.0	0.6700	177.47	0.9134	0.6696	177.52	0.9131	0.6692	177.57	0.9128
260.0	0.6878	187.03	0.9510	0.6873	187.07	0.9506	0.6868	187.12	0.9502
270.0	0.7077	196.86	0.9880	0.7071	196.89	0.9876	0.7065	196.92	0.9872
280.0	0.7302	207.00	1.0249	0.7294	207.01	1.0244	0.7286	207.02	1.0239
290.0	22.089	317.74	1.4158	0.7550	217.48	1.0612	0.7540	217.48	1.0606
300.0	23.415	325.93	1.4436	16.505	322.71	1.4198	0.7838	228.35	1.0975
310.0	24.666	334.06	1.4703	17.583	331.30	1.4480	13.252	328.20	1.4281
320.0	25.860	342.18	1.4961	18.587	339.77	1.4748	14.167	337.13	1.4565
330.0	27.011	350.32	1.5211	19.537	348.18	1.5007	15.013	345.88	1.4834
340.0	28.126	358.49	1.5455	20.444	356.58	1.5258	15.807	354.55	1.5093
350.0	29.213	366.71	1.5693	21.320	364.99	1.5502	16.562	363.18	1.5343

Chart R115.3: Thermodynamic properties–*continued*

p/MPa	0.60			0.80			1.00		
T K	v dm³/kg	h kJ/kg	s kJ/kgK	v dm³/kg	h kJ/kg	s kJ/kgK	v dm³/kg	h kJ/kg	s kJ/kgK
360.0	30.276	374.99	1.5926	22.168	373.43	1.5740	17.287	371.80	1.5586
370.0	31.320	383.34	1.6155	22.995	381.91	1.5972	17.988	380.43	1.5822
380.0	32.347	391.76	1.6380	23.804	390.45	1.6200	18.669	389.10	1.6053
390.0	33.360	400.25	1.6600	24.598	399.05	1.6423	19.334	397.81	1.6280
400.0	34.362	408.83	1.6818	25.379	407.72	1.6643	19.984	406.57	1.6502
410.0	35.353	417.49	1.7031	26.150	416.45	1.6858	20.623	415.39	1.6719
420.0	36.336	426.22	1.7242	26.910	425.26	1.7071	21.253	424.28	1.6933
430.0	37.311	435.04	1.7449	27.663	434.14	1.7280	21.873	433.22	1.7144
440.0	38.279	443.94	1.7654	28.409	443.10	1.7485	22.486	442.24	1.7351
450.0	39.241	452.92	1.7856	29.148	452.13	1.7688	23.093	451.32	1.7555
p/MPa	2.00			3.00			4.00		
210.0	0.6125	141.28	0.7519	0.6114	141.66	0.7508	0.6104	142.03	0.7497
220.0	0.6244	150.30	0.7939	0.6232	150.65	0.7926	0.6220	151.01	0.7914
230.0	0.6374	159.36	0.8342	0.6360	159.69	0.8328	0.6346	160.02	0.8315
240.0	0.6516	168.52	0.8731	0.6500	168.81	0.8716	0.6484	169.11	0.8702
250.0	0.6672	177.82	0.9111	0.6653	178.08	0.9095	0.6635	178.35	0.9079
260.0	0.6845	187.32	0.9484	0.6822	187.54	0.9466	0.6801	187.77	0.9448
270.0	0.7036	197.07	0.9851	0.7008	197.23	0.9832	0.6982	197.41	0.9812
280.0	0.7250	207.10	1.0216	0.7216	207.20	1.0194	0.7184	207.31	1.0172
290.0	0.7493	217.46	1.0580	0.7449	217.47	1.0554	0.7408	217.51	1.0530
300.0	0.7773	228.19	1.0943	0.7715	228.08	1.0914	0.7662	228.02	1.0886
310.0	0.8107	239.39	1.1311	0.8025	239.10	1.1275	0.7954	238.89	1.1243
320.0	0.8527	251.21	1.1686	0.8401	250.62	1.1641	0.8297	250.19	1.1601
330.0	0.9115	264.06	1.2081	0.8886	262.85	1.2017	0.8720	262.02	1.1966
340.0	6.1686	341.28	1.4409	0.9600	276.32	1.2419	0.9277	274.63	1.2342
350.0	6.8103	352.03	1.4721	1.1456	294.73	1.2952	1.0128	288.66	1.2748
360.0	7.3618	362.15	1.5006	3.6736	346.97	1.4436	1.2206	307.26	1.3272
370.0	7.8575	371.92	1.5274	4.2686	360.18	1.4798	2.0784	338.15	1.4118
380.0	8.3148	381.49	1.5529	4.7315	371.73	1.5106	2.7718	357.57	1.4637
390.0	8.7437	390.94	1.5774	5.1283	382.54	1.5387	3.2352	371.65	1.5002
400.0	9.1508	400.32	1.6012	5.4839	392.92	1.5650	3.6043	383.93	1.5314
410.0	9.5407	409.66	1.6242	5.8108	403.05	1.5900	3.9217	395.34	1.5595
420.0	9.9164	418.99	1.6467	6.1166	413.02	1.6140	4.2059	406.25	1.5858
430.0	10.281	428.33	1.6687	6.4060	422.89	1.6372	4.4668	416.85	1.6108
440.0	10.635	437.69	1.6902	6.6823	432.69	1.6598	4.7101	427.24	1.6346
450.0	10.981	447.08	1.7113	6.9480	442.47	1.6817	4.9399	437.49	1.6577
p/MPa	5.00			6.00			7.00		
210.0	0.6093	142.41	0.7486	0.6083	142.79	0.7475	0.6074	143.17	0.7464
220.0	0.6208	151.36	0.7902	0.6197	151.72	0.7890	0.6186	152.08	0.7879
230.0	0.6333	160.35	0.8302	0.6321	160.68	0.8289	0.6308	161.02	0.8276
240.0	0.6469	169.42	0.8688	0.6455	169.73	0.8674	0.6441	170.04	0.8660
250.0	0.6618	178.62	0.9063	0.6601	178.90	0.9048	0.6585	179.18	0.9033
260.0	0.6780	188.00	0.9431	0.6760	188.24	0.9414	0.6741	188.49	0.9398
270.0	0.6958	197.60	0.9793	0.6934	197.80	0.9775	0.6912	198.00	0.9757
280.0	0.7154	207.45	1.0151	0.7125	207.59	1.0131	0.7098	207.75	1.0111
290.0	0.7371	217.57	1.0507	0.7335	217.65	1.0484	0.7302	217.75	1.0462
300.0	0.7614	227.99	1.0860	0.7569	227.99	1.0835	0.7528	228.02	1.0810

Chart R115.3: Thermodynamic properties–*continued*

p/MPa	5.00			6.00			7.00		
T K	v dm³/kg	h kJ/kg	s kJ/kgK	v dm³/kg	h kJ/kg	s kJ/kgK	v dm³/kg	h kJ/kg	s kJ/kgK
310.0	0.7890	238.74	1.1212	0.7832	238.63	1.1184	0.7780	238.57	1.1156
320.0	0.8209	249.86	1.1565	0.8131	249.61	1.1532	0.8063	249.42	1.1501
330.0	0.8588	261.42	1.1921	0.8480	260.96	1.1881	0.8386	260.61	1.1845
340.0	0.9060	273.54	1.2283	0.8896	272.75	1.2233	0.8764	272.16	1.2190
350.0	0.9691	286.47	1.2658	0.9417	285.11	1.2591	0.9216	284.14	1.2537
360.0	1.0654	300.81	1.3061	1.0110	298.24	1.2961	0.9779	296.66	1.2890
370.0	1.2538	318.11	1.3535	1.1125	312.56	1.3353	1.0513	309.85	1.3251
380.0	1.6328	338.82	1.4087	1.2749	328.53	1.3779	1.1519	323.90	1.3626
390.0	2.0803	357.62	1.4576	1.5166	345.62	1.4223	1.2913	338.82	1.4013
400.0	2.4657	373.06	1.4967	1.8011	362.10	1.4640	1.4705	354.12	1.4401
410.0	2.7872	386.44	1.5297	2.0815	377.08	1.5011	1.6739	369.09	1.4770
420.0	3.0650	398.68	1.5592	2.3381	390.69	1.5339	1.8824	383.28	1.5112
430.0	3.3128	410.24	1.5865	2.5699	403.32	1.5636	2.0836	396.64	1.5427
440.0	3.5391	421.37	1.6120	2.7810	415.27	1.5910	2.2729	409.28	1.5717
450.0	3.7491	432.22	1.6364	2.9757	426.76	1.6169	2.4501	421.37	1.5989

References for R115

[Asto55] ASTON, J.G. ; WILLS, P.E. AND ZOLKI, T.P.: The Heatcapacities from 10.9K, Heats of Transition, Fusion and Vaporization, Vapor Pressures and Entropy of Pentafluorochloroethane, the Barrier Hindering Internal Rotation; J. Am. Chem. Soc. 77(1955) 3939/3941

[Barh66] BARHO, W.: Berechnung der Molwärme der Fluor-Chlor-Derivate von Methan und Ethan im Zustand idealer Gase; Dissertation Universität Karlsruhe (1966)

[Erns69] ERNST, G.: Zur universellen Darstellung des Realanteils der spezifischen Wärme von Gasen; Chemie-Ing.-Techn. 41(1969)9 544/551

[Erns70] ERNST, G. AND BÜSSER, J.: Ideal and real gas state heat capacities cp of C_3H_8, $i-C_4H_{10}$, C_2F_5Cl, CH_2ClCF_3, $CF_2ClCFCl_2$ and CHF_2Cl; J. Chem. Therm. (1970)2 787/791

[Hwan61] HWANG, Y.T.: The constant heat capacities of gaseous tetrafluoromethane, chlorodifluoromethane, dichlorotetrafluoroethane and chloropentafluoroethane; Ph.D. Dissertation University of Michigan (1961)

[KMR 81] - : Kältemaschinenregeln, 7. Auflage; Verlag C.F. Müller, Karlsruhe (1981)

[Matt66] MATTHIAS, H. AND LÖFFLER H.L.: Thermodynamische Eigenschaften von Pentafluormonochloräthan; Lehrstuhl und Institut für Thermodynamik und Kältetechnik TU Berlin (1966)

[Mear66] MEARS, W.H. ; ROSENTHAL, E. AND SINKA, J.V.: Pressure-Volume-Temperature Behavior of Pentafluoromonochloroethane; J. Chem. Eng. Data 11(1966)3 338/343

[Meye69] MEYER, K.J.: Über den Zusammenhang zwischen der Schallgeschwindigkeit und der Wärmeleitfähigkeit bei flüssigen Fluor-Chlor-Derivaten des Methans und Äthans; Kältetechnik-Klimatisierung 21(1969)9 270/275

R 115. ph

T=360
T=365
T=370
T=375
T=380
T=385
T=390
T=395
T=400
T=405
T=410
T=415
T=420
T=425
T=430
T=435
T=440
T=445
s=1.51
s=1.54
s=1.57
s=1.60
s=1.63
s=1.66
s=1.69
s=1.72
s=1.75
s=1.78
s=1.81
s=1.84
s=1.87
s=1.90
s=1.93
s=1.96
300
317
334
351
368
385
402
419
436
453
h/(kJ/kg)

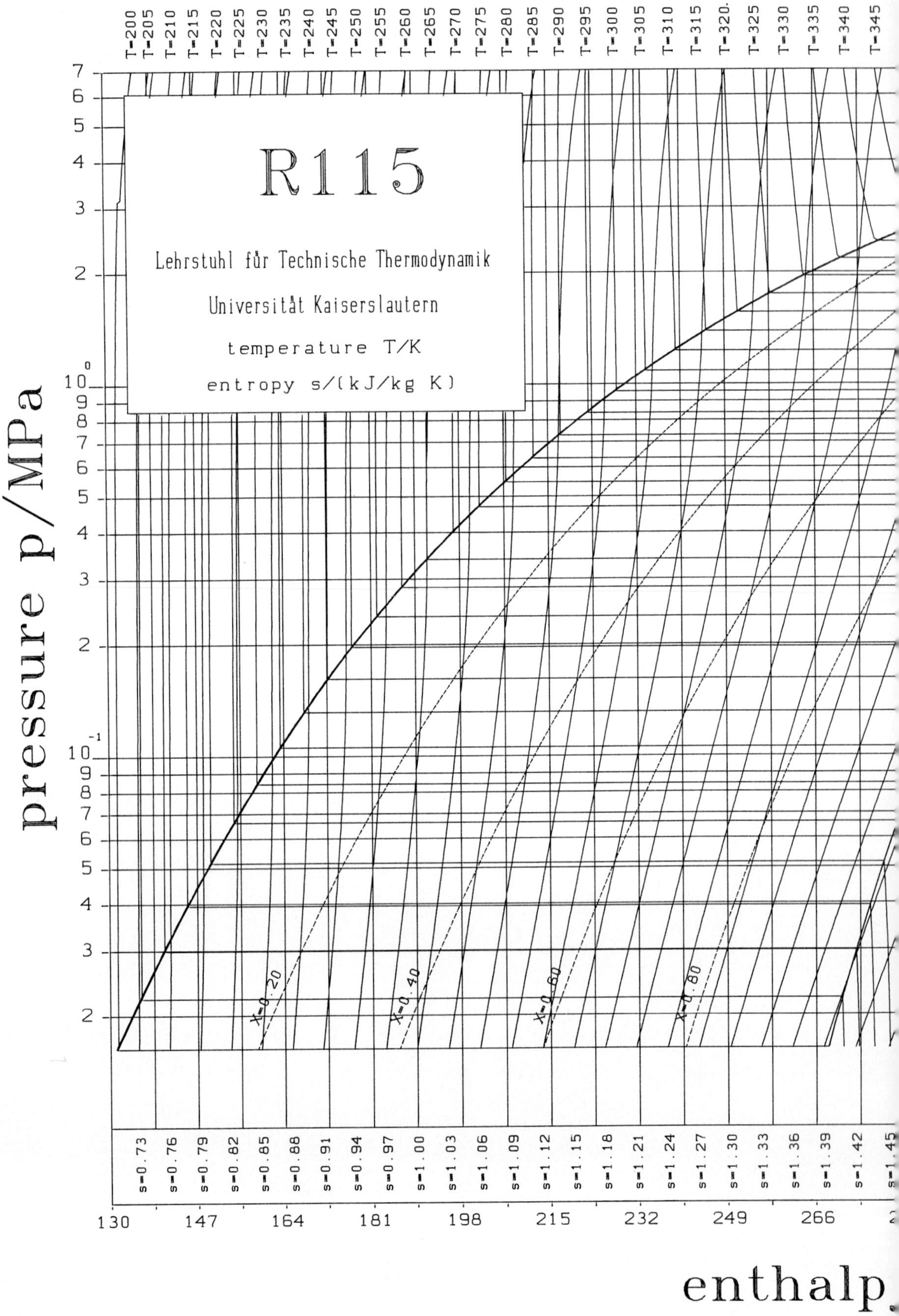
R115
Lehrstuhl für Technische Thermodynamik
Universität Kaiserslautern
temperature T/K
entropy s/(kJ/kg K)
pressure p/MPa
T=200
T=205
T=210
T=215
T=220
T=225
T=230
T=235
T=240
T=245
T=250
T=255
T=260
T=265
T=270
T=275
T=280
T=285
T=290
T=295
T=300
T=305
T=310
T=315
T=320.
T=325
T=330
T=335
T=340
T=345
X=0.20
X=0.40
X=0.60
X=0.80
s=0.73
s=0.76
s=0.79
s=0.82
s=0.85
s=0.88
s=0.91
s=0.94
s=0.97
s=1.00
s=1.03
s=1.06
s=1.09
s=1.12
s=1.15
s=1.18
s=1.21
s=1.24
s=1.27
s=1.30
s=1.33
s=1.36
s=1.39
s=1.42
s=1.45
130
147
164
181
198
215
232
249
266
enthalp

R 115. Ts

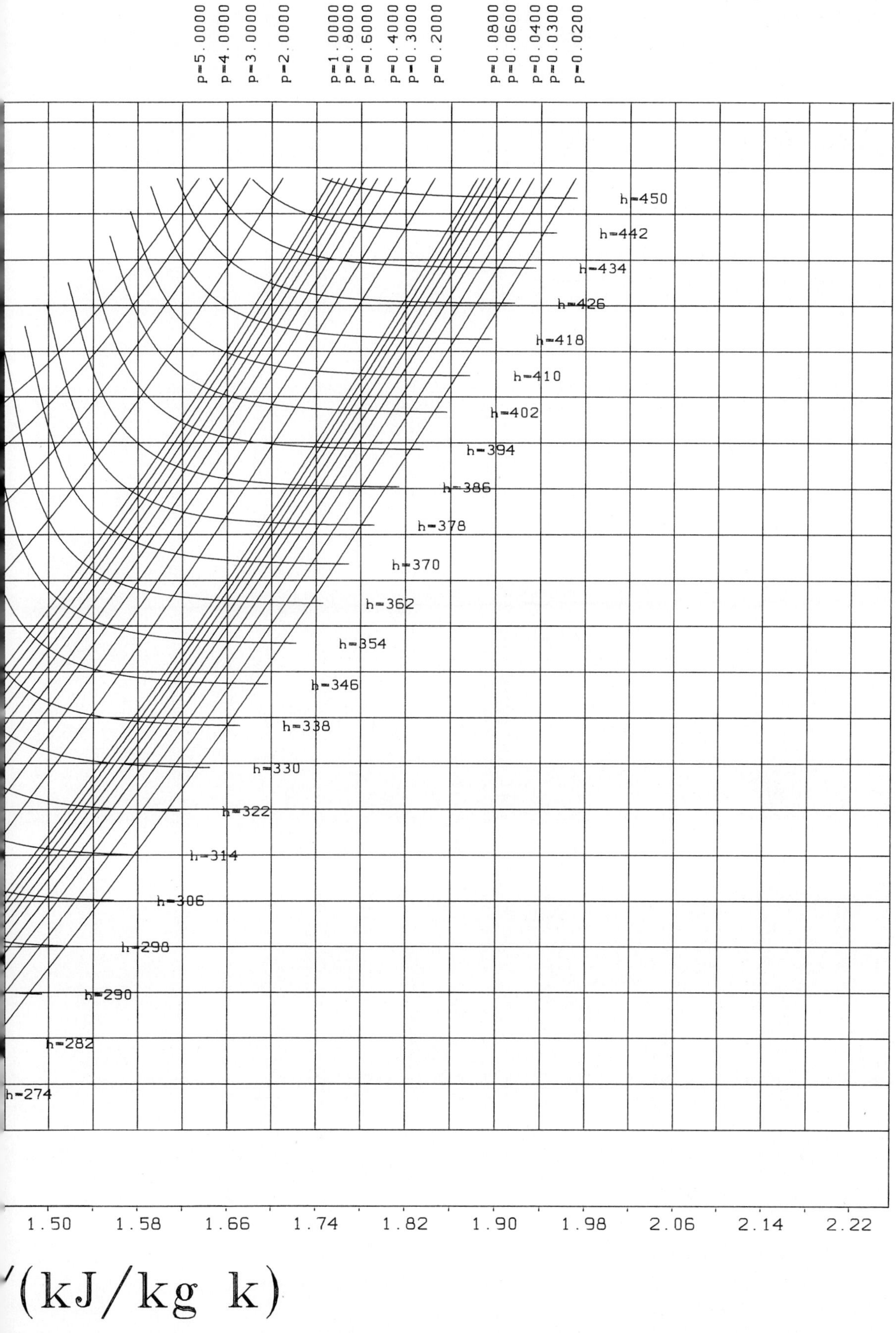

p=5.0000
p=4.0000
p=3.0000
p=2.0000
p=1.0000
p=0.8000
p=0.6000
p=0.4000
p=0.3000
p=0.2000
p=0.0800
p=0.0600
p=0.0400
p=0.0300
p=0.0200
h=450
h=442
h=434
h=426
h=418
h=410
h=402
h=394
h=386
h=378
h=370
h=362
h=354
h=346
h=338
h=330
h=322
h=314
h=306
h=298
h=290
h=282
h=274
1.50
1.58
1.66
1.74
1.82
1.90
1.98
2.06
2.14
2.22
'(kJ/kg k)

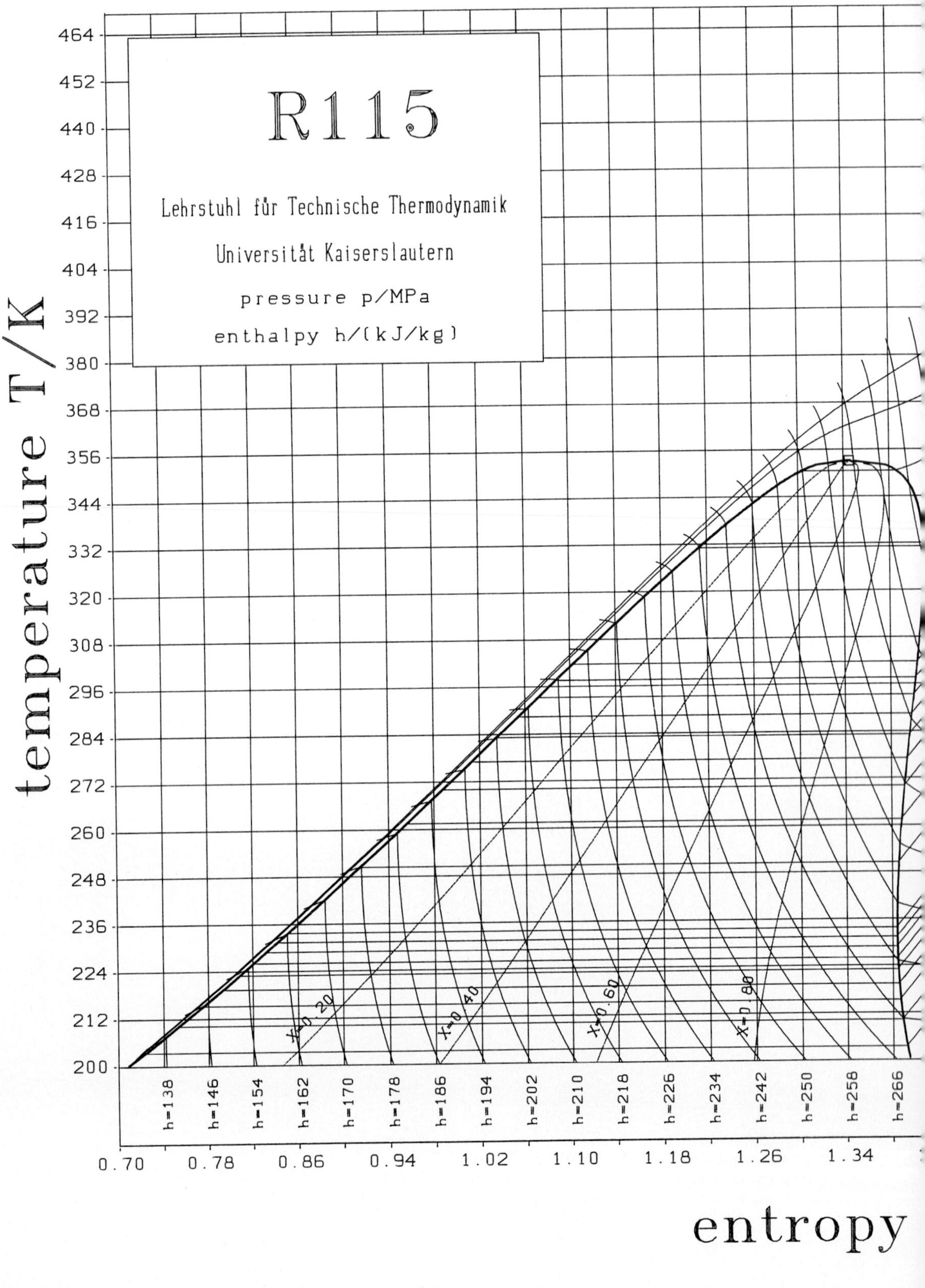
R115
Lehrstuhl für Technische Thermodynamik
Universität Kaiserslautern
pressure p/MPa
enthalpy h/(kJ/kg)
temperature T/K
entropy
464
452
440
428
416
404
392
380
368
356
344
332
320
308
296
284
272
260
248
236
224
212
200
X=0.20
X=0.40
X=0.60
X=0.80
h=138
h=146
h=154
h=162
h=170
h=178
h=186
h=194
h=202
h=210
h=218
h=226
h=234
h=242
h=250
h=258
h=266
0.70
0.78
0.86
0.94
1.02
1.10
1.18
1.26
1.34

R 115. hs

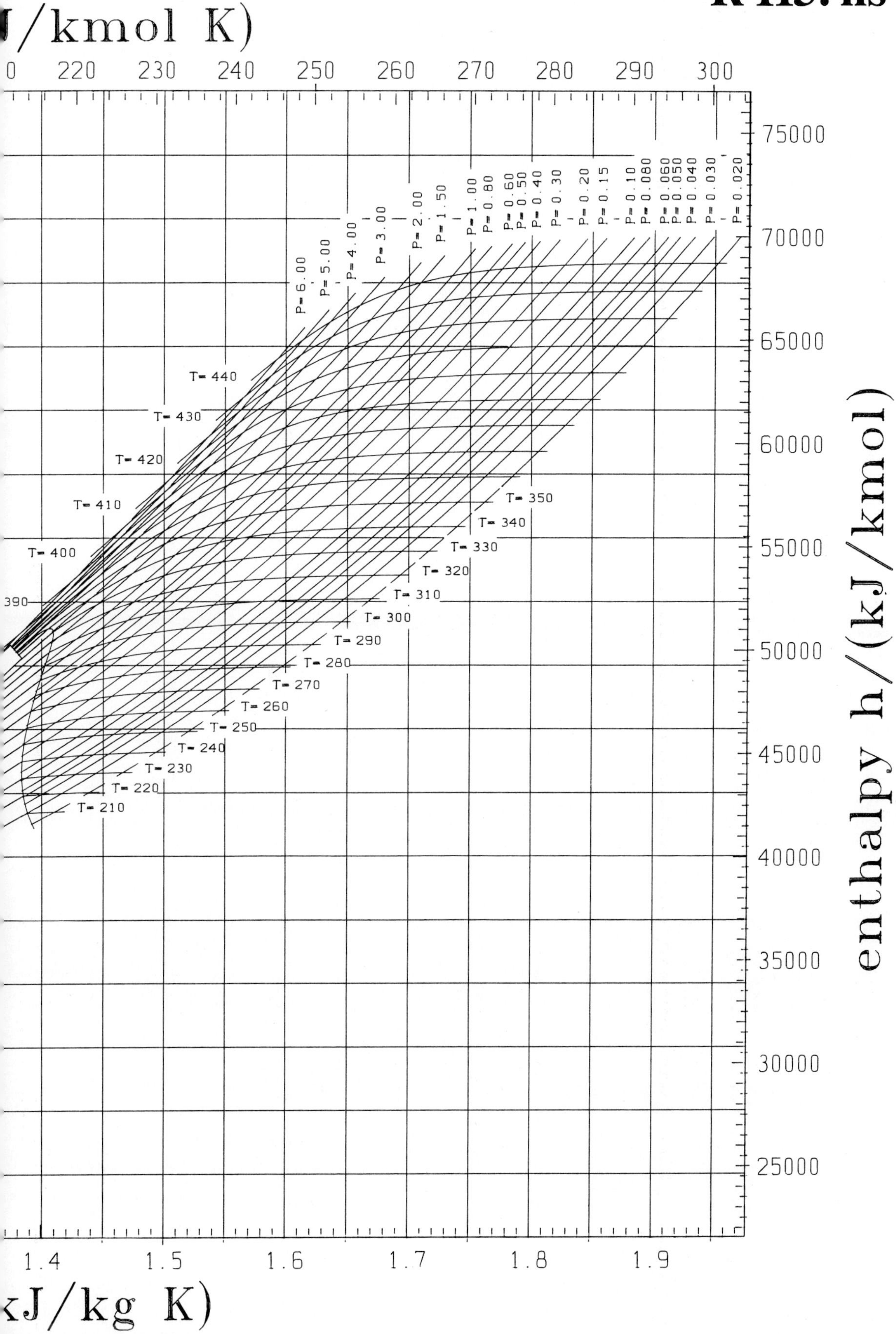

J/kmol K)
0
220
230
240
250
260
270
280
290
300
75000
70000
65000
60000
55000
50000
45000
40000
35000
30000
25000
enthalpy h/(kJ/kmol)
P= 6.00
P= 5.00
P= 4.00
P= 3.00
P= 2.00
P= 1.50
P= 1.00
P= 0.80
P= 0.60
P= 0.50
P= 0.40
P= 0.30
P= 0.20
P= 0.15
P= 0.10
P= 0.080
P= 0.060
P= 0.050
P= 0.040
P= 0.030
P= 0.020
T= 440
T= 430
T= 420
T= 410
T= 400
390
T= 350
T= 340
T= 330
T= 320
T= 310
T= 300
T= 290
T= 280
T= 270
T= 260
T= 250
T= 240
T= 230
T= 220
T= 210
1.4
1.5
1.6
1.7
1.8
1.9
kJ/kg K)

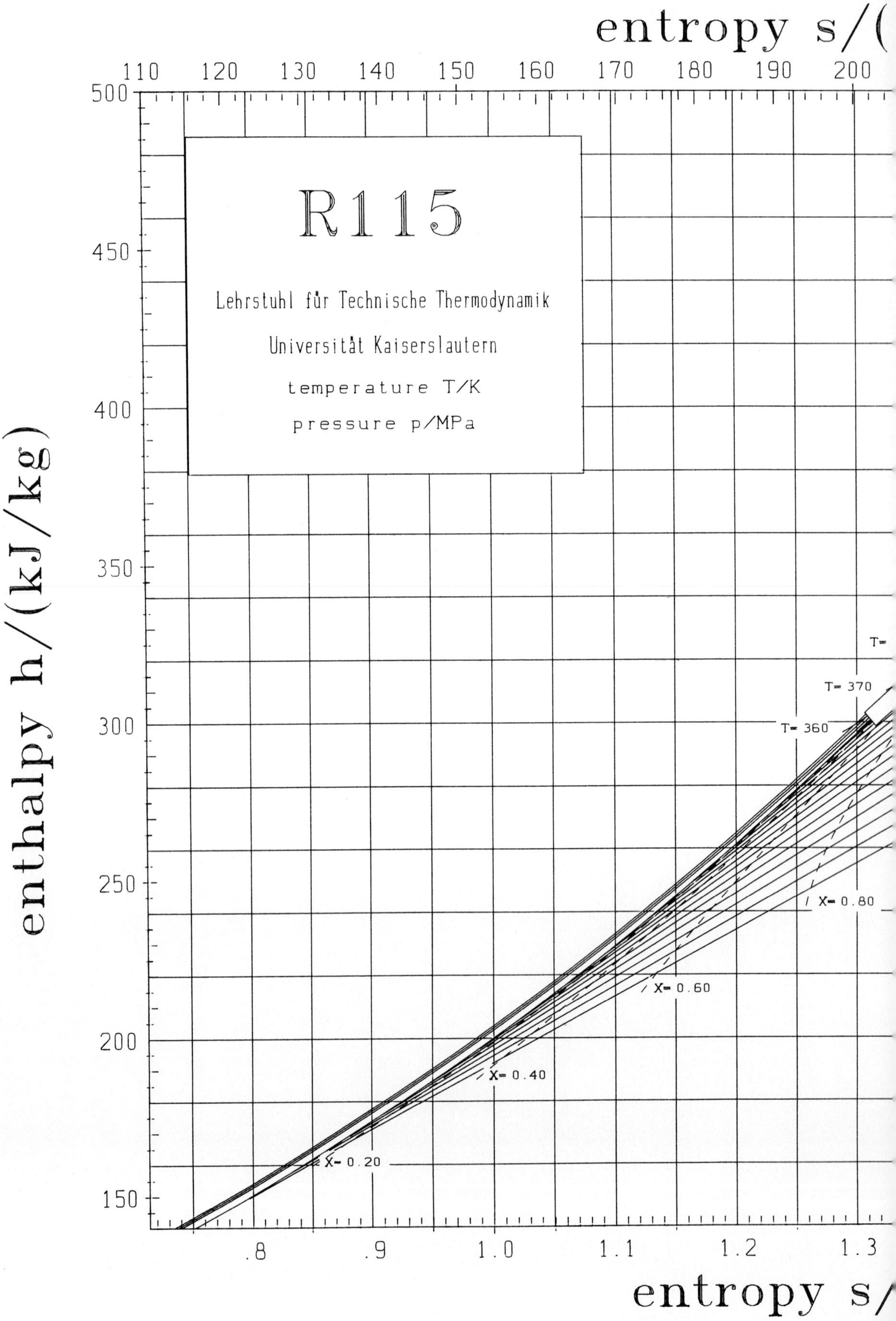

entropy s/(
110 120 130 140 150 160 170 180 190 200
500
450
400
350
300
250
200
150
R115
Lehrstuhl für Technische Thermodynamik
Universität Kaiserslautern
temperature T/K
pressure p/MPa
enthalpy h/(kJ/kg)
T=
T= 370
T= 360
X= 0.80
X= 0.60
X= 0.40
X= 0.20
.8 .9 1.0 1.1 1.2 1.3
entropy s/

Refrigerant RC318 (C_4F_8)

Table RC318.1: List of references for p, ρ, T data

Author	n	p/MPa		ρ/g cm^{-3}		T/K	
		min	max	min	max	min	max
BAMBACH, G. *[Bamb 56]*	46	0.02	2.87	0.0022	0.43	255	391
DOUSLIN, D.R. *[Dous 59]*	143	1.64	39.2	0.1500	1.40	373	623
GELLER, Z.I. *[Gell 75]*	205	0.25	59.02	0.3028	1.73	233	548
MARTIN, J.J. *[Mart 62]*	35	0.40	13.07	0.0344	0.89	318	486

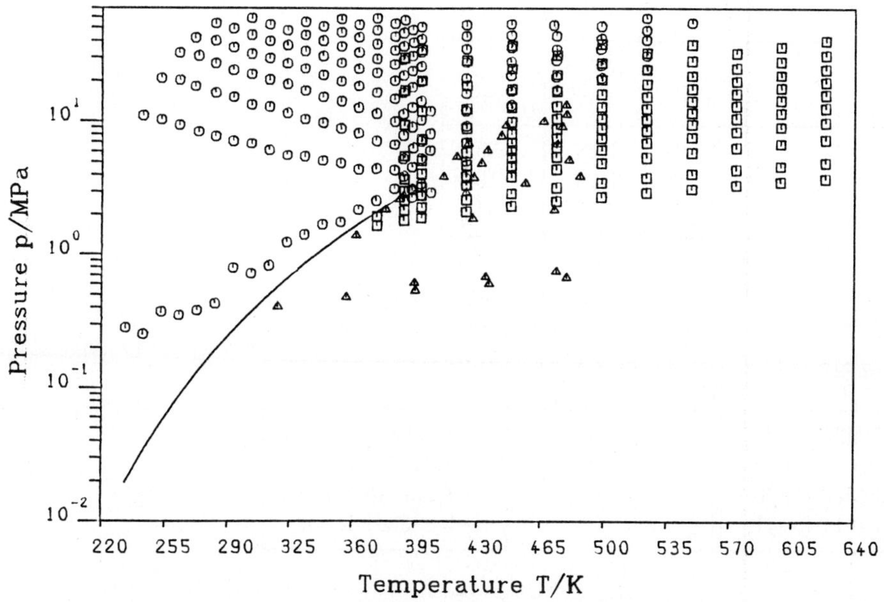

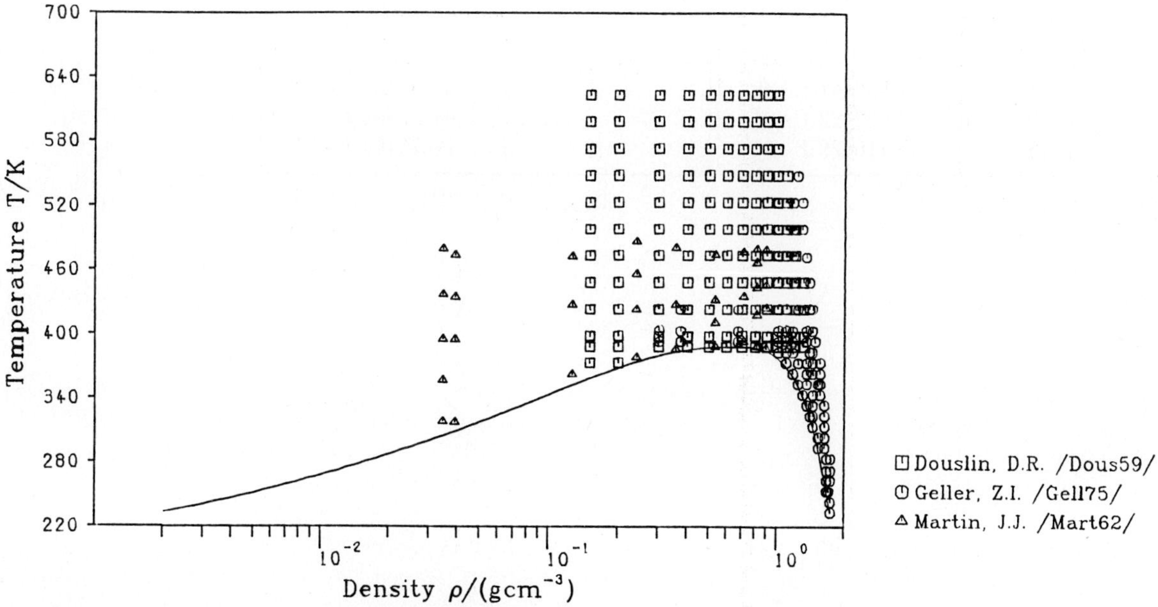

Fig.RC318.1: Distribution of p, ρ, T data projected in the p, T and in the T, ρ plane

Table RC318.2: Experimental vapor pressure data

Author	Δ / %		
	std	opt	*EOS*
BAMBACH, G. *[Bamb 56]*	3.80	3.76	3.60
DOUSLIN, D.R. *[Dous 59]*	0.01	0.01	0.03
FURUKAWA, G.T. *[Furu 54]*	0.02	0.02	0.02
KLETSKII, A.V. *[Klet 67]*	0.06	0.06	0.10
MARTIN, J.J. *[Mart 62]*	0.73	0.73	0.74
MATTHIAS, H. *[Matt 65]*	0.34	0.34	0.42
WHIPPLE, G.H. *[Whip 52]*	3.15	3.15	3.14

Table RC318.3: Experimental saturated liquid density data

Author	Δ / %		
	std	opt	*EOS*
BAMBACH, G. *[Bamb 56]*	1.14	1.18	0.95
DOUSLIN, D.R. *[Dous 59]*	1.03	1.18	1.10
MARTIN, J.J. *[Mart 62]*	0.35	0.15	0.91

Table RC318.4: Coefficients of the standard vapor pressure equation

Index i	b_i
5	−7.659423 E+00
6	1.390923 E+00
9	−4.254569 E+00
15	−8.534914 E−01

Table RC318.5: Coefficients of the quasi-optimal vapor pressure equation

Index i	b_i
5	−7.736884 E+00
6	2.043795 E+00
7	−1.577618 E+00
10	−3.893301 E+00

Table RC318.6: Coefficients of the standard saturated liquid density equation

Index i	c_i
8	1.832747 E+00
9	7.432422 E−01
16	5.716629 E−01

Table RC318.7: Coefficients of the quasi-optimal saturated liquid density equation

Index i	c_i
6	1.555136 E−03
8	1.756503 E+00
9	8.879105 E−01

Table RC318.8: Coefficients of the *Bender-EOS*

Range of validity: one-phase region

$$0.25 \leq p/\text{MPa} \leq 59.02$$
$$0.03441 \leq \rho/\text{g cm}^{-3} \leq 1.7333$$
$$233.35 \leq T/\text{K} \leq 623.23$$

	Specific units	Molar units
a_{01}	7.335120729 E−02	2.935222028 E+00
a_{02}	6.984726387 E+01	2.795008223 E+03
a_{03}	−1.753471145 E+04	−7.016690416 E+05
a_{04}	6.425573100 E+06	2.571257434 E+08
a_{05}	−4.649643840 E+08	−1.860601554 E+10
a_{06}	2.222487769 E−02	1.779057223 E−01
a_{07}	−1.951638818 E+00	−1.562248029 E+01
a_{08}	5.419829736 E+03	4.338465827 E+04
a_{09}	−1.308538610 E−02	−2.095337208 E−02
a_{10}	1.901300272 E+01	3.044514829 E+01
a_{11}	4.859880707 E−02	1.556717610 E−02
a_{12}	−5.831167355 E+01	−1.867840274 E+01
a_{13}	2.250284814 E+01	1.441911364 E+00
a_{14}	−1.972780465 E+07	−1.579171496 E+08
a_{15}	1.573508244 E+10	1.259562031 E+11
a_{16}	−3.284452012 E+12	−2.629138462 E+13
a_{17}	1.681489020 E+07	5.386147782 E+06
a_{18}	−2.326260811 E+10	−7.451481609 E+09
a_{19}	6.882888545 E+12	2.204727740 E+12
a_{20}	2.600000000 E+00	1.040416041 E−01

Table RC318.9: Deviations between experimental and predicted p, ρ, T data

Region	n	Δ / %	
		p	ρ
Gas	96	0.43	0.60
Critical	45	0.42	2.26
Liquid	236	1.92	0.08
Total	377	1.58	0.86

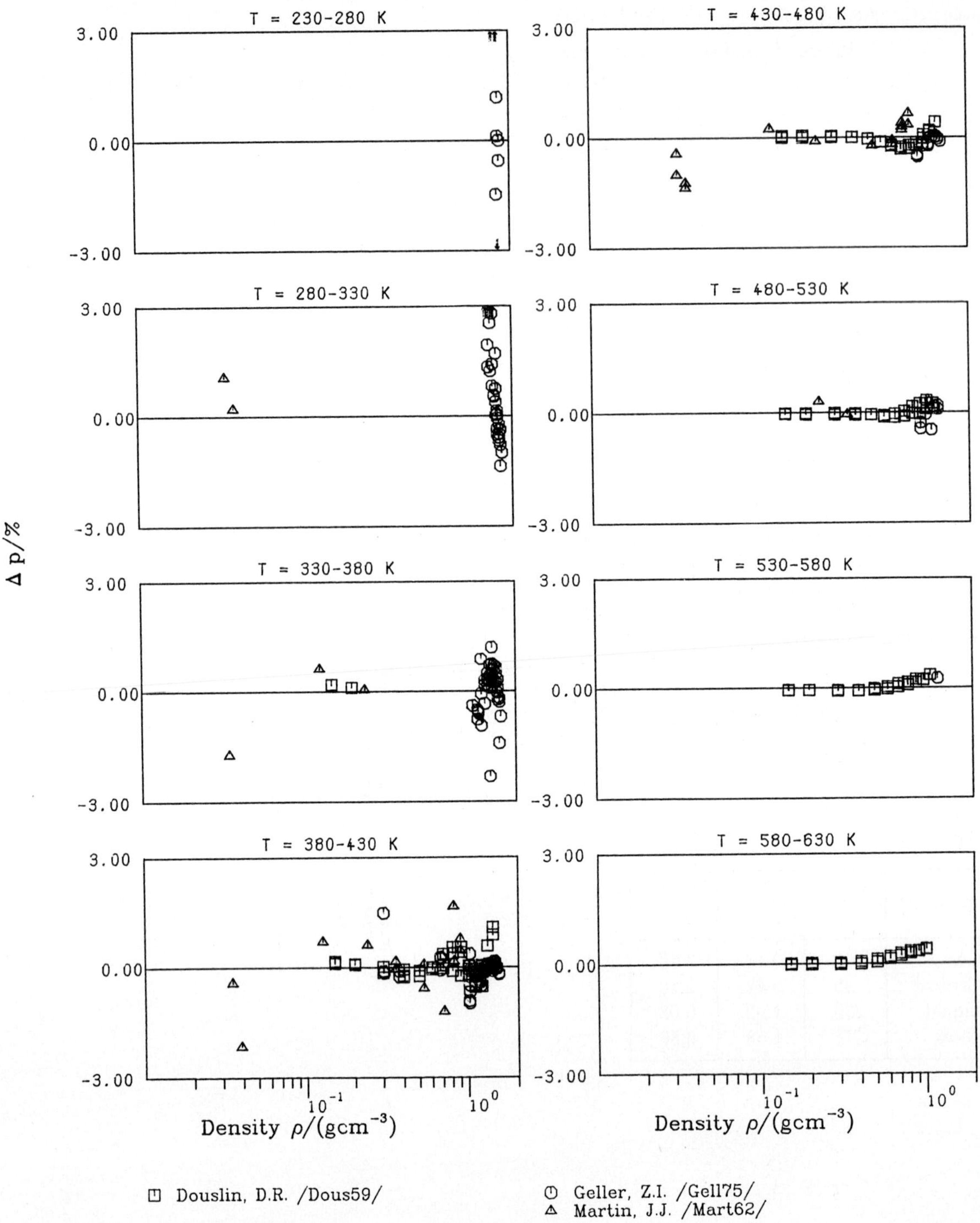

Fig.RC318.2: Deviations between experimental and calculated p, ρ, T data

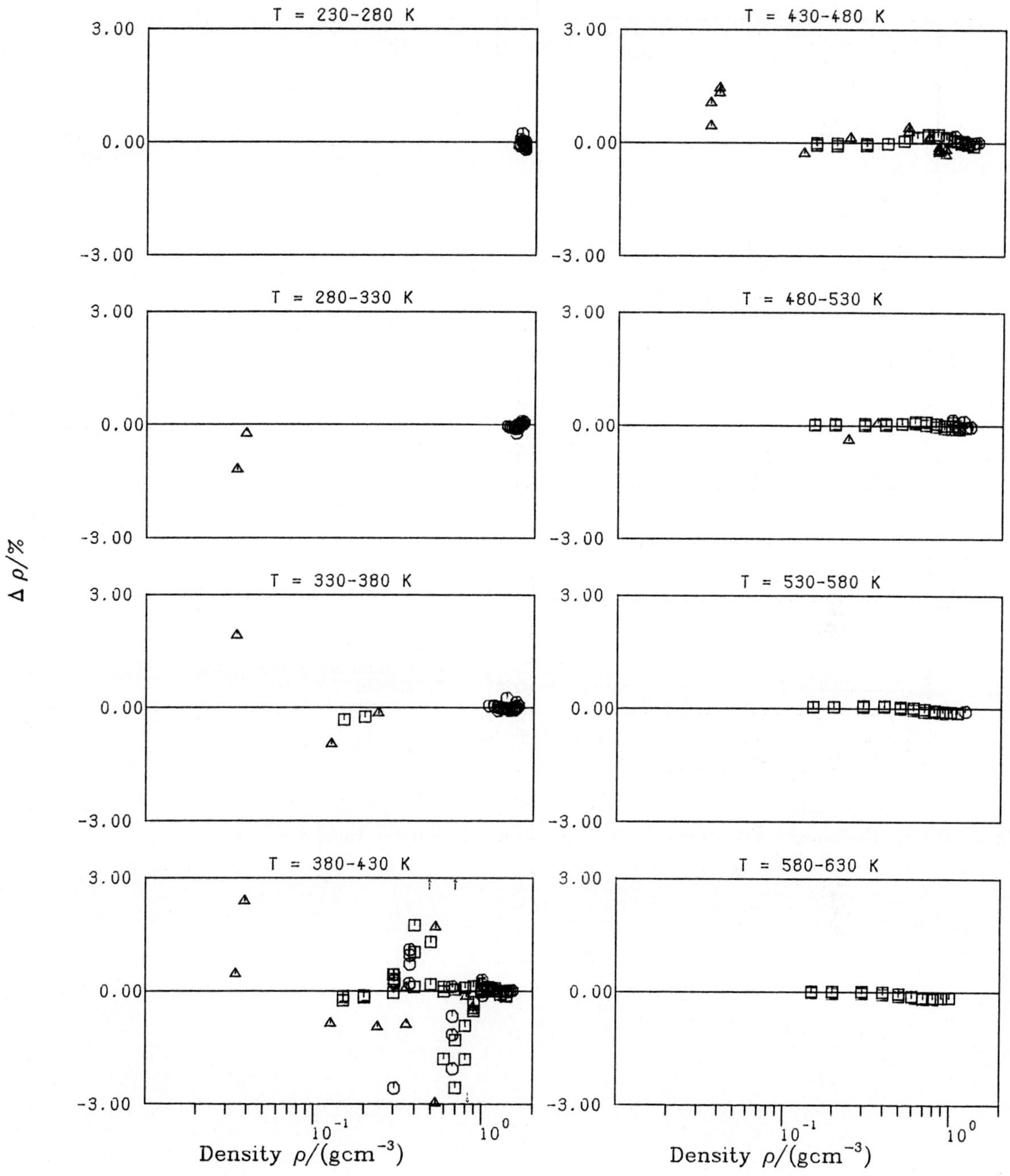

□ Douslin, D.R. /Dous59/
○ Geller, Z.I. /Gell75/
△ Martin, J.J. /Mart62/

Fig.RC318.2: Deviations between experimental and calculated p, ρ, T data *-continued*

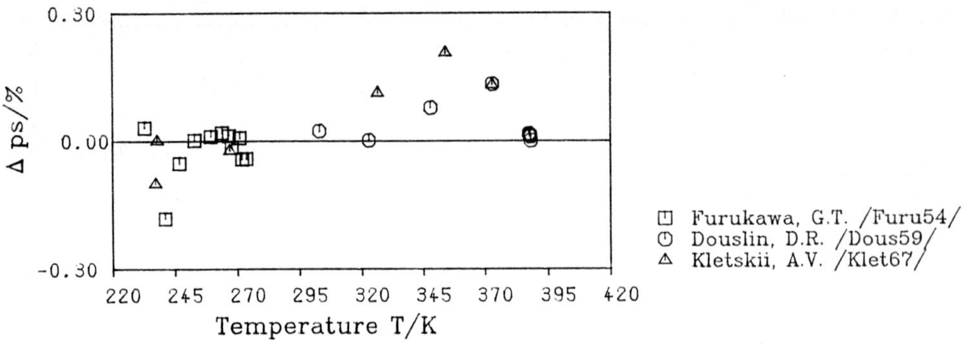

Fig.RC318.3: Deviations between experimental and calculated vapor pressures

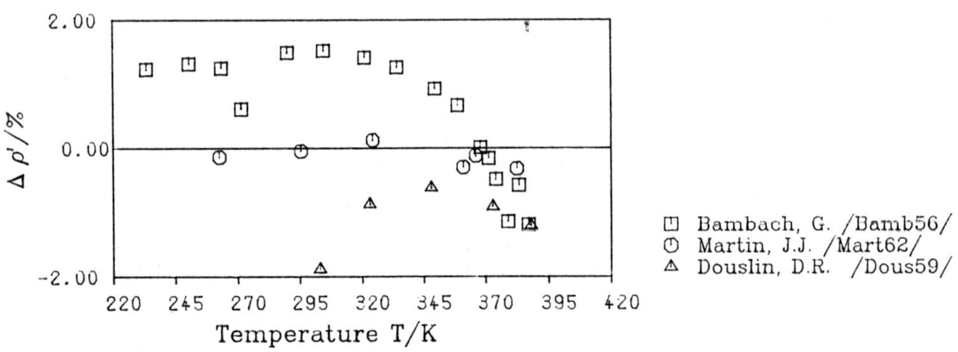

Fig.RC318.4: Deviations between experimental and calculated saturated liquid densities

Table RC318.10: Information on ideal gas state heat capacity equation

Author	T / K min	T / K max
KLETSKII, A.V. *[Klet 67a]*	200	700
Coefficients of equation (15)		
Index i	d_i	
1	1.210000 E−01	
2	2.903000 E−03	
3	−2.532700 E−06	
4	7.719100 E−10	
5	0.000000 E+00	

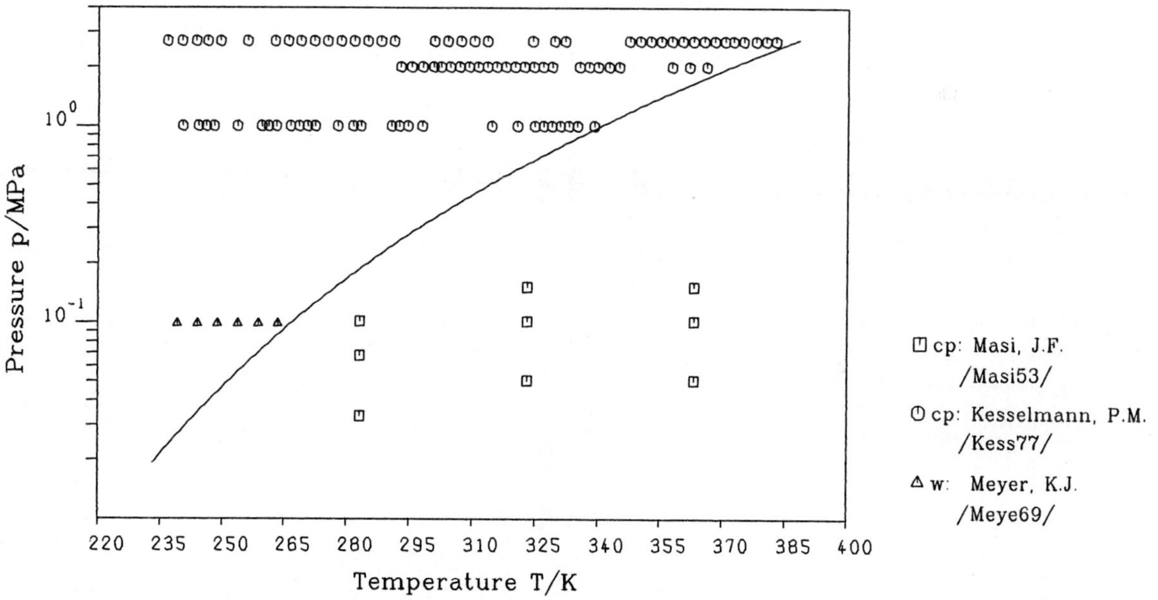

Fig.RC318.5: Distribution of $c_p(p,T)$ and $w(p,T)$ data

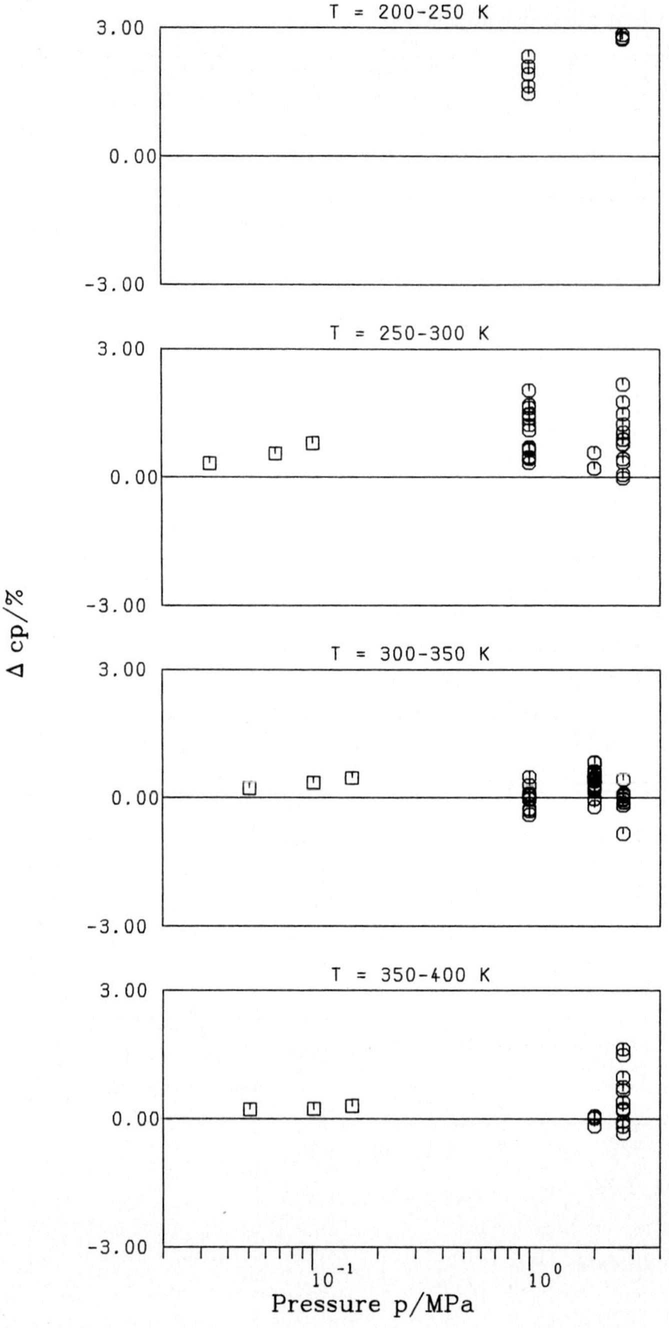

□ Masi, J.F. /Masi53/ ○ Kesselmann, P.M. /Kess77/

Fig.RC318.6: Deviations between experimental and calculated specific isobaric heat capacities c_p

Chart RC318.1: Saturation properties

T K	p MPa	v' dm^3/kg	v'' dm^3/kg	h' kJ/kg	h'' kJ/kg	r kJ/kg	s' kJ/kgK	s'' kJ/kgK
235.0	0.02137	0.5800	450.84	161.36	289.06	127.69	0.8481	1.3914
236.0	0.02261	0.5811	427.77	162.34	289.71	127.37	0.8522	1.3919
237.0	0.02390	0.5822	406.10	163.31	290.36	127.05	0.8563	1.3924
238.0	0.02525	0.5833	385.74	164.29	291.02	126.73	0.8604	1.3929
239.0	0.02667	0.5844	366.58	165.27	291.68	126.40	0.8645	1.3934
240.0	0.02814	0.5855	348.55	166.25	292.33	126.08	0.8686	1.3940
241.0	0.02969	0.5867	331.58	167.24	292.99	125.76	0.8727	1.3945
242.0	0.03130	0.5878	315.59	168.22	293.65	125.43	0.8768	1.3951
243.0	0.03298	0.5889	300.52	169.21	294.31	125.10	0.8809	1.3957
244.0	0.03473	0.5901	286.30	170.20	294.97	124.77	0.8849	1.3963
245.0	0.03656	0.5913	272.89	171.19	295.64	124.45	0.8890	1.3969
246.0	0.03846	0.5924	260.23	172.18	296.30	124.12	0.8930	1.3975
247.0	0.04045	0.5936	248.26	173.18	296.96	123.78	0.8970	1.3982
248.0	0.04251	0.5948	236.96	174.18	297.63	123.45	0.9011	1.3988
249.0	0.04466	0.5960	226.27	175.18	298.29	123.12	0.9051	1.3995
250.0	0.04689	0.5972	216.15	176.18	298.96	122.78	0.9091	1.4002
251.0	0.04922	0.5984	206.58	177.18	299.63	122.44	0.9131	1.4009
252.0	0.05163	0.5996	197.52	178.19	300.29	122.10	0.9171	1.4016
253.0	0.05414	0.6008	188.93	179.20	300.96	121.76	0.9211	1.4024
254.0	0.05674	0.6020	180.78	180.21	301.63	121.42	0.9251	1.4031
255.0	0.05945	0.6033	173.06	181.22	302.30	121.08	0.9290	1.4039
256.0	0.06225	0.6045	165.74	182.24	302.97	120.73	0.9330	1.4046
257.0	0.06516	0.6058	158.78	183.26	303.64	120.38	0.9370	1.4054
258.0	0.06817	0.6070	152.18	184.27	304.31	120.04	0.9409	1.4062
259.0	0.07130	0.6083	145.90	185.30	304.98	119.69	0.9449	1.4070
260.0	0.07454	0.6096	139.94	186.32	305.65	119.33	0.9488	1.4078
261.0	0.07789	0.6109	134.27	187.35	306.33	118.98	0.9527	1.4086
262.0	0.08136	0.6122	128.88	188.38	307.00	118.62	0.9567	1.4094
263.0	0.08495	0.6135	123.74	189.41	307.67	118.27	0.9606	1.4103
264.0	0.08866	0.6148	118.85	190.44	308.35	117.91	0.9645	1.4111
265.0	0.09250	0.6162	114.20	191.48	309.02	117.55	0.9684	1.4120
266.0	0.09648	0.6175	109.76	192.51	309.69	117.18	0.9723	1.4128
267.0	0.10058	0.6189	105.53	193.55	310.37	116.82	0.9762	1.4137
268.0	0.10482	0.6202	101.50	194.60	311.04	116.45	0.9801	1.4146
269.0	0.10919	0.6216	97.653	195.64	311.72	116.08	0.9840	1.4155
270.0	0.11371	0.6230	93.981	196.69	312.39	115.71	0.9878	1.4164
271.0	0.11838	0.6244	90.475	197.74	313.07	115.33	0.9917	1.4173
272.0	0.12319	0.6258	87.127	198.79	313.74	114.96	0.9956	1.4182
273.0	0.12815	0.6272	83.928	199.84	314.42	114.58	0.9994	1.4191
274.0	0.13326	0.6287	80.870	200.90	315.10	114.20	1.0033	1.4201
275.0	0.13854	0.6301	77.946	201.96	315.77	113.82	1.0071	1.4210
276.0	0.14397	0.6316	75.149	203.02	316.45	113.43	1.0109	1.4219
277.0	0.14957	0.6330	72.473	204.08	317.12	113.04	1.0148	1.4229
278.0	0.15533	0.6345	69.912	205.15	317.80	112.65	1.0186	1.4238
279.0	0.16127	0.6360	67.460	206.21	318.47	112.26	1.0224	1.4248
280.0	0.16738	0.6375	65.111	207.28	319.15	111.87	1.0262	1.4258
281.0	0.17366	0.6390	62.860	208.36	319.82	111.47	1.0301	1.4267
282.0	0.18013	0.6406	60.703	209.43	320.50	111.07	1.0339	1.4277
283.0	0.18678	0.6421	58.635	210.51	321.17	110.67	1.0377	1.4287
284.0	0.19362	0.6437	56.651	211.59	321.85	110.26	1.0414	1.4297

Chart RC318.1: Saturation properties–*continued*

T K	p MPa	v' dm³/kg	v'' dm³/kg	h' kJ/kg	h'' kJ/kg	r kJ/kg	s' kJ/kgK	s'' kJ/kgK
285.0	0.20064	0.6452	54.748	212.67	322.52	109.86	1.0452	1.4307
286.0	0.20787	0.6468	52.921	213.75	323.20	109.45	1.0490	1.4317
287.0	0.21529	0.6484	51.168	214.84	323.87	109.03	1.0528	1.4327
288.0	0.22291	0.6501	49.484	215.93	324.54	108.62	1.0566	1.4337
289.0	0.23073	0.6517	47.867	217.02	325.22	108.20	1.0603	1.4347
290.0	0.23877	0.6533	46.312	218.11	325.89	107.78	1.0641	1.4357
291.0	0.24701	0.6550	44.818	219.21	326.56	107.35	1.0678	1.4367
292.0	0.25548	0.6567	43.382	220.31	327.23	106.93	1.0716	1.4378
293.0	0.26416	0.6584	42.001	221.41	327.90	106.50	1.0753	1.4388
294.0	0.27306	0.6601	40.672	222.51	328.57	106.06	1.0791	1.4398
295.0	0.28219	0.6618	39.393	223.62	329.24	105.63	1.0828	1.4409
296.0	0.29155	0.6636	38.162	224.72	329.91	105.19	1.0865	1.4419
297.0	0.30115	0.6653	36.976	225.84	330.58	104.74	1.0903	1.4429
298.0	0.31098	0.6671	35.835	226.95	331.25	104.30	1.0940	1.4440
299.0	0.32105	0.6689	34.735	228.06	331.91	103.85	1.0977	1.4450
300.0	0.33137	0.6708	33.675	229.18	332.58	103.40	1.1014	1.4461
301.0	0.34194	0.6726	32.654	230.30	333.24	102.94	1.1051	1.4471
302.0	0.35276	0.6745	31.669	231.43	333.91	102.48	1.1088	1.4481
303.0	0.36384	0.6764	30.719	232.55	334.57	102.02	1.1125	1.4492
304.0	0.37518	0.6783	29.802	233.68	335.23	101.55	1.1162	1.4502
305.0	0.38678	0.6802	28.918	234.81	335.89	101.08	1.1199	1.4513
306.0	0.39865	0.6821	28.065	235.94	336.55	100.61	1.1236	1.4524
307.0	0.41079	0.6841	27.241	237.08	337.21	100.13	1.1272	1.4534
308.0	0.42321	0.6861	26.445	238.22	337.87	99.65	1.1309	1.4545
309.0	0.43592	0.6881	25.677	239.36	338.52	99.16	1.1346	1.4555
310.0	0.44890	0.6902	24.934	240.51	339.18	98.67	1.1383	1.4566
311.0	0.46218	0.6922	24.217	241.65	339.83	98.18	1.1419	1.4576
312.0	0.47574	0.6943	23.523	242.80	340.48	97.68	1.1456	1.4587
313.0	0.48961	0.6965	22.853	243.95	341.13	97.18	1.1492	1.4597
314.0	0.50377	0.6986	22.204	245.11	341.78	96.67	1.1529	1.4608
315.0	0.51824	0.7008	21.577	246.27	342.43	96.16	1.1565	1.4618
316.0	0.53302	0.7030	20.970	247.43	343.08	95.65	1.1602	1.4629
317.0	0.54811	0.7052	20.382	248.59	343.72	95.13	1.1638	1.4639
318.0	0.56352	0.7075	19.814	249.76	344.36	94.60	1.1675	1.4650
319.0	0.57925	0.7098	19.263	250.93	345.00	94.07	1.1711	1.4660
320.0	0.59531	0.7121	18.730	252.10	345.64	93.54	1.1747	1.4671
321.0	0.61170	0.7144	18.214	253.28	346.28	93.00	1.1784	1.4681
322.0	0.62842	0.7168	17.713	254.46	346.91	92.45	1.1820	1.4691
323.0	0.64549	0.7193	17.229	255.64	347.54	91.90	1.1856	1.4702
324.0	0.66290	0.7217	16.759	256.82	348.17	91.35	1.1893	1.4712
325.0	0.68065	0.7242	16.303	258.01	348.80	90.79	1.1929	1.4722
326.0	0.69876	0.7268	15.861	259.20	349.42	90.22	1.1965	1.4733
327.0	0.71723	0.7293	15.432	260.40	350.05	89.65	1.2001	1.4743
328.0	0.73605	0.7319	15.016	261.59	350.67	89.07	1.2037	1.4753
329.0	0.75525	0.7346	14.613	262.80	351.28	88.49	1.2074	1.4763
330.0	0.77481	0.7373	14.221	264.00	351.90	87.89	1.2110	1.4773
331.0	0.79476	0.7400	13.841	265.21	352.51	87.30	1.2146	1.4783
332.0	0.81508	0.7428	13.472	266.42	353.12	86.69	1.2182	1.4793
333.0	0.83579	0.7457	13.113	267.64	353.72	86.08	1.2218	1.4803
334.0	0.85689	0.7486	12.764	268.86	354.32	85.47	1.2254	1.4813

Chart RC318.1: Saturation properties—*continued*

T K	p MPa	v' dm³/kg	v'' dm³/kg	h' kJ/kg	h'' kJ/kg	r kJ/kg	s' kJ/kgK	s'' kJ/kgK
335.0	0.87838	0.7515	12.426	270.08	354.92	84.84	1.2290	1.4823
336.0	0.90028	0.7545	12.097	271.31	355.52	84.21	1.2326	1.4832
337.0	0.92258	0.7575	11.777	272.54	356.11	83.57	1.2362	1.4842
338.0	0.94529	0.7606	11.466	273.77	356.69	82.92	1.2398	1.4852
339.0	0.96841	0.7638	11.163	275.01	357.28	82.27	1.2434	1.4861
340.0	0.99196	0.7670	10.869	276.26	357.86	81.60	1.2471	1.4871
341.0	1.01593	0.7703	10.582	277.50	358.43	80.93	1.2507	1.4880
342.0	1.04033	0.7737	10.304	278.76	359.00	80.25	1.2543	1.4889
343.0	1.06517	0.7771	10.032	280.01	359.57	79.56	1.2579	1.4898
344.0	1.09046	0.7806	9.7682	281.27	360.13	78.86	1.2615	1.4907
345.0	1.11619	0.7842	9.5109	282.54	360.69	78.15	1.2651	1.4916
346.0	1.14237	0.7878	9.2602	283.81	361.24	77.43	1.2687	1.4925
347.0	1.16901	0.7916	9.0159	285.09	361.78	76.70	1.2724	1.4934
348.0	1.19612	0.7954	8.7778	286.37	362.32	75.95	1.2760	1.4943
349.0	1.22370	0.7993	8.5456	287.65	362.86	75.20	1.2796	1.4951
350.0	1.25176	0.8033	8.3193	288.95	363.38	74.44	1.2833	1.4959
351.0	1.28030	0.8074	8.0984	290.25	363.91	73.66	1.2869	1.4967
352.0	1.30933	0.8116	7.8829	291.55	364.42	72.87	1.2905	1.4976
353.0	1.33885	0.8160	7.6726	292.86	364.93	72.07	1.2942	1.4983
354.0	1.36888	0.8204	7.4673	294.18	365.43	71.25	1.2978	1.4991
355.0	1.39942	0.8250	7.2667	295.50	365.92	70.42	1.3015	1.4999
356.0	1.43047	0.8297	7.0708	296.83	366.40	69.57	1.3052	1.5006
357.0	1.46204	0.8345	6.8793	298.17	366.87	68.71	1.3088	1.5013
358.0	1.49415	0.8395	6.6921	299.51	367.34	67.83	1.3125	1.5020
359.0	1.52679	0.8447	6.5090	300.86	367.79	66.93	1.3162	1.5027
360.0	1.55998	0.8500	6.3298	302.22	368.24	66.02	1.3199	1.5033
361.0	1.59372	0.8555	6.1545	303.59	368.67	65.08	1.3237	1.5039
362.0	1.62803	0.8612	5.9828	304.97	369.10	64.12	1.3274	1.5045
363.0	1.66290	0.8671	5.8145	306.36	369.50	63.14	1.3311	1.5051
364.0	1.69835	0.8732	5.6496	307.76	369.90	62.14	1.3349	1.5056
365.0	1.73439	0.8796	5.4879	309.17	370.28	61.11	1.3387	1.5061
366.0	1.77102	0.8862	5.3292	310.59	370.65	60.06	1.3425	1.5066
367.0	1.80826	0.8931	5.1733	312.02	371.00	58.98	1.3463	1.5070
368.0	1.84611	0.9003	5.0202	313.47	371.33	57.87	1.3501	1.5074
369.0	1.88460	0.9079	4.8695	314.93	371.64	56.72	1.3540	1.5077
370.0	1.92372	0.9158	4.7213	316.40	371.94	55.54	1.3579	1.5080
371.0	1.96348	0.9241	4.5752	317.89	372.21	54.31	1.3618	1.5082
372.0	2.00392	0.9329	4.4312	319.40	372.45	53.05	1.3658	1.5084
373.0	2.04502	0.9423	4.2889	320.93	372.67	51.74	1.3698	1.5085
374.0	2.08681	0.9522	4.1482	322.48	372.85	50.37	1.3738	1.5085
375.0	2.12931	0.9627	4.0089	324.06	373.01	48.95	1.3779	1.5085
376.0	2.17253	0.9741	3.8706	325.66	373.12	47.46	1.3821	1.5083
377.0	2.21648	0.9863	3.7330	327.29	373.19	45.90	1.3863	1.5081
378.0	2.26119	0.9995	3.5958	328.96	373.21	44.26	1.3906	1.5077
379.0	2.30667	1.0139	3.4585	330.67	373.18	42.51	1.3950	1.5072
380.0	2.35296	1.0299	3.3206	332.42	373.07	40.65	1.3995	1.5065
381.0	2.40007	1.0477	3.1813	334.24	372.88	38.64	1.4041	1.5056
382.0	2.44804	1.0678	3.0397	336.13	372.59	36.46	1.4090	1.5044
383.0	2.49690	1.0911	2.8943	338.11	372.18	34.06	1.4140	1.5029
384.0	2.54670	1.1186	2.7431	340.22	371.59	31.37	1.4194	1.5010

Chart RC318.1: Saturation properties–*continued*

T K	p MPa	v' dm³/kg	v'' dm³/kg	h' kJ/kg	h'' kJ/kg	r kJ/kg	s' kJ/kgK	s'' kJ/kgK
385.0	2.59749	1.1525	2.5826	342.51	370.76	28.25	1.4252	1.4985
386.0	2.64934	1.1968	2.4066	345.09	369.56	24.47	1.4317	1.4951
387.0	2.71130	1.2374	1.4004	347.46	352.07	4.61	1.4376	1.4495

Chart RC318.2: Saturation properties

p MPa	T K	v' dm³/kg	v'' dm³/kg	h' kJ/kg	h'' kJ/kg	r kJ/kg	s' kJ/kgK	s'' kJ/kgK
0.0210	234.69	0.5797	458.36	161.06	288.85	127.79	0.8468	1.3913
0.0220	235.51	0.5806	438.85	161.86	289.39	127.53	0.8502	1.3917
0.0230	236.31	0.5814	420.98	162.64	289.91	127.27	0.8535	1.3920
0.0240	237.07	0.5823	404.55	163.39	290.41	127.03	0.8566	1.3924
0.0250	237.82	0.5831	389.40	164.11	290.90	126.79	0.8597	1.3928
0.0260	238.53	0.5839	375.37	164.81	291.37	126.56	0.8626	1.3932
0.0270	239.23	0.5847	362.36	165.50	291.83	126.33	0.8655	1.3935
0.0280	239.90	0.5854	350.24	166.16	292.27	126.11	0.8682	1.3939
0.0290	240.56	0.5862	338.93	166.80	292.70	125.90	0.8709	1.3943
0.0300	241.20	0.5869	328.35	167.43	293.12	125.69	0.8735	1.3946
0.0310	241.82	0.5876	318.43	168.04	293.53	125.49	0.8760	1.3950
0.0320	242.42	0.5883	309.11	168.64	293.93	125.29	0.8785	1.3953
0.0330	243.01	0.5890	300.34	169.22	294.32	125.10	0.8809	1.3957
0.0340	243.59	0.5896	292.07	169.79	294.70	124.91	0.8832	1.3960
0.0350	244.15	0.5903	284.25	170.35	295.07	124.73	0.8855	1.3964
0.0360	244.70	0.5909	276.86	170.89	295.44	124.55	0.8877	1.3967
0.0370	245.23	0.5915	269.85	171.42	295.79	124.37	0.8899	1.3971
0.0380	245.76	0.5921	263.20	171.95	296.14	124.19	0.8920	1.3974
0.0390	246.27	0.5927	256.88	172.46	296.48	124.02	0.8941	1.3977
0.0400	246.78	0.5933	250.86	172.96	296.82	123.86	0.8961	1.3980
0.0410	247.27	0.5939	245.13	173.45	297.14	123.69	0.8981	1.3984
0.0420	247.76	0.5945	239.66	173.93	297.46	123.53	0.9001	1.3987
0.0430	248.23	0.5950	234.44	174.41	297.78	123.37	0.9020	1.3990
0.0440	248.70	0.5956	229.44	174.87	298.09	123.22	0.9039	1.3993
0.0450	249.15	0.5961	224.67	175.33	298.40	123.06	0.9057	1.3996
0.0460	249.60	0.5967	220.09	175.78	298.70	122.91	0.9075	1.3999
0.0470	250.05	0.5972	215.70	176.23	298.99	122.76	0.9093	1.4002
0.0480	250.48	0.5977	211.49	176.66	299.28	122.62	0.9110	1.4005
0.0490	250.91	0.5982	207.44	177.09	299.56	122.47	0.9127	1.4009
0.0500	251.33	0.5988	203.55	177.51	299.84	122.33	0.9144	1.4011
0.0510	251.74	0.5993	199.81	177.93	300.12	122.19	0.9161	1.4014
0.0520	252.15	0.5998	196.20	178.34	300.39	122.05	0.9177	1.4017
0.0530	252.55	0.6002	192.73	178.74	300.66	121.92	0.9193	1.4020
0.0540	252.95	0.6007	189.38	179.14	300.92	121.78	0.9209	1.4023
0.0550	253.33	0.6012	186.15	179.54	301.19	121.65	0.9224	1.4026
0.0560	253.72	0.6017	183.03	179.92	301.44	121.52	0.9239	1.4029
0.0570	254.10	0.6021	180.02	180.31	301.69	121.39	0.9254	1.4032
0.0580	254.47	0.6026	177.11	180.68	301.94	121.26	0.9269	1.4035
0.0590	254.84	0.6031	174.29	181.06	302.19	121.13	0.9284	1.4037
0.0600	255.20	0.6035	171.56	181.43	302.43	121.01	0.9298	1.4040
0.0610	255.56	0.6040	168.92	181.79	302.67	120.88	0.9313	1.4043
0.0620	255.91	0.6044	166.37	182.15	302.91	120.76	0.9327	1.4045
0.0630	256.26	0.6048	163.89	182.50	303.14	120.64	0.9340	1.4048
0.0640	256.61	0.6053	161.48	182.85	303.38	120.52	0.9354	1.4051
0.0650	256.95	0.6057	159.15	183.20	303.60	120.40	0.9368	1.4053
0.0660	257.28	0.6061	156.88	183.54	303.83	120.29	0.9381	1.4056
0.0670	257.61	0.6065	154.68	183.88	304.05	120.17	0.9394	1.4059
0.0680	257.94	0.6070	152.55	184.22	304.27	120.06	0.9407	1.4061
0.0690	258.27	0.6074	150.47	184.55	304.49	119.94	0.9420	1.4064
0.0700	258.59	0.6078	148.45	184.88	304.71	119.83	0.9432	1.4066

Chart RC318.2: Saturation properties–*continued*

p MPa	T K	v' dm³/kg	v'' dm³/kg	h' kJ/kg	h'' kJ/kg	r kJ/kg	s' kJ/kgK	s'' kJ/kgK
0.0710	258.91	0.6082	146.48	185.20	304.92	119.72	0.9445	1.4069
0.0720	259.22	0.6086	144.57	185.52	305.13	119.61	0.9457	1.4071
0.0730	259.53	0.6090	142.71	185.84	305.34	119.50	0.9470	1.4074
0.0740	259.84	0.6094	140.89	186.15	305.54	119.39	0.9482	1.4076
0.0760	260.44	0.6102	137.41	186.77	305.95	119.18	0.9505	1.4081
0.0780	261.03	0.6109	134.09	187.38	306.35	118.97	0.9529	1.4086
0.0800	261.61	0.6117	130.93	187.98	306.74	118.76	0.9551	1.4091
0.0820	262.18	0.6124	127.93	188.56	307.12	118.56	0.9574	1.4096
0.0840	262.74	0.6132	125.06	189.14	307.50	118.36	0.9596	1.4100
0.0860	263.29	0.6139	122.32	189.70	307.87	118.16	0.9617	1.4105
0.0880	263.82	0.6146	119.70	190.26	308.23	117.97	0.9638	1.4110
0.0900	264.35	0.6153	117.19	190.80	308.58	117.78	0.9659	1.4114
0.0920	264.87	0.6160	114.79	191.34	308.93	117.59	0.9679	1.4119
0.0940	265.38	0.6167	112.49	191.87	309.28	117.41	0.9699	1.4123
0.0960	265.88	0.6174	110.27	192.39	309.62	117.22	0.9718	1.4127
0.0980	266.38	0.6180	108.15	192.90	309.95	117.04	0.9738	1.4132
0.1000	266.86	0.6187	106.11	193.41	310.28	116.87	0.9757	1.4136
0.1020	267.34	0.6193	104.14	193.91	310.60	116.69	0.9775	1.4140
0.1040	267.81	0.6200	102.25	194.40	310.92	116.52	0.9793	1.4144
0.1060	268.27	0.6206	100.43	194.88	311.23	116.35	0.9811	1.4148
0.1080	268.73	0.6212	98.672	195.36	311.54	116.18	0.9829	1.4152
0.1100	269.18	0.6219	96.976	195.83	311.84	116.01	0.9847	1.4156
0.1150	270.28	0.6234	92.985	196.98	312.58	115.60	0.9889	1.4166
0.1200	271.34	0.6249	89.315	198.10	313.30	115.21	0.9930	1.4176
0.1250	272.37	0.6263	85.929	199.18	313.99	114.82	0.9970	1.4185
0.1300	273.37	0.6277	82.794	200.23	314.67	114.44	1.0008	1.4195
0.1350	274.33	0.6291	79.883	201.25	315.32	114.07	1.0045	1.4204
0.1400	275.27	0.6305	77.173	202.24	315.96	113.71	1.0082	1.4212
0.1450	276.19	0.6318	74.643	203.21	316.57	113.36	1.0117	1.4221
0.1500	277.08	0.6331	72.275	204.16	317.17	113.01	1.0151	1.4230
0.1550	277.94	0.6344	70.055	205.08	317.76	112.68	1.0184	1.4238
0.1600	278.79	0.6357	67.969	205.99	318.33	112.34	1.0216	1.4246
0.1650	279.61	0.6369	66.005	206.87	318.89	112.02	1.0248	1.4254
0.1700	280.42	0.6381	64.152	207.73	319.43	111.70	1.0278	1.4262
0.1750	281.21	0.6393	62.402	208.58	319.97	111.39	1.0308	1.4269
0.1800	281.98	0.6405	60.745	209.41	320.49	111.08	1.0338	1.4277
0.1850	282.74	0.6417	59.174	210.22	321.00	110.77	1.0366	1.4284
0.1900	283.47	0.6428	57.683	211.02	321.49	110.48	1.0395	1.4292
0.1950	284.20	0.6440	56.266	211.80	321.98	110.18	1.0422	1.4299
0.2000	284.91	0.6451	54.917	212.57	322.46	109.89	1.0449	1.4306
0.2100	286.29	0.6473	52.405	214.07	323.39	109.33	1.0501	1.4320
0.2200	287.62	0.6494	50.113	215.52	324.29	108.77	1.0551	1.4333
0.2300	288.91	0.6515	48.014	216.92	325.15	108.24	1.0600	1.4346
0.2400	290.15	0.6536	46.083	218.28	325.99	107.71	1.0646	1.4359
0.2500	291.36	0.6556	44.301	219.60	326.80	107.20	1.0692	1.4371
0.2600	292.52	0.6576	42.651	220.88	327.58	106.70	1.0735	1.4383
0.2700	293.66	0.6595	41.119	222.13	328.35	106.21	1.0778	1.4395
0.2800	294.76	0.6614	39.692	223.35	329.08	105.73	1.0819	1.4406
0.2900	295.84	0.6633	38.360	224.54	329.80	105.26	1.0859	1.4417
0.3000	296.88	0.6651	37.114	225.70	330.50	104.80	1.0898	1.4428

Chart RC318.2: Saturation properties–*continued*

p MPa	T K	v' dm^3/kg	v'' dm^3/kg	h' kJ/kg	h'' kJ/kg	r kJ/kg	s' kJ/kgK	s'' kJ/kgK
0.3100	297.90	0.6670	35.945	226.84	331.18	104.34	1.0936	1.4439
0.3200	298.90	0.6688	34.847	227.95	331.85	103.90	1.0973	1.4449
0.3300	299.87	0.6705	33.812	229.04	332.49	103.46	1.1009	1.4459
0.3400	300.82	0.6723	32.836	230.10	333.12	103.02	1.1044	1.4469
0.3500	301.75	0.6740	31.914	231.14	333.74	102.60	1.1079	1.4479
0.3600	302.66	0.6757	31.041	232.16	334.34	102.18	1.1112	1.4488
0.3700	303.55	0.6774	30.214	233.17	334.93	101.76	1.1145	1.4498
0.3800	304.42	0.6791	29.428	234.15	335.51	101.36	1.1177	1.4507
0.3900	305.27	0.6807	28.682	235.12	336.07	100.95	1.1209	1.4516
0.4000	306.11	0.6824	27.971	236.07	336.63	100.56	1.1240	1.4525
0.4100	306.94	0.6840	27.293	237.01	337.17	100.16	1.1270	1.4533
0.4200	307.74	0.6856	26.647	237.93	337.70	99.77	1.1300	1.4542
0.4300	308.54	0.6872	26.029	238.83	338.22	99.39	1.1329	1.4550
0.4400	309.32	0.6888	25.439	239.72	338.73	99.01	1.1358	1.4558
0.4500	310.08	0.6903	24.873	240.60	339.23	98.63	1.1386	1.4567
0.4600	310.84	0.6919	24.332	241.47	339.73	98.26	1.1413	1.4574
0.4700	311.58	0.6934	23.812	242.32	340.21	97.89	1.1440	1.4582
0.4800	312.31	0.6950	23.313	243.16	340.69	97.53	1.1467	1.4590
0.4900	313.03	0.6965	22.834	243.99	341.15	97.17	1.1493	1.4598
0.5000	313.74	0.6980	22.373	244.80	341.61	96.81	1.1519	1.4605
0.5100	314.43	0.6995	21.930	245.61	342.06	96.45	1.1545	1.4612
0.5200	315.12	0.7010	21.503	246.41	342.51	96.10	1.1570	1.4620
0.5300	315.80	0.7025	21.091	247.19	342.95	95.75	1.1595	1.4627
0.5400	316.47	0.7040	20.694	247.97	343.38	95.41	1.1619	1.4634
0.5500	317.12	0.7055	20.311	248.74	343.80	95.06	1.1643	1.4641
0.5600	317.77	0.7070	19.941	249.49	344.22	94.72	1.1667	1.4647
0.5700	318.41	0.7084	19.584	250.24	344.63	94.38	1.1690	1.4654
0.5800	319.05	0.7099	19.238	250.98	345.03	94.05	1.1713	1.4661
0.5900	319.67	0.7113	18.904	251.72	345.43	93.72	1.1736	1.4667
0.6000	320.29	0.7128	18.580	252.44	345.82	93.38	1.1758	1.4674
0.6100	320.90	0.7142	18.266	253.16	346.21	93.06	1.1780	1.4680
0.6200	321.50	0.7156	17.962	253.86	346.59	92.73	1.1802	1.4686
0.6300	322.09	0.7171	17.668	254.57	346.97	92.40	1.1823	1.4692
0.6400	322.68	0.7185	17.382	255.26	347.34	92.08	1.1845	1.4698
0.6500	323.26	0.7199	17.104	255.95	347.71	91.76	1.1866	1.4704
0.6600	323.84	0.7213	16.835	256.63	348.07	91.44	1.1887	1.4710
0.6700	324.40	0.7227	16.573	257.30	348.42	91.12	1.1907	1.4716
0.6800	324.96	0.7241	16.319	257.97	348.78	90.81	1.1928	1.4722
0.6900	325.52	0.7255	16.072	258.63	349.12	90.49	1.1948	1.4728
0.7000	326.07	0.7269	15.831	259.28	349.47	90.18	1.1967	1.4733
0.7100	326.61	0.7283	15.597	259.93	349.80	89.87	1.1987	1.4739
0.7200	327.15	0.7297	15.370	260.57	350.14	89.56	1.2007	1.4744
0.7300	327.68	0.7311	15.148	261.21	350.47	89.26	1.2026	1.4750
0.7400	328.21	0.7325	14.932	261.84	350.79	88.95	1.2045	1.4755
0.7600	329.24	0.7353	14.516	263.09	351.43	88.34	1.2082	1.4766
0.7800	330.26	0.7380	14.120	264.32	352.06	87.74	1.2119	1.4776
0.8000	331.26	0.7408	13.744	265.52	352.67	87.14	1.2155	1.4786
0.8200	332.24	0.7435	13.385	266.71	353.26	86.55	1.2191	1.4796
0.8400	333.20	0.7462	13.042	267.88	353.84	85.96	1.2225	1.4805
0.8600	334.15	0.7490	12.714	269.04	354.41	85.37	1.2259	1.4814

Chart RC318.2: Saturation properties–*continued*

p MPa	T K	v' dm^3/kg	v'' dm^3/kg	h' kJ/kg	h'' kJ/kg	r kJ/kg	s' kJ/kgK	s'' kJ/kgK
0.8800	335.07	0.7517	12.401	270.17	354.97	84.79	1.2293	1.4823
0.9000	335.99	0.7544	12.101	271.29	355.51	84.22	1.2326	1.4832
0.9200	336.89	0.7572	11.813	272.40	356.04	83.64	1.2358	1.4841
0.9400	337.77	0.7599	11.537	273.49	356.56	83.07	1.2390	1.4849
0.9600	338.64	0.7626	11.272	274.56	357.07	82.50	1.2421	1.4858
0.9800	339.49	0.7654	11.017	275.63	357.56	81.94	1.2452	1.4866
1.0000	340.34	0.7681	10.771	276.68	358.05	81.38	1.2483	1.4874
1.0200	341.17	0.7709	10.535	277.71	358.53	80.82	1.2513	1.4882
1.0400	341.99	0.7736	10.307	278.74	359.00	80.26	1.2542	1.4889
1.0600	342.79	0.7764	10.088	279.75	359.45	79.70	1.2571	1.4896
1.0800	343.59	0.7791	9.8761	280.75	359.90	79.15	1.2600	1.4904
1.1000	344.37	0.7819	9.6714	281.75	360.34	78.59	1.2629	1.4911
1.1500	346.29	0.7889	9.1891	284.18	361.40	77.22	1.2698	1.4928
1.2000	348.14	0.7959	8.7445	286.55	362.40	75.85	1.2765	1.4944
1.2500	349.94	0.8031	8.3332	288.87	363.35	74.48	1.2830	1.4959
1.3000	351.68	0.8103	7.9512	291.13	364.26	73.12	1.2894	1.4973
1.3500	353.37	0.8176	7.5954	293.35	365.11	71.76	1.2955	1.4986
1.4000	355.02	0.8251	7.2629	295.52	365.93	70.40	1.3016	1.4999
1.4500	356.62	0.8327	6.9514	297.66	366.70	69.04	1.3074	1.5010
1.5000	358.18	0.8404	6.6587	299.75	367.42	67.67	1.3132	1.5021
1.5500	359.70	0.8484	6.3830	301.82	368.11	66.29	1.3188	1.5031
1.6000	361.18	0.8565	6.1226	303.85	368.75	64.90	1.3243	1.5040
1.6500	362.63	0.8649	5.8760	305.85	369.36	63.51	1.3297	1.5049
1.7000	364.05	0.8735	5.6421	307.82	369.92	62.10	1.3351	1.5056
1.7500	365.43	0.8824	5.4196	309.77	370.44	60.67	1.3403	1.5063
1.8000	366.78	0.8915	5.2074	311.70	370.92	59.22	1.3454	1.5069
1.8500	368.10	0.9011	5.0047	313.61	371.36	57.75	1.3505	1.5074
1.9000	369.40	0.9110	4.8106	315.51	371.76	56.26	1.3555	1.5078
1.9500	370.66	0.9213	4.6243	317.39	372.12	54.73	1.3605	1.5081
2.0000	371.90	0.9321	4.4449	319.26	372.43	53.17	1.3654	1.5084
2.1000	374.31	0.9554	4.1046	322.97	372.91	49.94	1.3751	1.5085
2.2000	376.63	0.9816	3.7842	326.68	373.17	46.49	1.3847	1.5082
2.3000	378.85	1.0118	3.4786	330.41	373.18	42.77	1.3943	1.5072
2.4000	381.00	1.0477	3.1815	334.24	372.88	38.65	1.4041	1.5056
2.5000	383.06	1.0927	2.8850	338.24	372.14	33.90	1.4143	1.5028
2.6000	385.05	1.1544	2.5744	342.63	370.71	28.08	1.4255	1.4984
2.7000	386.78	1.2445	2.2503	347.45	368.21	20.77	1.4376	1.4913

Chart RC318.3: Thermodynamic properties

p/MPa	0.25			0.40			0.50		
T K	v dm³/kg	h kJ/kg	s kJ/kgK	v dm³/kg	h kJ/kg	s kJ/kgK	v dm³/kg	h kJ/kg	s kJ/kgK
240.0	0.5853	166.32	0.8684	0.5851	166.37	0.8682	0.5850	166.40	0.8681
250.0	0.5969	176.24	0.9089	0.5967	176.28	0.9087	0.5966	176.31	0.9086
260.0	0.6093	186.37	0.9486	0.6091	186.41	0.9484	0.6090	186.44	0.9483
270.0	0.6227	196.72	0.9876	0.6225	196.76	0.9874	0.6223	196.78	0.9873
280.0	0.6373	207.30	1.0261	0.6370	207.33	1.0259	0.6368	207.35	1.0257
290.0	0.6533	218.11	1.0641	0.6529	218.14	1.0638	0.6526	218.16	1.0636
300.0	46.057	333.78	1.4607	0.6705	229.19	1.1013	0.6702	229.20	1.1011
310.0	48.041	341.97	1.4876	28.518	339.92	1.4632	0.6899	240.51	1.1382
320.0	49.984	350.28	1.5139	29.884	348.43	1.4902	23.118	347.07	1.4777
330.0	51.893	358.71	1.5399	31.205	357.04	1.5167	24.260	355.82	1.5046
340.0	53.775	367.27	1.5654	32.489	365.74	1.5426	25.357	364.64	1.5310
350.0	55.634	375.95	1.5906	33.743	374.54	1.5681	26.419	373.54	1.5568
360.0	57.474	384.75	1.6154	34.974	383.45	1.5932	27.453	382.54	1.5821
370.0	59.297	393.68	1.6399	36.185	392.47	1.6180	28.464	391.63	1.6070
380.0	61.105	402.74	1.6640	37.379	401.60	1.6423	29.457	400.82	1.6315
390.0	62.902	411.91	1.6878	38.559	410.85	1.6663	30.434	410.12	1.6557
400.0	64.687	421.20	1.7114	39.727	420.21	1.6900	31.398	419.52	1.6795
410.0	66.463	430.61	1.7346	40.884	429.67	1.7134	32.350	429.03	1.7030
420.0	68.231	440.13	1.7575	42.031	439.25	1.7365	33.293	438.64	1.7261
430.0	69.991	449.77	1.7802	43.171	448.93	1.7593	34.227	448.36	1.7490
440.0	71.744	459.52	1.8026	44.303	458.72	1.7818	35.153	458.18	1.7716
450.0	73.492	469.37	1.8248	45.429	468.61	1.8040	36.072	468.10	1.7939
460.0	75.234	479.33	1.8466	46.549	478.60	1.8259	36.986	478.12	1.8159
470.0	76.971	489.39	1.8683	47.664	488.70	1.8477	37.894	488.23	1.8376
480.0	78.703	499.55	1.8897	48.775	498.89	1.8691	38.797	498.44	1.8591
490.0	80.432	509.80	1.9108	49.881	509.17	1.8903	39.696	508.75	1.8804
500.0	82.157	520.16	1.9317	50.983	519.55	1.9113	40.591	519.15	1.9014
510.0	83.879	530.60	1.9524	52.082	530.02	1.9320	41.483	529.63	1.9222
520.0	85.597	541.13	1.9729	53.178	540.58	1.9525	42.371	540.21	1.9427
530.0	87.313	551.76	1.9931	54.271	551.22	1.9728	43.257	550.86	1.9630
540.0	89.026	562.46	2.0131	55.361	561.95	1.9928	44.139	561.60	1.9831
550.0	90.737	573.25	2.0329	56.449	572.76	2.0127	45.019	572.42	2.0029
560.0	92.446	584.12	2.0525	57.534	583.64	2.0323	45.897	583.32	2.0226
570.0	94.152	595.07	2.0719	58.617	594.61	2.0517	46.773	594.30	2.0420
580.0	95.857	606.09	2.0910	59.699	605.64	2.0709	47.646	605.35	2.0612
590.0	97.560	617.18	2.1100	60.778	616.75	2.0899	48.518	616.47	2.0802
600.0	99.261	628.35	2.1288	61.856	627.94	2.1087	49.388	627.66	2.0990
610.0	100.96	639.58	2.1473	62.932	639.18	2.1273	50.257	638.92	2.1176
620.0	102.66	650.89	2.1657	64.007	650.50	2.1457	51.124	650.24	2.1360

Chart RC318.3: Thermodynamic properties—*continued*

p/MPa	0.60			0.80			1.00		
T K	v dm³/kg	h kJ/kg	s kJ/kgK	v dm³/kg	h kJ/kg	s kJ/kgK	v dm³/kg	h kJ/kg	s kJ/kgK
240.0	0.5849	166.43	0.8680	0.5847	166.50	0.8678	0.5845	166.56	0.8675
250.0	0.5964	176.34	0.9084	0.5962	176.40	0.9082	0.5959	176.46	0.9080
260.0	0.6088	186.46	0.9481	0.6085	186.52	0.9479	0.6082	186.58	0.9476
270.0	0.6221	196.81	0.9872	0.6218	196.86	0.9869	0.6214	196.91	0.9866
280.0	0.6366	207.37	1.0256	0.6361	207.42	1.0253	0.6357	207.46	1.0250
290.0	0.6524	218.17	1.0635	0.6518	218.21	1.0631	0.6513	218.24	1.0628
300.0	0.6699	229.21	1.1009	0.6692	229.24	1.1005	0.6686	229.26	1.1002
310.0	0.6895	240.51	1.1379	0.6887	240.52	1.1375	0.6879	240.53	1.1371
320.0	0.7121	252.10	1.1747	0.7109	252.09	1.1743	0.7099	252.08	1.1738
330.0	19.589	354.51	1.4941	0.7371	264.00	1.2109	0.7356	263.96	1.2103
340.0	20.573	363.47	1.5208	14.509	360.86	1.5030	0.7669	276.25	1.2470
350.0	21.514	372.49	1.5470	15.324	370.19	1.5300	11.527	367.54	1.5149
360.0	22.422	381.58	1.5726	16.091	379.52	1.5563	12.235	377.22	1.5421
370.0	23.305	390.75	1.5977	16.823	388.89	1.5820	12.894	386.85	1.5685
380.0	24.166	400.01	1.6224	17.529	398.31	1.6071	13.516	396.47	1.5942
390.0	25.010	409.37	1.6467	18.212	407.80	1.6318	14.111	406.13	1.6193
400.0	25.840	418.82	1.6706	18.878	417.37	1.6560	14.684	415.84	1.6438
410.0	26.657	428.37	1.6942	19.530	427.02	1.6798	15.240	425.61	1.6680
420.0	27.464	438.03	1.7175	20.169	436.76	1.7033	15.782	435.45	1.6917
430.0	28.261	447.78	1.7404	20.798	446.59	1.7264	16.312	445.36	1.7150
440.0	29.051	457.63	1.7631	21.418	456.51	1.7492	16.832	455.36	1.7380
450.0	29.833	467.58	1.7854	22.030	466.52	1.7717	17.344	465.43	1.7606
460.0	30.609	477.62	1.8075	22.635	476.62	1.7939	17.848	475.60	1.7830
470.0	31.380	487.76	1.8293	23.234	486.81	1.8158	18.345	485.84	1.8050
480.0	32.145	498.00	1.8509	23.828	497.09	1.8375	18.837	496.17	1.8267
490.0	32.906	508.32	1.8722	24.418	507.46	1.8589	19.323	506.58	1.8482
500.0	33.663	518.74	1.8932	25.003	517.91	1.8800	19.806	517.08	1.8694
510.0	34.417	529.24	1.9140	25.584	528.45	1.9008	20.284	527.66	1.8904
520.0	35.167	539.83	1.9346	26.161	539.08	1.9215	20.758	538.31	1.9111
530.0	35.914	550.50	1.9549	26.736	549.78	1.9419	21.230	549.05	1.9315
540.0	36.658	561.26	1.9750	27.308	560.56	1.9620	21.698	559.86	1.9517
550.0	37.400	572.09	1.9949	27.877	571.42	1.9819	22.163	570.75	1.9717
560.0	38.139	583.00	2.0145	28.443	582.36	2.0017	22.626	581.71	1.9915
570.0	38.877	593.99	2.0340	29.008	593.37	2.0211	23.087	592.75	2.0110
580.0	39.612	605.05	2.0532	29.570	604.45	2.0404	23.546	603.86	2.0303
590.0	40.345	616.18	2.0722	30.130	615.61	2.0595	24.002	615.03	2.0494
600.0	41.077	627.38	2.0911	30.689	626.83	2.0783	24.457	626.27	2.0683
610.0	41.807	638.65	2.1097	31.246	638.11	2.0970	24.911	637.57	2.0870
620.0	42.536	649.98	2.1281	31.801	649.46	2.1154	25.362	648.94	2.1055

Chart RC318.3: Thermodynamic properties—*continued*

p/MPa	2.00			4.00			6.00		
T K	v dm³/kg	h kJ/kg	s kJ/kgK	v dm³/kg	h kJ/kg	s kJ/kgK	v dm³/kg	h kJ/kg	s kJ/kgK
240.0	0.5834	166.88	0.8664	0.5813	167.53	0.8643	0.5794	168.19	0.8622
250.0	0.5947	176.76	0.9068	0.5923	177.38	0.9045	0.5900	178.00	0.9023
260.0	0.6067	186.86	0.9464	0.6039	187.43	0.9439	0.6013	188.02	0.9415
270.0	0.6197	197.16	0.9852	0.6164	197.69	0.9826	0.6133	198.24	0.9801
280.0	0.6336	207.68	1.0235	0.6297	208.15	1.0207	0.6261	208.65	1.0180
290.0	0.6488	218.43	1.0612	0.6441	218.83	1.0581	0.6397	219.27	1.0552
300.0	0.6654	229.39	1.0984	0.6597	229.71	1.0950	0.6545	230.07	1.0918
310.0	0.6839	240.59	1.1351	0.6768	240.80	1.1314	0.6705	241.07	1.1279
320.0	0.7047	252.05	1.1715	0.6957	252.11	1.1673	0.6879	252.27	1.1635
330.0	0.7287	263.80	1.2076	0.7169	263.65	1.2028	0.7071	263.67	1.1985
340.0	0.7570	275.89	1.2437	0.7411	275.46	1.2381	0.7286	275.28	1.2332
350.0	0.7920	288.44	1.2801	0.7693	287.57	1.2732	0.7528	287.13	1.2675
360.0	0.8386	301.66	1.3173	0.8033	300.06	1.3083	0.7806	299.23	1.3016
370.0	0.9113	316.18	1.3571	0.8460	313.03	1.3439	0.8131	311.64	1.3356
380.0	5.0741	383.39	1.5375	0.9035	326.71	1.3803	0.8522	324.40	1.3697
390.0	5.6465	395.25	1.5684	0.9912	341.60	1.4190	0.9007	337.59	1.4039
400.0	6.1204	406.41	1.5966	1.1641	359.18	1.4635	0.9631	351.33	1.4387
410.0	6.5390	417.24	1.6234	1.6054	382.04	1.5199	1.0470	365.75	1.4743
420.0	6.9216	427.90	1.6490	2.1552	402.59	1.5695	1.1626	380.93	1.5109
430.0	7.2787	438.47	1.6739	2.5691	418.39	1.6067	1.3180	396.71	1.5480
440.0	7.6166	449.02	1.6982	2.8961	432.09	1.6382	1.5071	412.50	1.5843
450.0	7.9396	459.55	1.7218	3.1736	444.80	1.6667	1.7092	427.67	1.6184
460.0	8.2506	470.11	1.7450	3.4195	456.95	1.6935	1.9065	442.00	1.6499
470.0	8.5518	480.70	1.7678	3.6436	468.79	1.7189	2.0916	455.59	1.6791
480.0	8.8448	491.33	1.7902	3.8518	480.42	1.7434	2.2637	468.63	1.7066
490.0	9.1308	502.01	1.8122	4.0477	491.94	1.7671	2.4243	481.26	1.7326
500.0	9.4109	512.75	1.8339	4.2339	503.38	1.7903	2.5752	493.62	1.7576
510.0	9.6858	523.54	1.8553	4.4123	514.78	1.8128	2.7180	505.79	1.7817
520.0	9.9562	534.40	1.8764	4.5843	526.16	1.8349	2.8540	517.82	1.8051
530.0	10.223	545.32	1.8972	4.7507	537.54	1.8566	2.9843	529.76	1.8278
540.0	10.485	556.30	1.9177	4.9125	548.94	1.8779	3.1097	541.63	1.8500
550.0	10.745	567.34	1.9379	5.0702	560.36	1.8989	3.2310	553.47	1.8717
560.0	11.002	578.44	1.9580	5.2244	571.80	1.9195	3.3487	565.29	1.8930
570.0	11.256	589.61	1.9777	5.3755	583.27	1.9398	3.4633	577.10	1.9139
580.0	11.507	600.84	1.9973	5.5238	594.78	1.9598	3.5751	588.92	1.9345
590.0	11.757	612.13	2.0165	5.6697	606.33	1.9796	3.6844	600.74	1.9547
600.0	12.005	623.48	2.0356	5.8133	617.92	1.9990	3.7915	612.59	1.9746
610.0	12.250	634.88	2.0545	5.9549	629.55	2.0183	3.8966	624.45	1.9942
620.0	12.495	646.35	2.0731	6.0947	641.22	2.0372	4.0000	636.33	2.0135

Chart RC318.3: Thermodynamic properties–*continued*

p/MPa	8.00			10.00			15.00		
T K	v dm³/kg	h kJ/kg	s kJ/kgK	v dm³/kg	h kJ/kg	s kJ/kgK	v dm³/kg	h kJ/kg	s kJ/kgK
240.0	0.5774	168.85	0.8601	0.5756	169.53	0.8581	0.5713	171.24	0.8534
250.0	0.5878	178.64	0.9001	0.5857	179.29	0.8980	0.5809	180.95	0.8930
260.0	0.5988	188.63	0.9393	0.5964	189.24	0.9370	0.5908	190.84	0.9318
270.0	0.6104	198.80	0.9777	0.6076	199.39	0.9753	0.6013	200.91	0.9698
280.0	0.6227	209.18	1.0154	0.6195	209.72	1.0129	0.6122	211.17	1.0071
290.0	0.6357	219.74	1.0524	0.6320	220.24	1.0498	0.6237	221.59	1.0436
300.0	0.6497	230.48	1.0889	0.6454	230.93	1.0860	0.6358	232.17	1.0795
310.0	0.6648	241.41	1.1247	0.6597	241.79	1.1217	0.6485	242.91	1.1147
320.0	0.6811	252.51	1.1599	0.6749	252.82	1.1567	0.6619	253.80	1.1493
330.0	0.6988	263.80	1.1947	0.6914	264.01	1.1911	0.6762	264.82	1.1832
340.0	0.7182	275.27	1.2289	0.7093	275.37	1.2250	0.6913	275.99	1.2165
350.0	0.7396	286.93	1.2627	0.7287	286.90	1.2584	0.7073	287.28	1.2493
360.0	0.7636	298.80	1.2961	0.7500	298.60	1.2914	0.7245	298.71	1.2815
370.0	0.7907	310.88	1.3293	0.7736	310.47	1.3239	0.7429	310.27	1.3132
380.0	0.8216	323.21	1.3621	0.7998	322.53	1.3561	0.7625	321.96	1.3443
390.0	0.8575	335.80	1.3948	0.8291	334.78	1.3879	0.7837	333.77	1.3750
400.0	0.8998	348.67	1.4274	0.8622	347.23	1.4194	0.8065	345.70	1.4052
410.0	0.9500	361.85	1.4600	0.8997	359.88	1.4506	0.8310	357.75	1.4349
420.0	1.0105	375.34	1.4925	0.9424	372.72	1.4816	0.8575	369.91	1.4643
430.0	1.0831	389.14	1.5249	0.9910	385.75	1.5122	0.8860	382.18	1.4931
440.0	1.1692	403.15	1.5571	1.0461	398.93	1.5426	0.9166	394.54	1.5215
450.0	1.2684	417.26	1.5889	1.1078	412.23	1.5724	0.9494	406.99	1.5495
460.0	1.3780	431.31	1.6197	1.1760	425.59	1.6018	0.9844	419.51	1.5770
470.0	1.4941	445.15	1.6495	1.2497	438.95	1.6305	1.0216	432.09	1.6041
480.0	1.6124	458.70	1.6780	1.3278	452.25	1.6585	1.0609	444.70	1.6306
490.0	1.7300	471.96	1.7054	1.4088	465.43	1.6857	1.1019	457.35	1.6567
500.0	1.8449	484.94	1.7316	1.4914	478.49	1.7121	1.1447	470.00	1.6823
510.0	1.9565	497.68	1.7568	1.5743	491.40	1.7377	1.1888	482.65	1.7073
520.0	2.0643	510.24	1.7812	1.6568	504.18	1.7625	1.2340	495.29	1.7319
530.0	2.1685	522.65	1.8049	1.7383	516.83	1.7866	1.2801	507.91	1.7559
540.0	2.2694	534.95	1.8278	1.8184	529.38	1.8100	1.3267	520.51	1.7795
550.0	2.3671	547.17	1.8503	1.8970	541.85	1.8329	1.3737	533.08	1.8025
560.0	2.4621	559.33	1.8722	1.9741	554.24	1.8552	1.4209	545.63	1.8251
570.0	2.5545	571.45	1.8936	2.0496	566.58	1.8771	1.4681	558.16	1.8473
580.0	2.6447	583.54	1.9147	2.1236	578.88	1.8985	1.5152	570.66	1.8691
590.0	2.7328	595.62	1.9353	2.1962	591.16	1.9195	1.5620	583.15	1.8904
600.0	2.8190	607.70	1.9556	2.2675	603.42	1.9401	1.6087	595.63	1.9114
610.0	2.9036	619.77	1.9756	2.3375	615.66	1.9603	1.6550	608.10	1.9320
620.0	2.9865	631.85	1.9952	2.4064	627.91	1.9802	1.7009	620.56	1.9523

Chart RC318.3: Thermodynamic properties–*continued*

p/MPa	20.00			40.00			60.00		
T K	v dm³/kg	h kJ/kg	s kJ/kgK	v dm³/kg	h kJ/kg	s kJ/kgK	v dm³/kg	h kJ/kg	s kJ/kgK
240.0	0.5674	173.00	0.8488	0.5541	180.29	0.8325	0.5435	187.86	0.8183
250.0	0.5764	182.66	0.8882	0.5617	189.81	0.8714	0.5503	197.30	0.8569
260.0	0.5858	192.49	0.9268	0.5696	199.51	0.9094	0.5571	206.92	0.8946
270.0	0.5956	202.51	0.9646	0.5776	209.38	0.9466	0.5641	216.71	0.9315
280.0	0.6059	212.70	1.0017	0.5859	219.41	0.9831	0.5712	226.66	0.9677
290.0	0.6165	223.05	1.0380	0.5943	229.59	1.0188	0.5784	236.76	1.0031
300.0	0.6276	233.56	1.0736	0.6030	239.91	1.0538	0.5857	246.99	1.0378
310.0	0.6392	244.20	1.1085	0.6119	250.36	1.0881	0.5931	257.35	1.0718
320.0	0.6513	254.99	1.1427	0.6209	260.93	1.1216	0.6006	267.82	1.1050
330.0	0.6639	265.90	1.1763	0.6302	271.61	1.1545	0.6083	278.40	1.1376
340.0	0.6772	276.94	1.2093	0.6397	282.39	1.1867	0.6160	289.07	1.1694
350.0	0.6912	288.09	1.2416	0.6495	293.26	1.2182	0.6238	299.84	1.2007
360.0	0.7058	299.36	1.2733	0.6595	304.23	1.2491	0.6318	310.70	1.2312
370.0	0.7212	310.73	1.3045	0.6697	315.29	1.2794	0.6398	321.64	1.2612
380.0	0.7375	322.21	1.3351	0.6801	326.43	1.3091	0.6480	332.66	1.2906
390.0	0.7546	333.80	1.3652	0.6908	337.65	1.3383	0.6562	343.75	1.3194
400.0	0.7727	345.48	1.3948	0.7017	348.95	1.3669	0.6646	354.92	1.3477
410.0	0.7917	357.25	1.4239	0.7129	360.32	1.3949	0.6731	366.16	1.3755
420.0	0.8117	369.12	1.4525	0.7244	371.77	1.4225	0.6817	377.47	1.4027
430.0	0.8329	381.07	1.4806	0.7360	383.28	1.4496	0.6903	388.85	1.4295
440.0	0.8551	393.09	1.5082	0.7480	394.85	1.4762	0.6991	400.29	1.4558
450.0	0.8784	405.19	1.5354	0.7601	406.50	1.5024	0.7080	411.80	1.4816
460.0	0.9028	417.36	1.5622	0.7725	418.20	1.5281	0.7169	423.37	1.5071
470.0	0.9283	429.59	1.5884	0.7852	429.96	1.5534	0.7260	435.00	1.5321
480.0	0.9548	441.87	1.6143	0.7980	441.78	1.5783	0.7351	446.68	1.5567
490.0	0.9824	454.19	1.6397	0.8111	453.66	1.6028	0.7444	458.43	1.5809
500.0	1.0109	466.55	1.6647	0.8244	465.59	1.6269	0.7537	470.23	1.6047
510.0	1.0403	478.94	1.6892	0.8379	477.56	1.6506	0.7630	482.08	1.6282
520.0	1.0704	491.36	1.7133	0.8515	489.59	1.6739	0.7725	493.99	1.6513
530.0	1.1012	503.80	1.7370	0.8654	501.67	1.6969	0.7820	505.95	1.6741
540.0	1.1326	516.25	1.7603	0.8794	513.79	1.7196	0.7916	517.96	1.6966
550.0	1.1645	528.72	1.7832	0.8936	525.95	1.7419	0.8013	530.02	1.7187
560.0	1.1967	541.19	1.8057	0.9079	538.16	1.7639	0.8110	542.12	1.7405
570.0	1.2293	553.67	1.8277	0.9224	550.41	1.7856	0.8207	554.28	1.7620
580.0	1.2621	566.16	1.8495	0.9370	562.69	1.8070	0.8306	566.47	1.7832
590.0	1.2950	578.65	1.8708	0.9517	575.01	1.8280	0.8404	578.71	1.8041
600.0	1.3280	591.15	1.8918	0.9665	587.37	1.8488	0.8503	590.99	1.8248
610.0	1.3611	603.66	1.9125	0.9813	599.76	1.8693	0.8603	603.31	1.8452
620.0	1.3941	616.18	1.9329	0.9963	612.19	1.8895	0.8703	615.67	1.8653

References for RC318

[Bamb56] BAMBACH, G.: Octafluorcyclobutan C_4F_8; Kältetechnik 8(1956)11 334/339

[Dous59] DOUSLIN, D.R. ; MOORE, R.T. AND WADDINGTON, G.: The pressure-volume-temperature properties of perfluorocyclobutane: equations of state, virial coefficients and intermolecular potential energy functions; J. Phys. Chem. 63(1959)9 1959/1966

[Furu54] FURUKAWA, G.T. ; MCCOSKEY, R.E. AND REILLY, M.L.: Heat Capacity, Heats of Transition, Fusion and Vaporization, and Vapor Pressure of Octafluorocyclobutane; J. Res. NBS 52(1954)1 11/16

[Gell75] GELLER, Z.I. ; PORICHANSKIJ, E.G. AND ROMANOV, V.K.: Thermal properties of fluid Freon-C318; Thermophysical Properties of Matter and Substances 8(1975) 71/78

[Kess77] KESSELMAN, P.M. ; PORICHANSKIJ, E.G. ; ROMANOV, V.K. AND SCHUMASCHENKO, A.B.: Complex Study of Thermophysical Properties of Freon C318; Teplofiziceskie svojstva vescestv i materialov 11(1977) 4/10

[Klet67] KLETSKII, A.V. AND PETRIK,L.E.: Vapor Pressure Curve of Perfluorocyclobutane; Russian Journal of Physical Chemistry 41(1967)5 628/629

[Klet67a] KLETSKII, A.V.: Thermodynamic properties of Freon RC318; Kholodilnaja Tekhnika 44(1967)4 21/24

[Mart62] MARTIN, J.J.: Thermodynamic properties of perfluorocyclobutane; J. Chem. Eng. Data 7(1962)1 68/72

[Masi53] MASI, J.: Heat Capacity of Gaseous Perfluorocyclobutane; J. Am. Chem. Soc. 75(1953) 5082/5084

[Matt65] MATTHIAS, H. AND LÖFFLER, H.J.: Thermodynamische Eigenschaften von Octafluorcyclobutan C_4F_8 (RC318); Lehrstuhl und Institut für Thermodynamik und Kältetechnik TU Berlin (1965)

[Meye69] MEYER, K.J.: Über den Zusammenhang zwischen der Schallgeschwindigkeit und der Wärmeleitfähigkeit bei flüssigen Fluor-Chlor-Derivaten des Methans und Äthans; Kältetechnik-Klimatisierung 21(1969)9 270/275

[Neve60] DE NEVERS, N. AND JOSEPH, J.M.: The constant volume heat capacities of gaseous perfluorocyclobutane and propylene; AIChE J. 6(1960)1 43/49

[Wask66] WASKOW, E.T.: Thermodynamische Eigenschaften von Octafluorocyclobutan; Russian Journal Engineering Physics 10(1966)1 127/130

[Whip52] WHIPPLE, G.H.: Vapor-Liquid Equilibria of some Fluorinated Hydrocarbon Systems; Ind. Eng. Chem. 44(1952)7 1664/1667

RC 318. ph

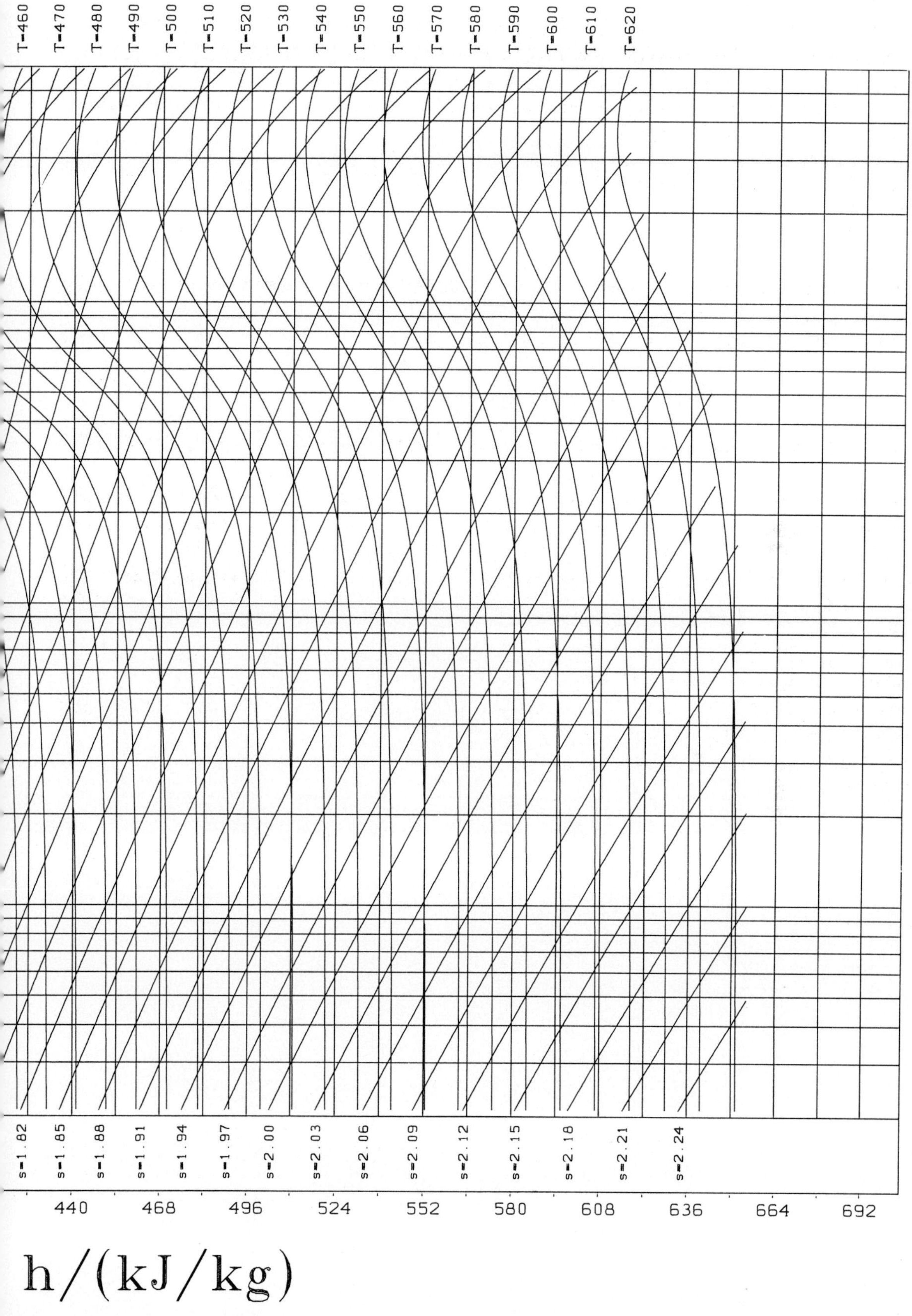

T=460
T=470
T=480
T=490
T=500
T=510
T=520
T=530
T=540
T=550
T=560
T=570
T=580
T=590
T=600
T=610
T=620
s=1.82
s=1.85
s=1.88
s=1.91
s=1.94
s=1.97
s=2.00
s=2.03
s=2.06
s=2.09
s=2.12
s=2.15
s=2.18
s=2.21
s=2.24
440
468
496
524
552
580
608
636
664
692
h/(kJ/kg)

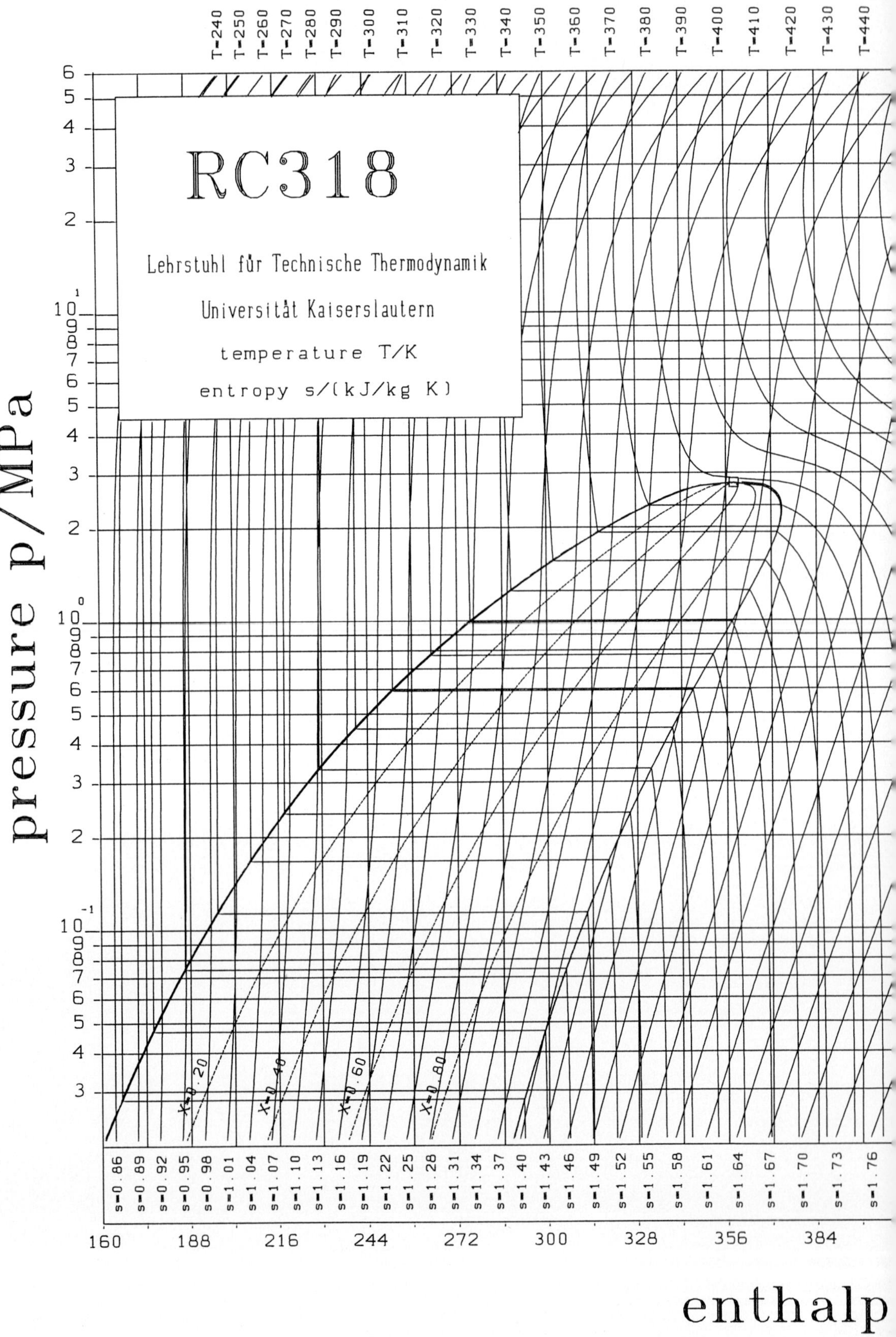
RC318
Lehrstuhl für Technische Thermodynamik
Universität Kaiserslautern
temperature T/K
entropy s/(kJ/kg K)
pressure p/MPa
enthalp
T=240 T=250 T=260 T=270 T=280 T=290 T=300 T=310 T=320 T=330 T=340 T=350 T=360 T=370 T=380 T=390 T=400 T=410 T=420 T=430 T=440
X=0.20 X=0.40 X=0.60 X=0.80
s=0.86 s=0.89 s=0.92 s=0.95 s=0.98 s=1.01 s=1.04 s=1.07 s=1.10 s=1.13 s=1.16 s=1.19 s=1.22 s=1.25 s=1.28 s=1.31 s=1.34 s=1.37 s=1.40 s=1.43 s=1.46 s=1.49 s=1.52 s=1.55 s=1.58 s=1.61 s=1.64 s=1.67 s=1.70 s=1.73 s=1.76
160 188 216 244 272 300 328 356 384

RC 318. Ts

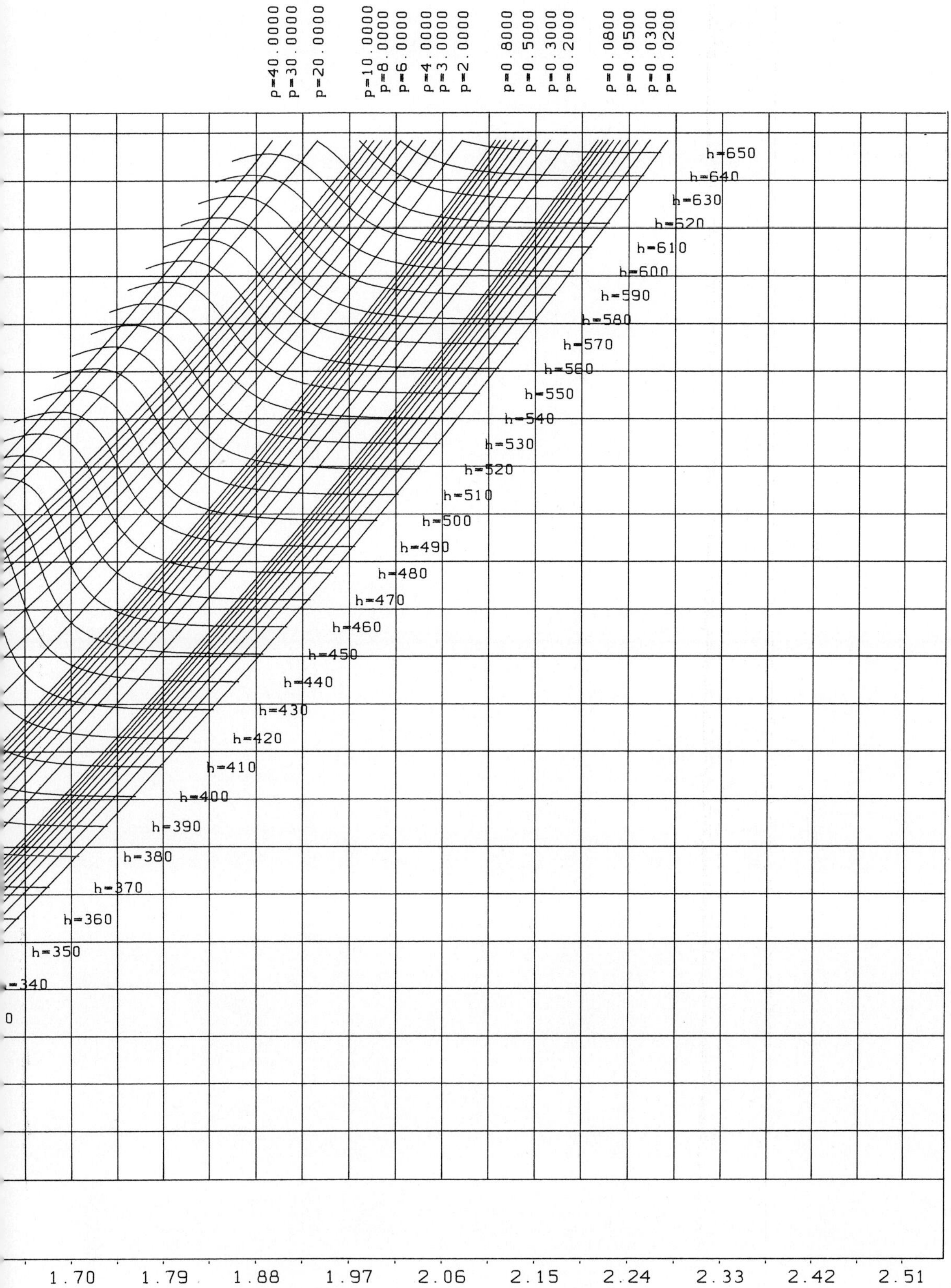

/(kJ/kg k)

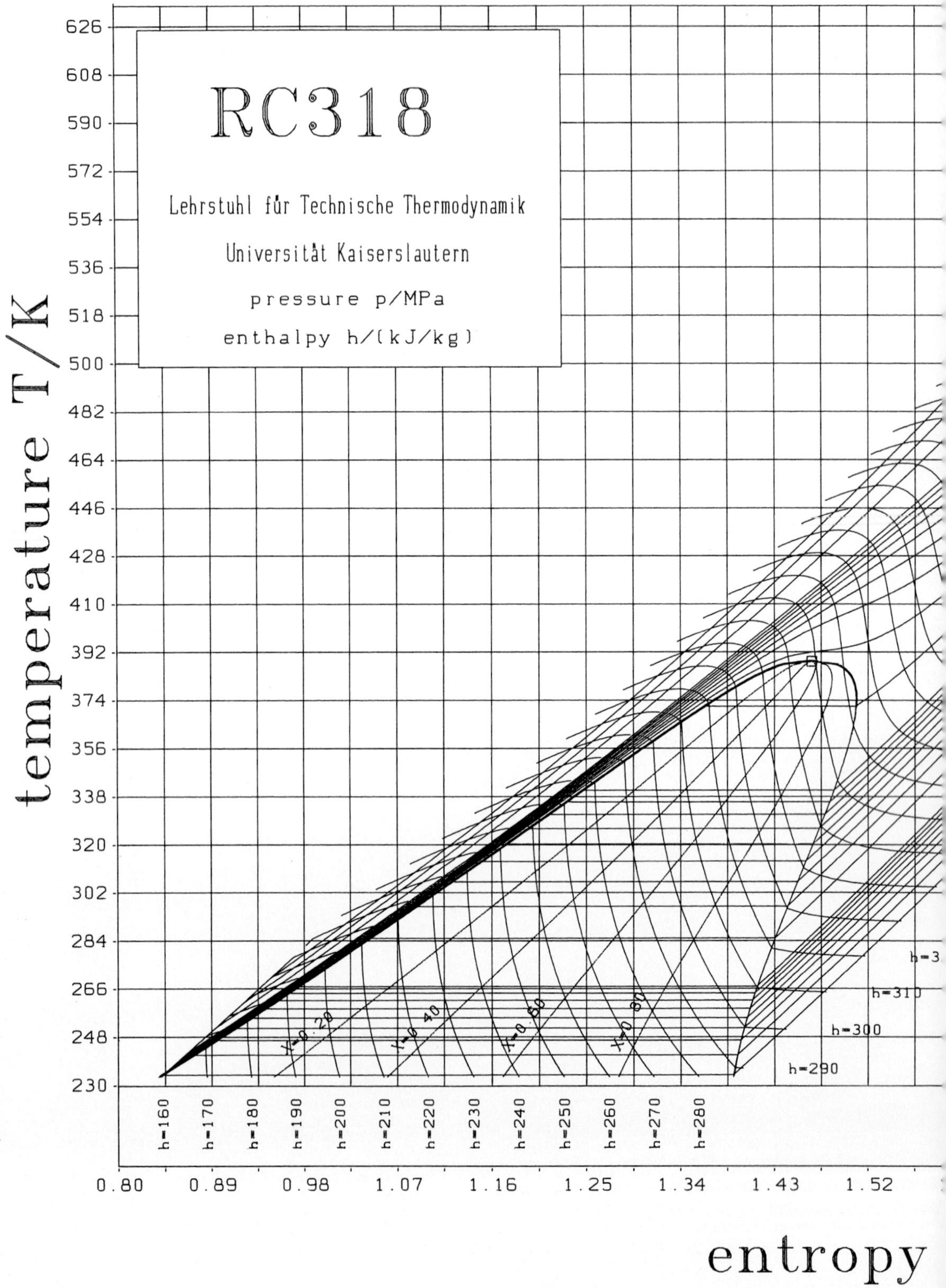

RC318
Lehrstuhl für Technische Thermodynamik
Universität Kaiserslautern
pressure p/MPa
enthalpy h/(kJ/kg)
temperature T/K
entropy
626
608
590
572
554
536
518
500
482
464
446
428
410
392
374
356
338
320
302
284
266
248
230
0.80
0.89
0.98
1.07
1.16
1.25
1.34
1.43
1.52
h=160
h=170
h=180
h=190
h=200
h=210
h=220
h=230
h=240
h=250
h=260
h=270
h=280
h=290
h=300
h=310
X=0.20
X=0.40
X=0.60
X=0.80

RC 318. hs

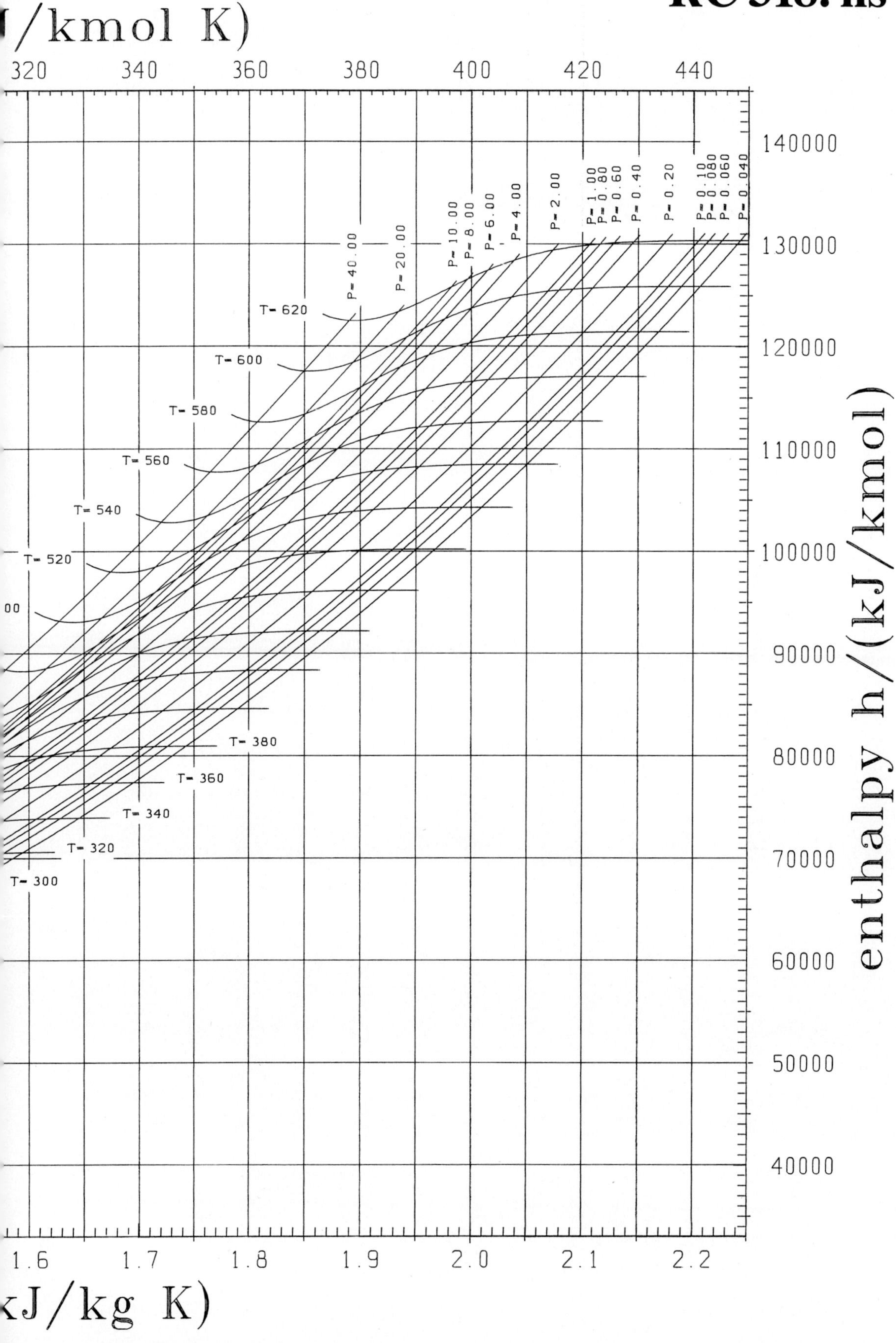

/kmol K)
320
340
360
380
400
420
440
kJ/kg K)
1.6
1.7
1.8
1.9
2.0
2.1
2.2
enthalpy h/(kJ/kmol)
140000
130000
120000
110000
100000
90000
80000
70000
60000
50000
40000
P= 40.00
P= 20.00
P= 10.00
P= 8.00
P= 6.00
P= 4.00
P= 2.00
P= 1.00
P= 0.80
P= 0.60
P= 0.40
P= 0.20
P= 0.10
P= 0.080
P= 0.060
P= 0.040
T= 620
T= 600
T= 580
T= 560
T= 540
T= 520
00
T= 380
T= 360
T= 340
T= 320
T= 300

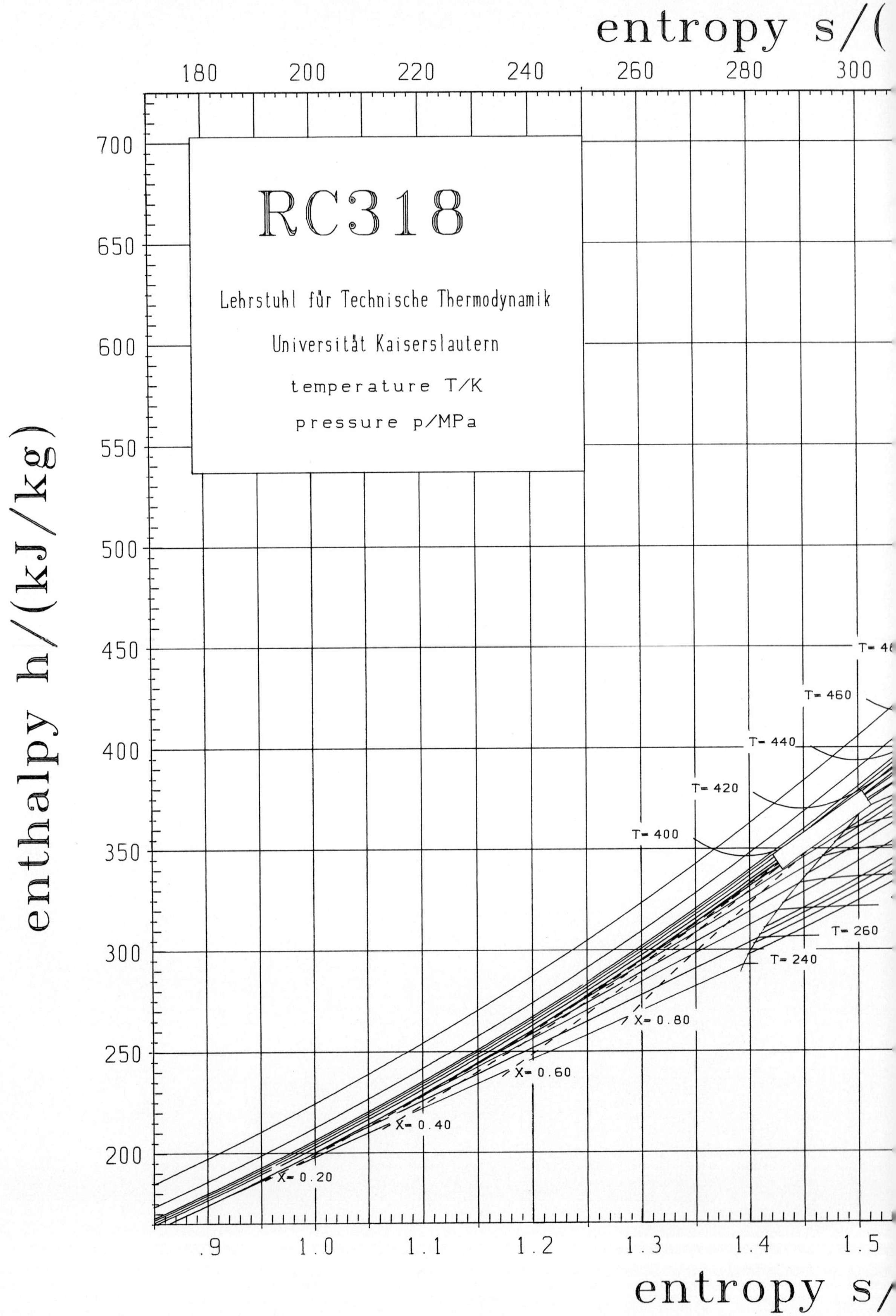

entropy s/(
180
200
220
240
260
280
300
RC318
Lehrstuhl für Technische Thermodynamik
Universität Kaiserslautern
temperature T/K
pressure p/MPa
enthalpy h/(kJ/kg)
700
650
600
550
500
450
400
350
300
250
200
T= 460
T= 440
T= 420
T= 400
T= 260
T= 240
X= 0.80
X= 0.60
X= 0.40
X= 0.20
.9
1.0
1.1
1.2
1.3
1.4
1.5
entropy s/

Refrigerant R500 ($R12/R152a$)

Table R500.1: List of references for p, ρ, T data

Author	n	p/MPa min	p/MPa max	ρ/g cm^{-3} min	ρ/g cm^{-3} max	T/K min	T/K max
PENNINGTON, W.A. *[Penn 55a]*	35	0.21	1.26	0.0086	0.06	298	391
PRASAD, M. *[Pras 78]*	160	0.02	0.74	0.0006	0.03	298	413
SINKA, J.V. *[Sink 67]*	28	1.39	5.85	0.0674	0.59	322	476

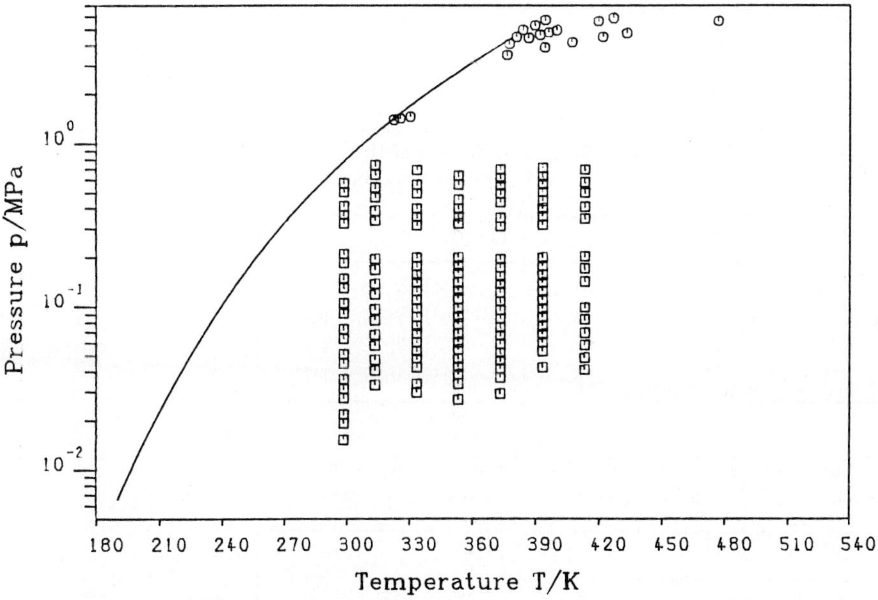

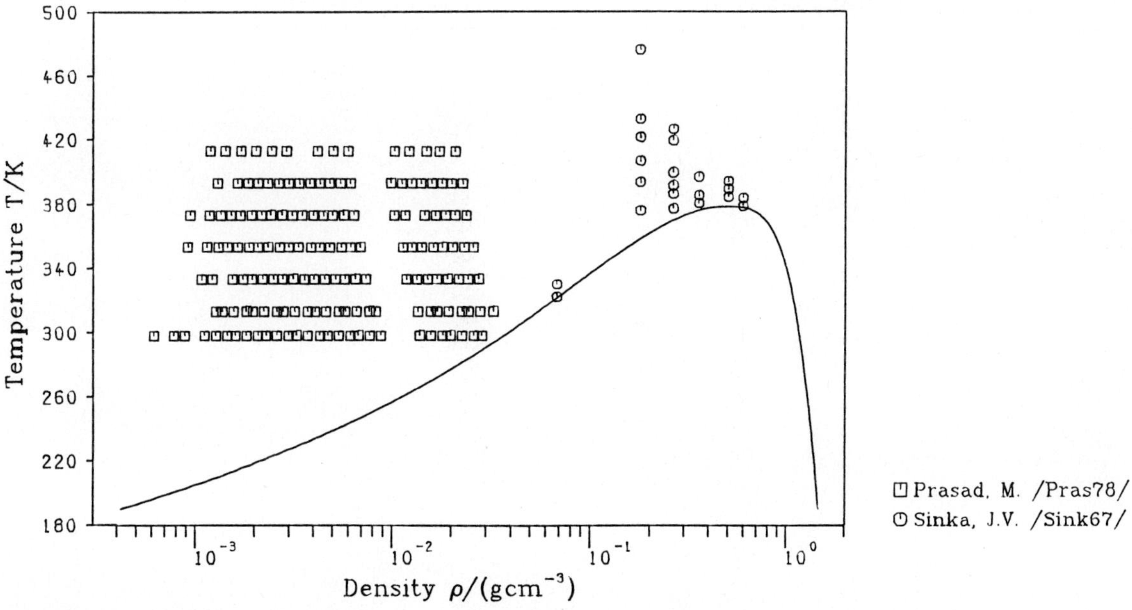

Fig.R500.1: Distribution of p, ρ, T data projected in the p, T and in the T, ρ plane

Table R500.2: Experimental vapor pressure data

Author	Δ / %		
	std	opt	*EOS*
PENNINGTON, W.A. *[Penn 50]*	0.33	0.32	0.33
PENNINGTON, W.A. *[Penn 55]*	0.07	0.06	0.07
SINKA, J.V. *[Sink 67]*	0.20	0.15	0.20

Table R500.3: Experimental saturated liquid density data

Author	Δ / %		
	std	opt	*EOS*
PENNINGTON, W.A. *[Penn 55a]*	0.15	0.13	0.10
SINKA, J.V. *[Sink 67]*	0.13	0.11	0.21

Table R500.4: Coefficients of the standard vapor pressure equation

Index i	b_i
5	−7.109686 E+00
6	1.781276 E+00
9	−4.089585 E+00
15	9.929842 E+00

Table R500.5: Coefficients of the quasi-optimal vapor pressure equation

Index i	b_i
4	−1.193158 E−01
5	−6.195464 E+00
14	−2.100277 E+01
18	7.255867 E+01

Table R500.6: Coefficients of the standard saturated liquid density equation

Index i	c_i
8	1.631446 E+00
9	9.970377 E−01
16	8.065858 E−02

Table R500.7: Coefficients of the quasi-optimal saturated liquid density equation

Index i	c_i
7	−9.119904 E−01
8	2.161232 E+01
10	6.170570 E+00

Table R500.8: Coefficients of the *Bender-EOS*

Range of validity: one-phase region

$0.02 \leq p/\text{MPa} \leq 5.85$

$0.00061 \leq \rho/\text{g cm}^{-3} \leq 0.5913$

$298.15 \leq T/\text{K} \leq 476.45$

	Specific units	Molar units
a_{01}	6.396207358 E−04	6.306973869 E−03
a_{02}	−6.282583192 E+01	−6.194934874 E+02
a_{03}	5.856348207 E+04	5.774646294 E+05
a_{04}	−1.238097651 E+06	−1.220824950 E+07
a_{05}	−2.167720996 E+08	−2.137479120 E+09
a_{06}	−2.724852502 E−01	−2.668030218 E−01
a_{07}	1.381217009 E+02	1.352414017 E+02
a_{08}	1.433604573 E+04	1.403709125 E+04
a_{09}	−8.088989248 E−02	−7.864864633 E−03
a_{10}	1.648831905 E+02	1.603147110 E+01
a_{11}	1.761800355 E−01	1.700994602 E−03
a_{12}	−3.391115169 E+02	−3.274076194 E+00
a_{13}	1.278110229 E+02	1.225360289 E−01
a_{14}	1.930097899 E+07	1.889848905 E+07
a_{15}	−1.804239536 E+10	−1.766615110 E+10
a_{16}	3.568797769 E+12	3.494376405 E+12
a_{17}	−1.140605067 E+08	−1.101238887 E+06
a_{18}	6.956411798 E+10	6.716322250 E+08
a_{19}	−1.144716448 E+13	−1.105208371 E+11
a_{20}	2.700000000 E+00	2.662332300 E−02

Table R500.9: Deviations between experimental and predicted p, ρ, T data

Region	n	Δ / %	
		p	ρ
Gas	176	0.20	0.36
Critical	11	0.33	1.97
Liquid	0	–	–
Total	187	0.22	0.62

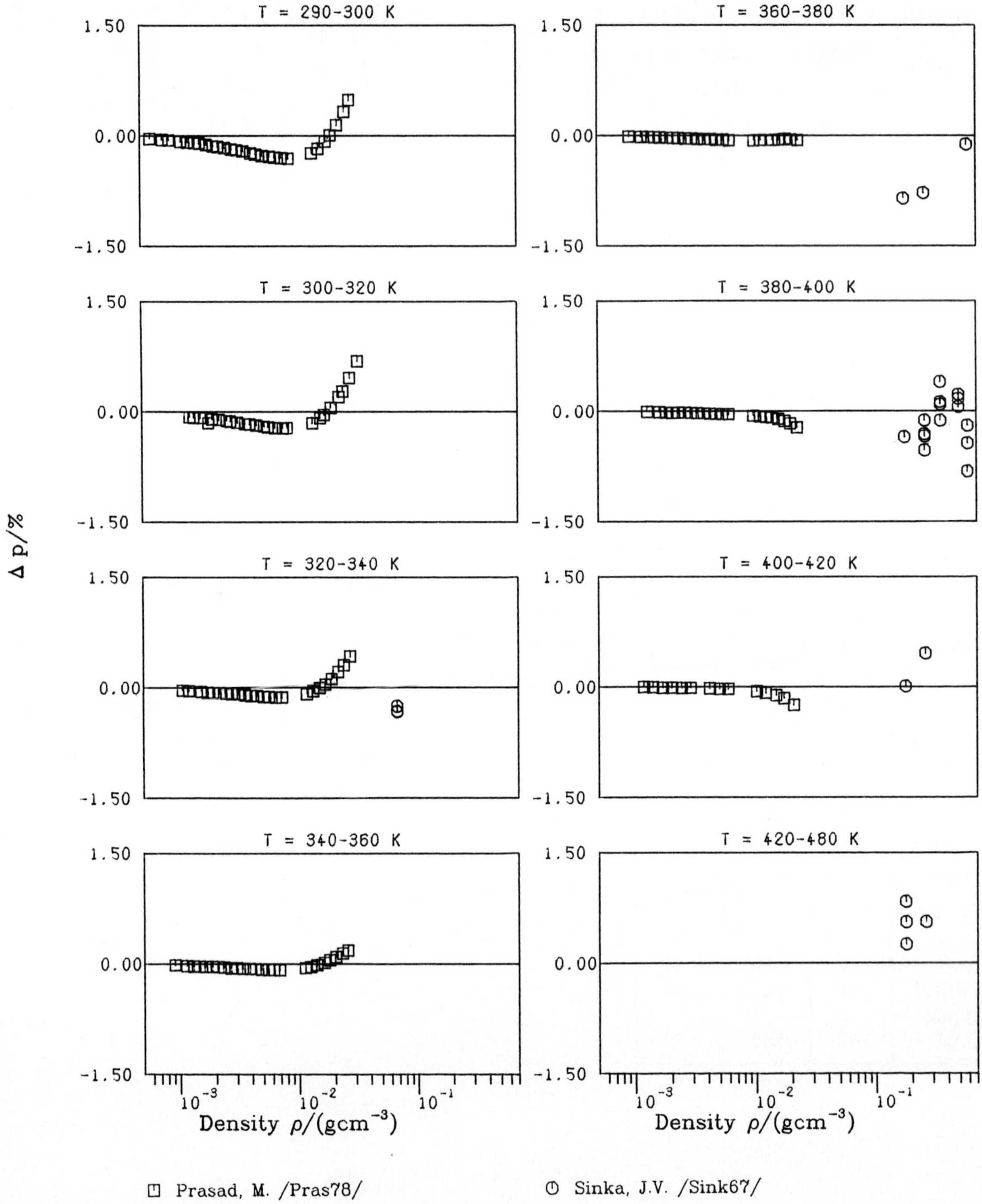

Fig.R500.2: Deviations between experimental and calculated p, ρ, T data

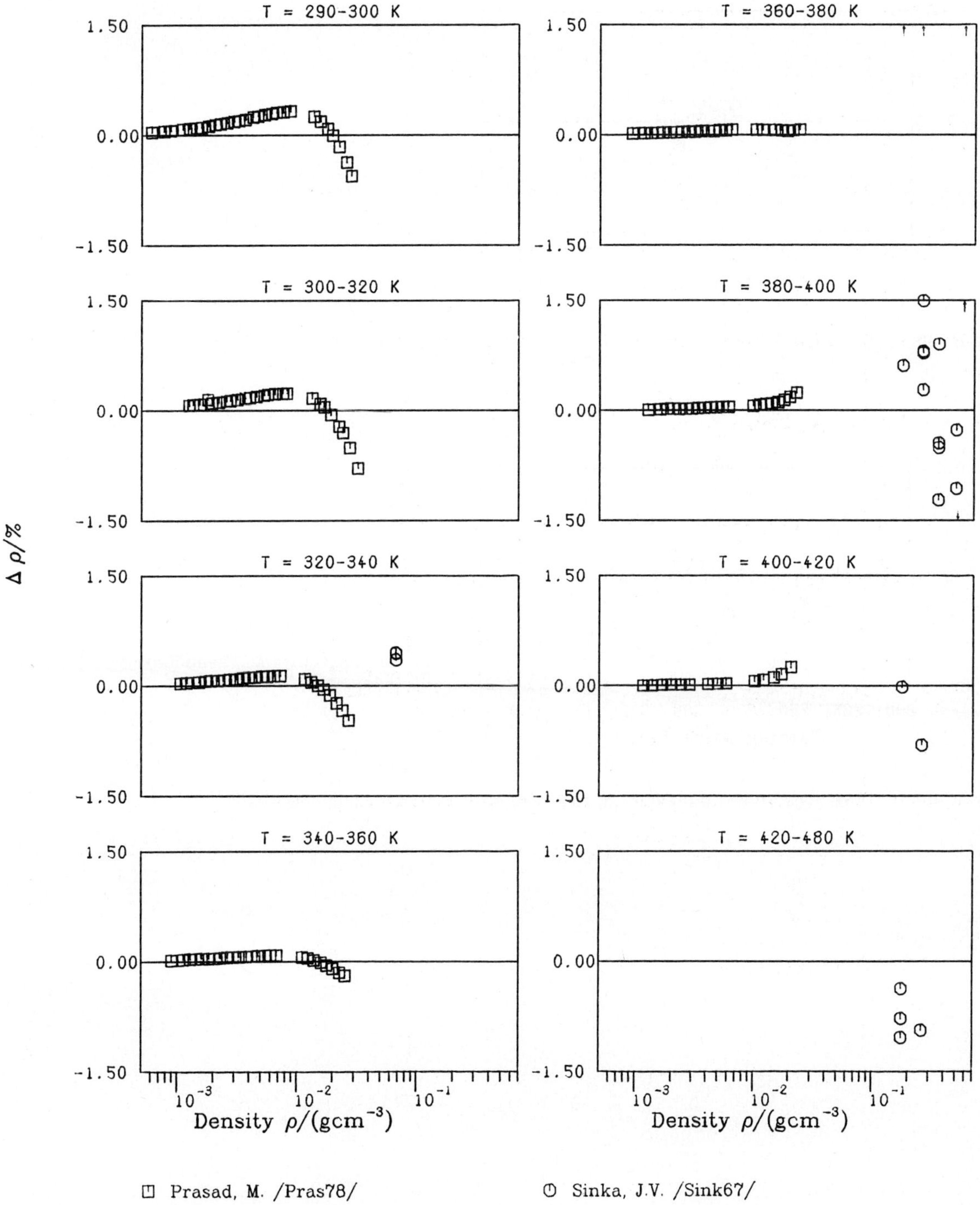

Fig.R500.2: Deviations between experimental and calculated p, ρ, T data *-continued*

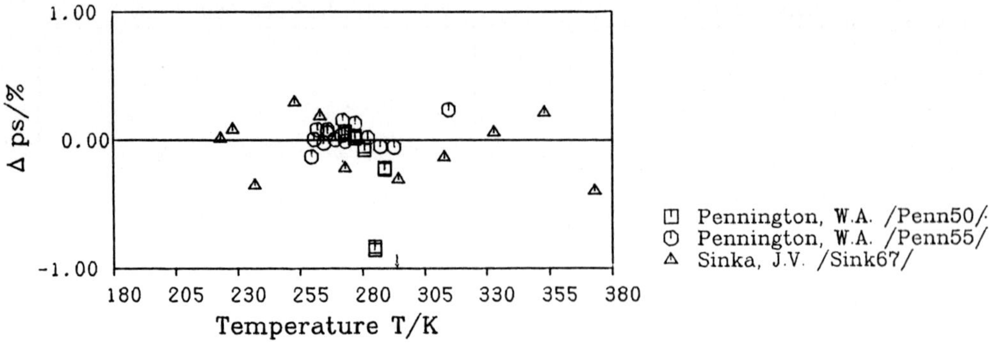

Fig.R500.3: Deviations between experimental and calculated vapor pressures

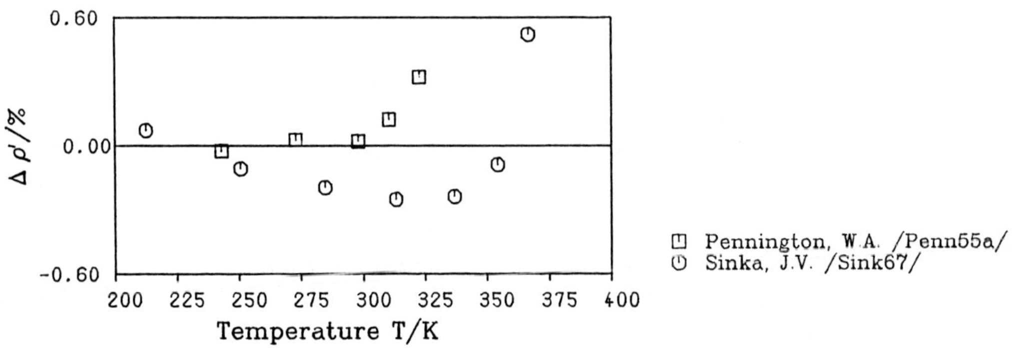

Fig.R500.4: Deviations between experimental and calculated saturated liquid densities

Table R500.10: Information on ideal gas state heat capacity equation

Author	T / K min	T / K max
BARHO, W. *[Barh 66]*	200	550

Coefficients of equation (15)	
Index i	d_i
1	2.060573 E−01
2	1.968933 E−03
3	−3.919962 E−07
4	−2.057727 E−09
5	1.586364 E−12

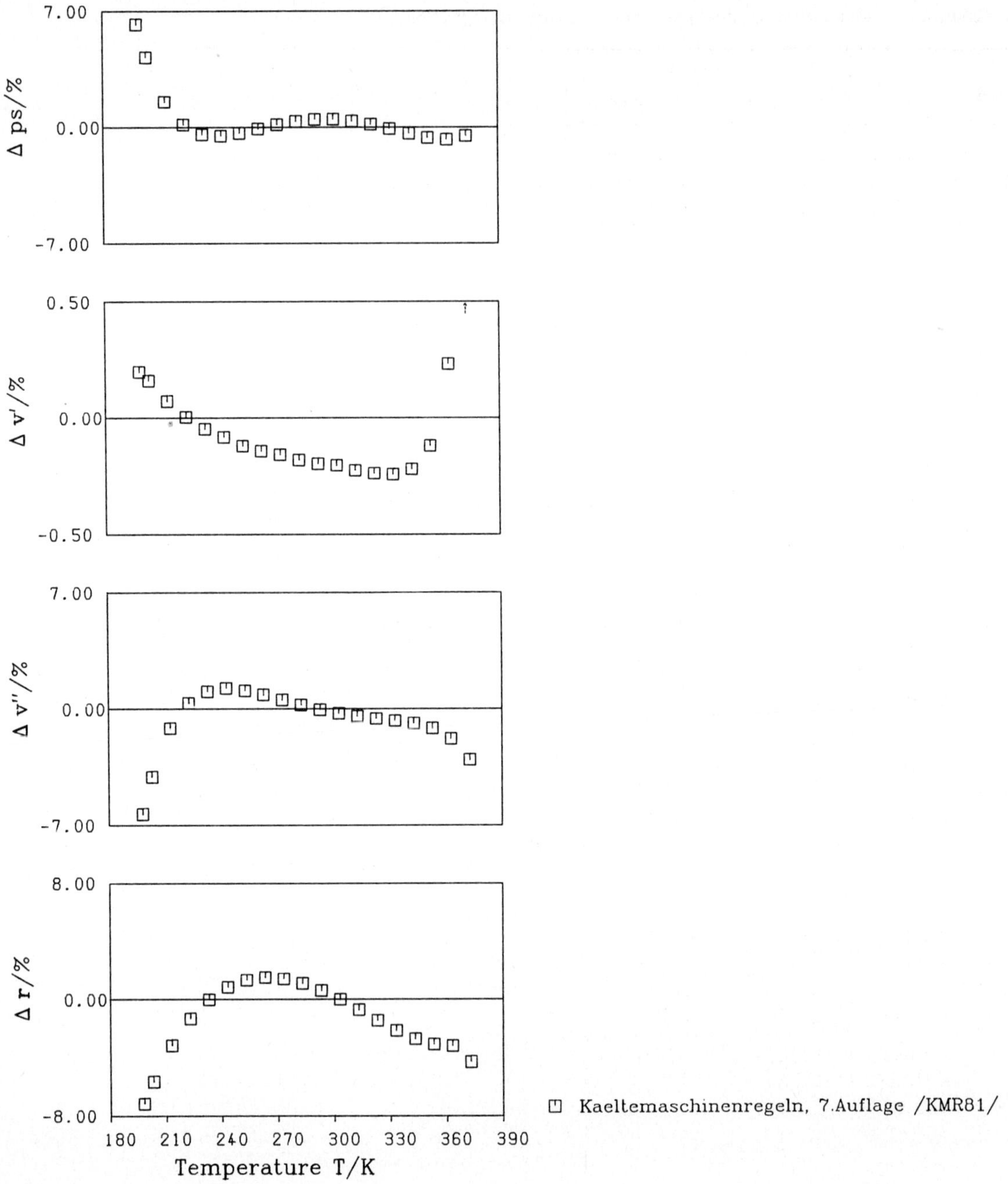

Fig.R500.5: Deviations between saturation properties from the *Kältemaschinenregeln* and those calculated by the *Bender-EOS*

Chart R500.1: Saturation properties

T K	p MPa	v' dm^3/kg	v'' dm^3/kg	h' kJ/kg	h'' kJ/kg	r kJ/kg	s' kJ/kgK	s'' kJ/kgK
225.0	0.05082	0.7290	362.40	152.88	359.50	206.62	0.8125	1.7308
226.0	0.05341	0.7304	346.04	153.63	360.06	206.42	0.8158	1.7292
227.0	0.05611	0.7317	330.56	154.40	360.61	206.22	0.8192	1.7276
228.0	0.05892	0.7331	315.89	155.17	361.17	205.99	0.8226	1.7260
229.0	0.06185	0.7345	301.99	155.96	361.72	205.76	0.8260	1.7245
230.0	0.06489	0.7359	288.81	156.76	362.27	205.51	0.8295	1.7230
231.0	0.06805	0.7373	276.31	157.57	362.83	205.26	0.8330	1.7215
232.0	0.07134	0.7387	264.45	158.39	363.38	204.99	0.8365	1.7201
233.0	0.07475	0.7401	253.19	159.22	363.93	204.71	0.8401	1.7186
234.0	0.07830	0.7416	242.50	160.07	364.48	204.41	0.8437	1.7172
235.0	0.08198	0.7430	232.35	160.92	365.03	204.11	0.8473	1.7159
236.0	0.08580	0.7445	222.70	161.79	365.58	203.79	0.8510	1.7145
237.0	0.08976	0.7459	213.52	162.66	366.13	203.47	0.8547	1.7132
238.0	0.09387	0.7474	204.80	163.55	366.68	203.13	0.8584	1.7119
239.0	0.09813	0.7489	196.50	164.45	367.22	202.78	0.8621	1.7106
240.0	0.10254	0.7504	188.60	165.35	367.77	202.42	0.8659	1.7093
241.0	0.10711	0.7519	181.07	166.27	368.32	202.05	0.8697	1.7081
242.0	0.11184	0.7534	173.91	167.19	368.86	201.67	0.8735	1.7068
243.0	0.11674	0.7550	167.08	168.13	369.40	201.28	0.8773	1.7056
244.0	0.12180	0.7565	160.57	169.07	369.95	200.87	0.8812	1.7045
245.0	0.12704	0.7581	154.37	170.03	370.49	200.46	0.8851	1.7033
246.0	0.13246	0.7596	148.45	170.99	371.03	200.04	0.8890	1.7022
247.0	0.13805	0.7612	142.80	171.96	371.57	199.61	0.8929	1.7010
248.0	0.14384	0.7628	137.41	172.94	372.10	199.17	0.8969	1.6999
249.0	0.14981	0.7644	132.27	173.93	372.64	198.71	0.9008	1.6989
250.0	0.15598	0.7660	127.35	174.92	373.17	198.25	0.9048	1.6978
251.0	0.16234	0.7676	122.66	175.93	373.71	197.78	0.9088	1.6968
252.0	0.16891	0.7693	118.17	176.94	374.24	197.30	0.9128	1.6957
253.0	0.17568	0.7709	113.88	177.96	374.77	196.81	0.9168	1.6947
254.0	0.18266	0.7726	109.77	178.99	375.30	196.31	0.9209	1.6937
255.0	0.18986	0.7743	105.84	180.03	375.83	195.80	0.9249	1.6927
256.0	0.19728	0.7760	102.08	181.07	376.35	195.28	0.9290	1.6918
257.0	0.20492	0.7777	98.485	182.12	376.88	194.75	0.9330	1.6908
258.0	0.21279	0.7794	95.039	183.18	377.40	194.22	0.9371	1.6899
259.0	0.22090	0.7811	91.738	184.24	377.92	193.67	0.9412	1.6890
260.0	0.22924	0.7829	88.574	185.32	378.44	193.12	0.9453	1.6881
261.0	0.23782	0.7846	85.542	186.39	378.95	192.56	0.9494	1.6872
262.0	0.24665	0.7864	82.634	187.48	379.47	191.99	0.9536	1.6863
263.0	0.25573	0.7882	79.846	188.57	379.98	191.41	0.9577	1.6855
264.0	0.26506	0.7900	77.170	189.67	380.49	190.82	0.9618	1.6846
265.0	0.27466	0.7918	74.602	190.78	381.00	190.22	0.9660	1.6838
266.0	0.28451	0.7937	72.137	191.89	381.51	189.62	0.9701	1.6830
267.0	0.29464	0.7955	69.770	193.00	382.01	189.01	0.9743	1.6822
268.0	0.30504	0.7974	67.496	194.13	382.51	188.39	0.9785	1.6814
269.0	0.31572	0.7993	65.311	195.26	383.01	187.76	0.9826	1.6806
270.0	0.32668	0.8012	63.211	196.39	383.51	187.12	0.9868	1.6799
271.0	0.33793	0.8031	61.192	197.53	384.00	186.47	0.9910	1.6791
272.0	0.34947	0.8051	59.251	198.68	384.50	185.82	0.9952	1.6784
273.0	0.36131	0.8070	57.383	199.83	384.99	185.16	0.9994	1.6776
274.0	0.37345	0.8090	55.587	200.98	385.48	184.49	1.0036	1.6769

Chart R500.1: Saturation properties–*continued*

T K	p MPa	v' dm³/kg	v'' dm³/kg	h' kJ/kg	h'' kJ/kg	r kJ/kg	s' kJ/kgK	s'' kJ/kgK
275.0	0.38590	0.8110	53.857	202.14	385.96	183.82	1.0078	1.6762
276.0	0.39865	0.8130	52.192	203.31	386.44	183.13	1.0120	1.6755
277.0	0.41173	0.8151	50.589	204.48	386.92	182.44	1.0162	1.6748
278.0	0.42512	0.8171	49.045	205.66	387.40	181.74	1.0204	1.6741
279.0	0.43884	0.8192	47.557	206.84	387.87	181.03	1.0246	1.6734
280.0	0.45289	0.8213	46.124	208.03	388.34	180.32	1.0288	1.6727
281.0	0.46727	0.8234	44.741	209.22	388.81	179.59	1.0330	1.6721
282.0	0.48199	0.8255	43.409	210.41	389.28	178.86	1.0372	1.6714
283.0	0.49706	0.8277	42.123	211.61	389.74	178.13	1.0414	1.6708
284.0	0.51248	0.8299	40.883	212.82	390.20	177.38	1.0456	1.6702
285.0	0.52825	0.8321	39.687	214.03	390.65	176.63	1.0498	1.6695
286.0	0.54438	0.8343	38.532	215.24	391.11	175.87	1.0540	1.6689
287.0	0.56088	0.8365	37.417	216.46	391.55	175.10	1.0582	1.6683
288.0	0.57774	0.8388	36.341	217.68	392.00	174.32	1.0624	1.6677
289.0	0.59498	0.8411	35.301	218.90	392.44	173.54	1.0666	1.6671
290.0	0.61259	0.8434	34.296	220.13	392.88	172.75	1.0708	1.6665
291.0	0.63059	0.8458	33.325	221.37	393.32	171.95	1.0750	1.6659
292.0	0.64898	0.8482	32.387	222.60	393.75	171.15	1.0792	1.6653
293.0	0.66776	0.8506	31.480	223.84	394.18	170.33	1.0833	1.6647
294.0	0.68694	0.8530	30.602	225.09	394.60	169.51	1.0875	1.6641
295.0	0.70652	0.8554	29.754	226.34	395.02	168.68	1.0917	1.6635
296.0	0.72652	0.8579	28.933	227.59	395.43	167.85	1.0959	1.6629
297.0	0.74692	0.8605	28.138	228.84	395.85	167.00	1.1001	1.6624
298.0	0.76775	0.8630	27.369	230.10	396.25	166.15	1.1042	1.6618
299.0	0.78899	0.8656	26.625	231.37	396.66	165.29	1.1084	1.6612
300.0	0.81067	0.8682	25.904	232.63	397.06	164.43	1.1126	1.6607
301.0	0.83278	0.8708	25.205	233.90	397.45	163.55	1.1167	1.6601
302.0	0.85533	0.8735	24.529	235.17	397.84	162.67	1.1209	1.6595
303.0	0.87832	0.8762	23.873	236.45	398.23	161.78	1.1250	1.6590
304.0	0.90176	0.8789	23.238	237.73	398.61	160.88	1.1292	1.6584
305.0	0.92566	0.8817	22.622	239.01	398.98	159.97	1.1333	1.6578
306.0	0.95002	0.8845	22.025	240.30	399.35	159.06	1.1375	1.6573
307.0	0.97484	0.8874	21.446	241.59	399.72	158.13	1.1416	1.6567
308.0	1.00013	0.8903	20.884	242.88	400.08	157.20	1.1457	1.6561
309.0	1.02589	0.8932	20.339	244.18	400.44	156.26	1.1499	1.6556
310.0	1.05214	0.8962	19.809	245.48	400.79	155.31	1.1540	1.6550
311.0	1.07887	0.8992	19.296	246.78	401.13	154.35	1.1581	1.6544
312.0	1.10610	0.9023	18.797	248.08	401.47	153.38	1.1622	1.6538
313.0	1.13382	0.9054	18.313	249.39	401.80	152.41	1.1663	1.6532
314.0	1.16204	0.9085	17.843	250.70	402.13	151.42	1.1704	1.6527
315.0	1.19078	0.9117	17.386	252.02	402.45	150.43	1.1745	1.6521
316.0	1.22002	0.9150	16.942	253.34	402.76	149.43	1.1786	1.6515
317.0	1.24979	0.9183	16.510	254.66	403.07	148.41	1.1827	1.6509
318.0	1.28008	0.9217	16.091	255.98	403.37	147.39	1.1868	1.6503
319.0	1.31090	0.9251	15.683	257.31	403.67	146.36	1.1909	1.6497
320.0	1.34226	0.9285	15.286	258.64	403.96	145.31	1.1949	1.6490
321.0	1.37416	0.9321	14.900	259.98	404.24	144.26	1.1990	1.6484
322.0	1.40660	0.9357	14.525	261.32	404.51	143.19	1.2031	1.6478
323.0	1.43961	0.9393	14.160	262.66	404.78	142.12	1.2072	1.6471
324.0	1.47317	0.9430	13.804	264.00	405.04	141.03	1.2112	1.6465

Chart R500.1: Saturation properties—*continued*

T K	p MPa	v' dm^3/kg	v'' dm^3/kg	h' kJ/kg	h'' kJ/kg	r kJ/kg	s' kJ/kgK	s'' kJ/kgK
325.0	1.50730	0.9468	13.458	265.35	405.29	139.93	1.2153	1.6458
326.0	1.54200	0.9507	13.120	266.70	405.53	138.82	1.2193	1.6452
327.0	1.57729	0.9546	12.792	268.06	405.76	137.70	1.2234	1.6445
328.0	1.61315	0.9586	12.472	269.42	405.99	136.57	1.2274	1.6438
329.0	1.64961	0.9627	12.160	270.78	406.20	135.42	1.2315	1.6431
330.0	1.68667	0.9668	11.856	272.15	406.41	134.26	1.2355	1.6423
331.0	1.72434	0.9711	11.560	273.52	406.61	133.08	1.2395	1.6416
332.0	1.76262	0.9754	11.271	274.90	406.79	131.89	1.2436	1.6409
333.0	1.80151	0.9798	10.989	276.28	406.97	130.69	1.2476	1.6401
334.0	1.84104	0.9844	10.714	277.66	407.14	129.47	1.2516	1.6393
335.0	1.88120	0.9890	10.446	279.05	407.29	128.24	1.2557	1.6385
336.0	1.92200	0.9937	10.184	280.44	407.43	126.99	1.2597	1.6377
337.0	1.96345	0.9986	9.9285	281.84	407.57	125.72	1.2637	1.6368
338.0	2.00556	1.0036	9.6788	283.25	407.69	124.44	1.2678	1.6359
339.0	2.04833	1.0086	9.4349	284.65	407.79	123.14	1.2718	1.6351
340.0	2.09177	1.0139	9.1966	286.07	407.89	121.82	1.2759	1.6341
341.0	2.13590	1.0192	8.9637	287.49	407.97	120.48	1.2799	1.6332
342.0	2.18072	1.0247	8.7359	288.91	408.03	119.12	1.2839	1.6322
343.0	2.22624	1.0304	8.5132	290.35	408.08	117.74	1.2880	1.6312
344.0	2.27247	1.0362	8.2954	291.79	408.12	116.33	1.2920	1.6302
345.0	2.31942	1.0422	8.0822	293.23	408.14	114.90	1.2961	1.6291
346.0	2.36709	1.0483	7.8735	294.69	408.14	113.45	1.3001	1.6280
347.0	2.41551	1.0547	7.6691	296.15	408.12	111.97	1.3042	1.6269
348.0	2.46467	1.0613	7.4689	297.62	408.09	110.47	1.3083	1.6257
349.0	2.51459	1.0680	7.2727	299.10	408.03	108.93	1.3124	1.6245
350.0	2.56528	1.0751	7.0804	300.58	407.95	107.37	1.3165	1.6233
351.0	2.61676	1.0823	6.8917	302.08	407.85	105.77	1.3206	1.6220
352.0	2.66903	1.0899	6.7066	303.59	407.73	104.14	1.3247	1.6206
353.0	2.72211	1.0977	6.5249	305.11	407.59	102.47	1.3289	1.6192
354.0	2.77601	1.1058	6.3465	306.65	407.41	100.76	1.3331	1.6177
355.0	2.83075	1.1143	6.1711	308.19	407.21	99.02	1.3373	1.6162
356.0	2.88633	1.1232	5.9987	309.76	406.98	97.22	1.3415	1.6146
357.0	2.94278	1.1324	5.8290	311.33	406.72	95.38	1.3457	1.6129
358.0	3.00011	1.1421	5.6620	312.93	406.42	93.49	1.3500	1.6111
359.0	3.05834	1.1523	5.4974	314.55	406.08	91.54	1.3543	1.6093
360.0	3.11748	1.1630	5.3351	316.18	405.71	89.53	1.3587	1.6074
361.0	3.17756	1.1743	5.1750	317.84	405.29	87.45	1.3631	1.6053
362.0	3.23859	1.1863	5.0167	319.53	404.83	85.30	1.3676	1.6032
363.0	3.30059	1.1990	4.8603	321.25	404.31	83.06	1.3721	1.6009
364.0	3.36359	1.2126	4.7053	322.99	403.74	80.74	1.3767	1.5985
365.0	3.42762	1.2271	4.5517	324.78	403.10	78.32	1.3814	1.5960
366.0	3.49269	1.2426	4.3990	326.61	402.40	75.79	1.3862	1.5932
367.0	3.55883	1.2595	4.2472	328.49	401.61	73.13	1.3911	1.5903
368.0	3.62608	1.2778	4.0957	330.42	400.74	70.32	1.3961	1.5872
369.0	3.69446	1.2978	3.9443	332.43	399.77	67.35	1.4013	1.5838
370.0	3.76402	1.3199	3.7923	334.51	398.69	64.18	1.4067	1.5801
371.0	3.83478	1.3445	3.6393	336.69	397.46	60.77	1.4123	1.5761
372.0	3.90680	1.3724	3.4845	338.99	396.08	57.08	1.4182	1.5717
373.0	3.98011	1.4043	3.3267	341.44	394.49	53.04	1.4246	1.5668
374.0	4.05477	1.4416	3.1644	344.10	392.65	48.55	1.4314	1.5612

Chart R500.1: Saturation properties—*continued*

T K	p MPa	v' dm^3/kg	v'' dm^3/kg	h' kJ/kg	h'' kJ/kg	r kJ/kg	s' kJ/kgK	s'' kJ/kgK
375.0	4.13083	1.4865	2.9952	347.03	390.48	43.44	1.4389	1.5548
376.0	4.20835	1.5428	2.8152	350.38	387.83	37.45	1.4475	1.5471
377.0	4.28742	1.6189	2.6156	354.41	384.42	30.01	1.4579	1.5375
378.0	4.36811	1.7396	2.3713	359.95	379.41	19.46	1.4722	1.5237

Chart R500.2: Saturation properties

p MPa	T K	v' dm^3/kg	v'' dm^3/kg	h' kJ/kg	h'' kJ/kg	r kJ/kg	s' kJ/kgK	s'' kJ/kgK
0.0510	225.07	0.7291	361.21	152.93	359.54	206.61	0.8127	1.7307
0.0520	225.46	0.7296	354.76	153.23	359.76	206.53	0.8140	1.7300
0.0530	225.84	0.7302	348.54	153.51	359.97	206.46	0.8153	1.7294
0.0540	226.22	0.7307	342.54	153.80	360.18	206.38	0.8165	1.7288
0.0550	226.59	0.7312	336.76	154.08	360.39	206.30	0.8178	1.7282
0.0560	226.96	0.7317	331.17	154.36	360.59	206.22	0.8190	1.7277
0.0570	227.32	0.7322	325.77	154.64	360.79	206.15	0.8202	1.7271
0.0580	227.68	0.7327	320.55	154.92	360.99	206.07	0.8214	1.7265
0.0590	228.03	0.7332	315.50	155.19	361.18	205.99	0.8226	1.7260
0.0600	228.37	0.7336	310.61	155.46	361.37	205.91	0.8238	1.7255
0.0610	228.71	0.7341	305.88	155.73	361.56	205.83	0.8250	1.7249
0.0620	229.05	0.7346	301.29	156.00	361.75	205.75	0.8262	1.7244
0.0630	229.38	0.7350	296.85	156.26	361.93	205.67	0.8273	1.7239
0.0640	229.71	0.7355	292.53	156.53	362.11	205.59	0.8285	1.7234
0.0650	230.04	0.7359	288.35	156.79	362.29	205.51	0.8296	1.7229
0.0660	230.36	0.7364	284.29	157.05	362.47	205.42	0.8307	1.7225
0.0670	230.67	0.7368	280.34	157.30	362.64	205.34	0.8318	1.7220
0.0680	230.98	0.7373	276.50	157.56	362.82	205.26	0.8329	1.7215
0.0690	231.29	0.7377	272.78	157.81	362.99	205.18	0.8340	1.7211
0.0700	231.60	0.7381	269.15	158.06	363.16	205.10	0.8351	1.7207
0.0710	231.90	0.7386	265.62	158.31	363.32	205.01	0.8361	1.7202
0.0720	232.20	0.7390	262.19	158.55	363.49	204.93	0.8372	1.7198
0.0730	232.49	0.7394	258.85	158.80	363.65	204.85	0.8383	1.7194
0.0740	232.78	0.7398	255.59	159.04	363.81	204.77	0.8393	1.7190
0.0760	233.36	0.7406	249.33	159.52	364.13	204.60	0.8414	1.7181
0.0780	233.92	0.7415	243.37	160.00	364.43	204.44	0.8434	1.7174
0.0800	234.47	0.7422	237.70	160.46	364.74	204.27	0.8454	1.7166
0.0820	235.01	0.7430	232.30	160.93	365.03	204.11	0.8473	1.7159
0.0840	235.53	0.7438	227.14	161.38	365.32	203.94	0.8493	1.7151
0.0860	236.05	0.7445	222.22	161.83	365.61	203.78	0.8512	1.7144
0.0880	236.56	0.7453	217.51	162.28	365.89	203.61	0.8530	1.7138
0.0900	237.06	0.7460	213.00	162.71	366.16	203.45	0.8549	1.7131
0.0920	237.55	0.7467	208.68	163.15	366.43	203.28	0.8567	1.7125
0.0940	238.03	0.7475	204.54	163.58	366.69	203.12	0.8585	1.7118
0.0960	238.50	0.7482	200.56	164.00	366.95	202.95	0.8603	1.7112
0.0980	238.97	0.7489	196.74	164.42	367.21	202.79	0.8620	1.7106
0.1000	239.43	0.7495	193.07	164.83	367.46	202.63	0.8637	1.7100
0.1020	239.88	0.7502	189.53	165.24	367.71	202.46	0.8654	1.7095
0.1040	240.32	0.7509	186.13	165.65	367.95	202.30	0.8671	1.7089
0.1060	240.76	0.7515	182.84	166.05	368.19	202.14	0.8688	1.7084
0.1080	241.19	0.7522	179.68	166.44	368.42	201.98	0.8704	1.7078
0.1100	241.61	0.7528	176.63	166.83	368.65	201.82	0.8720	1.7073
0.1150	242.65	0.7544	169.44	167.80	369.21	201.42	0.8760	1.7061
0.1200	243.65	0.7560	162.83	168.74	369.76	201.02	0.8798	1.7049
0.1250	244.61	0.7575	156.73	169.66	370.28	200.62	0.8836	1.7038
0.1300	245.55	0.7589	151.08	170.55	370.78	200.23	0.8872	1.7027
0.1350	246.46	0.7603	145.83	171.43	371.27	199.84	0.8908	1.7017
0.1400	247.34	0.7617	140.94	172.29	371.75	199.46	0.8943	1.7007
0.1450	248.20	0.7631	136.38	173.13	372.21	199.08	0.8976	1.6997
0.1500	249.03	0.7644	132.11	173.96	372.66	198.70	0.9009	1.6988

Chart R500.2: Saturation properties—*continued*

p MPa	T K	v' dm³/kg	v'' dm³/kg	h' kJ/kg	h'' kJ/kg	r kJ/kg	s' kJ/kgK	s'' kJ/kgK
0.1550	249.84	0.7658	128.11	174.77	373.09	198.32	0.9042	1.6980
0.1600	250.64	0.7670	124.34	175.56	373.51	197.95	0.9073	1.6971
0.1650	251.41	0.7683	120.80	176.34	373.93	197.58	0.9104	1.6963
0.1700	252.16	0.7695	117.45	177.11	374.33	197.22	0.9134	1.6956
0.1750	252.90	0.7708	114.29	177.86	374.72	196.86	0.9164	1.6948
0.1800	253.62	0.7720	111.30	178.60	375.10	196.50	0.9193	1.6941
0.1850	254.33	0.7731	108.46	179.33	375.47	196.14	0.9222	1.6934
0.1900	255.02	0.7743	105.77	180.05	375.84	195.79	0.9250	1.6927
0.1950	255.70	0.7754	103.21	180.75	376.19	195.44	0.9277	1.6921
0.2000	256.36	0.7766	100.77	181.45	376.54	195.09	0.9304	1.6914
0.2100	257.65	0.7788	96.234	182.81	377.21	194.41	0.9357	1.6902
0.2200	258.89	0.7809	92.092	184.13	377.86	193.73	0.9408	1.6891
0.2300	260.09	0.7830	88.297	185.41	378.48	193.07	0.9457	1.6880
0.2400	261.25	0.7851	84.805	186.66	379.08	192.42	0.9505	1.6870
0.2500	262.37	0.7871	81.582	187.89	379.66	191.77	0.9551	1.6860
0.2600	263.46	0.7890	78.598	189.08	380.22	191.14	0.9596	1.6851
0.2700	264.52	0.7910	75.826	190.24	380.75	190.51	0.9640	1.6842
0.2800	265.55	0.7928	73.245	191.38	381.28	189.90	0.9683	1.6834
0.2900	266.55	0.7947	70.835	192.49	381.78	189.29	0.9724	1.6826
0.3000	267.52	0.7965	68.579	193.59	382.27	188.69	0.9765	1.6818
0.3100	268.47	0.7983	66.463	194.65	382.75	188.09	0.9804	1.6810
0.3200	269.39	0.8000	64.475	195.70	383.21	187.51	0.9843	1.6803
0.3300	270.30	0.8018	62.602	196.73	383.66	186.93	0.9881	1.6796
0.3400	271.18	0.8035	60.835	197.74	384.09	186.36	0.9918	1.6790
0.3500	272.05	0.8052	59.165	198.73	384.52	185.79	0.9954	1.6783
0.3600	272.89	0.8068	57.584	199.70	384.93	185.23	0.9989	1.6777
0.3700	273.72	0.8085	56.086	200.66	385.34	184.68	1.0024	1.6771
0.3800	274.53	0.8101	54.663	201.60	385.73	184.13	1.0058	1.6765
0.3900	275.32	0.8117	53.310	202.52	386.12	183.59	1.0091	1.6760
0.4000	276.10	0.8132	52.023	203.43	386.49	183.06	1.0124	1.6754
0.4100	276.87	0.8148	50.795	204.33	386.86	182.53	1.0156	1.6749
0.4200	277.62	0.8163	49.624	205.21	387.22	182.01	1.0188	1.6744
0.4300	278.36	0.8178	48.505	206.08	387.57	181.49	1.0219	1.6739
0.4400	279.08	0.8194	47.435	206.94	387.91	180.97	1.0249	1.6734
0.4500	279.80	0.8208	46.411	207.79	388.25	180.46	1.0279	1.6729
0.4600	280.50	0.8223	45.430	208.62	388.58	179.96	1.0309	1.6724
0.4700	281.19	0.8238	44.488	209.44	388.90	179.46	1.0338	1.6720
0.4800	281.87	0.8252	43.585	210.25	389.21	178.96	1.0366	1.6715
0.4900	282.53	0.8267	42.716	211.05	389.52	178.47	1.0394	1.6711
0.5000	283.19	0.8281	41.881	211.84	389.83	177.98	1.0422	1.6707
0.5100	283.84	0.8295	41.078	212.63	390.12	177.50	1.0449	1.6703
0.5200	284.48	0.8309	40.304	213.40	390.42	177.02	1.0476	1.6698
0.5300	285.11	0.8323	39.558	214.16	390.70	176.54	1.0502	1.6695
0.5400	285.73	0.8337	38.839	214.91	390.98	176.07	1.0528	1.6691
0.5500	286.34	0.8351	38.145	215.66	391.26	175.60	1.0554	1.6687
0.5600	286.95	0.8364	37.475	216.39	391.53	175.14	1.0580	1.6683
0.5700	287.54	0.8378	36.827	217.12	391.80	174.68	1.0605	1.6679
0.5800	288.13	0.8391	36.201	217.84	392.06	174.22	1.0629	1.6676
0.5900	288.71	0.8405	35.595	218.55	392.32	173.76	1.0654	1.6672
0.6000	289.29	0.8418	35.009	219.26	392.57	173.31	1.0678	1.6669

Chart R500.2: Saturation properties–*continued*

p MPa	T K	v' dm^3/kg	v'' dm^3/kg	h' kJ/kg	h'' kJ/kg	r kJ/kg	s' kJ/kgK	s'' kJ/kgK
0.6100	289.85	0.8431	34.441	219.95	392.82	172.86	1.0702	1.6665
0.6200	290.41	0.8444	33.890	220.64	393.06	172.42	1.0725	1.6662
0.6300	290.97	0.8457	33.356	221.33	393.30	171.98	1.0748	1.6659
0.6400	291.51	0.8470	32.839	222.00	393.54	171.54	1.0771	1.6656
0.6500	292.05	0.8483	32.336	222.67	393.77	171.10	1.0794	1.6652
0.6600	292.59	0.8496	31.849	223.33	394.00	170.67	1.0816	1.6649
0.6700	293.12	0.8508	31.375	223.99	394.23	170.24	1.0838	1.6646
0.6800	293.64	0.8521	30.914	224.64	394.45	169.81	1.0860	1.6643
0.6900	294.16	0.8534	30.467	225.28	394.67	169.38	1.0882	1.6640
0.7000	294.67	0.8546	30.031	225.92	394.88	168.96	1.0903	1.6637
0.7100	295.18	0.8559	29.608	226.56	395.09	168.54	1.0924	1.6634
0.7200	295.68	0.8571	29.196	227.18	395.30	168.12	1.0945	1.6631
0.7300	296.17	0.8584	28.794	227.80	395.51	167.70	1.0966	1.6628
0.7400	296.66	0.8596	28.403	228.42	395.71	167.29	1.0987	1.6626
0.7600	297.63	0.8621	27.651	229.64	396.10	166.47	1.1027	1.6620
0.7800	298.58	0.8645	26.935	230.83	396.49	165.65	1.1066	1.6615
0.8000	299.51	0.8669	26.254	232.01	396.86	164.85	1.1105	1.6609
0.8200	300.42	0.8693	25.605	233.17	397.22	164.05	1.1143	1.6604
0.8400	301.32	0.8717	24.985	234.31	397.58	163.27	1.1181	1.6599
0.8600	302.20	0.8740	24.393	235.43	397.92	162.49	1.1217	1.6594
0.8800	303.07	0.8764	23.827	236.54	398.25	161.71	1.1253	1.6589
0.9000	303.93	0.8787	23.285	237.63	398.58	160.94	1.1289	1.6584
0.9200	304.76	0.8811	22.765	238.71	398.90	160.18	1.1324	1.6580
0.9400	305.59	0.8834	22.267	239.77	399.20	159.43	1.1358	1.6575
0.9600	306.40	0.8857	21.788	240.82	399.50	158.68	1.1391	1.6570
0.9800	307.21	0.8880	21.329	241.85	399.79	157.94	1.1425	1.6566
1.0000	308.00	0.8903	20.886	242.87	400.08	157.20	1.1457	1.6561
1.0200	308.77	0.8926	20.461	243.88	400.36	156.47	1.1489	1.6557
1.0400	309.54	0.8948	20.051	244.88	400.63	155.75	1.1521	1.6552
1.0600	310.30	0.8971	19.656	245.86	400.89	155.03	1.1552	1.6548
1.0800	311.04	0.8993	19.275	246.83	401.14	154.31	1.1583	1.6544
1.1000	311.78	0.9016	18.907	247.79	401.39	153.60	1.1613	1.6540
1.1500	313.58	0.9072	18.041	250.15	401.99	151.84	1.1687	1.6529
1.2000	315.32	0.9128	17.244	252.44	402.55	150.11	1.1758	1.6519
1.2500	317.01	0.9183	16.507	254.67	403.07	148.41	1.1827	1.6509
1.3000	318.65	0.9239	15.825	256.85	403.57	146.72	1.1894	1.6499
1.3500	320.24	0.9294	15.191	258.97	404.03	145.06	1.1959	1.6489
1.4000	321.80	0.9349	14.600	261.05	404.46	143.41	1.2023	1.6479
1.4500	323.31	0.9405	14.048	263.08	404.86	141.78	1.2084	1.6469
1.5000	324.79	0.9460	13.530	265.07	405.23	140.17	1.2144	1.6460
1.5500	326.23	0.9515	13.045	267.01	405.58	138.57	1.2202	1.6450
1.6000	327.64	0.9571	12.588	268.92	405.90	136.98	1.2259	1.6440
1.6500	329.01	0.9627	12.157	270.80	406.20	135.41	1.2315	1.6431
1.7000	330.36	0.9683	11.750	272.64	406.48	133.84	1.2369	1.6421
1.7500	331.67	0.9740	11.365	274.45	406.73	132.29	1.2423	1.6411
1.8000	332.96	0.9797	11.000	276.22	406.96	130.74	1.2475	1.6401
1.8500	334.22	0.9854	10.654	277.97	407.17	129.20	1.2526	1.6391
1.9000	335.46	0.9912	10.324	279.69	407.36	127.66	1.2575	1.6381
1.9500	336.68	0.9970	10.010	281.39	407.53	126.14	1.2624	1.6371
2.0000	337.87	1.0029	9.7112	283.06	407.67	124.61	1.2673	1.6361

Chart R500.2: Saturation properties—*continued*

p MPa	T K	v' dm³/kg	v'' dm³/kg	h' kJ/kg	h'' kJ/kg	r kJ/kg	s' kJ/kgK	s'' kJ/kgK
2.1000	340.19	1.0149	9.1525	286.33	407.90	121.57	1.2766	1.6340
2.2000	342.43	1.0271	8.6406	289.52	408.06	118.53	1.2857	1.6318
2.3000	344.59	1.0397	8.1694	292.64	408.13	115.50	1.2944	1.6296
2.4000	346.68	1.0526	7.7337	295.68	408.13	112.45	1.3029	1.6273
2.5000	348.71	1.0660	7.3293	298.66	408.05	109.38	1.3112	1.6249
2.6000	350.68	1.0799	6.9524	301.60	407.89	106.29	1.3193	1.6224
2.7000	352.59	1.0944	6.5999	304.48	407.65	103.17	1.3272	1.6198
2.8000	354.44	1.1095	6.2689	307.33	407.33	100.00	1.3349	1.6170
2.9000	356.24	1.1254	5.9571	310.14	406.92	96.78	1.3425	1.6142
3.0000	358.00	1.1421	5.6623	312.93	406.42	93.49	1.3500	1.6111
3.1000	359.71	1.1598	5.3826	315.70	405.82	90.12	1.3574	1.6079
3.2000	361.37	1.1787	5.1163	318.46	405.12	86.66	1.3647	1.6046
3.3000	362.99	1.1989	4.8617	321.23	404.31	83.09	1.3721	1.6009
3.4000	364.57	1.2207	4.6175	324.01	403.38	79.37	1.3794	1.5971
3.5000	366.11	1.2444	4.3821	326.82	402.31	75.50	1.3867	1.5929
3.6000	367.61	1.2705	4.1541	329.67	401.09	71.42	1.3941	1.5884
3.7000	369.08	1.2995	3.9321	332.59	399.69	67.10	1.4017	1.5835
3.8000	370.51	1.3321	3.7144	335.61	398.08	62.47	1.4095	1.5781
3.9000	371.91	1.3696	3.4991	338.77	396.21	57.44	1.4177	1.5721
4.0000	373.27	1.4137	3.2836	342.13	394.02	51.89	1.4263	1.5653
4.1000	374.60	1.4673	3.0644	345.81	391.40	45.59	1.4358	1.5575
4.2000	375.89	1.5361	2.8351	349.99	388.14	38.15	1.4465	1.5480
4.3000	377.16	1.6337	2.5814	355.14	383.77	28.64	1.4598	1.5357

Chart R500.3: Thermodynamic properties

p/MPa	0.02			0.04			0.07		
T K	v dm^3/kg	h kJ/kg	s kJ/kgK	v dm^3/kg	h kJ/kg	s kJ/kgK	v dm^3/kg	h kJ/kg	s kJ/kgK
300.0	1250.0	410.39	2.0027	623.55	410.10	1.9439	354.40	409.66	1.8961
310.0	1298.7	417.67	2.0265	644.81	417.39	1.9679	366.69	416.98	1.9201
320.0	1333.3	425.07	2.0500	666.04	424.81	1.9914	378.95	424.42	1.9437
330.0	1369.9	432.60	2.0732	687.24	432.36	2.0146	391.19	432.00	1.9670
340.0	1428.6	440.26	2.0961	708.42	440.03	2.0375	403.40	439.69	1.9900
350.0	1470.6	448.04	2.1186	729.58	447.83	2.0601	415.59	447.50	2.0126
360.0	1515.2	455.94	2.1409	750.72	455.74	2.0824	427.76	455.44	2.0350
370.0	1538.5	463.96	2.1628	771.84	463.77	2.1044	439.91	463.48	2.0570
380.0	1587.3	472.09	2.1845	792.95	471.91	2.1261	452.05	471.64	2.0788
390.0	1639.3	480.33	2.2059	814.05	480.16	2.1476	464.18	479.91	2.1002
400.0	1666.6	488.68	2.2271	835.13	488.53	2.1687	476.29	488.29	2.1215
410.0	1724.1	497.14	2.2480	856.20	496.99	2.1897	488.39	496.77	2.1424
420.0	1754.4	505.71	2.2686	877.26	505.57	2.2103	500.48	505.35	2.1631
430.0	1785.7	514.37	2.2890	898.32	514.24	2.2307	512.56	514.04	2.1835
440.0	1851.9	523.14	2.3091	919.36	523.01	2.2509	524.64	522.82	2.2037
450.0	1886.7	532.00	2.3290	940.40	531.88	2.2708	536.70	531.70	2.2237
460.0	1923.0	540.96	2.3487	961.43	540.84	2.2905	548.76	540.67	2.2434
470.0	1960.8	550.00	2.3682	982.45	549.89	2.3100	560.81	549.73	2.2629
p/MPa	0.10			0.15			0.20		
300.0	246.73	409.21	1.8652	162.96	408.46	1.8294	121.07	407.69	1.8036
310.0	255.43	416.56	1.8893	168.89	415.86	1.8537	125.60	415.14	1.8280
320.0	264.11	424.03	1.9130	174.78	423.38	1.8776	130.10	422.71	1.8520
330.0	272.76	431.63	1.9364	180.64	431.02	1.9011	134.57	430.39	1.8756
340.0	281.38	439.35	1.9594	186.47	438.77	1.9242	139.01	438.19	1.8989
350.0	289.99	447.18	1.9821	192.29	446.64	1.9470	143.43	446.09	1.9218
360.0	298.57	455.13	2.0045	198.08	454.62	1.9695	147.84	454.11	1.9444
370.0	307.14	463.20	2.0266	203.86	462.72	1.9917	152.22	462.24	1.9667
380.0	315.69	471.37	2.0484	209.62	470.92	2.0136	156.59	470.47	1.9886
390.0	324.23	479.66	2.0699	215.37	479.23	2.0352	160.94	478.80	2.0103
400.0	332.75	488.05	2.0911	221.11	487.64	2.0565	165.28	487.24	2.0316
410.0	341.26	496.54	2.1121	226.83	496.16	2.0775	169.61	495.78	2.0527
420.0	349.77	505.14	2.1328	232.54	504.78	2.0983	173.93	504.42	2.0735
430.0	358.26	513.83	2.1533	238.25	513.49	2.1188	178.24	513.15	2.0941
440.0	366.75	522.63	2.1735	243.94	522.30	2.1390	182.54	521.98	2.1144
450.0	375.22	531.51	2.1935	249.63	531.21	2.1590	186.83	530.90	2.1344
460.0	383.69	540.49	2.2132	255.31	540.20	2.1788	191.11	539.91	2.1542
470.0	392.16	549.56	2.2327	260.98	549.29	2.1983	195.39	549.01	2.1738

Chart R500.3: Thermodynamic properties—*continued*

p/MPa	0.30			0.40			0.50		
T K	v dm^3/kg	h kJ/kg	s kJ/kgK	v dm^3/kg	h kJ/kg	s kJ/kgK	v dm^3/kg	h kJ/kg	s kJ/kgK
300.0	79.127	406.12	1.7659	58.116	404.50	1.7379	45.472	402.81	1.7152
310.0	82.278	413.69	1.7907	60.584	412.18	1.7631	47.538	410.63	1.7408
320.0	85.392	421.35	1.8150	63.013	419.96	1.7878	49.562	418.52	1.7659
330.0	88.476	429.13	1.8390	65.409	427.83	1.8120	51.550	426.50	1.7904
340.0	91.533	437.00	1.8625	67.775	435.79	1.8358	53.506	434.56	1.8145
350.0	94.565	444.99	1.8856	70.116	443.86	1.8592	55.435	442.70	1.8381
360.0	97.576	453.07	1.9084	72.435	452.01	1.8822	57.340	450.94	1.8613
370.0	100.57	461.26	1.9308	74.734	460.27	1.9048	59.224	459.26	1.8841
380.0	103.54	469.55	1.9529	77.015	468.62	1.9271	61.091	467.68	1.9066
390.0	106.51	477.94	1.9747	79.280	477.07	1.9490	62.940	476.18	1.9286
400.0	109.45	486.43	1.9962	81.532	485.60	1.9706	64.775	484.77	1.9504
410.0	112.39	495.01	2.0174	83.771	494.24	1.9919	66.598	493.45	1.9718
420.0	115.31	503.69	2.0383	85.999	502.96	2.0129	68.408	502.22	1.9930
430.0	118.23	512.47	2.0590	88.216	511.78	2.0337	70.209	511.08	2.0138
440.0	121.13	521.33	2.0793	90.425	520.68	2.0542	71.999	520.02	2.0344
450.0	124.03	530.29	2.0995	92.625	529.67	2.0744	73.782	529.05	2.0546
460.0	126.92	539.33	2.1193	94.818	538.75	2.0943	75.556	538.16	2.0747
470.0	129.80	548.46	2.1390	97.003	547.91	2.1140	77.324	547.35	2.0944
p/MPa	0.60			0.80			1.00		
300.0	37.007	401.05	1.6957	26.329	397.27	1.6623	0.8667	232.64	1.1120
310.0	38.814	409.02	1.7218	27.836	405.59	1.6896	21.153	401.83	1.6618
320.0	40.574	417.04	1.7473	29.283	413.92	1.7160	22.438	410.55	1.6895
330.0	42.293	425.13	1.7722	30.681	422.27	1.7417	23.660	419.22	1.7161
340.0	43.980	433.29	1.7965	32.038	430.66	1.7667	24.834	427.87	1.7420
350.0	45.637	441.53	1.8204	33.362	439.09	1.7912	25.967	436.54	1.7671
360.0	47.268	449.84	1.8438	34.658	447.59	1.8151	27.067	445.23	1.7916
370.0	48.878	458.24	1.8668	35.929	456.14	1.8386	28.140	453.97	1.8155
380.0	50.469	466.72	1.8894	37.178	464.76	1.8616	29.189	462.74	1.8389
390.0	52.042	475.28	1.9117	38.409	473.46	1.8841	30.217	471.58	1.8619
400.0	53.601	483.93	1.9336	39.624	482.22	1.9063	31.228	480.47	1.8844
410.0	55.146	492.66	1.9551	40.824	491.06	1.9281	32.223	489.42	1.9065
420.0	56.679	501.48	1.9764	42.011	499.97	1.9496	33.205	498.43	1.9282
430.0	58.201	510.38	1.9973	43.188	508.96	1.9708	34.174	507.51	1.9496
440.0	59.714	519.36	2.0180	44.354	518.02	1.9916	35.133	516.66	1.9706
450.0	61.218	528.42	2.0383	45.511	527.16	2.0121	36.083	525.88	1.9913
460.0	62.714	537.56	2.0584	46.659	536.37	2.0324	37.024	535.16	2.0117
470.0	64.203	546.79	2.0783	47.801	545.66	2.0524	37.957	544.52	2.0318

Chart R500.3: Thermodynamic properties–*continued*

p/MPa	2.00			4.00			5.85		
T K	v dm³/kg	h kJ/kg	s kJ/kgK	v dm³/kg	h kJ/kg	s kJ/kgK	v dm³/kg	h kJ/kg	s kJ/kgK
300.0	0.8591	232.71	1.1094	0.8460	233.01	1.1047	0.8356	233.43	1.1009
310.0	0.8864	245.41	1.1510	0.8690	245.50	1.1457	0.8559	245.80	1.1415
320.0	0.9189	258.45	1.1925	0.8951	258.25	1.1861	0.8780	258.38	1.1814
330.0	0.9599	271.95	1.2340	0.9254	271.24	1.2261	0.9027	271.14	1.2207
340.0	9.9235	410.11	1.6433	0.9622	284.54	1.2658	0.9307	284.08	1.2593
350.0	10.831	421.06	1.6750	1.0102	298.29	1.3057	0.9637	297.21	1.2974
360.0	11.639	431.47	1.7043	1.0821	312.99	1.3471	1.0045	310.62	1.3352
370.0	12.379	441.57	1.7320	1.2415	331.28	1.3971	1.0597	324.54	1.3733
380.0	13.072	451.47	1.7584	4.2624	414.49	1.6198	1.1487	339.72	1.4138
390.0	13.728	461.25	1.7838	5.0417	431.90	1.6650	1.3899	360.72	1.4682
400.0	14.355	470.96	1.8084	5.6276	446.02	1.7008	2.3921	403.71	1.5771
410.0	14.958	480.62	1.8323	6.1239	458.77	1.7323	3.0966	428.89	1.6393
420.0	15.541	490.27	1.8555	6.5659	470.76	1.7612	3.5937	446.85	1.6826
430.0	16.109	499.90	1.8782	6.9707	482.28	1.7883	3.9989	462.09	1.7185
440.0	16.661	509.55	1.9004	7.3481	493.48	1.8140	4.3508	475.91	1.7503
450.0	17.202	519.22	1.9221	7.7042	504.46	1.8387	4.6671	488.87	1.7794
460.0	17.732	528.91	1.9434	8.0433	515.28	1.8625	4.9578	501.26	1.8066
470.0	18.253	538.64	1.9643	8.3685	525.99	1.8855	5.2290	513.25	1.8324

References for R500

[Barh66] BARHO, W.: Berechnung der Molwärme der Fluor-Chlor-Derivate von Methan und Ethan im Zustand idealer Gase; Dissertation Universität Karlsruhe (1966)

[KMR 81] - : Kältemaschinenregeln, 7. Auflage; Verlag C.F. Müller, Karlsruhe (1981)

[Penn50] PENNINGTON, W.A. AND REED, W.H.: Azeotrope of 1,1-Difluoroethane and Dichlorodifluoromethane as a Refrigerant; Chem. Eng. Progress 46(1950)9 464/466

[Penn55] PENNINGTON, W.A.: Thermodynamic Properties of Carrene-7, Part 1; World Refrigeration (1955) 255/258

[Penn55a] PENNINGTON, W.A.: Thermodynamic Properties of Carrene-7, Part 2; World Refrigeration (1955) 369/374

[Pras78] PRASAD, M. AND KUDCHADKER, A.P.: Pressure-volume-temperature relationship of refrigerant 500 gas; J. Chem. Eng. Data 23(1978)3 190/191

[Sink67] SINKA, J.V. AND MURPHY, K.P.: Pressure-Volume-Temperature Relationship for a Mixture of Difluorodichloromethane and 1,1-Difluoroethane; J. Chem. Eng. Data 12(1976)3 315/317

R 500. ph

T=390
T=395
T=400
T=405
T=410
T=415
T=420
T=425
T=430
T=435
T=440
T=445
T=450
T=455
T=460
T=465
T=470
T=475
s=1.90
s=1.94
s=1.98
s=2.02
s=2.06
s=2.10
s=2.14
s=2.18
s=2.22
s=2.26
s=2.30
s=2.34
s=2.38
s=2.42
370
394
418
442
466
490
514
538
562
586
h/(kJ/kg)

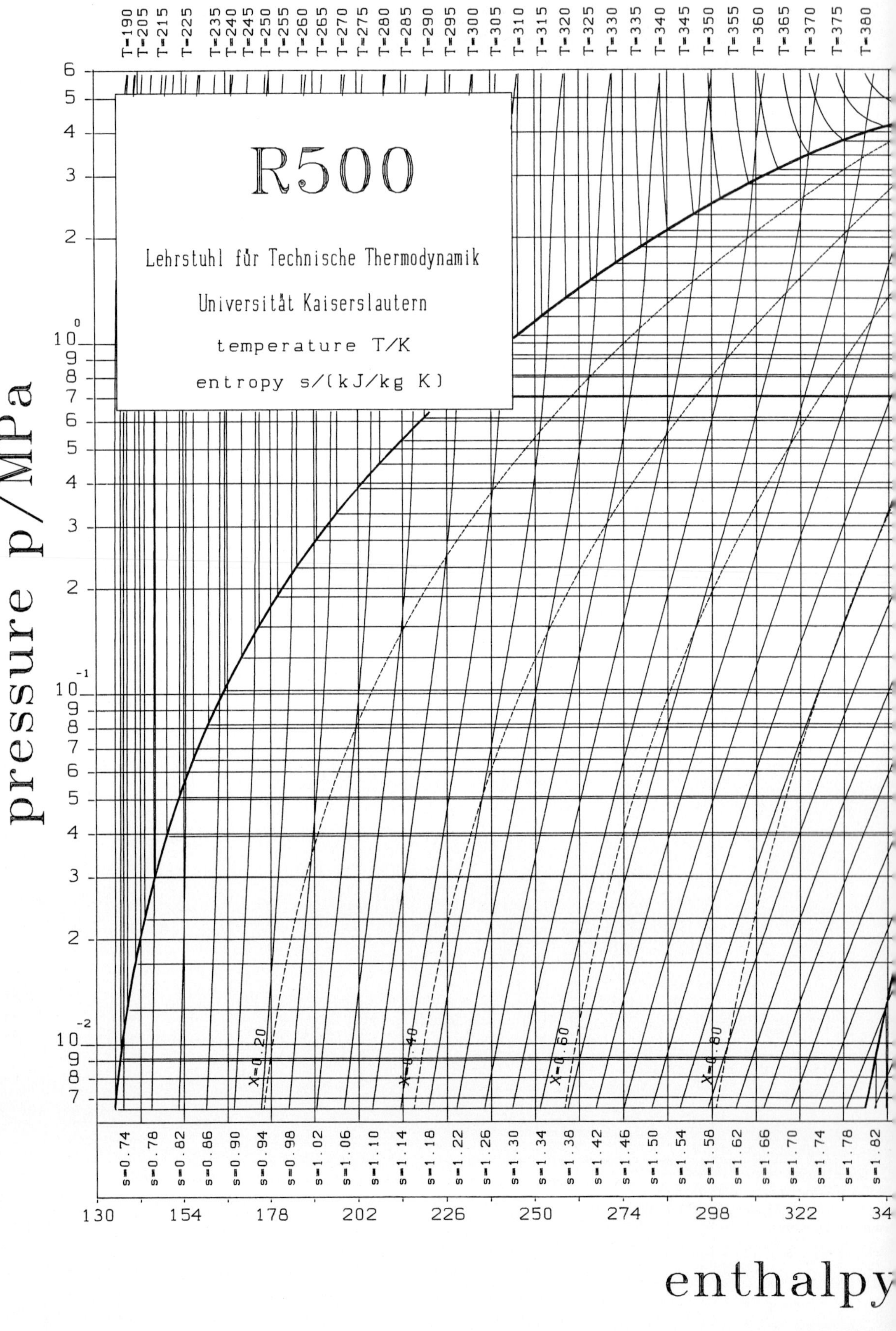

R500
Lehrstuhl für Technische Thermodynamik
Universität Kaiserslautern
temperature T/K
entropy s/(kJ/kg K)
pressure p/MPa
enthalpy

R 500. Ts

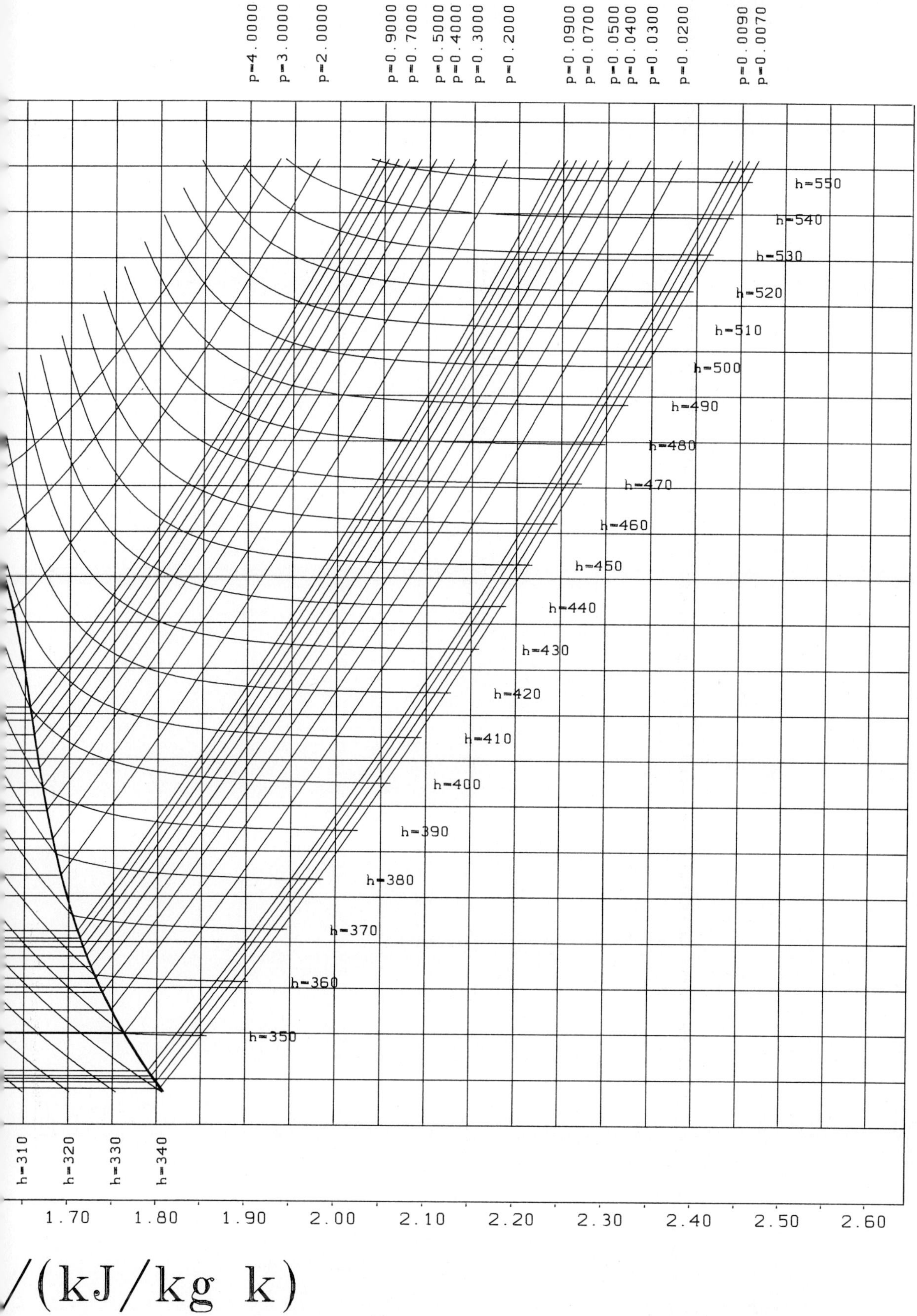

p=4.0000
p=3.0000
p=2.0000
p=0.9000
p=0.7000
p=0.5000
p=0.4000
p=0.3000
p=0.2000
p=0.0900
p=0.0700
p=0.0500
p=0.0400
p=0.0300
p=0.0200
p=0.0090
p=0.0070
h=550
h=540
h=530
h=520
h=510
h=500
h=490
h=480
h=470
h=460
h=450
h=440
h=430
h=420
h=410
h=400
h=390
h=380
h=370
h=360
h=350
h=310
h=320
h=330
h=340
1.70
1.80
1.90
2.00
2.10
2.20
2.30
2.40
2.50
2.60
/(kJ/kg k)

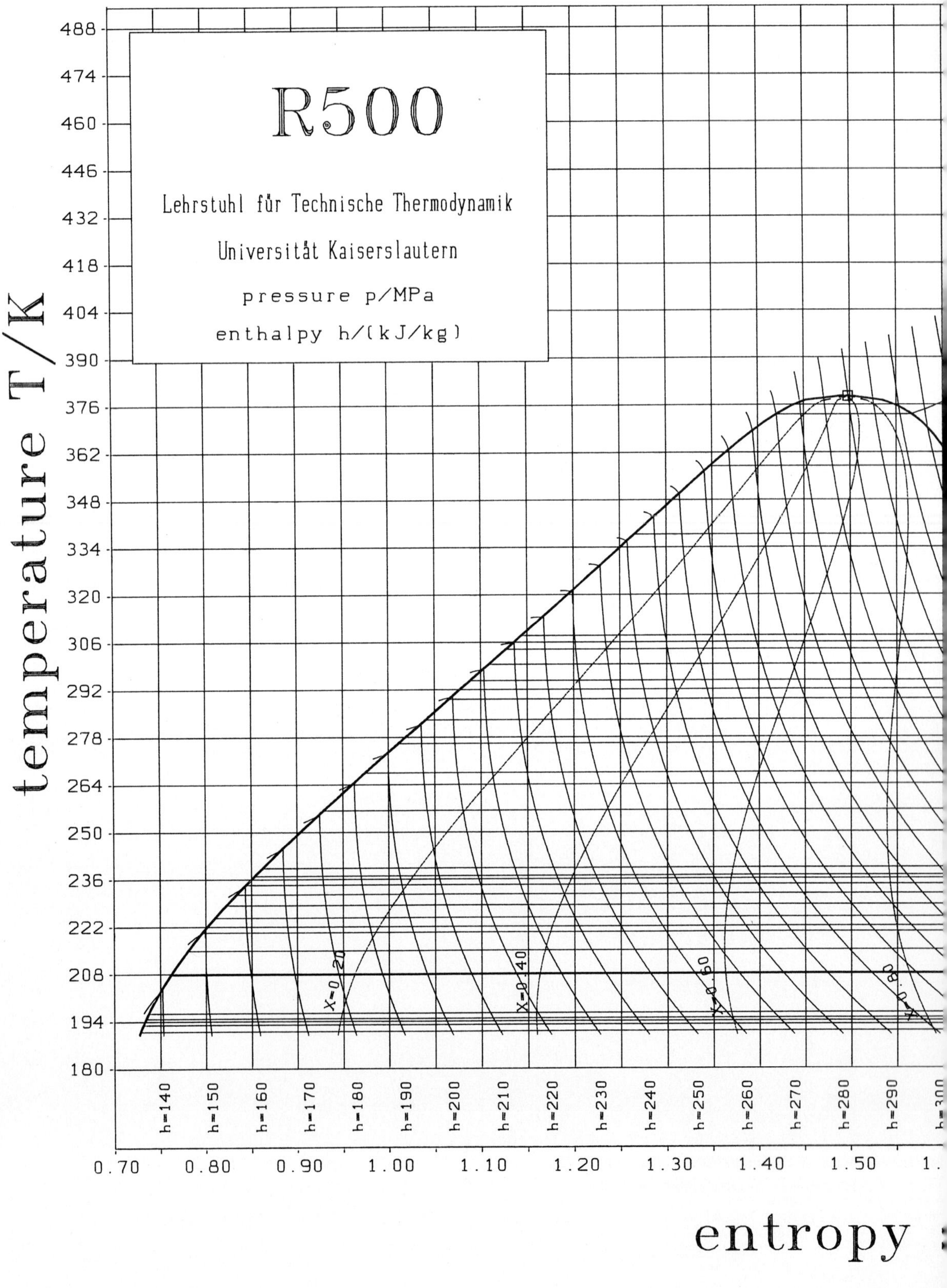

R500
Lehrstuhl für Technische Thermodynamik
Universität Kaiserslautern
pressure p/MPa
enthalpy h/(kJ/kg)
temperature T/K
488
474
460
446
432
418
404
390
376
362
348
334
320
306
292
278
264
250
236
222
208
194
180
X=0.20
X=0.40
X=0.60
X=0.80
h=140
h=150
h=160
h=170
h=180
h=190
h=200
h=210
h=220
h=230
h=240
h=250
h=260
h=270
h=280
h=290
0.70
0.80
0.90
1.00
1.10
1.20
1.30
1.40
1.50
entropy

R 500. hs

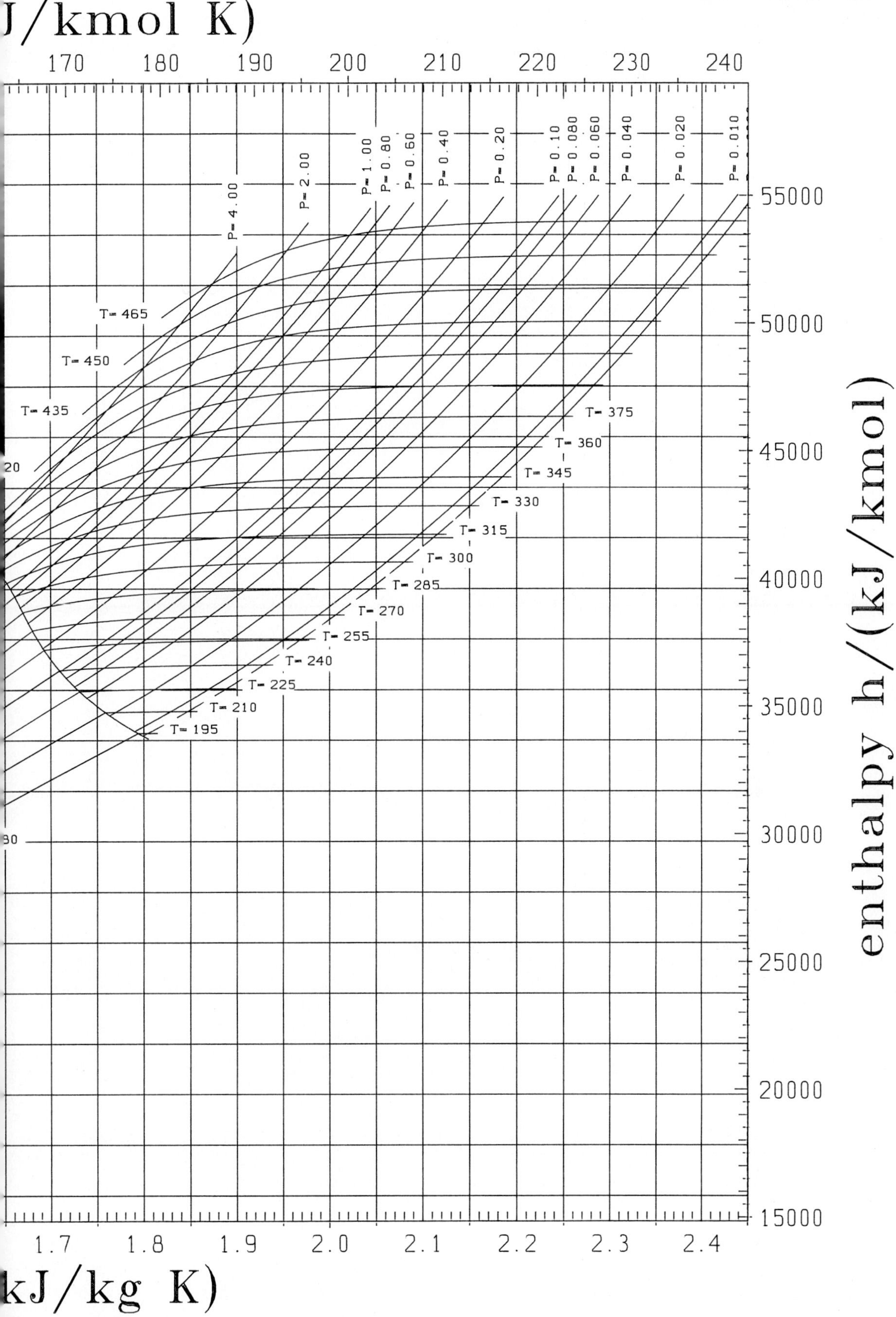
J/kmol K)
170
180
190
200
210
220
230
240
P= 4.00
P= 2.00
P= 1.00
P= 0.80
P= 0.60
P= 0.40
P= 0.20
P= 0.10
P= 0.080
P= 0.060
P= 0.040
P= 0.020
P= 0.010
T= 465
T= 450
T= 435
20
T= 375
T= 360
T= 345
T= 330
T= 315
T= 300
T= 285
T= 270
T= 255
T= 240
T= 225
T= 210
T= 195
30
55000
50000
45000
40000
35000
30000
25000
20000
15000
enthalpy h/(kJ/kmol)
1.7
1.8
1.9
2.0
2.1
2.2
2.3
2.4
kJ/kg K)

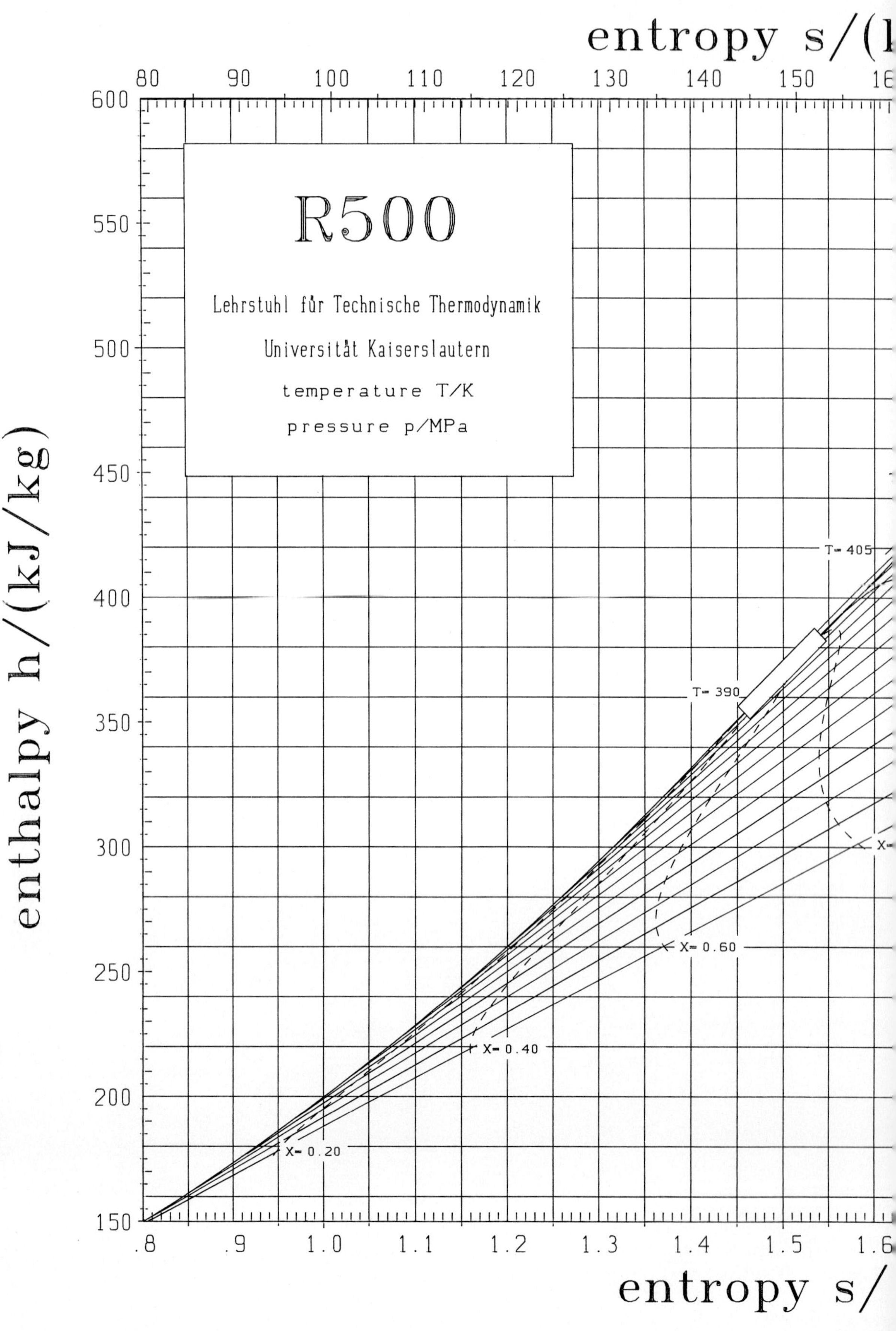

entropy s/(k
80
90
100
110
120
130
140
150
16
600
550
500
450
400
350
300
250
200
150
R500
Lehrstuhl für Technische Thermodynamik
Universität Kaiserslautern
temperature T/K
pressure p/MPa
enthalpy h/(kJ/kg)
T= 405
T= 390
X= 0.20
X= 0.40
X= 0.60
.8
.9
1.0
1.1
1.2
1.3
1.4
1.5
1.6
entropy s/

Refrigerant R502 (*R*22/*R*115)

Table R502.1: List of references for p, ρ, T data

Author	n	p/MPa min	p/MPa max	ρ/g cm^{-3} min	ρ/g cm^{-3} max	T/K min	T/K max
DÖRING, R. *[Doer 75]*	32	0.52	3.09	0.0288	0.16	275	367
KIMURA, T. *[Kimu 81]*	149	0.12	12.33	0.0044	1.04	273	383
MARTIN, J.J. *[Mart 70]*	64	0.57	13.62	0.0275	1.09	300	396
TODOROVIC, M. *[Todo 72]*	33	0.14	1.74	0.0065	0.08	293	362

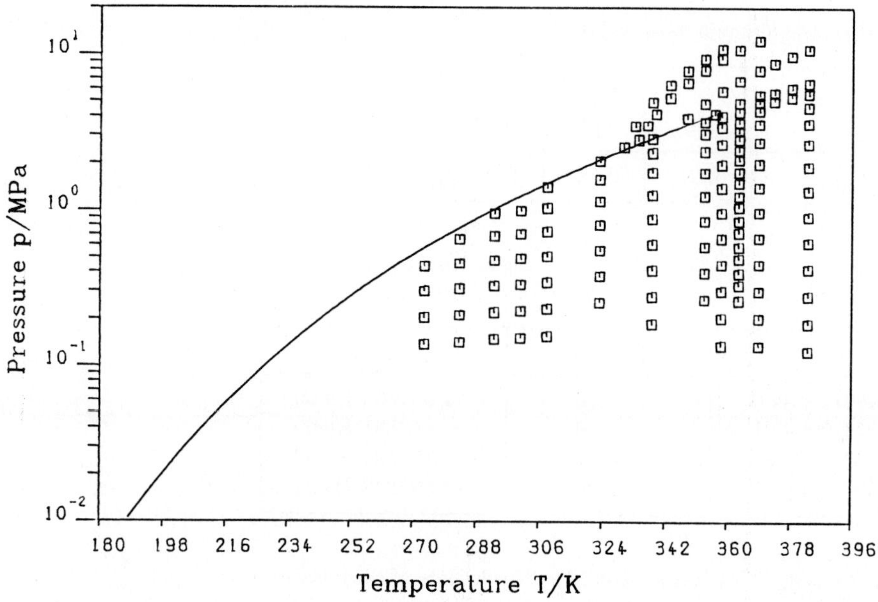

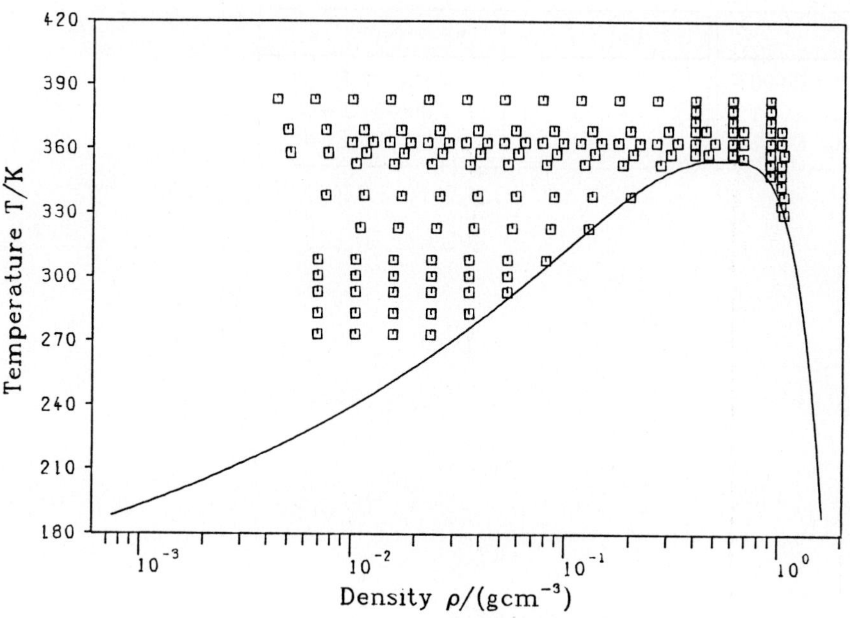

□ Kimura, T. /Kimu81/

Fig.R502.1: Distribution of p, ρ, T data projected in the p, T and in the T, ρ plane

Table R502.2: Experimental vapor pressure data

Author	Δ / %		
	std	opt	*EOS*
DÖRING, R. *[Doer 75]*	2.90	2.80	2.42
DUPONT *[DuPo 63]*	0.54	0.58	0.97
KIMURA, T. *[Kimu 81]*	0.08	0.03	0.03
LONG, L.J. *[Long 62]*	2.24	2.27	2.90
MARTIN, J.J. *[Mart 70]*	1.28	1.30	1.33
TODOROVIC, M. *[Todo 72]*	2.11	2.10	1.90

Table R502.3: Experimental saturated liquid density data

Author	Δ / %		
	std	opt	*EOS*
MARTIN, J.J. *[Mart 70]*	0.21	0.10	0.17

Table R502.4: Coefficients of the standard vapor pressure equation

Index i	b_i
5	−6.946679 E+00
6	9.497480 E−01
9	−6.253303 E−01
15	−1.890642 E+01

Table R502.5: Coefficients of the quasi-optimal vapor pressure equation

Index i	b_i
5	−7.087163 E+00
6	1.893801 E+00
7	−1.576190 E+00
21	−1.837848 E+02

Table R502.6: Coefficients of the standard saturated liquid density equation

Index i	c_i
8	1.651223 E+00
9	9.040326 E−01
16	2.614561 E−01

Table R502.7: Coefficients of the quasi-optimal saturated liquid density equation

Index i	c_i
7	−4.124142 E−02
8	2.006747 E+00
10	7.220296 E−01

Table R502.8: Coefficients of the *Bender-EOS*

Range of validity: one-phase region

$0.12 \leq p/\text{MPa} \leq 12.33$

$0.00439 \leq \rho/\text{g cm}^{-3} \leq 1.035$

$273.15 \leq T/\text{K} \leq 383.15$

	Specific units	Molar units
a_{01}	−3.118754640 E−02	−3.884271679 E−01
a_{02}	1.421951943 E+01	1.770978579 E+02
a_{03}	−7.402769780 E+03	−9.219824039 E+04
a_{04}	6.707259092 E+06	8.353596080 E+07
a_{05}	−1.869673925 E+07	−2.328596609 E+08
a_{06}	6.065258414 E−01	8.430277931 E−01
a_{07}	−3.347233358 E+02	−4.652416367 E+02
a_{08}	5.602170619 E+04	7.786618824 E+04
a_{09}	−5.682182617 E−01	−8.813978070 E−02
a_{10}	2.566848396 E+02	3.981594221 E+01
a_{11}	4.324693540 E−01	7.486456589 E−03
a_{12}	−2.767976817 E+02	−4.791631611 E+00
a_{13}	6.881401690 E+01	1.329419600 E−01
a_{14}	−1.761226038 E+07	−2.447978962 E+07
a_{15}	1.160666907 E+10	1.613244473 E+10
a_{16}	−2.064863049 E+12	−2.870012818 E+12
a_{17}	1.926800259 E+06	3.335474841 E+04
a_{18}	−2.131718031 E+10	−3.690207028 E+08
a_{19}	7.557095853 E+12	1.308205298 E+11
a_{20}	3.800000000 E+00	4.732732800 E−02

Table R502.9: Deviations between experimental and predicted p, ρ, T data

Region	n	Δ / %	
		p	ρ
Gas	104	0.16	0.20
Critical	21	1.22	0.81
Liquid	24	0.69	0.12
Total	149	0.33	0.37

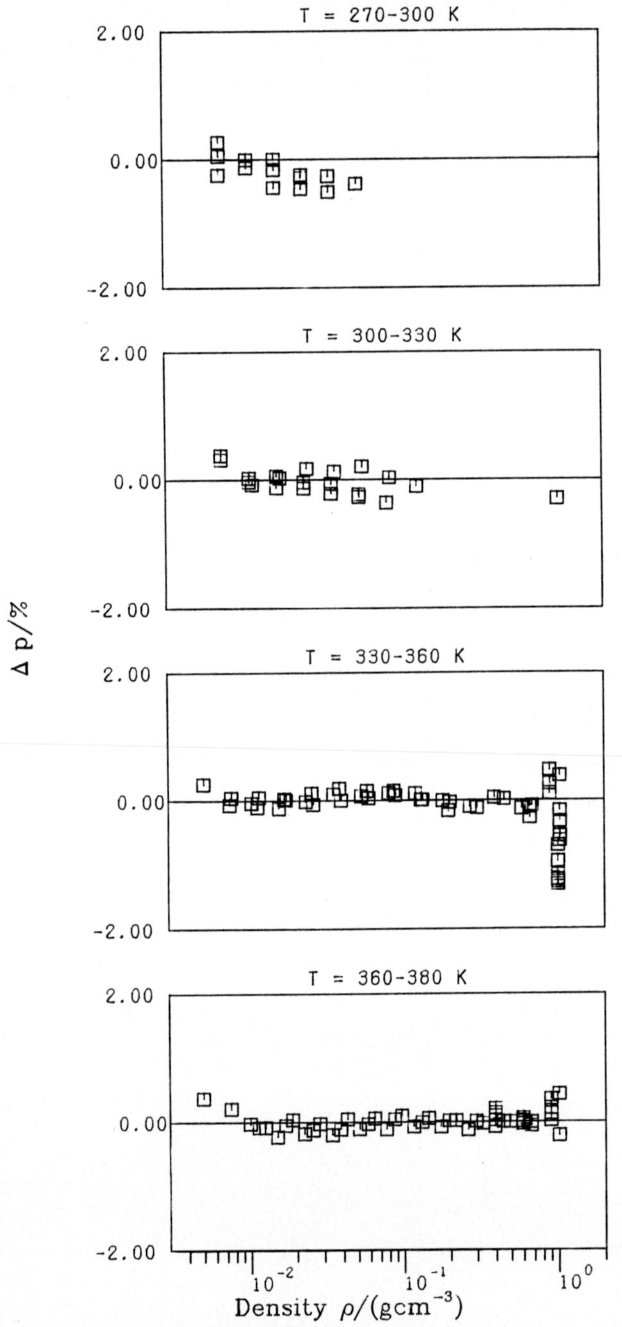

□ Kimura, T. /Kimu81/

Fig.R502.2: Deviations between experimental and calculated p, ρ, T data

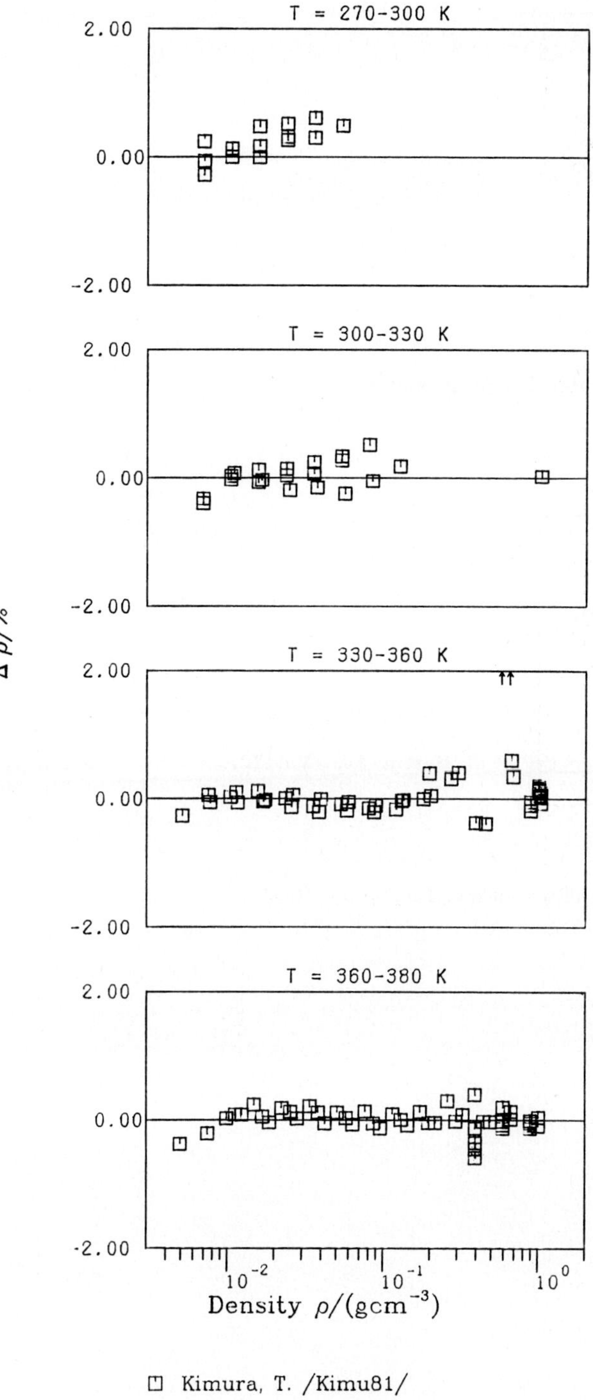

Fig.R502.2: Deviations between experimental and calculated p, ρ, T data *-continued*

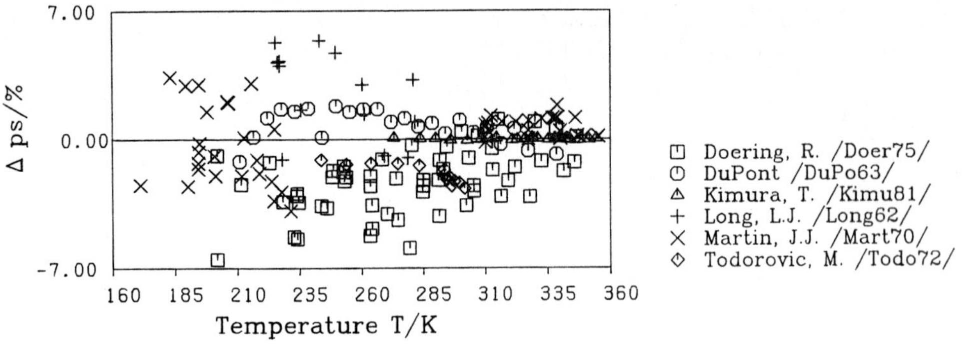

Fig.R502.3: Deviations between experimental and calculated vapor pressures

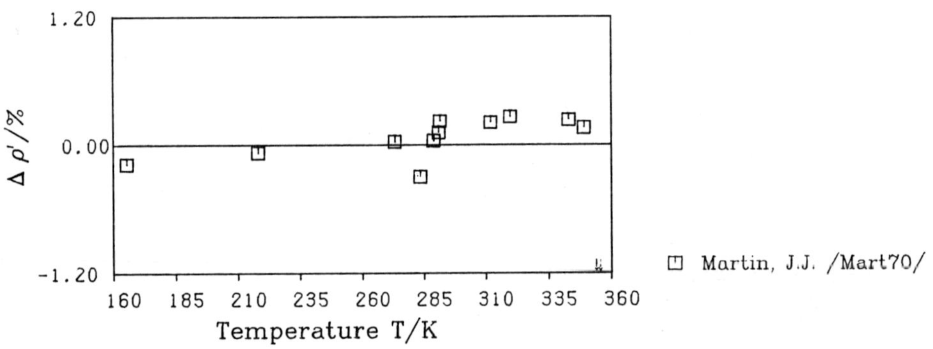

Fig.R502.4: Deviations between experimental and calculated saturated liquid densities

Table R502.10: Information on ideal gas state heat capacity equation

Author	T / K min	T / K max
BARHO ,W. *[Barh 66]*	200	550
Coefficients of equation (15)		
Index i	d_i	
1	2.192411 E−01	
2	1.832635 E−03	
3	−8.683826 E−07	
4	−7.944444 E−11	
5	−2.712121 E−13	

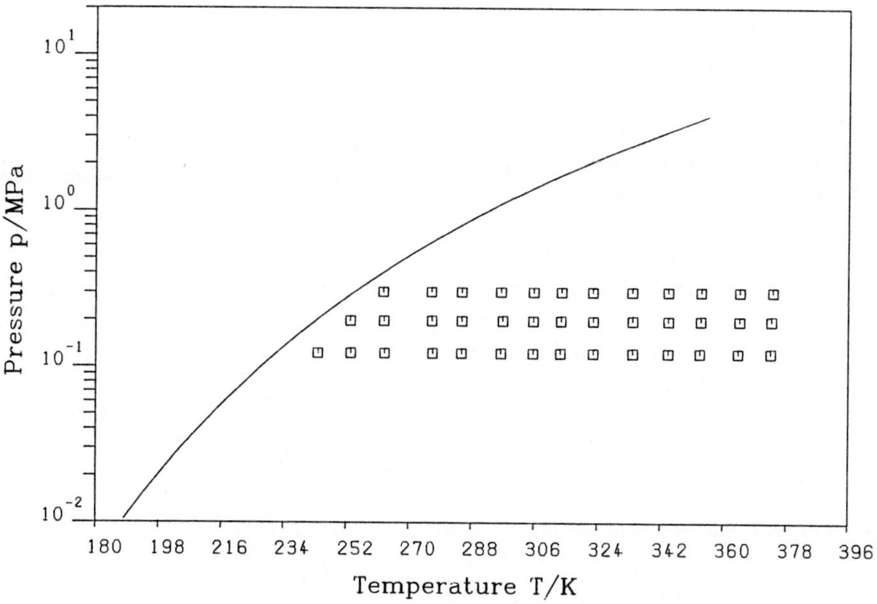

□ cp: Tanishita, I. /Tani72/

Fig.R502.5: Distribution of $c_p(p, T)$ data

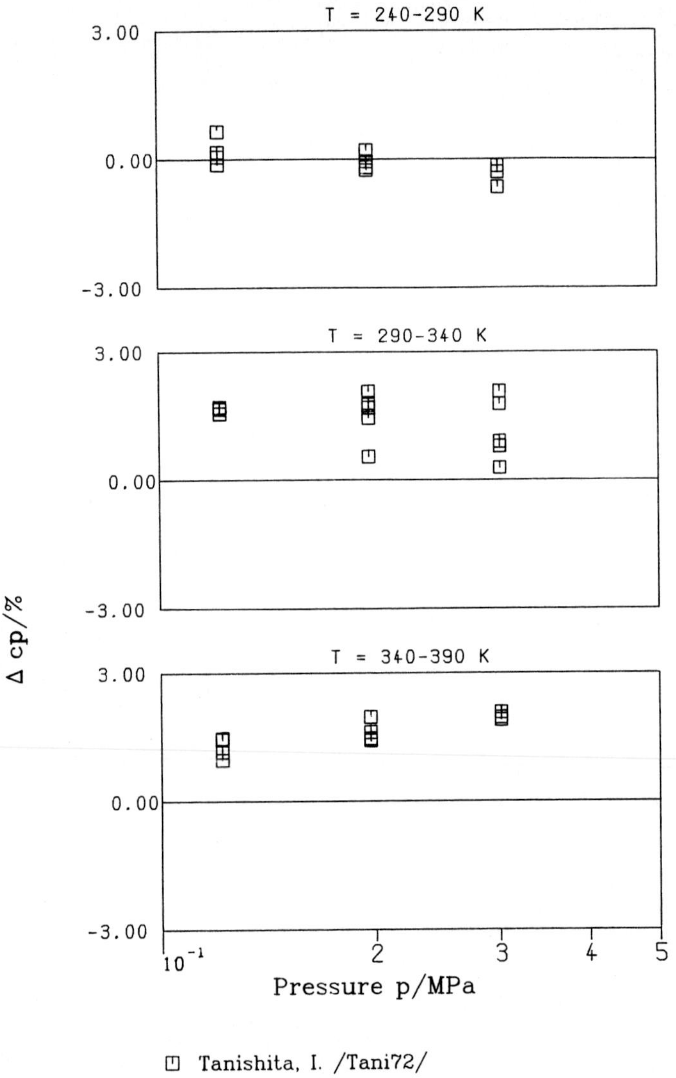

Fig.R502.6: Deviations between experimental and calculated specific isobaric heat capacities c_p

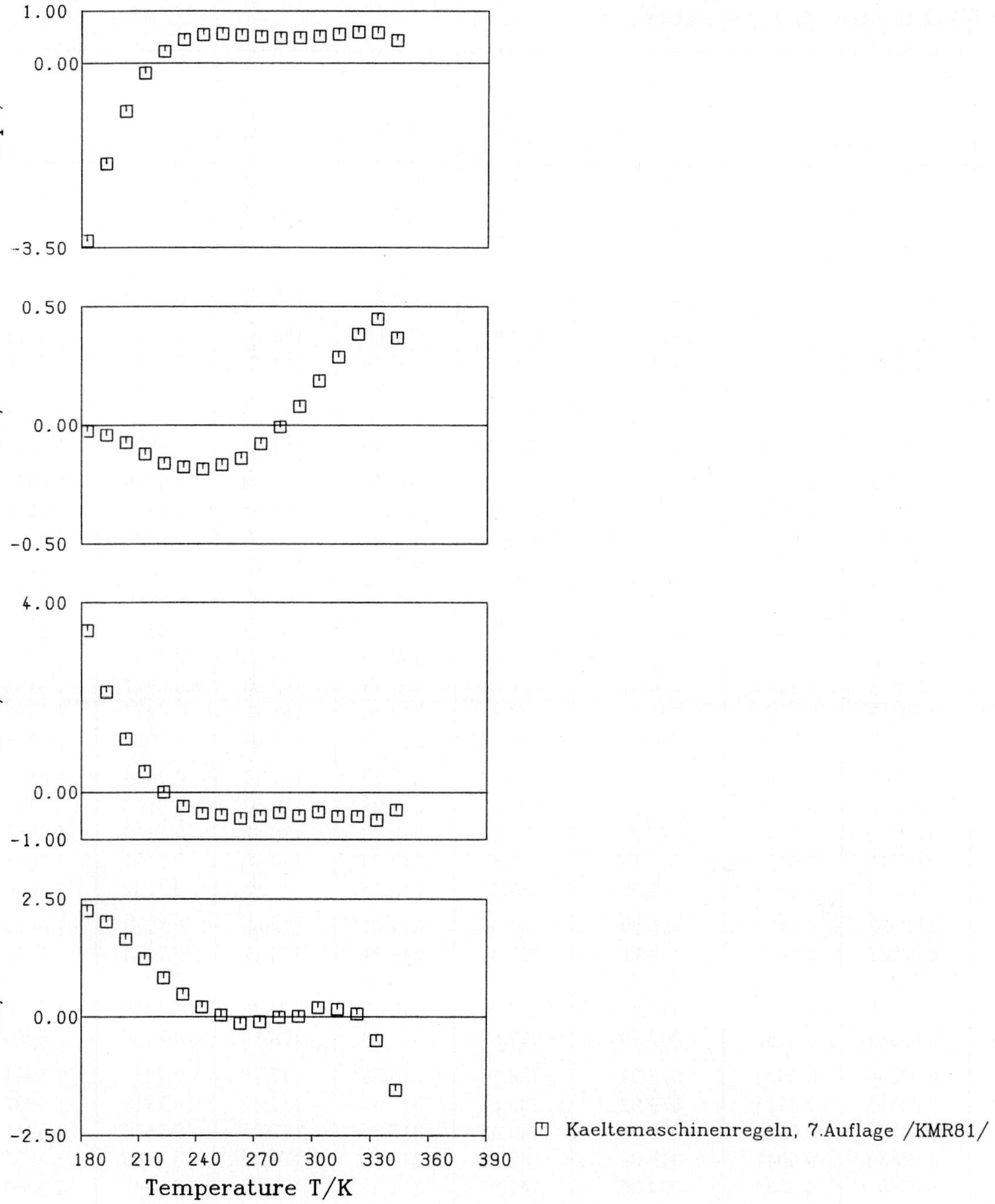

Fig.R502.7: Deviations between saturation properties from the *Kältemaschinenregeln* and those calculated by the *Bender-EOS*

Chart R502.1: Saturation properties

T K	p MPa	v' dm³/kg	v'' dm³/kg	h' kJ/kg	h'' kJ/kg	r kJ/kg	s' kJ/kgK	s'' kJ/kgK
180.0	0.00572	0.6143	2332.2	115.39	301.37	185.98	0.6282	1.6614
181.0	0.00617	0.6154	2170.5	116.06	301.87	185.81	0.6319	1.6585
182.0	0.00666	0.6165	2021.7	116.74	302.37	185.63	0.6356	1.6556
183.0	0.00718	0.6177	1884.7	117.43	302.88	185.44	0.6394	1.6528
184.0	0.00774	0.6188	1758.4	118.13	303.38	185.25	0.6432	1.6500
185.0	0.00833	0.6200	1641.8	118.84	303.88	185.04	0.6471	1.6473
186.0	0.00896	0.6211	1534.3	119.56	304.39	184.83	0.6509	1.6446
187.0	0.00963	0.6223	1434.9	120.29	304.89	184.60	0.6548	1.6420
188.0	0.01034	0.6234	1343.0	121.02	305.40	184.38	0.6587	1.6395
189.0	0.01109	0.6246	1257.9	121.76	305.90	184.14	0.6627	1.6370
190.0	0.01189	0.6257	1179.1	122.51	306.41	183.89	0.6666	1.6345
191.0	0.01273	0.6269	1106.1	123.27	306.91	183.64	0.6706	1.6321
192.0	0.01363	0.6281	1038.3	124.03	307.42	183.39	0.6746	1.6297
193.0	0.01457	0.6293	975.39	124.80	307.93	183.12	0.6786	1.6274
194.0	0.01557	0.6304	916.94	125.58	308.43	182.85	0.6826	1.6251
195.0	0.01663	0.6316	862.59	126.36	308.94	182.58	0.6866	1.6229
196.0	0.01775	0.6328	812.01	127.15	309.45	182.29	0.6907	1.6207
197.0	0.01893	0.6340	764.91	127.95	309.95	182.00	0.6947	1.6186
198.0	0.02017	0.6352	721.03	128.75	310.46	181.71	0.6987	1.6165
199.0	0.02147	0.6364	680.10	129.56	310.97	181.41	0.7028	1.6144
200.0	0.02285	0.6376	641.91	130.37	311.47	181.11	0.7069	1.6124
201.0	0.02430	0.6389	606.24	131.18	311.98	180.80	0.7109	1.6104
202.0	0.02582	0.6401	572.92	132.01	312.49	180.48	0.7150	1.6085
203.0	0.02742	0.6413	541.75	132.83	312.99	180.16	0.7191	1.6066
204.0	0.02910	0.6426	512.59	133.67	313.50	179.83	0.7232	1.6047
205.0	0.03087	0.6438	485.29	134.50	314.01	179.50	0.7273	1.6029
206.0	0.03272	0.6451	459.71	135.34	314.51	179.17	0.7314	1.6011
207.0	0.03466	0.6463	435.72	136.19	315.02	178.83	0.7354	1.5993
208.0	0.03669	0.6476	413.22	137.04	315.52	178.48	0.7395	1.5976
209.0	0.03882	0.6489	392.10	137.89	316.03	178.13	0.7436	1.5959
210.0	0.04105	0.6502	372.27	138.75	316.53	177.78	0.7477	1.5943
211.0	0.04338	0.6515	353.62	139.62	317.04	177.42	0.7518	1.5927
212.0	0.04581	0.6528	336.09	140.48	317.54	177.06	0.7559	1.5911
213.0	0.04836	0.6541	319.59	141.35	318.05	176.69	0.7600	1.5895
214.0	0.05101	0.6554	304.06	142.23	318.55	176.32	0.7641	1.5880
215.0	0.05379	0.6567	289.43	143.11	319.05	175.94	0.7682	1.5865
216.0	0.05668	0.6581	275.64	143.99	319.55	175.56	0.7723	1.5851
217.0	0.05970	0.6594	262.63	144.88	320.06	175.18	0.7763	1.5836
218.0	0.06285	0.6608	250.36	145.77	320.56	174.79	0.7804	1.5822
219.0	0.06613	0.6622	238.77	146.66	321.06	174.40	0.7845	1.5808
220.0	0.06954	0.6635	227.83	147.56	321.56	174.00	0.7886	1.5795
221.0	0.07309	0.6649	217.49	148.46	322.06	173.60	0.7926	1.5782
222.0	0.07678	0.6663	207.71	149.36	322.55	173.19	0.7967	1.5769
223.0	0.08063	0.6677	198.45	150.27	323.05	172.78	0.8008	1.5756
224.0	0.08462	0.6691	189.70	151.18	323.55	172.37	0.8049	1.5744
225.0	0.08877	0.6706	181.40	152.09	324.04	171.95	0.8089	1.5731
226.0	0.09308	0.6720	173.54	153.01	324.54	171.53	0.8130	1.5719
227.0	0.09755	0.6734	166.09	153.93	325.03	171.10	0.8170	1.5708
228.0	0.10219	0.6749	159.03	154.86	325.53	170.67	0.8211	1.5696
229.0	0.10701	0.6764	152.33	155.78	326.02	170.23	0.8251	1.5685

Chart R502.1: Saturation properties–*continued*

T K	p MPa	v' dm^3/kg	v'' dm^3/kg	h' kJ/kg	h'' kJ/kg	r kJ/kg	s' kJ/kgK	s'' kJ/kgK
230.0	0.11200	0.6778	145.96	156.71	326.51	169.79	0.8292	1.5674
231.0	0.11717	0.6793	139.92	157.65	327.00	169.35	0.8332	1.5663
232.0	0.12253	0.6808	134.18	158.59	327.49	168.90	0.8372	1.5653
233.0	0.12808	0.6824	128.72	159.53	327.98	168.45	0.8413	1.5642
234.0	0.13383	0.6839	123.53	160.47	328.46	167.99	0.8453	1.5632
235.0	0.13977	0.6854	118.59	161.42	328.95	167.53	0.8493	1.5622
236.0	0.14592	0.6870	113.89	162.37	329.44	167.07	0.8533	1.5612
237.0	0.15228	0.6886	109.42	163.32	329.92	166.60	0.8573	1.5603
238.0	0.15885	0.6901	105.16	164.28	330.40	166.12	0.8613	1.5593
239.0	0.16564	0.6917	101.09	165.24	330.88	165.64	0.8653	1.5584
240.0	0.17266	0.6933	97.220	166.20	331.36	165.16	0.8693	1.5575
241.0	0.17990	0.6950	93.526	167.17	331.84	164.67	0.8733	1.5566
242.0	0.18738	0.6966	90.001	168.14	332.32	164.18	0.8773	1.5558
243.0	0.19510	0.6983	86.637	169.11	332.79	163.68	0.8813	1.5549
244.0	0.20306	0.6999	83.425	170.09	333.27	163.18	0.8853	1.5541
245.0	0.21127	0.7016	80.356	171.06	333.74	162.68	0.8893	1.5533
246.0	0.21973	0.7033	77.424	172.05	334.21	162.17	0.8933	1.5525
247.0	0.22846	0.7050	74.621	173.03	334.68	161.65	0.8972	1.5517
248.0	0.23744	0.7067	71.940	174.02	335.15	161.13	0.9012	1.5509
249.0	0.24670	0.7085	69.376	175.01	335.62	160.61	0.9052	1.5502
250.0	0.25623	0.7102	66.922	176.01	336.08	160.08	0.9091	1.5494
251.0	0.26604	0.7120	64.572	177.01	336.55	159.54	0.9131	1.5487
252.0	0.27614	0.7138	62.323	178.01	337.01	159.00	0.9171	1.5480
253.0	0.28653	0.7156	60.167	179.01	337.47	158.46	0.9210	1.5473
254.0	0.29721	0.7175	58.102	180.02	337.93	157.91	0.9250	1.5466
255.0	0.30820	0.7193	56.122	181.03	338.39	157.35	0.9289	1.5460
256.0	0.31949	0.7212	54.223	182.05	338.84	156.79	0.9328	1.5453
257.0	0.33109	0.7230	52.402	183.07	339.30	156.23	0.9368	1.5447
258.0	0.34302	0.7250	50.654	184.09	339.75	155.66	0.9407	1.5440
259.0	0.35526	0.7269	48.976	185.11	340.20	155.08	0.9446	1.5434
260.0	0.36784	0.7288	47.365	186.14	340.64	154.50	0.9486	1.5428
261.0	0.38075	0.7308	45.818	187.18	341.09	153.91	0.9525	1.5422
262.0	0.39400	0.7328	44.331	188.21	341.53	153.32	0.9564	1.5416
263.0	0.40760	0.7348	42.902	189.25	341.97	152.72	0.9603	1.5410
264.0	0.42154	0.7368	41.529	190.29	342.41	152.12	0.9643	1.5405
265.0	0.43585	0.7388	40.208	191.34	342.85	151.51	0.9682	1.5399
266.0	0.45052	0.7409	38.937	192.39	343.29	150.90	0.9721	1.5394
267.0	0.46555	0.7430	37.714	193.44	343.72	150.28	0.9760	1.5388
268.0	0.48097	0.7451	36.537	194.50	344.15	149.65	0.9799	1.5383
269.0	0.49676	0.7473	35.404	195.56	344.58	149.02	0.9838	1.5378
270.0	0.51294	0.7494	34.313	196.62	345.00	148.38	0.9877	1.5373
271.0	0.52951	0.7516	33.261	197.69	345.43	147.73	0.9916	1.5368
272.0	0.54648	0.7538	32.248	198.76	345.85	147.08	0.9955	1.5363
273.0	0.56385	0.7561	31.272	199.84	346.26	146.43	0.9994	1.5358
274.0	0.58163	0.7583	30.330	200.92	346.68	145.76	1.0033	1.5353
275.0	0.59983	0.7606	29.422	202.00	347.09	145.09	1.0072	1.5348
276.0	0.61845	0.7630	28.546	203.09	347.50	144.41	1.0111	1.5343
277.0	0.63750	0.7653	27.701	204.18	347.91	143.73	1.0150	1.5339
278.0	0.65698	0.7677	26.885	205.27	348.31	143.04	1.0189	1.5334
279.0	0.67690	0.7701	26.098	206.37	348.72	142.34	1.0228	1.5330

Chart R502.1: Saturation properties—*continued*

T K	p MPa	v' dm^3/kg	v'' dm^3/kg	h' kJ/kg	h'' kJ/kg	r kJ/kg	s' kJ/kgK	s'' kJ/kgK
280.0	0.69727	0.7725	25.337	207.48	349.11	141.64	1.0267	1.5325
281.0	0.71809	0.7750	24.602	208.58	349.51	140.93	1.0306	1.5321
282.0	0.73937	0.7775	23.892	209.69	349.90	140.21	1.0344	1.5316
283.0	0.76111	0.7801	23.206	210.81	350.29	139.48	1.0383	1.5312
284.0	0.78332	0.7826	22.543	211.93	350.68	138.75	1.0422	1.5308
285.0	0.80601	0.7852	21.902	213.05	351.06	138.01	1.0461	1.5303
286.0	0.82919	0.7879	21.282	214.18	351.44	137.26	1.0500	1.5299
287.0	0.85285	0.7906	20.682	215.31	351.81	136.50	1.0539	1.5295
288.0	0.87701	0.7933	20.101	216.45	352.18	135.73	1.0578	1.5291
289.0	0.90167	0.7960	19.539	217.59	352.55	134.96	1.0617	1.5287
290.0	0.92684	0.7988	18.995	218.74	352.92	134.18	1.0656	1.5282
291.0	0.95253	0.8017	18.468	219.89	353.28	133.38	1.0695	1.5278
292.0	0.97874	0.8046	17.958	221.05	353.63	132.58	1.0733	1.5274
293.0	1.00547	0.8075	17.463	222.21	353.98	131.78	1.0772	1.5270
294.0	1.03274	0.8105	16.984	223.37	354.33	130.96	1.0811	1.5266
295.0	1.06056	0.8135	16.520	224.54	354.67	130.13	1.0850	1.5262
296.0	1.08892	0.8166	16.070	225.72	355.01	129.29	1.0889	1.5257
297.0	1.11783	0.8197	15.633	226.90	355.35	128.45	1.0928	1.5253
298.0	1.14731	0.8229	15.209	228.09	355.68	127.59	1.0967	1.5249
299.0	1.17736	0.8261	14.798	229.28	356.00	126.72	1.1007	1.5245
300.0	1.20798	0.8294	14.399	230.48	356.32	125.84	1.1046	1.5240
301.0	1.23918	0.8327	14.012	231.68	356.63	124.95	1.1085	1.5236
302.0	1.27098	0.8361	13.636	232.89	356.94	124.05	1.1124	1.5232
303.0	1.30337	0.8396	13.271	234.10	357.24	123.14	1.1163	1.5227
304.0	1.33636	0.8431	12.917	235.32	357.54	122.22	1.1203	1.5223
305.0	1.36997	0.8467	12.572	236.55	357.83	121.28	1.1242	1.5218
306.0	1.40419	0.8503	12.237	237.78	358.12	120.34	1.1281	1.5214
307.0	1.43904	0.8540	11.912	239.02	358.39	119.37	1.1321	1.5209
308.0	1.47453	0.8579	11.595	240.26	358.66	118.40	1.1360	1.5204
309.0	1.51065	0.8617	11.287	241.52	358.93	117.41	1.1400	1.5200
310.0	1.54742	0.8657	10.987	242.77	359.19	116.41	1.1439	1.5195
311.0	1.58485	0.8697	10.696	244.04	359.44	115.40	1.1479	1.5190
312.0	1.62295	0.8739	10.412	245.31	359.68	114.36	1.1519	1.5184
313.0	1.66171	0.8781	10.136	246.59	359.91	113.32	1.1559	1.5179
314.0	1.70116	0.8824	9.8670	247.88	360.14	112.26	1.1599	1.5174
315.0	1.74130	0.8868	9.6049	249.18	360.35	111.18	1.1639	1.5168
316.0	1.78213	0.8913	9.3495	250.48	360.56	110.08	1.1679	1.5163
317.0	1.82367	0.8960	9.1006	251.79	360.76	108.97	1.1719	1.5157
318.0	1.86593	0.9007	8.8580	253.11	360.94	107.83	1.1760	1.5151
319.0	1.90891	0.9056	8.6213	254.44	361.12	106.68	1.1800	1.5144
320.0	1.95262	0.9106	8.3904	255.78	361.29	105.51	1.1841	1.5138
321.0	1.99707	0.9157	8.1651	257.13	361.44	104.31	1.1882	1.5131
322.0	2.04227	0.9210	7.9452	258.49	361.59	103.10	1.1923	1.5124
323.0	2.08824	0.9264	7.7305	259.85	361.72	101.86	1.1964	1.5117
324.0	2.13497	0.9320	7.5207	261.23	361.83	100.60	1.2005	1.5110
325.0	2.18249	0.9378	7.3157	262.62	361.94	99.31	1.2046	1.5102
326.0	2.23080	0.9437	7.1154	264.03	362.02	98.00	1.2088	1.5094
327.0	2.27991	0.9498	6.9194	265.44	362.10	96.66	1.2130	1.5086
328.0	2.32983	0.9562	6.7277	266.87	362.15	95.29	1.2172	1.5077
329.0	2.38058	0.9627	6.5401	268.31	362.19	93.88	1.2215	1.5068

Chart R502.1: Saturation properties—*continued*

T K	p MPa	v' dm³/kg	v'' dm³/kg	h' kJ/kg	h'' kJ/kg	r kJ/kg	s' kJ/kgK	s'' kJ/kgK
330.0	2.43217	0.9695	6.3564	269.76	362.21	92.45	1.2257	1.5059
331.0	2.48460	0.9766	6.1765	271.23	362.21	90.98	1.2300	1.5049
332.0	2.53789	0.9839	6.0001	272.72	362.19	89.47	1.2343	1.5038
333.0	2.59205	0.9915	5.8271	274.22	362.15	87.93	1.2387	1.5027
334.0	2.64710	0.9994	5.6574	275.74	362.08	86.34	1.2431	1.5016
335.0	2.70305	1.0077	5.4908	277.28	361.99	84.71	1.2475	1.5004
336.0	2.75992	1.0163	5.3270	278.84	361.87	83.03	1.2520	1.4991
337.0	2.81771	1.0254	5.1661	280.42	361.72	81.30	1.2565	1.4978
338.0	2.87644	1.0349	5.0076	282.02	361.53	79.51	1.2611	1.4963
339.0	2.93614	1.0449	4.8516	283.65	361.31	77.66	1.2657	1.4948
340.0	2.99681	1.0555	4.6978	285.31	361.06	75.75	1.2704	1.4932
341.0	3.05848	1.0668	4.5459	287.00	360.76	73.76	1.2752	1.4915
342.0	3.12116	1.0787	4.3958	288.72	360.41	71.69	1.2800	1.4897
343.0	3.18487	1.0914	4.2472	290.49	360.02	69.53	1.2850	1.4877
344.0	3.24964	1.1051	4.0999	292.29	359.56	67.27	1.2900	1.4856
345.0	3.31548	1.1198	3.9535	294.14	359.04	64.90	1.2952	1.4833
346.0	3.38242	1.1358	3.8078	296.04	358.45	62.40	1.3005	1.4808
347.0	3.45048	1.1533	3.6622	298.01	357.77	59.75	1.3059	1.4781
348.0	3.51970	1.1725	3.5163	300.05	356.99	56.93	1.3116	1.4752
349.0	3.59010	1.1940	3.3694	302.18	356.09	53.90	1.3174	1.4719
350.0	3.66171	1.2182	3.2207	304.42	355.05	50.63	1.3236	1.4683
351.0	3.73457	1.2460	3.0690	306.80	353.83	47.03	1.3301	1.4641
352.0	3.80872	1.2786	2.9126	309.36	352.39	43.03	1.3371	1.4594
353.0	3.88422	1.3182	2.7489	312.16	350.64	38.47	1.3448	1.4538
354.0	3.96110	1.3686	2.5732	315.34	348.44	33.09	1.3535	1.4470
355.0	4.03945	1.4384	2.3758	319.17	345.50	26.32	1.3640	1.4382

Chart R502.2: Saturation properties

p MPa	T K	v' dm³/kg	v'' dm³/kg	h' kJ/kg	h'' kJ/kg	r kJ/kg	s' kJ/kgK	s'' kJ/kgK
0.0071	182.84	0.6175	1905.5	117.32	302.80	185.47	0.6388	1.6532
0.0072	183.03	0.6177	1880.8	117.45	302.89	185.44	0.6395	1.6527
0.0073	183.21	0.6179	1856.8	117.58	302.98	185.40	0.6402	1.6522
0.0074	183.40	0.6181	1833.4	117.71	303.08	185.37	0.6409	1.6517
0.0076	183.75	0.6185	1788.4	117.96	303.26	185.30	0.6423	1.6507
0.0078	184.11	0.6189	1745.6	118.21	303.43	185.22	0.6436	1.6497
0.0080	184.45	0.6193	1704.9	118.45	303.61	185.15	0.6450	1.6488
0.0082	184.78	0.6197	1666.1	118.69	303.78	185.09	0.6462	1.6479
0.0084	185.11	0.6201	1629.2	118.92	303.94	185.02	0.6475	1.6470
0.0086	185.44	0.6205	1593.8	119.16	304.10	184.95	0.6488	1.6461
0.0088	185.75	0.6208	1560.1	119.38	304.26	184.88	0.6500	1.6453
0.0090	186.06	0.6212	1527.8	119.61	304.42	184.81	0.6512	1.6445
0.0092	186.37	0.6215	1496.8	119.83	304.57	184.75	0.6524	1.6437
0.0094	186.67	0.6219	1467.1	120.04	304.72	184.68	0.6535	1.6429
0.0096	186.96	0.6222	1438.6	120.26	304.87	184.61	0.6547	1.6421
0.0098	187.25	0.6225	1411.3	120.47	305.02	184.55	0.6558	1.6414
0.0100	187.53	0.6229	1385.0	120.68	305.16	184.48	0.6569	1.6407
0.0102	187.81	0.6232	1359.7	120.88	305.30	184.42	0.6580	1.6399
0.0104	188.09	0.6235	1335.3	121.09	305.44	184.35	0.6591	1.6392
0.0106	188.36	0.6238	1311.8	121.29	305.58	184.29	0.6602	1.6386
0.0108	188.62	0.6241	1289.2	121.48	305.71	184.23	0.6612	1.6379
0.0110	188.89	0.6244	1267.3	121.68	305.84	184.17	0.6622	1.6372
0.0115	189.52	0.6252	1216.0	122.15	306.17	184.01	0.6647	1.6357
0.0120	190.14	0.6259	1168.7	122.62	306.48	183.86	0.6672	1.6342
0.0125	190.73	0.6266	1125.2	123.07	306.78	183.71	0.6695	1.6327
0.0130	191.31	0.6273	1084.8	123.50	307.07	183.57	0.6718	1.6314
0.0135	191.86	0.6279	1047.4	123.93	307.35	183.42	0.6740	1.6300
0.0140	192.40	0.6286	1012.5	124.34	307.62	183.28	0.6762	1.6288
0.0145	192.92	0.6292	980.00	124.75	307.89	183.14	0.6783	1.6276
0.0150	193.43	0.6298	949.57	125.14	308.15	183.01	0.6803	1.6264
0.0155	193.93	0.6304	921.04	125.52	308.40	182.87	0.6823	1.6253
0.0160	194.41	0.6309	894.23	125.90	308.64	182.74	0.6842	1.6242
0.0165	194.88	0.6315	868.99	126.27	308.88	182.61	0.6861	1.6232
0.0170	195.34	0.6320	845.19	126.63	309.11	182.48	0.6880	1.6222
0.0175	195.78	0.6326	822.70	126.98	309.34	182.35	0.6898	1.6212
0.0180	196.22	0.6331	801.42	127.33	309.56	182.23	0.6915	1.6202
0.0185	196.65	0.6336	781.25	127.67	309.77	182.11	0.6933	1.6193
0.0190	197.06	0.6341	762.11	128.00	309.98	181.99	0.6949	1.6184
0.0195	197.47	0.6346	743.91	128.32	310.19	181.87	0.6966	1.6176
0.0200	197.87	0.6351	726.60	128.64	310.39	181.75	0.6982	1.6168
0.0210	198.64	0.6360	694.36	129.27	310.79	181.52	0.7014	1.6151
0.0220	199.39	0.6369	664.96	129.87	311.16	181.29	0.7044	1.6136
0.0230	200.11	0.6378	638.02	130.45	311.53	181.07	0.7073	1.6122
0.0240	200.80	0.6386	613.26	131.02	311.88	180.86	0.7101	1.6108
0.0250	201.47	0.6394	590.41	131.57	312.22	180.65	0.7128	1.6095
0.0260	202.11	0.6402	569.26	132.10	312.54	180.44	0.7155	1.6083
0.0270	202.74	0.6410	549.63	132.62	312.86	180.24	0.7180	1.6071
0.0280	203.35	0.6418	531.35	133.12	313.17	180.05	0.7205	1.6059
0.0290	203.94	0.6425	514.29	133.62	313.47	179.85	0.7229	1.6048
0.0300	204.51	0.6432	498.32	134.10	313.76	179.66	0.7253	1.6038

Chart R502.2: Saturation properties–*continued*

p MPa	T K	v' dm^3/kg	v'' dm^3/kg	h' kJ/kg	h'' kJ/kg	r kJ/kg	s' kJ/kgK	s'' kJ/kgK
0.0310	205.07	0.6439	483.35	134.56	314.04	179.48	0.7276	1.6028
0.0320	205.62	0.6446	469.29	135.02	314.32	179.30	0.7298	1.6018
0.0330	206.15	0.6453	456.05	135.47	314.59	179.12	0.7320	1.6008
0.0340	206.67	0.6459	443.56	135.91	314.85	178.94	0.7341	1.5999
0.0350	207.17	0.6466	431.75	136.34	315.10	178.77	0.7361	1.5991
0.0360	207.67	0.6472	420.59	136.76	315.35	178.60	0.7382	1.5982
0.0370	208.15	0.6478	410.00	137.17	315.60	178.43	0.7401	1.5974
0.0380	208.62	0.6484	399.95	137.57	315.84	178.27	0.7421	1.5966
0.0390	209.08	0.6490	390.40	137.97	316.07	178.11	0.7440	1.5958
0.0400	209.54	0.6496	381.31	138.35	316.30	177.95	0.7458	1.5951
0.0410	209.98	0.6502	372.65	138.74	316.52	177.79	0.7476	1.5943
0.0420	210.42	0.6507	364.39	139.11	316.74	177.63	0.7494	1.5936
0.0430	210.84	0.6513	356.50	139.48	316.96	177.48	0.7512	1.5929
0.0440	211.26	0.6518	348.95	139.84	317.17	177.33	0.7529	1.5923
0.0450	211.67	0.6524	341.73	140.20	317.38	177.18	0.7546	1.5916
0.0460	212.08	0.6529	334.81	140.55	317.58	177.03	0.7562	1.5910
0.0470	212.47	0.6534	328.17	140.89	317.78	176.89	0.7578	1.5903
0.0480	212.86	0.6539	321.80	141.23	317.98	176.74	0.7594	1.5897
0.0490	213.25	0.6544	315.68	141.57	318.17	176.60	0.7610	1.5892
0.0500	213.62	0.6549	309.80	141.90	318.36	176.46	0.7625	1.5886
0.0510	213.99	0.6554	304.14	142.22	318.55	176.32	0.7641	1.5880
0.0520	214.36	0.6559	298.69	142.54	318.73	176.19	0.7656	1.5875
0.0530	214.72	0.6564	293.44	142.86	318.91	176.05	0.7670	1.5869
0.0540	215.07	0.6568	288.37	143.17	319.09	175.92	0.7685	1.5864
0.0550	215.42	0.6573	283.49	143.48	319.26	175.78	0.7699	1.5859
0.0560	215.77	0.6578	278.77	143.78	319.44	175.65	0.7713	1.5854
0.0570	216.11	0.6582	274.21	144.08	319.61	175.52	0.7727	1.5849
0.0580	216.44	0.6587	269.81	144.38	319.78	175.40	0.7741	1.5844
0.0590	216.77	0.6591	265.54	144.67	319.94	175.27	0.7754	1.5839
0.0600	217.10	0.6596	261.42	144.96	320.10	175.14	0.7767	1.5835
0.0610	217.42	0.6600	257.42	145.25	320.26	175.02	0.7780	1.5830
0.0620	217.73	0.6604	253.55	145.53	320.42	174.89	0.7793	1.5826
0.0630	218.05	0.6609	249.80	145.81	320.58	174.77	0.7806	1.5821
0.0640	218.36	0.6613	246.16	146.08	320.73	174.65	0.7819	1.5817
0.0650	218.66	0.6617	242.63	146.36	320.89	174.53	0.7831	1.5813
0.0660	218.96	0.6621	239.20	146.63	321.04	174.41	0.7843	1.5809
0.0670	219.26	0.6625	235.87	146.89	321.19	174.29	0.7856	1.5805
0.0680	219.55	0.6629	232.63	147.16	321.33	174.18	0.7868	1.5801
0.0690	219.85	0.6633	229.49	147.42	321.48	174.06	0.7879	1.5797
0.0700	220.13	0.6637	226.43	147.67	321.62	173.95	0.7891	1.5793
0.0710	220.42	0.6641	223.45	147.93	321.76	173.83	0.7903	1.5789
0.0720	220.70	0.6645	220.56	148.18	321.90	173.72	0.7914	1.5786
0.0730	220.98	0.6649	217.74	148.43	322.04	173.61	0.7925	1.5782
0.0740	221.25	0.6653	214.99	148.68	322.18	173.50	0.7937	1.5778
0.0760	221.79	0.6660	209.71	149.17	322.45	173.28	0.7959	1.5771
0.0780	222.32	0.6668	204.69	149.65	322.71	173.06	0.7980	1.5765
0.0800	222.84	0.6675	199.91	150.12	322.97	172.85	0.8001	1.5758
0.0820	223.35	0.6682	195.35	150.58	323.22	172.64	0.8022	1.5752
0.0840	223.85	0.6689	191.01	151.04	323.47	172.43	0.8042	1.5745
0.0860	224.34	0.6696	186.85	151.49	323.71	172.23	0.8062	1.5739

Chart R502.2: Saturation properties—*continued*

p MPa	T K	v' dm³/kg	v'' dm³/kg	h' kJ/kg	h'' kJ/kg	r kJ/kg	s' kJ/kgK	s'' kJ/kgK
0.0880	224.82	0.6703	182.89	151.93	323.95	172.03	0.8082	1.5734
0.0900	225.29	0.6710	179.09	152.36	324.19	171.83	0.8101	1.5728
0.0920	225.75	0.6716	175.45	152.78	324.42	171.63	0.8120	1.5722
0.0940	226.21	0.6723	171.96	153.20	324.64	171.44	0.8138	1.5717
0.0960	226.66	0.6729	168.60	153.62	324.86	171.25	0.8156	1.5712
0.0980	227.10	0.6736	165.39	154.02	325.08	171.06	0.8174	1.5707
0.1000	227.53	0.6742	162.29	154.42	325.29	170.87	0.8192	1.5702
0.1020	227.96	0.6748	159.31	154.82	325.51	170.69	0.8209	1.5697
0.1040	228.38	0.6755	156.45	155.21	325.71	170.51	0.8226	1.5692
0.1060	228.79	0.6761	153.68	155.59	325.92	170.32	0.8243	1.5687
0.1080	229.20	0.6767	151.02	155.97	326.12	170.15	0.8259	1.5683
0.1100	229.60	0.6773	148.45	156.35	326.31	169.97	0.8276	1.5678
0.1150	230.58	0.6787	142.39	157.26	326.80	169.54	0.8315	1.5668
0.1200	231.53	0.6801	136.83	158.15	327.26	169.11	0.8353	1.5657
0.1250	232.45	0.6815	131.69	159.01	327.71	168.70	0.8390	1.5648
0.1300	233.34	0.6829	126.94	159.85	328.14	168.30	0.8426	1.5639
0.1350	234.20	0.6842	122.52	160.66	328.56	167.90	0.8461	1.5630
0.1400	235.04	0.6855	118.41	161.45	328.97	167.52	0.8495	1.5622
0.1450	235.85	0.6868	114.57	162.23	329.36	167.14	0.8527	1.5614
0.1500	236.65	0.6880	110.98	162.98	329.75	166.76	0.8559	1.5606
0.1550	237.42	0.6892	107.61	163.72	330.12	166.40	0.8590	1.5599
0.1600	238.17	0.6904	104.44	164.44	330.48	166.04	0.8620	1.5592
0.1650	238.91	0.6916	101.46	165.15	330.84	165.69	0.8650	1.5585
0.1700	239.62	0.6927	98.651	165.84	331.18	165.34	0.8678	1.5579
0.1750	240.33	0.6939	95.994	166.52	331.52	165.00	0.8707	1.5572
0.1800	241.01	0.6950	93.478	167.18	331.85	164.67	0.8734	1.5566
0.1850	241.68	0.6961	91.094	167.83	332.17	164.34	0.8761	1.5560
0.1900	242.34	0.6972	88.830	168.47	332.48	164.01	0.8787	1.5555
0.1950	242.99	0.6982	86.679	169.10	332.79	163.69	0.8813	1.5549
0.2000	243.62	0.6993	84.630	169.71	333.09	163.37	0.8838	1.5544
0.2100	244.85	0.7013	80.816	170.92	333.67	162.75	0.8887	1.5534
0.2200	246.03	0.7034	77.335	172.08	334.23	162.15	0.8934	1.5525
0.2300	247.17	0.7053	74.146	173.20	334.77	161.56	0.8979	1.5516
0.2400	248.28	0.7072	71.213	174.30	335.28	160.98	0.9023	1.5507
0.2500	249.35	0.7091	68.506	175.36	335.78	160.42	0.9066	1.5499
0.2600	250.39	0.7109	65.999	176.40	336.26	159.87	0.9107	1.5492
0.2700	251.40	0.7127	63.671	177.40	336.73	159.33	0.9147	1.5484
0.2800	252.37	0.7145	61.504	178.39	337.18	158.80	0.9185	1.5478
0.2900	253.33	0.7162	59.480	179.34	337.62	158.28	0.9223	1.5471
0.3000	254.26	0.7179	57.586	180.28	338.05	157.77	0.9260	1.5465
0.3100	255.16	0.7196	55.810	181.20	338.46	157.26	0.9295	1.5459
0.3200	256.04	0.7212	54.140	182.09	338.86	156.77	0.9330	1.5453
0.3300	256.91	0.7229	52.568	182.97	339.25	156.28	0.9364	1.5447
0.3400	257.75	0.7245	51.085	183.83	339.63	155.80	0.9397	1.5442
0.3500	258.57	0.7261	49.683	184.68	340.00	155.33	0.9430	1.5437
0.3600	259.38	0.7276	48.356	185.51	340.37	154.86	0.9461	1.5432
0.3700	260.17	0.7291	47.099	186.32	340.72	154.40	0.9492	1.5427
0.3800	260.94	0.7307	45.905	187.12	341.06	153.95	0.9523	1.5422
0.3900	261.70	0.7322	44.770	187.90	341.40	153.50	0.9552	1.5418
0.4000	262.44	0.7336	43.689	188.67	341.73	153.06	0.9582	1.5414

Chart R502.2: Saturation properties—*continued*

p MPa	T K	v' dm^3/kg	v'' dm^3/kg	h' kJ/kg	h'' kJ/kg	r kJ/kg	s' kJ/kgK	s'' kJ/kgK
0.4100	263.17	0.7351	42.659	189.43	342.05	152.62	0.9610	1.5409
0.4200	263.89	0.7366	41.676	190.18	342.37	152.19	0.9638	1.5405
0.4300	264.59	0.7380	40.738	190.91	342.67	151.76	0.9666	1.5401
0.4400	265.29	0.7394	39.840	191.64	342.98	151.34	0.9693	1.5398
0.4500	265.97	0.7408	38.980	192.35	343.27	150.92	0.9719	1.5394
0.4600	266.63	0.7422	38.157	193.06	343.56	150.50	0.9746	1.5390
0.4700	267.29	0.7436	37.367	193.75	343.84	150.10	0.9771	1.5387
0.4800	267.94	0.7450	36.609	194.43	344.12	149.69	0.9797	1.5383
0.4900	268.57	0.7463	35.880	195.11	344.40	149.29	0.9822	1.5380
0.5000	269.20	0.7477	35.180	195.77	344.66	148.89	0.9846	1.5377
0.5100	269.82	0.7490	34.506	196.43	344.93	148.49	0.9870	1.5374
0.5200	270.43	0.7504	33.857	197.08	345.18	148.10	0.9894	1.5370
0.5300	271.03	0.7517	33.231	197.72	345.44	147.72	0.9917	1.5367
0.5400	271.62	0.7530	32.628	198.36	345.69	147.33	0.9940	1.5365
0.5500	272.20	0.7543	32.045	198.98	345.93	146.95	0.9963	1.5362
0.5600	272.78	0.7556	31.483	199.60	346.17	146.57	0.9986	1.5359
0.5700	273.35	0.7569	30.940	200.21	346.41	146.20	1.0008	1.5356
0.5800	273.91	0.7581	30.414	200.82	346.64	145.82	1.0030	1.5353
0.5900	274.46	0.7594	29.906	201.42	346.87	145.45	1.0051	1.5351
0.6000	275.01	0.7606	29.414	202.01	347.10	145.09	1.0072	1.5348
0.6100	275.55	0.7619	28.937	202.60	347.32	144.72	1.0093	1.5346
0.6200	276.08	0.7631	28.476	203.18	347.54	144.36	1.0114	1.5343
0.6300	276.61	0.7644	28.028	203.75	347.75	144.00	1.0135	1.5341
0.6400	277.13	0.7656	27.594	204.32	347.96	143.64	1.0155	1.5338
0.6500	277.64	0.7668	27.172	204.88	348.17	143.29	1.0175	1.5336
0.6600	278.15	0.7681	26.763	205.44	348.38	142.93	1.0195	1.5333
0.6700	278.66	0.7693	26.365	205.99	348.58	142.58	1.0214	1.5331
0.6800	279.15	0.7705	25.979	206.54	348.78	142.24	1.0234	1.5329
0.6900	279.65	0.7717	25.603	207.08	348.97	141.89	1.0253	1.5327
0.7000	280.13	0.7729	25.238	207.62	349.17	141.54	1.0272	1.5325
0.7100	280.61	0.7741	24.883	208.15	349.36	141.20	1.0291	1.5322
0.7200	281.09	0.7752	24.537	208.68	349.54	140.86	1.0309	1.5320
0.7300	281.56	0.7764	24.200	209.21	349.73	140.52	1.0327	1.5318
0.7400	282.03	0.7776	23.872	209.73	349.91	140.19	1.0346	1.5316
0.7600	282.95	0.7799	23.240	210.75	350.27	139.52	1.0381	1.5312
0.7800	283.85	0.7822	22.640	211.76	350.62	138.86	1.0416	1.5308
0.8000	284.74	0.7846	22.068	212.76	350.96	138.20	1.0451	1.5305
0.8200	285.61	0.7868	21.524	213.74	351.29	137.55	1.0485	1.5301
0.8400	286.46	0.7891	21.003	214.70	351.61	136.91	1.0518	1.5297
0.8600	287.30	0.7914	20.507	215.65	351.92	136.27	1.0550	1.5294
0.8800	288.12	0.7936	20.031	216.59	352.23	135.64	1.0583	1.5290
0.9000	288.93	0.7959	19.576	217.52	352.53	135.01	1.0614	1.5287
0.9200	289.73	0.7981	19.140	218.43	352.82	134.39	1.0645	1.5283
0.9400	290.51	0.8003	18.722	219.33	353.10	133.77	1.0676	1.5280
0.9600	291.29	0.8025	18.320	220.22	353.38	133.16	1.0706	1.5277
0.9800	292.05	0.8047	17.934	221.10	353.65	132.55	1.0735	1.5274
1.0000	292.80	0.8069	17.563	221.97	353.91	131.94	1.0764	1.5271
1.0200	293.54	0.8091	17.205	222.83	354.17	131.34	1.0793	1.5268
1.0400	294.26	0.8113	16.861	223.68	354.42	130.74	1.0822	1.5265
1.0600	294.98	0.8134	16.529	224.52	354.67	130.15	1.0850	1.5262

Chart R502.2: Saturation properties—*continued*

p MPa	T K	v' dm^3/kg	v'' dm^3/kg	h' kJ/kg	h'' kJ/kg	r kJ/kg	s' kJ/kgK	s'' kJ/kgK
1.0800	295.69	0.8156	16.209	225.35	354.91	129.56	1.0877	1.5259
1.1000	296.39	0.8178	15.900	226.18	355.14	128.97	1.0904	1.5256
1.1500	298.09	0.8231	15.172	228.20	355.71	127.51	1.0971	1.5249
1.2000	299.74	0.8285	14.501	230.17	356.24	126.07	1.1036	1.5242
1.2500	301.34	0.8339	13.882	232.09	356.74	124.65	1.1098	1.5235
1.3000	302.90	0.8392	13.308	233.98	357.21	123.24	1.1159	1.5228
1.3500	304.41	0.8445	12.775	235.82	357.66	121.84	1.1219	1.5221
1.4000	305.88	0.8499	12.277	237.63	358.08	120.45	1.1276	1.5214
1.4500	307.31	0.8552	11.812	239.41	358.48	119.07	1.1333	1.5208
1.5000	308.71	0.8606	11.376	241.15	358.85	117.70	1.1388	1.5201
1.5500	310.07	0.8660	10.967	242.86	359.20	116.34	1.1442	1.5194
1.6000	311.40	0.8714	10.582	244.55	359.53	114.99	1.1495	1.5188
1.6500	312.70	0.8768	10.218	246.21	359.84	113.63	1.1547	1.5181
1.7000	313.97	0.8823	9.8747	247.84	360.13	112.29	1.1598	1.5174
1.7500	315.21	0.8878	9.5496	249.45	360.40	110.94	1.1647	1.5167
1.8000	316.43	0.8933	9.2412	251.05	360.65	109.60	1.1696	1.5160
1.8500	317.63	0.8989	8.9482	252.62	360.88	108.26	1.1745	1.5153
1.9000	318.79	0.9046	8.6695	254.17	361.09	106.92	1.1792	1.5146
1.9500	319.94	0.9103	8.4040	255.70	361.28	105.58	1.1838	1.5138
2.0000	321.07	0.9161	8.1506	257.22	361.45	104.24	1.1884	1.5131
2.1000	323.25	0.9278	7.6769	260.20	361.75	101.54	1.1974	1.5115
2.2000	325.36	0.9399	7.2422	263.13	361.97	98.84	1.2062	1.5099
2.3000	327.40	0.9524	6.8414	266.01	362.12	96.11	1.2147	1.5082
2.4000	329.38	0.9653	6.4702	268.86	362.20	93.34	1.2231	1.5065
2.5000	331.29	0.9787	6.1248	271.66	362.21	90.55	1.2313	1.5046
2.6000	333.15	0.9926	5.8023	274.44	362.14	87.70	1.2393	1.5026
2.7000	334.95	1.0072	5.4997	277.19	361.99	84.80	1.2473	1.5004
2.8000	336.70	1.0226	5.2148	279.93	361.77	81.83	1.2551	1.4982
2.9000	338.40	1.0388	4.9455	282.67	361.45	78.78	1.2629	1.4957
3.0000	340.05	1.0561	4.6898	285.40	361.04	75.64	1.2707	1.4931
3.1000	341.66	1.0746	4.4460	288.14	360.54	72.39	1.2784	1.4903
3.2000	343.24	1.0946	4.2125	290.91	359.91	69.01	1.2862	1.4872
3.3000	344.77	1.1163	3.9877	293.70	359.17	65.47	1.2940	1.4838
3.4000	346.26	1.1402	3.7700	296.55	358.28	61.73	1.3019	1.4801
3.5000	347.72	1.1669	3.5576	299.47	357.22	57.75	1.3100	1.4760
3.6000	349.14	1.1972	3.3488	302.49	355.95	53.46	1.3183	1.4714
3.7000	350.53	1.2323	3.1411	305.66	354.43	48.77	1.3270	1.4661
3.8000	351.88	1.2745	2.9312	309.05	352.57	43.52	1.3363	1.4600
3.9000	353.21	1.3275	2.7138	312.78	350.23	37.44	1.3465	1.4525
4.0000	354.50	1.4001	2.4785	317.15	347.10	29.95	1.3585	1.4429

Chart R502.3: Thermodynamic properties

p/MPa	0.12			0.15			0.20		
T K	v dm^3/kg	h kJ/kg	s kJ/kgK	v dm^3/kg	h kJ/kg	s kJ/kgK	v dm^3/kg	h kJ/kg	s kJ/kgK
270.0	163.22	352.41	1.6662	129.66	351.90	1.6482	96.085	351.04	1.6244
280.0	169.85	359.16	1.6907	135.06	358.71	1.6729	100.25	357.95	1.6495
290.0	176.42	366.01	1.7148	140.40	365.61	1.6971	104.35	364.93	1.6740
300.0	182.94	372.97	1.7383	145.68	372.60	1.7208	108.40	371.99	1.6979
310.0	189.42	380.03	1.7615	150.92	379.70	1.7441	112.41	379.15	1.7214
320.0	195.87	387.20	1.7843	156.12	386.90	1.7670	116.37	386.40	1.7444
330.0	202.28	394.47	1.8066	161.30	394.20	1.7894	120.31	393.74	1.7670
340.0	208.67	401.85	1.8287	166.45	401.60	1.8115	124.22	401.19	1.7892
350.0	215.04	409.34	1.8504	171.58	409.11	1.8333	128.11	408.73	1.8111
360.0	221.39	416.94	1.8718	176.69	416.73	1.8547	131.98	416.38	1.8326
370.0	227.73	424.64	1.8929	181.78	424.44	1.8759	135.83	424.12	1.8538
380.0	234.05	432.44	1.9137	186.86	432.26	1.8967	139.67	431.96	1.8748
p/MPa	0.30			0.40			0.50		
270.0	62.443	349.24	1.5892	45.551	347.33	1.5624	35.348	345.28	1.5400
280.0	65.399	356.36	1.6151	47.920	354.70	1.5892	37.384	352.94	1.5678
290.0	68.279	363.52	1.6402	50.203	362.06	1.6150	39.323	360.52	1.5944
300.0	71.101	370.74	1.6647	52.421	369.44	1.6400	41.187	368.09	1.6201
310.0	73.874	378.02	1.6885	54.585	376.86	1.6644	42.993	375.66	1.6449
320.0	76.608	385.38	1.7119	56.708	384.33	1.6881	44.753	383.26	1.6690
330.0	79.309	392.82	1.7348	58.795	391.87	1.7113	46.475	390.90	1.6925
340.0	81.983	400.35	1.7573	60.854	399.48	1.7340	48.167	398.60	1.7155
350.0	84.634	407.96	1.7793	62.888	407.17	1.7563	49.833	406.37	1.7381
360.0	87.266	415.67	1.8010	64.902	414.94	1.7782	51.479	414.21	1.7601
370.0	89.881	423.46	1.8224	66.899	422.80	1.7997	53.106	422.12	1.7818
380.0	92.482	431.35	1.8435	68.881	430.74	1.8209	54.717	430.12	1.8031
p/MPa	0.60			0.80			1.00		
270.0	0.7491	196.64	0.9875	0.7485	196.67	0.9871	0.7478	196.70	0.9866
280.0	30.314	351.06	1.5491	0.7721	207.48	1.0264	0.7713	207.49	1.0259
290.0	32.037	358.91	1.5766	22.839	355.40	1.5459	0.7984	218.74	1.0653
300.0	33.674	366.68	1.6030	24.220	363.66	1.5739	18.465	360.30	1.5486
310.0	35.247	374.41	1.6283	25.519	371.78	1.6005	19.625	368.91	1.5769
320.0	36.769	382.15	1.6529	26.757	379.82	1.6261	20.708	377.33	1.6036
330.0	38.251	389.91	1.6768	27.947	387.83	1.6507	21.734	385.63	1.6291
340.0	39.701	397.70	1.7000	29.100	395.84	1.6746	22.716	393.88	1.6538
350.0	41.124	405.55	1.7228	30.222	403.87	1.6979	23.664	402.11	1.6776
360.0	42.524	413.46	1.7451	31.320	411.93	1.7206	24.584	410.33	1.7008
370.0	43.906	421.44	1.7669	32.398	420.04	1.7428	25.482	418.58	1.7234
380.0	45.272	429.49	1.7884	33.458	428.20	1.7646	26.362	426.87	1.7455

Chart R502.3: Thermodynamic properties—*continued*

p/MPa	2.00			4.00			6.00		
T K	v dm^3/kg	h kJ/kg	s kJ/kgK	v dm^3/kg	h kJ/kg	s kJ/kgK	v dm^3/kg	h kJ/kg	s kJ/kgK
270.0	0.7446	196.86	0.9845	0.7386	197.23	0.9804	0.7332	197.65	0.9765
280.0	0.7672	207.57	1.0234	0.7598	207.80	1.0188	0.7532	208.10	1.0145
290.0	0.7931	218.70	1.0625	0.7836	218.73	1.0571	0.7754	218.87	1.0523
300.0	0.8236	230.31	1.1018	0.8110	230.07	1.0956	0.8004	230.00	1.0900
310.0	0.8609	242.56	1.1420	0.8432	241.89	1.1343	0.8291	241.52	1.1278
320.0	0.9098	255.74	1.1838	0.8824	254.33	1.1738	0.8627	253.52	1.1658
330.0	8.9512	371.65	1.5444	0.9331	267.64	1.2148	0.9033	266.09	1.2045
340.0	9.7075	382.02	1.5754	1.0062	282.39	1.2588	0.9547	279.43	1.2443
350.0	10.377	391.81	1.6038	1.1490	300.88	1.3123	1.0246	293.91	1.2863
360.0	10.990	401.25	1.6304	3.4487	368.31	1.5024	1.1331	310.37	1.3327
370.0	11.561	410.47	1.6556	4.2005	385.92	1.5507	1.3517	331.27	1.3899
380.0	12.103	419.55	1.6798	4.7258	399.39	1.5867	1.8876	360.11	1.4668
p/MPa	8.00			10.00			12.00		
270.0	0.7282	198.11	0.9728	0.7236	198.61	0.9692	0.7193	199.13	0.9658
280.0	0.7472	208.46	1.0104	0.7417	208.86	1.0065	0.7367	209.31	1.0028
290.0	0.7681	219.10	1.0477	0.7615	219.39	1.0435	0.7555	219.74	1.0394
300.0	0.7913	230.06	1.0849	0.7832	230.21	1.0801	0.7760	230.44	1.0757
310.0	0.8174	241.36	1.1219	0.8074	241.33	1.1166	0.7986	241.41	1.1117
320.0	0.8472	253.04	1.1590	0.8345	252.78	1.1529	0.8235	252.68	1.1474
330.0	0.8819	265.16	1.1963	0.8652	264.59	1.1893	0.8514	264.25	1.1831
340.0	0.9233	277.81	1.2341	0.9006	276.80	1.2257	0.8828	276.16	1.2186
350.0	0.9744	291.11	1.2726	0.9423	289.48	1.2625	0.9186	288.42	1.2541
360.0	1.0402	305.27	1.3125	0.9922	302.69	1.2997	0.9600	301.08	1.2898
370.0	1.1303	320.62	1.3545	1.0538	316.53	1.3376	1.0084	314.17	1.3256
380.0	1.2622	337.61	1.3998	1.1316	331.13	1.3765	1.0658	327.73	1.3618

References for R502

[Barh66] BARHO, W.: Berechnung der Molwärme der Fluor-Chlor-Derivate von Methan und Ethan im Zustand idealer Gase; Dissertation Universität Karlsruhe (1966)

[Doer75] DÖRING, R.: Thermodynamic properties of azeotropic refrigerant compound R502 determined as a result of new experimental investigations; 14th International Congress of Refrigeration, Moskau (1975) 196/207

[DuPo63] - : Thermodynamic properties of Freon R502 Refrigerant, Freon Products Division; E.I. DuPont de Nemours & Co,Wilmington Delaware (1963)

[Kimu81] KIMURA, T. ; UEMATSU, M. AND WATANABE, K.: Measurements of compressibility factor and vapor pressure for refrigerant 502; J. Chem. Eng. Data 26(1981) 109/112

[KMR 81] - : Kältemaschinenregeln, 7. Auflage; Verlag C.F. Müller, Karlsruhe (1981)

[Long62] LONG, L.J.: The Effect of Pressure and Temperature on Azeotropic Composition in the Monochlorodifluoromethane/Monochloropentafluoroethane and Monochlorodifluoromethane/Propane Systems; MS Thesis in Chemistry, University of Delaware (1962)

[Mart70] MARTIN, J.J. AND DOWNING, R.C.: Thermodynamic properties of refrigerant 502; ASHRAE 76(1970)2 129/139

[Tani72] TANISHITA, I.; WATANABE, K.; KIJIMA, J.; OGUCHI, K. AND KONDO, H.: Status report on the thermodynamic properties research of fluorocarbon Refrigerants; Second report, Keio University, Tokyo (1972)

[Todo72] TODOROVIC, M.: Termicne i kaloricne velicne stanja freona-502, azeotropske mesavine freona-22 i freona-115 (I); Klimat. Grej.-Hlad. 1(1972)1 2/9

R 502. ph

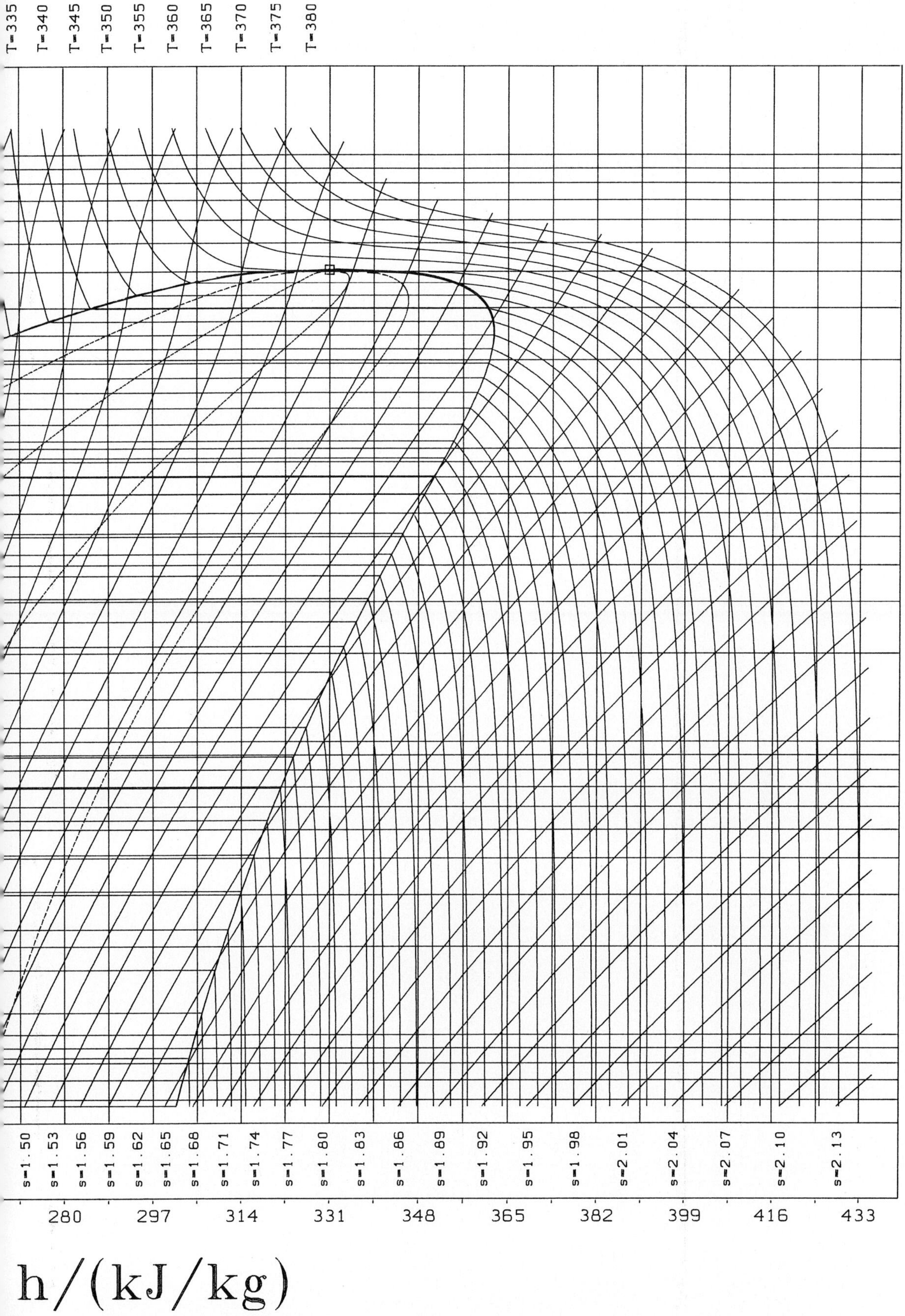

T=335
T=340
T=345
T=350
T=355
T=360
T=365
T=370
T=375
T=380
s=1.50
s=1.53
s=1.56
s=1.59
s=1.62
s=1.65
s=1.68
s=1.71
s=1.74
s=1.77
s=1.80
s=1.83
s=1.86
s=1.89
s=1.92
s=1.95
s=1.98
s=2.01
s=2.04
s=2.07
s=2.10
s=2.13
280
297
314
331
348
365
382
399
416
433
h/(kJ/kg)

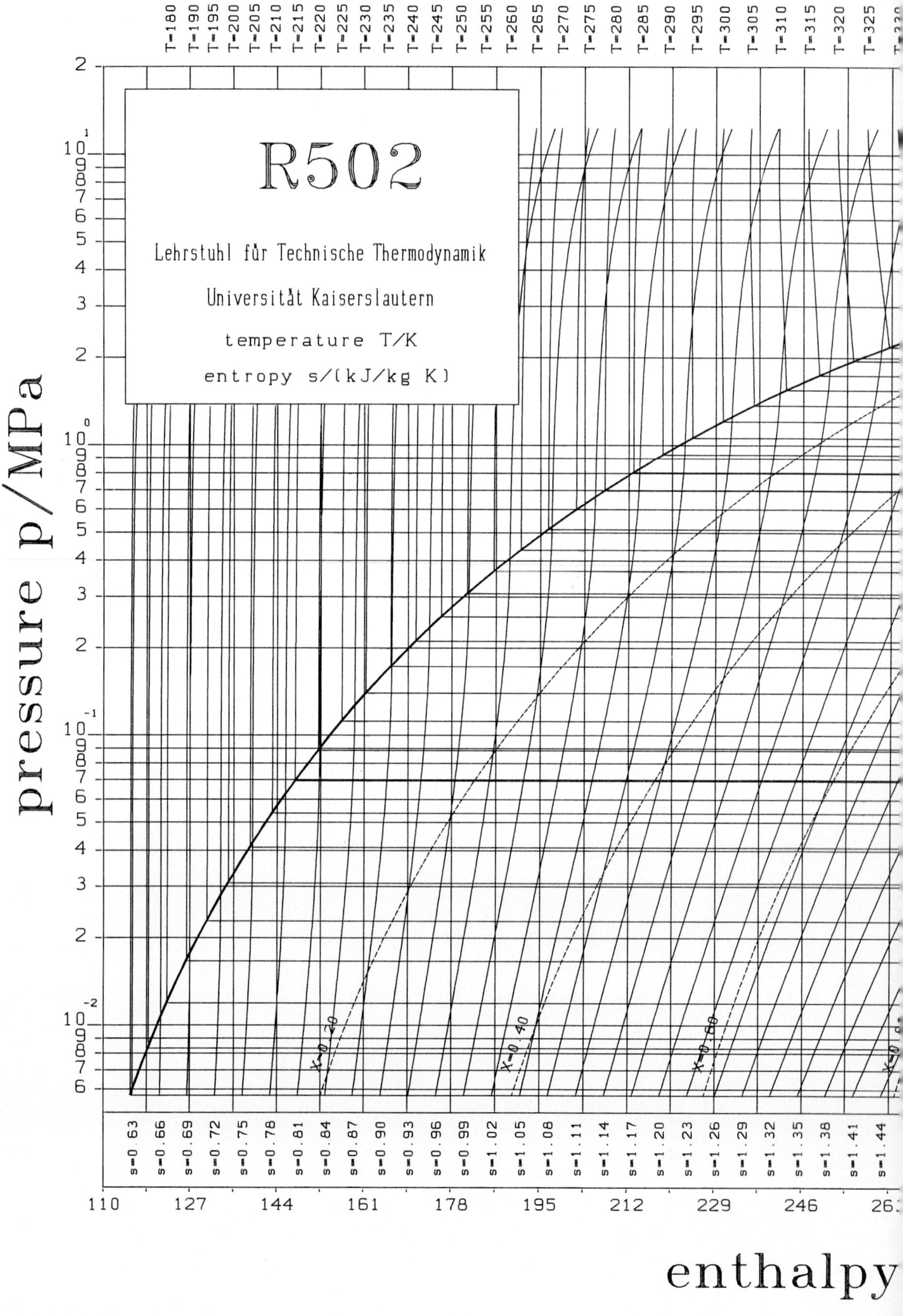
R502
Lehrstuhl für Technische Thermodynamik
Universität Kaiserslautern
temperature T/K
entropy s/(kJ/kg K)
pressure p/MPa
enthalpy
T=180 T=190 T=195 T=200 T=205 T=210 T=215 T=220 T=225 T=230 T=235 T=240 T=245 T=250 T=255 T=260 T=265 T=270 T=275 T=280 T=285 T=290 T=295 T=300 T=305 T=310 T=315 T=320 T=325
s=0.63 s=0.66 s=0.69 s=0.72 s=0.75 s=0.78 s=0.81 s=0.84 s=0.87 s=0.90 s=0.93 s=0.96 s=0.99 s=1.02 s=1.05 s=1.08 s=1.11 s=1.14 s=1.17 s=1.20 s=1.23 s=1.26 s=1.29 s=1.32 s=1.35 s=1.38 s=1.41 s=1.44
X=0.20 X=0.40 X=0.60
110 127 144 161 178 195 212 229 246

R 502. Ts

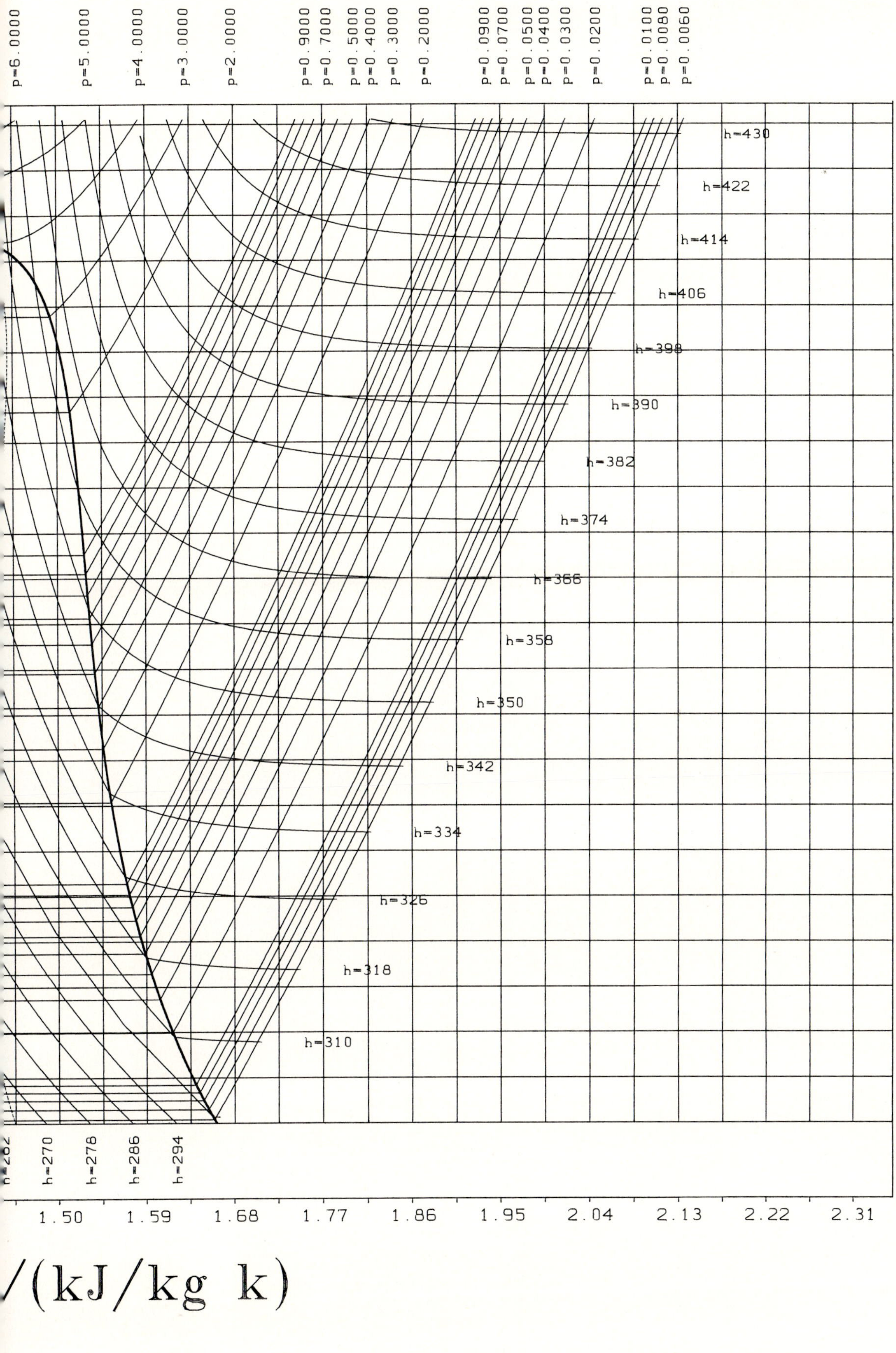

p=6.0000
p=5.0000
p=4.0000
p=3.0000
p=2.0000
p=0.9000
p=0.7000
p=0.5000
p=0.4000
p=0.3000
p=0.2000
p=0.0900
p=0.0700
p=0.0500
p=0.0400
p=0.0300
p=0.0200
p=0.0100
p=0.0080
p=0.0060
h=430
h=422
h=414
h=406
h=398
h=390
h=382
h=374
h=366
h=358
h=350
h=342
h=334
h=326
h=318
h=310
h=270
h=278
h=286
h=294
1.50
1.59
1.68
1.77
1.86
1.95
2.04
2.13
2.22
2.31
/(kJ/kg k)

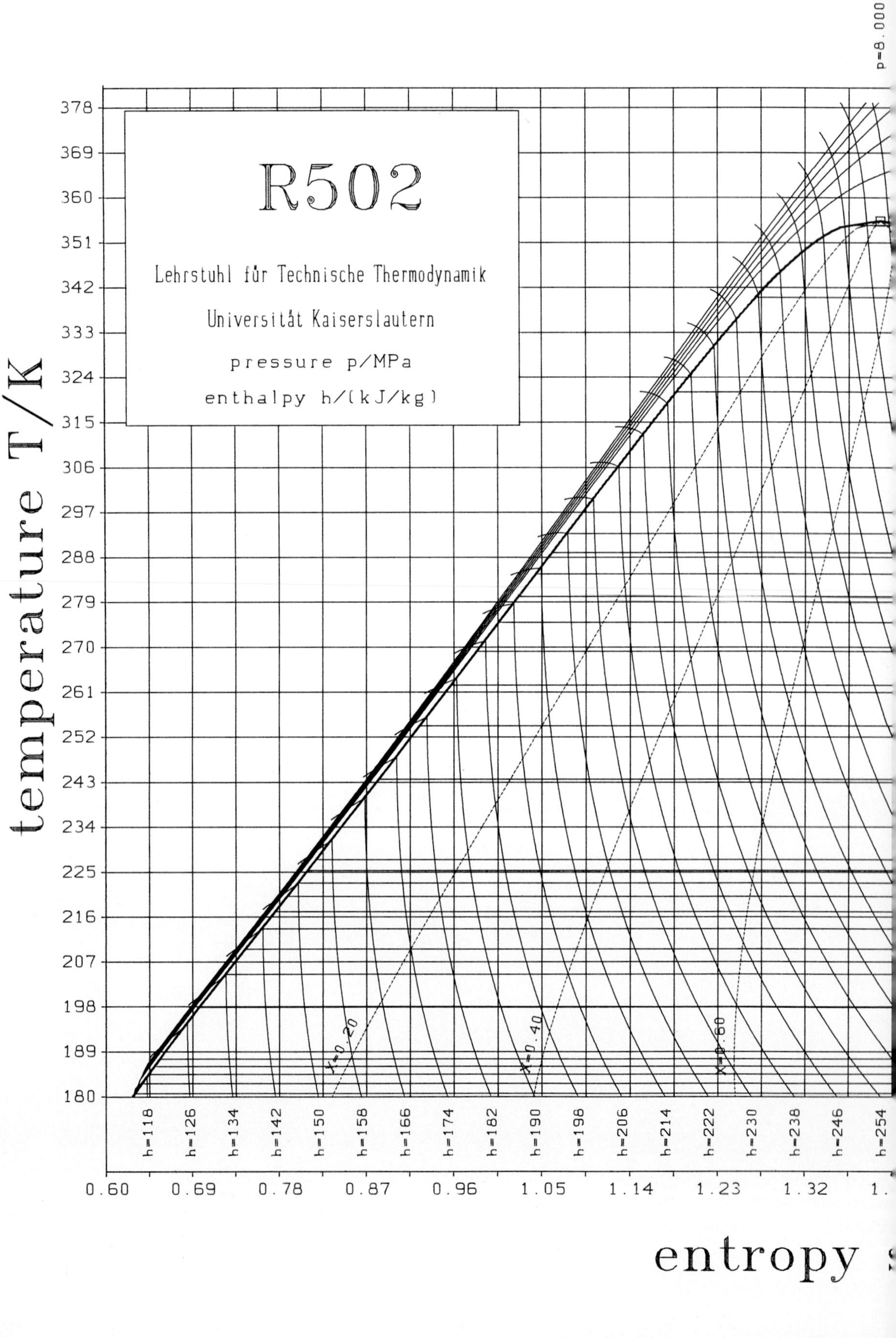

R502
Lehrstuhl für Technische Thermodynamik
Universität Kaiserslautern
pressure p/MPa
enthalpy h/(kJ/kg)
temperature T/K
p=8.0000
378
369
360
351
342
333
324
315
306
297
288
279
270
261
252
243
234
225
216
207
198
189
180
x=0.20
x=0.40
x=0.60
h=118
h=126
h=134
h=142
h=150
h=158
h=166
h=174
h=182
h=190
h=198
h=206
h=214
h=222
h=230
h=238
h=246
h=254
0.60
0.69
0.78
0.87
0.96
1.05
1.14
1.23
1.32
entropy

R 502. hs

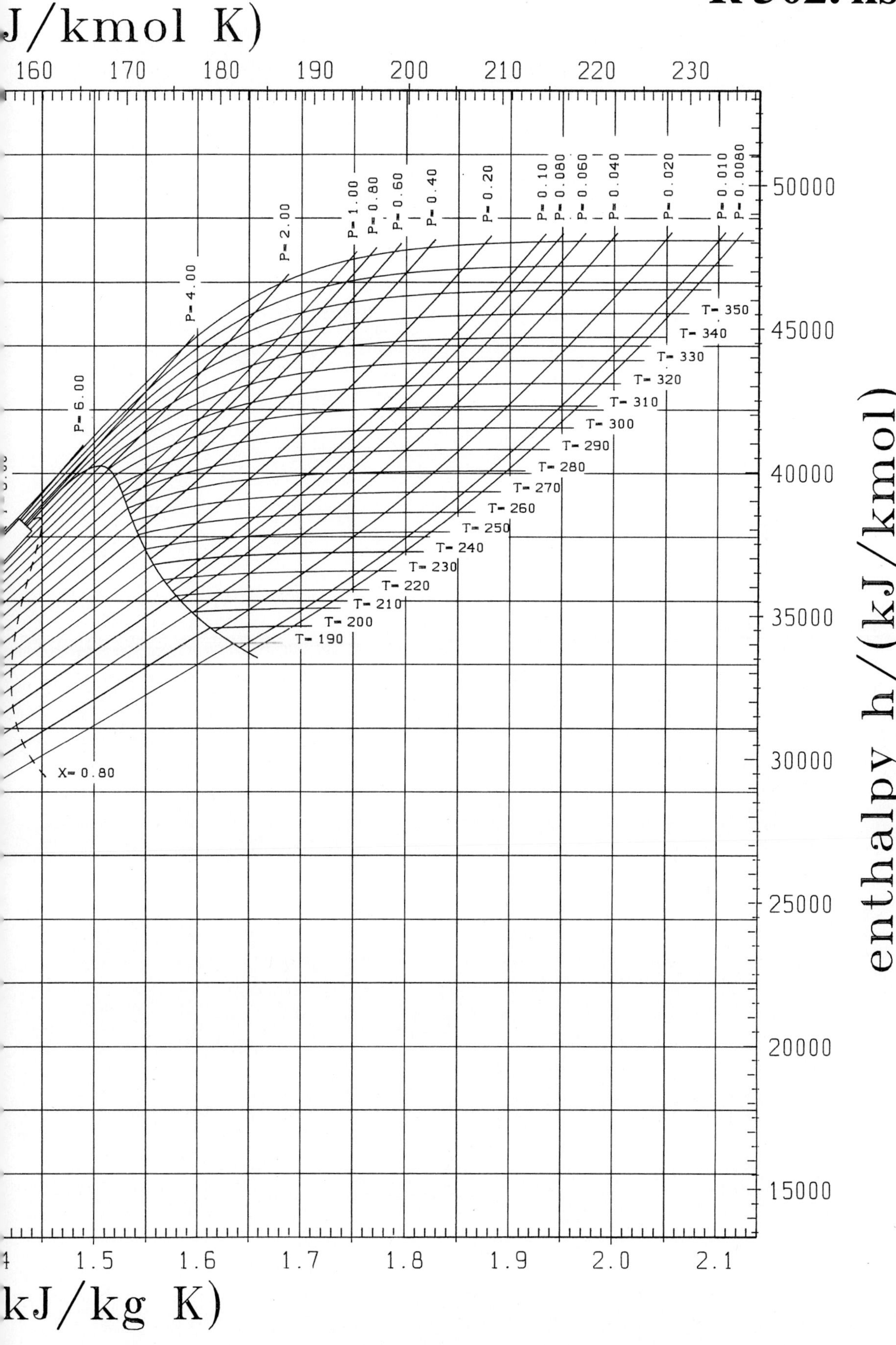
J/kmol K)
160
170
180
190
200
210
220
230
50000
45000
40000
35000
30000
25000
20000
15000
enthalpy h/(kJ/kmol)
P= 6.00
P= 4.00
P= 2.00
P= 1.00
P= 0.80
P= 0.60
P= 0.40
P= 0.20
P= 0.10
P= 0.080
P= 0.060
P= 0.040
P= 0.020
P= 0.010
P= 0.0080
T= 350
T= 340
T= 330
T= 320
T= 310
T= 300
T= 290
T= 280
T= 270
T= 260
T= 250
T= 240
T= 230
T= 220
T= 210
T= 200
T= 190
X= 0.80
1.5
1.6
1.7
1.8
1.9
2.0
2.1
kJ/kg K)

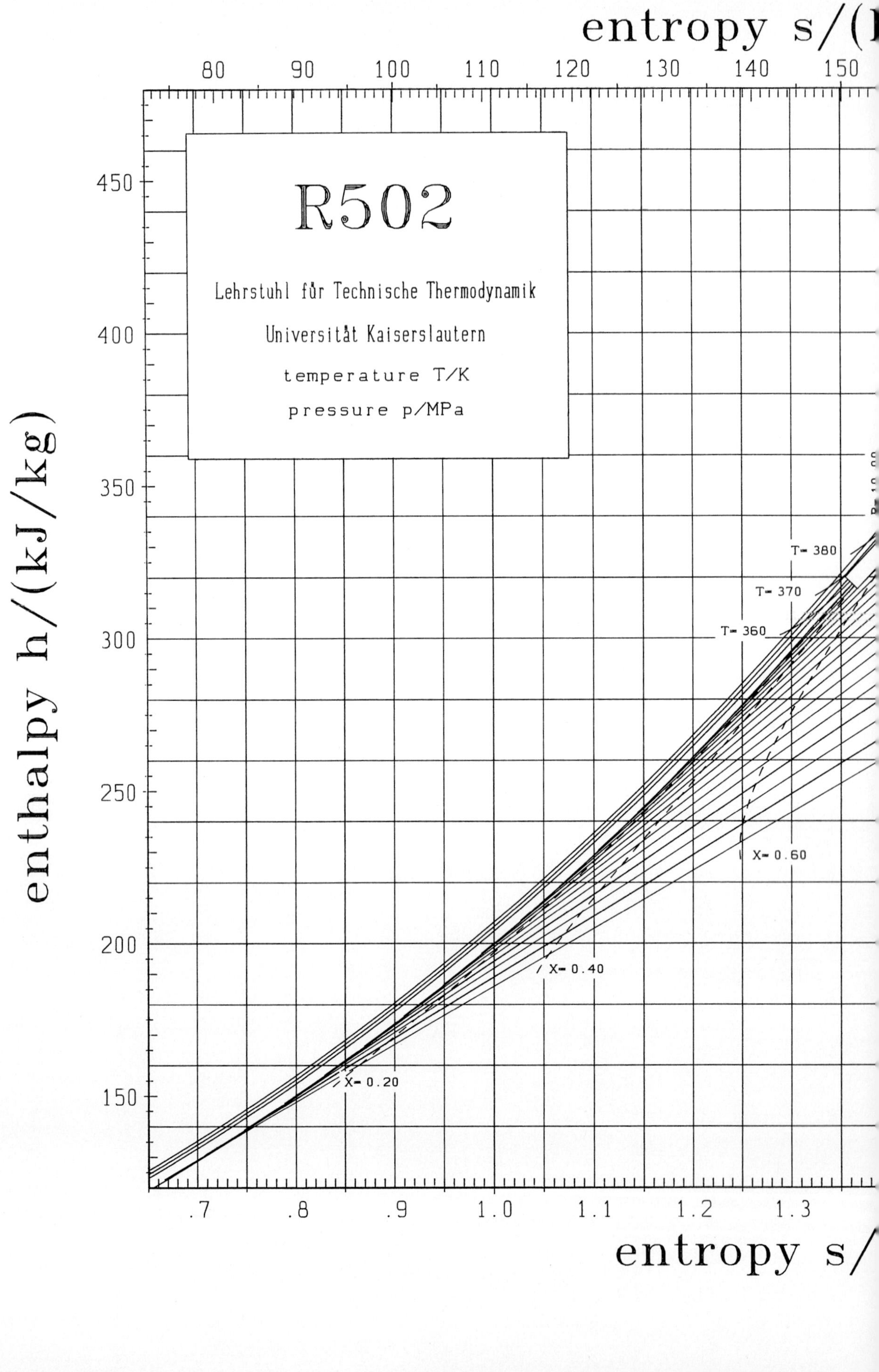
entropy s/(
80 90 100 110 120 130 140 150
R502
Lehrstuhl für Technische Thermodynamik
Universität Kaiserslautern
temperature T/K
pressure p/MPa
enthalpy h/(kJ/kg)
450
400
350
300
250
200
150
T= 380
T= 370
T= 360
X= 0.60
X= 0.40
X= 0.20
.7 .8 .9 1.0 1.1 1.2 1.3
entropy s/

Refrigerant R503 ($R13/R23$)

Table R503.1: List of references for p, ρ, T data

Author	n	p/MPa min	p/MPa max	ρ/g cm^{-3} min	ρ/g cm^{-3} max	T/K min	T/K max
Oguchi, K. *[Oguc 73]*	169	0.29	10.69	0.0155	0.76	213	372
Sinka, J.V. *[Sink 70]*	33	1.49	7.57	0.0523	0.61	298	492

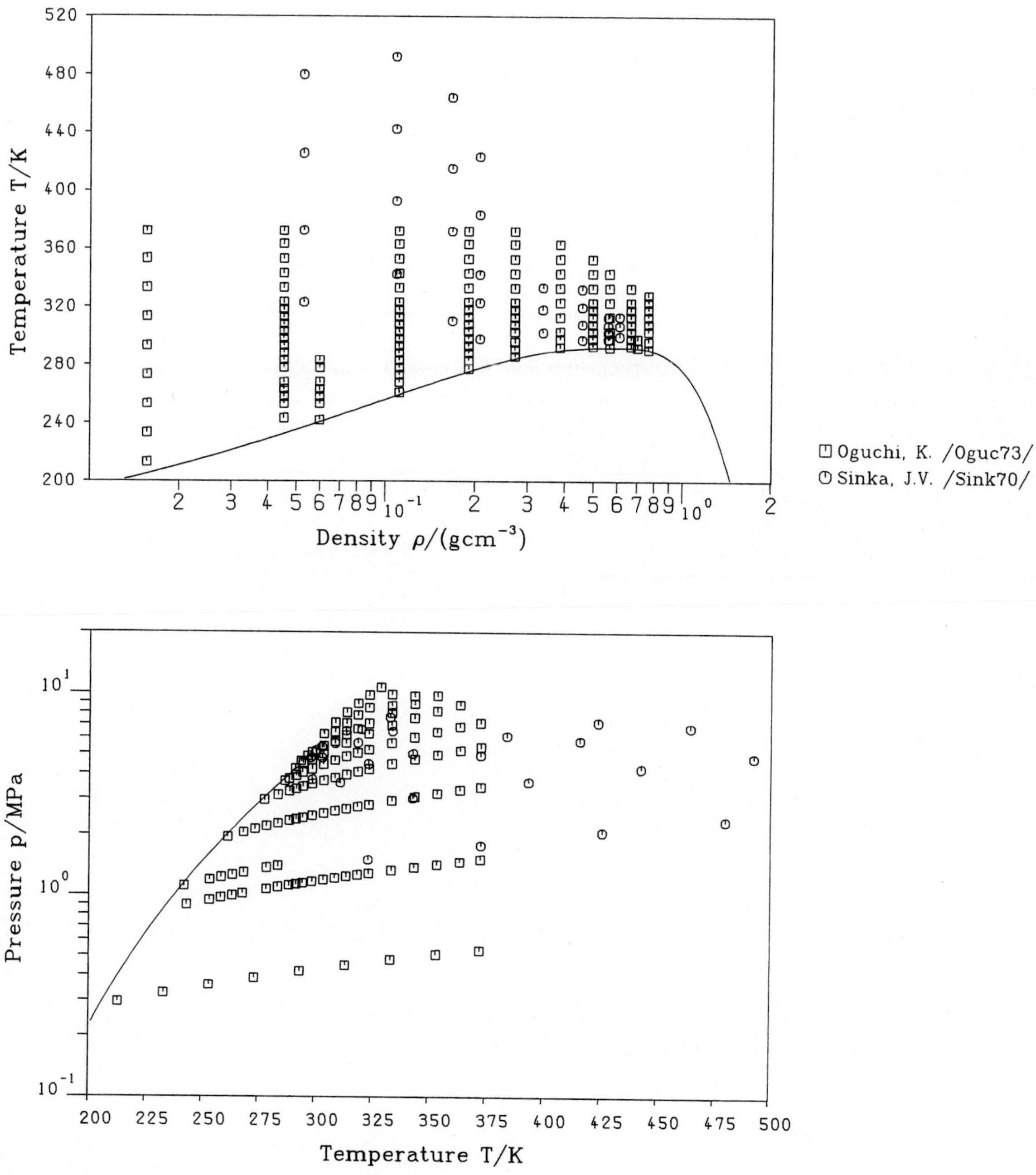

Fig.R503.1: Distribution of p, ρ, T data projected in the p, T and in the T, ρ plane

Table R503.2: Experimental vapor pressure data

Author	Δ / %		
	std	opt	*EOS*
SINKA, H.V. *[Sink 70]*	0.05	0.04	0.05

Table R503.3: Experimental saturated liquid density data

Author	Δ / %		
	std	opt	*EOS*
SINKA, H.V. *[Sink 70]*	0.11	0.02	0.11

Table R503.4: Coefficients of the standard vapor pressure equation

Index i	b_i
5	−6.922355 E+00
6	1.345650 E+00
9	−2.908168 E+00
15	1.902114 E−01

Table R503.5: Coefficients of the quasi-optimal vapor pressure equation

Index i	b_i
2	5.218030 E−03
3	−4.150070 E−02
5	−6.275198 E+00
13	−7.966459 E+00

Table R503.6: Coefficients of the standard saturated liquid density equation

Index i	c_i
8	1.788127 E+00
9	6.462856 E−01
16	2.296617 E+00

Table R503.7: Coefficients of the quasi-optimal saturated liquid density equation

Index i	c_i
1	1.159015 E−03
8	1.672146 E+00
9	8.998816 E−01

Table R503.8: Coefficients of the *Bender-EOS*

Range of validity: one-phase region

$0.29 \leq p/\mathrm{MPa} \leq 10.69$

$0.0155 \leq \rho/\mathrm{g\ cm^{-3}} \leq 0.76237$

$213.02 \leq T/\mathrm{K} \leq 492.3$

	Specific units	Molar units
a_{01}	6.572961414 E−02	5.007149495 E−01
a_{02}	4.606933847 E+01	3.509469329 E+02
a_{03}	1.083259296 E+03	8.252050935 E+03
a_{04}	2.868837561 E+06	2.185422618 E+07
a_{05}	−3.359364701 E+07	−2.559096305 E+08
a_{06}	1.923920925 E−01	1.279179323 E−01
a_{07}	−1.538325333 E+00	−1.022803969 E+00
a_{08}	6.724380393 E+03	4.470915748 E+03
a_{09}	−3.544978203 E−01	−2.057181071 E−02
a_{10}	4.817216001 E+01	2.795471510 E+00
a_{11}	1.850461543 E−01	9.372463369 E−04
a_{12}	−5.386616122 E+01	−2.728284869 E−01
a_{13}	2.005338534 E+01	8.864946618 E−03
a_{14}	−4.072049164 E+07	−2.707429930 E+07
a_{15}	2.090676551 E+10	1.390052045 E+10
a_{16}	−2.763915047 E+12	−1.837675829 E+12
a_{17}	7.759309200 E+07	3.930037970 E+05
a_{18}	−5.644474575 E+10	−2.858888443 E+08
a_{19}	1.010234139 E+13	5.116768031 E+10
a_{20}	4.000000000 E+00	3.047119360 E−02

Table R503.9: Deviations between experimental and predicted p, ρ, T data

Region	n	Δ / %	
		p	ρ
Gas	122	0.30	0.36
Critical	79	0.26	1.17
Liquid	0	–	–
Total	201	0.29	0.82

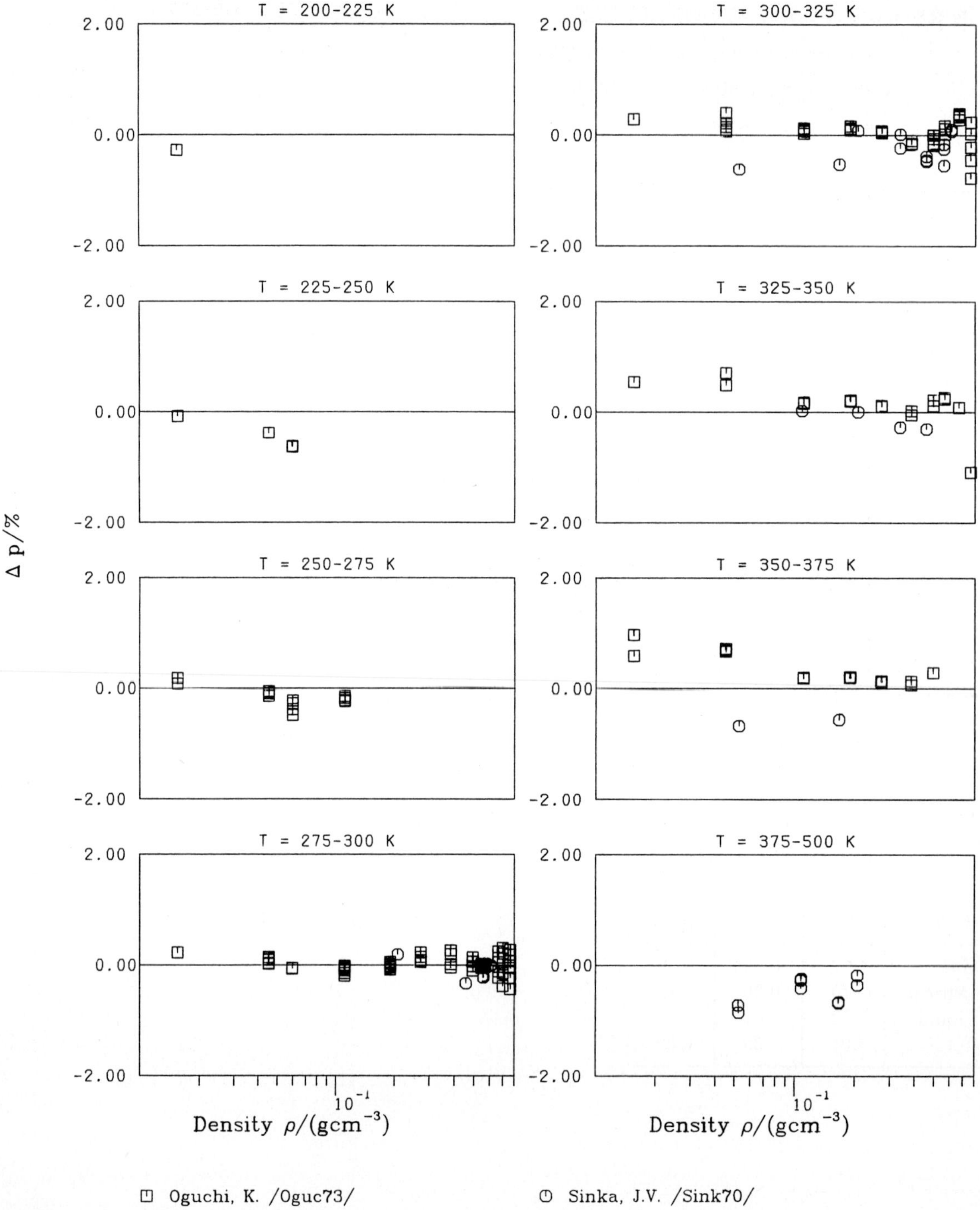

Fig.R503.2: Deviations between experimental and calculated p, ρ, T data

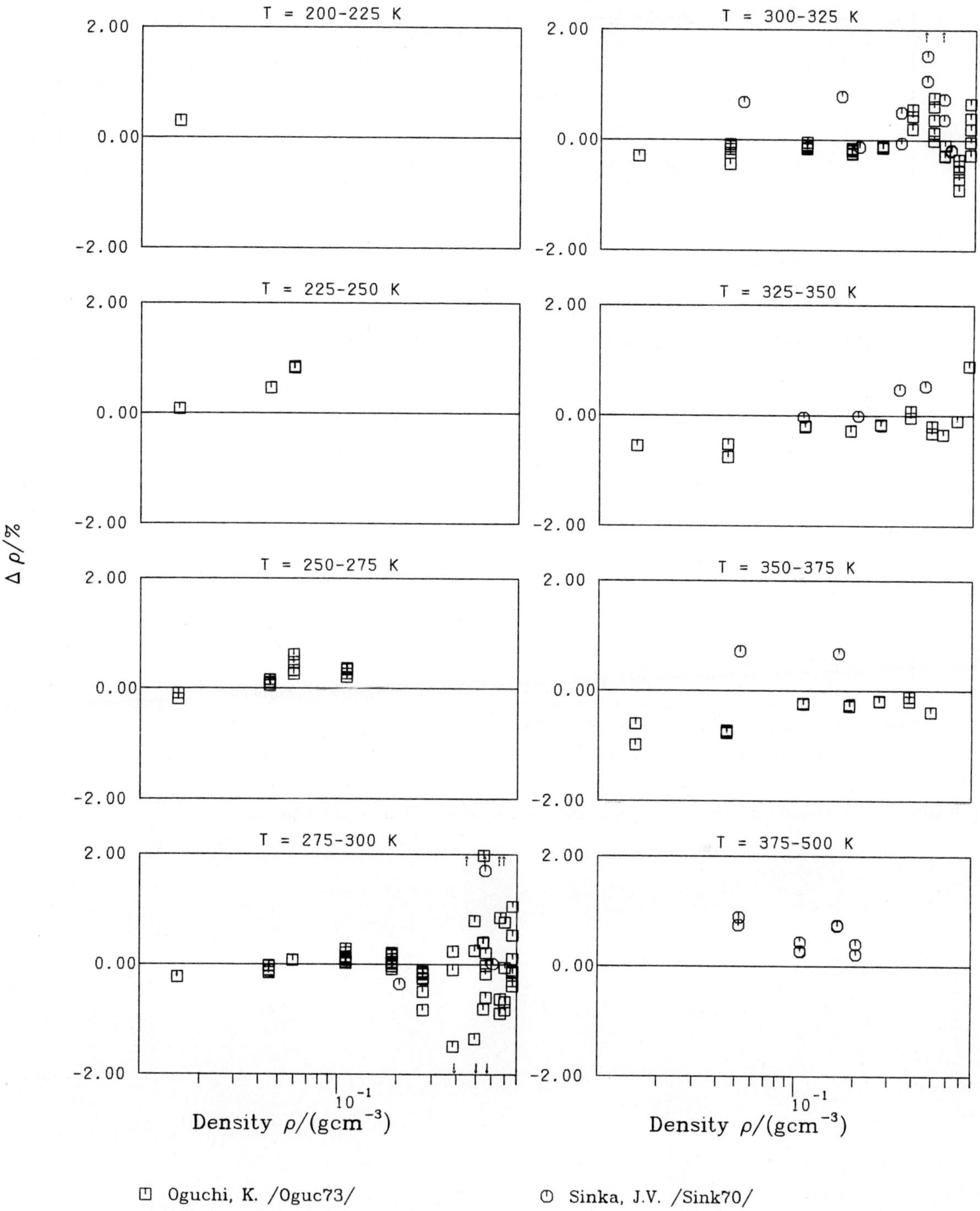

Fig.R503.2: Deviations between experimental and calculated p, ρ, T data *-continued*

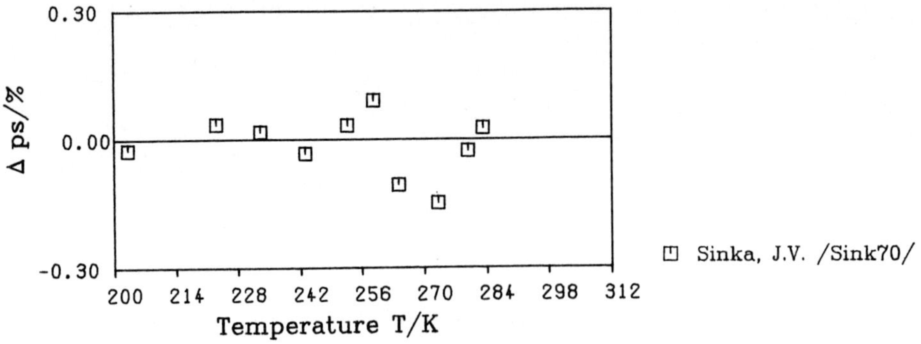

Fig.R503.3: Deviations between experimental and calculated vapor pressures

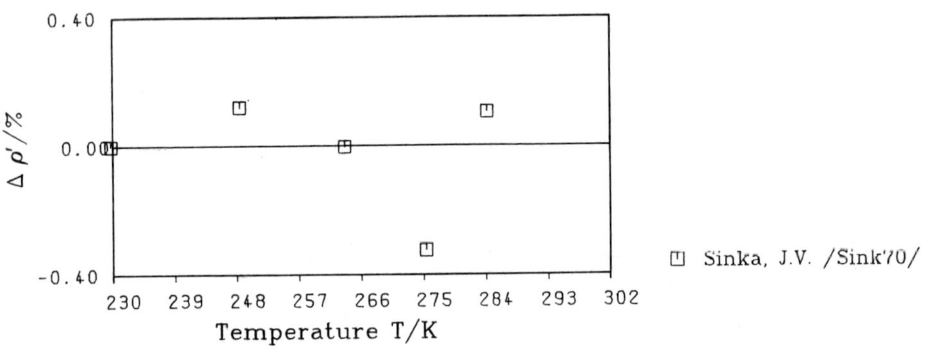

Fig.R503.4: Deviations between experimental and calculated saturated liquid densities

Table R503.10: Information on ideal gas state heat capacity equation

Author	T / K min	T / K max
CHEN, S.S. *[Chen 76]* RODGERS, A.S. *[Rodg 74]*	150	600

Coefficients of equation (15)	
Index i	d_i
1	2.600758 E−01
2	1.172641 E−03
3	2.185403 E−06
4	−5.909009 E−09
5	3.703543 E−12

Chart R503.1: Saturation properties

T K	p MPa	v' dm³/kg	v'' dm³/kg	h' kJ/kg	h'' kJ/kg	r kJ/kg	s' kJ/kgK	s'' kJ/kgK
205.0	0.27746	0.7007	64.651	116.24	281.94	165.70	0.6579	1.4662
206.0	0.29032	0.7031	61.915	117.34	282.29	164.95	0.6632	1.4639
207.0	0.30363	0.7056	59.318	118.44	282.63	164.19	0.6685	1.4617
208.0	0.31739	0.7081	56.853	119.55	282.97	163.42	0.6738	1.4595
209.0	0.33162	0.7106	54.512	120.65	283.30	162.65	0.6791	1.4573
210.0	0.34632	0.7131	52.287	121.77	283.63	161.87	0.6843	1.4551
211.0	0.36151	0.7157	50.172	122.88	283.96	161.08	0.6895	1.4530
212.0	0.37720	0.7182	48.159	124.00	284.28	160.29	0.6948	1.4508
213.0	0.39339	0.7208	46.244	125.11	284.60	159.49	0.7000	1.4488
214.0	0.41010	0.7234	44.421	126.24	284.92	158.68	0.7052	1.4467
215.0	0.42733	0.7261	42.684	127.36	285.23	157.87	0.7104	1.4446
216.0	0.44510	0.7287	41.028	128.49	285.53	157.04	0.7155	1.4426
217.0	0.46342	0.7314	39.449	129.62	285.83	156.21	0.7207	1.4406
218.0	0.48229	0.7341	37.943	130.75	286.13	155.38	0.7258	1.4386
219.0	0.50173	0.7368	36.506	131.88	286.42	154.54	0.7310	1.4366
220.0	0.52175	0.7395	35.134	133.02	286.71	153.68	0.7361	1.4346
221.0	0.54235	0.7423	33.823	134.16	286.99	152.83	0.7412	1.4327
222.0	0.56356	0.7451	32.571	135.30	287.26	151.96	0.7463	1.4308
223.0	0.58537	0.7480	31.373	136.45	287.54	151.09	0.7513	1.4289
224.0	0.60781	0.7508	30.228	137.60	287.80	150.20	0.7564	1.4270
225.0	0.63087	0.7537	29.132	138.75	288.06	149.31	0.7615	1.4251
226.0	0.65458	0.7566	28.084	139.90	288.32	148.42	0.7665	1.4232
227.0	0.67894	0.7596	27.079	141.06	288.57	147.51	0.7715	1.4213
228.0	0.70396	0.7626	26.117	142.22	288.81	146.59	0.7765	1.4195
229.0	0.72966	0.7656	25.196	143.38	289.05	145.67	0.7815	1.4176
230.0	0.75604	0.7686	24.312	144.55	289.29	144.74	0.7865	1.4158
231.0	0.78312	0.7717	23.464	145.72	289.51	143.80	0.7915	1.4140
232.0	0.81091	0.7749	22.651	146.89	289.73	142.85	0.7965	1.4122
233.0	0.83942	0.7780	21.871	148.06	289.95	141.89	0.8014	1.4104
234.0	0.86866	0.7813	21.121	149.24	290.16	140.92	0.8064	1.4086
235.0	0.89863	0.7845	20.402	150.42	290.36	139.94	0.8113	1.4068
236.0	0.92937	0.7878	19.710	151.61	290.55	138.95	0.8162	1.4050
237.0	0.96087	0.7912	19.046	152.79	290.74	137.94	0.8212	1.4032
238.0	0.99314	0.7946	18.407	153.99	290.92	136.93	0.8261	1.4014
239.0	1.02621	0.7980	17.792	155.18	291.09	135.91	0.8310	1.3996
240.0	1.06007	0.8015	17.201	156.38	291.26	134.88	0.8359	1.3979
241.0	1.09475	0.8051	16.631	157.59	291.42	133.83	0.8408	1.3961
242.0	1.13025	0.8087	16.083	158.79	291.57	132.78	0.8456	1.3943
243.0	1.16659	0.8124	15.556	160.01	291.71	131.71	0.8505	1.3925
244.0	1.20378	0.8161	15.047	161.22	291.85	130.62	0.8554	1.3907
245.0	1.24183	0.8199	14.557	162.44	291.97	129.53	0.8603	1.3889
246.0	1.28076	0.8238	14.084	163.67	292.09	128.42	0.8651	1.3872
247.0	1.32057	0.8277	13.629	164.90	292.20	127.30	0.8700	1.3854
248.0	1.36129	0.8317	13.189	166.14	292.30	126.16	0.8748	1.3836
249.0	1.40291	0.8358	12.764	167.38	292.39	125.01	0.8797	1.3817
250.0	1.44547	0.8400	12.355	168.62	292.47	123.84	0.8845	1.3799
251.0	1.48896	0.8442	11.959	169.88	292.54	122.66	0.8894	1.3781
252.0	1.53341	0.8486	11.577	171.14	292.60	121.46	0.8943	1.3763
253.0	1.57882	0.8530	11.207	172.40	292.65	120.25	0.8991	1.3744
254.0	1.62522	0.8575	10.850	173.67	292.69	119.01	0.9040	1.3725

Chart R503.1: Saturation properties—*continued*

T K	p MPa	v' dm^3/kg	v'' dm^3/kg	h' kJ/kg	h'' kJ/kg	r kJ/kg	s' kJ/kgK	s'' kJ/kgK
255.0	1.67261	0.8622	10.505	174.95	292.71	117.76	0.9088	1.3706
256.0	1.72101	0.8669	10.171	176.24	292.73	116.49	0.9137	1.3687
257.0	1.77043	0.8718	9.8472	177.53	292.73	115.20	0.9186	1.3668
258.0	1.82089	0.8768	9.5342	178.84	292.72	113.88	0.9235	1.3649
259.0	1.87241	0.8819	9.2310	180.15	292.69	112.55	0.9284	1.3629
260.0	1.92499	0.8872	8.9372	181.47	292.66	111.19	0.9333	1.3609
261.0	1.97866	0.8926	8.6525	182.80	292.60	109.81	0.9382	1.3589
262.0	2.03343	0.8982	8.3764	184.14	292.53	108.40	0.9431	1.3569
263.0	2.08931	0.9039	8.1086	185.49	292.45	106.96	0.9481	1.3548
264.0	2.14633	0.9098	7.8488	186.85	292.35	105.50	0.9531	1.3527
265.0	2.20450	0.9159	7.5965	188.22	292.23	104.01	0.9581	1.3505
266.0	2.26384	0.9222	7.3514	189.61	292.09	102.48	0.9631	1.3484
267.0	2.32436	0.9288	7.1133	191.01	291.93	100.92	0.9681	1.3461
268.0	2.38608	0.9356	6.8817	192.43	291.76	99.33	0.9732	1.3438
269.0	2.44903	0.9426	6.6565	193.86	291.56	97.70	0.9783	1.3415
270.0	2.51322	0.9499	6.4372	195.31	291.33	96.03	0.9835	1.3391
271.0	2.57867	0.9576	6.2237	196.77	291.09	94.31	0.9887	1.3367
272.0	2.64539	0.9655	6.0155	198.26	290.81	92.55	0.9939	1.3342
273.0	2.71343	0.9739	5.8125	199.77	290.51	90.74	0.9992	1.3316
274.0	2.78278	0.9826	5.6143	201.30	290.17	88.87	1.0046	1.3289
275.0	2.85349	0.9919	5.4207	202.86	289.80	86.94	1.0100	1.3261
276.0	2.92556	1.0016	5.2313	204.45	289.40	84.95	1.0155	1.3233
277.0	2.99903	1.0118	5.0458	206.07	288.95	82.88	1.0211	1.3203
278.0	3.07392	1.0228	4.8640	207.73	288.47	80.74	1.0268	1.3172
279.0	3.15026	1.0344	4.6855	209.42	287.93	78.51	1.0326	1.3140
280.0	3.22807	1.0469	4.5099	211.16	287.34	76.18	1.0385	1.3106
281.0	3.30739	1.0604	4.3370	212.95	286.68	73.74	1.0446	1.3070
282.0	3.38825	1.0749	4.1661	214.80	285.96	71.17	1.0508	1.3032
283.0	3.47069	1.0909	3.9968	216.71	285.16	68.45	1.0573	1.2992
284.0	3.55473	1.1084	3.8286	218.71	284.26	65.56	1.0640	1.2948
285.0	3.64043	1.1280	3.6607	220.80	283.26	62.46	1.0710	1.2902
286.0	3.72781	1.1501	3.4921	223.01	282.12	59.11	1.0784	1.2851
287.0	3.81694	1.1754	3.3217	225.36	280.81	55.45	1.0863	1.2795
288.0	3.90787	1.2051	3.1476	227.91	279.29	51.38	1.0948	1.2732
289.0	4.00064	1.2411	2.9673	230.72	277.48	46.76	1.1041	1.2659
290.0	4.09534	1.2866	2.7762	233.92	275.26	41.33	1.1148	1.2573
291.0	4.19204	1.3487	2.5657	237.75	272.36	34.61	1.1275	1.2464
292.0	4.29082	1.4468	2.3135	242.88	268.15	25.27	1.1446	1.2311

Chart R503.2: Saturation properties

p MPa	T K	v' dm³/kg	v'' dm³/kg	h' kJ/kg	h'' kJ/kg	r kJ/kg	s' kJ/kgK	s'' kJ/kgK
0.2700	204.40	0.6992	66.353	115.58	281.73	166.15	0.6547	1.4676
0.2800	205.20	0.7012	64.091	116.46	282.01	165.55	0.6590	1.4658
0.2900	205.98	0.7031	61.980	117.31	282.28	164.97	0.6631	1.4640
0.3000	206.73	0.7049	60.004	118.14	282.54	164.39	0.6671	1.4623
0.3100	207.47	0.7068	58.151	118.96	282.79	163.83	0.6710	1.4607
0.3200	208.19	0.7086	56.409	119.75	283.03	163.28	0.6748	1.4591
0.3300	208.89	0.7103	54.768	120.53	283.27	162.74	0.6785	1.4575
0.3400	209.57	0.7121	53.221	121.29	283.49	162.20	0.6821	1.4560
0.3500	210.25	0.7138	51.759	122.04	283.71	161.68	0.6856	1.4546
0.3600	210.90	0.7154	50.375	122.77	283.93	161.16	0.6890	1.4532
0.3700	211.55	0.7171	49.062	123.49	284.14	160.65	0.6924	1.4518
0.3800	212.18	0.7187	47.817	124.19	284.34	160.15	0.6957	1.4505
0.3900	212.79	0.7203	46.632	124.88	284.54	159.65	0.6989	1.4492
0.4000	213.40	0.7219	45.505	125.56	284.73	159.17	0.7021	1.4479
0.4100	213.99	0.7234	44.431	126.23	284.91	158.68	0.7052	1.4467
0.4200	214.58	0.7249	43.406	126.89	285.10	158.21	0.7082	1.4455
0.4300	215.15	0.7265	42.426	127.53	285.27	157.74	0.7112	1.4443
0.4400	215.72	0.7280	41.490	128.17	285.44	157.28	0.7141	1.4432
0.4500	216.27	0.7294	40.593	128.79	285.61	156.82	0.7169	1.4420
0.4600	216.82	0.7309	39.734	129.41	285.78	156.37	0.7197	1.4409
0.4700	217.35	0.7323	38.911	130.02	285.94	155.92	0.7225	1.4399
0.4800	217.88	0.7338	38.120	130.61	286.09	155.48	0.7252	1.4388
0.4900	218.40	0.7352	37.360	131.20	286.24	155.04	0.7279	1.4378
0.5000	218.91	0.7366	36.630	131.78	286.39	154.61	0.7305	1.4368
0.5100	219.42	0.7379	35.927	132.36	286.54	154.18	0.7331	1.4358
0.5200	219.91	0.7393	35.250	132.92	286.68	153.76	0.7356	1.4348
0.5300	220.40	0.7407	34.597	133.48	286.82	153.34	0.7381	1.4339
0.5400	220.89	0.7420	33.968	134.03	286.96	152.92	0.7406	1.4329
0.5500	221.36	0.7433	33.361	134.58	287.09	152.51	0.7430	1.4320
0.5600	221.83	0.7447	32.774	135.11	287.22	152.10	0.7454	1.4311
0.5700	222.30	0.7460	32.208	135.65	287.35	151.70	0.7478	1.4302
0.5800	222.76	0.7473	31.660	136.17	287.47	151.30	0.7501	1.4293
0.5900	223.21	0.7485	31.130	136.69	287.59	150.90	0.7524	1.4285
0.6000	223.66	0.7498	30.617	137.20	287.71	150.51	0.7547	1.4276
0.6100	224.10	0.7511	30.121	137.71	287.83	150.12	0.7569	1.4268
0.6200	224.53	0.7523	29.639	138.21	287.94	149.73	0.7591	1.4260
0.6300	224.96	0.7536	29.172	138.71	288.05	149.35	0.7613	1.4251
0.6400	225.39	0.7548	28.720	139.20	288.16	148.97	0.7634	1.4243
0.6500	225.81	0.7561	28.281	139.68	288.27	148.59	0.7655	1.4236
0.6600	226.22	0.7573	27.854	140.16	288.38	148.21	0.7676	1.4228
0.6700	226.64	0.7585	27.440	140.64	288.48	147.84	0.7697	1.4220
0.6800	227.04	0.7597	27.037	141.11	288.58	147.47	0.7717	1.4213
0.6900	227.45	0.7609	26.646	141.58	288.68	147.10	0.7737	1.4205
0.7000	227.84	0.7621	26.265	142.04	288.78	146.74	0.7757	1.4198
0.7100	228.24	0.7633	25.895	142.49	288.87	146.38	0.7777	1.4190
0.7200	228.63	0.7645	25.535	142.95	288.96	146.02	0.7797	1.4183
0.7300	229.01	0.7656	25.184	143.40	289.06	145.66	0.7816	1.4176
0.7400	229.40	0.7668	24.842	143.84	289.14	145.30	0.7835	1.4169
0.7600	230.15	0.7691	24.184	144.72	289.32	144.60	0.7873	1.4155
0.7800	230.89	0.7714	23.559	145.58	289.49	143.90	0.7909	1.4142

Chart R503.2: Saturation properties—*continued*

p MPa	T K	v' dm^3/kg	v'' dm^3/kg	h' kJ/kg	h'' kJ/kg	r kJ/kg	s' kJ/kgK	s'' kJ/kgK
0.8000	231.61	0.7736	22.964	146.43	289.65	143.22	0.7945	1.4129
0.8200	232.32	0.7759	22.397	147.26	289.80	142.54	0.7981	1.4116
0.8400	233.02	0.7781	21.855	148.09	289.95	141.87	0.8015	1.4103
0.8600	233.71	0.7803	21.338	148.89	290.09	141.20	0.8049	1.4091
0.8800	234.38	0.7825	20.843	149.69	290.23	140.54	0.8083	1.4079
0.9000	235.04	0.7847	20.370	150.47	290.37	139.89	0.8115	1.4067
0.9200	235.70	0.7868	19.916	151.25	290.49	139.25	0.8147	1.4055
0.9400	236.34	0.7890	19.481	152.01	290.62	138.61	0.8179	1.4044
0.9600	236.97	0.7911	19.063	152.76	290.73	137.97	0.8210	1.4033
0.9800	237.60	0.7932	18.662	153.50	290.85	137.34	0.8241	1.4021
1.0000	238.21	0.7953	18.276	154.24	290.96	136.72	0.8271	1.4010
1.0200	238.81	0.7974	17.904	154.96	291.06	136.10	0.8301	1.4000
1.0400	239.41	0.7994	17.547	155.67	291.16	135.49	0.8330	1.3989
1.0600	240.00	0.8015	17.202	156.38	291.26	134.88	0.8359	1.3979
1.0800	240.58	0.8036	16.869	157.08	291.35	134.28	0.8387	1.3968
1.1000	241.15	0.8056	16.548	157.77	291.44	133.68	0.8415	1.3958
1.1500	242.55	0.8107	15.793	159.46	291.65	132.19	0.8483	1.3933
1.2000	243.90	0.8157	15.097	161.10	291.83	130.73	0.8549	1.3909
1.2500	245.21	0.8207	14.455	162.70	292.00	129.30	0.8613	1.3886
1.3000	246.49	0.8257	13.861	164.27	292.14	127.88	0.8675	1.3863
1.3500	247.72	0.8306	13.308	165.80	292.27	126.48	0.8735	1.3841
1.4000	248.93	0.8355	12.793	167.29	292.38	125.09	0.8794	1.3819
1.4500	250.11	0.8404	12.312	168.76	292.48	123.72	0.8851	1.3797
1.5000	251.25	0.8453	11.862	170.19	292.56	122.36	0.8906	1.3776
1.5500	252.37	0.8502	11.439	171.60	292.62	121.02	0.8960	1.3756
1.6000	253.46	0.8551	11.042	172.98	292.67	119.68	0.9013	1.3735
1.6500	254.53	0.8600	10.667	174.34	292.70	118.36	0.9065	1.3715
1.7000	255.57	0.8649	10.313	175.68	292.72	117.04	0.9116	1.3696
1.7500	256.59	0.8698	9.9788	177.00	292.73	115.73	0.9166	1.3676
1.8000	257.59	0.8747	9.6618	178.30	292.72	114.43	0.9215	1.3657
1.8500	258.57	0.8797	9.3609	179.58	292.71	113.13	0.9263	1.3638
1.9000	259.53	0.8847	9.0749	180.84	292.68	111.83	0.9310	1.3619
1.9500	260.47	0.8897	8.8027	182.09	292.63	110.54	0.9356	1.3600
2.0000	261.39	0.8948	8.5432	183.32	292.58	109.26	0.9401	1.3581
2.1000	263.19	0.9050	8.0589	185.74	292.43	106.69	0.9490	1.3544
2.2000	264.92	0.9155	7.6156	188.12	292.24	104.12	0.9577	1.3507
2.3000	266.60	0.9261	7.2077	190.45	292.00	101.55	0.9661	1.3470
2.4000	268.22	0.9371	6.8310	192.74	291.71	98.97	0.9743	1.3433
2.5000	269.80	0.9484	6.4816	195.01	291.38	96.37	0.9824	1.3396
2.6000	271.32	0.9601	6.1561	197.25	291.00	93.75	0.9903	1.3359
2.7000	272.80	0.9722	5.8519	199.47	290.57	91.09	0.9982	1.3321
2.8000	274.25	0.9849	5.5664	201.68	290.09	88.40	1.0059	1.3282
2.9000	275.65	0.9981	5.2975	203.89	289.55	85.66	1.0135	1.3243
3.0000	277.01	1.0120	5.0434	206.09	288.95	82.86	1.0211	1.3203
3.1000	278.34	1.0267	4.8023	208.30	288.29	79.98	1.0287	1.3161
3.2000	279.64	1.0423	4.5726	210.53	287.56	77.03	1.0364	1.3118
3.3000	280.91	1.0591	4.3528	212.78	286.75	73.97	1.0440	1.3073
3.4000	282.14	1.0771	4.1417	215.07	285.85	70.79	1.0517	1.3026
3.5000	283.35	1.0968	3.9377	217.40	284.86	67.46	1.0596	1.2977
3.6000	284.53	1.1185	3.7395	219.80	283.75	63.94	1.0677	1.2924

Chart R503.2: Saturation properties–*continued*

p MPa	T K	v' dm^3/kg	v'' dm^3/kg	h' kJ/kg	h'' kJ/kg	r kJ/kg	s' kJ/kgK	s'' kJ/kgK
3.7000	285.68	1.1428	3.5455	222.29	282.50	60.20	1.0760	1.2868
3.8000	286.81	1.1703	3.3540	224.90	281.07	56.17	1.0847	1.2806
3.9000	287.91	1.2024	3.1627	227.68	279.43	51.75	1.0940	1.2737
4.0000	288.99	1.2408	2.9685	230.70	277.49	46.79	1.1041	1.2660
4.1000	290.05	1.2892	2.7665	234.09	275.13	41.04	1.1153	1.2568
4.2000	291.08	1.3549	2.5472	238.11	272.08	33.98	1.1287	1.2454
4.3000	292.09	1.4594	2.2862	243.47	267.63	24.16	1.1466	1.2293

Chart R503.3: Thermodynamic properties

p/MPa	0.29			0.35			0.40		
T K	v dm^3/kg	h kJ/kg	s kJ/kgK	v dm^3/kg	h kJ/kg	s kJ/kgK	v dm^3/kg	h kJ/kg	s kJ/kgK
210.0	63.609	284.83	1.4762	0.7131	121.77	0.6843	0.7130	121.77	0.6842
220.0	67.546	291.13	1.5056	55.085	290.05	1.4841	47.535	289.12	1.4682
230.0	71.363	297.45	1.5337	58.361	296.52	1.5128	50.492	295.72	1.4975
240.0	75.089	303.81	1.5607	61.539	302.99	1.5404	53.343	302.30	1.5255
250.0	78.746	310.24	1.5870	64.643	309.51	1.5670	56.116	308.90	1.5525
260.0	82.349	316.74	1.6125	67.690	316.10	1.5928	58.829	315.55	1.5785
270.0	85.909	323.34	1.6374	70.691	322.76	1.6179	61.494	322.26	1.6039
280.0	89.433	330.04	1.6617	73.656	329.51	1.6425	64.121	329.06	1.6286
290.0	92.928	336.84	1.6856	76.591	336.35	1.6665	66.718	335.94	1.6528
300.0	96.399	343.74	1.7090	79.500	343.30	1.6900	69.289	342.92	1.6764
310.0	99.850	350.75	1.7320	82.389	350.34	1.7131	71.839	350.00	1.6996
320.0	103.28	357.87	1.7546	85.260	357.49	1.7358	74.371	357.17	1.7224
330.0	106.70	365.10	1.7768	88.117	364.75	1.7582	76.888	364.45	1.7448
340.0	110.11	372.44	1.7987	90.960	372.11	1.7801	79.391	371.84	1.7668
350.0	113.50	379.89	1.8203	93.792	379.58	1.8018	81.883	379.32	1.7885
360.0	116.89	387.44	1.8416	96.614	387.15	1.8231	84.366	386.91	1.8099
370.0	120.26	395.10	1.8626	99.427	394.83	1.8442	86.839	394.60	1.8310
380.0	123.63	402.87	1.8833	102.23	402.61	1.8649	89.305	402.40	1.8518
390.0	126.99	410.73	1.9037	105.03	410.49	1.8854	91.764	410.29	1.8723
400.0	130.35	418.70	1.9239	107.83	418.47	1.9056	94.217	418.28	1.8925
410.0	133.70	426.76	1.9438	110.61	426.54	1.9255	96.665	426.36	1.9125
420.0	137.05	434.92	1.9635	113.40	434.71	1.9452	99.108	434.54	1.9322
430.0	140.39	443.17	1.9829	116.18	442.98	1.9646	101.55	442.82	1.9516
440.0	143.73	451.52	2.0021	118.95	451.33	1.9838	103.98	451.18	1.9709
450.0	147.06	459.95	2.0210	121.72	459.78	2.0028	106.41	459.63	1.9899
460.0	150.39	468.47	2.0398	124.49	468.31	2.0216	108.84	468.17	2.0086
470.0	153.72	477.08	2.0583	127.25	476.92	2.0401	111.27	476.79	2.0272
480.0	157.05	485.77	2.0766	130.02	485.61	2.0584	113.69	485.49	2.0455
490.0	160.37	494.53	2.0946	132.78	494.39	2.0765	116.11	494.27	2.0636

p/MPa	0.50			0.60			0.80		
210.0	0.7128	121.79	0.6839	0.7125	121.81	0.6837	0.7121	121.85	0.6832
220.0	36.914	287.15	1.4402	0.7393	133.03	0.7359	0.7387	133.05	0.7353
230.0	39.443	294.06	1.4709	32.036	292.29	1.4478	0.7685	144.55	0.7864
240.0	41.848	300.86	1.4999	34.157	299.36	1.4779	24.472	296.11	1.4403
250.0	44.163	307.64	1.5276	36.177	306.33	1.5064	26.147	303.57	1.4707
260.0	46.412	314.43	1.5542	38.122	313.28	1.5336	27.728	310.87	1.4994
270.0	48.610	321.26	1.5800	40.012	320.23	1.5598	29.243	318.11	1.5267
280.0	50.768	328.15	1.6050	41.859	327.23	1.5853	30.707	325.32	1.5529
290.0	52.893	335.12	1.6295	43.671	334.28	1.6100	32.133	332.56	1.5783
300.0	54.991	342.17	1.6534	45.455	341.40	1.6342	33.527	339.84	1.6030
310.0	57.067	349.30	1.6768	47.216	348.60	1.6578	34.897	347.17	1.6271
320.0	59.125	356.53	1.6997	48.958	355.88	1.6809	36.246	354.57	1.6505
330.0	61.166	363.86	1.7223	50.684	363.26	1.7036	37.578	362.04	1.6735
340.0	63.195	371.28	1.7444	52.396	370.72	1.7259	38.896	369.60	1.6961
350.0	65.211	378.81	1.7662	54.096	378.29	1.7478	40.201	377.24	1.7182

Chart R503.3: Thermodynamic properties—*continued*

p/MPa	0.50			0.60			0.80		
T K	v dm^3/kg	h kJ/kg	s kJ/kgK	v dm^3/kg	h kJ/kg	s kJ/kgK	v dm^3/kg	h kJ/kg	s kJ/kgK
360.0	67.218	386.43	1.7877	55.786	385.94	1.7694	41.495	384.96	1.7400
370.0	69.216	394.15	1.8089	57.467	393.69	1.7906	42.781	392.78	1.7614
380.0	71.206	401.97	1.8297	59.140	401.54	1.8115	44.058	400.68	1.7825
390.0	73.189	409.89	1.8503	60.806	409.48	1.8322	45.328	408.67	1.8032
400.0	75.166	417.90	1.8706	62.466	417.52	1.8525	46.591	416.75	1.8237
410.0	77.138	426.00	1.8906	64.120	425.64	1.8726	47.849	424.92	1.8439
420.0	79.105	434.20	1.9103	65.770	433.86	1.8924	49.102	433.18	1.8638
430.0	81.067	442.49	1.9298	67.415	442.17	1.9119	50.351	441.52	1.8834
440.0	83.026	450.87	1.9491	69.056	450.56	1.9312	51.596	449.95	1.9028
450.0	84.981	459.34	1.9681	70.694	459.05	1.9503	52.837	458.46	1.9219
460.0	86.933	467.89	1.9869	72.328	467.61	1.9691	54.074	467.05	1.9408
470.0	88.882	476.52	2.0055	73.960	476.26	1.9877	55.309	475.72	1.9594
480.0	90.828	485.23	2.0238	75.589	484.98	2.0061	56.541	484.47	1.9779
490.0	92.772	494.03	2.0420	77.215	493.78	2.0242	57.771	493.30	1.9961
p/MPa	1.00			2.00			4.00		
210.0	0.7116	121.88	0.6826	0.7094	122.07	0.6801	0.7052	122.47	0.6753
220.0	0.7381	133.08	0.7347	0.7352	133.21	0.7320	0.7296	133.53	0.7268
230.0	0.7676	144.56	0.7857	0.7636	144.61	0.7827	0.7563	144.79	0.7768
240.0	18.563	292.46	1.4073	0.7962	156.32	0.8325	0.7860	156.29	0.8258
250.0	20.069	300.55	1.4404	0.8353	168.47	0.8821	0.8205	168.10	0.8740
260.0	21.453	308.30	1.4707	0.8861	181.41	0.9328	0.8625	180.40	0.9222
270.0	22.755	315.86	1.4993	9.4773	302.04	1.3937	0.9183	193.58	0.9719
280.0	23.998	323.34	1.5265	10.391	311.75	1.4291	1.0066	208.77	1.0271
290.0	25.197	330.79	1.5526	11.201	320.77	1.4607	3.2062	282.08	1.2818
300.0	26.362	338.23	1.5779	11.946	329.40	1.4900	4.3130	303.59	1.3549
310.0	27.499	345.71	1.6024	12.645	337.81	1.5176	4.9768	317.14	1.3994
320.0	28.615	353.23	1.6263	13.311	346.09	1.5438	5.5114	328.59	1.4357
330.0	29.712	360.81	1.6496	13.951	354.30	1.5691	5.9782	339.03	1.4679
340.0	30.794	368.45	1.6724	14.571	362.48	1.5935	6.4024	348.91	1.4973
350.0	31.863	376.17	1.6948	15.175	370.66	1.6172	6.7969	358.43	1.5249
360.0	32.920	383.97	1.7168	15.765	378.86	1.6403	7.1694	367.72	1.5511
370.0	33.969	391.85	1.7383	16.343	387.09	1.6629	7.5249	376.87	1.5762
380.0	35.008	399.81	1.7596	16.912	395.36	1.6850	7.8669	385.93	1.6004
390.0	36.041	407.85	1.7805	17.472	403.69	1.7066	8.1978	394.94	1.6237
400.0	37.067	415.98	1.8010	18.025	412.07	1.7278	8.5195	403.91	1.6465
410.0	38.088	424.19	1.8213	18.572	420.51	1.7486	8.8334	412.88	1.6686
420.0	39.103	432.49	1.8413	19.113	429.01	1.7691	9.1406	421.86	1.6903
430.0	40.114	440.87	1.8610	19.649	437.58	1.7893	9.4421	430.86	1.7114
440.0	41.121	449.33	1.8805	20.181	446.22	1.8091	9.7385	439.88	1.7322
450.0	42.124	457.87	1.8997	20.709	454.92	1.8287	10.030	448.93	1.7525
460.0	43.124	466.49	1.9186	21.234	463.69	1.8480	10.318	458.02	1.7725
470.0	44.120	475.19	1.9373	21.755	472.52	1.8670	10.603	467.15	1.7921
480.0	45.114	483.97	1.9558	22.274	481.42	1.8857	10.884	476.32	1.8114
490.0	46.106	492.81	1.9740	22.789	490.39	1.9042	11.163	485.54	1.8304

Chart R503.3: Thermodynamic properties–*continued*

p/MPa	6.00			8.00			10.00		
T K	v dm^3/kg	h kJ/kg	s kJ/kgK	v dm^3/kg	h kJ/kg	s kJ/kgK	v dm^3/kg	h kJ/kg	s kJ/kgK
210.0	0.7012	122.91	0.6707	0.6974	123.37	0.6663	0.6939	123.87	0.6620
220.0	0.7245	133.89	0.7218	0.7198	134.30	0.7171	0.7153	134.75	0.7126
230.0	0.7496	145.06	0.7714	0.7435	145.38	0.7664	0.7379	145.76	0.7616
240.0	0.7771	156.40	0.8197	0.7692	156.60	0.8141	0.7620	156.89	0.8089
250.0	0.8082	167.96	0.8669	0.7976	167.99	0.8606	0.7882	168.13	0.8548
260.0	0.8445	179.85	0.9135	0.8298	179.60	0.9061	0.8174	179.54	0.8996
270.0	0.8892	192.26	0.9604	0.8678	191.54	0.9512	0.8507	191.17	0.9435
280.0	0.9492	205.60	1.0089	0.9148	204.02	0.9966	0.8900	203.15	0.9870
290.0	1.0426	220.89	1.0625	0.9775	217.42	1.0436	0.9386	215.65	1.0309
300.0	1.2501	241.83	1.1334	1.0710	232.50	1.0947	1.0024	229.00	1.0761
310.0	1.9287	276.29	1.2463	1.2338	250.80	1.1546	1.0924	243.70	1.1243
320.0	2.6801	302.03	1.3281	1.5213	272.83	1.2246	1.2267	260.31	1.1770
330.0	3.2036	318.75	1.3796	1.8945	294.38	1.2909	1.4170	278.44	1.2328
340.0	3.6184	332.20	1.4198	2.2622	312.58	1.3453	1.6459	296.40	1.2864
350.0	3.9743	344.11	1.4543	2.5899	327.87	1.3896	1.8856	312.99	1.3345
360.0	4.2931	355.14	1.4854	2.8804	341.29	1.4274	2.1190	328.01	1.3768
370.0	4.5863	365.62	1.5141	3.1422	353.53	1.4609	2.3394	341.71	1.4144
380.0	4.8607	375.75	1.5411	3.3827	365.00	1.4915	2.5459	354.43	1.4483
390.0	5.1206	385.63	1.5668	3.6069	375.95	1.5200	2.7398	366.42	1.4795
400.0	5.3690	395.35	1.5914	3.8182	386.54	1.5468	2.9228	377.87	1.5085
410.0	5.6080	404.96	1.6151	4.0192	396.88	1.5723	3.0967	388.94	1.5358
420.0	5.8391	414.49	1.6381	4.2117	407.03	1.5968	3.2628	399.72	1.5618
430.0	6.0636	423.98	1.6604	4.3972	417.05	1.6204	3.4223	410.28	1.5866
440.0	6.2825	433.43	1.6822	4.5768	426.98	1.6432	3.5761	420.68	1.6105
450.0	6.4964	442.88	1.7034	4.7512	436.85	1.6654	3.7251	430.96	1.6336
460.0	6.7060	452.32	1.7242	4.9211	446.66	1.6870	3.8697	441.15	1.6560
470.0	6.9119	461.77	1.7445	5.0871	456.45	1.7080	4.0106	451.28	1.6778
480.0	7.1144	471.24	1.7644	5.2497	466.22	1.7286	4.1482	461.36	1.6990
490.0	7.3139	480.73	1.7840	5.4092	475.99	1.7487	4.2829	471.40	1.7197

References for R503

[Chen76] CHEN, S.S.; WILHOIT, R.C. AND ZWOLINSKI, B.J.: Ideal Gas Thermodynamic Properties of Six Chlorofluoromethanes; J. Phys. Chem. Ref. Data 5(1976)3 571/580

[Oguc73] OGUCHI, K. ; TANISHITA, I. ; WATANABE, K. ; YAMAGUCHI, T. ; SASAYAMA, A. ; MATSUSHITA, Y. AND OTAKI, S.: Pressure-Volume-Temperature Properties of R-503; Institut International dux Froid Bulletin, Commission B1 (1973/1974) 25/31

[Oguc77] OGUCHI, K.: Physical properties of refrigerants. PVT relationship; Reito 52(1977) 869/889

[Rodg74] RODGERS, A.S.; CHAO, J.; WILHOIT, R.C. AND ZWOLINSKI, B.J.: Ideal Gas Thermodynamic Properties of Eight Chloro- and Fluoromethanes; J. Phys. Chem. Ref. Data 3(1974)1 117/140

[Sink70] SINKA, J.V. ; ROSENTHAL, E. AND DIXON, R.P.: Pressure-Volume-Temperature Relationship for a Mixture of Monochlorotrifluoromethane and Trifluoromethane; J. Chem. Eng. Data 15(1970)1 73/74

R 503. ph

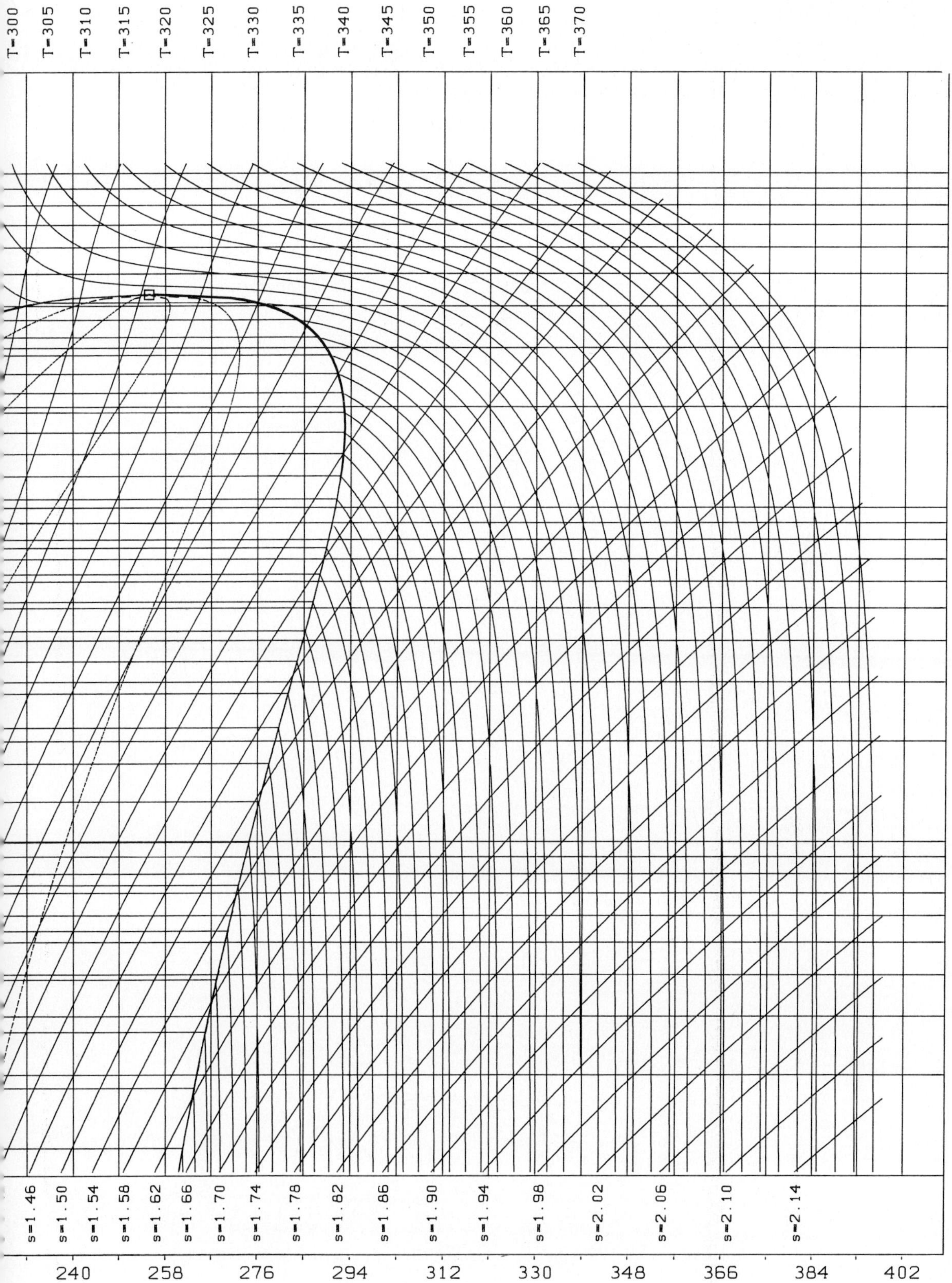

h/(kJ/kg)

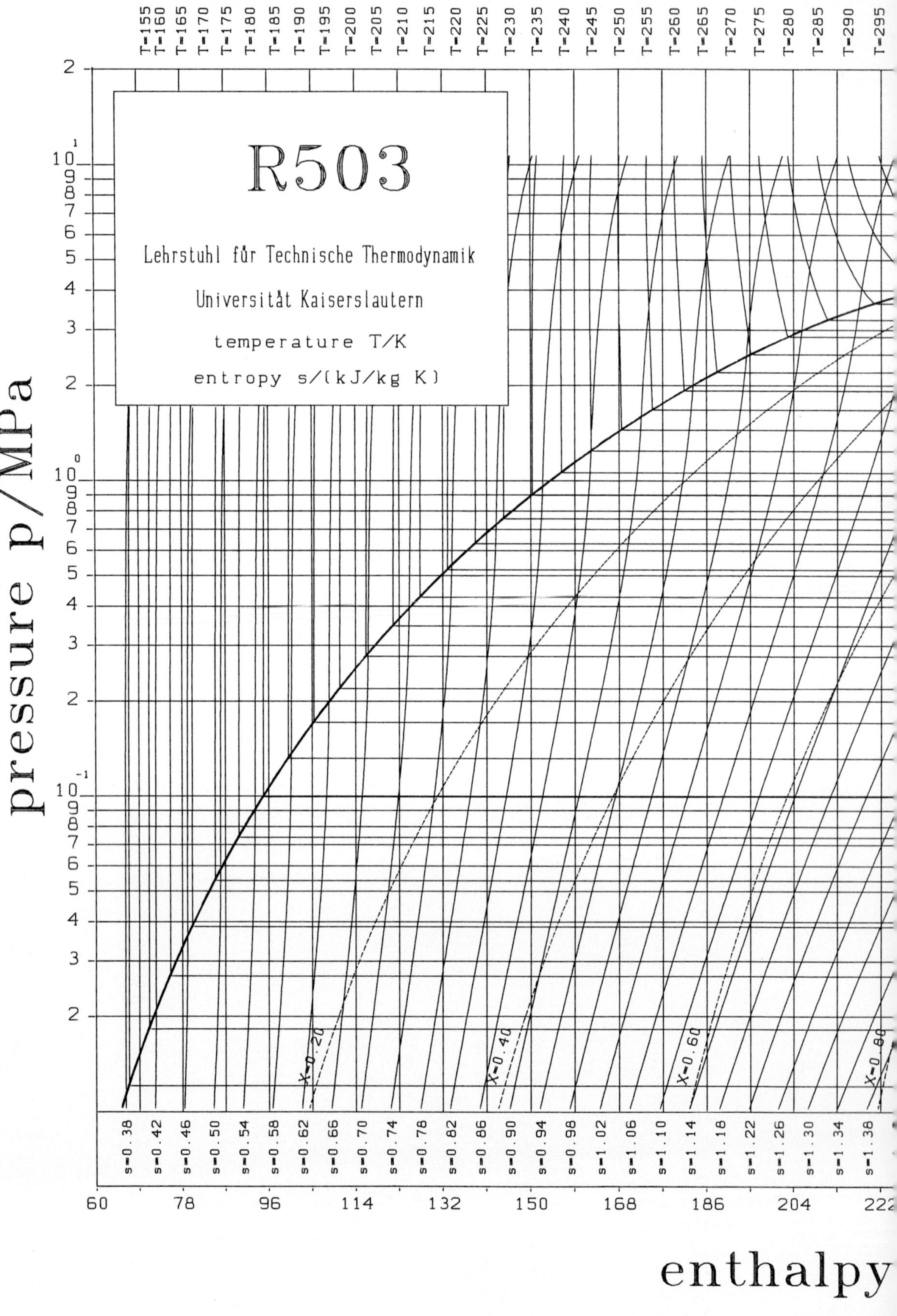

R503
Lehrstuhl für Technische Thermodynamik
Universität Kaiserslautern
temperature T/K
entropy s/(kJ/kg K)
pressure p/MPa
enthalpy
T=155 T=160 T=165 T=170 T=175 T=180 T=185 T=190 T=195 T=200 T=205 T=210 T=215 T=220 T=225 T=230 T=235 T=240 T=245 T=250 T=255 T=260 T=265 T=270 T=275 T=280 T=285 T=290 T=295
s=0.38 s=0.42 s=0.46 s=0.50 s=0.54 s=0.58 s=0.62 s=0.66 s=0.70 s=0.74 s=0.78 s=0.82 s=0.86 s=0.90 s=0.94 s=0.98 s=1.02 s=1.06 s=1.10 s=1.14 s=1.18 s=1.22 s=1.26 s=1.30 s=1.34 s=1.38
X=0.20 X=0.40 X=0.60 X=0.80
60 78 96 114 132 150 168 186 204 222

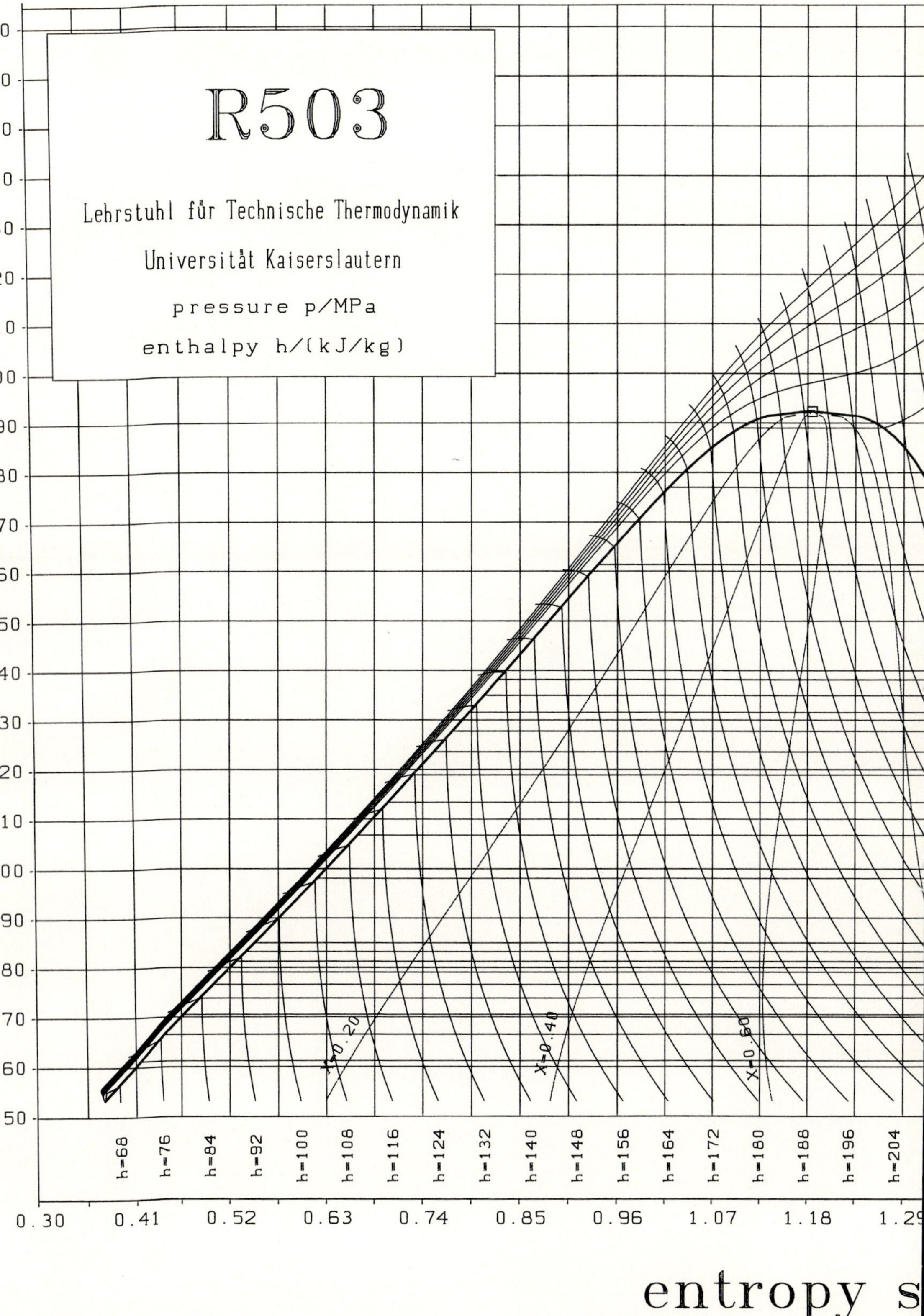

R503
Lehrstuhl für Technische Thermodynamik
Universität Kaiserslautern
pressure p/MPa
enthalpy h/(kJ/kg)
X=0.20
X=0.40
X=0.60
h=68
h=76
h=84
h=92
h=100
h=108
h=116
h=124
h=132
h=140
h=148
h=156
h=164
h=172
h=180
h=188
h=196
h=204
0.30
0.41
0.52
0.63
0.74
0.85
0.96
1.07
1.18
entropy s

R 503. Ts

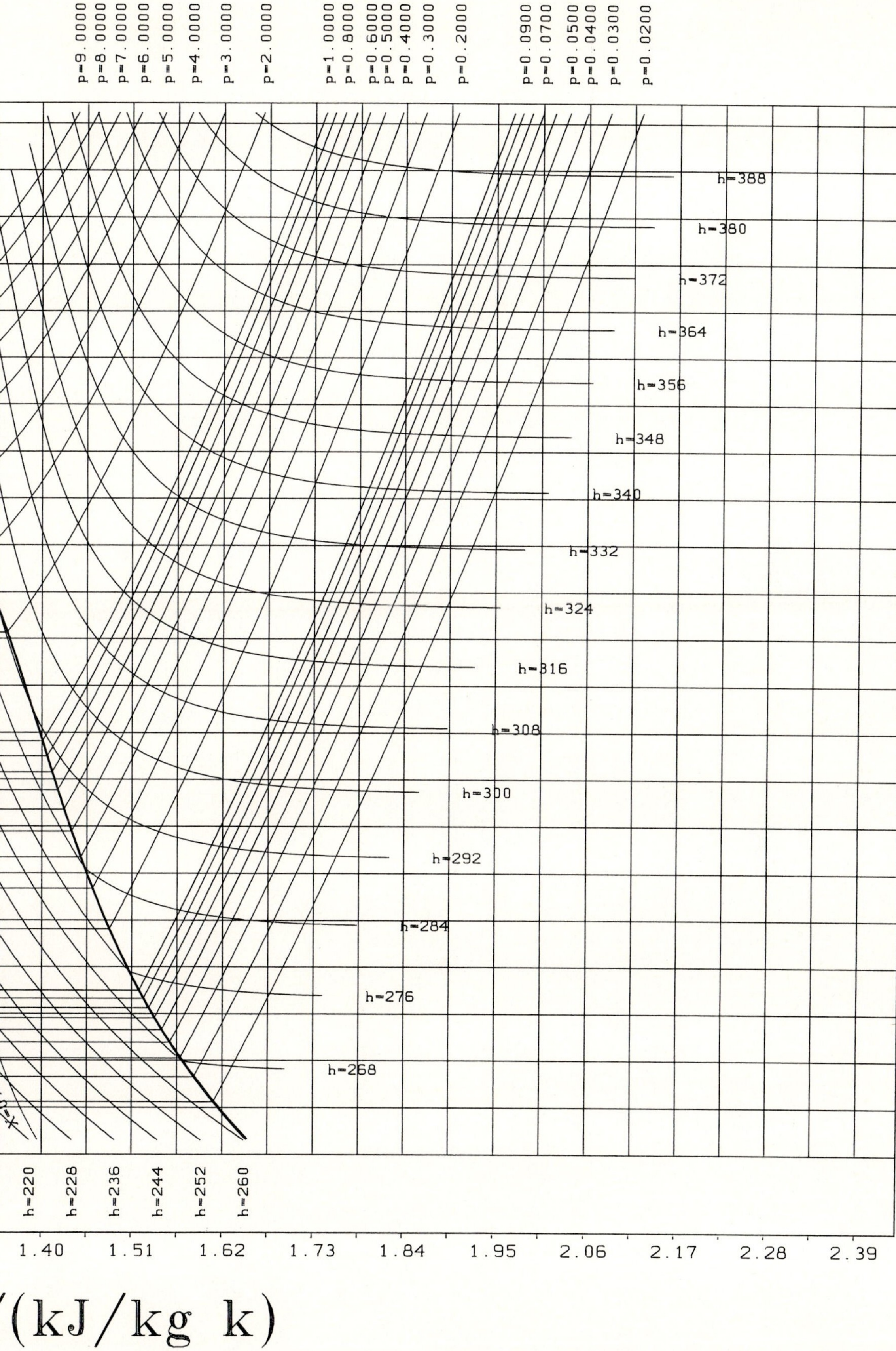

p=9.0000
p=8.0000
p=7.0000
p=6.0000
p=5.0000
p=4.0000
p=3.0000
p=2.0000
p=1.0000
p=0.8000
p=0.6000
p=0.5000
p=0.4000
p=0.3000
p=0.2000
p=0.0900
p=0.0700
p=0.0500
p=0.0400
p=0.0300
p=0.0200
h=388
h=380
h=372
h=364
h=356
h=348
h=340
h=332
h=324
h=316
h=308
h=300
h=292
h=284
h=276
h=268
h=220
h=228
h=236
h=244
h=252
h=260
1.40
1.51
1.62
1.73
1.84
1.95
2.06
2.17
2.28
2.39
/(kJ/kg k)

R 503. hs

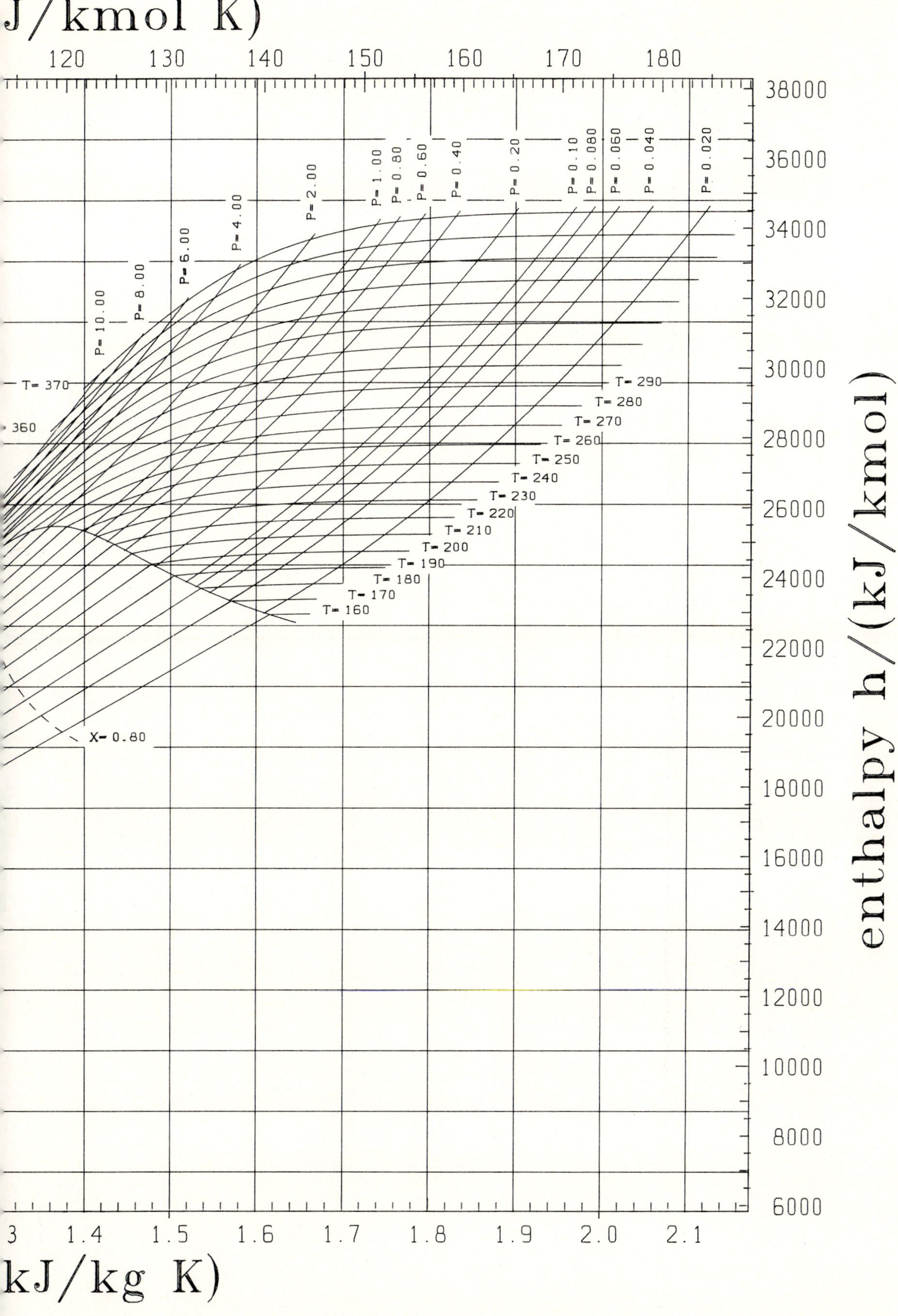

J/kmol K)
120 130 140 150 160 170 180
P= 10.00
P= 8.00
P= 6.00
P= 4.00
P= 2.00
P= 1.00
P= 0.80
P= 0.60
P= 0.40
P= 0.20
P= 0.10
P= 0.080
P= 0.060
P= 0.040
P= 0.020
T= 370
360
T= 290
T= 280
T= 270
T= 260
T= 250
T= 240
T= 230
T= 220
T= 210
T= 200
T= 190
T= 180
T= 170
T= 160
X= 0.80
38000 36000 34000 32000 30000 28000 26000 24000 22000 20000 18000 16000 14000 12000 10000 8000 6000
enthalpy h/(kJ/kmol)
3 1.4 1.5 1.6 1.7 1.8 1.9 2.0 2.1
kJ/kg K)

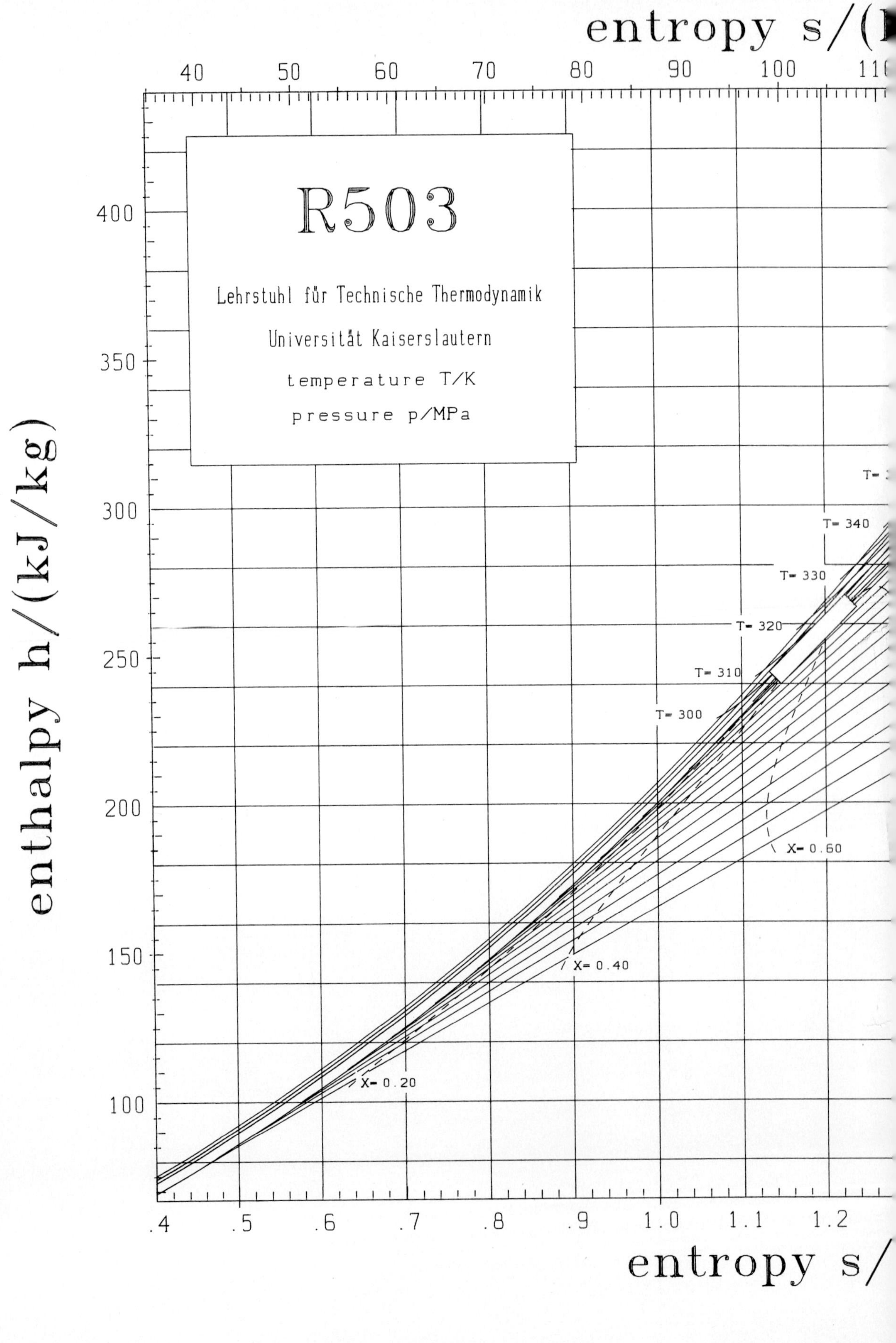

entropy s/(
40
50
60
70
80
90
100
R503
Lehrstuhl für Technische Thermodynamik
Universität Kaiserslautern
temperature T/K
pressure p/MPa
enthalpy h/(kJ/kg)
400
350
300
250
200
150
100
T= 340
T= 330
T= 320
T= 310
T= 300
X= 0.60
X= 0.40
X= 0.20
.4
.5
.6
.7
.8
.9
1.0
1.1
1.2
entropy s/

R 503. Ts

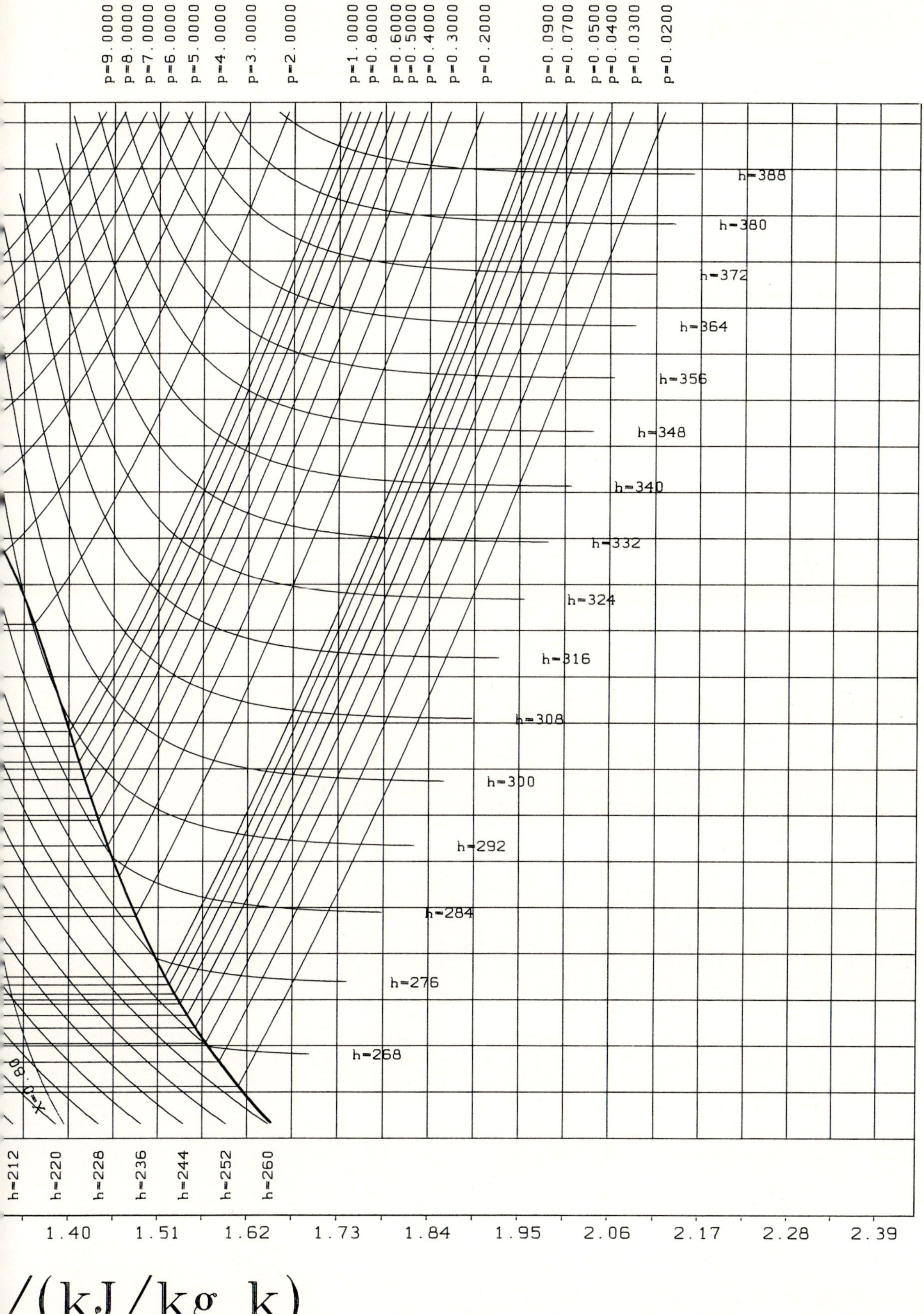

/(kJ/kg k)

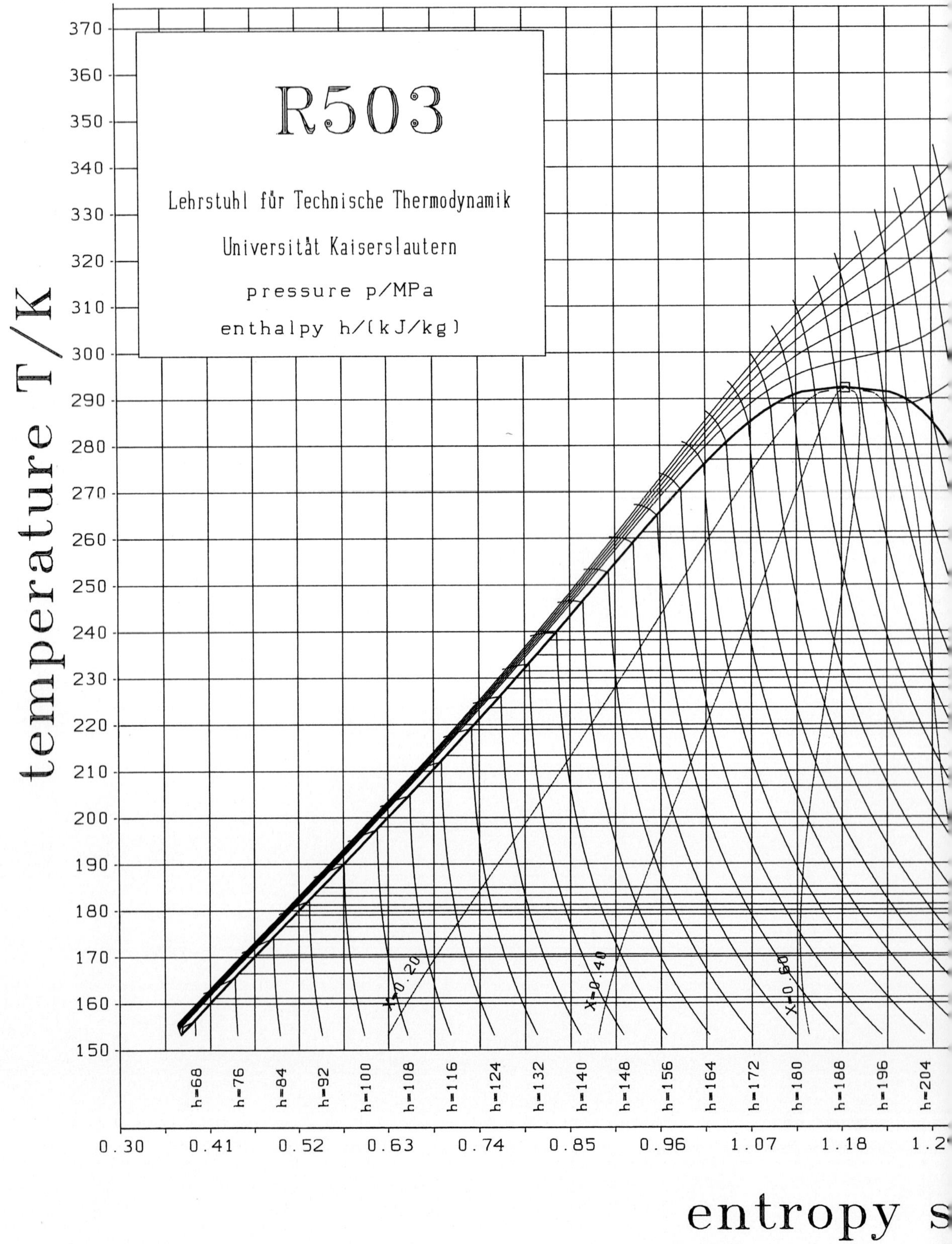

R503
Lehrstuhl für Technische Thermodynamik
Universität Kaiserslautern
pressure p/MPa
enthalpy h/(kJ/kg)
temperature T/K
entropy s
X=0.20
X=0.40
X=0.60
h=68
h=76
h=84
h=92
h=100
h=108
h=116
h=124
h=132
h=140
h=148
h=156
h=164
h=172
h=180
h=188
h=196
h=204

U. Grigull, H. Sandner

Heat Conduction

1984. XX, 187 pp. (International Series in Heat and Mass Transfer). Hardcover DM 98,–
ISBN 3-540-13186-8

In Cooperation with Hemisphere Publishing Corporation

Contents: Introductory Remarks. – Transport Properties. – One-Dimensional Steady-State Heat Conduction. – Steady-State Conduction in the Presence of Heat Sources. – Steady-State Conduction in Several Dimensions. – Nonsteady One-Dimensional Heat Conduction. – Thermal Explosions. – Continuously Operating Heat Sources. – Moving Sources. – Nonsteady Heat Conduction in Several Dimensions. – Nonsteady Heat Conduction with Phase Transition. – References. – Appendices A–G. – Index.

The book is a simple yet surprisingly complete introduction to the subject of heat conduction. The merit of the book is that it presents the subject to the novice in a simple and direct manner. In addition to its clarity, the book is characterized by a felicitous selection of topics, both in what it contains and in what it omits. The emphasis is not only on simple analytic methods, but also on the development of an intuitive "feeling" for numerical values, so important for success in design. The cumulative effect of the exposition and of the worked examples is to implant in the student a visual appreciation of rates of heat transferred by conduction, of size, of the effects of varying and changing properties and, above all, of the importance of thinking in numbers and not only in symbols.

Springer-Verlag
Berlin Heidelberg New York
London Paris Tokyo Hong Kong

Distribution rights for USA, Canada and Mexico: Hemisphere Publishing Corporation Washington–New York–London

Springer